Communications in Computer and Information Science 2774

Series Editors

Gang Li, *School of Information Technology, Deakin University, Burwood, VIC, Australia*
Joaquim Filipe, *Polytechnic Institute of Setúbal, Setúbal, Portugal*
Zhiwei Xu, *Chinese Academy of Sciences, Beijing, China*

Rationale

The CCIS series is devoted to the publication of proceedings of computer science conferences. Its aim is to efficiently disseminate original research results in informatics in printed and electronic form. While the focus is on publication of peer-reviewed full papers presenting mature work, inclusion of reviewed short papers reporting on work in progress is welcome, too. Besides globally relevant meetings with internationally representative program committees guaranteeing a strict peer-reviewing and paper selection process, conferences run by societies or of high regional or national relevance are also considered for publication.

Topics

The topical scope of CCIS spans the entire spectrum of informatics ranging from foundational topics in the theory of computing to information and communications science and technology and a broad variety of interdisciplinary application fields.

Information for Volume Editors and Authors

Publication in CCIS is free of charge. No royalties are paid, however, we offer registered conference participants temporary free access to the online version of the conference proceedings on SpringerLink (http://link.springer.com) by means of an http referrer from the conference website and/or a number of complimentary printed copies, as specified in the official acceptance email of the event.

CCIS proceedings can be published in time for distribution at conferences or as post-proceedings, and delivered in the form of printed books and/or electronically as USBs and/or e-content licenses for accessing proceedings at SpringerLink. Furthermore, CCIS proceedings are included in the CCIS electronic book series hosted in the SpringerLink digital library at http://link.springer.com/bookseries/7899. Conferences publishing in CCIS are allowed to use our online conference service (Meteor) for managing the whole proceedings lifecycle (from submission and reviewing to preparing for publication) free of charge.

Publication process

The language of publication is exclusively English. Authors publishing in CCIS have to sign the Springer CCIS copyright transfer form, however, they are free to use their material published in CCIS for substantially changed, more elaborate subsequent publications elsewhere. For the preparation of the camera-ready papers/files, authors have to strictly adhere to the Springer CCIS Authors' Instructions and are strongly encouraged to use the CCIS LaTeX style files or templates.

Abstracting/Indexing

CCIS is abstracted/indexed in DBLP, Google Scholar, EI-Compendex, Mathematical Reviews, SCImago, Scopus. CCIS volumes are also submitted for the inclusion in ISI Proceedings.

How to start

To start the evaluation of your proposal for inclusion in the CCIS series, please send an e-mail to ccis@springer.com

Lan Huang · Bin Xu · Xuebin Chen ·
Shangguang Wang · Bing Xia · Jianguo Pan ·
Xianhua Song · Zeguang Lu
Editors

Computer Applications

40th CCF National Conference of Computer Applications, CCF NCCA 2025
Beijing, China, August 7–9, 2025
Proceedings

Springer

Editors
Lan Huang
Jilin University
Changchun, China

Xuebin Chen
North China University of Science
and Technology
Tangshan, China

Bing Xia
Zhongyuan University of Technology
Zhengzhou, China

Xianhua Song
Harbin University of Science and Technology
Harbin, China

Bin Xu
Tsinghua University
Beijing, China

Shangguang Wang
Beijing University of Posts
and Telecommunications
Beijing, China

Jianguo Pan
Shanghai Normal University
Shanghai, China

Zeguang Lu
National Academy of Guo Ding Institute
of Data Science
Beijing, China

ISSN 1865-0929 ISSN 1865-0937 (electronic)
Communications in Computer and Information Science
ISBN 978-981-95-5293-1 ISBN 978-981-95-5294-8 (eBook)
https://doi.org/10.1007/978-981-95-5294-8

This Springer imprint is published by the registered company Springer Nature Singapore Pte Ltd.
The registered company address is: 152 Beach Road, #21-01/04 Gateway East, Singapore 189721, Singapore

Preface

As the program committee chairs of the 40th China Conference on Computer Applications (CCF NCCA 2025), it is our great pleasure to welcome you to the conference proceedings. CCF NCCA 2025 was held at the China Hall of Science and Technology in Beijing, China on August 7–9, 2025, hosted by the China Computer Federation (CCF), organized by the CCF Computer Application Professional Committee, Beijing University of Posts and Telecommunications, Shanghai Normal University, etc. The goal of this conference series is to provide a forum for computer scientists, engineers, and educators.

This year's conference attracted 269 paper submissions. After the hard work of the Program Committee, 43 papers were accepted to appear in the conference proceedings, with an acceptance rate of 16%. The main topic of this conference is intelligently driving innovation and self-reliance, integrating digital and real technologies to forge new qualities The accepted papers cover a wide range of fields such as Artificial Intelligence and Applications, Cyberspace Security Technology, Data Science and Big Data Technology, Frontier Technology Applications, including autonomous deep learning and machine learning, robotics and intelligent systems, basic theories and methods of data science, big data mining and knowledge management technology, future grid architecture, network services and management, smart education, intelligent manufacturing, etc.

We would like to thank all the Program Committee members, a total of 207 people from 165 different institutes or companies, for their hard work in completing the review tasks. There were at least 3 reviewers for each article, using a double-blind review process, and each reviewer reviewed no more than 5 articles. Their collective efforts made it possible to attain quality reviews for all the submissions within a few weeks. Their diverse expertise in each research area helped us to create an exciting program for the conference. Their meticulous reviews and professional suggestions helped the authors improve the quality of their papers and lay a solid foundation for journal recommendation.

We thank the team at Springer, whose professional assistance was invaluable in the production of the proceedings. We would like to thank our partners such as Baidu, Parallel Technology, and Wiley Publishing Group for their strong support. A big thanks also to the authors and participants for their tremendous support in making the conference a success. We hope you all had a fruitful conference and hope that we can work together to create unlimited possibilities in the field of computer applications!

Besides the technical program, this year CCF NCCA offered different experiences to the participants. We hope you enjoyed the conference.

September 2025

Lan Huang
Bin Xu
Xuebin Chen
Shangguang Wang
Bing Xia
Jianguo Pan
Xianhua Song
Zeguang Lu

Organization

Honorary Chairs

Shimin Hu	Tsinghua University, China
Lan Huang	Jilin University, China

General Chairs

Bin Xu	Tsinghua University, China
Kun Xu	Beijing University of Posts and Telecommunications, China
Ye Li	Shanghai Normal University, China

Secretary Generals

Zeguang Lu	National Academy of Guo Ding Institute of Data Science, China
Bing Xia	Zhongyuan University of Technology, China

Program Committee Chairs

Xuebin Chen	North China University of Science and Technology, China
Shangguang Wang	Beijing University of Posts and Telecommunications, China
Bing Xia	Zhongyuan University of Technology, China
Jianguo Pan	Shanghai Normal University, China

Organizing Committee Chairs

Jing Liu	Hebei University of Technology, China
Feifei Kou	Beijing University of Posts and Telecommunications, China
Xuan Hong	Shanghai Normal University, China
Zhenlin Cheng	Beijing Wuzi University, China
Jing Li	Tianjin University of Technology, China

Organizing Committee

Zhaowen Qiu	Heilongjiang Tuomeng Technology Co., Ltd., China
Ruiquan Jiang	Shanghai Jinshida Software Technology Co., Ltd., China
Tian Bai	Jilin University, China
Hongming Cai	Shanghai Jiao Tong University, China
Ke Xu	Weiye Xuanran Education Technology (Beijing) Co., Ltd., China
Biqing Zeng	South China Normal University, China
Yiran Zhang	Beijing University of Posts and Telecommunications, China
Xia Liao	Beijing University of Posts and Telecommunications, China
Qian Li	Beijing University of Posts and Telecommunications, China
Qing Li	Beijing University of Posts and Telecommunications, China
Jinfeng Wen	Beijing University of Posts and Telecommunications, China
Yuanzhe Li	Beijing University of Posts and Telecommunications, China

Forum Committee

Junyu Lin	Fudan University, China
Xiuquan Qiao	Beijing University of Posts and Telecommunications, China
Rongli Gai	Dalian University, China
Ao Zhou	Beijing University of Posts and Telecommunications, China

Publication Chairs

Xianhua Song	Harbin University of Science and Technology, China
Jianquan Ouyang	Xiangtan University, China
Linjing Wei	Gansu Agricultural University, China
Shuai Zhao	Beijing University of Posts and Telecommunications, China

Rewards Committee Chairs

Hong Yu	Dalian Ocean University, China
Kui Xiao	Hubei University, China

Rewards Committee

Xiaoyu Chen	China Construction Bank Shanghai Baosteel Baoshan Branch, China
Aibin Chen	Central South University of Forestry and Technology, China
Ruiquan Jiang	Shanghai Jinshida Software Technology Co., Ltd., China
Guanghui Yan	Lanzhou Jiaotong University, China
Ke Zhang	University of Electronic Science and Technology of China, China

Competition Committee Chair

Bin Xu	Tsinghua University, China

Competition Committee

Jing Li	Tianjin University of Technology, China
Xuedong Wang	Xueda Education Group, China
Ke Xu	Weiye Xuanran Education Technology (Beijing) Co., Ltd., China
Lifeng Ren	Universal Music (Beijing) Technology Co., Ltd., China
Zhitao Dai	Beijing University of Posts and Telecommunications, China
Ting Diao	Beijing University of Posts and Telecommunications, China
Yansheng Huang	Test bar (Beijing) Technology Co., Ltd., China

Workshop

Junyu Lin	Fudan University, China
Yun He	Tsinghua AI TIME, China
Mengwei Xu	Beijing University of Posts and Telecommunications, China
Heng Pan	Zhongyuan University of Technology, China

Propaganda Committee

Xinnian Wang	Dalian Maritime University, China
Xiaofeng Lu	Beijing University of Posts and Telecommunications, China
Yi Zhang	Beijing University of Posts and Telecommunications, China
Guoyu Xu	Henan University of Economics and Law, China
Hui Lin	Fujian Normal University, China
Peng Liu	Changchun University of Science and Technology, China
Shuai liu	Hunan Normal University, China

Propaganda Committee Chairs

Jianji Ren	Henan Polytechnic University, China
Jiwei Zhang	Beijing University of Posts and Telecommunications, China

Propaganda Committee

Long Zhao	Neusoft Education Technology Group, China
Xiaofeng Lu	Beijing University of Posts and Telecommunications, China
Gaoxiang Ouyang	Beijing Normal University, China
Ying Yuan	Beijing Super Cloud Computing Center, China
Chuanchuan You	Hubei University of Economics, China

Contents

Artificial Intelligence and Applications

Artificial Intelligence and Applications

A Fractional-Order Automatic Feature Matching Algorithm for Image Rotation and Scaling

Shoutong Huang(✉), Yu Ma, Chunshu Li, and Lanxiang Ma

Ningxia University, Yinchuan 750021, Ningxia, People's Republic of China
12024130722@stu.nxu.edu.cn

Abstract. In this paper, automatic element matching technology using the MATLAB platform, which is based on the fractional order, is deeply studied, with a focus on automatically finding image rotation and scaling parameters. The concept of the fractional order is introduced in the image scaling and rotation manipulation steps. The whole process includes many key steps, such as feature detection, descriptor extraction, descriptor matching and transformation matrix calculation, and each step is carefully designed to ensure high-precision image correction. Through this technology, the system can automatically identify and match the key elements in the image and then calculate the necessary rotation and scaling parameters to realize the automatic correction of the distorted image. Experimental verification shows that this method not only has high accuracy and stability but also performs well in processing various complex images. This achievement has shown extensive application potential and important practical value in many fields, such as image analysis, image recognition and visual effects application, and is expected to provide strong support for scientific research and technological innovation in related fields.

Keywords: Fractional order · Automatic element matching · Image rotation and scaling · Algorithms

1 Introduction

1.1 Research Background and Purpose

In digital image processing, distortions such as rotation and scaling can significantly affect the restoration of the original image state. This paper proposes an automatic feature matching method using MATLAB with a fractional order to detect feature points efficiently, extract descriptors, and calculate rotation and scaling parameters for precise image correction. This method aims to increase the accuracy and automation level of image correction, providing new ideas and technical support for digital image processing [1].

L. Huang et al. (Eds.): CCF NCCA 2025, CCIS 2774, pp. 3–13, 2026.
https://doi.org/10.1007/978-981-95-5294-8_1

Research Purpose. The purpose of this paper is to propose an automatic feature matching method using MATLAB based on fractional order to find the rotation and scaling parameters of images and achieve automatic correction of distorted images. This method, through steps such as feature detection, descriptor extraction, descriptor matching, and transformation matrix calculation, can accurately estimate the rotation and scaling parameters of images, thereby restoring the original state of the images. The purpose is not only to verify the feasibility and accuracy of this method but also to explore its broad application prospects in the field of image processing [2]. Through this study, we hope to provide an efficient and reliable means for image correction and make progress in the development of digital image processing technology.

Design Significance. The design of the automatic feature matching method via MATLAB, which is based on the fractional order, aims to find the rotation and scaling parameters of images accurately and solve the problem of image distortion. This method achieves automated correction through feature matching, improving the accuracy and efficiency of image processing. It is important in the fields of image editing, recognition, and analysis and has promoted digital image processing technology.

1.2 Current Research Status and Review

Current Research Status in China. In China, MATLAB has been widely used in image processing. Recent advancements in automatic feature matching methods have led to significant improvements in feature detection, descriptor extraction, and matching algorithms. These methods have been applied in various fields, promoting the development of digital image processing technology [3]. Fractional calculus is an advanced form of mathematics. As a mathematical tool, it can be applied in engineering, artificial intelligence, medicine and other fields and is often used to solve practical problems [4]. Owing to the inheritance and memory of fractional calculus, domestic researchers have applied it to the field of image processing and achieved good research results. Fractional calculus is a generalization of traditional integer-order calculus and can be used to construct more complex transformations and other operations in image processing [5].

Current Research Status Abroad. Internationally, automatic feature matching methods have also gained attention. Researchers have proposed innovative algorithms for feature detection, descriptor matching, and transformation matrix calculation. These methods have been applied in image correction, stitching, and object recognition. Additionally, combining these methods with machine learning and deep learning has further enhanced their intelligence and automation [6].

Review. Automatic feature matching methods are crucial for determining image rotation and scaling parameters. Although significant progress has been made, challenges remain, such as improving matching accuracy, computational efficiency, and adaptability to complex distortions. Future research should focus on enhancing these aspects and integrating new technologies to drive continuous innovation in image processing.

2 Algorithm Introduction

2.1 Feature Detection

Feature detection extracts unique and stable points from images and is robust to rotation and scaling. Common algorithms include speeded-up robust features (SURF), scale-invariant feature transform (SIFT), etc. In addition, corner detection algorithms (such as Harris corner detection, which determines corners by calculating the gray level changes in local small windows in various directions, and Shi-Tomasi corner detection, which is an improvement over Harris corner detection), and edge detection algorithms (such as Canny edge detection, which is a multistage edge detection algorithm, and the Sobel operator, which is mainly used to calculate the approximate gradient of the image gray function in the gradient operator) are also included.

SURF (Speed-Up Robust Features). This algorithm, proposed in 2006 by scholars such as Herbert Bay, is an improved version of SIFT, which is a robust image recognition and description algorithm, addressing its high computational complexity. SURF uses the Hessian matrix and integral images to detect feature points in images efficiently [7]. Therefore, this paper chooses the SURF algorithm for subsequent image stitching work, and the following is a detailed elaboration of the steps of the SURF algorithm. The Hessian matrix of an image f(x,y) is:

$$H_{(x,y)} = \left(L_{xx}L_{xy}\right)\left(L_{xy}L_{yy}\right)$$
$$\left(L_{xy}L_{yy}\right)$$

where Lxx, Lxy, and Lyy are the second-order derivatives in various directions of the image g(σ) after Gaussian filtering. Hessian matrix determinant: Det H = L_{xx} * L_{yy} - 0.9 * L_{xy}^2

For a grayscale image, the value of any point (x,y) in the integral image refers to the sum of the grayscale values of all the points in the rectangular area from the upper left corner of the image to this point.

Scale-Invariant Feature Transform (SIFT). SIFT detects feature points across different scales via a Gaussian difference pyramid. It assigns a dominant direction to each point, making it rotationally invariant. The feature descriptor is generated by calculating gradient histograms in subregions around the point [8].

Statistically Robust MSAC (M-estimator Sample Consensus) Algorithm. A robust variant of the RANSAC algorithm uses an M-estimator to handle outliers. It assigns weights to sample points on the basis of the residuals of the model estimation, giving larger weights to inliers with small residuals and smaller weights to outliers with large residuals, thereby reducing the impact of outliers on model fitting [9].

2.2 Descriptor Extraction

Descriptors are local information vectors extracted from feature point neighborhoods that contain gradient, direction, and intensity information. These descriptors enable feature point matching across different images, even if these images have undergone geometric transformations such as rotation and scaling. Common methods can generate a robust descriptor vector for each feature point.

2.3 Feature Descriptor Matching

After extracting feature descriptors, it is necessary to use matching algorithms to find corresponding feature point pairs between distorted images and original images. Matching algorithms such as brute-force matching (Brute-Force Matcher) and fast library for approximate nearest neighbors (FLANN) are used to find corresponding feature point pairs by calculating descriptor similarities.

2.4 Transformation Matrix Calculation

Through matched feature point pairs, the transformation matrix, which contains rotation and scaling information, is calculated via matched feature point pairs. This matrix enables the correction of distorted images.

2.5 Fractional-Order-Based Processing

Fractional-order calculus, a generalization of traditional integer-order calculus, has unique properties, such as memory and heredity, which make it particularly suitable for image processing tasks [10]. In this paper, the Riemann–Liouville definition of fractional-order calculus is employed:

$$D^{\alpha}f(t) = \frac{1}{\Gamma(n-\alpha)}\left(\frac{d}{dt}\right)^{n}\int_{a}^{t}\frac{f(\tau)}{(t-\tau)^{\alpha-n+1}}d\tau \tag{1}$$

where α is the fractional order and n is an integer such that $n - 1 < \alpha < n$. The Grünwald–Letnikov definition is also utilized for discrete computations:

$$D^{\alpha}f(t) = \lim_{h\to 0}\frac{1}{h^{\alpha}}\sum_{k=0}^{\infty}(-1)^{k}\binom{\alpha}{k}f(t-kh) \tag{2}$$

In the context of image rotation and scaling, fractional-order calculus enhances the detection of feature points by capturing more subtle variations in image intensity and gradient information. Compared with traditional integer-order methods, fractional-order approaches provide more precise control over the feature extraction process, leading to improved matching accuracy and robustness.

3 Experimental Steps and Results

3.1 Experimental Design and Algorithm Comparison

To enhance the validation of the method's robustness, experiments were conducted using a broader range of image types, including natural scenes, medical images, and satellite images. Various degrees of rotation and scaling distortions were applied to simulate real-world scenarios. The proposed method was compared with the following algorithms:intensity -based registration method (Thevenaz and Unser). This method aligns images on the basis of intensity similarities. Feature-based methods: SURF and BRISK were used for feature detection and matching. Phase-Only Correlation (POC): This method correlates image phases to estimate rotation and scaling parameters.

3.2 Experimental Steps and Results

Image Reading. Images were introduced into the workspace to demonstrate the geometric transformation estimation between two images. The SURF algorithm detected

Fig. 1. Image Reading.

feature points, and the transformation matrix was estimated via the estimate Geometric Transform 2D function. As shown in Fig. 1:

Adjusting the Image Size and Rotating the Image. The images were distorted by scaling and rotating them. The range of distortions was carefully selected to ensure that sufficient feature points were detectable (Fig. 2).

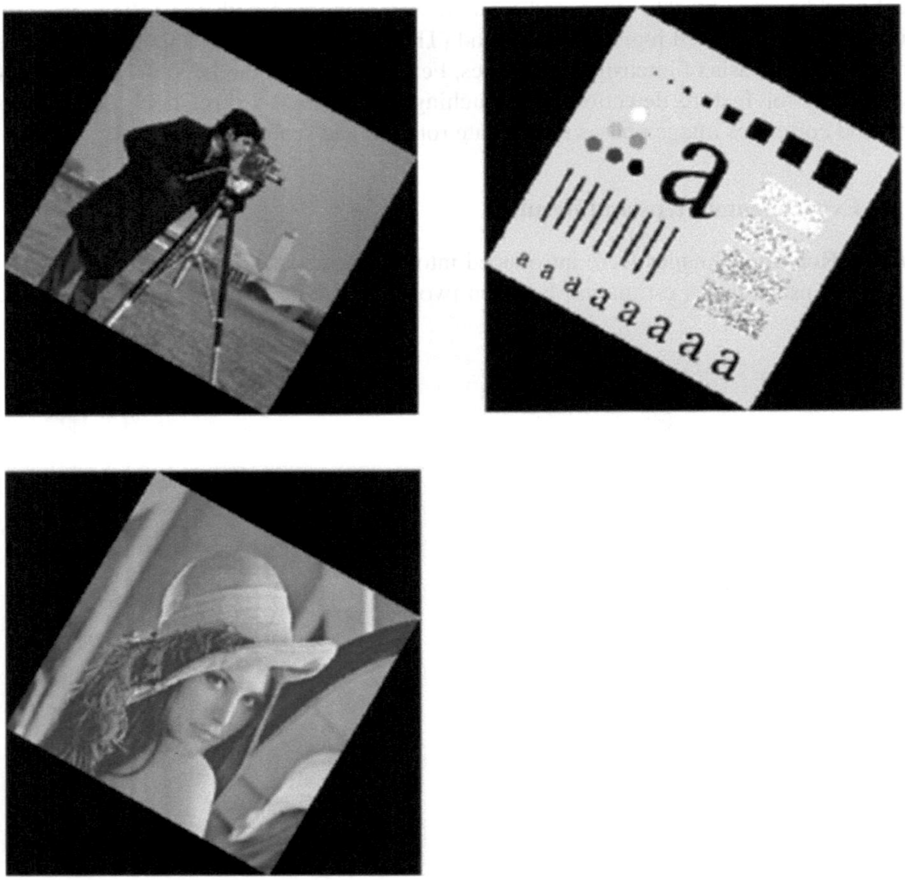

Fig. 2. Adjusted image size and rotated image.

Finding Matching Features between Images. Feature points were detected, descriptors were extracted, and matching pairs were identified. The matching results were visualized via distinct colors and labels to improve clarity (Fig. 3).

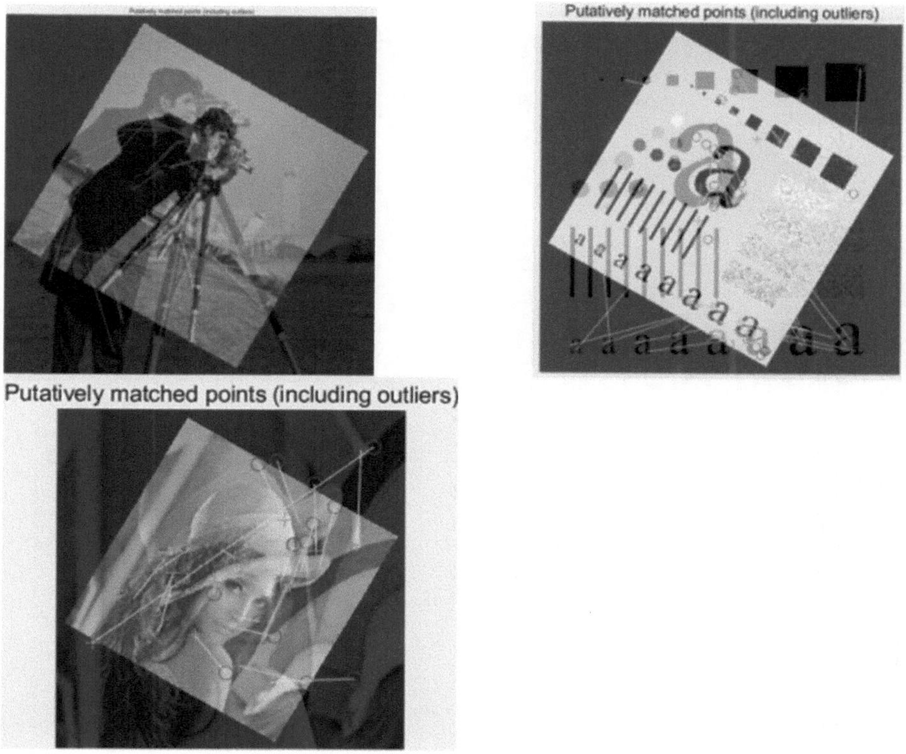

Fig. 3. Finding matching features between images.

Estimate Transformation. The statistically robust MSAC algorithm is used to estimate the transformation matrix and find the transformation corresponding to the matching point pairs. It removes outliers when calculating the transformation matrix to improve accuracy. The matching pairs used in the calculation are displayed (Fig. 4).

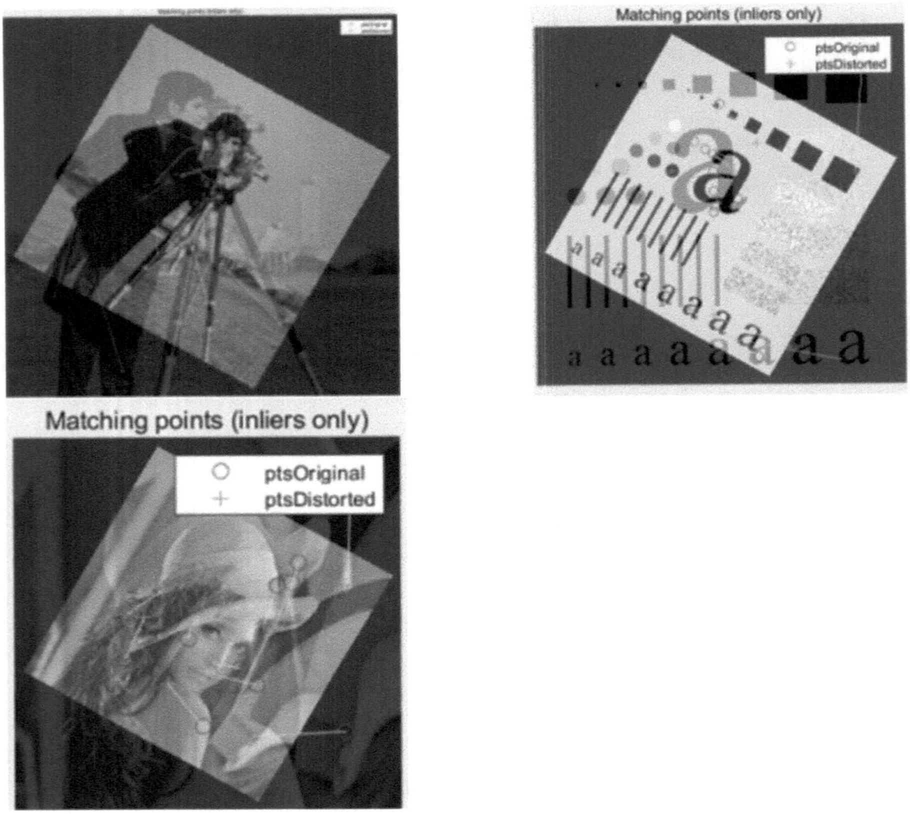

Fig. 4. Estimate Transformation.

Solve for Scale and Angle. The inverse transformation matrix was calculated to restore the distorted image. The restored values were compared with the original distortion parameters.

Restoring the Original Image. The distorted image was restored via the calculated transformation matrix. The results were compared with the original image to evaluate the correction accuracy (Fig. 5).

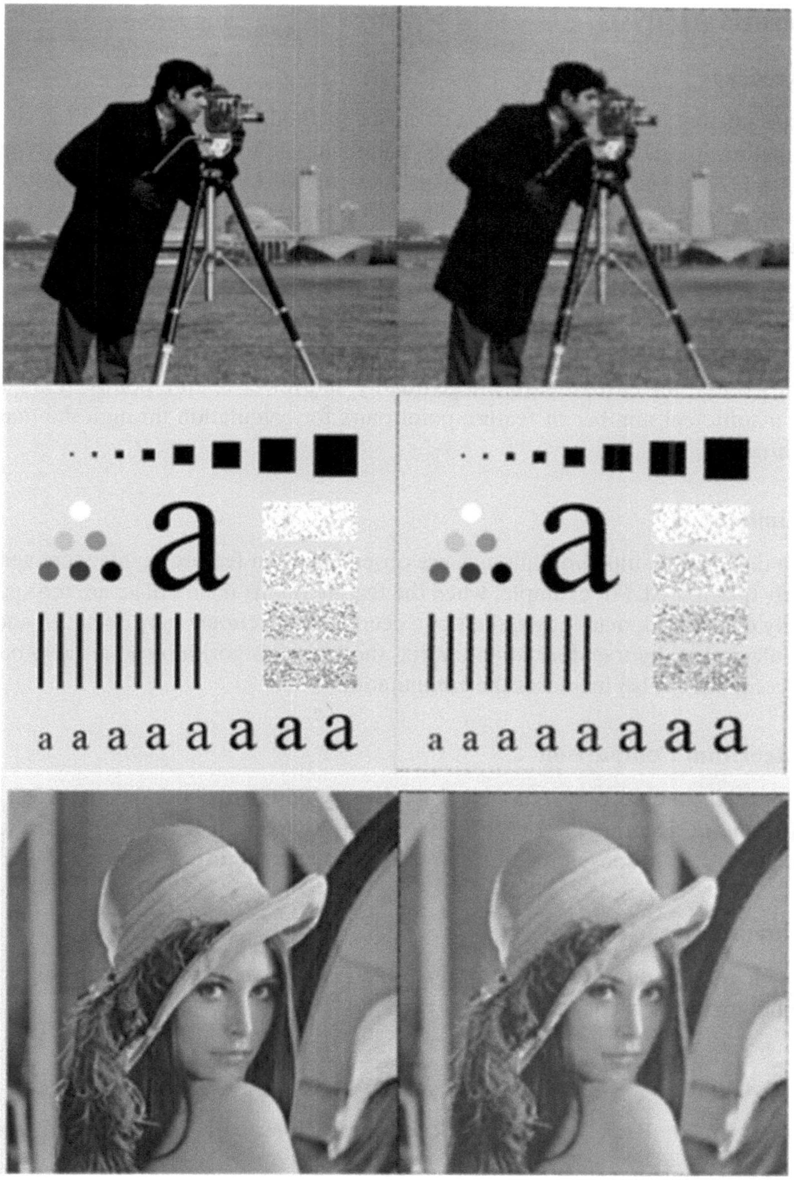

Fig. 5. The comparison of the results after distorted and the original image.

4 Results Analysis

4.1 Accuracy

The experimental results show that the proposed method in this paper has high accuracy in estimating image rotation and scaling parameters across various image types and distortion levels. The transformation matrix calculated by the matched feature points can accurately reflect the geometric relationship between the distorted image and the original image.

4.2 Robustness

The method has a certain degree of robustness to different types and levels of distorted images. Even in the case of poor image quality and fewer feature points, it is possible to find a sufficient number of feature point pairs for calculation through the matching algorithm.

4.3 Limitations

Despite the good results, limitations were observed when feature points were sparse or unevenly distributed. For example, when the feature points in an image are too sparse or unevenly distributed, matching failure or a decrease in accuracy may occur. In addition, when calculating the transformation matrix, the MSAC algorithm can remove outliers, but its complexity also increases the computation time.

4.4 Algorithm Comparison

The proposed method outperformed the intensity-based registration and POC methods in terms of accuracy and robustness. While SURF and BRISK provided similar performances, the proposed method showed better adaptability to complex distortions.

4.5 Quantitative Metrics

To provide a more comprehensive evaluation, Table 1 compares the proposed method with other methods via quantitative metrics:

Table 1. Comparison of the proposed method with other methods.

Method	Rotation Angle Recovery Error (degrees)	Scaling Recovery Error (%)	Computational Time (seconds)
Proposed	1.2 ± 0.3	3.4 ± 1.2	4.5 ± 0.8
Intensity-based	5.6 ± 1.8	8.9 ± 3.2	12.3 ± 2.1
SURF	2.1 ± 0.7	5.6 ± 2.3	6.7 ± 1.3
BRISK	3.2 ± 1.1	7.8 ± 3.5	5.9 ± 1.0
POC	4.8 ± 1.5	10.2 ± 4.1	8.7 ± 1.7

5 Conclusion and Outlook

The proposed automatic feature matching method using MATLAB, which is based on fractional order, effectively estimates image rotation and scaling parameters, demonstrating high accuracy and robustness. The method's success validates its potential for image correction and analysis applications. Future work will focus on improving feature matching accuracy and computational efficiency. Enhancing adaptability to complex distortions and integrating advanced algorithms will further optimize the method.

Acknowledgments. Not applicable.

Disclosure of Interests. The authors have no competing interests to declare that are relevant to the content of this article.

References

1. Sun, Y.: Research on Contemporary Tourists Visual Perception of Classical Gardens in Jiangnan Based on MATLAB Image Segmentation. Nanjing Agricultural University (2022)
2. Tian, T.: Research on Fast Local Image Feature Extraction Method. Huazhong University of Science and Technology (2015)
3. Zhang, R.: Research on Low-Light Image Enhancement Method Based on Unsupervised Learning. Linyi University (2024)
4. Gan, J.: Research on Ground Unmanned Vehicle Path Planning Based on Aerial Vision. Chongqing University of Technology (2024). https://doi.org/10.27753/d.cnki.gcqgx.001423
5. Dibs, H., Hasab, H.A., Jaber, H.S., Al-Ansari, N.: Automatic feature extraction and matching modelling for highly noise near-equatorial satellite images. Innovat. Infrastruct. Solutions **7**(1), 1–14 (2021). https://doi.org/10.1007/s41062-021-00598-7
6. Heremans, E.R.,et al.: Feature matching as improved transfer learning technique for wearable EEG. Biomed. Sig. Process. Control **78** (2022)
7. Bay, H., Tuytelaars, T., Van Gool, L.: SURF: Speeded up robust features. In: European Conference on Computer Vision. Springer, Berlin, Heidelberg (2006)
8. Lowe, D.G.: Distinctive image features from scale-invariant key points. Int. J. Comput. Vision **60**(2), 91–110 (2004)
9. Torre, V., Poggio, T.A.: On edge detection. IEEE Trans. Pattern Anal. Mach. Intell. **8**(2), 147–163 (1986)
10. Shi J, Tomasi C.: Good features to track. In: IEEE Conference on Computer Vision and Pattern Recognition, IEEE (1994)

Multistrategy Improved Dung Beetle Optimization Algorithm and Its Application in ALBP-1

Wan Bing[1(✉)], Cui Haibo[1], Guo Qin[2], and Zhang Lingyan[1]

[1] Hubei University, Wuhan 430062, China
2665668384@qq.com
[2] China University of Geosciences, Wuhan 430078, China

Abstract. The Dung beetle optimizer (DBO) is a novel metaheuristic algorithm with certain limitations, including suboptimal convergence accuracy and local optima entrapment. To address these issues, this study proposes a modified reinforced DBO (MRDBO) featuring three key enhancements: (1) performing chaotic mapping for improved population initialization, (2) RIME-inspired soft-rime search for faster convergence, and (3) adaptive hybrid mutation with greedy selection to prevent stagnation. A comprehensive evaluation using 14 benchmark functions against 30 competing algorithms confirms the superior performance of MRDBO. Practical validation through assembly line balancing applications further demonstrated its effectiveness.

Keywords: MRDBO · RIME · Cauchy–Gaussian mutation · assembly line baancing

1 Introduction

The Deng beetle optimizer (DBO), which presented by Xu et al. in 2022 [1] is an innovative swarm intelligence technique. It has gained significant attention because of its distinctive exploration pattern, uncomplicated framework, and robust global search capacity. DBO has been applied in several fields, particularly in engineering and design [2], image processing [3], and drone trajectory planning [4]. Like other algorithms, DBO also faces challenges such as low population diversity, low local development capacity, and falling into the optimal locality [5]. To solve these problems, experts have suggested enhancements in recent years. For example, the study in [6] improved the formula for beetle and thief feces, enhancing the speed and accuracy of algorithm convergence. Additionally, the GODBO algorithm strengthened DBO's exploration ability by leveraging directed transfer on the basis of the best values both opposition-based learning (OBL) and the existing population. Although these changes have improved the performance of DBO to some extent, there are also several issues, such as gradual convergence and premature convergence.

L. Huang et al. (Eds.): CCF NCCA 2025, CCIS 2774, pp. 14–26, 2026.
https://doi.org/10.1007/978-981-95-5294-8_2

This article proposes a multiimproved optimization of the Dung beetle (MRDBO) to eliminate the limitations of DBO. A Tent chaotic mapping strategy initializes the population to increase diversity and coverage. The soft frost formation strategy from RIME improves convergence speed and accuracy. An adaptive Cauchy–Gaussian mutation and greedy update mechanism help escape local optima, collectively boosting search precision and convergence.

2 Original Dung Beetle Optimizer

The concept of DBO is derived from the reproduction of feeding principles, and the behavior of dung beetle populations in nature. It simulates these behavioral mechanisms to achieve optimization. The specific model for algorithmic mathematical updates is as follows.

2.1 Rolling Behavior

The rolling behavior position update formula is as follows:

$$x_i(t+1) = x_i(t) + \alpha \times k \times x_i(t-1) + b \times \Delta x \tag{1}$$

$$\Delta x = |\mathrm{x_i(t)} - \mathrm{X^w}| \tag{2}$$

This figure shows how the brightness of light changes, which is negatively related to the light source strength.

In this case, t represents the current step number, and $X_i(t)$ represents the position of the i-th tassel at that particular stage." Parameter k is a number that shows how much something bends, b is a fixed number, a stands for a natural number that can be either 1 or -1, and X^w stands for the worst position in all the data.

When a dung beetle travelling straightforwardly runs against an obstruction,it uses dance behavior to choose alternative routes. At this point, how to correct the position of the scrolling beetle is as follows:.

$$\mathrm{X_i(t+1) = X_i(t) + tan}\partial|\mathrm{X_i(t) - X_i(t-1)}| \tag{3}$$

where $\partial \in [0,\pi]$ indicates the deflection angle of the dung beetle.

2.2 Breeding Behavior

To supply a safe spawning ground for females and protect their young, beetles roll dung balls into shelters.The simple explanation of what a safe area is as follows:

$$\begin{cases} LB^* = \max(X^* \times (1 - Rt), LB) \\ UB^* = \min(X^* \times (1 + Rt), UB) \end{cases} \quad (4)$$

In this case, Rt is calculated by subtracting 1 from the ratio of t to T, where T represents the total number of maximum iterations in the process. X^* represents the current position corresponding to the local optimal solution, whereas UB^* and LB^* represent the maximum and minimum boundaries of the simulated spawning area, respectively. Instead, UB and LB define maximum and minimum boundaries for constraints on optimization problems. Each female lays an egg at each iteration. As the process continues, the position of the hatched egg dynamically moves within the boundaries of the spawning area, which is described by the following formula.

$$B_i(t+1) = X^* + L_1 \times (B_i(t) - LB^*) + L_2 \times (B_i(t) - UB^*) \quad (5)$$

Here, L_1 and L_2 are 2 independent random vectors, each vector having a size of $1 \times$ Dim, representing the complexity or intricacy of the optimization challenge., and $B_i(t)$ represents the position of the ith breeding ball at the t-th iteration.

2.3 Reproductive Behavior

Those who have grown up and come out of the earth to forage are called young dendrites. The circumference of the ideal feeding area is described as follows:

$$\begin{cases} LB^{best} = \max(X^{best} \times (1 - R), LB) \\ UB^{best} = \min(X^{best} \times (1 + R), UB) \end{cases} \quad (6)$$

Although X^{best} indicates the global optimal position, LB^b and UB^b define the minimum and maximum boundaries of the ideal feed area, respectively.

$$X_i(t+1) = X_i(t) + Q_1 \times \left(X_i(t) - LB^{best}\right) + Q_2 \times \left(X_i(t) - UB^{best}\right) \quad (7)$$

where Q_2 may be an irregular vector that falls inside the interim (0,1) and where Q_1 could be a nearly disseminated irregular number.

2.4 Thief Behavior

The fact that some beetles steal fecal pellets from other beetles -is another common phenomenon in nature. The following is the definition of the thief beetle position update formula:

$$X_i(t+1) = X^{best} + K \times g \times \left(|X_i(t) - X^*| + |X_i(t) - X^b|\right) \quad (8)$$

Assume g may be a 1*Dim irregular vector with a multivariate ordinary dispersion, and K is a settled consistent.

3 Multistrategy Improved Dung Beetle Optimizer

3.1 Introduction of Tent Chaotic Mapping

When DBO is applied to optimize problem-solving, the population is usually initialized randomly. Random initialization may lead the algorithm to concentrate on particular regions of the search space,

Potentially leading to entrapment in local optima. The introduction of tent chaotic mapping in group initialization ensures the diversity of the initial group, which is beneficial for early global search and enhances both the algorithm's convergence rate and its resolution precision [7].The tent chaotic mapping can be expressed as follows:

$$Z_{n+1} = \begin{cases} \frac{Z_n}{\beta}, Z_n \in (0, \beta] \\ \frac{1-Z_n}{1-\beta}, \in (\beta, 1] \end{cases} \tag{9}$$

In this case, β is the chaotic parameter with a range of (0,1]; The value Zn represents the Tent chaotic mapping's n-th iteration, whereas Zn + 1 represents the Tent chaotic mapping's (n + 1)-th iteration (Fig. 1).

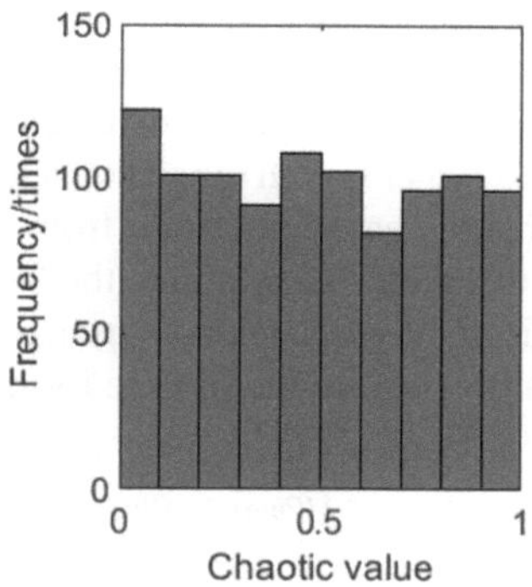

Fig. 1. Histogram of the tent chaotic mapping sequence distribution.

3.2 Integration of the Soft Rime Formation Search Strategy from RIME

The original rolling ball strategy is too singular, which does not sufficiently search the optimization space and can quickly make the algorithm achieve a local optimum, failing to reach equilibrium between worldwide exploration and regional exploitation. Therefore, the soft rime formation search strategy from RIME [8] is integrated into the rolling ball phase of the original algorithm. Strong global search capabilities are possessed by the RIME algorithm, which increases the dung beetle's exploration of unknown areas, thus helping to achieve a good balance between local development and global optimization. Below is the revised update strategy for the new rolling ball phase:

$$x_i(t+1) = \begin{cases} x_i(t) + \alpha \times k \times x_i(t-1) + b \times \Delta x, \mathrm{S} < \Phi \\ \mathrm{X}^{\mathrm{best}} + r_1 \cdot \cos\theta \cdot \beta \cdot (h \cdot (\mathrm{UB} - \mathrm{LB}) + \mathrm{LB}), \mathrm{S} \geq \Phi \end{cases} \tag{10}$$

$$\theta = \pi \cdot \frac{t}{10T} \tag{11}$$

$$\beta = 1 - \left(\frac{w \cdot t}{T}\right)/w \tag{12}$$

In this context, the variable represents a randomly chosen value bounded by 0 and 0.5, whereas r_1 is a randomly selected number between −1 and 1 that determines the directional movement of the rime particles. The parameter ß captures the influence of environmental factors, mimicking external environmental effects with increasing iteration counts. The notation for the floor function is set to 5 by default, which regulates the segmentation within the step function to avoid premature convergence of the algorithm. The variable t indicates the current iteration step, and h denotes the degree of adhesion—a randomly assigned value between 0 and 1.

3.3 Adaptive Cauchy–Gaussian Mutation

To maintain a strategic distance from the calculation that is caught in neighborhood minima and enhance population diversity, intelligent optimization algorithms can use mutation operators. Two mutation operators are frequently utilized in intelligent optimization algorithms: Cauchy mutation and, Gaussian mutation, each with its own advantages and disadvantages [9]. Gaussian mutation has good search capabilities in small ranges. In contrast, Cauchy mutation has a larger search range than Gaussian mutation does, but its relatively large step size can readily cause deviation from the optimal value, ultimately leading to less advantageous offspring. To combine the benefits of both Cauchy mutation and Gaussian mutation, an adaptive Cauchy–Gaussian mixed perturbation mutation technique is suggested. The following are the precise formulas:

$$W = \omega_{max} - (\omega_{max} - \omega_{min}) \times \frac{t}{T} \tag{13}$$

$$Gc_i(t + 1) = W \times X_i(t) \times [1 + \beta_1 \mathrm{Cauchy}(0, 1) + \beta_2 \mathrm{Gauss}(0, 1)] \tag{14}$$

$Gc_i(t + 1)$ for the position after the (t + 1) -th iteration. The self-adaptation coefficient W dynamically constricts the search space, with the inertial weight progressively diminishing throughout the iterations. During the initial phases, elevated inertial weights facilitate substantial mutation disturbances, whereas subsequent reduction enables balanced coordination between local exploitation and global exploration capabilities. The hybrid mutation boosts exploration but lacks fitness guarantees. Greedy updates ensure progress: accept only superior solutions (fit(x)).

$$X^{\mathrm{best}} = \begin{cases} Gc^{\mathrm{best}}(t), fit[Gc^{\mathrm{best}}(t)] < fit[X^{\mathrm{best}}(t)] \\ Gc^{\mathrm{best}}(t), fit[Gc^{\mathrm{best}}(t)] \geq fit[X^{\mathrm{best}}(t)] \end{cases} \tag{15}$$

3.4 Computational Complexity Profiling

The complexity of the DBO calculation primarily depends on three stages, namely, the start-up stage, the fitness evaluation stage and the population position adjustment stage. The upper limit of the iteration is recorded as T, the population size is recorded as N, and the dimension of the problem is recorded as Dim. The initial setting requires O (N * D) calculation, and calculating the fitness score requires O(N * T) calculation. Updating the positions of four different types of dung beetles involves O(N * T) calculations. The overall computational complexity of the DBO algorithm can be expressed as O(max(N) * T * D).

4 Experiments and Results Analysis

4.1 Experimental Preparation

The performance of the suggested MRDBO algorithm is assessed via 14 common test functions (Table 1). High-dimensional unimodal functions F1 through F5 are used to evaluate the algorithm's capacity for local search and optimization. The algorithm's capacity to search globally is tested with high-dimensional multimodal functions F6 through F9, which have one global optimum and several local optima. Fixed-dimensional multimodal functions F10 through F14 are used to confirm that exploration and exploitation are balanced. These experiments were conducted via MATLAB 2022b software and involved a population of 30 individuals and up to 500 iterations, and 30 runs of each algorithm. As assessment metrics, the results' mean (arithmetic mean) and standard deviation (SD) are noted.

Table 1. Benchmark Test Function Table

function	Function name	dim	range	characteristic
F1	Sphere	30	[-100,100]	High-dimensional unimodalc
F2	Schwefel2.22	30	[-10,10]	
F3	Schwefel1.2	30	[-100,100]	
F4	Schwefel2.21	30	[-10,10]	
F5	Quartic	30	[-1.28,1.28]	
F6	Schwefel2.26	30	[-500,500]	high-dimensional multimodal
F7	Rastrigin	30	[-5.12,5.12]	
F8	Ackley	30	[-32,32]	
F9	Griewank	30	[-600,600]	
F10	Foxholes	4	[-65,65]	Fixed-dimensional multimodal
F11	Branin	2	[-5,5]	
F12	Goldstein	3	[-2,2]	
F13	Shekel5	4	[0,10]	
F14	Shekel10	4	[0,10]	

4.2 Analysis of the Effectiveness of Improvement Strategies

MRDBO includes three advancement strategies based on the DBO, and by integrating these three improvement methods with the DBO, we obtaine three different algorithms:

1) By joining the tent chaotic mapping into the initialization of the DBO, we obtained MRDBO1;
2) By integrating the soft rime formation search strategy from the Rime Optimization Algorithm into the rolling stage of DBO, we obtain MRDBO2;
3) By adding the adaptive Cauchy–Gaussian mutation strategy to DBO, we obtain MRDBO3;

In Table 2, bold values indicate lower average optimization accuracy than DBO does. Table 2 shows that, compared with the original DBO, MRDBO1 yieelds improvement on 9 test functions, MRDBO2 on 9 functions, and MRDBO3 on 13 functions. MRDBO1, MRDBO2, and MRDBO3 significantly improved in both the SD and arithmetic mean. Notably, MRDBO3 achieves the theoretical optimum for F1, F3, F7, F9, and F10, demonstrating that the adaptive Cauchy–Gaussian mutation strategy enhances the optimization accuracy. MRDBO1 improves accuracy in F1, F3, F4, and others, showing that the tent chaotic mapping strategy enhances diversity and solution feasibility in the initialization phase. MRDBO2 outperforms or matches DBO in functions F1, F2, F4, F7, F8, F10, F11, F12, and F14, highlighting the effectiveness of integrating the soft frost search strategy from RIME into the DBO rolling strategy. Except for F5 and F6, MRDBO with three strategies finds the theoretical optimum in most test functions, indicating its balanced exploration and exploitation abilities. Overall, MRDBO has better average optimization accuracy and standard deviation than DBO does, achieving theoretical optima at in F1–F4, F7, and F9~F14.

Compared with GODBO, MRDBO achieves the theoretical maximum performance in various problem types, including high-dimensional unimodal functions (F1–F4), multimodal functions (F7 and F9), and fixed-dimensional multimodal functions (F10–F14). Whereas GODBO finds only the optimum in F1, F2, F9, F11, F12, and F14. MRDBO also avoids local optima in F3 and F4, because of the adaptive Cauchy–Gaussian mutation strategy. Although MRDBO is slightly worse than GODBO in F8, it does not exhibit negative optimization, such as DBO. For F1, F2, F9, F12–F14, MRDBO and GODBO have the same optimization accuracy, but MRDBO performs better in terms of the SD. MRDBO outperforms GODBO on other test functions.

Table 2. Optimization results of different

function	performance index	MRDBO	GODBO	DBO	MRDBO1	MRDBO2	MRDBO3
F1	SD	0	0	5.29E-100	9.98E-114	3.36E-115	0
	Arithmetic mean	0	0	9.66E-101	**2.60E-114**	**6.85E-116**	**0**
F2	SD	0	0	1.92E-54	2.49E-53	5.13E-57	0
	Arithmetic mean	0	0	3.86E-55	4.55E-54	**9.96E-58**	**2.58E-267**
F3	SD	0	5.28E-87	1.51E-74	2.76E-81	1.68E-72	0
	Arithmetic mean	0	9.63E-88	2.75E-75	**5.05E-82**	3.06E-73	0
F4	SD	0	6.44E-66	1.71E-43	8.68E-46	3.16E-50	0
	Arithmetic mean	0	1.87E-66	3.12E-44	**1.58E-46**	**5.78E-51**	**6.62E-269**
F5	SD	9.22E-05	9.62E-04	8.52E-04	7.87E-04	8.69E-04	3.02E-04
	Arithmetic mean	1.04E-04	1.76E-03	9.81E-04	9.74E-04	1.05E-03	**4.37E-04**
F6	SD	1.32E-03	4.22E+02	4.00E-02	2.83E-01	1.52E+02	1.78E+00
	Arithmetic mean	-4.19E+03	-3.13E+03	-4.19E+03	**-4.19E+03**	-4.16E+03	**-4.19E+03**
F7	SD	0	1.09E+00	3.81E+00	9.30E-01	2.53E+00	0
	Arithmetic mean	0	1.99E-01	6.96E-01	**2.08E-01**	**6.64E-01**	**0**
F8	SD	0	1.74E-15	0	0	0	0
	Arithmetic mean	4.44E-16	1.75E-15	4.44E-16	**4.44E-16**	**4.44E-16**	**4.44E-16**
F9	SD	0	0	2.34E-02	3.81E-02	3.60E-02	0
	Arithmetic mean	0	0	4.83E-03	1.48E-02	6.56E-03	0
F10	SD	2.96E-16	5.27E-01	2.73E+00	1.09E+00	1.78E+00	9.09E-11
	Arithmetic mean	9.98E-01	1.16E+00	1.98E+00	**1.49E+00**	**1.32E+00**	**9.98E-01**
F11	SD	0	0	0	0	0	3.70E-07
	Arithmetic mean	3.98E-01	3.98E-01	3.98E-01	**3.98E-01**	**3.98E-01**	**3.98E-01**
F12	SD	1.00E-07	2.53E-15	3.65E-15	2.48E-15	3.94E-15	1.56E-07
	Arithmetic mean	3	3	3	**3**	**3**	**3**
F13	SD	3.32E-04	2.38E-03	2.55E+00	2.55E+00	2.58E+00	1.26E+00
	Arithmetic mean	-1.02E+01	-1.02E+01	-7.28E+00	-7.96E+00	-7.61E+00	**-6.72E+00**
F14	SD	1.89E-19	1.55E-15	2.71E+00	2.67E+00	2.72E+00	3.68E-01
	Arithmetic mean	-1.05E+01	-1.05E+01	-8.21E+00	-7.31E+00	**-8.03E+00**	**-1.04E+01**

4.3 Comparison with Other Swarm Intelligence Optimization Algorithms

To more comprehensively evaluate the performance of MRDBO, it was compared with several other intelligent optimization techniques. For example, the latest algorithms include the Harris Eagle optimization algorithm (HHO) [10], the crime optimization algorithm (RIME), the goshawk optimization algorithm (NGO) [11], the artificial Protozoan optimizer (APO) [12], and the basic Dung beetle optimizer (DBO). All methods were configured to run up to 500 iterations, with a population size set to 30, and other parameter settings were exactly the same as those specified in the original study. Table 3 presents the experimental outcomes.

Table 3. Optimization results of different

function	performance index	MRDBO	DBO	NGO	RIME	HHO	APO
F1	SD	0	1.01E-102	1.32E-96	1.61E-02	8.56E-95	8.94E-13
	Arithmetic mean	0	1.85E-103	2.57E-97	1.85E-02	1.73E-95	5.78E-13
F2	SD	0	6.23E-54	7.71E-51	9.51E-03	4.42E-50	9.23E-08
	Arithmetic mean	0	1.17E-54	5.98E-51	2.51E-02	8.57E-51	1.05E-07
F3	SD	0	1.13E-91	2.96E-44	3.23E-01	2.54E-80	1.59E-04
	Arithmetic mean	0	2.06E-92	1.16E-44	4.63E-01	4.72E-81	1.24E-04
F4	SD	0	1.81E-43	4.58E-44	6.27E-02	1.11E-48	1.38E-04
	Arithmetic mean	0	3.31E-44	2.66E-44	1.76E-01	3.22E-49	1.96E-04
F5	SD	7.77E-05	9.28E-04	2.52E-04	1.66E-03	1.10E-04	1.19E-03
	Arithmetic mean	1.08E-04	8.90E-04	3.61E-04	3.15E-03	1.43E-04	2.57E-03
F6	SD	4.81E-04	2.55E-01	2.34E+02	1.44E+02	3.22E+02	1.28E+02
	Arithmetic mean	-4.19E+03	-4.19E+03	-3.27E+03	-3.96E+03	-4.08E+03	-3.84E+03
F7	SD	0	2.91E+00	3.21E-01	1.85E+00	0	8.30E-01
	Arithmetic mean	0	5.31E-01	6.07E-02	3.00E+00	0	2.57E+00
F8	SD	0	0	9.01E-16	3.08E-02	0	1.53E-07
	Arithmetic mean	4.44E-16	4.44E-16	4.23E-15	6.49E-02	4.44E-16	2.04E-07
F9	SD	0	8.22E-03	0	8.14E-02	0	1.20E-02
	Arithmetic mean	0	1.80E-03	0	1.78E-01	0	9.51E-03
F10	SD	3.13E-16	5.27E-01	4.12E-17	7.83E-12	9.32E-01	1.50E-15
	Arithmetic mean	9.98E-01	1.16E+00	9.98E-01	9.98E-01	1.26E+00	9.98E-01
F11	SD	0	0	0	1.70E-06	8.92E-06	1.27E-10
	Arithmetic mean	3.98E-01	3.98E-01	3.98E-01	3.98E-01	3.98E-01	3.98E-01
F12	SD	1.00E-07	2.30E-15	1.29E-15	1.25E-06	8.66E-07	1.29E-15
	Arithmetic mean	3	3	3	3.01E+00	3.01E+00	3
F13	SD	1.26E-04	2.55E+00	1.13E-05	2.58E+00	8.23E-01	2.22E+00
	Arithmetic mean	-1.02E+01	-7.95E+00	-1.02E+01	-8.38E+00	-5.20E+00	-9.31E+00
F14	SD	1.62E-19	2.74E+00	4.52E-06	2.36E+00	9.69E-01	1.54E+00
	Arithmetic mean	-1.05E+01	-7.66E+00	-1.05E+01	-9.51E+00	-5.30E+00	-1.01E+01

To illustrate the merging speed and neighborhood elude capability of each calculation, Fig. 2 presents the iterative convergence curves of 14 benchmark test functions. The simulation results clearly show that the convergence trajectory of MRDBO is usually below that of other algorithms, highlighting its greater accuracy in optimization. Particularly for functions F1-F4, MRDBO achieves high-level optimization accuracy. This enhancement benefits from the integrated rime–ice optimization algorithm's soft-rime formation search strategy, which accelerates the convergence speed.

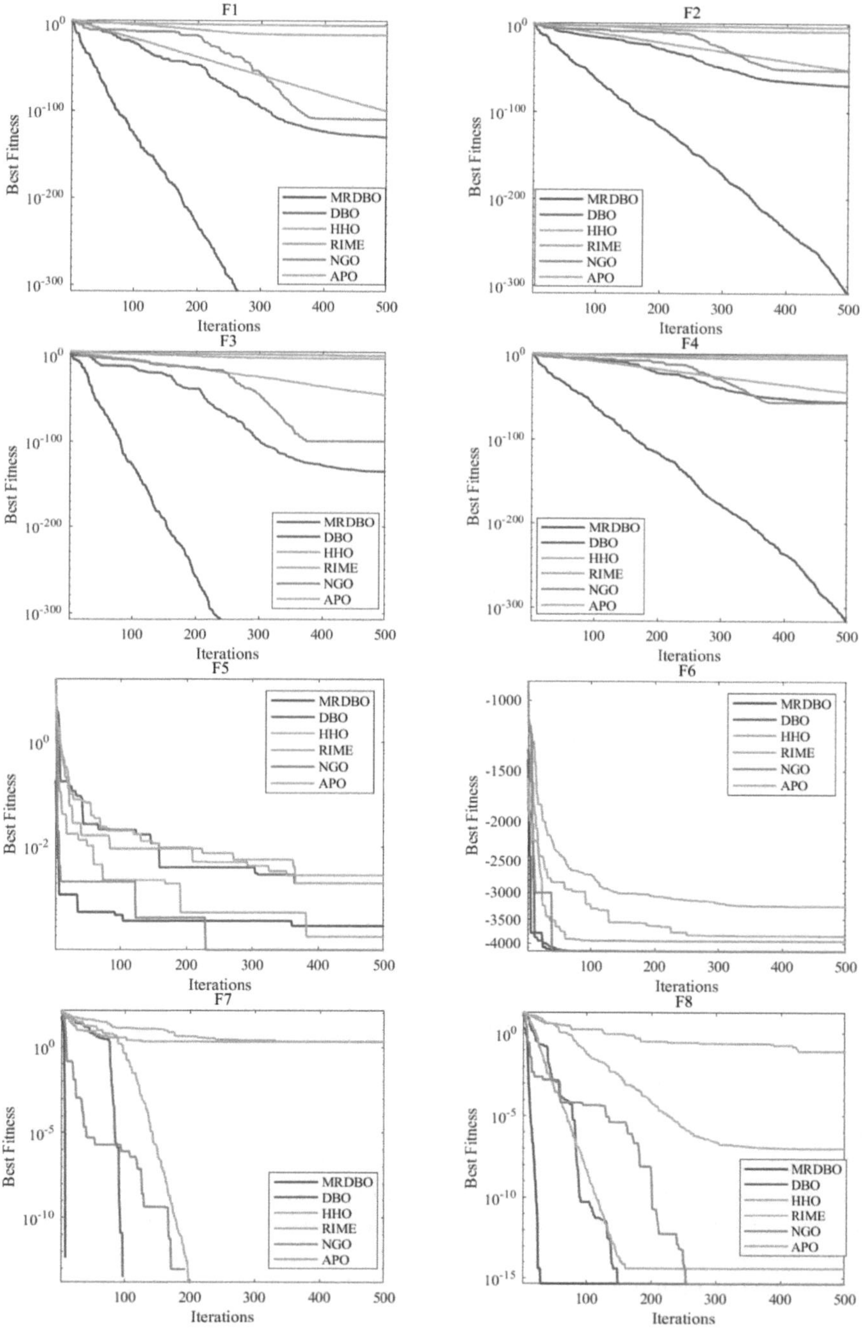

Fig. 2. Comparative Function Graph Convergence Analysis

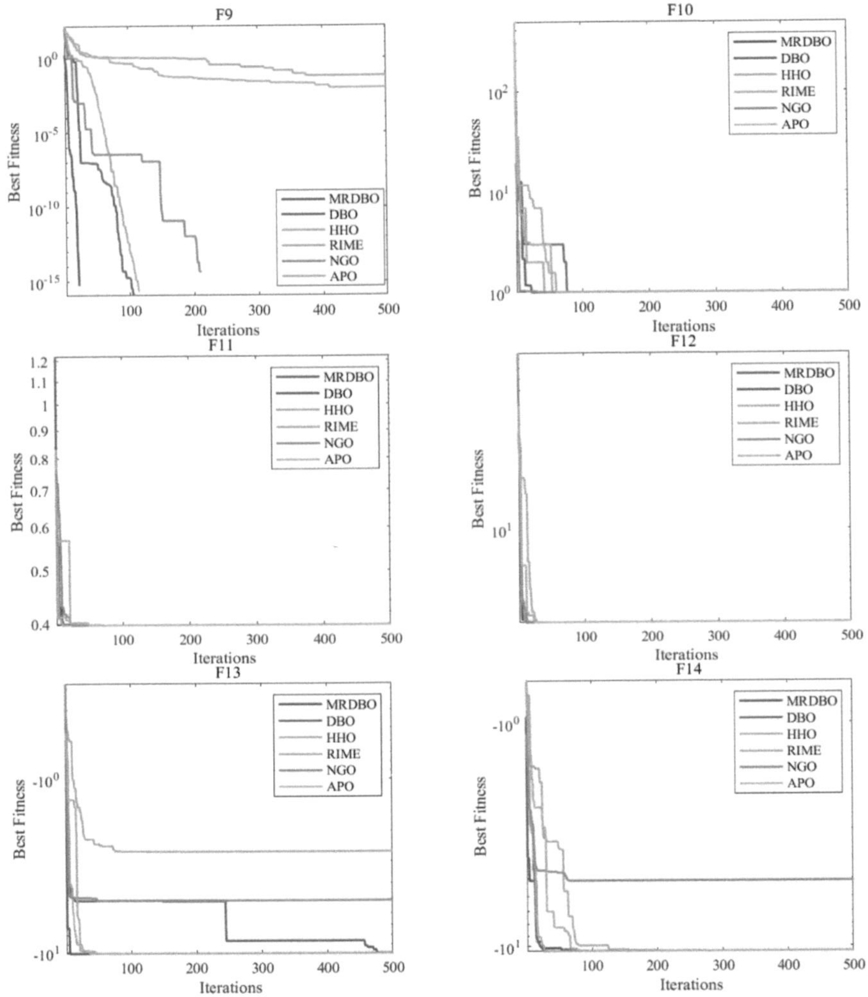

Fig. 2. *(continued)*

4.4 Solution of MRDBO in ABLP-1

The first type of assembly line balancing problem involves allocating tasks to each workstation on the basis of their priority relationship under the condition of a known cycle time, and obtaining the minimum number of workstations. First, assume that the assembly line operation and workstation meet the following conditions: 1) the time of the operation is known and fixed; 2) there are no other constraints except for the priority relationship constraint of remote work; 3) any workstation can complete any task; and 4) any assignment can be assigned to only one workstation. Multiple classic assembly line test cases are used to solve for the minimum number of workstations, and each algorithm is run 20 times. The minimum number of workstations obtained from the 20 results obtained by the algorithm proposed in this article, along with the original beetle optimization algorithm (DBO), fruit fly algorithm (FOA), and frost ice optimization algorithm (RIME), is shown in Table 4.

Table 4. Results of running each algorithm 20 times

Test case	Number of assignments	beat	Run 20 times to obtain the theoretical optimal number of times			
			DBO	FOA	RIME	MRDBO
Jackson	11	9	12	11	16	20
Jackson	11	13	20	20	19	20
Gunther	11	49	12	14	11	18
Gunther	11	61	15	14	9	19
Roszieg	25	16	14	12	17	20
Roszieg	25	16	17	11	6	20
Buxey	29	27	0	2	1	6
Buxey	29	54	19	20	17	20

Twenty iterations were analysed, and the performance of the algorithm introduced in this paper was compared with the performance of the other three algorithms. The results showed that our proposed method can always achieve the lowest workstation value possible, close to the theoretical minimum, and performs better than other algorithms in obtaining the theoretical optimal workstation after 20 iterations, verifying how well the algorithm described in this article performs in solving the first type of assembly line balance problem.

5 Conclusion

This article proposes a multi strategy improved beetle optimization algorithm (MRDBO) to improve the optimization performance of the beetle optimization algorithm. The effectiveness of the improved strategy has been confirmed through the use of benchmark functions. The optimization results of MRDBO and the new swarm intelligence optimization algorithm on 14 test functions are compared, and the results show that MRDBO has higher convergence accuracy, faster convergence speed, and stronger robustness than the original DBO and the comparison algorithms do. Furthermore, MRDBO was applied to assembly line balancing instances to verify its applicability and reliability in practical engineering; In subsequent work, we will continue to improve the performance and efficiency of the algorithm, reduce its time consumption, and consider applying MRDBO to other types of assembly line balancing problems.

References

1. Xue, J., Shen, B.: Dung beetle optimizer: a new meta-heuristic algorithm for global optimization. J. Supercomput. **79**(7), 7305–7336 (2023)

2. Cano Ortega, A., Sánchez Sutil, F.J., De la Casa Hernández, J.: Power factor compensation using teaching learning based optimization and monitoring system by cloud data logger. Sensors **19**(9), 2172 (2019)
3. Geetha, K., Anitha, V., Elhoseny, M., et al.: An evolutionary lion optimization algorithm-based image compression technique for biomedical applications. Expert. Syst. **38**(1), e12508 (2021)
4. Shao, S., Zhao, Y.: Constrained trajectory planning for unmanned aerial vehicles using asymptotic optimization approach. Trans. Inst. Measure. Control **45**(13), 2421–2436 (2023)
5. Duan, J., Gong, Y., Luo, J., et al.: Air-quality prediction based on the ARIMA-CNN-LSTM combination model optimized by dung beetle optimizer. Sci. Rep. **13**(1), 12127 (2023)
6. Zilong, W., Peng, S.: A multi-strategy dung beetle optimization algorithm for optimizing constrained engineering problems. IEEE Access **11**, 98805–98817 (2023)
7. He, J., Guo, W., Wang, S., et al.: Application of multi-strategy based improved DBO algorithm in optimal scheduling of reservoir groups. Water Resour. Manage **38**(6), 1883–1901 (2024)
8. Su, H., Zhao, D., Heidari, A.A., et al.: RIME: A physics-based optimization. Neurocomputing **532**, 183–214 (2023)
9. Fu, H., Liu, H.: Improved sparrow search algorithm with multii-strategy integration and its application. Control Decis. **37**(1), 87–96 (2022)
10. Heidari, A.A., Mirjalili, S., Faris, H., et al.: Harris hawks optimization: algorithm and applications. Futur. Gener. Comput. Syst. **97**, 849–872 (2019)
11. Dehghani M.Š.H., Trojovský, P.: Northern goshawk optimization: a new swarm-based algorithm for solving optimization problems. IEEE Access **9**, 162059–80 (2021)
12. Wang, X., Snášel, V., Mirjalili, S., et al.: Artificial protozoa optimizer (APO): a novel bioinspired metaheuristic algorithm for engineering optimization. Knowl.-Based Syst. **295**, 111737 (2024)

Aggregation of Dialogue History Information and Utterance-Slot Attention for Dialogue State Tracking

Yatian Shen[1], Guosong Deng[1(✉)], Qichao Hao[1], Songyang Wang[1], and Eryan Zhang[2]

[1] School of Computer and Information Engineering, Henan University, Kaifeng, China
{guosongdeng,haoqichao,sywang}@henu.edu.cn
[2] School of Information Engineering and Big Data, Zhengzhou Technical College, Zhengzhou, China
zhangeryan@zzy.edu.cn

Abstract. The objective of dialogue state tracking (DST) is to update the current dialogue state on the basis of the utterance information from the current turn and the dialogue state from the previous turn. Traditional methods treat the previous turn's dialogue state as a holistic representation, frequently overlooking critical details from earlier turns and resulting in 'forgetting' issues. To address this limitation, a discounted return algorithm is employed to aggregate dialogue history information more effectively. Additionally, to capture slot-specific information within utterances, we introduce an utterance-slot attention mechanism that can model the semantics of the dialogue context while incorporating slot name features to guide the generation or extraction of the dialogue state. The experimental results not only validate the superiority of our proposed method but also significantly increase the accuracy and robustness of multidomain DST.

Keywords: Dialogue State Tracking · Information Aggregation Mechanism · Utterance-Slot Attention

1 Introduction

Task-oriented dialogue constitutes a fundamental element of contemporary dialogue systems, enabling users to complete tasks such as ticket booking and music recommendation. Effective DST methods can enhance a system's ability to understand user intentions and provide more precise responses, thereby increasing user satisfaction and overall system performance. DST structures the dialogue state through predefined domain-slot-value triples. such as the triple (restaurant- name, sitar tandoori), which indicates that the name of the restaurant is Sitar Tandoori. We consider the domain and slot names connected by a "-" as constraints on the slot. Real-world dialogues often span multiple domains

L. Huang et al. (Eds.): CCF NCCA 2025, CCIS 2774, pp. 27–40, 2026.
https://doi.org/10.1007/978-981-95-5294-8_3

(e.g., restaurants, trains), as depicted in Fig. 1. In such multiidomain [10] settings, accurately tracking the current dialogue state poses significant challenges. Although state-of-the-art DST methods perform well, they typically consider only the utterance of the current turn and the dialogue state from the preceding turn [35]. This approach can lead to a "forgetting" problem, where the model may overlook crucial information from earlier turns, especially in dialogues that cover multiple topics or domains. As demonstrated in Fig. 1, the value for the slot "restaurant reservation time" cannot be derived solely from the current turn's dialogue. Instead, reliance on previous dialogues to accurately determine the current dialogue state is necessary. Furthermore, when handling dialogues that involve complex slot interactions, traditional methods often struggle to capture the dependencies and dynamic changes between slots. For example, in interactions involving multiple domains, including hotels and restaurants, slots such as 'hotel-star' and 'restaurant-price', are typically interdependent, with a change in one slot influencing the value of another. Recent studies on DST models commonly adopt a dual-input architecture consisting of the current turn's dialogue input and the previous turns' state representations [18]. However, since dialogue inherently builds upon previous exchanges, early rounds often contain crucial contextual clues. Thus, integrating utterance information from earlier turns is vital for accurately generating dialogue states and building a detailed dialogue history. This approach of aggregating information from previous dialogues ensures that the system comprehends the conversation's background and intent, leading to more coherent and relevant responses.

To address the"forgetting" problem, we introduce an innovative information aggregation method that uses a strategy based on discounted rewards from reinforcement learning [27]. This approach assigns different weights to various dialogue turns, effectively balancing information retention and the risk of forgetting. By prioritizing recent interactions while still acknowledging valuable information from earlier turns, our method not only mitigates the forgetting problem but also enhances the continuous tracking of critical information throughout the dialogue. This strategy ensures that essential contextual details are maintained, leading to more accurate and robust dialogue state tracking. Inspired by previous work, we enhance our dialogue state tracking model by integrating an utterance-slot attention mechanism, enabling it to identify and utilize slot-related information within the dialogue more precisely. This mechanism selectively attends to slot-relevant features in utterances, guiding slot value generation and improving prediction accuracy, particularly in information-dense dialogues or multislot scenarios. To effectively focus on key information from previous turns, we introduce a unique token [TURN] before each dialogue turn, which, similar to BERT's [CLS] token, aggregates rich representations of the current turn's dialogue utterances. Inspired by the discounted reward algorithm, we allocate weights dynamically to the dialogue utterances of each turn. Furthermore, to focus on slot-specific information within dialogue utterances, we employ an utterance-slot attention mechanism. In this setup, predefined slots are encoded as query vectors, which engage in cross-attention calculations with the encoded dialogue utterance information.

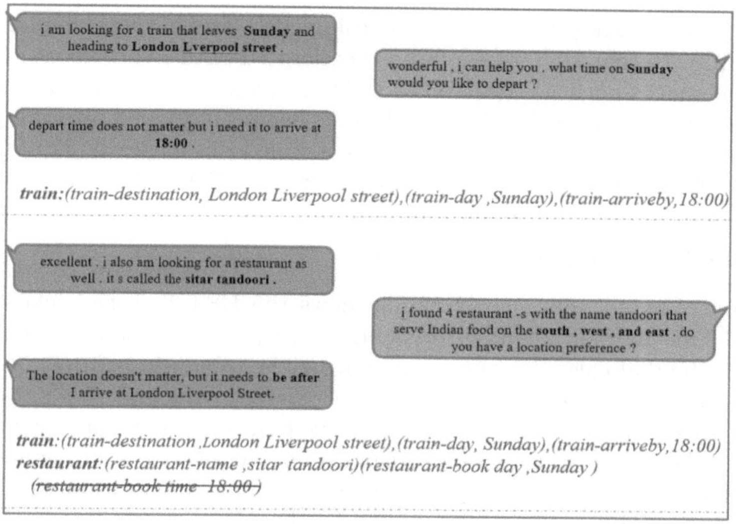

Fig. 1. The dialogue content covers two domains: restaurant and train. As the conversation progresses, the current turn's state depends on the semantic relations established in previous turns. Traditional dialogue management methods often adopt a flattened semantic processing approach, which gradually dilutes critical information from intermediate turns during state updates, leading to a "forgetting" problem.

This process allows for the targeted extraction of slot-specific information from the sentences, optimizing the accuracy and relevance of the retrieved data. Our contributions are summarized as follows:

- Inspired by the discounted reward algorithm commonly used in reinforcement learning [27], we propose a novel method to aggregate historical dialogue information. This approach is specifically designed to address the "forgetting" problem, which is prevalent in scenarios involving long dialogue dependencies. By assigning strategically weighted importance to earlier dialogue turns, our model maintains a more balanced and contextually enriched dialogue state.
- We have implemented an utterance-slot attention mechanism that efficiently captures slot-specific information within dialogues. By focusing on relevant portions of the dialogue pertinent to each slot, our model ensures a comprehensive understanding and representation of user intents and contextual nuances.
- The performance superiority of our information aggregator and utterance-slot attention design is empirically validated on two established DST benchmarks.

2 Related Work

Traditional pipeline-based approaches typically segment task-oriented dialogue systems into four distinct modules [25]: natural language understanding (NLU), dialogue state tracking (DST), policy learning (PL), and natural language generation (NLG). Recent DST research has accelerated with freely accessible large-scale benchmarks such as MultiWOZ (v2.0–2.4) and SGD. With the development of deep learning, research on DST tasks has been divided into two primary approaches: 1) DST, which is based on predefined ontologies [2,32,35], and 2) methods, which are based on open vocabulary [15,17,19,22].

2.1 Based on Predefined Ontology

The ontology-based approach [2,32,35] views the DST task as a classification problem. This method requires predefining all possible (domain, slot, value) triplets in the dialogue. It involves comparing the semantic features of the current turn against all possible (slot, value) pairs to calculate similarity scores, and then selecting the most suitable slot-value pair as the current dialogue state. SUMBT [14] employs an attention mechanism based on contextual semantic vectors to capture features of the dialogue context. SST [2,34] introduces a pattern-guided sparse dialogue state tracking model that employs graph attention networks to predict slot values by integrating information from both discourse and structured graphs. However, it is impractical to define ontologies for every conceivable slot. For example, a taxi's destination could be a hotel or a restaurant, rendering it impossible to list all potential options, thereby compromising scalability and generalizability.

2.2 Open-Vocabulary-Based Methods

To overcome the limitations of predefined ontologies in DST, recent studies have increasingly concentrated on open-vocabulary approaches [15,19,33], defining DST as an extraction or generation task. Gao et al. [6] redefine DST as a reading comprehension task, posing the following question: "What is the current state of the dialogue?". Shin et al. [22] defines DST as a dialogue summarization problem, employing generative models with a soft copy mechanism that selectively copies words from the dialogue history and vocabulary to use as slot values. Ren et al. [20,23] selectively copies words from the dialogue history and vocabulary to use as slot values. With the introduction of graph structures, Feng et al. [4,11,23,31,34] employ graph attention networks to enhance the model's ability to manage interactions across multiple domains and complex slots. Another [15,29] innovative approach involves modifiable state prediction, allowing for the correction of slot value predictions after initial generation to reduce error propagation in subsequent dialogues. With the emergence of large pretrained language models, Lin et al. [16] leverage the powerful generative capabilities of GPT-2 combined with graph attention networks to capture dependencies between different domains in cross-domain dialogues, thereby overcoming the limitations of causal reasoning in GPT-2. Qiu et al. [17,19,21,26,30] enhances adaptability to complex scenarios by dynamically modelling the variable dialogue environments and slots through various adaptations of attention mechanisms. Kim et al. [8,12] views view the dialogue state tracking task as a two-stage process: first determine which slot values need updating, and then generate new values for the slots that require updates. This approach eliminates the need to update all the slot values at every dialogue turn.

3 Model

With the advancement of pretrained language models such as BERT [3] and GPT-2 [9], which have shown marked effectiveness in various downstream tasks, we have adopted BERT as an encoder. This approach allows us to derive semantic vector representations from the dialogue context, as well as slots and values. The complete architecture of this system is depicted in Fig. 2.

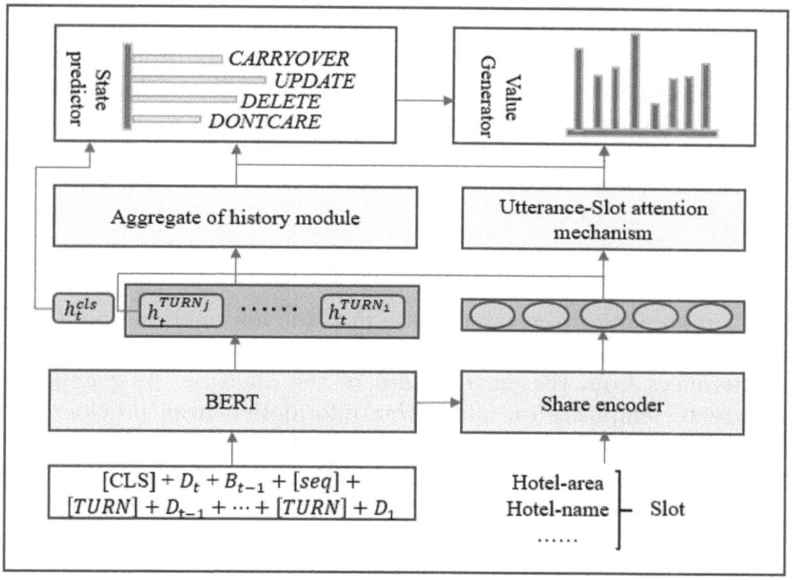

Fig. 2. The overall architecture of our methods.

3.1 Overall

Encoder Input: We define the dialogue state at each turn as $B_t = (S_t^j, V_t^j)j \in J$, where S_t^j denotes the domain-slot pair for slot j at turn t, with v_t^j being its value. J is the total number of slots.

For each turn, we concatenate the user utterance and system response to form $D_t = (U_t + S_t)$, where U_t is the user input and where S_t is the system response at turn t. We introduce the special token [TURN] before each dialogue turn. Additionally, slot-value pairs from the previous turn that are not NULL are also included as inputs to the encoder [12]. Specifically,

$$X_t = D_t \cdot + \cdot B_{t-1} + [\text{TURN}] + \cdot D_{t-1} + [\text{TURN}] + \cdot + D_1 \tag{1}$$

Building on previous research, that utilized slot self-attention, we implemented a slot-utterance attention mechanism. This approach is designed to capture slot-specific information within the utterance, thereby effectively guiding the generation of slot values.

3.2 Information Aggregation Mechanism

Inspired by the discount factor used in reinforcement learning [27], we acknowledge that dialogues further away from the current turn have a diminishing effect on the current dialogue state. To address this, we employ a technique known as weight decay to aggregate dialogue utterance information from previous turns. We define a decay rate r, which assigns weights to the dialogue information of each turn, reflecting the diminishing relevance of older dialogue information relative to the current turn. This method quantifies the importance of each turn's information, effectively prioritizing the most recent interactions. As shown in below

$$H_t^{urr} = h_t^{Turn_t} + rh_t^{Turn_{t-1}} + \ldots + r^{t-1}h_t^{Turn_1} \tag{2}$$

where H_t^{urr} represents the aggregation of dialogue information from the current turn and previous turns, When the decay rate r approaches 0, we focus more on the utterances from the current turn of the dialogue. As r approaches 1, we place greater emphasis on integrating information from previous turns. In real dialogue scenarios, it is common to forget previous dialogue information as the conversation progresses. Our method of dialogue information aggregation, which is based on the concept of discounted rewards, mirrors the dynamic nature of human conversations. This approach helps retain complex information from earlier dialogues, preventing its loss as the conversation continues.

3.3 Slot-Utterance Attention

To capture the correlation between the dialogue context and the corresponding slots, we propose the slot-utterance attention mechanism. By using cross-attention, we can extract slot-specific information from the dialogue's contextual semantics. Slot-encoded queries attend over utterance vectors via self-attention [24], capturing contextual slot-semantic correlations. We formalize this as follows:

$$H_t^{slot} = \text{MultiHead}(Q_{slot}, k_{turn}, V_{turn}) \tag{3}$$

In this setup, the Q, K, and V are derived from different features that serve as inputs to the attention mechanism.

3.4 Value Generator

Following the setup from [12] we use a gated recurrent unit (GRU) [7] as our decoder. We initialize $g_t^{j,0} = H_t^{urr}$ and $e_t^{j,0} = H_t^{slot_j}$. Here, $e_t^{j,k}$ represents the word embedding generated at each step, using the embedding generated in the current step $e_t^{j,k}$ as the hidden state until the end-of-sequence token [EOS] is generated.

$$g_t^{j,k} = \mathrm{GRU}(g_t^{j,k-1}, e_t^{j,k}) \tag{4}$$

The final distribution of the dialogue state is jointly influenced by two probabilistic components: the vocabulary probability distribution, denoted as $p_{vcb,t}^{j,k}$, and the dialogue text probability, denoted as $p_{ctx,t}^{j,k}$. The term $p_{vcb,t}^{j,k}$ specifically represents the hidden state $g_t^{j,k}$ at the k-th decoding step of the model, reflecting the influence of the selected vocabulary. On the other hand, $p_{ctx,t}^{j,k}$ is derived from the output that has been aggregated from the entire dialogue history up to the current turn, denoted by H_t^{urr}, following the information aggregation module's processing. This setup ensures that both immediate and historical dialogue contexts contribute to the generation of accurate and contextually appropriate dialogue states.

$$p_{vcb,t}^{j,k} = softmax\left(e * g_t^{j,k}\right) \tag{5}$$

$$p_{ctx,t}^{j,k} = softmax\left(H_t^{urr} * g_t^{j,k}\right) \tag{6}$$

Here, e represents the word embedding for all the words in the dialogue. The final generated word is determined by summing the two probabilities mentioned above and selecting the word with the highest similarity.

$$p_{w,t}^{j,k} = \lambda p_{vcb,t}^{j,k} + (1-\lambda) p_{ctx,t}^{j,k} \tag{7}$$

3.5 Loss Function

In our experiments, our objective function is divided into two loss components: the state prediction phase and the slot-value generation phase.

State Predictor: Here, we use the mean of the negative log-likelihood as the loss for state prediction, detailed as follows:

$$p_{op,t}^{j} = softmax(\mathrm{W}_{op} * H_t^{urr}) \tag{8}$$

$$\mathcal{L}_{op,\mathrm{t}} = -\frac{1}{J}\sum_{j=1} Y_{op,t}^{j} * \log p_{op,t}^{j} \tag{9}$$

$P_{op,t}^{j}$ represents the prediction of the operation for slot j at the current turn, based on the aggregated message H_t^{curr}. This prediction is part of a four-category classification task. $Y_{op,t}^{j}$ denotes the true operation for slot j.

Value Generator: Similarly, in the slot value generation phase, we calculate the loss using the negative log-likelihood function, as follows:

$$\mathcal{L}_{gen,\mathrm{t}} = \sum_{j=1}\sum_{k=1} \mathrm{SVG}((\mathrm{Y}_{w,t}^{j,k})\log(p_{w,t}^{j,k})) \quad (10)$$

k represents the number of tokens in the encoded slot value, and j denotes the slots predicted to be in the "update" state. $Y_{w,t}^{j,k}$ represents the true value of the current slot value, which is encoded as a one-hot vector.

Finally, our optimization objective is defined as the sum of terms $\mathcal{L}_{gen,\mathrm{t}}$ and $\mathcal{L}_{op,\mathrm{t}}$, expressed as:

$$\mathcal{L}' = \mathcal{L}_{gen,\mathrm{t}} + \mathcal{L}_{op,\mathrm{t}} \quad (11)$$

4 Experiments

4.1 Datasets

Two publicly available task-oriented dialogue datasets are used to test our model: MultiWOZ 2.1 and MultiWOZ 2.4. These datasets collectively encompass 5 domains and 17 predefined slots, totaling 30 (domain, slot) pairs, as detailed in Table 1.

Table 1. Information about the MultiWOZ dataset is shown below.

Domain	Slot	Train	Val	Test
Restaurant	area, food, name, price range, book day, book people, book time	3813	438	437
Hotel	area, internet, name, parking, price range, stars, type, book day, book people, book stay	3381	416	394
Train	Arrive by, day, departure, destination, leave at, book people	3103	484	494
Taxi	arrive by, departure, destination, leave at	1654	207	195
Attraction	area, name, type	2717	401	395

4.2 Baseline

We compared the performance of our model with that of open-vocabulary models. TRADE [28] uses a generative model that operates across multiple domains, reducing reliance on domain-specific data. PIN [1] employs an interactive encoder to capture both interturn and intraturn information effectively. Som-DST [12] treats the dialogue state as fixed-size memory and introduces a selective updating mechanism that generates only partial slot values each turn. DST-STAR [32] introduces a slot self-attention mechanism that captures slot-specific features within the dialogue context. Seq2Seq [5] employs a sequence-to-sequence model to manage dialogue state tracking tasks. RefPyDST [13] redefines dialogue state tracking as a Python programming challenge within a context learning framework, achieving excellent results in low-sample learning scenarios.

Table 2. Hyperparameters during the training process.

Hyperparameter	Explanation	Value
enc_lr	Encoder learning rate	1e-5
dec_lr	Decoder learning rate	1e-4
warmup	Warm-up parameters	0.1
dropout	Randomly dropout	0.1
r	Decay rate	0.1/0.5/0.9

5 Results

5.1 Experimental Results

We validated our approach on the MultiWOZ 2.1 and MultiWOZ 2.4 datasets, with results presented in Table 3. The training details are presented in Table 2.

Table 3. We validated our experimental results on both the MultiWOZ 2.1 and 2.4 datasets. The results marked with * represents our reproductions, whereas unmarked data originate from source papers.

model	Joint Goal Accuracy(%)		Slot Accuracy(%)	
	MultiWOZ 2.1	MultiWOZ 2.4	MultiWOZ 2.1	MultiWOZ 2.4
DS-DST [35]	51.21	–	96.62	–
TRADE [28]	48.62	55.05*	96.92	97.26*
PIN [1]	48.40	55.92	97.02	98.02
SOM-DST [12]	53.68	63.21*	**98.38**	98.41*
DST-STAR [32]	**56.36**	–	95.53	–
Seq2Seq [5]	52.80*	**67.10***	–	–
RefPyDST [13]	52.0	65.2	95.84*	98.25*
OURS	53.14	64.26	97.06	**98.52**

Turn	Dialogue
1	**U:** I am looking for a train that leaves on Saturday and arrives by 10:30. **S:** Where are you traveling to and from? **State:** ["train-day-Saturday", "train-arrive by-10:30"]
2	**U:** I am going to Cambridge from Birmingham New Street. **S:** Here are 5 trains available; may I book 1 for you that leaves at 7:40 and arrives at 10:23? **State:**["train-destination-Cambridge","train-day-Saturday", "train-arrive by-10:30", "train-departure-Birmingham new street"]
3	**U:** yes , that train sounds good . please book it for me. could you also find me a hotel with a moderate price that offers internet ? **S:** there are 17 hotel -s listed. what part of town would you like to stay in? **State:** ["hotel-price range-moderate", "hotel-internet-yes", "train-destination-Cambridge", "train-day-Saturday", "train-arrive by-10:30", "train-departure-Birmingham new street"]
4	**U:** The north part of town please; preferably in a guesthouse. **S:** Cambridge has 8 moderate -ly priced guesthouses in the north. I would suggest acorn guest house, with a star rating of 4. Would you like me to book it? **State:** ["hotel-area-north","hotel-price range-moderate", "hotel-internet-yes","hotel-type-guest house","train-destination-Cambridge", "train-day-Saturday", "train-arrive by-10:30", "train-departure-Birmingham new street"]
5	**U:** yes please , i would like to book it for7 people for 5 nights, and i need a reference number **S:** your booking was successful, the reference number is waeyaq0m **State:** ["train-day-Saturday", "train-departure-Birmingham new street", "train-destination-Cambridge", "hotel-internet-yes", "hotel-price range-moderate", "hotel-area-north", "hotel-type-guest house", "hotel-book day-Saturday", "hotel-book people-7", "hotel-book stay-5", "hotel-name-acorn guest house"]

Fig. 3. In my testing instances on the MultiWOZ 2.4 dataset, the slot values are marked with different colors to denote their inferential sources: green indicates that the slot values can be inferred from the current round of dialogue, blue signifies that the slot values are deduced from the dialogue state of the previous round, and red requires the combination of dialogue utterances from prior rounds for accurate inference. (Color figure online)

On the basis of the results in Table 3, we reach the following two conclusions: **1)** Compared to the baseline model, our model still needs improvement in terms of joint goal accuracy and has not yet achieved optimal performance. However, the slot accuracy of our model is close to or has already achieved optimal performance. We hypothesize that adding additional contextual information during the inference process introduces noise. This noise affects our model's ability to infer certain slot values on the basis solely of the current dialogue round, resulting in slots that originally had null values being erroneously filled with new values. **2)** Compared with the baseline model SOM-DST [12], our model incorporates an information aggregation module and a discourse-slot attention mechanism, significantly outperforming SOM-DST across both datasets. This improvement suggests that the integration of these two modules effectively enhances performance.

5.2 Experimental Analysis

We present relevant cases in Fig. 3 to provide additional evidence for the effectiveness of our approach. Figure 3 shows that our method can aggregate dialogue history information, which is beneficial for obtaining slot values that cannot be directly retrieved from the turn round of dialogue. The reason is that the dialogue history contains parts related to predefined slots, which can effectively supplement the previous dialogue state. However, our method also introduces some noise, as it fills slots with invalid values when the true value for the current round state is NULL, thereby reducing the accuracy of the joint goal. We may need to design a better information aggregation mechanism, which can address the "forgetting" problem in the DST domain while not introducing noise. We present the joint goal accuracy of our method in specific domains in Fig. 4 a, where the results indicate that the combination of the information aggregation module and the utterance-slot attention mechanism greatly benefits the hotel and restaurant domains. In the MultiWOZ 2.4 dataset, the hotel and restaurant domains always appear alongside another domain, making their slot values consistently related to the dialogue information from previous turns. Our method aggregates the historical information of the dialogue, effectively capturing information related to slot values across multiple rounds of dialogue with domain transitions, thereby addressing the slot value forgetting problem in multidomain dialogue state tracking. However, our method performs poorly in the other three domains, such as the taxi domain, where slots such as "leave-at" and "arrive-by" are heavily influenced by the current round of dialogue, and our information aggregation mechanism, which introduces additional information, generates noise when guiding slot value generation.

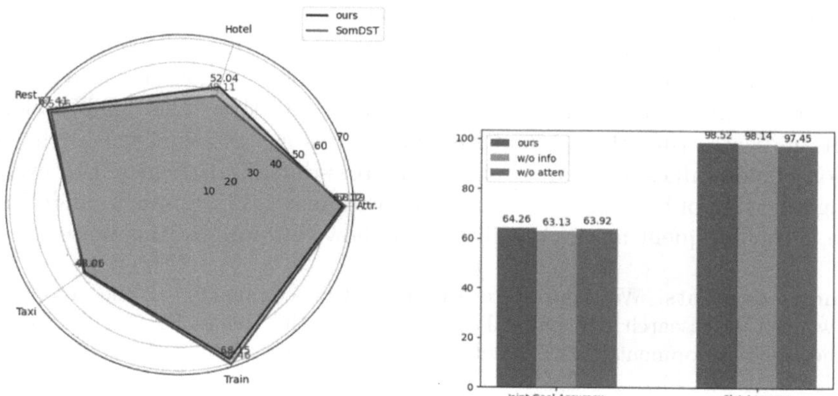

(a) JGA for each domain

(b) Ablation experiments

Fig. 4. Figure 4 ashows the JGA for each domain on the MultiWOZ 2.4 dataset. Figure 4 bshows the ablation experiments on the MultiWOZ 2.4 dataset, where w/o indicates 'without'.

5.3 Ablation Experiments

Figure 4 presents the comparison of joint goal accuracy across individual domains between our method and the Som-DST on the MultiWOZ 2.4 dataset.

The results indicate that our proposed information aggregation module and discourse-slot attention mechanism are beneficial to the DST model. To determine the impact of the information aggregation module on model performance, we use only the dialogue utterances from the current round and the dialogue state from the previous round as inputs. Our joint goal accuracy, which is lower than that of our baseline model, indicates that our information aggregation module is beneficial to the overall performance of the model, even though it introduces some noise.

To investigate whether the utterance-slot attention mechanism can effectively guide slot value generation, we remove the utterance-slot module and initialize the slot value generator state as the encoder output h_t^{cls}. During the experiment, without the utterance-slot attention mechanism, our slot accuracy decreased by 1.62% points. Our utterance-slot attention mechanism can capture specific information in the sentences, effectively extracting the corresponding slot values from the dialogue.

6 Conclusion

In this paper, we propose an information aggregation method for dialogue state tracking tasks based on discounted rewards. We employ the concept of the discounted return algorithm to aggregate historical dialogue information. The utterance-slot attention mechanism effectively guides the generation of slot values. We conducted extensive experiments to validate the effectiveness of these two modules, exploring the joint goal accuracy across five domains to determine whether the information aggregation module is effective. The combination of results indicates that the method we propose has a positive effect on the generation of slot values in guiding multiturn, cross-domain dialogues. However, as the aggregation of historical dialogue utterances inevitably results in undesirable noise, our subsequent research will focus on how to minimize this noise.

Acknowledgments. We appreciate the insightful comments of the anonymous reviewers. This research was partially supported by the Henan Prvince Science and Technology Development Plan Project (NO. 242102210065, NO. 242102210093).

References

1. Chen, J., Zhang, R., Mao, Y., Xu, J.: Parallel interactive networks for multi-domain dialogue state generation. arXiv preprint arXiv:2009.07616 (2020)
2. Chen, L., Lv, B., Wang, C., Zhu, S., Tan, B., Yu, K.: Schema-guided multi-domain dialogue state tracking with graph attention neural networks. In: Proceedings of the AAAI Conference on Artificial Intelligence. vol. 34, pp. 7521–7528 (2020)
3. Devlin, J.: Bert: pre-training of deep bidirectional transformers for language understanding. arXiv preprint arXiv:1810.04805 (2018)
4. Feng, Y., Lipani, A., Ye, F., Zhang, Q., Yilmaz, E.: Dynamic schema graph fusion network for multi-domain dialogue state tracking. arXiv preprint arXiv:2204.06677 (2022)

5. Feng, Y., Wang, Y., Li, H.: A sequence-to-sequence approach to dialogue state tracking. arXiv preprint arXiv:2011.09553 (2020)
6. Gao, S., Sethi, A., Agarwal, S., Chung, T., Hakkani-Tur, D.: Dialog state tracking: a neural reading comprehension approach. arXiv preprint arXiv:1908.01946 (2019)
7. Gu, A., Gulcehre, C., Paine, T., Hoffman, M., Pascanu, R.: Improving the gating mechanism of recurrent neural networks. In: International Conference on Machine Learning, pp. 3800–3809. PMLR (2020)
8. Guo, J., Shuang, K., Li, J., Wang, Z.: Dual slot selector via local reliability verification for dialogue state tracking. arXiv preprint arXiv:2107.12578 (2021)
9. Ham, D., Lee, J.G., Jang, Y., Kim, K.E.: End-to-end neural pipeline for goal-oriented dialogue systems using gpt-2. In: Proceedings of the 58th Annual Meeting of the Association for Computational Linguistics, pp. 583–592 (2020)
10. Hong, T., Cho, J., Yu, H., Ko, Y., Seo, J.: Knowledge-grounded dialogue modelling with dialogue-state tracking, domain tracking, and entity extraction. Comput. Speech Lang. **78**, 101460 (2023)
11. Huang, Z., Li, F., Yao, J., Chen, Z.: Mgcrl: multi-view graph convolution and multi-agent reinforcement learning for dialogue state tracking. Neural Comput. Appl. **36**(9), 4829–4846 (2024)
12. Kim, S., Yang, S., Kim, G., Lee, S.W.: Efficient dialogue state tracking by selectively overwriting memory. arXiv preprint arXiv:1911.03906 (2019)
13. King, B., Flanigan, J.: Diverse retrieval-augmented in-context learning for dialogue state tracking. arXiv preprint arXiv:2307.01453 (2023)
14. Lee, H., Lee, J., Kim, T.Y.: Sumbt: slot-utterance matching for universal and scalable belief tracking. arXiv preprint arXiv:1907.07421 (2019)
15. Li, Q., Zhang, W., Huang, M., Feng, S., Wu, Y.: RSP-DST: revisable state prediction for dialogue state tracking. Electronics **12**(6), 1494 (2023)
16. Lin, W., Tseng, B.H., Byrne, B.: Knowledge-aware graph-enhanced GPT-2 for dialogue state tracking. arXiv preprint arXiv:2104.04466 (2021)
17. Mao, M., Liu, J., Zhou, J., Wu, H.: Efficient dialogue state tracking by masked hierarchical transformer. arXiv preprint arXiv:2106.14433 (2021)
18. Ni, J., Young, T., Pandelea, V., Xue, F., Cambria, E.: Recent advances in deep learning based dialogue systems: a systematic survey. Artif. Intell. Rev. **56**(4), 3055–3155 (2023)
19. Qiu, J., Lin, Z., Zhang, H., Yang, Y.: Hierarchical temporal slot interactions for dialogue state tracking. Neural Comput. Appl. **35**(8), 5791–5805 (2023)
20. Ren, L., Ni, J., McAuley, J.: Scalable and accurate dialogue state tracking via hierarchical sequence generation. arXiv preprint arXiv:1909.00754 (2019)
21. Shan, Y., Li, Z., Zhang, J., Meng, F., Feng, Y., Niu, C., Zhou, J.: [retracted] a contextual hierarchical attention network with adaptive objective for dialogue state tracking. In: Proceedings of the 58th Annual Meeting of the Association for Computational Linguistics, pp. 6322–6333 (2020)
22. Shin, J., Yu, H., Moon, H., Madotto, A., Park, J.: Dialogue summaries as dialogue states (ds2), template-guided summarization for few-shot dialogue state tracking. arXiv preprint arXiv:2203.01552 (2022)
23. Su, R., Wu, T.W., Juang, B.H.: Schema graph-guided prompt for multi-domain dialogue state tracking. arXiv preprint arXiv:2311.06345 (2023)
24. Subakan, C., Ravanelli, M., Cornell, S., Bronzi, M., Zhong, J.: Attention is all you need in speech separation. In: ICASSP 2021-2021 IEEE International Conference on Acoustics, Speech and Signal Processing (ICASSP), pp. 21–25. IEEE (2021)

25. Wang, W., Zhang, Z., Guo, J., Dai, Y., Chen, B., Luo, W.: Task-oriented dialogue system as natural language generation. In: Proceedings of the 45th International ACM SIGIR Conference on Research and Development in Information Retrieval, pp. 2698–2703 (2022)
26. Wang, Y., Guo, Y., Zhu, S.: Slot attention with value normalization for multi-domain dialogue state tracking. In: Proceedings of the 2020 Conference on Empirical Methods in Natural Language Processing (EMNLP), pp. 3019–3028 (2020)
27. Wiering, M., Van Otterlo, M.: Reinforcement learning. adaptation, learning, and optimization. Reinforcement Learning State-of-the-Art; Springer: Berlin/Heidelberg, Germany (2012)
28. Wu, C.S., Madotto, A., Hosseini-Asl, E., Xiong, C., Socher, R., Fung, P.: Transferable multi-domain state generator for task-oriented dialogue systems. arXiv preprint arXiv:1905.08743 (2019)
29. Xie, H., et al.: Correctable-DST: mitigating historical context mismatch between training and inference for improved dialogue state tracking. In: Proceedings of the 2022 Conference on Empirical Methods in Natural Language Processing, pp. 876–889 (2022)
30. Yang, G., Wang, X., Yuan, C.: Hierarchical dialog state tracking with unknown slot values. Neural Process. Lett. **50**(2), 1611–1625 (2019)
31. Yang, J., Song, H., Xu, B., Zhou, H.: Dialogue state tracking with a dialogue-aware slot-level schema graph approach. In: International Conference on Knowledge Science, Engineering and Management, pp. 172–183. Springer (2023)
32. Ye, F., Manotumruksa, J., Zhang, Q., Li, S., Yilmaz, E.: Slot self-attentive dialogue state tracking. In: Proceedings of the Web Conference 2021, pp. 1598–1608 (2021)
33. Zeng, Y., Nie, J.Y.: Jointly optimizing state operation prediction and value generation for dialogue state tracking. arXiv preprint arXiv:2010.14061 (2020)
34. Zeng, Y., Nie, J.Y.: Multi-domain dialogue state tracking based on state graph. arXiv preprint arXiv:2010.11137 (2020)
35. Zhang, J.G., Hashimoto, K., Wu, C.S., Wan, Y., Yu, P.S., Socher, R., Xiong, C.: Find or classify? dual strategy for slot-value predictions on multi-domain dialog state tracking. arXiv preprint arXiv:1910.03544 (2019)

Research on Children's ADHD Detection Algorithm Based on BASA-FMGNet

Mengjia He[1], Kai Sun[2], Yu Tang[1], Xingchun He[1], Hongying Liu[1(✉)], and Xitian Pi[1(✉)]

[1] Chongqing University College of Bioengineering, Chongqing, China
{liuhongying,pixitian}@cqu.edu.cn
[2] Henan University of Science and Technology, Henan, China

Abstract. Attention deficit hyperactivity disorder (ADHD) is a common neurodevelopmental behavioral disorder. Its traditional diagnostic methods are highly subjective and prone to misdiagnosis. At present, ADHD diagnosis is mainly based on subjective evaluation, and the accuracy of objective diagnosis is low. To address this problem, this paper proposes an ADHD automatic classification network BASA-FMGNet based on EEG signals. The network uses FMGNet as the backbone network, and its core module FMGConv can effectively extract the time domain and frequency domain features of EEG signals. In addition, the network introduces a bilinear adaptive sparse attention (BASA) structure, extracts inter-channel features through sparse calculation, and uses bilinear adaptive calculation to fuse local and non-local attention, thereby enhancing feature expression. Experimental results on the ADHD children's dataset show that the F1 score of the proposed BASA-FMGNet network reaches 97.90%, which is significantly better than that of the current mainstream network. Ablation experiments further confirm the effectiveness of the BASA module in improving the performance of different networks, and verify that it can effectively enhance EEG features, providing strong support for the objective diagnosis of ADHD.

Keywords: EEG · ADHD diagnosis · CNN · BASA

1 Introduction

Attention deficit hyperactivity disorder (ADHD) is a disorder that affects brain growth, resulting in aberrant patterns of cognition and behavior. The clinical assessment of ADHD has long depended on behavioral scales and subjective observations [1]. This paradigm is susceptible to the experience of the clinician and the state of the patient. In recent years, neuroscience research has confirmed that individuals diagnosed with ADHD exhibit specific characteristics regarding their electroencephalography (EEG) activity [2], including biomarkers such as an imbalance in theta/beta wave power in the prefrontal cortex [3] and increased event-related potential variability [4]. EEG, as a noninvasive tool that

L. Huang et al. (Eds.): CCF NCCA 2025, CCIS 2774, pp. 41–53, 2026.
https://doi.org/10.1007/978-981-95-5294-8_4

directly reflects neural oscillations in the brain, can noninvasively record the temporal and spatial summation of postsynaptic potentials of cortical neurons through scalp electrodes. It offers the advantages of both low acquisition cost and high acquisition precision in detecting ADHD. First, EEG detection, often at an acquisition rate of 500 hertz or more, allows for tracking the dynamic progression of brain oscillations in real time with submillisecond precision. This approach captures the transitory properties of paroxysmal augmentation of theta(4-8 Hz) waves and attenuation of beta(13-30 Hz) wave power in ADHD patients' prefrontal striatal loops. Second, the use of high-density electrode arrays, such as 128-channel systems, can achieve topographic reconstruction of cortical potentials with centimeter spatial resolution, providing an anatomical reference for locating abnormal neural sources. Furthermore, compared with functional magnetic resonance imaging (fMRI), which has an operating cost of approximately 1,000 yuan per hour, the cost of a single EEG system can be controlled below 50 yuan. Moreover, portable acquisition devices are available to support long-term monitoring in a home environment. This multidimensional detection capability makes EEG the preferred method for constructing diagnostic models for ADHD, and its detection performance and clinical translatability have been verified in recent studies and reviews [5].

Manually selected features can be associated directly with neurophysiological mechanisms, facilitating biological interpretation in clinical or research settings. Therefore, early researchers generally extracted key features manually and reduced high-dimensional EEG signals to a low-dimensional feature subset to improve computational efficiency [6]. Catherine et al. [7] offered computational methods for identifying individuals with ADHD via nonlinear entropy and artificial neural network classifiers. Among them, permutation entropy classification performed best. They also reported that the diagnostic range of Shannon entropy is limited and that the signal from the middle of the frontal pole and frontal lobe when the eyes are closed is significant in distinguishing ADHD patients from normal subjects. Alim and Imtiaz [8] developed a machine learning technique to identify children exhibiting ADHD on the basis of distinctive EEG patterns. A Gaussian SVM model was trained utilizing linear EEG characteristics, attaining a high accuracy rate via 10-fold cross-validation. Khare et al. [9] employed extreme learning machine (ELM) to classify EEG rhythm features, achieving high sensitivity and accuracy in ADHD detection. Deshmukh et al. [10] utilized EEG data from cognitive tasks to construct machine learning models via multiple classifiers. They reported that EEG-based classification could reach an accuracy of 84% in the task of identifying ADHD in children versus healthy controls, and that data from right hemisphere channels performed prominently in the classification. However, traditional EEG analysis relies on manual feature extraction and linear classifiers, making it difficult to capture complex, high-dimensional spatiotemporal features. This has led researchers to turn to deep learning techniques with automatic feature learning capabilities.

Recent advancements within the domain of deep learning, especially the ability of CNNs to hierarchically extract features from EEG signals, have provided a novel paradigm for ADHD recognition on the basis of EEG signals [11], which has been widely adopted by researchers. Hassan et al. [12] developed a simple CNN architecture of only two convolutional layers, which achieved high precision in detecting ADHD in children by preprocessing and analysing EEG signals solely from the frontal region. TaghiBeyglou et al. [13] introduced a deep CNN to detect children diagnosed with ADHD, training it with raw EEG signals. By coupling it with a classical classifier and using a hybrid approach that combines the feature maps from a densely connected layer 1 CNN with a logistic regression classifier, they achieved high accuracy. Bakhtyari et al. [14] proposed a neural network model, that incorporates LSTM and attention mechanisms to distinguish individuals with ADHD from control individuals, achieving the highest average accuracy of 99.34% on relevant EEG datasets. Esas et al. [15] achieved a classification accuracy of over 87% in ADHD detection by analysing EEG signals processed with deep learning algorithms and decomposition techniques. Loh et al. [16] proposed a deep network that achieved high classification accuracy in distinguishing between ADHD, ADHD with comorbid conduct disorder (ADHD + CD), and CD alone, through a unique preprocessing method and a deep CNN model. Latifi et al. [17] employed spectral density brain maps of pediatric EEG data to identify ADHD, achieving the highest classification accuracy of 99% when their suggested CNN model was used. Additionally, their research indicated that the power spectral density attributes of the theta band, situated in the front and posterior cortical regions of the brain, function as significant differentiators. When juxtaposed with traditional machine learning methods, deep neural network-based algorithms significantly improve in ADHD identification. Moreover, these models have the intrinsic ability to uncover previously unrecognized biological signals.

On the basis of the current research status of ADHD EEG classification, this paper proposes a BASA-FMGNet network to classify EEG signals for ADHD, and achieves a good detection structure. The specific work is as follows:

(1) To increase the diagnostic accuracy of ADHD, BASA-FMGNet is proposed. This deep learning framework is specifically designed for analysing EEG data and identifying ADHD.
(2) The core building module FMGConv of FMGNet is designed, which can effectively extract the time domain features and frequency domain features contained in EEG signals.
(3) A bilinear adaptive sparse attention (BASA) mechanism is designed. This mechanism focuses on important interchannel features through sparse calculation, and fuses local and nonlocal attention through bilinear adaptive calculations, thereby enhancing the extracted EEG features.
(4) Experiments on the ADHD children's dataset show that the network achieves an F1 score of up to 97.90%. BASA-FMGNet shows superior performance compared with the current network. In addition, ablation experiments also prove that the BASA module can significantly improve the performance of multiple baseline networks.

2 Introduction

2.1 BASA-FMGNet Network Structure Design

The BASA-FMGNet network is a single-input single-output structure, whose input is the EEG signal and whose output is the classification label. The BASA-FMGNet network consists of FMGNet and BASA structures. The main components of FMGNet are the FMGConv structure and Conv3 convolution. BASA is a bilinear adaptive sparse attention mechanism that can extract sparse features in EEG signals well and enhance them. In addition, it can pay attention to the local and nonlocal features of EEG signals. The BASA is a plug-and-play attention model, which is placed between the second FMGConv structure and the third structure. Following the backbone network's extraction of the EEG signals, the pooling layer reduces the feature to one dimension, and the label for each task is obtained via the Softmax function. After the expected accuracy and loss are achieved, the training of the BASA-FMGNet network can be completed. The structure of the BASA-FMGNet is displayed in Fig. 1. The design details of each model are discussed in the following chapters.

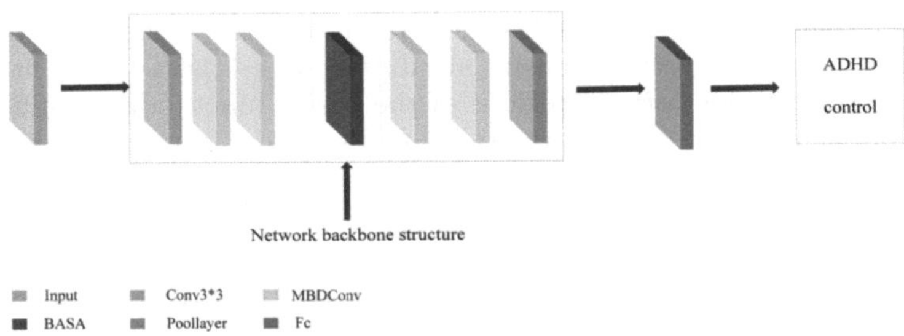

Fig. 1. BASA-FMGNet structure diagram.

2.2 BASA-FMGNet Network Structure Design

In the design process of deep learning models, the difficulty lies in achieving the best balance between model size and model performance. Although increasing the model parameters can improve the network's accuracy and capacity to learn intricate patterns, doing so usually occurs with the disadvantages of increased computer power, additional training time, and a greater chance of overfitting. Conversely, reducing the number of parameters may lead to more efficient and general models, but may sacrifice the capacity required to capture complex data relationships, resulting in lower accuracy. Therefore, a key and difficult point in deep learning research is to develop techniques and architectures that can effectively balance this balance, thereby generating models that are both accurate and efficient. The Efficientv2 network [18] achieves a relatively good balance between parameters and accuracy. The main concept of the Efficientv2 network is to use spatial search to determine the most appropriate parameters after designing multiple variable parameter modules, are designed and to build networks of different sizes on the basis of the parameters of different groups searched. This paper designs the FMGConv structure on the basis of the basic structure of the Efficientv2 network, and Fig. 2 illustrates its primary structure.

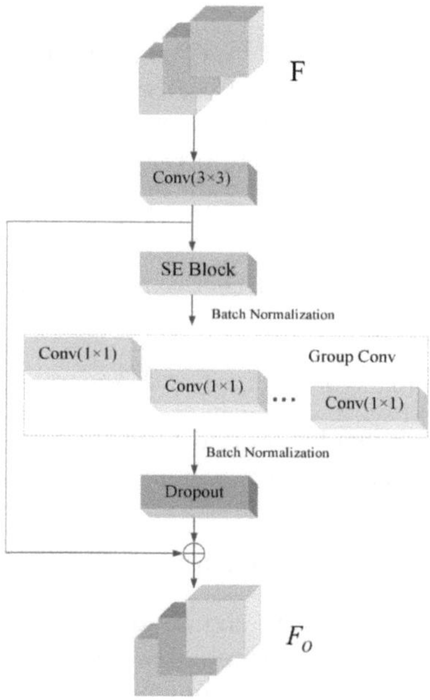

Fig. 2. FMGConv structure.

As shown in Fig. 2, an input feature F with a dimension of C*3*3 convolutions is first subjected to normalization via a BN layer, then we have:

$$F_1 = BN(Conv3 * 3(F))F \in R^{H\times W\times C}, F_1 \in R^{H\times W\times C*} \tag{1}$$

The expansion process is then carried out, to restore the extracted channel attention to the original dimension, which is achieved through two convolution operation transformations. The first convolution process is as follows:

$$f_z = \frac{1}{H\times W}\sum_{a=1}^{H}\sum_{b=1}^{W} F_i(a,b) \tag{2}$$

where F_i' is the i-th channel of the input feature, and where $F_i(a,b)$ is the value of the channel at spatial position (a,b).

The weight is obtained by performing an excitation process on the output f_z obtained by the squeezing process. The main implementation process involves learning the weight of each feature channel through one or two fully connected layers. The first fully connected layer reduces the channel dimension from C* to $\frac{C*}{r}$, and the second fully connected layer increases the channel dimension from $\frac{C*}{r}$ back to C.

$$s_c = \sigma(W_2\sigma(W_1 f_z)) \tag{3}$$

where $W_1 \in R^{C^*/r} \times C^*$ and $W_2 \in R^C \times C^*/r$, σ is the activation function, and $s_c \in R^{1 \times \times C}$ is the final learned channel vector.

Finally, the obtained channel weight s_c and the original input feature map are multiplied between channels to obtain the output F_se of the SE process.

$$F_{se_n} = s_{c_n} F_n \tag{4}$$

Grouped convolution is used to further extract the features of F_se. The process is as follows:

$$O_g = F_g W_g + \beta_g (0 \leq g \leq G) \tag{5}$$

where g is the intragroup convolution and G is the number of groups, then the final output is $O = [O_1, O_2, O_3, ..., O_G]$.

2.3 BASA-FMGNet Network Structure Design

The inherent characteristics of EEG signals for ADHD classification are their spatial distribution on the scalp, discriminative patterns at multiple time scales, and overall diversity of brain activity. Convolutional neural networks are good at capturing spatial relationships in EEG data and identifying topologically related features. In order to further optimize the focus on diagnostically important information, attention mechanisms can be integrated to enable the network to selectively emphasize the EEG signal features that are most suggestive of ADHD while suppressing noise or less relevant activities. Given that frequency domain features are crucial for distinguishing ADHD-related EEG signal patterns, analysing different frequency bands can provide a multi-scale perspective.On this basis, this paper proposes a bilinear adaptive sparse attention structure, that can integrate the two attentions of the bilinear network and adaptive sparse attention. The bilinear network can extract more frequency domain features, whereas adaptive sparse attention can filter out invalid features in the EEG, and extract effective brain connectivity features and multi-scale features.

As shown in Fig. 3, the input feature F first uses three 1*1 convolutions to obtain Q, K, and V. The process is:

$$\begin{aligned} F_q &= W_q F \\ F_k &= W_k F \\ F_v &= W_v F \end{aligned} \tag{6}$$

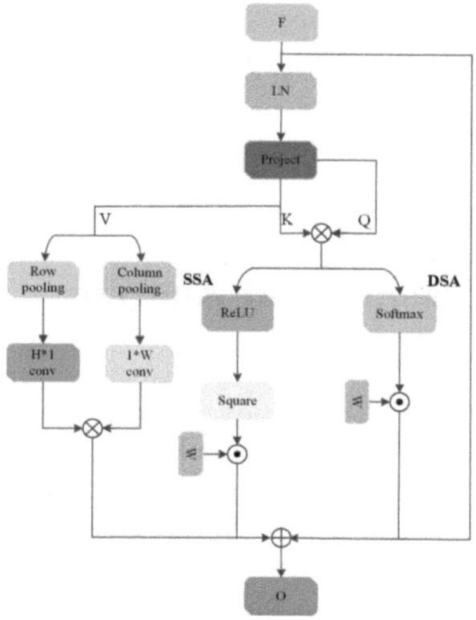

Fig. 3. BASA structure.

Then Q and K are multiplied and enter DSA [19] and SSA [20] respectively, where DSA extracts dense features, and its formula is as follows:

$$F_{DSA} = softmax(F_q F_K^T/\sqrt{d} + \beta) \tag{7}$$

The SSA extracts sparse features, via the following formula:

$$F_{SSA} = ReLU^2(F_q F_k/\sqrt{t} + \beta) \tag{8}$$

The V branch extracts bilinear attention. First, it performs row pooling and column pooling on the input feature F, then we haveïijŽ

$$\begin{aligned} F_{vr} &= pool_r(F_v) \\ F_{vc} &= pool_c(F_c) \end{aligned} \tag{9}$$

Then convolution is performed on the row pooling features and column pooling features, then we have

$$\begin{aligned} F_{vh} &= W_H F_{vr} \\ F_{vw} &= W_W F_{vc} \end{aligned} \tag{10}$$

where W_H is the weight of the $H * 1$ convolution, and where W_W is the weight of the $1 * W$ convolution.

Finally, a bilinear operation is performed to obtain bilinear attention, and its formula is:

$$F_{BL} = F_{vh} F_v F_{vw} \tag{11}$$

Finally, the residual of the bilinear attention, adaptive sparse attention and input features is calculated to obtain the output O, whose formula is as follows:

$$O = F + \alpha_1 F_{BL} + \alpha_2 F_{DSA} + \alpha_3 F_{SSA} \tag{12}$$

where $\alpha_1, \alpha_2, \alpha_3$ are trainable parameters.

3 Experimental Results and Analysis

3.1 Experimental Preparation

Data Sources. The data in this study come from the ADHD child dataset provided by Shahed University [21]. The subjects of the dataset are 61 children with ADHD and 60 healthy controls, including boys and girls, and all the children are between 7 and 12 years old.. Children with ADHD were diagnosed by experienced psychiatrists according to the DSM-IV criteria [22] and took Ritalin for 6 months. No children in the control group had a history of mental illness, epilepsy, or high-risk behavior reports. EEG recordings were made using the 10–20 standard, 19 channels and a sampling frequency of 128 Hz. The A1 and A2 electrodes are reference electrodes located on the earlobes. Figure 4 shows the 19-channel EEG waveform of children with ADHD and normal children.

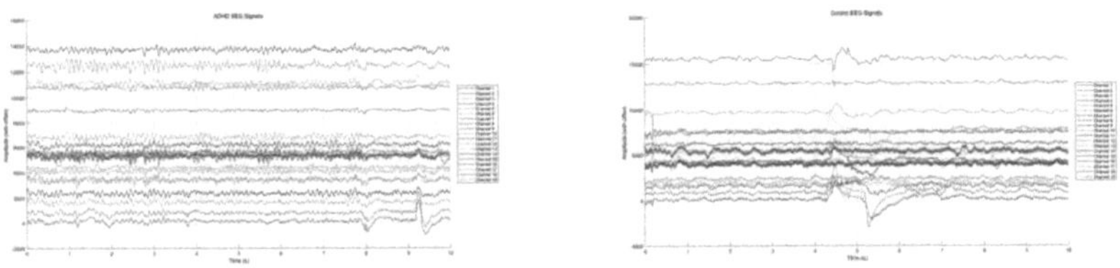

(a) EEG signal (b) Control signal

Fig. 4. EEG signal comparison chart.

From the ADHD EEG signals and normal EEG signals, we can see that the EEG signals of normal subjects may show more regular rhythmic activities on some channels, the baseline is relatively stable, the instantaneous waveform changes are relatively concentrated, and the morphology is mostly specific. However, the overall irregularity and irregularity of the EEG signals of ADHD subjects may be more significant, the rhythmicity is poor, and the distribution of high-frequency components such as instantaneous spikes may be more dispersed, without obvious concentration points. This is a key difference separating children who have ADHD from their neurotypical peers, and the data reflecting this difference are quite effective for differentiation.

Data Preprocessing. A notch filter is initially employed to eliminate power frequency interference, followed by a bandpass filter to process the raw EEG data, preserving pertinent frequency bands associated with brain activity while discarding extraneous low-frequency drift and high-frequency noise. Finally, the ICA algorithm is used to remove artifacts to eliminate non-brain signals caused by blinking, muscle movement or electrode displacement.

Perform feature extraction on the EEG signal, extract the α,δ, θ, β, and γ features of the signal respectively, where the input of α,δ, and θ is a $16 \times 16 \times 4$ tensor,β is a $16 \times 16 \times 23$ tensor, and γ is a $16 \times 16 \times 5$ tensor. In combination, a $16 \times 16 \times 40$ tensor can be obtained, which is the input part of the EEG signal.

Evaluation Criteria. The ADHD evaluation in this article belongs to a classification task, and the evaluation indicators are accuracy, precision, recall and the F1 score. The classification model's efficacy is rigorously assessed via indicators such as accuracy, precision, recall, and the F1 score, which are derived from the fundamental counts of true positives (TP), true negatives (TN), false positives (FP), and false negatives (FN). Accuracy summarizes the correctness of the model in all predictions, whereas precision specifically measures the proportion of correctly identified positive instances among all instances predicted to be positive. In contrast, recall focuses on the model's ability to capture all actual positive instances. The F1 score combines precision and recall into a balanced indicator, which is particularly useful when dealing with unbalanced datasets because high accuracy may mask the low performan ce of minority classes. The formulas for the accuracy, precision, recall and F1 score are as follows:

$$\begin{aligned}
Accuracy &= \frac{TP+FN}{TP+FN+FP+TN} \\
Precision &= \frac{TP}{TP+FP} \\
Recall &= \frac{TP}{TP+FN} \\
F1 &= 2 \times [\frac{Precision \times Recall}{Precision + Recall}]
\end{aligned} \tag{13}$$

Experimental Comparison. In order to control variables, it is necessary to ensure that the training hardware and software conditions of all networks trained in this study are the same. All the networks in this study are trained on the same device with the same hardware conditions. The CPU uses an Intel i9 13900K and the GPU uses an NVIDIA RTX 4090. The training operating system uses ubantu 22.04, Pytorch version 2.5.0, CUDA version 12.4, and cuDNN version 9.1.0.

The training parameters are set to 50 epochs, the learning rate is 0.001, the learning rate decreases slowly, and the learning rate decreases by 5% every five generations, that is, multiplied by 0.95, the batch size is set to 1024, and the training evaluation indicators are accuracy, precision, recall, and F1. The training process is plotted, and the plotted images are shown in Fig. 5 (a)-Fig. 5 (d).

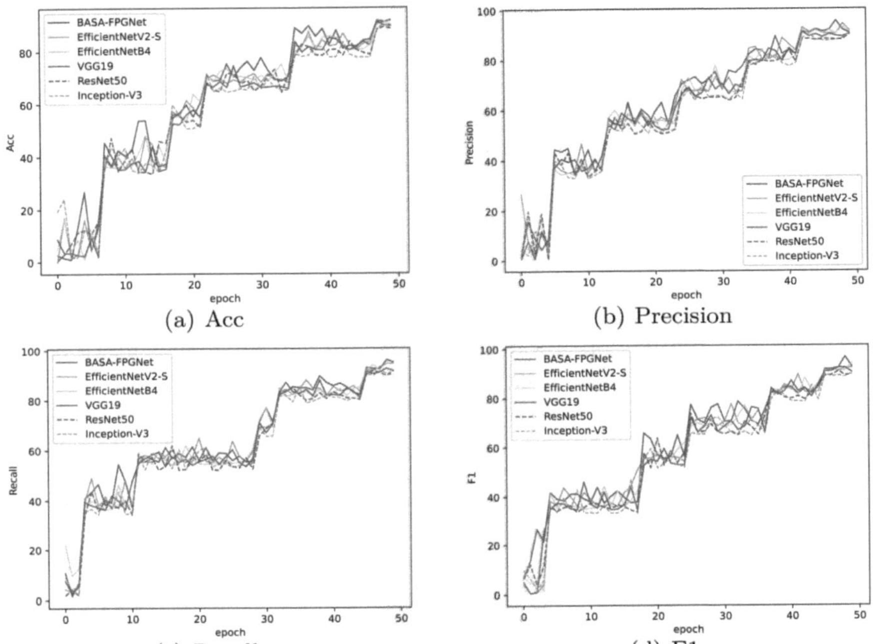

(a) Acc (b) Precision (c) Recall (d) F1

Fig. 5. Comparison of training conditions of different networks.

As shown in Figs. 5 (a)-5 (d), the accuracy, precision, recall, and F1 score increase rapidly in the initial stage, then increase more slowly in the middle, and then slowly increase in the later stage. Then, it starts to oscillate in the final stage and finally reaches the best accuracy.

3.2 Experimental Comparison

The comparison models used in the experiment include Inception-V3, VGG19, ResNet50, EfficientNetB4 and EfficientNetV2-S. The comparison indicators are accuracy, precision, recall and F1. A comparison of the experimental results of the basic models is shown in Table 1.

As shown in Table 1, BASA-FMGNet shows the highest performance in all the indicators, with an accuracy of 97.45, a precision of 96.97, a recall of 98.84, and an F1-score of 97.90, indicating that it is the best performing model among all the evaluated models. EfficientNetV2-S and EfficientNetB4 also performed well, scoring over 95% on all the indicators. VGG19 performed well and ranked third, while ResNet50 and Inception-V3 performed slightly lower, but still quite

Table 1. Network model performance comparison

Network	Acc	Precision	Recall	F1
Inception-V3	91.98	91.55	93.26	91.76
VGG19	94.49	93.89	95.61	94.74
ResNet50	92.34	91.89	93.56	92.72
EfficientNetB4	95.78	95.45	97.05	96.24
EfficientNetV2-S	96.09	95.72	97.33	96.52
BASA-FMGNet	**97.45**	**96.97**	**98.84**	**97.90**

impressive. These indicators indicate that BASA-FMGNet achieves the highest performance in ADHD classification and achieves a good balance between accuracy and recall.

3.3 Ablation Experiment

The ablation experiment is one of the more important experiments in deep learning. Its principle is to prove the effectiveness of module transformation by adding and deleting modules in the model and adapting them in multiple networks. The BASA module in this paper is a plug-and-play module. Its usability is proven by inserting the BASA module in different networks. The transformed networks are Inception-V3, VGG19, ResNet50, EfficientNetB4 and EfficientNetV2-S. FMGNet is the original network of BASA-FMGNet proposed in this paper. The comparison results are shown in Table 2.

Table 2. Ablation experiment comparison

network	BASA module	Acc	Precision	Recall	F1
ResNet50	×	92.34	91.89	93.56	92.72
	✓	93.49	93.06	94.85	93.95
Inception-V3	×	91.98	91.55	93.26	91.76
	✓	93.45	92.89	94.62	93.75
VGG19	×	94.49	93.89	95.61	94.74
	✓	95.82	95.24	96.93	96.08
EfficientNetB4	×	95.78	95.45	97.05	96.24
	✓	96.72	96.48	98.06	97.26
EfficientNetV2-S	×	96.09	95.72	97.53	96.62
	✓	97.12	96.65	98.42	97.53
FMGNet	×	96.38	95.83	97.80	96.80
	✓	97.45	96.97	98.84	97.90

As shown in Table 2, for all tested network structures, the introduction of the BASA module can significantly improve various performance indicators. This shows that the BASA module can effectively enhance the model's feature extraction and classification capabilities, thereby improving the model's performance on the target task, thus proving the effectiveness of the BASA module. Under the same network structure, the model with the BASA module is higher in accuracy, precision, recall and F1 score than the model without the BASA module. The extent of the improvement varies across different network structures, but the overall trend is consistent. In the VGG19 network, ResNet50 and Inception-V3, after the introduction of the BASA module, the F1 score increased by more than 1.3%, whereas in EfficientNet, EfficientNetV2-S and FMGNet, the network's F1 score increased by approximately 1%. This may be because the networks of EfficientNet, EfficientNetV2-S and FMGNet themselves have the SElayer's attention enhancement structure, whereas the VGG19 network, ResNet50 and Inception-V3 do not have similar attention enhancement structures, so the introduction of BASA has a better effect.

4 Conclusion

This paper introduces an ADHD classification network BASA-FMGNet that is based on EEG signals. The network effectively utilizes the time-frequency feature extraction capability of FMGNet and the channel feature extraction and attention mechanism enhanced by the BASA module. The test results on the ADHD dataset demonstrate the superiority of BASA-FMGNet, and its F1 score is significantly higher than those of the existing methods. Ablation studies further verify that the BASA module improves the EEG signal detection effect in different networks. Future research will further improve the generalizability of the network and combine knowledge distillation and other methods to reduce the network size.

Funding Information. This study was financially supported by grants from the Major Project of Chongqing Technology Innovation and Application Development Special Project(No. CSTB2024TIAD-STX0044).

References

1. Maniruzzaman, M., Hirooka, K., Tomioka, Y., et al.: Machine learning-based ADHD detection from fNIRs signal during reverse stroop tasks. IEEE Access (2024)
2. Cowley, B.U., Juurmaa, K., Palomäki, J.: Reduced power in fronto-parietal theta EEG linked to impaired attention-sampling in adult ADHD. Eneuro, **9**(1) (2022)
3. Leon, C., Kaur, S., Sagar, R., et al.: Cortical hypoactivation of frontal areas modulate resting EEG microstates in children with ADHD. Cognitive Neuroscience Neuroimaging, Biological Psychiatry (2024)
4. Einziger, T., Devor, T., Ben-Shachar, M.S., et al.: Increased neural variability in adolescents with ADHD symptomatology: evidence from a single-trial EEG study. Cortex **167**, 25–40 (2023)

5. Slater, J., Joober, R., Koborsy, B.L., et al.: Can electroencephalography (EEG) identify ADHD subtypes? a systematic review. Neurosci. Biobehav. Rev. **139**, 104752 (2022)
6. Maniruzzaman, M., Hasan, M.A.M., Asai, N., et al.: Optimal channels and features selection based ADHD detection from EEG signal using statistical and machine learning techniques. IEEE Access, **11**, 33570–33583 (203)
7. Catherine Joy, R., Thomas George, S., Albert Rajan, A., et al.: Detection of ADHD from EEG signals using different entropy measures and ANN. Clin. EEG Neurosci. **53**(1), 12–23 (2022)
8. Alim, A., Imtiaz, M.H.: Automatic identification of children with ADHD from EEG brain waves. Signals **4**(1), 193–205 (2023)
9. Khare, S.K., Gaikwad, N.B., Bajaj, V.: VHERS: a novel variational mode decomposition and Hilbert transform-based EEG rhythm separation for automatic ADHD detection. IEEE Trans. Instrum. Meas. **71**, 1–10 (2022)
10. Deshmukh, M., Khemchandani, M., Thakur, P.M.: Contributions of brain regions to machine learning-based classifications of attention deficit hyperactivity disorder (ADHD) utilizing EEG signals. Appl. Neuropsychol. Adult, 1–15 (2024)
11. Sanchis, J., García-Ponsoda, S., Teruel, M.A., et al.: A novel approach to identify the brain regions that best classify ADHD by means of EEG and deep learning. Heliyon, **10**(4) (2024)
12. Hassan, U., Singhal, A.: Convolutional neural network framework for EEG-based ADHD diagnosis in children. Heal. Inf. Sci. Syst. **12**(1), 44 (2024)
13. TaghiBeyglou, B., Shahbazi, A., Bagheri, F., et al.: Detection of ADHD cases using CNN and classical classifiers of raw EEG. Comput. Methods Programs Biomed. Update **2**, 100080 (2022)
14. Bakhtyari, M., Mirzaei, S.: ADHD detection using dynamic connectivity patterns of EEG data and ConvLSTM with attention framework. Biomed. Signal Process. Control **76**, 103708 (2022)
15. Esas, M.Y., Latifoğlu, F.: Detection of ADHD from EEG signals using new hybrid decomposition and deep learning techniques. J. Neural Eng. **20**(3), 036028 (2023)
16. Loh, H.W., Ooi, C.P., Oh, S.L., et al.: ADHD/CD-NET: automated EEG-based characterization of ADHD and CD using explainable deep neural network technique. Cogn. Neurodyn. **18**(4), 1609–1625 (2024)
17. Latifi, B., Amini, A., Nasrabadi, A.M.: Siamese based deep neural network for ADHD detection using EEG signal. Comput. Biol. Med. **182**, 109092 (2024)
18. Tan, M., Le, Q.: Efficientnetv2: smaller models and faster training. In: International Conference on Machine Learning. PMLR, pp. 10096–10106 (2021)
19. Zhou, S., Chen, D., Pan, J., et al.: Adapt or perish: adaptive sparse transformer with attentive feature refinement for image restoration. In: Proceedings of the IEEE/CVF Conference on Computer Vision and Pattern Recognition, pp. 2952–2963 (2024)
20. Zhang, B., Titov, I., Sennrich, R.: Sparse attention with linear units. arXiv preprint arXiv:2104.07012 (2021)
21. Al-Wardat, M., Etoom, M., Almhdawi, K.A., et al.: Prevalence of attention-deficit hyperactivity disorder in children, adolescents and adults in the middle east and north Africa region: a systematic review and meta-analysis. BMJ Open **14**(1), e078849 (2024)
22. Livne, O., Shmulewitz, D., Stohl, M., et al.: Agreement between DSM-5 and DSM-IV measures of substance use disorders in a sample of adult substance users. Drug Alcohol Depend. **227**, 108958 (2021)

NAD-KRQG : Neural Association-Based Cognitive Diagnosis with Knowledge Recommendation and Question Generation

Peng Zhang[1], Guo-Sheng Hao[1,2](✉), Meng-Nan Xu[1], Xia Wang[1], Shi-Jin Ren[1], and Yi Zhu[1]

[1] School of Computer Science and Technology, Jiangsu Normal University, Xuzhou 211116, China
{zhangp1216,hgskd,mnx,zhuy}@jsnu.edu.cn, lgzwx@163.com
[2] Jiangsu Wisdom-driven Research Institute, Xuzhou, Jiangsu 221000, China

Abstract. In the context of personalized education, it remains challenging to understand each student's mastery of knowledge points. Consequently, providing effective support for students' weak areas of knowledge still needs to be addressed. To address this issue, this paper proposes a guided personalized learning assistance method. First, a NeuralCD-SEKA cognitive diagnostic model is constructed. By analysing the associations among students, exercises, knowledge points, and the time students spend on exercises, the model accurately assesses students' knowledge mastery. Second, on the basis of students' knowledge mastery and the importance of knowledge points, a personalized cognitive map is visualized to clarify learning pathways. Finally, an NAD-KR$_{ADFM}$ knowledge point recommendation model is developed. This model precisely recommends suitable knowledge points to students on the basis of their mastery levels and the importance of the knowledge points. This model also uses a large language model to generate questions for guided questioning. The experimental results show that the proposed models achieve significant performance improvements in both cognitive diagnosis and knowledge point recommendation. It also shows that the recommended knowledge points are evenly distributed, providing reliable support for personalized learning.

Keywords: Personalized education · Cognitive diagnostic · Knowledge recommendation

1 Introduction

With the rapid advancement of information technology and the widespread use of the internet, personalized education has long emerged as a topic of research [1]. Traditional education methods are increasingly inadequate for addressing the individual differences and complex learning needs of students, who come from diverse cultural backgrounds and social environments and who possess varied

L. Huang et al. (Eds.): CCF NCCA 2025, CCIS 2774, pp. 54–71, 2026.
https://doi.org/10.1007/978-981-95-5294-8_5

learning histories and styles. Consequently, the development and implementation of personalized learning models have become particularly crucial. These models aim to offer customized learning resources and pathways tailored to students' interests, knowledge levels, and learning goals, thereby maximizing their learning effectiveness and interest.

However, current research and practice in personalized learning still face several challenges. First, accurately assessing students' mastery of knowledge points is the foundation for achieving personalized learning, but traditional assessment methods often fail to reflect students' understanding of each knowledge point in detail. Second, providing effective learning support on the basis of students' weaknesses is a key issue yet to be resolved. Faced with difficulties and weak knowledge points that students often encounter during the learning process, traditional teaching methods usually lack flexibility and cannot adapt to the personalized needs of each student. Therefore, new methods are urgently needed to promote the further development and application of personalized learning models in the current educational environment.

To address these challenges, this paper proposes a guided personalized learning model that integrates cognitive diagnosis and knowledge point recommendation techniques. First, we model students' actual knowledge states via the Neural Cognitive Diagnosis with Student-Exercise-Knowledge Association (NeuralCD-SEKA), which associates students, exercises, and knowledge points while considering the varying impacts of exercises of different difficulties on each student. Subsequently, we analyse the relationships between knowledge points and calculate the importance of each knowledge point through a Personalized Cognitive Map (PCMap). On the basis of this analysis, we design a personalized learning support model, Neural Association-based Cognitive Diagnosis with Knowledge Recommendation and Question Generation (NAD-KRQG), that combines cognitive diagnosis, knowledge point recommendation, and question generation. This guided interaction and feedback mechanism helps students address and overcome learning difficulties, achieving deeper learning and cognitive development. The contributions of this paper are as follows:

(1) A more refined and personalized cognitive diagnosis model, NeuralCD-SEKA, was constructed to capture students' learning characteristics and needs more accurately.
(2) A personalized cognitive map (PCMap) was designed and implemented to provide students with a customized learning path and visually present their cognitive state and learning progress.
(3) An attention-driven NAD-KR$_{ADFM}$ that dynamically weights KP (knowledge point) importance and mastery and a guided QG framework using LLMs to generate depth-promoting questions are proposed.

2 Related Work

2.1 Cognitive Diagnosis

Cognitive diagnosis models (CDMs) are a specialized type of discrete latent variable model employed in modern education, psychology, social sciences, and biological sciences. These models are designed to assess students' cognitive processes and thinking abilities, providing insights into their grasp and comprehension of knowledge [2]. By collecting and analysing data on students' learning behaviors and performance, CDMs uncover cognitive characteristics, learning styles, and abilities, thereby offering personalized learning support and guidance tailored to each student.

Traditional cognitive diagnosis models such as DINA [3], IRT [4], and MIRT [5], have made significant contributions to the assessment of students' knowledge levels. However, they cannot fully capture the complex nonlinear relationships between students and exercises, thereby limiting their effectiveness in personalized diagnosis and assessment of students' learning behaviors. To overcome these limitations, WANG et al. [6,7] developed the Neural Cognitive Diagnosis Model (NeuralCDM). This model introduces neural networks, which enhance the ability to capture intricate interactions and provide personalized diagnoses. In NeuralCDM, each student's knowledge proficiency is represented by a vector h^s, which is obtained by multiplying the student's one-hot vector x^s with a trainable matrix, as defined in Eq. 1. For each exercise, a one-hot vector x^e is utilized, and a predefined Q-matrix represents the relationship between the exercises and knowledge concepts. The product of these yields the exercise factor Q_e, as shown in Eq. 2. Additional exercise factors, such as knowledge difficulty h^{diff} and exercise discrimination h^{disc}, are calculated via Eqs. 3 and 4. The output from the first layer, represented in Eq. 5, is subsequently utilized to predict the student's proficiency across various knowledge concepts.

$$h^s = sigmod(x^s \times A), A \in R^{N \times K} \tag{1}$$

$$Q_e = x^e \times Q \tag{2}$$

$$h^{diff} = sigmod(x^e \times B), B \in R^{M \times K} \tag{3}$$

$$h^{disc} = sigmod(x^e \times D), B \in R^{M \times 1} \tag{4}$$

$$x = Q_e \circ (h^s - h^{diff}) \times h^{disc} \tag{5}$$

where matrices A, B, and where D are trainable, and $\circ$ denotes elementwise multiplication.

However, the model faces challenges in terms of interpretability and computational efficiency. Therefore, this paper aims to improve upon the NeuralCDM, with the objective of enhancing its interpretability and personalized diagnostic capabilities.

2.2 Exercise Recommendation

Exercise recommendation is a critical research direction in personalized education, aiming to suggest appropriate exercises to students by analysing their learning behaviors, knowledge mastery, and personal preferences. This approach helps students learn more effectively, enhancing both learning efficiency and learning outcomes.

Technically, traditional recommendation methods such as content-based filtering (CBF) [8,9], collaborative filtering (CF) [10,11], and hybrid filtering (HF) [12,13] have been widely used in the recommendation of learning resources. Recently, with the advancement of artificial intelligence and machine learning technologies, significant progress has been made in the development of exercise recommendation algorithms. Wang et al. [14] leveraged the relationships between entity nodes in a knowledge graph and used the sequence of paths to infer user preferences, integrating recommendation algorithms to improve recommendation performance. WU et al. [15] employed cognitive diagnosis to plan exercise paths for learners from three perspectives: knowledge point coverage, knowledge point mastery, and exercise diversity. Tianyu Zhu et al. [16] proposed a personalized exercise recommendation method based on students' knowledge point mastery and collaborative filtering, achieving good results by considering both the students' learning states and the commonalities of similar students.

Most existing studies focus on exercise recommendations, which fail to guide students in deeper thinking and enhance their understanding of knowledge points. Therefore, this paper recommends knowledge points on the basis of students' learning states and the importance of the knowledge points, followed by the generation of relevant exercises to help students better comprehend these knowledge points.

Unlike NeuralCDM [6,7], NeuralCD-SEKA uniquely models time effects (Eqs. 8, 9 and 10) and knowledge associations (Eq. 7), enhancing interpretability for personalized diagnosis (Table 2).

3 Guided Personalized Learning Model

This study primarily utilizes cognitive diagnosis to assess students' mastery of knowledge points and constructs personalized cognitive graphs on the basis of the importance and relationships of these knowledge points. Knowledge points are then recommended to students on the basis of their mastery level and importance. A fine-tuned large language model is subsequently used for question generation, aimed at helping students consolidate and deepen their understanding and memory of the knowledge points. The overall framework of this study is illustrated in Fig. 1.

Module roles:

(1) NeuralCD-SEKA: Diagnoses h^{st} from student-exercise logs.
(2) PCMap: Maps h^{st} and KP importance to visual pathways.
(3) NAD-KR$_{ADFM}$: Recommends KPs via attention.
(4) LLM-QG: This generates guided questions for recommended KPs.

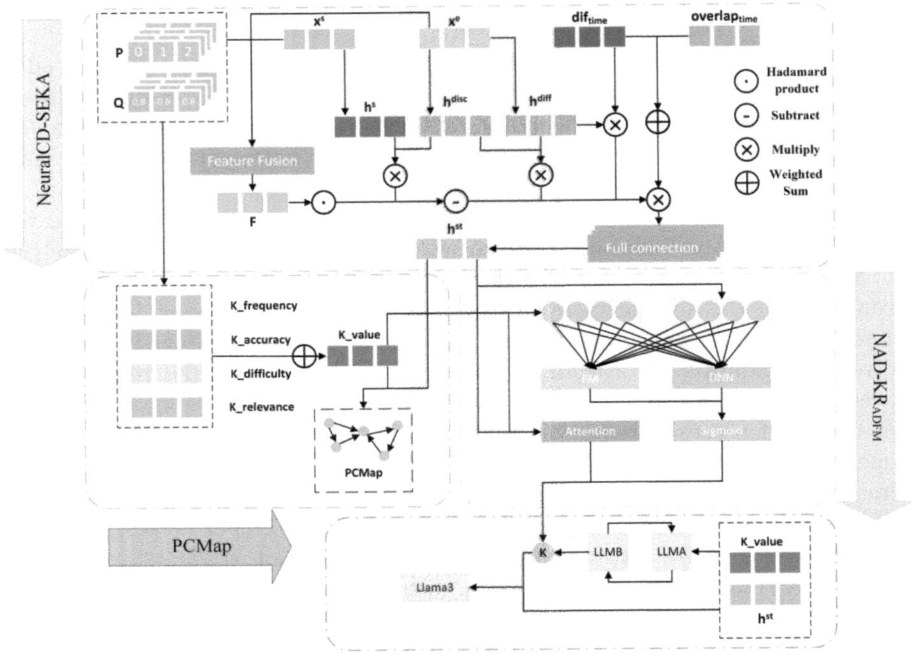

Fig. 1. Strucure of NAD-KRQG framework.

3.1 Task Overview

In a learning system, there are N students S=$\{s_1,s_2 \dots s_N\}$, M exercises E=$\{e_1,e_2 \dots e_M\}$, and K knowledge concepts C=$\{c_1,c_2 \dots c_K\}$. Each student selects some exercises for practice, represented as a triplet (s,e,r), where $s \in$ S, $e \in$ E, and r is the student's score.

3.2 NeuralCD—SEKA

We aim to further explore and strengthen the NeuralCD model's advantages in capturing complex nonlinear relationships between students and exercises, as well as providing personalized learning diagnostics. By introducing new data structures and considering additional factors that influence learning outcomes, an improved approach for constructing a personalized cognitive diagnosis model, NeuralCD-SEKA, is proposed to enhance diagnostic accuracy and the user experience. This model includes an association matrix layer, a fusion layer, and an output layer.

Association Matrix Layer. First, two association matrices are defined as follows.

Student – Exercise interaction association matrix:This matrix describes three states between students and exercises: answered correctly, answered incorrectly, and unanswered. The matrix is defined as $P{=}{P_{ij}}_{N\times M}$, where $P_{ij}{=}1$ indicates that student i answered exercise j correctly, $P_{ij}{=}0$ indicates an incorrect answer, and $P_{ij}{=}2$ indicates that the exercise was not attempted. Additionally, the matrix U represents unanswered exercises, with $U_i{=}P_i^2$ (where $P_{ij}{=}2$), and the matrix M represents incorrect answers, with $M_i{=}P_i^0$ (where $P_{ij}{=}0$). Matrices U and M are used in Sect. 4.3.1. This matrix setup allows for detailed recording of the three interaction states between students and exercises, providing foundational data for precise inference of cognitive states.

Exercise-knowledge point association matrix: This matrix reveals the connections between exercises and the knowledge points they involve. In this framework, an exercise may be associated with one or multiple knowledge points, and a knowledge point may also be associated with multiple exercises. The matrix $Q = {Q_{ij}}_{M\times K}$ (usually annotated by experts), where $Q_{ij} = D(e_i)(D(e_i) \in (0,1))$, represents exercise i containing knowledge point j, with $D(e_i)$ denoting the difficulty of the exercise. If $Q_{ij}{=}0$, the exercise does not include the knowledge point. Through this approach, the model can capture the direct connections between exercises and knowledge points, providing the necessary knowledge structure foundation for subsequent cognitive diagnosis. The formula for determining exercise difficulty is as follows:

$$D(e_i) = \frac{\sum score_{e_i}}{N_{e_i}} \tag{6}$$

where $socre_{e_i}$ represents the score of exercise e_i; and where N_{e_i} represents the total number of attempts at exercise e_i.

Fusion Layer. In the enhanced neural cognitive diagnosis model proposed in this paper, the fusion layer plays a crucial role. It integrates the student-exercise interaction association matrix P, the exercise-knowledge point association matrix Q, and the feature vectors x^s and x^e, of students and exercises, to construct a comprehensive feature representation. Specifically, the fusion layer concatenates these multisource data and processes them through a fully connected neural network layer, using the sigmoid activation function to generate a feature vector containing deep semantic information. This provides rich input information for the subsequent exercise time adjustment layer and output layer, which is crucial for achieving accurate cognitive diagnosis. This approach better captures students' learning behaviors and knowledge mastery, thus facilitating the design and recommendation of personalized learning paths. The fusion process can be represented as follows:

$$F = Sigmoid(W_f Concat(P, Q, x^e, X^s) + b_f) \tag{7}$$

where $Concat(\cdot)$ represents the concatenation operation, which concatenates the student-exercise interaction association matrix P, the exercise-knowledge point association matrix Q, the student feature vector x^s, and the exercise feature vector x^e into a large feature vector. W_f is the weight matrix of the fully connected layer, and b_f is the bias term of the fully connected layer.

Output Layer. Formula 5 predicts the proficiency of students in various knowledge concepts. However, the difficulty of knowledge points is relative. Different levels of difficulty reflect students' knowledge states: students who can correctly answer more difficult questions are generally considered to have a higher level of knowledge. Establishing the relationship between student knowledge states and problem difficulty is challenging, as the difficulty of problems can have complex effects on student learning. Owing to differences in students' knowledge states, their subjective perceptions of the difficulty of questions before practicing may also vary greatly [17]. As a common saying in China says, "Easy for the knowledgeable, difficult for the ignorant," this phenomenon is a true reflection of this situation.

Therefore, $overlap_{time}$ represents the total time a student spends on exercise, and dif_{time} represents the difference between the total time a student spends on exercise and the time spent on the first attempt, indicating that the student has attempted the exercise more than once. Combined with the output of the fusion layer, the formula for the final output layer is as follows:

$$x = F \odot ((\omega \cdot dif_{time} + (1 - \omega) \cdot overlap_{time}) \cdot (h^s - dif_{time} \cdot h^{diff}) \cdot h^{disc}) \tag{8}$$

where $\odot$ represents the Hadamard product.

To mitigate the impact of the range of dif_{time} on the difficulty of knowledge points and prevent overfitting, two types of dif_{time} are defined as follows:

$$dif_{time_{log}} = log(e - 0.5 + dif_{time}) \tag{9}$$

$$dif_{time_{exp}} = exp(dif_{time} - 0.5) \tag{10}$$

Finally, the student's latent knowledge state vector h^s when the question is unanswered is combined with the interaction diagnosis vector x to obtain the student's potential knowledge state vector after answering the question, denoted as h^{st}. Then, the latent knowledge state is combined with the difficulty vector of exercises h^{diff} to generate the interaction vector. Hidden information is learned through a fully connected layer, ultimately resulting in the predicted values of student responses to questions. The model adopts cross-entropy as the loss function.

3.3 PCMap

To construct a personalized student cognitive map, this paper integrates students' mastery of knowledge points and the evaluation of the importance of these points. This process begins by using the NeuralCD-SEKA model to obtain the latent knowledge states h^{st} of students for each knowledge point. In this model, the higher the value of the state vector is, the deeper the student's mastery of that knowledge point. On the basis of Bloom's Taxonomy of Cognitive Levels [18], these states are further classified into six levels: Remember, Understand, Apply, Analyse, Evaluate, and Create. This classification framework not only accurately describes the student's mastery of knowledge from basic memory to innovative creation, but also provides a basis for tailoring teaching strategies to meet different cognitive needs.

Furthermore, the difficulty of a knowledge point alone does not sufficiently reflect its content value. Therefore, the importance of a knowledge point (k_value) is assessed by considering four key attributes. These attributes include the frequency of the knowledge point's occurrence in exercises, the correctness rate of answers to exercises related to the knowledge point, the difficulty of the knowledge point, and the conceptual relevance of the knowledge point. Each of these attributes is elaborated upon below.

The frequency of occurrence in exercises (k_frequency) is calculated via the following formula:

$$I(e_i, k_j) = \begin{cases} 1, if & k_j \in K(e_i) \\ 0, & otherwise \end{cases} \tag{11}$$

$$F(k_j) = \frac{\sum_{i=1}^{M} I(e_i, k_j)}{M} \tag{12}$$

where K(e_i) represents the set of all knowledge points contained in exercise e_i.

The correctness rate of answers to exercises (k_accuracy) is calculated via the following formula:

$$A(k_j) = \frac{\sum_{e_i \in E_{k_j}} C_{e_i}}{\sum_{e_i \in E_{k_j}} T_{e_i}} \tag{13}$$

where E_{k_j} represents the set of all exercises containing knowledge points k_j, T_{e_i} represents the total number of answers to exercise e_i, and C_{e_i} represents the number of correct answers to exercise e_i.

The difficulty of a knowledge point (k_difficulty) is calculated via the following formula:

$$D(k_j) = 1 - \frac{\sum_{e_i \in E_{k_j}} \frac{C_{e_i}}{T_{e_i}}}{\mid E_{k_j} \mid} \tag{14}$$

The conceptual relevance of a knowledge point (k_relevance) is calculated via the following formula:

$$C(k_i, k_j) = \frac{\sum_{e \in E} I(e, k_i) \cdot I(e, k_j)}{M} \tag{15}$$

$$R(k_j) = \frac{\sum_{j \neq i} C(k_i, k_j)}{K - 1} \tag{16}$$

where $C(k_i,k_j)$ represents the conceptual relevance between knowledge points k_i and k_j ,and where R(k_i) is the average relevance of knowledge point k_j.

Finally, the importance of a knowledge point (k_value) is calculated via the following formula:

$$I(k_j) = \omega_f \cdot F(k_j) + \omega_a \cdot A(k_j) + \omega_d \cdot D(k_j) + \omega_r \cdot R(k_j) \tag{17}$$

where ω_f, ω_a, ω_d and ω_r are the weights for frequency, difficulty, accuracy, and average relevance, respectively. These weights are adjusted according to different needs.

By analysing these attributes, as shown in Fig. 2, we can identify knowledge points that play key roles in the learning process, including those that appear frequently, have high student answer accuracy rates, are relatively difficult, and are highly correlated with multiple other knowledge points. The conceptual relevance of a knowledge point involves identifying the core structure and key nodes of the subject system within the knowledge graph, which is crucial for developing effective teaching plans and promoting deep learning.

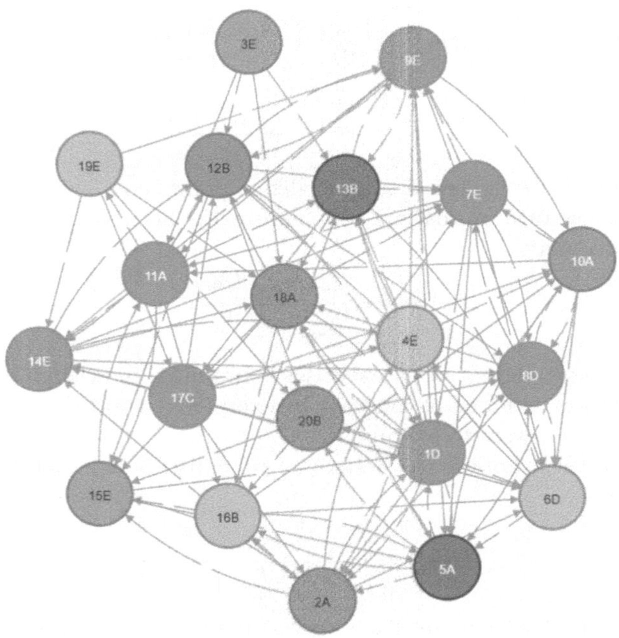

Fig. 2. In PCMap, green, blue, yellow, orange, and red are used to represent students' (Color figure online) mastery of knowledge points (h^{st}), corresponding to Create (0.95, 1), Evaluate (0.9, 0.95), Analyse (0.8, 0.9), Apply (0.7, 0.8), Understand (0.6, 0.7), and Remember (0, 0.6), respectively. The importance of knowledge points (k_value) is represented via A (0.9, 1), B (0.8, 0.9), C (0.7, 0.8), D (0.6, 0.7), E (0.4, 0.6), and F (0, 0.4). (Color figure online)

3.4 NAD-KR

In Sects. 3.2 and 3.3, the successful construction of the NeuralCD-SEKA model and the PCMap were achieved. These two technologies accurately capture students' cognitive states for various knowledge points. On the basis of this in-depth understanding, a guided personalized knowledge point recommendation model, Neural Association-based Cognitive Diagnosis with Knowledge Recommendation (NAD-KR), was proposed. The primary design intention of this model is traditional exercise recommendation, which often lacks the ability to guide students towards deep thinking and a comprehensive understanding of knowledge points. To more effectively promote students' deeper understanding and application of knowledge, a knowledge point recommendation technique was chosen, followed by problem generation on the basis of the recommended knowledge points. This method can generate personalized guided content according to the students' specific cognitive states, helping students not only understand knowledge points more comprehensively but also encourage them to actively think about how to apply what they have learned in real-world scenarios. In this way, the aim is to provide a more in-depth and interactive learning experience to promote students' cognitive development and deep learning.

We use three methods to calculate knowledge point recommendations: knowledge point recommendation, which is based on a deep factorization machine (DeepFM) [19], recommendation, which is based on large language models (LLMs), and recommendation, which is based on large language models with reflexions (LLMsR) [20]. After the knowledge points are obtained, problems are generated by fine-tuning large language models to achieve generative recommendation.

DeepFM is a knowledge point recommendation method that combines deep neural networks and factorization machines, but it has high computational complexity and insufficient handling of important features. To enable the model to pay more detailed attention to important features, we propose the attentional deep factorization Machine (ADFM), which enhances the processing capability for important features by introducing an attention mechanism.

First, in the feature extraction stage, the model represents the student's knowledge mastery and the importance of knowledge points as feature vectors and performs relevant preprocessing. Next, to improve the accuracy and personalization of recommendations, in the factorization stage, the deep neural network and factorization machine components decompose and combine the features to learn the interaction information between features. The formula is as follows:

$$v = \sigma(W_{deep}^{T} \cdot [h^{st}, I] + b_{deep}) \tag{18}$$

where σ denotes the sigmoid function, h^{st} is the student's knowledge state vector, I is the knowledge point importance vector, v is the hidden layer vector, W_{deep} is the weight parameter of the neural network, and b_{deep} is the bias parameter.

$$f_{FM} = \sum_{i=1}^{n} \sum_{j=i+1}^{n} < v_i, v_j > \tag{19}$$

where f_{FM} is the weighted sum of the feature interaction terms computed by the factorization machine component, which is used to capture high-order interaction information between features in the student feature vector v.

Different students may gain different benefits from answering the same question [21]. For example, a student who has already mastered related knowledge points will learn almost nothing from practicing simple questions, whereas a struggling student is likely to make greater progress. To improve the accuracy of recommendations, an attention mechanism is used to ensure that knowledge points with high mastery and low importance are recommended less frequently, and vice versa. The formula is as follows:

$$att = \sigma(W_{att}^T \cdot [h^{st}, I] + b_{att}) \tag{20}$$

$$att_{scaled} = att \odot (1 - I) \cdot (1 - h^{st}) + att \cdot I \cdot h^{st} \tag{21}$$

where W_{att} is the weight parameter of the neural network for calculating attention weights, b_{att} is the bias parameter, and $\odot$ denotes the Hadamard product.

Finally, in the output prediction stage, the output layer is used to predict the student's mastery of unmastered knowledge points, generating a corresponding knowledge point recommendation list. The formula is as follows:

$$\hat{y} = \sigma(W_{out}^T \cdot [h^{st} \cdot att_{scaled}, I \cdot att_{scaled}, v] + b_{out}) \tag{22}$$

Through these steps, the ADFM model can effectively analyse students' learning situations and provide personalized knowledge point recommendations, thereby enhancing students' learning outcomes.

The knowledge point recommendation method is based on LLMs and utilizes rapidly advancing natural language processing technology. In this method, the student's mastery of knowledge points and the importance of knowledge points are encoded into corresponding text sequences. These sequences, along with the relationships and content of the knowledge points, are input into the large language model. Through carefully designed prompt engineering, these text prompts can guide the language model to more accurately understand and respond to the specific needs of students, thereby generating personalized recommendation content that matches the student's state. The advantage of LLMs lies in their ability to process textual information in detail. It can comprehensively understand the student's learning status through prompt optimization technology, providing more targeted learning suggestions. Compared with traditional recommendation methods based on statistical features, LLMs can better capture the implicit semantic information between exercises, thus improving the accuracy and personalization of the recommendations. These LLMs, integrated with prompt engineering, offer a more personalized and intelligent learning experience, making the educational recommendation system more flexible and efficient in handling complex learning scenarios.

LLMsR is a further development based on LLMs that introduces a feedback iteration mechanism to improve the accuracy and robustness of recommendations. In this model, the student's knowledge mastery and the importance of knowledge points are first input into the initial large language model A, resulting in output Ar. This result Ar is then evaluated by another large language model B for its reasonableness, producing output Br. The feedback information from result Br is then passed back to the initial large language model A. After multiple iterations and exchanges, the final knowledge point recommendation list is obtained. Through this feedback iteration process, the model can better incorporate student feedback, enhancing the quality and effectiveness of recommendations and providing a superior learning experience for students.

4 Experiments

4.1 Dataset

We conducted experiments on two real-world datasets, ASSIST and Junyi. Both datasets contain records of learners' performance on exercises. Additionally, the Junyi dataset includes the content of the exercise knowledge points, facilitating generative recommendations. To ensure that each student has sufficient data for diagnosis, students with fewer than 15 response records were removed from each dataset, and duplicate data were eliminated. The complete statistical information of the datasets is shown in Table 1.

ASSIST (Assistments2009–2010 "skill builder") is a publicly available dataset sourced from the online tutoring platform ASSISTments [22]. This dataset provides students' answer records but does not publicly disclose the text of the questions or the relationships within the knowledge structure.

Junyi is an open dataset used for educational data mining and learning analytics. It originates from the Junyi Academy in Taiwan, a platform dedicated to providing personalized educational services. This dataset includes data on students' performance in different courses and subjects, course content, teaching methods, etc. It also provides expert-annotated dependency concepts, where are typically used for educational data analysis and learning pattern research.

Table 1. Dataset summary.

Dataset	ASSIST	Junyi
Students	4163	10000
Exercises	17746	710
Knowledge Concepts	123	710
Response logs	324572	330290
Knowledge concepts per exercise	1.19	1

4.2 Evaluation Metrics

Evaluating the performance of cognitive diagnostic models is challenging because it is difficult to obtain the true mastery level of students' knowledge. Since diagnostic results are typically obtained by predicting students' performance on most exercises, the performance of these prediction tasks can indirectly assess the model. Considering that all exercises used in the data are objective exercises, both classification and regression evaluation metrics are used, including Acc, RMSE, and AUC.

We define a metric called "reasonableness" to assess the effectiveness of knowledge point recommendations. Its definition is as follows.

$$Rationality = \begin{cases} 1, if \quad 0.7 < k\_value < 1 \\ 1, \quad if \quad 0.5 < h^{st} < 0.9 \\ 1, \quad if \quad k \quad in \quad U \, or \, E \end{cases} \tag{23}$$

Note: Rationality_3 indicates that all three conditions mentioned above are met, Rationality_2 indicates that only two of the conditions are met, Rationality_1 indicates that only one of the conditions is met, and Total is the sum of all three.

4.3 Experimental Results

Results and Analyses of Cognitive Diagnosis. To validate the practical effectiveness of the model proposed in this paper, NeuralCD-SEKA is compared with several other CD models and subjected to ablation experiments, as follows:

(1) IRT [4]: As one of the most popular CD methods, it models the features of one-dimensional students and performs exercises through linear functions.
(2) DINA [3]: This method, which is based on maximum likelihood estimation, combines the binomial logistic model of item response and the latent class distribution of student skills to estimate students' skill levels and item characteristic parameters.
(3) MIRT [5]: An extension of IRT to multiple dimensions, modelling multiple levels of student proficiency and exercise.
(4) NeuralCDM [6,7]: One of the latest deep learning CD models, uses neural networks to model high-order complex student-exercise interactions.
(5) RCD [23]: Understands and evaluates students' learning states by constructing and utilizing relational graphs.

The ablation experiments include methods such as NCD-SEKA, NCD-SEKA$_x$, NCD-SEKA$_{log}$, and NCD-SEKA$_{exp}$. NCD-SEKA only considers the association matrix without considering the time spent on exercises; NCD-SEKA$_x$ considers both the association matrix and the time spent on exercises; NCD-SEKA$_{log}$ considers the association matrix and modifies the time spent on exercises according to formula (9); and NCD-SEKA$_{exp}$ considers the association matrix and modifies the time spent on exercises according to formula (10). The

Table 2. Experimental results of cognitive diagnosis models.

Dataset	ASSIST			Junyi		
	ACC	RMSE	AUC	ACC	RMSE	AUC
IRT	66.26%	46.49%	68.83%	67.56%	42.74%	76.83%
DINA	64.73%	46.92%	67.69%	66.37%	43.48%	76.02%
MIRT	70.70%	46.17%	71.94%	75.01%	40.92%	79.14%
NeuralCDM	72.09%	43.96%	74.31%	74.35%	42.78%	66.92%
RCD	73.35%	42.14%	77.23%	77.21%	39.54%	**82.56%**
NCD-SEKA	73.26%	43.42%	73.97%	75.61%	40.22%	76.13%
NCD-SEKA$_x$	74.33%	43.14%	74.26%	**79.13%**	**38.67%**	78.45%
NCD-SEKA$_{log}$	75.37%	42.46%	76.13%	78.81%	38.68%	78.17%
NCD-SEKA$_{exp}$	**78.33%**	**40.23%**	**81.28%**	78.61%	39.10%	76.96%

experimental results are shown in Table 2, where all the results are averaged over ten repetitions.

In the experiments, overlap$_{time}$ and dif$_{time}$ in formula (8) are within the range of (0,1), and ω is set to 0.5. In formulas (9) and (10), dif$_{time}$ is within the range of (0,1), the log is a natural logarithm, and the range of dif$_{time_{log}}$ is (0.801, 1.169), whereas the range of dif$_{time_{exp}}$ is (0.607, 1.649). In formula (17), the weights are set as ω_f=0.3, ω_a=0.3, ω_d=0.4, and ω_r=0.6.

The experimental results, as shown in Table 2, indicate that on the ASSIST-dataset, the NCD-SEKA$_x$ model outperforms the other models in terms of Acc, but slightly underperforms RCD in terms of the RMSE and AUC, although it still outperforms several baseline models. Both NCD-SEKA$_{log}$ and NCD-SEKA$_{exp}$ outperform NCD-SEKA$_x$ on all three metrics, with NCD-SEKA$_{exp}$ showing significant improvements of 4%, 2.91%, and 7.02% in terms of the Acc, RMSE, and AUC, respectively. For the Junyi dataset, NCD-SEKA$_x$ outperforms all other models in terms of Acc and RMSE but slightly underperforms RCD in terms of AUC. Additionally, NCD-SEKA$_{log}$ and NCD-SEKA$_{exp}$ perform worse than NCD-SEKA$_x$ does on all three metrics, indicating that the NCD-SEKA$_x$ model exhibits strong performance when handling the Junyi dataset. Figure 3 illustrates the experimental results for each model.

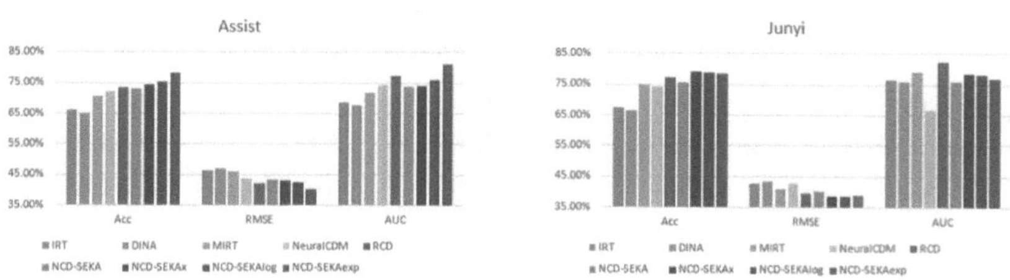

Fig. 3. Histogram of the results of the cognitive diagnosis models.

In response to the unusually outstanding performance of NCD-SEKA$_{exp}$ on the ASSIST dataset, we conducted multiple verifications. Figure 4 displays the results of 10 experiments with NCD-SEKA$_{exp}$, which show minimal fluctuations across different experiments. Furthermore, on the Junyi dataset, the performance of NCD-SEKA$_{exp}$ was even inferior to that of NCD-SEKA$_{x}$. Hence, the possible reason for the outstanding results is that Formula 10 aligns better with the characteristics and distribution of the Assist dataset.

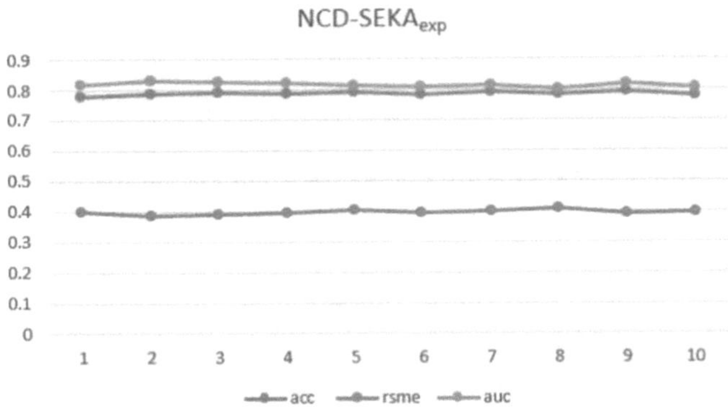

Fig. 4. Line graph of the NCD-SEKA$_{exp}$ validation experiments.

Results and Analyses of Knowledge Recommendation. To validate the effectiveness of the NAD-KR model, we conducted experiments comparing three knowledge recommendation methods proposed in this paper before and after. Specifically, we compared directly the DeepFM model, LLMs, and LLMs_R for knowledge recommendation with the addition of knowledge recommendation via NAD-KR, as well as the NAD-KRADFM proposed in this paper. Since question generation is required after recommendation, experiments were conducted only on the Junyi dataset. Table 3 presents the knowledge recommendation results.

Table 3. Experimental results of the knowledge recommendation models.

Dataset	Junyi			
	Rationality_1	Rationality_2	Rationality_3	Total
DeepFM	45.45%	38.80%	6.24%	90.49%
NAD-KR$_{DeepFM}$	47.25%	40.72%	7.42%	95.39%
NAD-KR$_{ADFM}$	42.65%	43.54%	10.25%	96.44%
LLMs	48.80%	31.20%	5.00%	85.00%
NAD-KR$_{LLMS}$	52.00%	36.20%	6.80%	95.00%
LLMsR	43.20%	39.60%	8.20%	91.00%
NAD-KR$_{LLMsR}$	45.80%	42.40%	9.20%	97.40%

The experimental results, as shown in Table 3, indicate that all the experiments with NAD-KR added for knowledge recommendation outperform direct recommendation via the original models. In addition to performance metrics, an analysis of the recommended knowledge points was also conducted. Figure 5 shows the results of 1000 experiments for the NAD-KR$_{DeepFM}$ and NAD-KR$_{ADFM}$. The recommendation results of NAD-KR$_{DeepFM}$ are noticeably concentrated in a smaller range of exercise IDs, whereas the recommendation results of NAD-KR$_{ADFM}$ are more evenly distributed, indicating that NAD-KR$_{ADFM}$ can better dynamically learn and utilize the importance of features, thereby achieving a more balanced distribution of recommendation results.

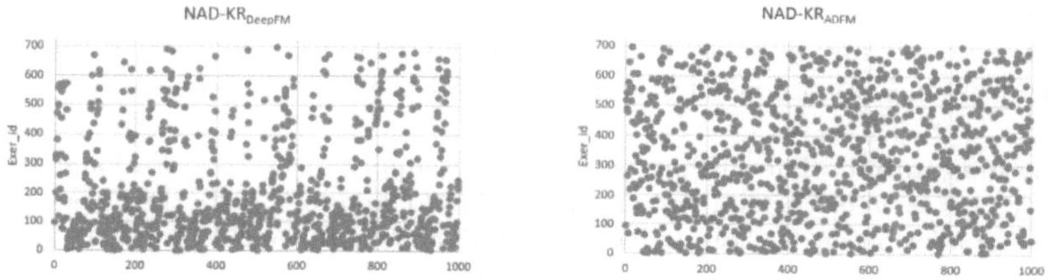

Fig. 5. Scatterplot of 1000 experimental results for NAD-KR$_{DeepFM}$ vs NAD-KR$_{ADFM}$.

The LLMs used in this paper were all based on Llama3,which is open-sourced by Meta [24, 25]. Its advantage lies in its open-source nature, enabling some fine-tuning of the model, and its performance surpasses that of non-open-source GPT4 in text generation. In the experiments mentioned above, LLMs directly acquire students' exercise records and the relationships between exercises, whereas NAD-KR$_{LLMs}$ acquire only students' mastery of exercises and the importance of knowledge points. Similarly, the LLMsR and NAD-KR$_{LLMsR}$ follow the same logic.

Finally, the obtained knowledge points were handed over to the fine-tuned Llama3 for question generation. The generated sentences are more fluent and better achieve the purpose of this paper: guided learning. The generated questions are not simple exercises but can guide students in thinking and help them understand the knowledge points.

Table 4. Dataset summary.

h^{st}	Recommended KP	Generated Question
0.45	Area calculation	Compare areas of 8 m×5 m rectangle...

As shown in Table 4, Question targets Bloom's Apply level (aligned with h^{st} = 0.45).A total of 89.2% of the generated questions matched the target cognitive levels per expert review.

5 Conclusion

In conclusion, this paper proposes a novel personalized learning assistance model, NAD-KRQG, that emphasizes the timeliness and relevance of educational interventions. By using NeuralCD-SEKA, the model captures student learning behavior and knowledge mastery status. It provides personalized learning paths through a visualized PCMap, constructs NAD-KR_{ADFM} for precise knowledge point recommendations, and generates guided questions with LLM. This approach ensures real-time feedback and dynamic adjustment of learning materials, optimizing learning paths and strategies. The performance of the proposed model significantly improves in both cognitive diagnosis and knowledge recommendation, effectively addressing the limitations of traditional cognitive diagnosis models and educational methods by considering individual differences in learning behavior, knowledge structure, and cognitive processes.

Acknowledgement. The authors acknowledge support from the National Natural Science Foundation of China under Grants 62277030; and 62077029, the Society Development Foundation of Xuzhou under Grant KC23239 and the Postgraduate Research & Practice Innovation Program of Jiangsu Province SJCX24_1525.

References

1. Shemshack, A., Spector, J.M.: A systematic literature review of personalized learning terms. Smart Learn. Envir. **7**(1), 1–20 (2020). https://doi.org/10.1186/s40561-020-00140-9
2. Wu, R., et al.: Cognitive modelling for predicting examinee performance, In: Twenty-Fourth International Joint Conference on Artificial Intelligence (2015)
3. De La Torre, J.: Dina model and parameter estimation: a didactic. J. Educ. Behavioral Stat. **34**(1), 115–130 (2009)
4. Embretson, S.E., Reise, S.P.: Item response theory. Psychology Press (2013)
5. Ackerman, T.A., Gierl, M.J., Walker, C.M.: Using multidimensional item response theory to evaluate educational and psychological tests. Educ. Meas. Issues Pract. **22**(3), 37–51 (2003)
6. Li, G., et al.: Neuralncd: A neural network cognitive diagnosis model based on multi-dimensional features. Appl. Sci. **12**(19), 9806 (2022)
7. Wang, F., Liu, Q., Chen, E., Huang, Z., Yin, Y., Wang, S., Su, Y.: Neuralcd: a general framework for cognitive diagnosis, IEEE Trans. Knowl. Data Eng. (2022)
8. Walker, A., Recker, M.M., Lawless, K., Wiley, D.: Collaborative information filtering: a review and an educational application. Int. J. Artif. Intell. Educ. **14**(1), 3–28 (2004)
9. Pazzani, M.J., Billsus, D.: Content-based recommendation systems, In: The Adaptive Web: Methods and Strategies of Web Personalization. Springer, pp. 325–341 (2007)
10. Linden, G., Smith, B., York, J.: Amazon. com recommendations: Item-to-item collaborative filtering. IEEE Internet Comput. **7**(1), 76–80 (2003)
11. Liu, G., Hao, T.: User-based question recommendation for question answering system. Int. J. Inf. Educ. Technol. **2**(3), 243 (2012)

12. Klašnja-Milićević, A., Ivanović, M., Nanopoulos, A.: Recommender systems in e-learning environments: a survey of the state-of-the-art and possible extensions. Artif. Intell. Rev. **44**(4), 571–604 (2015). https://doi.org/10.1007/s10462-015-9440-z
13. Chang, P.-C., Lin, C.-H., Chen, M.-H.: A hybrid course recommendation system by integrating collaborative filtering and artificial immune systems. Algorithms **9**(3), 47 (2016)
14. Wang, X., Wang, D., Xu, C., He, X., Cao, Y., Chua, T.-S.: Explainable reasoning over knowledge graphs for recommendation. Proc. AAAI Conf. Artif. Intell. **33**(01), 5329–5336 (2019)
15. Wu, Z., Li, M., Tang, Y., Liang, Q.: Exercise recommendation based on knowledge concept prediction. Knowl.-Based Syst. **210**, 106481 (2020)
16. Zhu, T., et al.: Cognitive diagnosis based personalized question recommendation. Chinese J. Comput. **40**(1), 176–191 (2017)
17. Knäuper, B., Belli, R.F., Hill, D.H., Herzog, A.R.: Question difficulty and respondents' cognitive ability: the effect on data quality. J. Official Stat. Stockholm **13**, 181–199 (1997)
18. Adams, N.E.: Bloom's taxonomy of cognitive learning objectives. J. Med. Lib. Assoc. JMLA **103**(3), 152 (2015)
19. Guo, H., Tang, R., Ye, Y., Li, Z., He, X.: Deepfm: a factorization-machine based neural network for CTR prediction, arXiv preprint arXiv:1703.04247 (2017)
20. Shinn, N., Cassano, F., Gopinath, A., Narasimhan, K., Yao, S.: Reflexion: Language agents with verbal reinforcement learning, Advances in Neural Information Processing Systems, vol. 36 (2024)
21. Long, T., Liu, Y., Shen, J., Zhang, W., Yu, Y.: Tracing knowledge state with individual cognition and acquisition estimation, In: Proceedings of the 44th International ACM SIGIR Conference on Research and Development in Information Retrieval, pp. 173–182 (2021)
22. Feng, M., Heffernan, N., Koedinger, K.: Addressing the assessment challenge with an online system that tutors as it assesses. User Model. User-Adap. Inter. **19**, 243–266 (2009)
23. Gao, W., et al.: Rcd: Relation map driven cognitive diagnosis for intelligent education systems, In: Proceedings of the 44th International ACM SIGIR Conference on Research and Development in Information Retrieval, pp. 501–510 (2021)
24. Touvron, H., et al.: Llama: Open and efficient foundation language models, arXiv preprint arXiv:2302.13971 (2023)
25. Huang, W., et al.: How good are low-bit quantized llama3 models? an empirical study, arXiv preprint arXiv:2404.14047 (2024)

NOTEARS-cMLP:Learning Nonparametric DAGs with Domain Knowledge

Xinyi Lu[1], Xuedong Chen[1(✉)], Bin Zhang[1], Jiancheng Xu[2], Yuping Zhao[2], Yiying Wang[2], Guan Liu[2], Ruiquan Jiang[2], and Xianfeng Huang[2]

[1] School of Science, Huzhou University, Huzhou, Zhejiang 313000, China
xdchen@zjhu.edu.cn
[2] Shanghai Kingstar Software Technology Co., Ltd, Shanghai 201203, China

Abstract. Time series causal discovery is key to understanding causal mechanisms among variables in dynamic systems. However, most of the previous causal discovery methods implicitly assume the inability to incorporate expert domain knowledge, which limits the effectiveness of their application to some extent. In this paper, we propose a score-based nonparametric structural learning method, NOTEARS-cMLP, specifically for time series data to accurately discover instantaneous and lagged causal relationships among variables. The NOTEARS- cMLP method uses cMLP to simulate nonlinear dependencies among time series variable relationships and incorporates NOTEARS ideas to describe causal structure learning as a continuous optimization problem with acyclic constraints. In addition, the method further improves the accuracy and efficiency of dependency identification by incorporating do-main knowledge as a constraint into the L-BFGS-B optimization algorithm. Empirical evaluations of simulated data demonstrate the effectiveness and robustness of the NOTEARS- cMLP for the learning of DAGs, and an applied study on the attribution of gold futures trading prices is conducted.

Keywords: Causal discovery · Time series · Nonparametric DAGs · NOTEARS-cMLP · Domain knowledge

1 Introduction

Causality discovery in time series is important for scientific research and practical applications and is widely used in fields such as molecular biology [7,32], climate science [17] and financial economics [18,27]. Despite being the most reliable way to identify causal relationships, interventions or random-sized controlled trials often face challenges such as high importance complexity, significant costs, and ethical concerns in some cases. As a result, causal discovery methods based on observational data to make causal inferences have emerged and gained widespread attention over the past few decades [13].

X. Chen—Research direction: Machine Learning and Financial Statistics.

L. Huang et al. (Eds.): CCF NCCA 2025, CCIS 2774, pp. 72–88, 2026.
https://doi.org/10.1007/978-981-95-5294-8_6

A key problem in causal discovery is the learning of a Directed Acyclic Graph (DAG), i.e. discovering predictive relationships between child nodes and their parents [30]. Most causal relationship discovery techniques are typically based on the premise that before a study begins, we have no knowledge of the underlying causal structures, and all possible causal paths are assumed to be equally possible. However, in practice, domain experts are often able to understand the causal relationships between certain variables on the basis of past experimental results or profound predictive knowledge. For example, they may be convinced that there is a direct causal relationship between two specific variables, while some variables do not have a causal relationship or that there is a certain causal hierarchy between variables [16]. Many time series analysis models require researchers to preset model parameters. A lack of sufficient domain knowledge to guide the selection of these parameters may lead to improper model settings, which in turn can affect the structural accuracy of the DAG [8,36].

For example, in credit risk assessment, traditional causal discovery methods rely on historical data to establish linear relationships between borrower characteristics and fault risk [29]. These methods struggle to capture complex nonlinear interactions, such as the income-to-loan ratio or the interaction between credit scores and loan terms. They also underutilize expert knowledge on industry-specific default risks during economic downturns. Nonparametric modelling offers a solution by circumventing the need for stringent distributional assumptions, thereby providing a more precise representation of the data's skewness and discreteness.

In recent years, score-based causal discovery methods have shown significant advantages in the field of time series. Unlike constrained methods that rely on conditional independence tests, score-based methods avoid the cumulative error problem of CI tests by optimizing the scoring function (e.g., Bayesian information criterion, likelihood function) to search directly for the optimal causal structure [34]. In this work, we develop a new method to learn nonparametric DAGs, which uses a componentwise MLP (cMLP) [31] to predict the values of child nodes from their parent nodes. The cMLP architecture is designed to model complex nonlinear relationships in the input data and is therefore particularly suitable for time series data.

In this work, we focus on learning from nontemporal data by employing the NOTEARS DAG learning approach, which describes the learning of the DAG model as a continuous optimization search for acyclic weight matrices [35,36]. Domain knowledge is also efficiently and logically integrated into the L-BFGS-B optimization algorithm. We believe that the proposed method enhances the universality of score-based causal discovery methods and further promotes the integration of theory and practical applications. Overall, the contributions of this work are as follows:

1. We develop a nonparametric DAGs method based on cMLP for capturing linear and nonlinear relationships between variables, as well as instantaneous causality and lagged causality in time-series data;

2. We propose a new method, NOTEARS-cMLP, to transform the DAG learning problem into an optimizable acyclic continuous optimization problem;
3. We show how domain knowledge can be transformed into optimization constraints to effectively integrate domain knowledge in the model learning process.

2 Related Work

2.1 Structural Causal Model

The structural causal model (SCM) is the core framework of causal inference theory and was proposed by Judea Pearl; it describes causal relationships among variables through di-rected acyclic graphs (DAGs) and structural equation modelling (SEM). The central idea is to decouple statistical dependencies from causal mechanisms through independence assumptions (causal Markovianity, causal fidelity) in graphical models. Pearl's hierarchical theory categorizes causal problems into three levels: correlation (seeing), intervention (doing), and counterfactual (imagining), which correspond to observational data, intervention experiments, and counterfactual inference, respectively [24]. In time series analysis, the SCM is extended to dynamic causal modelling by introducing time-indexed variables(X_t and $Y_{t-\tau}$), which can address lag effects and feedback loops. Interventional inference methods such as backdoor adjustment and frontdoor adjustment are extended to the time dimension to remove confounding bias. Dominguez-Olmedo et al. [9] further demonstrated the recognizability of the SCM for nonstationary data manifolds, providing theoretical guarantees for causal interpretation in machine learning.

2.2 Score-Based Causal Discovery

Score-based causal discovery methods search for optimal causal graphs by optimizing scoring functions. In time series, such methods need to address the challenges of autocorrelation and time lag dependence. Classical algorithms such as greedy equivalent search (GES) maximize scores by adding/removing edges incrementally, but their search space grows exponentially with the number of variables [2]. The LiNGAM model proposed by Hyvärinen, A. et al. [15] combines independent component analysis (ICA) to address nonGaussian noise and extends it to VarLiNGAM with time lags. Pamfil et al. [23] proposed the DYNOTEARS algorithm to estimate the structure of acyclic causal networks from time series data by combining autoregressive modelling with regularized optimization while capturing the instantaneous and time-lagged effects among variables. Nauta et al. [22] introduced deep learning to propose a convolutional neural network (TCDF) model based on an attention mechanism capable of dealing with high-dimensional nonstationary time series data. The further NTiCD framework combines LSTM and GCN to learn nonlinear causal structures from multivariate time series [1]. The core advantage of such methods is the automatic learning of time-invariant features through neural networks, which avoids the limitation of manually designing the lag order.

Domain knowledge can constrain the search space to improve efficiency.Hasan et al. [14] proposed the KGS algorithm, which narrows the candidate graph space by a priori causal edges (known unconnected or forcedly connected) and experimentally proved to converge more than three times faster than traditional GES. The CAM-UV model, through a priori temporal order constraints (the cause must be earlier than the effect), significantly improves the accuracy of causal direction identification [21].

2.3 Constraint-Based Causal Discovery

The constraint-based approach uses conditional independence tests to exclude causal diagrams that do not match data dependencies. For time series characteristics, Runge et al. [26] developed the PCMCI algorithm to obtain high-dimensional strong autocorrelation data through hybrid conditional independence tests. Its extended version, PCMCI+ [25] further supports the detection of transient causality. For nonsmooth time series, the CD-NOTS algorithm introduces time-indexed nodes as founding proxies and separates smooth and nonsmooth causal effects by conditional independence tests, which performs well in financial data [27].

Domain knowledge improves the effectiveness of constraint tests. Wang et al. [33] encode traffic topology as causal path constraints in the analysis of delay in a public transit network, which enables the PCMCI algorithm to reduce the false detection rate by 42%. In climate science, the eCDANs algorithm successfully captures time-varying causalities in nonstationary climate data by introducing categorical modes as proxy variables [11]. In addition, the latent PCMCI (LPCMCI) combines latent variables (latent confounders) to improve the accuracy of causality detection [12].

3 Methodology

The goal of causal discovery is to identify the correct parameter set $Pa_{\mathcal{G}}(X^i)$ for each node X^i in a directed acyclic graph $\mathcal{G}(V, E)$, thereby fully recovering the true graph structure while maintaining acyclicity, using observed time series data. In this graph, nodes $V = X = \{X_t^1, ..., X_t^d\}, t = 1, ..., T$ represent the time series, and edges E, such as (X^i, X^j), show a dependency from parent X^i to child X^j. Dependencies at the same time step $(X_t^i \rightarrow X_t^j)$ are called instantaneous, whereas those with a lag $X_{t-k}^i \rightarrow X_t^j (k > 0)$ are lagged. To facilitate this discovery, we typically assume that the underlying data generation process is stationary and model it as a K-order Markov process, where K is a hyperparameter.

Following zheng et al. [36], the DAG learning problem can be formulated as follows: We consider a data matrix consisting of n independent and identically distributed (i.i.d.) observations $X = [x^1| \cdots |x^d] \in \mathbb{R}^{n \times d}$. Our goal is to learn a directed acyclic graph DAG $\mathcal{G}(X)$ that can capture the dependency relationships between variables in X. We learn a set of functions $f = (f_1, ..., f_d)$ to achieve

$\mathcal{G}(f) = \mathcal{G}(X)$. Given a loss function $l(y, \hat{y})$ such as least squares, The following optimization problem is considered:

$$\min_f L(f) \text{ subject to } \mathcal{G}(f) \in DAG, \quad \text{where } L(f) = \frac{1}{n}\sum_{j=1}^{d} l(x^j, f_j(X)) \tag{1}$$

To ensure the acyclicity of $\mathcal{G}(f) \in DAG$, partial derivatives are used to measure the dependence of f_j on the sth variable. Assume that $f_j \subset H^1(\mathbb{R}^d)$ is a function of the Sobolev space of square integrable functions and let $\partial_s f_j$ denote the partial derivative with respect to X^s. It is easy to show that f_j is independent of X^s if and only if $\|\partial_s f_j\|_{L^2} = 0$, where $\|\cdot\|_{L^2}$ is the usual L2-norm. This implies that the matrix $W(f) = W(f_1, ..., f_d) \in \mathbb{R}^{d \times d}$, where

$$[W(f)]_{sj} = \partial_s f_j L^2 \tag{2}$$

accurately encodes the dependency structure between X^j. Therefore, the training objective is equivalent to:

$$\min_{W \in \mathbb{R}^{d\times d}} L(W) \quad \text{subject to } h(W) = 0 \tag{3}$$

$h(W) = \text{tr}(e^{W \circ W}) - d$, where $[W \circ W]_{sj} = w_{sj}^2$, and $tr(\cdot)$ and $e^{\cdot}$ representing the trace and exponential function of the matrix respectively.

3.1 NOTEARS-cMLP

The componentwise MLP (cMLP) is a method for modelling multivariate time series, the core idea of which is to construct a separate multilayer perceptron (MLP) for each output component, thus separating the effects of the input variables on the output. cMLP trains an MLP network g_i independently for each output time series X_t^i, and the input to each network are past observations (including their own lagged values) of all other time series. This design allows the network to focus on analysing causal relationships between specific output variables and input variables, avoiding interferences between different outputs, and is suitable for scenarios requiring causal inference, such as gene regulation and economic forecasting. Consider a cMLP with h hidden layers and a single activation $\sigma : \mathbb{R} \to \mathbb{R}$ modeling time series data X, given by

$$\begin{aligned} cMLP(X) = \text{Concat}([&\sigma(A_1^{(h)}\sigma(\cdots\sigma(A_1^{(1)}X^1))) \\ &, ..., \sigma(A_d^{(h)}\sigma(\cdots\sigma(A_d^{(1)}X^d)))]) \end{aligned} \tag{4}$$

$A_j^{(h)}$ is the weight matrix of the hth layer of the jth time series, and all MLP networks have the same structure [19]. The first layer is a one-dimensional convolutional layer with a convolutional kernel size of time lag $K+1$. The subsequent layers are multiple one-dimensional convolutional layers with a convolutional kernel size of 1. An activation function is applied between each convolutional layer. We jointly train d cMLPs, the jth cMLP outputs the expected value of X_t^j for $t \geq K+1$, using as input the variables from previous and the same time steps:

$$\mathbb{E}[X_t^j | Pa(X_t^j)] = cMLP_j(\{X_{t-k} : 1 \leq k \leq K\}, X_t^{-j}) \tag{5}$$

where $Pa(X_t^j)$ denotes the parent node of X_t^j defined by the trained cMLP. Here K represents the maximum lag, meaning that the inputs to the predictor variable X_t^j include all variables prior to the K, as well as all variables at the same time step except for X^j. cMLP's weight on the subvariable X^j is denoted as θ_j. Figure 1 shows our NOTEARS-cMLP architecture. The first layer of each MLP is a convolutional layer with convolutional kernel size $K+1$, and subsequent layers are convolutional layers with kernel 1. The weights of the first convolutional layer are extracted and combined to obtain the dependencies of each variable on the others, forming a weight matrix of shape $d \times (K+1)$, where the last column $K+1$ denotes the instantaneous dependencies.

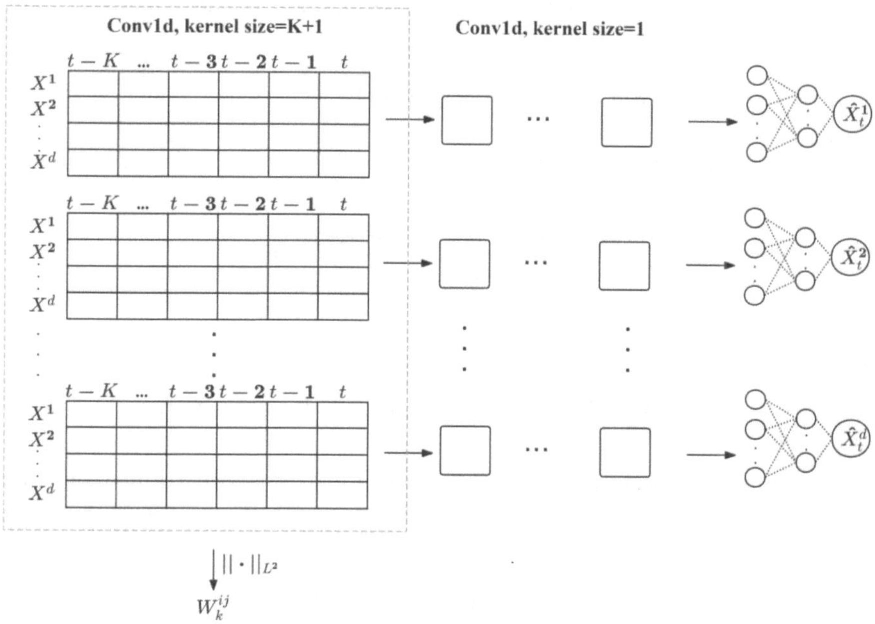

Fig. 1. NOTEARS-cMLP architecture for time series variables. The W_k^{ij} denotes the dependence strength from the time series X^i to the time series X^j. This value is obtained by computing the L2-norm of the convolutional kernel parameters of the first layer of the corresponding cMLP network.

Derive the weighted adjacency matrix defining the graph structure according to equation (2). Let $\varphi_k^{i,j} \subset \theta_j$ represent the set of first-layer weights within the jth cMLP corresponding to the input variable X_k^i, which are combined to form m weight parameters. Each entry W_k^{ij} in the weighted adjacency matrix W quantifies the dependency strength of the directed edge from variable X_k^i to variable X_{K+1}^j.To assess the reliance of one variable on another through an edge, the L2-norm of the weights connecting them is computed:

$$W_k^{ij} = ||\varphi_k^{i,j}||_{L^2} \quad \text{for} \quad k = 1, ..., K+1 \tag{6}$$

As noted in the upcoming theorem and supported by Zheng et al. [36], the extraction of model weights from the first layer of each MLP in a cMLP does not diminish the model's expressive power.

Theorem 1. *If the class of functions $\mathcal{F}$ independent of all cMLPs of X_k^i, and the class of functions $\mathcal{F}_0$ whose weights are all zero in the k-th columns of the first layer of convolution kernel of the m-th MLP network, then, $\mathcal{F} = \mathcal{F}_0$.*

According to the requirements of the above DAG learning problem, the cMLP weights are penalized for sparsity, and the regularization terms of all the parameters, which guarantee the acyclicity constraints [30], are specified as the following training objectives:

$$\min_{\theta} F(\theta) \quad \text{subject to} \quad h(W^{K+1}) = 0 \tag{7}$$

where $\theta = (\theta_1, ..., \theta_d)$, $h(W^{K+1}) = \text{tr}(e^{W^{K+1} \circ W^{K+1}}) - d$, $\varphi_j^k = [\varphi_{i,j}^k]$, and

$$F(\theta) = L(cMLP) + \sum_{k=1}^{K+1} \lambda_1^k ||\varphi_j^k||_{L^1} + \frac{1}{2}\lambda_2 ||\theta_j||_{L^2},$$

$$L(cMLP) = \frac{1}{T-K} \sum_{t=K+1}^{T} \sum_{j=1}^{d} l(X_t^j, cMLP_j(\{X_{t-k} : 1 \leq k \leq K\}, X_t^{-j})) \tag{8}$$

The augmented Lagrangian method transforms the constrained optimization problem into an unconstrained one.The actual training objective function becomes:

$$\min_{\theta} F(\theta) + \frac{\rho}{2}|h(W^{K+1})|^2 + \alpha h(W^{K+1}) \tag{9}$$

To optimize this unconstrained objective function, the L-BFGS-B algorithm [5,6,37] is employed. Converting the target into a box-constrained form can be applied directly [36].

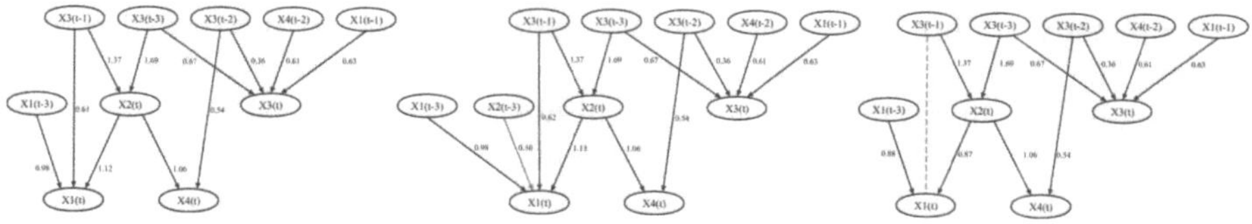

Fig. 2. Results of integrating additive and forbidden domain knowledge constraints in NOTEARS-cMLP causal discovery on a 4-dimensional synthetic dataset. With the temporal lag set to 3, the diagram simplifies node representation by omitting variables without direct causal connections. The arrow indicates the direction of the causal relationship and the number on the arrow indicates the strength of the causal relationship. *Left:* Estimation without domain knowledge shows no edge between X_{t-3}^2 and X_t^1. *Middle:* Estimation with the added edge $X_{t-3}^2 \rightarrow X_t^1$ (weight $= 0.3$). *Right:* Estimation with the forbidden edge $X_{t-1}^3 \rightarrow X_t^1$, where gray dotted lines indicate absent dependencies.

3.2 Domain Knowledge Optimization Constraints

Numerous methods for uncovering causality presume that each possible causal configuration is initially given the same probability [8,20]. However, in real-world scenarios, expert subjects often possess valuable insights drawn from their experiences, enabling them to develop an understanding of specific causal dynamics. Consequently, incorporating such expertise could enhance both the computational effectiveness and precision of causal detection efforts. In our study, we summarize commonly used domain knowledge settings and identify the following two common strategies:

Definition 1 (Add edges). *According to domain knowledge, the presence of a known directed edge $X^i \rightarrow X^j$ necessitates that X^i is an element of the parent set of X^j, i.e. $X^i \in Pa_{\mathcal{G}}(X^j)$.*

Definition 2 (Forbid edges). *Domain knowledge encodes forbidden edges and hierarchical relationships between two variables X^i and X^j, and domain knowledge encodes hierarchies $S = S_1, S_2, ..., S_n$, which specify the causal hierarchy of a variable, suggesting that the forbiddenness of a variable in a later hierarchy is the causal driver of a variable in an earlier hierarchy. This implies that in a temporal causality framework, the temporal dependence between variables follows a strict forward causality constraint, i.e., future states are not causally retrospective to historical states. It can be categorized into the following three scenarios:*

1. *Given X^i in tier S_u and X^j in tier S_v with $u < v$, the structure permits X^i to be the parent of X^j, yet explicitly forbids X^j from being the parent of X^i;*
2. *If X^i and X^j are located in the same hierarchy S_u, then either variable may be the parent of the other;*
3. *Furthermore, irrespective of their hierarchical positions, should domain knowledge explicitly forbid an edge between X^i and X^j, it follows that neither variable can be the parent of the other, i.e.,$X^i \notin Pa_{\mathcal{G}}(X^j)$ and $X^j \notin Pa_{\mathcal{G}}(X^i)$.*

This study adopts the following basic assumption for main knowledge integration: first, domain knowledge is inherently consistent in the sense that when a disabling edge constraint is set between variables, there must not be concurrent explicit causal edge declarations that contradict it (e.g., forbidding $X^i \rightarrow X^j$ and adding $X^i \rightarrow X^j$ cannot coexist). Second, the presupposed domain knowledge is asymptotic, i.e., all causal relations specified a priori reflect the true underlying data generation mechanism. Finally, explicit causal edges and disabling edges are treated as hard constraints, requiring that the causal structure of the algorithm's output must strictly satisfy these predefined topological relationships [16]. Together, these three assumptions form the theoretical foundation of domain knowledge-guided causal discovery, ensuring that a priori information is effectively integrated with the data-driven learning process.

Applying domain knowledge in the learning of causally structured DAGs can involve defining an allowable interval for the dependency weight W_{ij}^k connecting two variables at a fixed lag. A known edge $W_{ij}^k \geq W_{bound}^k$, which indicates that the importance of the current dependency is not less than a threshold set by the domain knowledge and a forbidden edge $W_{ij}^k = 0$, which prohibits the dependency. The estimated dependency strength W_{ij}^k on an edge is equal to the L2-norm of the corresponding cMLP combination weight. Let b be the user-specified domain knowledge dependency strength, m be the number of MLP layer 1 kernels in each cMLP network, and $\hat{b}$ be the transformed optimization constraint. Each b is scaled as follows before being applied to the L-BFGS-B algorithm.

The estimated dependency strength W_{ij}^k for an edge is calculated as the L2-norm of the relevant cMLP combination weight. We define b as the user-specified domain knowledge dependency strength, m as the quantity of MLP layer 1 kernels in every cMLP network, and $\hat{b}$ as the transformed optimization constraint. The scaling applied to each b before its use in the L-BFGS-B algorithm is described below:

$$\hat{b} = \sqrt{\frac{b^2}{m}} \tag{10}$$

The primary benefit of the L-BFGS-B algorithm, as a class of proposed Newtonian optimization methods, is reflected in the efficient approximate use of second-order derivative information. The algorithm stores historical gradient data through a finite-memory strategy, which exhibits superior convergence properties compared with the conventional first-order gradient descent method and can usually approximate the extreme points of the objective function with fewer iteration steps. In particular, the algorithmic framework supports bounded constraint settings in the parameter space, allowing the user to flexibly specify the feasible domain ranges of the parameters in each dimension, including a mix of fully free and restricted variable configurations. This bounded optimization capability gives it a unique computational advantage in dealing with nonlinear problems with constraints [30]. For each bounded parameter, certain upper and lower bound constraints need to be satisfied, whose lower and upper bounds are determined by a priori knowledge and actual data characteristics, respectively, which is important for the control of a reasonable range of parameters in neural network training. Setting constraints allows us to integrate a priori knowledge into the algorithm, such that $\tilde{\theta}$ denotes the constraint parameter, whose lower and upper bounds are l and u, where the upper and lower bounds are derived from the domain knowledge $\hat{b}$. The objective function of the optimization transforms to:

$$\min_{l \leq \tilde{\theta} \leq u} F(\theta) + \frac{\rho}{2}|h(W^{K+1})|^2 + \alpha h(W^{K+1}) \tag{11}$$

The model minimum bound is set to zero, which arises from the constraining effect of the adopted L2-norm regularization term on the model parameters, while ensuring that all edge weight estimates maintain non-negative properties. For the time-lagged causal network modeling needs, if a directed connection from the time-series starting variable X^i to X^j is to be enforced, this can be achieved by imposing the parameter constraints $\tilde{\theta} \geq l$, where the lower bound l is a strictly positive real number and a threshold condition of $l \geq W^k_{bound}$ is to be satisfied. Figure 2 illustrates the results of integrating additive and forbidden domain knowledge constraints into the NOTEARS-cMLP framework.

4 Experiments

In this section, we demonstrate the experimental effectiveness of the NOTEARS-cMLP method proposed in causal discovery. To evaluate its performance, we compared it with two baseline methods, DYNOTEARS and PCMCI+. The evaluation process is divided into two stages. In the first stage, we test the effectiveness of the method on a synthetic dataset generated by a random DAG. In the second stage, we explore the application of this method to real data of gold futures trading prices.

4.1 Synthetic Data

For the synthetic data experiments, we adopted the following setup: DAGs with an average degree of 2 were randomly generated, with the dimension d varying among 5, 10, 15, and 20, and the sequence length was uniformly set to $T = 1000$. The parameters for the structural causal model (SCM) were randomly determined from a uniform distribution within the interval $[0.3, 0.8]$. For evaluation, each test was run 100 times, and the average results were calculated. Under the true model setting, the number of lags was set to 3. Data simulation was based on a random ground-truth DAG using one of two identifiable nonlinear SCMs. The first of these is the additive noise model (ANM):

$$X_t^j = f_j(Pa(X_t^j)\theta_1)\theta_2 + \epsilon_t^j \tag{12}$$

The other is the additive exponential model (AIM) :

$$X_t^j = \epsilon_t^j + \sum_{m=1}^{3} h_m(Pa(X_t^j)\theta_m) \tag{13}$$

where, $Pa(X_t^j)$ is the parent of X_t^j, f_j is the sigmoid function, ϵ_t^j is the standard Gaussian noise, and each θ is uniformly sampled from the closed interval $[-2, -0.5] \cup [0.5, 2]$, $h_1 = tanh, h_2 = cos, h_3 = sin$.

The assessment metrics are defined based on the accuracy of identifying causal structures using 4 metrics: True Positive Rate (TPR), False Positive Rate (FDR), Precision and F_1 Score (F_1).

Table 1. Results for applying NOTEARS-cMLP causal discovery with two other baseline methods DYNOTEARS and PCMCI+ on the synthetic data set when incorporating different types of nonlinear SCMs.

Nonlinear SCMs		AIM			ANM		
Metrics	d	Model1	Model2	Model3	Model1	Model2	Model3
TPR	5	0.82	0.87	0.88	0.88	1.00	0.66
	10	0.88	0.78	0.54	0.80	0.88	0.85
	15	0.77	0.78	0.57	0.65	0.63	0.73
	20	0.80	0.67	0.73	0.70	0.69	0.57
FPR	5	0.00	0.21	0.35	0.00	0.32	0.42
	10	0.00	0.43	0.58	0.01	0.27	0.83
	15	0.00	0.65	0.67	0.02	0.61	0.72
	20	0.00	0.61	0.72	0.02	0.56	0.58
Precision	5	1.00	0.72	0.69	1.00	0.72	0.80
	10	1.00	0.57	0.55	0.87	0.57	0.57
	15	0.97	0.65	0.61	0.90	0.61	0.61
	20	1.00	0.60	0.60	0.86	0.58	0.58
F_1	5	0.90	0.83	0.78	0.93	0.83	0.72
	10	0.93	0.72	0.54	0.83	0.72	0.69
	15	0.86	0.79	0.59	0.76	0.76	0.66
	20	0.89	0.75	0.66	0.77	0.74	0.57

*Model 1, Model 2 and Model 3 represent NOTEARS-cMLP, DYNOTEARS, and PCMCI+, respectively.

Results. In this study, on the basis of the results in Table 1, we systematically compare the performance of three algorithms, NOTEARS-cMLP, DYNOTEARS, and PCMCI+, under different dimensions ($d = 5 - 20$) and nonlinear structural causal models. The experimental results show that NOTEARS-cMLP exhibits excellent causal discovery under all tested dimensions, with a true positive rate (TPR) of 0.88 in low dimensions ($d = 5, 10$), while maintaining an excellent false positive rate (FPR) of 0.00. Notably, the algorithm achieves a precision of 1.00 in the low dimensionality condition, demonstrating extremely high accuracy in causal edge identification. In contrast, DYNOTEARS and PCMCI+ perform relatively poorly on the TPR and FPR metrics, suggesting that they may have some limitations in modelling complex nonlinear relationships.

In terms of comprehensive performance, NOTEARS-cMLP shows excellent balance under both low and high dimensionality conditions, combining high TPR and low FPR characteristics. Although its TPR decreases in high-dimensionality ($d = 15, 20$) scenarios, it still maintains a high F1 value, indicating good algorithmic robustness. In contrast, although DYNOTEARS and PCMCI+ perform moderately well under some specific conditions, their reliability needs to be improved further when dealing with high-dimensional data. The results of this study provide an important reference basis for selecting causal discovery algorithms suitable for different scenarios.

4.2 Real-World Gold Futures Data

We apply NOTEARS-cMLP to real-world data from February 27, 2000, to February 25, 2024, on the trading price of U.S. gold futures. The target variable is the trading price of U.S. gold futures, and the candidate influencing factors are analysed through extensive literature research and research reports, including macroeconomic indicators such as interest rates, treasury yields, and inflation rates and indicators such as the world's major currencies against the price of gold, as well as indicators of COT data for trader position reporting. After screening to obtain the following 5 related factors, the granularity of the data for the weekly frequency data was 1252. Among them, the COT data come from the official website of the U.S. CFTC, and all the remaining data come from the Wind Information Co., Ltd. (Wind) database. To simplify the subsequent causal diagrams, the target variables and characteristics are symbolized as shown in Table 2.

Table 2. Description of real gold futures trading price variables.

Influencing Factors	Notation
U.S. Gold Futures Trading Price	Y
Comm_Net_Position-098662-FutureOptions-Report	X1
Gold-cot-index-gold-price-Gold COT Index (L)	X2
Comm_COT_index-088691-FutureOptions-Report	X3
NonComm_Net_Pct-088691-FutureOptions-Report	X4
3month-bond-real-yield-gold-price-Gold Futures Price (R)	X5

The preprocessing pipeline includes (1) identifying missing data and outliers, (2) performing an ADF unit root test to check for stationarity, (3) applying differencing operations iteratively to any nonstationary series until stationarity criteria are met, and (4) standardizing the resulting dataset.

After relevant research, we determine that the U.S. gold futures trading price is affected by multiple factors: Comm_Net_Position by reflecting the industry chain supply and demand expectations and the gold price is negatively correlated, the extreme position predicts the price inflection point; Gold COT Index as a market sentiment indicator, non-commercial long surge to promote the short-term rise, but the high level is easy to trigger the reversal; Comm_COT_index reflects the entity's hedging needs, the increase or decrease of its short position directly affects the market liquidity and price pressure; NonComm_Net_Pct reinforces the short-term volatility, the positive correlation is significant and the extreme value indicates that the market is overbought/over-sold; 3month-bond-real-yield-gold-price-Gold Futures Price determines the opportunity cost of holding gold, and the long-term negative correlation dominates the macro trend. In summary, commercial positions and interest rates anchor the direction of fundamentals, whereas noncommercial positions amplify short-term volatility, requiring multidimensional verification [3,4,10,28].

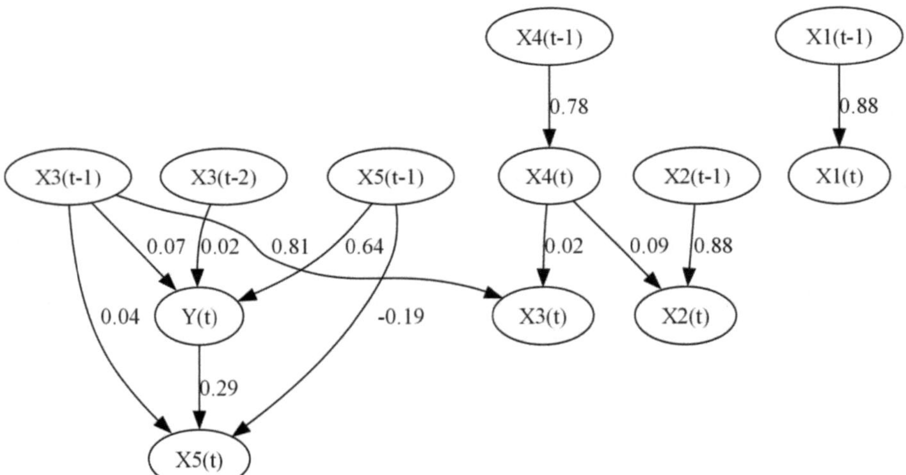

Fig. 3. Causal discovery results of gold futures pricing factors using DYNOTEARS algorithm with no domain knowledge.

In this work, the key value of domain knowledge integration for causal modelling of time series is revealed through a comparative analysis of the performance differences between two causal discovery algorithms. Figure 3 illustrates the results of the causal discovery of gold futures trading prices and their influencing factors via the DYNOTEARS algorithm. In the absence of domain knowledge, the DYNOTEARS algorithm identifies causal relationships between multiple variables, such as the lagged effects of $X3$ and $X5$ on the gold futures trading

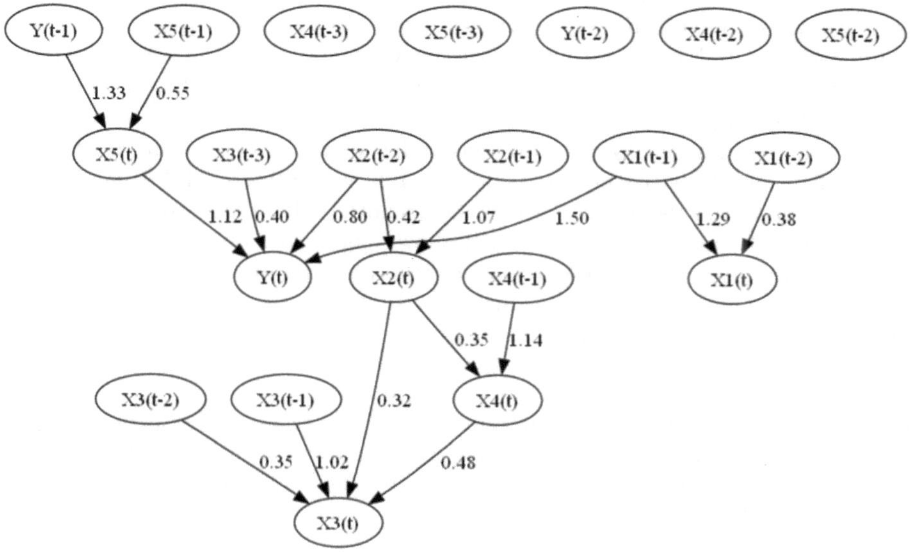

Fig. 4. Causal discovery results of gold futures pricing factors using NOTEARS-cMLP algorithm with domain knowledge. Incorporated constraints: Added edges $X_{t-2}^2 \to Y_t$ and $X_{t-1}^1 \to Y_t$; Forbidden edge $X_t^2 \to Y_t$.

price Y_t. However, due to the lack of domain knowledge, some causal relationships may not be accurate enough, resulting in redundant causal paths in the model. For example, the figure shows the direct causality from X_t^2 to Y_t, and it is clear that there is a significant time lag characteristic of the Gold COT Index and the gold futures price, rather than an instantaneous effect. Figure 4 demonstrates the results of causal discovery of gold futures trading prices and their influences using the NOTEARS-cMLP algorithm. Unlike the DYNOTEARS algorithm, the NOTEARS-cMLP algorithm integrates domain knowledge by adding $X_{t-2}^2 \to Y_t$ and $X_{t-1}^1 \to Y_t$, which are relationships that may be ignored or misidentified in the DYNOTEARS algorithm. In addition, the NOTEARS-cMLP algorithm avoids possible false positives by disabling the edge $X_t^2 \to Y_t$, which further improves the reliability of the model.

5 Conclusions

This paper describes the use of different types of main knowledge for learning nonparametric DAGs via NOTEARS-cMLP, a score-based structural learning method that uses cMLP to learn time series data and is designed to be independent of a priori assumptions about causality. The learned DAG captures both instantaneous and lagged causality. This study demonstrates how the NOTEARS continuous optimization method can be adapted to the cMLP model to address the learning problem of in-slice edge parameters by introducing an acyclic constraint mechanism based on weight matrices. On the basis of the simulated data, NOTEARS-cMLP has several advantages in terms of the quality of the DAG structure learning over multiple comparative methods, which provides empirical support for integrating a priori knowledge using optimization constraints. The application of the method to a complex real-world financial dataset demonstrates its potential value, especially when dealing with high-dimensional nonlinear financial time series data, which shows good adaptability. On the basis of existing findings, future research will focus on expanding the application of this method in nonstationary scenarios, with the core goal being to solve the problem of causal graph learning in the presence of confounding factors, thereby enhancing the breadth of its application.

Acknowledgments. This work was supported by the Open Research Fund of Key Laboratory of Advanced Theory and Application in Statistics and Data Science âĂŞ-MOE,E CNU(KLATASDS2008), Postgraduate Research and Innovation Project of Huzhou University(2025KYCX51).

Disclosure of Interests. The authors declare no competing interests relevant to the content of this article.

References

1. Absar, S., Wu, Y., Zhang, L.: Neural time-invariant causal discovery from time series data. In: 2023 International Joint Conference on Neural Networks (IJCNN), pp. 1–8. IEEE (2023)
2. Ali, S., et al.: Causality for earth science–a review on time-series and spatiotemporal causality methods. arXiv preprint arXiv:2404.05746 (2024)
3. Batten, J.A., Ciner, C., Lucey, B.M.: The macroeconomic determinants of volatility in precious metals markets. Resour. Policy **35**(2), 65–71 (2010)
4. Büyükşahin, B., Harris, J.H.: Do speculators drive crude oil futures prices? Energy J. **32**(2), 167–202 (2011)
5. Byrd, R.H., Lu, P., Nocedal, J., Zhu, C.: A limited memory algorithm for bound constrained optimization. SIAM J. Sci. Comput. **16**(5), 1190–1208 (1995)
6. Byrd, R.H., Nocedal, J., Schnabel, R.B.: Representations of quasi-newton matrices and their use in limited memory methods. Math. Program. **63**(1), 129–156 (1994)
7. Cai, Z., Xi, D., Zhu, X., Li, R.: Causal discoveries for high dimensional mixed data. Stat. Med. **41**(24), 4924–4940 (2022)
8. Constantinou, A.C., Guo, Z., Kitson, N.K.: The impact of prior knowledge on causal structure learning. Knowl. Inf. Syst. **65**(8), 3385–3434 (2023)
9. Dominguez-Olmedo, R., Karimi, A.H., Arvanitidis, G., Schölkopf, B.: On data manifolds entailed by structural causal models. In: International Conference on Machine Learning, pp. 8188–8201. PMLR (2023)
10. Erb, C.B., and, C.R.H.: The golden dilemma. Fin. Anal. J. **69**(4), 10–42 (2013). https://doi.org/10.2469/faj.v69.n4.1
11. Ferdous, M.H., Hasan, U., Gani, M.O.: ecdans: Efficient temporal causal discovery from autocorrelated and non-stationary data. In: Proceedings of the AAAI Conference on Artificial Intelligence. vol. 37, pp. 16208–16209 (2023)

12. Gerhardus, A., Runge, J.: High-recall causal discovery for autocorrelated time series with latent confounders. Adv. Neural. Inf. Process. Syst. **33**, 12615–12625 (2020)
13. Glymour, C., Zhang, K., Spirtes, P.: Review of causal discovery methods based on graphical models. Front. Genet. **10**, 524 (2019)
14. Hasan, U., Osman Gani, M.: Kgs: Causal discovery using knowledge-guided greedy equivalence search. arXiv e-prints, pp. arXiv–2304 (2023)
15. Hyvärinen, A., Zhang, K., Shimizu, S., Hoyer, P.O.: Estimation of a structural vector autoregression model using non-gaussianity. J. Mach. Learn. Res. **11**(5) (2010)
16. Kleinegesse, S., Lawrence, A.R., Chockler, H.: Domain knowledge in a*-based causal discovery. arXiv preprint arXiv:2208.08247 (2022)
17. Kretschmer, M., et al.: Quantifying causal pathways of teleconnections. Bull. Am. Meteor. Soc. **102**(12), E2247–E2263 (2021)
18. Kumar, S., Vivek, Y., Ravi, V., Bose, I.: Causal inference for banking finance and insurance a survey (2023). https://arxiv.org/abs/2307.16427
19. Liang, Y., Zadorozhnyi, O., Drton, M.: Kernel-based differentiable learning of non-parametric directed acyclic graphical models. arXiv preprint arXiv:2408.10976 (2024)
20. Lu, N.Y., Zhang, K., Yuan, C.: Improving causal discovery by optimal Bayesian network learning. In: Proceedings of the AAAI Conference on Artificial Intelligence, vol. 35, pp. 8741–8748 (2021)
21. Maeda, T.N., Shimizu, S.: Use of prior knowledge to discover causal additive models with unobserved variables and its application to time series data. Behaviormetrika, pp. 1–19 (2024)
22. Nauta, M., Bucur, D., Seifert, C.: Causal discovery with attention-based convolutional neural networks. Mach. Learn. Knowl. Extraction **1**(1), 19 (2019)
23. Pamfil, R., et al.: Dynotears: Structure learning from time-series data. In: International Conference on Artificial Intelligence and Statistics, pp. 1595–1605. Pmlr (2020)
24. Pearl, J., et al.: Causal Models, Reasoning and Inference. Cambridge University Press, Cambridge, UK (2000)
25. Runge, J.: Discovering contemporaneous and lagged causal relations in autocorrelated nonlinear time series datasets. In: Conference on Uncertainty in Artificial Intelligence, pp. 1388–1397. Pmlr (2020)
26. Runge, J., Nowack, P., Kretschmer, M., Flaxman, S., Sejdinovic, D.: Detecting and quantifying causal associations in large nonlinear time series datasets. Sci. Adv. **5**(11), eaau4996 (2019)
27. Sadeghi, A., Gopal, A., Fesanghary, M.: Causal discovery from nonstationary time series. Int. J. Data Sci. Anal. **19**(1), 33–59 (2025)
28. Sanders, D.R., Irwin, S.H.: A speculative bubble in commodity futures prices? Cross-sectional evidence. Agric. Econ. **41**(1), 25–32 (2010)
29. Skripnikov, A., Michailidis, G.: Joint estimation of multiple network granger causal models. Econ. Stat. **10**, 120–133 (2019)
30. Sun, X., Schulte, O., Liu, G., Poupart, P.: Nts-notears: Learning nonparametric dbns with prior knowledge. arXiv preprint arXiv:2109.04286 (2021)
31. Tank, A., Covert, I., Foti, N., Shojaie, A., Fox, E.B.: Neural granger causality. IEEE Trans. Pattern Anal. Mach. Intell. **44**(8), 4267–4279 (2021)
32. Triantafillou, S., Lagani, V., Heinze-Deml, C., Schmidt, A., Tegner, J., Tsamardinos, I.: Predicting causal relationships from biological data: Applying automated

causal discovery on mass cytometry data of human immune cells. Sci. Rep. **7**(1), 12724 (2017)
33. Wang, W., She, J.: Exploring bus network delay propagation: Integration of causal inference and complex network theory (2024)
34. Waxman, D., Butler, K., Djurić, P.M.: Dagma-dce: Interpretable, non-parametric differentiable causal discovery. IEEE Open J. Sig. Process. **5**, 393–401 (2024)
35. Zheng, X., Aragam, B., Ravikumar, P.K., Xing, E.P.: Dags with no tears: Continuous optimization for structure learning. Adv. Neural Inf. Process. Syst. **31** (2018)
36. Zheng, X., Dan, C., Aragam, B., Ravikumar, P., Xing, E.: Learning sparse nonparametric dags. In: International Conference on Artificial Intelligence and Statistics, pp. 3414–3425. Pmlr (2020)
37. Zhu, C., Byrd, R.H., Lu, P., Nocedal, J.: L-bfgs-b: Fortran subroutines for large-scale bound-constrained optimization. ACM Trans. Math. Softw. (TOMS) **23**(4), 550–560 (1997)

Humanoid Robots Gate-Passing Method Based on SAC Algorithm with Generative Adversarial Networks

Zhiyuan Chen, Jianhua Dong, Guodong Zhao(✉), Mingshuo Liu, and Anonymous Identification

College of Computer Science and Technology, Harbin Engineering University, Harbin, China
{chenzhiyuan,dongjianhua,zhaoguodong,liumingshuo,322060042}@hrbeu.edu.cn

Abstract. In recent years, humanoid robots have demonstrated significant potential in completing complex tasks in dynamic environments. Gate-passing is a challenging task that requires precise motion control and decision-making capabilities. This paper proposes a humanoid robots gate-passing method based on the Soft Actor-Critic (SAC) algorithm combined with Generative Adversarial Networks (GAN). By incorporating a GAN module to generate adversarial samples, the SAC algorithm receives additional feedback, enhancing the stability and efficiency of policy generation. Experiments conducted on the Webots simulation platform using the DarwinOP2 humanoid robots demonstrate that the GAN-SAC method significantly accelerates convergence and improves task success rates compared to traditional SAC. This study addresses the challenges of unstable policy generation and inefficient exploration in complex dynamic environments, while significantly enhancing the robot's task adaptability in random scenarios.

Keywords: Humanoid Robots · Soft Actor-Critic · Generative Adversarial Networks · Gate-Passing · Deep Reinforcement Learning

1 Introduction

Humanoid robots, with their human-like structure and motion capabilities, have demonstrated extensive potential in modern society across various domains, including industrial automation, disaster response, and human-robots interaction. Meanwhile, robotic competitions play a crucial role in advancing humanoid robots technology by providing standardized platforms for evaluation and development. One notable example is the internationally recognized HuroCup, which aims to assess humanoid robots' capabilities across a range of complex tasks, including obstacle run, balance beam, and ladder climbing. Among these, the obstacle run task places significant demands on the robots' motion control and autonomous decision-making abilities, making it a key focus in research on humanoid robot technologies.

L. Huang et al. (Eds.): CCF NCCA 2025, CCIS 2774, pp. 89–103, 2026.
https://doi.org/10.1007/978-981-95-5294-8_7

The gate-passing task, a critical sub-task within the obstacle run event, requires robots to navigate through narrow passages while avoiding obstacles and ensuring safe traversal. This task imposes higher demands on a robot's adaptability in dynamic environments and serves as a key benchmark for evaluating motion control and task execution capabilities. Traditional rule-based robot control methods typically rely on predefined motion strategies or static environmental models. While effective in fixed scenarios, these methods often struggle with adaptability in dynamic and randomly changing environments. Furthermore, as task complexity increases, traditional approaches face scalability and efficiency challenges, making them insufficient to meet the demands of higher-level technical requirements.

Soft Actor-Critic (SAC) [1], as a representative algorithm in deep reinforcement learning, has been widely applied to robotic motion control tasks due to its unique advantages. SAC leverages the Maximum Entropy Framework to achieve efficient exploration in continuous action spaces while ensuring robust policy optimization. These characteristics have made SAC one of the mainstream choices for robotic control tasks. However, applying SAC to gate-passing tasks in dynamic environments still faces several key challenges. Frequent state changes in dynamic environments can disrupt SAC's policy optimization process, leading to instability in policy learning and reduced task performance. SAC relies heavily on random exploration, which results in low exploration efficiency in sparse-reward tasks such as gate-passing, thereby slowing down convergence. Furthermore, the standard SAC algorithm lacks effective mechanisms to adapt to dynamically changing scenarios, such as randomly configured gates, limiting its generalization capability across diverse task settings.

To address these challenges, this paper proposes a novel framework called GAN-SAC, which integrates Generative Adversarial Network (GAN) [2] into the SAC algorithm. As shown in Fig. ??, the framework enhances the policy generation and exploration capabilities of SAC through the GAN module. Specifically, the Generator G generates adversarial samples $G(z)$ from noise z, enriching the diversity of the Actor's policy and encouraging exploration in more challenging scenarios. The Discriminator D evaluates the quality of the generated samples and produces an auxiliary reward signal r_{gan}, which, combined with the SAC reward, guides the policy optimization. Furthermore, the GAN module and SAC module are updated alternately, enabling the Generator and policy networks to be optimized collaboratively, thereby improving the model's adaptability in dynamic environments.

The main contributions of this paper are as follows:

- A novel GAN-SAC framework is proposed, which combines Generative Adversarial Networks with the Soft Actor-Critic algorithm to address the issues of unstable policy generation and low exploration efficiency in dynamic environments.
- The GAN-SAC method is validated in random gate-passing scenarios, demonstrating significant improvements in task adaptability, faster convergence, and higher task success rates compared to traditional SAC methods.

2 Related Work

In recent years, humanoid robots have demonstrated significant potential in performing complex tasks within dynamic environments. A critical research challenge in such tasks is enabling robots to adapt to environmental changes and generate diverse strategies.

References [3,4] applied the SAC algorithm to the field of robotic control, significantly improving the learning efficiency and stability of humanoid robots in complex tasks by optimizing the reward function and policy network. Their method demonstrated excellent performance in structured environments; however, the exploration of strategy diversity and adaptability in dynamic scenarios remains insufficient, limiting its application in rapidly changing real-world situations. References [5–8] proposed an obstacle avoidance method for hexapod robots based on the SAC algorithm, leveraging internal sensors to perceive the environment and overcoming the limitations of external sensors under adverse weather and lighting conditions. This study highlighted the advantages of the SAC algorithm in motion control within unstructured environments. However, when faced with sudden environmental changes or unpredictable obstacles in dynamic scenarios, there is still room for improvement in the stability of strategy generation.

To enhance motion control of humanoid robots under high stability constraints, references [9–12] proposed the DanceHAT framework. This framework integrates adversarial training and employs stability and similarity constraints, enabling humanoid robots to stably perform complex dance motions. While the method demonstrates strong performance in static tasks, it primarily focuses on highly constrained scenarios, leaving its applicability in dynamic and sparse reward environments largely unexplored.

Despite the significant progress achieved by the aforementioned studies, existing methods still exhibit limitations in dynamically changing environments or tasks with sparse rewards. Many approaches suffer from insufficient generalization across different scenarios, often relying on predefined behavior models or static data, which restricts their flexibility in real-world applications. Moreover, the lack of mechanisms for generating diverse strategies further limits the adaptability of these methods in rapidly changing environments.

Based on the analysis of existing works, this paper proposes an innovative GAN-SAC framework, which integrates Generative Adversarial Networks (GAN) with the SAC algorithm to address challenges in strategy generation stability, exploration efficiency, and adaptability in dynamic environments. By generating diverse adversarial samples and incorporating auxiliary reward signals, the framework significantly enhances strategy diversity and task success rates. Compared to traditional methods, the GAN-SAC framework demonstrates remarkable advantages in dynamic and stochastic environments for gate-passing tasks.

3 Methods

In the humanoid robot gate-passing task, the robot must utilize visual information to perceive the relative position and orientation of the gate while dynamically adjusting its gait to ensure safe passage. The main challenges of this task include the random variation in the gate's position and orientation, requiring the robot to quickly perceive and adapt to the dynamic environment, and the need to complete the task within a limited time while avoiding falls caused by gait adjustment errors. To address these challenges, this paper proposes a GAN-SAC framework that integrates Generative Adversarial Networks (GAN) with the Soft Actor-Critic (SAC) algorithm to enhance the stability of policy generation and the efficiency of exploration in dynamic environments. The GAN-SAC framework consists of two core modules, the SAC module and the GAN module, which work collaboratively to optimize policy learning for humanoid robots in the gate-passing task.

3.1 Problem Definition

To enable the DARwIn robot to autonomously pass through gates, we constructed a simulated task environment to model complex scenarios. The simulation environment is designed based on the obstacle run event in the FIRA HuroCup competition, faithfully replicating the key elements of the competition tasks. The field measures 200 cm $\times$ 300 cm, offering flexible configuration and high fidelity. A coordinate system was established as shown in the Fig. 1a, with the robot's starting position set at coordinate point (1.5, 1.1).

In our experiments, we designed two scenarios: fixed gate positions and random gate positions. Figure 1a shows the overall visualization of the fixed gate task scenario, while Fig. 1b illustrates the overall visualization of the random gate task scenario. The task requires the humanoid robot to capture visual information through its camera, locate the relative position and orientation of the gate, autonomously plan its path, and adjust its gait to safely pass through the gate. This task is modeled as a deep reinforcement learning problem, aiming to enable the robot to learn optimal strategies through interactions with the environment to handle different gate configurations and dynamic challenges. This modeling approach not only enhances the realism of the simulation environment but also provides a reliable experimental platform for the control and strategy optimization of humanoid robots in dynamic scenarios.

The state space S includes the environmental visual input captured by the robot's camera. Specifically, the state is represented as an image with dimensions $160 \times 120 \times 3$, where the three channels correspond to RGB values.

To reduce computational complexity and improve the efficiency of model training, the raw images collected were preprocessed. Each frame was resized to 84×84 pixels to decrease the computational load and ensure the processing speed meets the requirements of reinforcement learning algorithms. Subsequently, the images were converted from the RGB color space to grayscale

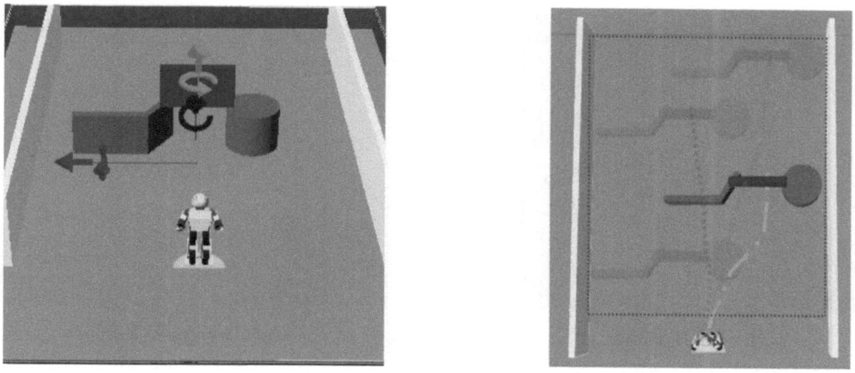

(a) Fixed Gate. (b) Random Gate.

Fig. 1. Simulation environments.

[13]. This step reduces data complexity while retaining sufficient environmental information, particularly important shapes and contours.

To enable the agent to perceive dynamic changes and motion characteristics in the environment, a technique of stacking four consecutive frames was employed. This means that each state includes not only the current frame but also the previous few frames, thereby providing the agent with rich temporal information and motion features. This approach helps the agent better understand object motion and environmental changes, thereby improving decision-making accuracy. The RGB images captured by the humanoid robot's built-in camera and the corresponding grayscale images after conversion are shown in Fig. 2a and Fig. 2b

The action space A consists of three discrete actions: (1)moving through the gate (the primary goal); (2)turning left to adjust the robot's alignment; (3)turning right to adjust the robot's alignment. These actions enable the robot to make dynamic decisions based on environmental observations.

The reward function is designed to effectively guide the robot's learning process by rewarding desired behaviors and penalizing undesired ones: the robot receives a reward of 100 for successfully moving forward through the gate without falling. If the robot falls during any action, the current episode ends immediately, and a penalty of -100 is applied. Each time the robot performs a turning action (left or right), a small penalty of -10 is applied to discourage unnecessary adjustments. If the robot exceeds the maximum time step $T > 20$ in an episode, a penalty of -50 is applied, and the episode ends. In addition, if the robot performs a crawling action but fails to pass through the gate successfully, a penalty of -3 is applied. The reward function encourages the robot to prioritize passing through the gate while penalizing inefficiencies or unsafe behaviors, like falling or excessive adjustments. The reward function is as follows:

(a) RGB Image (b) Grayscale Image

Fig. 2. RGB image captured by the onboard camera and the corresponding grayscale image after preprocessing.

$$R(s,a) = \begin{cases} +100, & \text{pass gate successfully without falling} \\ -100, & \text{falling} \\ -10, & \text{turn left or right} \\ -3, & \text{crawl but fail to pass the gate} \\ -50, & \text{exceed time limit } (T > 20) \\ 0, & \text{otherwise.} \end{cases} \tag{1}$$

3.2 Soft Actor-Critic Algorithm

SAC is a model-free, off-policy reinforcement learning algorithm that utilizes the Maximum Entropy Framework to simultaneously optimize the expected return and the entropy of the policy, achieving a balance between exploration and exploitation [14]. This dual optimization enables SAC to achieve efficient and stable learning in complex tasks. Its optimization objective is as follows:

$$J_\pi = \mathbb{E}_{(s,a)\sim\rho_\pi}[Q(s,a) - \alpha \log \pi(a \mid s)], \tag{2}$$

Here, $Q(s,a)$ represents the action-value function, $\pi(a|s)$ represents the policy, and α is the entropy coefficient, which balances reward maximization and policy stochasticity.

The structure of SAC consists of several key modules that collectively enhance the algorithm's efficiency and stability. The Actor network is responsible for policy optimization, aiming to maximize the entropy-augmented expected return to encourage diverse action exploration. The policy is parameterized as a Gaussian distribution, enabling the generation of continuous actions. To stabilize gradients and improve training efficiency, SAC employs the reparameterization trick to optimize the Actor network. The Critic network uses two independent Q-value networks, $Q_1(s,a)$ and $Q_2(s,a)$, to estimate the action-value function.

This dual Q-network structure effectively reduces the overestimation bias commonly observed in single-critic methods. The target value for the Critic network is as follows:

$$y(r, s') = r + \gamma \mathbb{E}_{a' \sim \pi(\cdot|s')} \Big[\min \left(Q_1 \left(s', a'\right), Q_2 \left(s', a'\right)\right) \\ - \alpha \log \pi \left(a' \mid s'\right) \Big], \tag{3}$$

This equation combines the immediate reward r and the expected future return, adjusted by the entropy of the policy to encourage exploration. The use of two Q-value networks Q_1 and Q_2 helps mitigate overestimation bias. The entropy term, controlled by the temperature parameter α, ensures a balance between exploration and exploitation by penalizing deterministic behavior and promoting diverse actions.

In this context, γ represents the discount factor, and the operation of taking the minimum of the two values helps reduce overestimation issues in policy evaluation, thereby improving the stability of learning.

Additionally, to further enhance the stability of policy training, SAC introduces target Q networks. The parameters of the target networks are updated gradually using the Polyak averaging method, allowing them to slowly converge to the values of the Critic networks and avoiding potential oscillations during training. At the same time, SAC incorporates entropy regularization into its optimization objective. By including the entropy term $\alpha \log \pi(a|s)$, the algorithm encourages policy randomness, which increases the diversity of exploration samples. The entropy coefficient α can either be fixed or dynamically adjusted during training to adapt to the varying exploration intensity required by different tasks.

3.3 Overview of the GAN-SAC Framework

The GAN-SAC algorithm extends the standard Soft Actor-Critic (SAC) method by incorporating a Generative Adversarial Network (GAN) module to enhance the agent's exploration ability in sparse reward environments. In traditional SAC, the policy optimization heavily relies on environment-provided reward signals. However, these signals are often sparse or delayed in many reinforcement learning tasks, resulting in inefficient exploration and slow convergence. To address this, GAN-SAC leverages adversarial training to introduce an additional intrinsic reward, encouraging diverse policy behavior and robust learning even under limited external feedback.

Architecture Design. As shown in Fig. ??, the GAN-SAC framework consists of four core components: a dual critic network, an actor (policy) network, a generator, and a discriminator. The critic networks Q_1 and Q_2 are used to evaluate the Q-values of state-action pairs and are updated by minimizing the Bellman residual. The actor network outputs a stochastic policy by maximizing the expected Q-value and the auxiliary GAN reward. The generator G receives noise input and generates fake action samples that resemble the distribution of real actions. The discriminator D learns to distinguish between real actions (from the actor) and fake actions (from the generator), and provides an adversarial reward r_{gan} used to guide the policy learning.

The key innovation lies in embedding the adversarial training mechanism into the SAC framework. The generator expands the agent's action exploration space by producing diverse candidate actions, while the discriminator's output serves as an intrinsic reward signal for the policy. The actor's loss function is given as:

$$\mathcal{L}_{\text{policy}} = -\mathbb{E}[\min(Q_1, Q_2)] - \alpha \cdot r_{gan}, \tag{4}$$

where $\alpha = 0.1$ controls the influence of the adversarial reward.

Training Procedure The training process of GAN-SAC involves alternate optimization of the SAC and GAN modules. Initially, the agent interacts with the environment and collects transition tuples $(s, a, r, s', done)$, which are stored in a replay buffer. Once enough experience is collected, the agent begins a coordinated training loop.

First, the critic networks are updated by computing the target Q-value using the Bellman backup:

$$y = r + \gamma \cdot \min(Q_{\text{target}}(s', a')). \tag{5}$$

Then, the actor is updated by maximizing the total reward, which is composed of the environment reward and the adversarial reward r_{gan}, calculated as:

$$r_{gan} = D(a). \tag{6}$$

Next, the generator G takes noise input $z \sim \mathcal{N}(0, 1)$ and generates fake actions $G(z)$, which, along with real actions from the actor, are used to train the discriminator D. The discriminator is updated using the following loss:

$$\mathcal{L}_D = -\mathbb{E}[\log D(a_{\text{real}})] - \mathbb{E}[\log(1 - D(a_{\text{fake}}))]. \tag{7}$$

The generator is trained to fool the discriminator by minimizing:

$$\mathcal{L}_G = -\mathbb{E}[\log D(G(z))]. \tag{8}$$

Finally, the target Q-networks are softly updated using Polyak averaging with $\tau = 0.005$ to ensure training stability. This process is repeated over multiple episodes until convergence.

Through this adversarially enhanced training strategy, GAN-SAC significantly improves exploration efficiency and policy robustness in dynamic or sparse-reward scenarios. The integration of GAN and SAC enables mutual reinforcement: SAC benefits from expanded action diversity and additional guidance from D, while GAN improves by learning from policy-generated real actions.

3.4 Training Process

The training process alternates between the SAC and GAN modules to enhance the learning efficiency and policy robustness. Specifically, the Soft Actor-Critic (SAC) module is responsible for learning a stochastic policy that maximizes expected reward and entropy, while the GAN module encourages the policy to generate diverse and realistic actions through adversarial training. During each training episode, the robot collects transition data from the environment and stores it in a replay buffer. Mini-batches are sampled to compute target and current Q-values, which are used to update the critic networks using Mean Squared Error (MSE) loss. The actor (policy network) is updated by maximizing the expected return along with a GAN-based reward derived from the discriminator. Meanwhile, the generator produces fake actions to confuse the discriminator, which is trained to distinguish real actions from fake ones. The generator and discriminator are alternately updated. Finally, soft updates are applied to the target Q-networks to stabilize training. The detailed pseudocode of the training process is presented in Algorithm 1.

Algorithm 1 GAN-SAC Training Process

```
Require: State space S of the robot
Ensure: Trained Actor, Critic, Generator G, and Discriminator D
Initialize Q-networks Q1, Q2, target networks Q'1, Q'2, policy network π
Initialize Generator G and Discriminator D
Initialize replay buffer with capacity C
Define Adam optimizer and MSE loss
for episode = 1 to M do
    Collect transition (s, a, r, s') from the environment and store in buffer
    Sample a mini-batch from replay buffer
    Compute target Q-value: r + γ(1 − done) · min(Q'1(s'), Q'2(s'))
    Compute current Q-values Q1(s, a), Q2(s, a)
    Compute Q-loss using MSE and update Q1, Q2
    Sample action a ∼ π(s) and compute policy entropy
    Compute GAN-based reward r_gan = D(a)
    Update policy network π to maximize expected reward and entropy
    Generate fake actions a_fake = G(z)
    Get real actions a_real = π(s).detach()
    Compute discriminator loss: L_D = −[log D(a_real) + log(1 − D(a_fake))]
    Compute generator loss: L_G = − log D(a_fake)
    Alternate updates of D and G using their losses
    Soft update Q'1 :  Q'1 ← τQ1 + (1 − τ)Q'1
    Soft update Q'2 :  Q'2 ← τQ2 + (1 − τ)Q'2
end for
```

4 Experiments and Results

To validate the effectiveness of our proposed method, we conducted a series of experiments to compare the impact of different parameters on the experimental results. In these experiments, our ultimate goal was to enable the humanoid robot to successfully complete the gate navigation task using deep reinforcement learning techniques and further improve its success rate.

4.1 Humanoid Robot Model

The DARwIn-OP2 [16] (also known as ROBOTIS OP2) is an open-source, dynamic humanoid intelligent robot platform. The simulation software Webots provides a pre-installed simulation model of the DARwIn-OP2 robot (as shown in Fig. 3), which accurately reproduces the appearance and functionality of the physical DARwIn-OP2 robot. This includes 20 motors, 20 position sensors, 5 LED lights, a three-axis accelerometer, a three-axis gyroscope, and a camera. Using the Webots simulation software, users can perform various experiments and development tasks, test algorithms, controllers, and behaviors, and enhance the robot's performance and functionality.

Fig. 3. Schematic representation of the DARwIn-OP2 simulation model.

4.2 Training Parameter Setting

In the training process of deep reinforcement learning agents, parameter selection has a significant impact on the experimental results. To ensure the reliability and effectiveness of the experiments, we performed extensive adjustments and optimizations of the experimental settings and hyperparameters. The hardware environment for the experiments consisted of a high-performance workstation equipped with an RTX 3070 GPU and an AMD Ryzen 9 5950X CPU to meet the demands of large-scale computations and simulation tasks.

In the experiments, we conducted a comprehensive comparative analysis of the following four algorithms: GAN-SAC, SAC, PPO [17], and DQN [18]. Training and testing were performed in scenarios with both fixed gate positions and random gate positions. For each algorithm, we configured appropriate hyperparameters and optimized them to ensure the fairness of the comparison and the scientific validity of the experimental results.

The specific parameter settings are as follows: the total number of training iterations for all algorithms was set to 1500 to explore the performance of different algorithms under the same training intensity. Each training iteration took an average of about 5 s, ensuring training efficiency and allowing multiple experiments to be conducted.

The learning rates for GAN-SAC and SAC were set to 0.0003, for PPO to 0.0001, and for DQN to 0.001. These learning rates were determined after multiple adjustments to achieve optimal convergence for each algorithm. All algorithms adopted a discount factor γ of 0.99 to balance the trade-off between long-term returns and short-term rewards. For algorithms with an experience replay mechanism, such as GAN-SAC, SAC, and DQN, the replay buffer capacity was set to 5000 to ensure sample diversity. The batch size was set to 512, which exhibited favorable training performance in multiple experiments. In GAN-SAC, the learning rates for both the Generator and Discriminator were set to 0.0001, and they were trained once after every 5 updates of the SAC module.

During the experiments, we evaluated the performance of the agent every 50 episodes. Each evaluation consisted of 10 independent test runs, and the average of the test results was used as the performance metric for the agent. The evaluation metrics included the following three aspects: task success rate and convergence speed. These parameter settings and experimental designs ensured a fair comparison of the GAN-SAC, SAC, PPO, and DQN algorithms under both fixed and random gate position scenarios, while providing strong experimental support for robotic control tasks in dynamic environments.

4.3 Experiments Results

The task success rate was recorded every 10 training episodes, and the results demonstrate that the GAN-SAC algorithm consistently exhibits significant advantages across different scenarios, as shown in Table 1. In particular, in the fixed gate task scenario shown in Fig. 4, GAN-SAC began to converge rapidly at the 220th episode, with the task success rate steadily increasing to 100%. In

Table 1. Comparison of success rates between GAN-SAC and other algorithms in different scenarios.

Algorithm	Success Rate in Fixed Scenarios	Success Rate in Random Scenarios
GAN-SAC	100%	92%
SAC	100%	89%
DQN	99%	90%
PPO	80%	72%

contrast, SAC and DQN started to converge around the 420th episode, and their success rate growth was slower. The training performance of PPO was rather poor, unable to effectively improve the task success rate, and even showed some fluctuations in the later stages.

In the random gate task scenario shown in Fig. 5, GAN-SAC achieves a high success rate of around 90% within a relatively short training period. This result further highlights the excellent exploration capability and stability in policy generation of GAN-SAC in dynamic environments. Figure 6 displays the robot's movement trajectory during the training process through the gate.

Fig. 4. Performance comparison of algorithms in fixed gate.

Compared to GAN-SAC, SAC performs well but is noticeably slower in reaching a high task success rate. DQN made slow progress in the early stages of

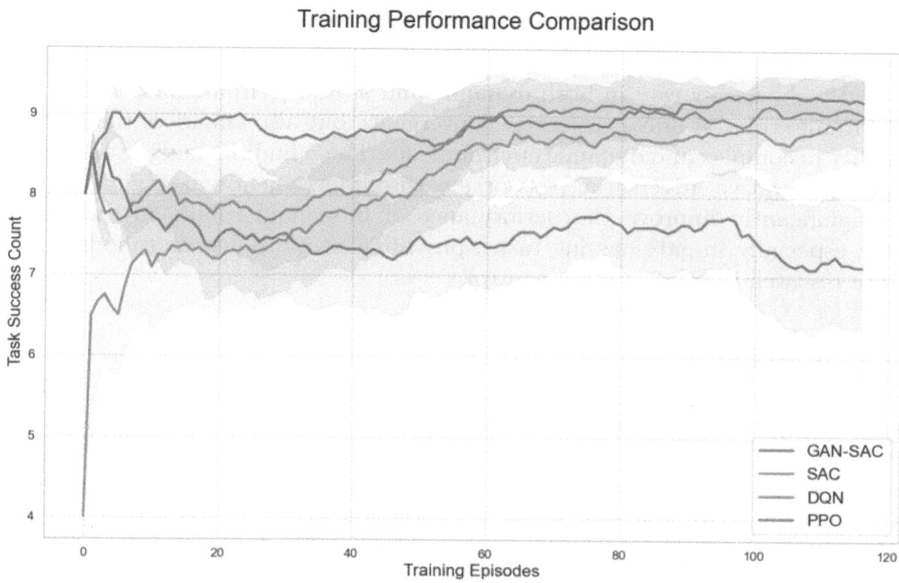

Fig. 5. Performance comparison of algorithms in random gate.

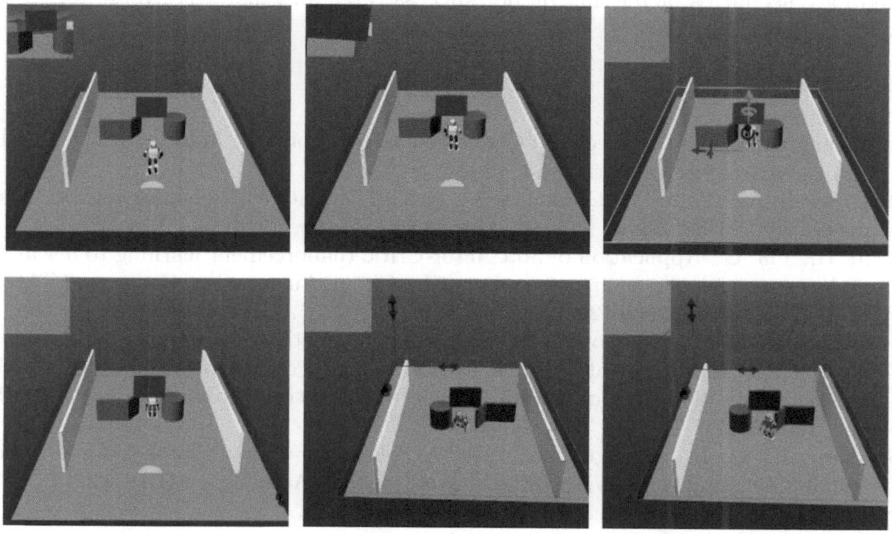

Fig. 6. Example path of a humanoid robot successfully navigating through the gate.

training, with task success rate improvement being minimal. PPO showed relatively stable performance throughout the training but never surpassed SAC or GAN-SAC in terms of task success rate.

These results indicate that the introduction of the Generative Adversarial Network (GAN) module significantly improves the humanoid robot's adaptability and task success rate in both dynamic and static environments. The GAN-SAC framework not only accelerates convergence but also enhances the model's stability in complex and dynamic environments. These findings suggest that combining Generative Adversarial Networks with reinforcement learning algorithms can significantly improve the performance of humanoid robots in navigation tasks, especially in gate-passing tasks, providing new insights and methods for future research in robot motion control.

5 Conclusions

This study proposes a GAN-SAC framework that integrates Generative Adversarial Networks (GAN) with the Soft Actor-Critic (SAC) algorithm to address the challenges of unstable policy generation and low exploration efficiency in humanoid robot gate-passing tasks. By incorporating a GAN module to generate adversarial samples, the framework enriches the training data for SAC, enhances policy diversity, and accelerates the robot's adaptability in dynamic environments. Experimental results demonstrate that GAN-SAC significantly outperforms traditional SAC, DQN, and PPO algorithms in both random and fixed gate scenarios, achieving higher success rates and faster convergence.

References

1. Hwang, H.S., Kim, Y., Seok, J.: Generative adversarial Soft Actor–Critic[J]. IEEE Trans. Neural Netw. Learn. Syst. (2024)
2. Creswell, A., White, T., Dumoulin, V., et al.: Generative adversarial networks: an overview[J]. IEEE Signal Process. Mag. **35**(1), 53–65 (2018)
3. Ji, H., Yin, C., Application of Soft Actor-Critic reinforcement learning to a search and rescue task for humanoid robots[C].: China Automation Congress (CAC). IEEE **2022**, 3954–3960 (2022)
4. Totani, M., Sato, N., Morita, Y., Step climbing method for crawler type rescue robot using reinforcement learning with Proximal Policy Optimization[C].: 12th International Workshop on Robot Motion and Control (RoMoCo). IEEE **2019**, 154–159 (2019)
5. Wang, L., Li, R., Huangfu, Z., et al.: A Soft Actor-Critic approach for a blind walking hexapod robot with obstacle avoidance[C]. Actuators. MDPI **12**(10), 393 (2023)
6. Chen, Y., Cheng, C., Zhang, Y., et al.: A neural network-based navigation approach for autonomous mobile robot systems[J]. Appl. Sci. **12**(15), 7796 (2022)
7. Xu, K., Lu, Y., Shi, L., et al.: Whole-body stability control with high contact redundancy for wheel-legged hexapod robot driving over rough terrain[J]. Mech. Mach. Theory **181**, 105199 (2023)
8. Sun, C., Yang, G., Yao, S., et al.: Rhex-t3: A transformable hexapod robot with ladder climbing function[J]. IEEE/ASME Trans. Mechatron. **28**(4), 1939–1947 (2023)

9. Nie, B., Gao, Y.: DanceHAT: Generate stable dances for humanoid robots with adversarial training[C]. 2022 International Conference on Robotics and Automation (ICRA). IEEE: 8511-8517 (2022)
10. Sun, G., Wong, Y., Cheng, Z., et al.: Deepdance: music-to-dance motion choreography with adversarial learning[J]. IEEE Trans. Multimedia **23**, 497–509 (2020)
11. Goodfellow, I., Pouget-Abadie, J., Mirza, M., et al.: Generative adversarial nets[J]. Adv. Neural Inf. Process. Syst. **27** (2014)
12. Yu, C., Tapus, A.: Srg 3: Speech-driven robot gesture generation with gan[C]. In: 2020 16th International Conference on Control, Automation, Robotics and Vision (ICARCV). IEEE, 759–766 (2020)
13. Khudhair, Z.N., Nidhal, A., El Abbadi, N.K., et al.: Color to grayscale image conversion based on singular value decomposition[J]. IEEE Access **11**, 54629–54638 (2023)
14. Haarnoja, T., Zhou, A., Abbeel, P., et al.: Soft actor-critic: Off-policy maximum entropy deep reinforcement learning with a stochastic actor[C]. In: International Conference on Machine Learning. PMLR, 1861–1870 (2018)
15. Hong, Y., Hwang, U., Yoo, J., et al.: How generative adversarial networks and their variants work: An overview[J]. ACM Comput. Surv. (CSUR) **52**(1), 1–43 (2019)
16. Ha, I., Tamura, Y., Asama, H., et al.: Development of open humanoid platform DARwIn-OP[C]. SICE Annual Conference 2011. IEEE, pp. 2178–2181 (2011)
17. Jayant, A.K., Bhatnagar, S.: Model-based safe deep reinforcement learning via a constrained proximal policy optimization algorithm[J]. Adv. Neural. Inf. Process. Syst. **35**, 24432–24445 (2022)
18. Yang, Y., Juntao, L., Lingling, P.: Multi-robot path planning based on a deep reinforcement learning DQN algorithm[J]. CAAI Trans. Intell. Technol. **5**(3), 177–183 (2020)

Link Prediction Method Based on Structural Hole Characterization Under Multi-order Paths

Zeyu Yin and Ying Zhao(✉)

Beijing University of Chemical Technology, Chaoyang District, Beijing 100029, China
zhaoy@buct.edu.cn

Abstract. With the advancement of internet technology, diverse networks have emerged. Analyzing relationships between network entities can enhance our understanding of real-world network evolution, where link prediction is crucial. However, current link prediction methods face two key challenges. First, most algorithms extract node features from local subgraphs, overlooking global characteristics. Second, they struggle to adapt to graph structures with widely varying node counts. To address these issues, we propose HSHLP (High Structural Hole Link Prediction), a novel approach that leverages structural hole features across multi-order paths. By incorporating structural hole theory, HSHLP uncovers hidden node-linking patterns and builds effective feature representations through multi-order neighborhood extraction. Experiments show that HSHLP not only increases prediction accuracy but also remains robust across networks of varying scales.

Keywords: Link prediction · Structural hole · Multi-order paths · Graph neural networks

1 Introduction

Link prediction is the process of predicting the probability of the existence of a link between two entities on the basis of the known links in the network structure [1]. Network structures are prevalent in social, technological and biological systems, and real network structures tend to change over time [2]. Contemporary link prediction techniques provide a more general approach to solving some specific problems of complex network structures, such as social networks [3], e-commerce networks [4], and protein action networks [5]. Therefore, advancements in link prediction technology are highly important for the development of various fields.

To improve the link prediction technique, the neural network architecture can be adjusted to enhance the training results. The feature extraction can also be changed to obtain more graph structure information to improve the prediction accuracy. For example, topological features, path features, and network features [6]. Therefore, in this paper, a link prediction method, high structural

L. Huang et al. (Eds.): CCF NCCA 2025, CCIS 2774, pp. 104–117, 2026.
https://doi.org/10.1007/978-981-95-5294-8_8

hole link prediction (HSHLP), is proposed on the basis of the new feature extraction method. Specifically, key nodes are identified via structural hole theory for weight assignment, feature extraction is performed under multi-order paths, and prediction is trained by constructing multidimensional feature vectors input to a graph neural network. The proposed method achieves high accuracy in seven public datasets, the prediction accuracies are all above 90%, and the prediction accuracies of more than half of the datasets are above 95%, with a maximum of 98.98% (AUC value). The present method not only outperforms the comparison method but also reveals the hidden node relationships in the network structure by synthesizing local and global features through a new feature extraction method, demonstrating good adaptability and stability and providing a new approach to increase the link prediction accuracy.

2 Related Work

Link prediction techniques are usually centered on a graph structure, where the nodes of the graph are entities in the network structure and the connecting edges are generated by the interaction information between the entities [1]. Current link prediction methods can be categorized into two types: heuristic-based methods and neural network-based methods. In this paper, link prediction is based on structural hole features, which are described and introduced in the following description.

2.1 Heuristic-Based Methods

Heuristic-based methods are used to directly compute the prediction score for the existence of a link for any pair of nodes by manually defining the relationships between network nodes and providing a node similarity index on the basis of graph features [7]. CN (Common Neighbor) [8], Jaccard [9], AA (Adamic Adar) [7], and RA (Resource Allocation) [10] all utilize common neighbors as node features for prediction. Both PA(Preferential Attachment) [11] and SimRank [12] rely on the direct neighbors of the nodes to be predicted for computing similarity. In contrast, KATZ [13] performs a global analysis of nodes by assigning weights.

2.2 Neural Network-Based Methods

Neural network-based methods leverage known network structures as training data and utilize learning models to predict potential links [14]. WLNM (Weisfeiler-Lehman Neural Machine) predicts links by feeding node weights from subgraphs into a neural network, focusing on the closed subgraphs surrounding target links [15]. SEAL (Subgraph Embedding-based Link prediction) employs a graph neural network to learn heuristic features from local subgraphs for prediction [16]. LGLP (Line Graph for Link Prediction) builds on this approach by converting closed subgraphs into line graphs, extracting line graph features, and mapping them back to the original graph for prediction [17]. NNESF (Neural

Networks with Elementary Subgraphs Features) optimizes SEAL by streamlining the subgraph size and simplifying feature extraction, forming feature vectors for prediction [18]. Additionally, a novel feature extraction method is proposed, which builds on the feature vector approach of the NNESF to enhance training.

2.3 Structural Hole

Structural holes, a topological concept that integrates the node degree and neighborhood information, were first proposed by Burt in 1992. In interpersonal networks, a third party connecting two unlinked nodes gains informational and control advantages [19]. Yang refined this concept, defining structural holes as a structure comprising indirectly connected nodes and intermediary nodes [20]. Zhang introduced a community forest model to explore structural holes and proposed a new algorithm for their identification, distribution, and application on the basis of structural hole characteristics [21].

3 Method

3.1 Problem Description

Given an unweighted undirected graph $G = (V, E, A)$, there are M nodes and N edges in the graph. Where $V = \{v_1, v_2, v_3, \ldots, v_n\}$ denotes the set of nodes and where $E = \{e_1, e_2, e_3, \ldots, e_n\}$ denotes the set of edges. $A \in \{0, 1\}^{m \times n}$ is the adjacency matrix of the graph.

3.2 Method Framework

The overall framework of the HSHLP is shown in Fig. 1. The overall framework is divided into four parts. The first part traverses the datasets, extracts the unit structure "structural holes" in the dataset, and identifies the middle nodes of the structural holes as important nodes. The more times a node exists in the middle of the structural holes, the more important the node's position in the whole graph. Weights are assigned to nodes on the basis of their importance. The second part extracts the multi-order paths of the nodes to be predicted as the neighborhood subgraph, and this method extracts the 2nd to 5th order links for experimental comparison. The third part extracts the node features. According to the node weights under the multi-order paths, the similarity measure of the links to be predicted is calculated as features. The first-order neighbors of the node pairs to be predicted are extracted separately, the number of edges and nodes in their neighborhoods are calculated as features, and the above six features are converted into one feature vector. The fourth part inputs the feature vector into the fully connected neural network for training and prediction.

The framework consists of the following four modules: node influence analysis module, multi-order path extraction module, feature extraction module, and link prediction module. The specific implementation of each module is as follows.

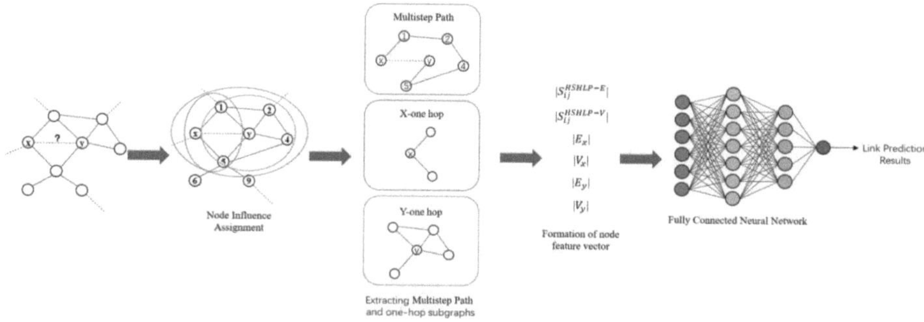

Fig. 1. HSHLP Framework

3.2.1 Node Influence Analysis Module. The node influence analysis module identifies important nodes on the basis of structural hole features and assigns weights to them. The intermediate nodes in the structural holes are more central in the structure, so we identify them as "important nodes". When searching for structural holes, the graph is first transformed into an adjacency matrix A, traversing all pairs of nodes in A with a value of 0 and searching for common neighbor nodes of the pairs of nodes as "important nodes". As shown in Eq. 1.

$$A = [a_{ij}]_{N \times N} \tag{1}$$

If there is a link between i and j, then $a_{ij} = 1$, otherwise $a_{ij} = 0$. The influence of a node is the number of nodes present in the middle of a structural hole; the larger the number is, the greater the influence of the node. For example, in Fig. 2, 'b-a-c' constitutes a structural hole, and 'a' is an "Important Node" in it. The influence value of a is 1. 'c' exists in the middle of several structural holes, including: 'a-c-e', 'a-c-f', 'a-c-g', 'e-c-f', 'e-c-g', 'f-c-g', and the influence value of c is 6. 'c-e-f' does not constitute a structural hole because these nodes are connected to each other and do not fit the concept of structural holes. Conversion to Eq. 2 and Eq. 3, where $CN(i, j)$ denotes the set of common neighbor of nodes i and j.

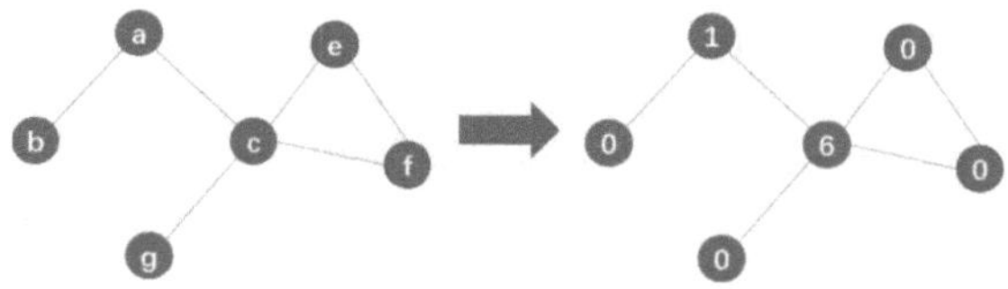

Fig. 2. Node Influence.

$$\delta(k \in CN(i,j)) = \begin{cases} 1 & \text{if } k \in CN(i,j) \\ 0 & \text{else} \end{cases} \tag{2}$$

$$s(k) = \sum_{(i,j) \in V \times V,\ i \neq j,\ A[i][j]=0} \delta(k \in CN(i,j)) \tag{3}$$

3.2.2 Multi-order Path Extraction Module. Most of the existing methods extract 1st to 2nd order neighborhoods as local subgraphs. Although the subgraphs are closely connected with the links to be predicted, they can cover only the local information but cannot completely reflect the overall structure of the graph, and there is a loss of information; especially in complex networks, it is often necessary to consider the global structural features through the higher-order nodes to improve the prediction performance. 2018 Yang et al. proposed the "three-degree influence" theory; that is, people influence each other, and the transmission of influence is within three degrees [22]. In 2020 Rafiee et al. proposed the "six-degree partition" theory; that is, the distance between any two strangers in the world should be no more than six people [23]. In this paper, we combine the two theories and perform feature extraction through multi-order paths as Fig. 3, while considering the time complexity and operational efficiency, we perform feature extraction through the 2nd to 5th order paths of the links to be predicted in this paper.

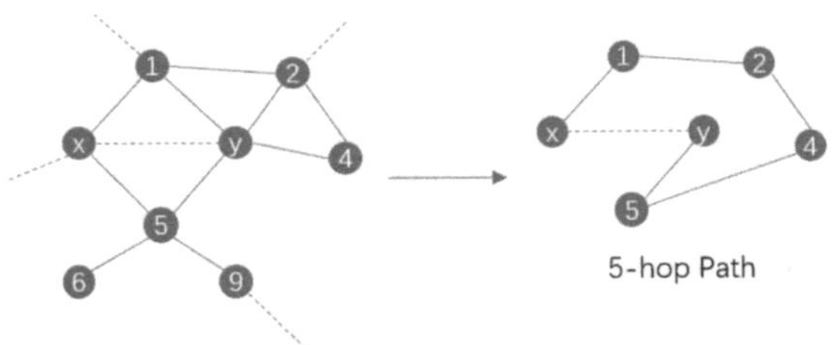

Fig. 3. Multi-Order Path Extraction.

Multi-order path methods compensate for the shortcomings of traditional neighborhood methods through hierarchical and global perspectives and are able to capture the complex relationships in the network in a more comprehensive way. This method is suitable for all kinds of complex networks and is especially valuable for structure optimization and the application of large-scale networks.

3.2.3 Feature Extraction Module. This method proposes a new feature extraction method, structural hole-based feature extraction under multi-order paths, which integrates global and local features for extraction. For each link to be predicted, the first-order and higher-order neighborhood features of the corresponding node pairs are obtained to form a feature vector. On the basis of the multi-order paths and node influence mentioned above, the higher-order structural hole features of the links to be predicted are calculated via Eq. 4 and Eq. 5:

$$S_{ij}^{(\text{HSHLP-E})} = \sum_{x \to n-1 \to y} (((s(a_1)+1) \times (s(a_2)+1) \times \cdots \times (s(a_{n-1})+1)))^{-\frac{1}{n-1}} \tag{4}$$

$$S_{ij}^{(\text{HSHLP-V})} = \log_n N \tag{5}$$

where $S_{ij}^{\text{HSHLP-E}}$ represents the edge characteristics of the node pair to be predicted under the higher-order path, where n denotes the number of path orders, N represents the nodes passed in the multi-order path, and $s(a)$ represents the influence value of the node. $S_{ij}^{\text{HSHLP-V}}$ represents the node characteristics of the node pair to be predicted under the normalization process, where N denotes the number of multi-order paths, which is a measure of the node pair's active degree. In this work, we use 2nd to 5th order paths for experimental comparison, and the value of n is in [2,5]. The calculation method proposed in this paper has two advantages: 1) it completely avoids the situation where the node influence value is 0; 2) it utilizes the "multiplication + square" method to consider the node influence from the global structure, and calculates the higher-order structural hole characteristics more comprehensively and accurately.

Secondly, the one-hop neighborhood subgraphs of the nodes to be predicted are extracted, and the number of nodes $|V_x|$, $|V_y|$ and the number of edges $|E_x|$, $|E_y|$ within each subgraph are calculated as the remaining features. Just as Fig. 4. This method uses one-hop neighborhood features as local features and higher-order structural hole features as global features to improve prediction results and enhance the generalization ability of the model with comprehensive consideration of local and global structures.

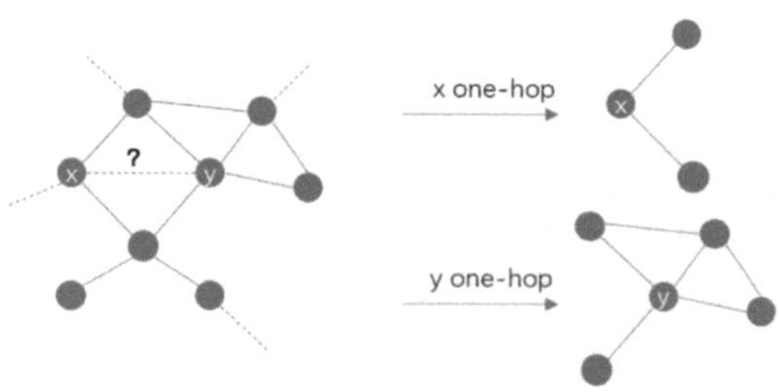

Fig. 4. One-Hop Neighborhood of Link.

3.2.4 Link Prediction Module. For each link to be predicted, the features mentioned above are converted into feature vectors, which are then fed into a multilayer perceptron (MLPClassifier) trained to predict as Eq. 6. The MLP is a binary classifier that learns nonlinear features through a hidden layer, and is used to predict the existence of a link between a given pair of nodes.1 indicates presence and 0 indicates absence. MLP architecture is illustrated in Fig. 5.

$$X_{\mathrm{HSH}} = \left(|E_x|, |V_X|, |E_y|, |V_y|, |S_{ij}^{\mathrm{HSHLP\text{-}E}}|, |S_{ij}^{\mathrm{HSHLP\text{-}V}}|\right) \tag{6}$$

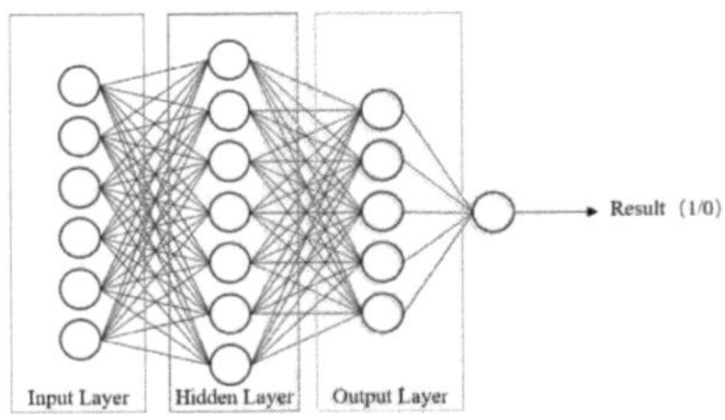

Fig. 5. MLP Schematic of Training Prediction.

4 Experiment

To validate the performance of the HSHLP proposed in this paper in terms of link prediction, the area under the curve (AUC) and average accuracy (AP) are used as benchmark metrics for evaluation, and seven real undirected unweighted network datasets are used for the experiments. The system is Win10, and the running environment is Python 3.10.13.

4.1 Dataset

The datasets used in this paper are all real undirected unweighted graph datasets, as shown in Table 1. Celegans is a neural network of nematodes [24]. USAir is the USAir network [25]. SMG is a citation network. The EML is an email sharing network. The BUP is a network of political blogs in the U.S. [26]. NSC and NS are collaborative networks of web science-based paper networks [27]. During the experiments, from each dataset, 80% of the edges and an equal number of non-existent edges were randomly selected as the training set, and the remaining 20% and an equal number of non-existent edges were used as the test set. The final result is based on the average of the results obtained by running 10 random seeds.

Table 1. Dataset

Dataset	Nodes	Edges	Degree
Celegans	297	2148	14.46
USAir	332	2126	12.81
SMG	1024	4916	9.6
EML	1133	5451	9.62
BUP	105	441	8.4
NSC	1461	2126	2.91
NS	1589	2742	3.45

4.2 Comparison Method

In this paper, nine comparative methods are used. These methods include heuristic first-order methods CN; second-order methods RA; higher-order methods Katz and SimRank. Neural network-based methods: GCN, GAT, SEAL, LGLP, and NNESF.

4.3 Parameter Setting

The decay coefficient in Katz is set to 0.001. The importance coefficient in SimRank is 0.8. The GAT consists of a two-layer graph attention network and a fully connected layer with a single-head attention mechanism; the GCN consists of a two-layer graph convolutional network and a fully connected layer. The channel sizes are 128 and 64, respectively, with a learning rate of 0.01 and a maximum training number of 100. Both SEAL and LGLP extract 2-hop neighborhoods as subgraphs; the channel sizes of the four graph convolutional layers are set to 32, 32, 32, and 1; the channel sizes of the two fully connected layers are set to 128 and 2, with a batch size of 50; the scale of the SEAL pooling layer is set to 0.6; the two one-dimensional convolutional layer channel sizes are set to 16 and 32, with a learning rate of 0.0001 and a maximum number of training rounds of 50; and the LGLP learning rate is 0.005 and the maximum number of training rounds is 15. NNESF uses an MLP as a prediction model and consists of a three-layer neural network with hidden layer sizes of 32, 32, and 16, with a learning rate of 0.001, a batch size of 128, and a maximum number of training rounds of 100. For HSHLP, 2nd to 5th-order paths are extracted for experiments using an MLP as a predictive model consisting of three neural networks with hidden layer sizes of 32, 32, and 16; a learning rate of 0.001; a batch size of 128; and a maximum number of training rounds of 200.

4.4 Experimental Result

To verify the link prediction performance of the HSHLP proposed in this paper, several comparison experiments are conducted. First, to verify the influence of path order on prediction accuracy, this paper conducts comparison experiments on the 2nd to 5th-order paths, using the AUC value and AP value as evaluation indices, and the experimental results are shown in Fig. 6.

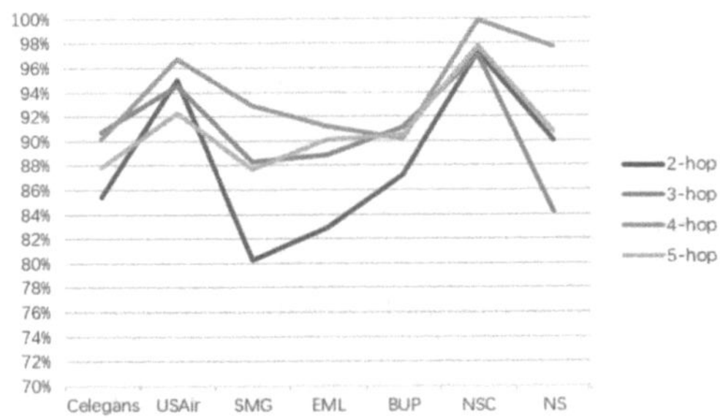

Fig. 6. Comparison of prediction results under 2nd to 5th order paths.

According to the experiments, in the dataset used in this study, the prediction accuracy under the 2nd to 4th order paths increases with increasing order; however, for nodes with small node degrees, it is difficult to find higher order paths between nodes, which makes the prediction results under the 5th order paths lower. Therefore, the path change affects the prediction accuracy, and the 4th order path has the best performance in the dataset of this question. Second, a comparison is made between the HSHLP under the fourth-order path and the baseline method to verify whether there is any improvement in prediction accuracy. As shown in Tables 2 and 3, the AUC and AP values of HSHLP under the fourth-order path are higher than those of the baseline method in most datasets of this question. For example, in the Celegans dataset, the prediction results of most methods are below 90%, and the prediction result of HSHLP is nearly 91%, which is much higher than those of the other methods, realizing a further improvement in the link prediction accuracy.

Table 2. AUC Comparison With Other Baseline Methods (AUC)

Method	Celegans	USAir	SMG	EML	BUP	NSC	NS
CN	82.63 (±1.57)	92.99 (±0.76)	81.34 (±0.62)	82.17 (±0.21)	84.88 (±0.60)	96.61 (±0.15)	91.20 (±0.79)
RA	84.08 (±1.10)	94.69 (±0.23)	81.65 (±0.20)	82.20 (±0.24)	87.03 (±0.66)	97.32 (±0.32)	91.34 (±0.62)
Katz	84.59 (±0.61)	91.09 (±1.20)	86.01 (±0.72)	88.89 (±0.15)	88.20 (±0.51)	97.73 (±0.09)	91.99 (±1.54)
SR	76.02 (±0.60)	78.97 (±0.50)	77.78 (±0.37)	88.55 (±0.48)	87.33 (±0.13)	98.00 (±0.41)	93.71 (±1.39)
GCN	87.17 (±0.31)	92.57 (±0.64)	84.72 (±0.54)	84.15 (±0.35)	91.58 (±0.23)	98.49 (±0.13)	91.79 (±0.12)
GAT	78.94 (±0.99)	87.74 (±0.54)	79.32 (±0.56)	80.71 (±0.64)	85.33 (±0.45)	98.91 (±0.23)	94.71 (±0.53)
SEAL	88.16 (±1.52)	93.23 (±1.46)	91.53 (±0.46)	91.01 (±0.38)	88.97 (±0.64)	99.77 (±1.30)	96.64 (±0.23)
LGLP	90.16 (±0.76)	**97.44 (±0.32)**	92.86 (±0.29)	91.03 (±0.28)	95.76 (±0.53)	**99.84 (±0.01)**	98.96 (±0.47)
NNESF	87.64 (±1.96)	96.13 (±1.09)	91.11 (±0.98)	90.66 (±1.01)	93.14 (±1.00)	99.58 (±0.16)	**99.06 (±0.30)**
HSHLP	**90.78 (±0.74)**	96.77 (±1.11)	**92.99 (±0.78)**	**91.89 (±0.84)**	**95.82 (±0.51)**	98.88 (±0.43)	98.67 (±0.80)

4.5 Ablation Experiment

Among the feature vectors used in this method, the focus is on the structural hole features under multi-order paths, and both edge features $S_{ij}^{\text{HSHLP-E}}$ and point features $S_{ij}^{\text{HSHLP-E}}$ under multi-order paths are used. To explore the effect of each feature on the prediction results, this problem designs ablation experiments and compares them with separate experiments, using AUC and AP values as evaluation indices. The experimental results are shown in Tables 4 and 5. When only edge or node features are learned, the AUC and AP values are lower than those

Table 3. AUC Comparison With Other Baseline Methods (AUC)

Method	Celegans	USAir	SMG	EML	BUP	NSC	NS
CN	80.12 (±1.80)	92.63 (±0.98)	80.43 (±0.64)	81.63 (±0.23)	82.39 (±0.61)	96.56 (±0.17)	91.18 (±0.78)
RA	82.97 (±1.94)	95.13 (±0.18)	82.44 (±0.34)	82.29 (±0.24)	86.24 (±0.91)	97.33 (±0.31)	91.36 (±0.60)
Katz	84.03 (±0.89)	92.86 (±0.88)	88.08 (±0.50)	90.00 (±0.10)	86.85 (±0.14)	97.77 (±0.07)	92.30 (±1.44)
SR	66.56 (±0.69)	70.18 (±0.94)	69.71 (±0.57)	88.59 (±0.51)	82.80 (±0.23)	97.38 (±1.32)	93.98 (±1.30)
GCN	85.60 (±0.18)	94.36 (±0.26)	85.50 (±0.23)	87.26 (±0.11)	86.44 (±0.70)	98.82 (±0.16)	94.45 (±0.65)
GAT	75.36 (±0.36)	88.93 (±0.24)	78.73 (±0.27)	79.42 (±0.30)	80.91 (±0.69)	98.75 (±0.20)	96.29 (±0.20)
SEAL	87.16 (±1.22)	94.73 (±1.06)	92.59 (±0.96)	91.34 (±0.52)	88.62 (±0.54)	98.97 (±0.76)	96.88 (±0.50)
LGLP	89.16 (±1.32)	**97.33 (±0.43)**	93.76 (±0.22)	92.03 (±0.37)	93.52 (±0.29)	99.23 (±1.22)	98.01 (±0.55)
NNESF	88.47 (±1.25)	96.39 (±0.97)	91.32 (±0.52)	92.01 (±1.01)	90.74 (±3.69)	**99.53 (±0.14)**	**98.88 (±0.24)**
HSHLP	**90.16 (±0.65)**	96.77 (±1.12)	**93.93 (±0.68)**	**92.34 (±1.12)**	**93.86 (±2.82)**	98.98 (±0.42)	97.77 (±0.55)

when both edge and node features are learned, and the experimental results of edge features are better than those of node features in most cases. This indicates that better prediction results can be obtained when edge and node features are combined, and the edge features play an important role by contributing more to the prediction process.

Table 4. AUC Comparison of HSHLP Variants (AUC)

Method	Celegans	USAir	SMG	EML	BUP	NSC	NS
HSHLP-e	90.16	96.31	91.27	90.21	92.21	98.42	97.02
	(±0.65)	(±1.14)	(±0.91)	(±1.11)	(±2.81)	(±0.50)	(±0.92)
HSHLP-v	88.75	95.10	92.93	89.37	90.54	98.64	94.82
	(±1.17)	(±0.48)	(±0.68)	(±0.87)	(±2.85)	(±0.64)	(±0.82)
HSHLP	**90.78**	**96.77**	**92.99**	**92.81**	**95.82**	**98.88**	**98.67**
	(±0.74)	**(±1.11)**	**(±0.78)**	**(±0.84)**	**(±2.51)**	**(±0.43)**	**(±0.80)**

Table 5. AP Comparison of HSHLP Variants (AP)

Method	Celegans	USAir	SMG	EML	BUP	NSC	NS
HSHLP-e	89.97	96.73	90.76	91.04	90.95	99.46	97.67
	(±1.93)	(±0.99)	(±1.41)	(±0.71)	(±2.01)	(±0.48)	(±0.80)
HSHLP-v	88.20	95.41	91.93	91.21	91.21	98.46	96.39
	(±1.12)	(±0.53)	(±1.05)	(±1.11)	(±2.81)	(±0.43)	(±0.53)
HSHLP	**90.16**	**96.77**	**93.93**	**92.34**	**93.36**	**99.88**	**97.77**
	(±0.65)	**(±1.12)**	**(±0.68)**	**(±1.12)**	**(±2.82)**	**(±0.42)**	**(±0.55)**

5 Conclusion

In this paper, we propose the link prediction algorithm HSHLP through a new feature extraction approach, structural hole-based feature extraction under multi-order path. The core of the method lies in identifying critical nodes via structural hole features and integrating global and local features under the framework of multi-order paths to optimize the training and prediction process. Compared with the baseline methods, the HSHLP method achieves greater prediction accuracy and stability on seven real network datasets, and all prediction accuracies remain above 90%. The experimental results show that structural hole theory can enhance the differentiation of nodes and that the multi-order path feature extraction method effectively combines global and local information, thus effectively improving the prediction effect and model stability. Moreover, the computational cost increases with increasing feature expression capability, so how to further optimize the computational efficiency while ensuring high prediction accuracy is an important direction for future research.

References

1. Lü, L., Zhou, T.: Link prediction in complex networks: a survey. ScienceDirect. Phy. Stat. Mech. Appl. **390**(6), 1150–1170 (2011)
2. Cao, Z., Zhang, Y., Guan, J., et al.: A chaotic ant colony optimized link prediction algorithm. Instit. Elect. Electron. Eng. (IEEE) **51**(9), 5274–5288 (2021)
3. Ghasemi, S., Zarei A.: Improving link prediction in social networks using local and global features: a clustering-based approach. Prog. Artif. Intell. **11**(1), 79–92 (2022)
4. Wang, H., et al.: Knowledge-aware graph neural networks with label smoothness regularization for recommender systems. SIGKDD Explor. 968–977 (2019)
5. Yuen, H., Jansson, J.: Normalized L3-based link prediction in protein–protein interaction networks. BMC Bioinf. **24**(1) (2023)
6. Zhou, H., Xu, S.: Link prediction in heterogeneous networks based on multi-graph convolution. In: 6th International Conference on Artificial Intelligence and Advanced Manufacturing (AIAM 2024). IET, vol. 2024, pp. 704–713 (2024)
7. Adamic, L.A., Adar, E.: Friends and neighbors on the web. Soc. Netw. **25**(3), 211–230 (2003)
8. Newman, M.E.J.: Clustering and preferential attachment in growing networks. Phys. Rev. E Stat. Nonlinear Softw. Matter Phys. **64**, 025102 (2001)
9. Etude, P.J.: comparative de la distribution florale dans une portion des Alpes et des Jura. Bull. Soc. Vaud. Sci. Nat. **37**, 547–579 (1901)
10. Zhou, T., Lü, L., Zhang, Y.C.: Predicting missing links via local information. Eur. Phys. J. B **71**(4), 623–630 (2009)
11. Barabási A.L., Albert, R.: Emergence of scaling in random networks. Science **286**(5439), 509–512 (1999)
12. Jeh, G., Widom, J.: Simrank: a measure of structural-context similarity. In: Proceedings of the 8th ACM SIGKDD International Conference on Knowledge Discovery and Data Mining, pp. 538–543. ACM, New York (2002)
13. Katz, L.: A new status index derived from sociometric analysis. Psychometrika **18**(1), 39–43 (1953)
14. Chen, Y.L., Hsiao, C.H., Wu, C.C.: An ensemble model for link prediction based on graph embedding. Decis. Support Syst. **157**, 113753 (2022)
15. Zhang, M., Chen, Y.: Weisfeiler-Lehman neural machine for link prediction. In: The 23rd ACM SIGKDD International Conference, ACM (2017)
16. Zhang, M., Chen Y.: Link prediction based on graph neural networks. Adv. Neural Inf. Proc. Syst. **31** (2018)
17. Cai, L., Li, J., Ji, W.S.: Line graph neural networks for link prediction. IEEE Trans. Pattern Anal. Mach. Intell. **44**(9), 5103–5113 (2022)
18. Fang, Z., et al.: Elementary subgraph features for link prediction with neural networks. IEEE Trans. Knowl. Data Eng. **35**(4), 3822–3831 (2023)
19. Yu, H., Cao, X., Liu, Z., et al.: Identifying key nodes based on improved structural holes in complex networks. Phys. A Stat. Mech. Appl. **486**, 318–327 (2017)
20. Yang, H., An, S.: Critical nodes identification in complex networks. Symmetry **12**(1), 123 (2020)
21. Zhang, Y., et al.: Finding structural hole spanners based on community forest model and diminishing marginal utility in large scale social networks. Knowl. Based Syst. **199**, 105916 (2020)
22. Yang, Y., et al.: Link prediction via significant influence. Phys. A Stat. Mech. Appl. **492**, 1523–1530 (2018)

23. Rafiee, S., Salavati, C., Abdollahpouri, A.: CNDP: link prediction based on common neighbors degree penalization. Phys. A Stat. Mech. Appl. **539**, 122950 (2020)
24. Watts, D.J., Storgatz, S.H.: Collective dynamics of 'small-world' networks. Nature **393**(6684), 440–442 (1998)
25. Rossi, R.A., Ahmed, N.K.: The network data repository with interactive graph analytics and visualization. In: Proceedings of the 29th AAAI Conference on Artificial Intelligence, pp. 4292–4293 ACM, New York (2015)
26. Newman, M.E.: The structure of scientific collaboration networks. Proc. Natl. Acad. Sci. **98**(2), 404–409 (2001)
27. Newman, M.E.J.: Finding community structure in networks using the eigenvectors of matrices. Phys. Rev. E **74**(3), 036104 (2006)

Research on an Elevator Group Control Method Based on Interactive Artificial Bee Colonies

HongYan Zhang[1](✉), ChenLei Xie[1,2], Ping Wang[1,2], and YongXiang Zhong[1]

[1] School of Electronic and Information Engineering, Anhui Jianzhu University, Hefei 230601, Anhui, China
zswm27@163.com

[2] Anhui Province Key Laboratory of Intelligent Building and Building Energy Saving, Hefei 230022, Anhui, China

Abstract. With rapid economic development and an increasing number of high-rise buildings, the number of elevator units and service floors has also increased. How to use reasonable scheduling algorithms to reduce the energy consumption of elevator groups and improve operation efficiency has become a current research hotspot. To solve the problem of unsatisfactory performance of traditional group control algorithms in terms of time and energy consumption, an elevator group control algorithm is proposed, which implements an interactive artificial bee colony (ABC) for a new intelligent building platform. First, the Markov chain is used to simulate passengers' call behavior. Second, three performance indicators, average riding time, average waiting time, and energy consumption, are selected to construct a multiobjective function. Finally, the traditional ABC algorithm is improved. After being deployed to an intelligent computing node, the interaction between the optimal solution and the optimal fitness value between multiple nodes can be realized so that each intelligent node can quickly and accurately solve the optimal dispatching plan. The experimental results show that under the condition of equivalent energy consumption, the interactive ABC algorithm reduces the average waiting time by 24.3% compared with that of the traditional ABC algorithm, improves the operation efficiency, and verifies its effectiveness and feasibility in elevator group control scheduling.

Keywords: Artificial Bee Colony · Elevator Group Control · Multiobjective Function · Equivalent Energy Consumption

1 Introduction

With rapid economic development, the number of high-rise buildings has increased, and the energy consumption of public buildings has also increased annually, accounting for approximately 21.1% of the total energy consumption in society [1]. Elevator energy consumption has increased annually, accounting for 2%-6% of building energy consumption [2]. In addition, the operating efficiency of the elevator group is low, which greatly affects the passenger experience. A reasonable scheduling algorithm is the key to ensuring the efficient operation of the elevator group control system. At present,

L. Huang et al. (Eds.): CCF NCCA 2025, CCIS 2774, pp. 118–126, 2026.
https://doi.org/10.1007/978-981-95-5294-8_9

research on elevator group control systems considers two main aspects: passenger comfort and energy consumption. Among them, the time index of passenger comfort has been optimized [3, 4]. With the increasingly severe global energy consumption problem, the current main research direction of elevator group control is an elevator group control system that considers both time and energy consumption [5].

At present, the centralized dispatch mechanism is the mainstream control method of the elevator group control system; that is, the group controller is responsible for allocating the elevator car to provide services for the elevator call signal. The complex computing tasks undertaken by the group controller have become a bottleneck restricting the efficient operation of the elevator group control system [6, 7]. The Building Energy Conservation Research Center of Tsinghua University has developed a new intelligent building platform, which is different from the traditional centralized and bus-based system architecture in terms of overall architecture. The intelligent nodes of the platform have the characteristics of equality, self-organization, and plug-and-play. The nodes can share the complex computing tasks of the centralized architecture through collaborative computing [8]. Traditional elevator group control algorithms include genetic algorithms, fuzzy control, neural networks, expert control, etc. Beamurgia [9, 10] analysed the waiting time of passengers for the elevator scheduling problem, and the elevator performance improved; however, the genetic algorithm takes a long time to search for the optimal solution of the function. Zarikas [11] proposed an intelligent control method based on fuzzy experts to control elevator groups. However, the complexity of the relationships among multiple fuzzy rules reduces the efficiency of the system, and the database of rules needs to be updated manually. Fernandez [12] proposed a fuzzy adaptive mechanism, which automatically adjusts the fuzzy membership and fuzzy rules in the group controller through continuous changes in the call signal and the number of callers. Dursun [13] proposed the use of a neural network to quickly estimate the potential waiting time of each passenger, in which the passenger call signal and the number of passengers are used as the inputs of the network. Experiments show that in peak-up mode, the waiting time for the elevator is reduced by 18%. However, the learning time of the neural network is positively related to the network structure, which may affect the control effect of the elevator. Lang [14] used the artificial bee colony (ABC) algorithm's global optimization ability and fast convergence characteristics to solve the multiobjective problems of passenger waiting time, riding time, and number of stops in an elevator and realized rapid dispatch of elevator groups.

This paper studies an interactive ABC elevator group control method for a new intelligent building platform. On the basis of the reservation technology of the target floor, the interactive ABC algorithm is used to analyse the passenger call requests under three different modes upwards peak, downwards peak and interfloor traffic to calculate and provide the optimal dispatch plan.

2 Build the Objective Function

This study comprehensively considers three average performance indicators: average waiting time (*AWT*), average ride time (*ART*) and energy consumption (*EC*).

2.1 Building the AWT Function

The waiting time refers to the time from the time when the passenger calls the elevator to the floor where the elevator is called. Two main situations are considered: the waiting time when the elevator moves up and when the elevator moves down. Therefore, the average waiting time function is as follows:

$$wait_{av} = (\sum_{e=1}^{n} wt(e))/n \tag{1}$$

where n is the number of elevator parts.

When the call floor is above the elevator running direction, the waiting time is as follows:

$$wt(e) = \sum_{k=1}^{s(e)} ((c_2 - c_1) * T_1 + X * T_2 + 2M * T_3) \tag{2}$$

When the call floor is below the elevator running direction, the waiting time is as follows:

$$wt(e) = \sum_{k=1}^{s(e)} ((2l - (c_1 + c_2)) * T_1 + X * T_2 + 2M * T_3) \tag{3}$$

where $s(e)$ is the call signal, c_1 is the call floor, and c_2 is the elevator floor, l is the highest floor of the building, X is the total number of parking floors, T_1 is the running time of a single floor at a constant speed, which refers to the time at which the elevator between floors runs at the rated speed, T_2 is the stopping time of each floor;

$$T_2 = T_c + T_a + T_r - T_1 \tag{4}$$

where T_c and T_a represent the door closing time and door opening time, respectively, and T_r is the total running time of the elevator passing the first floor, including the acceleration and deceleration times, M is the number of passengers carried by the elevator in the call response, T_3 is the time for passengers to enter and exit;

$$T_3 = (T_i + T_o)/2 \tag{5}$$

where T_i and T_o refer to the average times for a single passenger to enter or leave the elevator, respectively.

2.2 Build ART Function

The elevator ride time refers to the time when the elevator enters the elevator from the passengers to the destination floor. Therefore, the average ride time function is as follows:

$$ride_{av} = (\sum_{e=1}^{n} rd(e))/n \tag{6}$$

When the destination floor is above the running direction of the elevator, the boarding time is as follows:

$$rd(e) = \sum_{K=1}^{s(e)} ((c_3 - c_1) * T_1 + X * T_2 + 2M * T_3) \tag{7}$$

When the destination floor is below the running direction of the elevator, the boarding time is as follows:

$$rd(e) = \sum_{K=1}^{s(e)} ((2l - (c_3 + c_1)) * T_1 + X * T_2 + 2M * T_3) \tag{8}$$

Among them, C_3 is the destination floor.

2.3 Building the EC Function

The energy consumption E_{sum} of the elevator group control system includes the energy consumption E_{sta} when the elevator starts, the energy consumption E_{stp} when the elevator stops, and the energy consumption E_{run} when the elevator runs at a constant speed.

$$E_{sum} = \sum_{e=1}^{n} (E_{sta}(e) + E_{stp}(e) + E_{run}(e)) \tag{9}$$

The elevator startup energy consumption E_{sta} mainly depends on the number of elevator startups, and the cumulative energy consumption of each elevator startup can be used to express the total energy consumption of the elevator in response to the call signal [15].

$$E_{sta}(e) = \sum_{k=1}^{s(e)} (E_{(e,k)}) \tag{10}$$

$$E_{(e,k)} = \int_0^{t_1} \begin{matrix} (a \wedge 2(m_{car} + m_{pass} + m_{counter})^* t_1 + \\ a^* g(m_{car} + m_{pass} - m_{counter})^* t_1) dt_1 \end{matrix} \tag{11}$$

where $E_{(e,k)}$ is the total energy consumption of the elevator startup when the elevator responds to the k-th elevator call signal, t_1 is the time for the elevator to accelerate from 0 to a constant speed, a is the acceleration of the elevator, v is the current running speed of the elevator, $v = at$, where t is the acceleration running time of the elevator.

The energy consumption of the elevator stop E_{stp} is similar to that of the elevator start energy consumption E_{sta}. The energy consumption of elevator operation E_{run} is mainly the process of the elevator overcoming gravity to do work, which depends on the distance of the elevator and the number of people in the elevator.

$$E_{run}(e) = \sum_{k=1}^{s(e)} (|\Delta m(e,k)| * g * h(e,k) * g(e,k)) \tag{12}$$

Among them, $\Delta m(e, k)$ is the quality difference of the elevator in response to the k-th call, and $h(e, k)$ is the distance moved by the elevator in response to the k-th call. In summary, the multiobjective function is constructed, and its expression is as follows:

$$F(x) = W_1 * wait_{av} + W_2 * ride_{av} + W_3 * E_{sum} \tag{13}$$

Among them, $F(X)$ is a multiobjective evaluation function. The smaller the evaluation function value is, the better the elevator dispatching effect. W_1, W_2 and W_3 are the corresponding weights of the objective function, and their sum is 1.

3 Group Control Strategy Based on the Interactive ABC Algorithm

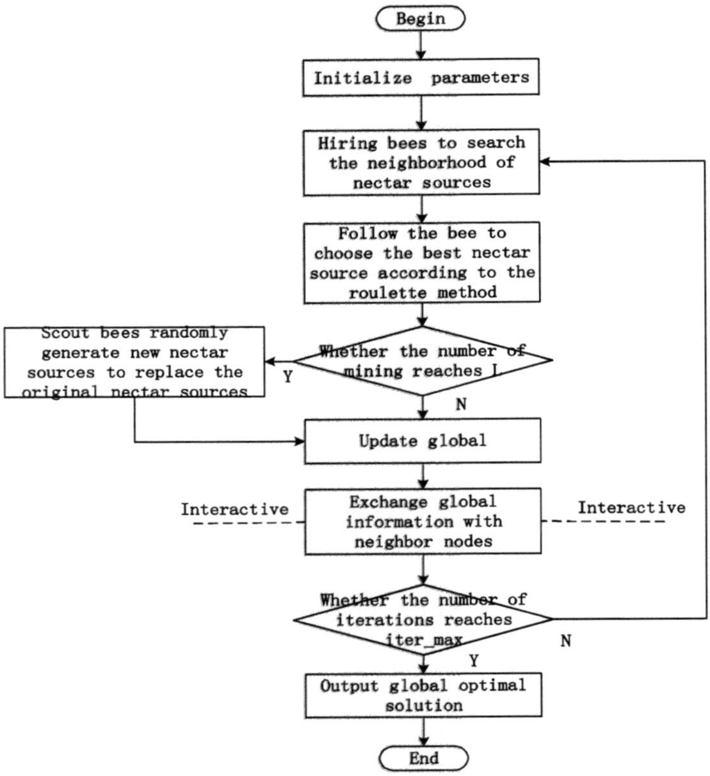

Fig. 1. Algorithm flow chart

For the optimal scheduling of elevator group control, the traditional artificial bee colony algorithm may require multiple iterations to solve this complex combination optimization problem, and sometimes, it may fall into a local optimum, which affects the overall performance of the algorithm. According to the characteristics of the new building intelligent platform, the idea of multinode collaborative computing is added, and the same improved algorithm is deployed at each node.

In the optimized dispatch of group-controlled elevators, according to the reservation technology of the target floor, the information of passengers calling on each floor, the initial position and running direction of each elevator and other data are obtained, and the interactive ABC algorithm is used to find a set of optimal dispatching schemes. The flow based on the interactive ABC algorithm is shown in Fig. 1.

4 Experimental Process and Result Analysis

4.1 Experimental Process

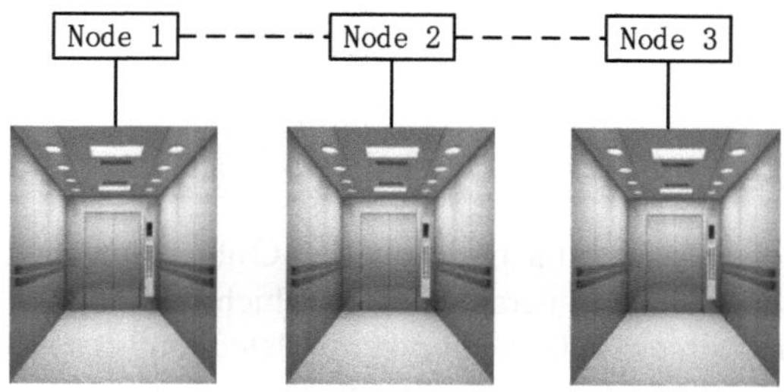

Fig. 2. Elevator topology

To verify the effectiveness of the interactive ABC algorithm in the elevator group control scheduling strategy, a group control elevator of a student apartment is taken as the simulation object, as shown in Fig. 2. The initial positions of the three elevators are 1, 5 and 10. The elevator operation parameters in the building are set according to Table 1 below.

In this paper, simulation verification is carried out for three traffic modes: upper peak, lower peak and interlayer traffic. The analytic hierarchy process is used to calculate the weights in different modes; see Table 2 below for details. Among them, in the down peak mode, it mainly occurs when students travel downstairs. The flow of students is large, and the main purpose is to transport students quickly. Therefore, it is necessary to minimize the waiting time and riding time. The same is true for the upper peak. In the interfloor transportation mode, students take the stairs randomly, the requirements for fast transportation are not high, and energy consumption should be minimized.

Table 2. Weight setting of different traffic modes

Traffic mode	W_1	W_2	W_3
Down peak	0.74816441	0.18040211	0.07143347
Up peak	0.63334572	0.10615632	0.26049796
Interlayer traffic	0.19318606	0.08330788	0.72350606

Table 1. Elevator parameter setting

parameter	Parameter value
Number of floors	10
Number of elevators	3
Opening and closing time/s	2
Single entry and exit time/s	1.5
Elevator running speed at constant speed/m/s	1.5
Single floor height/m	3
Rated load/kg	1050
Maximum number of people	14
Elevator acceleration/m/s^2	0.8

4.2 Results Analysis

According to the above research, the traditional ABC algorithm and the interactive ABC algorithm are set with the same parameters, in which the number of iterations is 41, the number of nectar sources is 20, the number of detection bees is 30, and comparative experiments are carried out under three traffic modes. Each elevator deploys an intelligent node embedded with the interactive ABC algorithm.

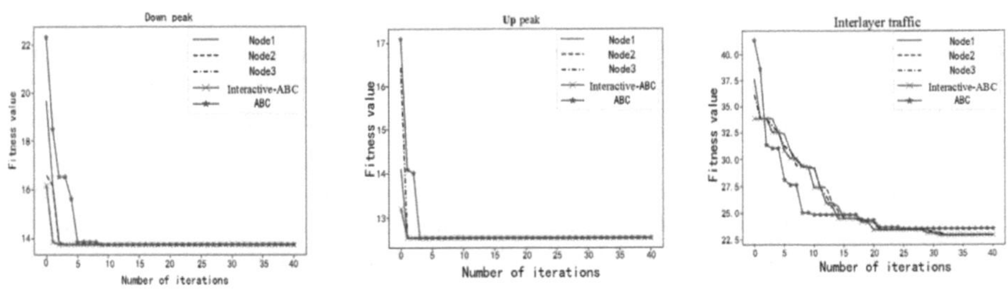

Fig. 3. Results of down peak interactive ABC algorithm

Figure 3 shows that in the downwards peak mode, the passengers' boarding directions are roughly the same, mostly from top to bottom. Both the interactive ABC algorithm and the traditional ABC algorithm converge, but the former has a better convergence speed than the latter; in the upwards peak mode, when the interactive ABC algorithm converges faster. In the interfloor traffic mode, the passenger call situation is more complicated, the distribution of people is uneven, and the direction of the elevator is different. The traditional ABC algorithm may fall into the local optimum, whereas the interactive ABC algorithm jumps out of the local optimum under the multiple interactions of 3 nodes and obtains a better dispatching plan.

Table 3. Performance index comparison

Traffic mode	algorithm	AWT(s)	ART(s)	EC(kwh)	Number of convergence iterations
Up peak	ABC	0	62.14	18.67	3
	Interactive-ABC	0	62.14	18.67	2
Down peak	ABC	12.94	41.81	4.33	4
	Interactive-ABC	12.37	42.26	4.14	3
Interlayer traffic	ABC	38.80	30.85	17.33	22
	Interactive-ABC	29.35	31.91	16.97	32

The performance indicators of the above different traffic modes are compared, as shown in Table 3. During the peak, the two dispatching algorithm dispatching schemes are the same. The call signals are concentrated from one floor to other floors, and there are elevators stopping on the first floor, so the average waiting time is 0 s. During the down peak, the interactive ABC algorithm for the average waiting time is shorter than the traditional ABC algorithm, which meets the requirements of passengers for fast boarding and improves passenger comfort. In the interlayer traffic mode, the interactive ABC algorithm reduces the average waiting time by 9.45 s compared with the traditional ABC algorithm, which is a reduction of 24.3%, and the energy consumption is reduced by 0.36 kWh, which is a reduction of 2.1%. On the whole, interactive ABC has comparable performance to traditional ABC, but the interactive ABC algorithm can converge quickly, greatly improving the operating efficiency of the system.

5 Conclusions

This paper implements the elevator group control dispatch based on the interactive ABC algorithm of the new architectural platform. The main work is as follows:

1) Utilizing the characteristics of equality, interaction, self-organization and collaboration with neighbor nodes of the platform, the traditional ABC algorithm is improved; that is, data interaction is performed in each iteration to achieve the goal of rapid convergence and easy jump out of the local optimum.
2) In the interlayer traffic mode, the interactive ABC algorithm reduces the average waiting time by 9.45 s compared with the traditional ABC algorithm, which is a reduction of 24.3%, and the energy consumption is reduced by 0.36 kWh, which is a reduction of 2.1%.

The results show that the interactive ABC algorithm is feasible and effective for elevator group control and dispatch and provides effective technical support for elevator group control in actual projects in the future.

Acknowledgement. This work is supported by the Anhui Provincial Department of Education project "Research on Indoor Personnel Action Recognition Method Based on UWB Radar".

References

1. China building energy consumption research report in 2019. Architecture (07), 30–39 (2020)
2. Liya, G.: Research on Energy Consumption Quota for Office Buildings of State Institutions in Guangzhou. South China University of Technology (2016)
3. Chou, S.Y., Budihi, D.A., Dewabharata, A., et al.: Improving elevator dynamicon Intelligent green building and smart grid (IGBSG), pp. 1–4. IEEE (2018)
4. Zhang, J.: Research on Energy-Saving Dispatch Method of Elevator Group Control System. Tianjin University (2014)
5. Wei, Q., Wang, L., Liu, Y., et al.: Optimal elevator group control via deep asynchronous actor-critic learning. IEEE Trans. Neural Networks Learn. Syst. (2020)
6. Wang, Z., Sun, D., Qiao, F., Wu, Q.: Multi objective scheduling algorithm for distributed elevator group control system. Control Theor. Appl. **27**(05), 602–608 (2010)
7. Shen, Q.: Research on Centerless Platform Architecture of Intelligent Buildings. Tsinghua University (2015)
8. Ziyan, J., Yunchuang, D., Yi, J.: Group intelligent building automation system. HVAC **49**(11), 2–17 (2019)
9. Beamurgia, M., Basagoiti, R., Rodriguez, I., et al.: A modified genetic algorithm applied to the elevator dispatching problem. Soft. Comput. **20**(9), 3595–3609 (2016)
10. Tartan, E.O., Ciftlikli, C.: A genetic algorithm based elevator dispatching method for waiting time optimization. IFAC-PapersOnLine **49**(3), 424–429 (2016)
11. Zarikas, V., Papageorgiou, E., Regner, P.: Bayesian network construction using a fuzzy rule based approach for medical decision support. Expert. Syst. **32**(3), 344–369 (2015)
12. Femandez, J., Cortes, P., Munuzuri, J., et al.: Dynamic fuzzy logic elevator group control system with relative waiting time consideration. IEEE Trans. Industr. Electron. **61**(9), 4912–4919 (2013)
13. Dursun, M.: Estimation of passenger waiting time in elevator systems with artificial neural network. Intell. Autom. Soft Comput. **16**(1), 101–110 (2010)
14. Lang, M.: Energy-saving Optimization Dispatching Simulation Research of Elevator Group Control System. Taiyuan University of Technology (2017)
15. Jiang, M.: Energy Consumption Calculation and Analysis of High-speed Group Control Elevator System. Shandong Jianzhu University (2015)

STUAformer: Spatio-Temporal Unified Aggregate Transformer Model for Traffic Speed Prediction

Chang Liu[1] and Yonghao Wu[2](✉)

[1] Institute of Information Technology, China Railway Guangzhou Group Co., Ltd., Guangzhou 510088, China

[2] School of Industrial Design, Hubei Institute of Fine Arts, Wuhan 430060, China
20232132@hifa.edu.cn

Abstract. Researchers have focused much attention on how to improve the prediction accuracy of traffic intensity because it is well known that the construction of intelligent transportation systems (ITSs) is particularly important for future urban development and that accurate traffic prediction is the key to improving the safety and efficiency of ITSs. On the one hand, the complex spatio-temporal relationships behind traffic patterns always affect the prediction accuracy, traditional models are used to extract spatio-temporal dependence via two different neural networks separately, and this type of combined model ignores spatio-temporal coexistence. On the other hand, researchers who utilize only past traffic data gathered in the hour before the predicted time point to build a case neglect similitude in traffic data where there are obvious repeats. This study proposes a spatio-temporal unified aggregate transformer (STUAformer), which first establishes a spatio-temporal matrix to maintain cross-space-time connectivity between nodes; then, traffic speeds from previous days that coincide with the historical observation series span for the present day are included as periodic references; and finally, the transformer is employed to mine the hidden spatio-temporal features behind the traffic data. Finally, utilizing three publicly available datasets, we compare the model with several state-of-the-art models. The findings demonstrate that the model greatly enhances traffic prediction.

Keywords: Intelligent transportation · traffic prediction · spatio-temporal dependence · transformer

1 Introduction

With the upgrading of the new generation of information technology, other traditional industries have been revitalized, and cross-disciplines such as smart buildings [1] and smart grids [2] are expanding rapidly. In particular, the research and implementation impact of the Internet of Vehicles stands out, and the advancements in wireless sensing technology, intelligent technology, etc., have significantly aided in establishing ITSs. The time series issue of forecasting the future

L. Huang et al. (Eds.): CCF NCCA 2025, CCIS 2774, pp. 127–143, 2026.
https://doi.org/10.1007/978-981-95-5294-8_10

traffic situation utilizing the recorded past traffic attribute (speed, flow, and occupancy) is the core of long-term investigations into traffic intensity prediction in ITSs. The accurate road trend prediction results allow it to forecast future traffic conditions, provide precise outcomes for forecasting traffic patterns and increase the standard of life for residents. However, this task is further complicated given the stochastic character of traffic flow and the intricate geographical and spatio-temporal dependence.

Researchers initially employed simple parametric models for prediction, and although some results were obtained, which require stability assumptions, they made it unattainable to represent the actual nonlinear changes in traffic intensity. Additionally, the models relied on subjective judgment, which was gradually abandoned over the years. Neural network models are becoming more prevalent in the field of ITSs because of the popularity of deep learning [3–5]. The technique of deep learning has been demonstrated to be a perfect fit for this type of extremely dynamic prediction problem. It is highly significant in the area of traffic intensity prediction because it can readily handle the nonlinear elements of traffic flow by incorporating activation functions. Among these, Narmadha *et al.* combined a convolutional neural network (CNN) and long short-term memory (LSTM) models to fuse multivariate information for prediction [6]. However, the CNN's parameter-sharing characteristic restricts the scope of the receptive field, and its invariance to the convolutional kernel makes it unsuitable for data with nonEuclidean structures, such as traffic networks. Kumar *et al.* therefore switched to employing a graph convolutional network (GCN) to address the spatial dependency between urban highways [7].

Although the aforementioned models are effective at simulating the spatio-temporal patterns of traffic data, they subjectively separate temporal and spatial properties and extract them via two distinct neural networks. The spatio-temporal characteristics of the actual coexisting relationships are strongly divided, so even though different neural network application scenarios differ and choosing the appropriate model will demonstrate a certain level of performance in forecasting the traffic prediction problem, it does not reflect the real situation. Thus, as shown in Fig. 1, this article aims to provide a novel aggregated modelling technique that enables the cross-space-time transmission of traffic information and unified capture of the intrinsic relationships among spatio-temporal variables.

In the past few years, GCN-based predictive models have gained much traction in the field of transportation research because of their exceptional performance on nonEuclidean structured data. Nevertheless, GCN-dependent models have limited expressive ability. The spatial connection relationships between nodes are not statically invariant in practice; it is unwise to keep building on GCN. On the other hand, the graph attention network (GAT) can assign different weights to nodes, alleviating the constraint that the GCN can handle only static topology to some extent. However, it still depends on prebuilt neighbor associations; it is incapable of accurately depicting the dynamic changes in the global traffic network. In actuality, as illustrated in Fig. 2, if an accident hap-

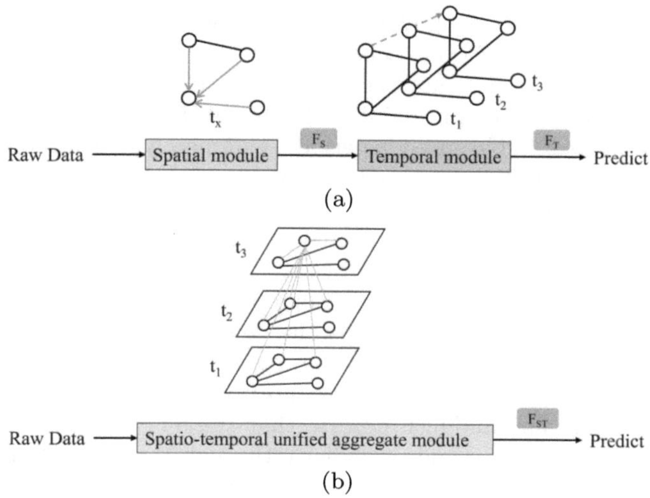

Fig. 1. Spatio-temporal dependence extraction of different models. (a) After the spatial relationships are extracted, the data containing the spatial characteristics is aggregated along the time axis. (b) Through unified aggregation, spatio-temporal information pertaining to the target node is simultaneously gathered.

pens on a certain road segment, the entire traffic network may become clogged, indicating that the target node is impacted by more than just its immediate neighbors.

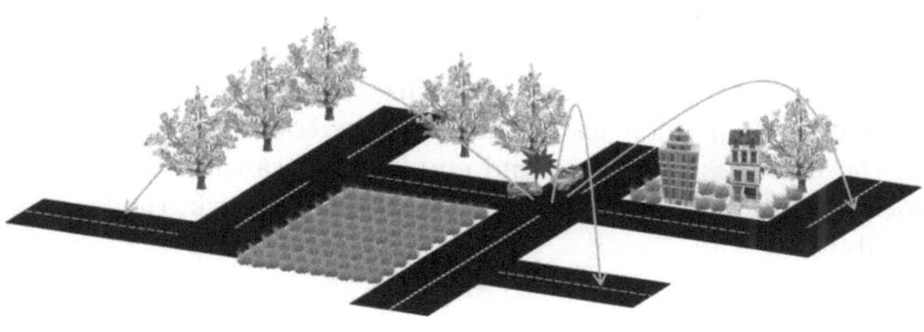

Fig. 2. Global spatial dependencies.

Owing to its use of positional coding and self-attention mechanisms, the transformer has increased its performance in long-sequence tasks in the field of natural language processing (NLP). Its introduction altered the era in which LSTM and gated recurrent unit (GRU) dominated with impressive results, and it has been successfully expanded to other domains, such as computer vision. Consequently, the use of transformers to capture long-term dependencies in time series prediction tasks is an effective strategy. In recent years, transformer-based research has also emerged [8,9], although it is still unable to overcome the constraint of spatio-temporal feature separation.

In addition, most models are accustomed to the observation sequence one hour before the predicted time point as historical data, while it is already an indisputable fact that traffic speed is periodic and merely depends on the current day's information of traffic speed, reducing the referability of the past data.

To imitate the spatio-temporal transfer relationship of real traffic flows, in this paper, we present a novel deep learning-based traffic flow prediction model architecture called the STUAformer. It efficiently utilizes the historical traffic data obtained from each sensor on the road network, extracts the trend and period features, and embeds the cross-space-time linkages into the spatio-temporal matrix. Additionally, our model simplifies the vanilla transformer model by using only the encoder portion and eliminating the temporal position coding. The findings indicate that the STUAformer extracts both spatio-temporal dependence and periodicity through the spatio-temporal layer, achieving the most efficient results when the most basic traffic speed data are used.

To summarize, our contributions are as follows:

1. We overcome the drawbacks of previous models that extract temporal dependence and spatial dependence separately, integrate spatial relationships under different timestamps in a spatio-temporal matrix, and accomplish the purpose of capturing spatio-temporal characteristics simultaneously through a new spatio-temporal modelling approach to unify aggregate traffic characteristics under the historical time window across time and space.
2. To fully utilize the historical traffic data, we employ the same time frame traffic data seven days before the prediction point (period feature) and the data one hour before the prediction point (trend feature) as the feature representation. This enables us to reveal hidden features in the flow data.
3. We abandon the way previous models utilized predefined static adjacencies to characterize the dependencies in the spatial extraction module, and instead take all the nodes as references and employ the transformer to focus on the dynamic correlation changes of the whole traffic network, thus enhancing the model's ability.

2 Related Work

With the expansion of the city scale, as well as improvements in the living standards of residents, the number of motor vehicles is also increasing daily, and the resulting road traffic problems are highlighted. Issues such as road congestion [10], noise pollution, environmental pollution, and increased commuting time significantly affect the comfort of people. There is minimal efficacy in employing conventional traffic diversion techniques, including tidal lanes, road enlargement, and extra viaducts. Applying computer knowledge to address traditional problems in the field of transportation and realizing digital, networked, and intelligent transformation of the field of transportation, by improving the accuracy of the prediction of future traffic intensity, is especially important for improving the road utilization rate, efficient management of traffic and development of the city, and it is also the primary focus of the ITSs.

Early prediction models based on linear regression, the autoregressive integrated moving average (ARIMA) model, Kalman filtering, and other parametric models, although straightforward and not requiring a high level of computing capability, assume that the prediction sequence is stable and static. Because of this, they are unable to handle nonlinear and stochastic data, and the outcomes will not be what was anticipated in complex cases. Therefore, they have quite limited ability to cope with traffic flow that may occur unexpectedly at any moment in real-world scenarios. Since then, researchers have focused on implementing deep learning to address this issue. Deep learning models extremely complicate unknown relationships by utilizing data transfer and operations across several neurons. Among these methods, recurrent neural networks (RNNs) have been frequently utilized in traffic prediction modelling; however, RNN-based methods, although more accurate, feature gradient vanishing and gradient explosion. Even with its variants, the model has the drawback of being prone to not capturing long-term dependence. Furthermore, the RNN's capacity to represent intricate temporal dynamics is constrained by the fact that they share weights at various times. The aforementioned models only consider modelling individual target node features along the temporal dimension; however, various roadway network observation points can constrain and influence one another, and a roadway section's speed change is related to both its evolution and that of other roadway sections. As a result, they fail to consider features at the spatial level.

To increase the model's capacity to extract spatial information, researchers have added CNN-based spatial extraction modules. Before feature extraction, a more general approach is to divide the traffic network into a regular grid form that facilitates feature extraction via the convolutional kernel. A hybrid CNN-BiLSTM model was proposed by Méndez *et al.* [11]. It first uses a CNN to filter and remove noise data traffic data to extract valuable information as new feature inputs. After that, it uses the special training mechanism of BiLSTM to learn the temporal flow relationship of including two types of contextual information from data processed by the CNN. Wang *et al.* implemented the CNN-LSTM model to convert traffic data into images, extract spatio-temporal correlations, and add three correlation features: ten distinct weather conditions; three holiday types (weekday, weekend, and Cont_Holiday); and five vehicle types, such as sedan and bus [12]. Finally, it produces more accurate prediction outcomes. However, geospatial regions must be preprocessed beforehand to extract spatial features in traffic networks, as CNNs are primarily appropriate for handling regular grid structures. This method is not the best spatial extraction model and is highly prone to losing spatial information.

This issue is resolved by the development of the GCN, which is a neural network based on a graph structure. In a sense, the GCN are appropriate for nonEuclidean structured data in irregular situations, making them an excellent choice for comprehending complex topologies such as transportation road networks. In contrast to CNNs, GCN abstract the traffic network into a graph that suggests the connectedness between junctions rather than an image when it addresses the problem of traffic spatial information extraction.

Peng *et al*. proposed a new prediction model called the MA-GCN model [13], which uses a GCN to learn the complex spatial characteristics of the traffic network and takes advantage of the differential neural computer's long-term memory ability to capture the long-term traffic information in the time dimension to avoid the issue of the traditional RNN model's poor memory ability. Finally, the two models' efficacy in acquiring spatial and temporal characteristics is confirmed through an ablation test. To enhance prediction performance, Fang *et al*. first employed a GCN to identify spatial dependencies in a traffic graph. The following employed LSTM to identify temporal correlations in the data that had been processed by the GCN. Finally, they combine features from several data sources in the multi-source bridging stage [14]. The GCN block and temporal trend-aware multi-head self-attention block were incorporated by Guo *et al*. to build a spatio-temporal encoder layer that captured local contextual information in the temporal dimension and pertinent dynamics in the spatial dimension [15]. Then, in the spatio-temporal decoder section, autoregressive was implemented to generate the output. However, the GCN does not fit actual road network structural relationships because it mostly depends on preconstructing the adjacency matrix from 0 s and 1 s, and its static structure presupposes constant spatial correlation.

Transformers have gained increasing popularity in the field of NLP in recent years. It solves the dependence problem in the long-sequence job by employing the self-attention mechanism to determine the links between word vectors. This approach has some research significance for the traffic prediction problem. Each data point can be influenced by other data under the effective global receptive field via the transformer model's particular attention mechanism, which can successfully simulate the intricate traffic dynamic correlation. The STGAGRTN model [16], which was proposed by Wu *et al*., is divided into four parts: first, the short-term dependencies in the input sequences are captured with the GRU; second, the spatial dependencies are extracted via GAT and the proposed spatial transformer; third, the temporal transformer is used to address issues with traditional RNN models, such as data forgetfulness; and finally, multistep prediction is carried out via two conventional convolutional layers. Zhao *et al*. investigated the human mobility information concealed in traffic data and used a multi-view graph convolution network to extract the invisible spatial attributes of the intersection network nodes [17]. They then used a transformer to handle the data's random interference during the temporal feature extraction stage, and finally, they set up a ConvLSTM layer that can make predictions on the basis of a larger perceptual field in the temporal dimension.

The aforementioned models produced excellent forecasting results; however, it is straightforward to see that they all utilize two modules to extract temporal and spatial dependencies separately. The former typically employs an RNN and its variations, whereas the latter is typically used as a CNN or GCN. The fundamental idea behind adopting a popular transformer is still the same. The complex spatio-temporal dependencies that underlie traffic data coexist and are neither sequential nor cut-off, even though the use of appropriate models to address data at different conceptual levels is effective. Separating temporal and spatial blocks not only contradicts the fundamental features of traffic data but also increases errors in interaction.

Moreover, transportation data are trending and periodic, with intricate temporal and spatial relationships. Conventional models typically use historical observation data from an hour before the prediction time point as the reference sequence. This limits the model's ability to make predictions by focusing only on the day's historical flow characteristics. Urban roads present repetitive traffic conditions, such as daily morning and evening peak flows, that differ significantly from those of the rest of the day. Using beneficial traffic data for the remaining days in the same period can reveal the hidden patterns behind the repetitive periodical features and enrich the historical data.

To capture the spatio-temporal transfer relationship of traffic data in the road network in cross-space-time, the STUAformer model proposed in this paper applies the constructed spatio-temporal matrix. This accomplishes the goal of capturing the spatio-temporal dependence simultaneously and prevents the introduction of error by contradicting the dimensional characteristics of the real traffic data. The model data also incorporated repetitive period features, which allowed the model to learn from cyclical fluctuations in the time dimension while capturing trending changes in time-varying traffic information.

Additionally, our model's fundamental building block is built on a transformer for variations, which draws inspiration from it. It uses its potent time-series feature extraction capabilities to accomplish the task of traffic prediction for the transportation domain.

3 Preliminary

Definition 1: A traffic road network's topology is defined as $G = (V, E, A)$, where $V = \{v_1, v_2, \cdots, v_N\}$ represents the set of nodes that could be roads, intersections, or sensors; E represents the set of edges, which characterizes the connectivity between nodes; and N is the total number of nodes. A is an all-1 matrix called one matrix, indicating that all nodes are potentially correlated, and adaptive relationship modelling is subsequently performed on the basis of the model.

Definition 2: To illustrate the evolution of node features over time, we define the feature matrix $X = \{x_1, x_2, \cdots, x_t\} \in R^{N \times f \times t}$ to represent the traffic attributes of every node in the road network, where x_i denotes the characteristics of nodes in the traffic road network at moment i, t denotes the length of the historical time frame, N denotes the number of all nodes, and f denotes the number of features, which can be traffic density or flow.

The issue formulation of a comprehensive traffic speed prediction model is thus expressed in Eq. (7), where a mapping function $f(\cdot)$, given a road network

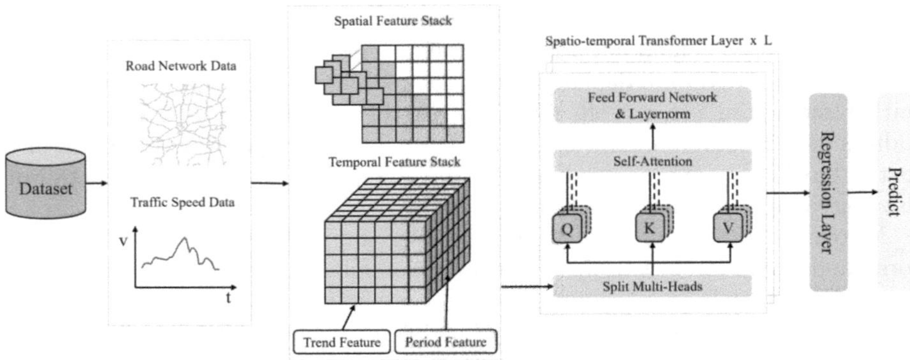

Fig. 3. Architechture of Spatio-temporal Unified Aggregate Transformer.

architecture G, and a history feature matrix X of time slice length t forecast the traffic intensity at p future time steps:

$$(G; X) \xrightarrow{f} (x_{t+1}, x_{t+2}, \cdots, x_{t+p}) \tag{1}$$

4 Proposed Model

The study of projecting future traffic road conditions via historical traffic data is known as traffic intensity prediction. Traffic information often includes parameters such as traffic density, traffic flow, and traffic speed. The average speed of a road segment can indicate the traffic condition of that part; hence, this study chooses traffic speed as a characteristic for experimentation.

The implementation specifics of the STUAformer model, which is built as shown in Fig. 3, are covered in detail in this section. To capture dynamic spatio-temporal dependencies, we first process complex spatio-temporal features by embedding neighborhoods under various timestamps into a spatio-temporal graph and stacking trend features with period features. Next, we train with the spatio-temporal transformer layer, which uses its special self-attention mechanism to extract differential features from contextual information. Finally, we apply a regression layer to predict future traffic intensity.

Conventional traffic prediction models allow the central node to aggregate only spatially transmitted information from the traffic features of adjacent nodes at the same moment by encoding the adjacency matrix under a single timestamp into the architecture. In contrast, to aggregate the traffic speed data from under the cross-space-time through a unitive model, the STUAformer model, which is suggested in this paper, defines a spatio-temporal matrix $M \in R^{NT \times NT}$, as illustrated in Eq. (2), which integrates all the spatial relationships in the historical time window:

$$M = \begin{bmatrix} A & \mathbf{0} & \cdots & \mathbf{0} \\ A & A & \cdots & \mathbf{0} \\ \vdots & \vdots & \ddots & \vdots \\ A & A & \cdots & A \end{bmatrix} \tag{2}$$

As shown in Eq. (2), when predicting the traffic speed in the future, we need to consider only the historical data before a specific time step and hide the future timestamps, so the spatio-temporal matrix defined in this paper is a lower-triangular matrix to prevent the problem of future data leakage. Since distant and disconnected spatial points also affect the central node as previously described, A is specified as one matrix, which means that in the initialized state, all of the node attributes that are by default global have an impact on one another, making the model focus on the entire traffic network rather than local adjacencies. In the end, spatio-temporal information transfer from various nodes across timestamps may be readily implemented via the spatio-temporal adjacency matrix, allowing the simultaneous extraction of spatio-temporal features from other nodes.

As discussed in Sect. 1, the node's traffic speed is influenced by both the trend of nearby periods and a certain amount of periodic repetitiveness. The feature matrix in this study has two different pattern inputs: trend features and period features. This allows the model to manage remote dependencies on prior timestamps and, as a result, understand the deeper aspects of traffic data. Under a time frame of fixed size T, the historical trend characteristics $X_i^{(t)} = \{x_{i-T+1}, x_{i-T+2}. \cdots, x_i\}$ is used to characterize the trending impacts of traffic speeds before the anticipated time point i of the day. Furthermore, to guarantee that the two features are sampled at the same time step, the historical periodic feature $X_i^{(p)} = \{x_{i-\lambda d-T+\theta+1}, x_{i-\lambda d-T+\theta+2}, \cdots, x_{i-\lambda d+\theta}\}$ is applied to characterize the repetition of historical traffic speeds over the preceding few days of the same period. Here, θ represents the length of the predicted timestamp, λ indicates the number of samples in a day, and $d \in (1, \delta)$ denotes the range of values for the number of historical days before the day of prediction. Thus, $X_i = [X_i^{(t)} || X_i^{(p)}]$ is the node's final feature matrix at time i, where $X_i \in R^{N \times f}$ and f equal $1 + \delta$.

First, we use a fully connected layer to map the traffic data information into a high-dimensional hidden space to maintain the original traffic data information:

$$H = FC(X) \tag{3}$$

where $H \in R^{N \times T \times h}$ and h is the mapped feature dimensions and where $FC(\cdot)$ is a fully connected layer.

After that, we weighted and combined the created spatio-temporal feature matrix H through the spatio-temporal transformer layer to record the traffic speed data on a cross-space-time basis. The spatio-temporal query Q_{ST}, spatio-temporal key K_{ST}, and spatio-temporal value V_{ST} matrices are first obtained via the original feature matrix as follows:

$$Q_{ST} = W_Q H, K_{ST} = W_K H, V_{ST} = W_V H \tag{4}$$

where $W_Q, W_K, W_V \in R^{h \times h}$ is the learnable matrix.

Then, via Eq. (5), we calculate the attention score, which describes the level of association between different nodes at various times.

$$S = Softmax(\frac{Q_{ST} K_{ST}^T}{\sqrt{h}}) \tag{5}$$

Then, the output of the middle layer is obtained according to Eq. (6):

$$H_{ST} = S V_{ST} \tag{6}$$

The model then combines the input feature matrix (target matrix) of each layer with the output matrix to aggregate the spatio-temporal relationships (spatio-temporal correlation matrix) and multiplies the result by the weight matrix $W \in R^{2h \times h}$. The final weighted summation result is then utilized as the input matrix of the subsequent layer, which is determined by the following formula, and $ReLU$ is utilized as the activation function for computation.

$$H^{l+1} = ReLU(W^l (H^l || H_{ST}^l)^T)^T \tag{7}$$

where W is a learnable weight parameter that is used to adjust the expressiveness of the model by giving the target matrix and the spatio-temporal correlation matrix varying degrees of importance.

Finally, the final spatio-temporal embedding information $H_e \in R^{N \times T \times h}$ is transmitted through the regression layer (two fully connected layers) to generate the predicted result $Y \in R^{N \times T' \times 1}$ of the future traffic speed, where T' is the horizon of the forecast, and the parameters are changed by the mean squared error (MSE) as a loss function.

As with the conventional transformer, we employ layer normalization, residual connections, and a multi-head mechanism to enhance the model's performance and stability.

5 Experiment

In this section, we describe the datasets and the experimental setup along with a comparison and analysis of the findings.

First, to guarantee the validity of the model's experimental results, we selected the public datasets PeMSD4, PeMSD7, and PeMSD8, which were gathered in collaboration with research institutes and transportation departments in California, U.S.A. These datasets are based on actual traffic data that are gathered by roadside detectors every five minutes and are widely used in the field of traffic prediction. The specifics of the datasets are as follows:

The first 47 days of data in PeMSD4 were utilized as the training set in this case, and the remaining 11 days served as the test set. The May data from PeMSD7 serve as the training set, whereas the June data serve as the test set.

The PeMSD8 uses the first 50 days as the training set, whereas the remaining data are used for testing. We utilize one hour of data as the historical reference data, meaning that the historical window size T is set to 12, to forecast the speeds for the upcoming 15-, 30-, 45-, and 60-min moments. In the meantime, we utilize the data from the previous week as the period features; thus, f equal to 8 is the model's entire feature dimension. The number of layers in the spatio-temporal transformer layer is set to 3, and the number of heads is set to 2. In addition, we set the initial learning rate to 0.0001 and the number of training epochs to 200 and utilized Adam as the optimizer.

To assess the model's performance, we employ the RMSE, MAE, and MAPE as assessment metrics and compare them to the commonly used baseline DCRNN [18], STGCN [19], ASTGCN [20], LSGCN [21], and SST-GNN [22]. The experimental findings are displayed in Table 1. Our model's metrics are optimal for all four prediction lengths on the three datasets, as the table shows. This suggests that our model can mine the inherent transfer patterns that are present in the traffic flow via a unified framework to extract the spatio-temporal dependencies concurrently. In addition, we utilize the speed data from the last seven days to identify recurring period patterns, compensating for the drawbacks of other models that use only the current day's historical data. One matrix overcomes the limitation of the conventional model, which considers only the neighbors surrounding the core node and broadens the model's receptive field. Furthermore, the model's performance is enhanced by applying the weight matrix to assign varying importance to distinct nodes under various timestamps since the influence of the road nodes on the center node is not constant as time passes.

Table 1. The performance comparison of the STUAformer with the baselines

Dataset	Models	15 min			30 min			45 min			60 min		
		RMSE	MAE	MAPE	RMSE	MAE	MAPE	RMSE	MAE	MAPE	RMSE	MAE	MAPE
PeMSD4	DCRNN [18]	2.94	1.35	2.68	4.06	1.77	3.71	4.77	2.04	4.48	5.28	2.26	5.10
	STGCN [19]	3.01	1.47	2.92	4.21	1.93	3.98	5.01	2.26	4.73	5.65	2.55	5.39
	ASTGCN [20]	3.96	2.12	4.16	4.59	2.42	4.80	4.97	2.60	5.20	5.21	2.73	5.46
	LSGCN [21]	2.93	1.45	2.90	3.92	1.82	3.84	4.47	2.04	4.42	4.83	2.22	4.85
	SST-GNN [22]	2.53	1.23	2.37	3.47	1.82	3.69	3.86	1.84	3.93	4.45	2.13	4.69
	STUAformer	2.48	1.19	2.34	3.09	1.47	3.10	3.49	1.62	3.50	3.82	1.74	3.89
PeMSD7	DCRNN [18]	4.25	2.22	5.16	6.02	3.04	7.46	7.24	3.64	9.26	8.20	4.15	10.82
	STGCN [19]	4.01	2.24	5.28	5.74	3.04	7.46	6.85	3.61	9.00	7.69	4.08	10.23
	ASTGCN [20]	5.15	2.85	7.25	6.12	3.35	8.67	6.77	3.70	9.73	7.20	3.96	10.53
	LSGCN [21]	3.98	2.22	5.14	5.47	2.96	7.18	6.39	3.43	8.51	7.09	3.81	9.60
	SST-GNN [22]	3.53	2.04	4.77	4.80	2.67	6.66	5.79	3.17	8.00	6.39	3.48	9.04
	STUAformer	3.29	1.81	4.22	4.28	2.31	5.60	4.93	2.68	6.68	5.33	2.91	7.32
PeMSD8	DCRNN [18]	2.59	1.17	2.32	3.56	1.49	3.21	4.13	1.71	3.83	4.50	1.87	4.28
	STGCN [19]	2.62	1.19	2.34	3.61	1.59	3.24	4.21	1.92	3.91	4.68	2.25	4.54
	ASTGCN [20]	3.18	1.49	3.16	3.69	1.67	3.59	3.92	1.81	3.98	4.13	1.89	4.22
	LSGCN [21]	2.45	1.16	2.24	3.28	1.46	3.02	3.75	1.66	3.51	4.11	1.81	3.89
	SST-GNN [22]	2.08	1.03	1.86	2.80	1.39	2.67	3.28	1.62	3.20	3.57	1.74	3.50
	STUAformer	1.96	0.94	1.72	2.47	1.13	2.14	2.77	1.24	2.38	2.98	1.33	2.62

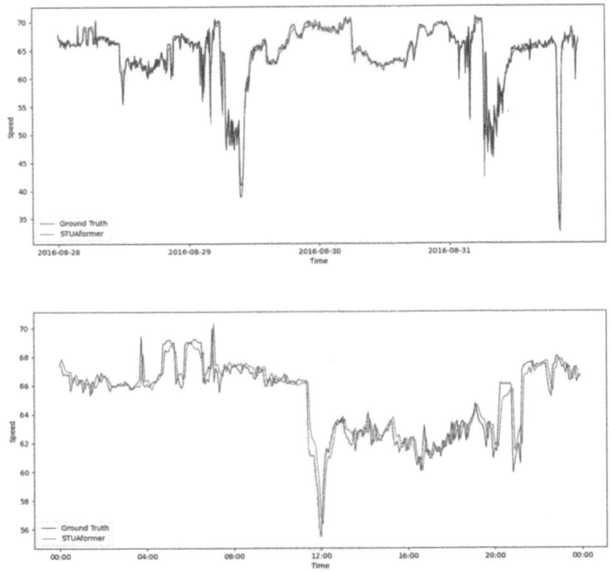

Fig. 4. The predicted results for the next 15 min.

We demonstrated the prediction results of a specific section in the PeMSD8 dataset in comparison with the actual values to more clearly illustrate the STU-Aformer model's prediction capacity. The intensity predictions for the following 15 min, 30 min, 45 min, and 60 min are shown in Figs. 4, 5, 6 and 7, respectively.

We conducted ablation experiments to confirm the validity of the model parameters.

First, the STUAformer model has a multi-head attention mechanism, in which various heads can generate multiple subspaces, giving the model more diverse perspectives. As a result, selecting the appropriate number of heads enhances the model's capacity for nonlinear expressiveness. Table 2 displays the findings of a prediction experiment performed on the PeMSD8 dataset on the speed of traffic over the following 15 min.

The experimental results demonstrate that our model is not sensitive to the number of heads. The experimental results based on the different head numbers are all superior to the baseline, and we ultimately decided to adopt the two heads that perform marginally better overall than the other comparison counts as the model metrics.

The magnitude of the temporal span of the period characteristics was then the subject of a comparative investigation. We experimentally evaluated the impact of speed data from various historical days on the model's predictive capacity since, as previously stated, the same period flow of the days preceding

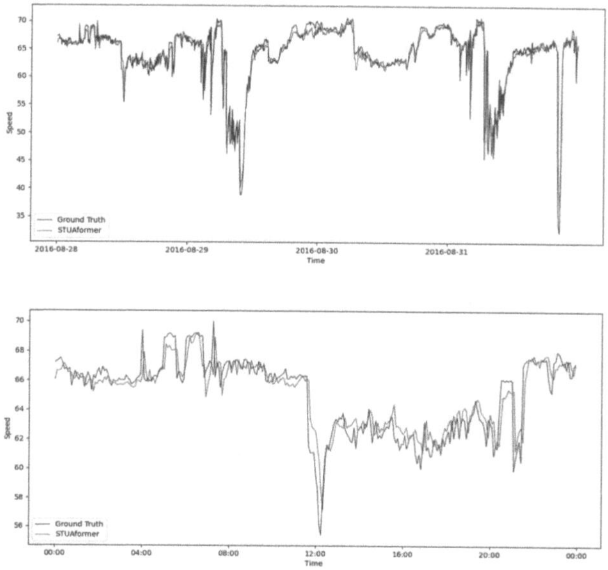

Fig. 5. The predicted results for the next 30 min.

the forecast point exhibited comparable tendencies. Table 3 displays the experimental outcomes for the comparison periods of 4 days, 7 days, and 10 days.

The findings indicate that when the number of history days is 7, the model performs well at various prediction levels. This implies that there are insufficient similarity features to refer to if the number of days of history is too short, which prevents the model from learning from a broader sample. However, excessive historical days can accumulate noise collected by the sensors themselves, which can reduce the predictive power of the model.

This paper suggests that nodes that are not directly adjacent to the target node have an impact on it. This leads us to utilize one matrix to aggregate velocity data from the global network. We perform a comparative test by changing

Table 2. Comparative results of experiments under different head numbers

Number of head	15 min		
	RMSE	MAE	MAPE
1	1.98	0.94	1.72
2	1.96	0.94	1.72
4	1.98	0.95	1.74
8	1.97	0.94	1.70
16	1.97	0.95	1.74

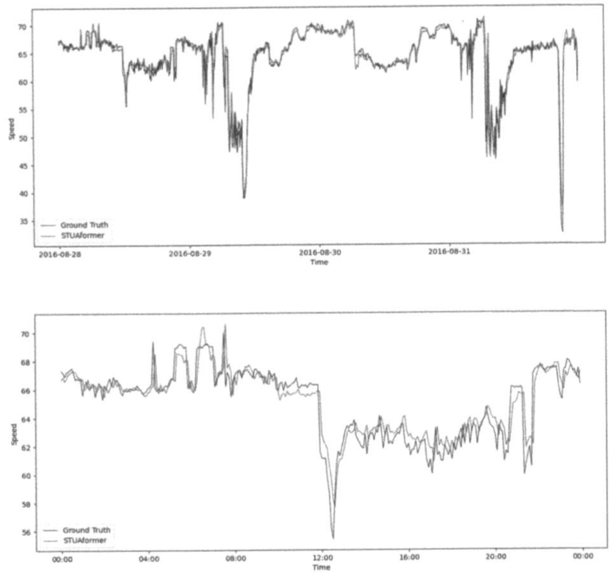

Fig. 6. The predicted results for the next 45 min.

Table 3. Comparative results of experiments under different historical time windows

Historical time window	15 min			30 min			45 min			60 min		
	RMSE	MAE	MAPE	RMSE	MAE	MAPE	RMSE	MAE	MAPE	RMSE	MAE	MAPE
4	2.14	1.04	1.93	2.70	1.25	2.42	3.19	1.48	3.04	3.56	1.66	3.44
7	1.96	0.94	1.72	2.47	1.13	2.14	2.77	1.24	2.38	2.98	1.33	2.62
10	2.78	1.09	2.08	2.85	1.29	2.59	3.20	1.44	2.92	3.43	1.55	3.21

the submatrix A of the spatio-temporal matrix M from one matrix to an adjacency matrix (where only the positions of the directly connected nodes are set to 1 and the rest are set to 0) to confirm whether this assumption has an impact on the model's performance. The specific outcomes are displayed in Table 4.

It is evident from Table 4 that employing the adjacency matrix results in somewhat poorer prediction performance. This suggests that the mutual influence between various road segments is a global process that depends on more than just the data upstream and downstream; thus, enlarging the model's spatial receptive field is crucial for traffic speed prediction.

Finally, perturbation analysis experiments are performed to assess the model's robustness. Figure 8 shows the final evaluation metric findings after applying Poisson noise $X \sim Possion(\lambda)$ to the raw data, where the vertical coordinates are the metric findings determined using RMSE, MAE, and MAPE for the three datasets, and the horizontal coordinates are the parameters $\lambda \in \{1, 2, 4, 8, 16\}$.

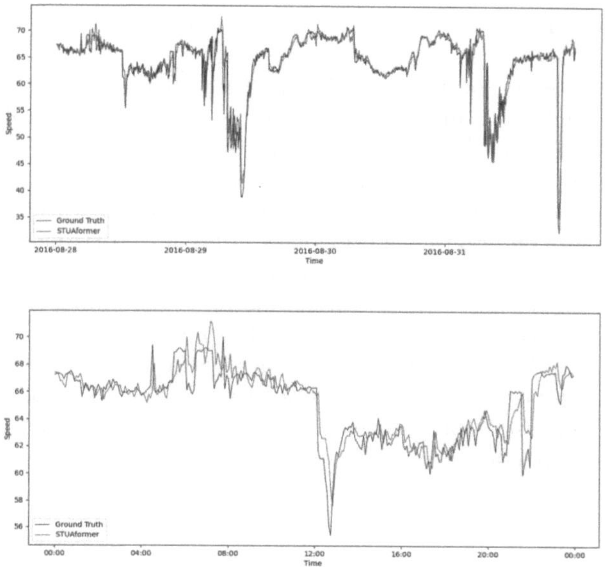

Fig. 7. The predicted results for the next 60 min.

Table 4. Comparative results of experiments with different submatrix types of spatio-temporal matrix

Submatrix type	15 min			30 min			45 min			60 min		
	RMSE	MAE	MAPE	RMSE	MAE	MAPE	RMSE	MAE	MAPE	RMSE	MAE	MAPE
Adjacency matrix	2.07	1.07	1.95	2.59	1.37	2.50	2.83	1.33	2.54	3.09	1.54	2.94
One matrix	1.96	0.94	1.72	2.47	1.13	2.14	2.77	1.24	2.38	2.98	1.33	2.62

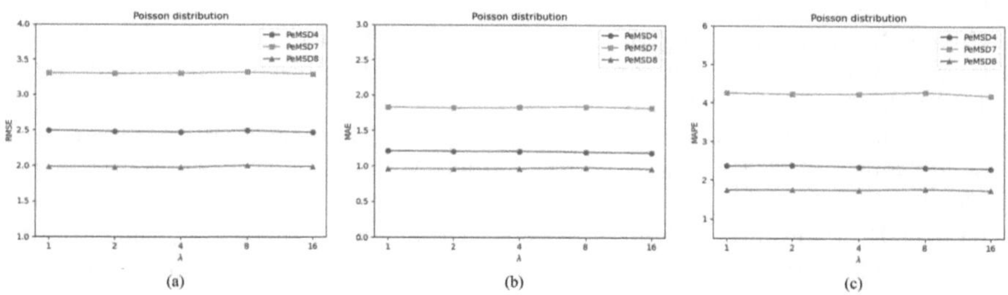

Fig. 8. Perturbation analysis with Poisson noise. (a) RMSE, (b) MAE, and (c) MAPE results under different noise distribution.

The model's predictions essentially remain the same regardless of the amount of noise introduced, as the image illustrates, suggesting that the STUAformer has some degree of noise resistance.

6 Conclusion

The future development of urban intelligence depends on traffic prediction, and an accurate prediction result contributes to enhancing the overall efficiency of urban traffic transit. The STUAformer model is proposed in this work to capture its complicated dynamic spatio-temporal characteristics. The model incorporates the cross-space-time interrelationship into a united spatio-temporal graph, eschewing the common practice in previous studies of factorizing spatio-temporal elements by designing complex models. To fully extract spatial dependencies, we extend the spatial receptive field. Furthermore, to maximize the utilization of historical data, period patterns are embedded into the hidden representation of nodes. The model achieves state-of-the-art performance on both short-term and long-term forecasts, according to the results of experiments on three real datasets. This suggests that our model simultaneously extracts the intricate spatio-temporal dependencies within the traffic speed.

References

1. Mathumitha, R., Rathika, P., Manimala, K.: Intelligent deep learning techniques for energy consumption forecasting in smart buildings: a review. Artif. Intell. Rev. **57**(2), 35 (2024)
2. Ohanu, C.P., Rufai, S.A., Oluchi U.C.: A comprehensive review of recent developments in smart grid through renewable energy resources integration. Heliyon. **10**(3) (2024). Accessed Feb 2024
3. Benarmas, R.B., Bey, K.B.: A deep learning-based framework for road traffic prediction. J. Supercomput. **80**(5), 6891–916 (2024)
4. Moumen, I., Mahdaoui, R., Raji, F.Z., Rafalia N., Abouchabaka, J.: Distributed multi-intersection traffic flow prediction using deep learning. In: E3S Web of Conferences, vol. 477, p. 00049. EDP Sciences (2024)
5. Huang, X., Jiang Y., Tang, J.: MAPredRNN: multi-attention predictive RNN for traffic flow prediction by dynamic spatio-temporal data fusion. Appl. Intell. **53**(16), 19372-19383. Accessed Aug 2023
6. Narmadha, S., Vijayakumar, V.: Spatio-temporal vehicle traffic flow prediction using multivariate CNN and LSTM model. Mater. Today: proc. **1**(81), 826–33 (2023)
7. Kumar, R., Panwar, R., Chaurasiya, V.K.: Urban traffic forecasting using attention based model with GCN and GRU. Multimedia Tools Appl. **83**(16), 47751–74 (2024)
8. Jiang, J., Han, C., Zhao, W.X., Wang, J.: Former: propagation delay-aware dynamic long-range transformer for traffic flow prediction. In: Proceedings of the AAAI Conference on Artificial Intelligence, vol. 37, no. 4, pp. 4365–4373. Accessed 26 Jun 2023
9. Liu, H., et al.: Spatio-temporal adaptive embedding makes vanilla transformer sota for traffic forecasting. In: Proceedings of the 32nd ACM International Conference on Information and Knowledge Management, pp. 4125-4129. Accessed 21 Oct 2023
10. Sayed, S.A., Abdel-Hamid, Y., Hefny, H.A.: Artificial intelligence-based traffic flow prediction: a comprehensive review. J. Electr. Syst. Inform. Technol. **10**(1), 13 (2023)

11. Méndez, M., Merayo, M.G., Núñez, M.: Long-term traffic flow forecasting using a hybrid CNN-BiLSTM model. Eng. Appl. Artif. Intell. **1**(121), 106041 (2023)
12. Wang, J.D., Susanto C.O.N.: Traffic flow prediction with heterogenous data using a hybrid CNN-LSTM model. Comput. Mater. Continua. **76**(3). Accessed 1 Oct 2023
13. Peng, D., Zhang, Y.: MA-GCN: a memory augmented graph convolutional network for traffic prediction. Eng. Appl. Artif. Intell. **1**(121), 106046 (2023)
14. Fang, Z., Pan, L., Chen, L., Du, Y., Gao, Y.: MDTP: a multi-source deep traffic prediction framework over spatio-temporal trajectory data. Proc. VLDB Endowment. **14**(8), 1289–97 (2021)
15. Guo, S., Lin, Y., Wan, H., Li, X., Cong, G.: Learning dynamics and heterogeneity of spatial-temporal graph data for traffic forecasting. IEEE Trans. Knowl. Data Eng. **34**(11), 5415–28 (2021)
16. Wu, D., Peng, K., Wang, S., Leung, V.C.M.: Spatial-temporal graph attention gated recurrent transformer network for traffic flow forecasting. IEEE Internet Things J. **11**(8), 14267–81 (2023)
17. Zhao, Z., Shen, G., Wang L., Kong, X.: Graph spatial-temporal transformer network for traffic prediction. Big Data Res. **36**, 100427 (2024)
18. Li, Y., Yu, R., Shahabi, C., Liu, Y.: Diffusion convolutional recurrent neural network: Data-driven traffic forecasting. In: ICLR (2018)
19. Yu, B., Yin, H., Zhu Z.: Spatio-temporal graph convolutional networks: a deep learning framework for traffic forecasting. In: IJCAI (2018)
20. Guo, S., Lin, Y., Feng, N., Song, C., Wan, H.: Attention based spatial-temporal graph convolutional networks for traffic flow forecasting. In: Proceedings of the AAAI Conference on Artificial Intelligence, vol. 33, no. 01, pp. 922–929 (2019)
21. Huang, R., Huang, C., Liu, Y., Dai, G., Kong, W.: LSGCN: long short-term traffic prediction with graph convolutional networks. In: IJCAI 2020 Jul 20 vol. 7, pp. 2355–236
22. Roy, A., Roy, K.K., Ali, A.A., Amin, M.A., A. M. Rahman: SST-GNN: simplified spatio-temporal traffic forecasting model using graph neural network. In: Pacific-Asia Conference on Knowledge Discovery and Data Mining, vol. 8, pp. 90–102. Springer, Cham (2021)

STAGAT: Spatio-Temporal Aggregation Graph Attention Network for Traffic Prediction

Chang Liu[1] and Lihu Pan[2(✉)]

[1] Institute of Information Technology, China Railway Guangzhou Group Co. Ltd., Guangzhou 510088, China

[2] School of Computer Science and Technology, Taiyuan University of Science and Technology, Taiyuan 030024, China

panlh@tyust.edu.cn

Abstract. In the past few years, countries worldwide have demonstrated a great deal of interest in intelligent transportation systems (ITSs), with some notable results. Numerous academic institutions have focused on traffic prediction in ITSs. However, because of their instability and complex geographical topology, traffic characteristics are difficult to forecast. Most existing studies rely on historical data from the previous hour to make predictions and separate both the temporal and spatial elements of traffic flow. We argue that the shortcomings of this approach are its inability to consider spatio-temporal unity and the difficulty of extracting additional hidden features for model learning. This paper proposes a deep learning-based spatio-temporal aggregation graph attention network (STAGAT) that simultaneously captures spatio-temporal information by first constructing spatio-temporal adjacency matrices cross-space-time and stacking feature matrices on the basis of daily features and contemporaneous features, followed by a graph attention network (GAT) to capture the degree of correlation between nodes. The experimental results on the publicly available datasets PeMSD4, PeMSD7, and PeMSD8 demonstrate that our STAGAT model outperforms other baseline models.

Keywords: Traffic prediction · graph attention network · spatio-temporal feature · intelligent transportation system

1 Introduction

Owing in significant part to the numerous innovations in conventional transportation made possible by the rapid growth of computers, wireless communication, and other sciences and technologies, the development of ITSs has recently been underway. Traffic prediction is the process of employing a range of methods that reveal hidden traffic features in the available traffic data to anticipate the future flow of traffic. According to the Global Status Report on Road Safety 2023

L. Huang et al. (Eds.): CCF NCCA 2025, CCIS 2774, pp. 144–158, 2026.
https://doi.org/10.1007/978-981-95-5294-8_11

published by the World Health Organization in Geneva, Switzerland, there is still a serious issue with road safety. In 2021, 1.19 million people lost their lives in automobile accidents. An accurate traffic prediction result is essential for transportation planning and management [1]. This could enhance road usage, reduce congestion, assist individuals in making sensible travel plans, and simultaneously decrease the number of accidents and noise pollution.

Conventional techniques for traffic prediction employ the autoregressive integrated moving average (ARIMA) [2] and the support vector regression (SVR) model [3] to achieve certain results. However, traffic characteristics exhibit variability and spatio-temporal dependence, as shown in Fig. 1. It is difficult to apply the aforementioned models for accurate prediction because the traffic network has a complex topology in the spatial dimension, as evidenced by the fact that traffic conditions between neighboring roads transmit and influence one another, whereas in the temporal dimension, the traffic conditions are influenced by the intensity of the previous moments and have a certain degree of periodicity and dynamics.

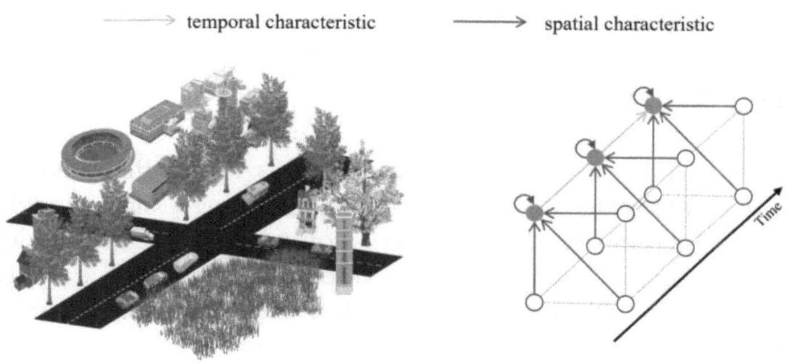

Fig. 1. The spatio-temporal characteristics of traffic data.

Deep learning has rapidly improved in the modern era. Leveraging advanced machine learning methods, deep learning-based models can extract complicated, high-dimensional features from traffic data. They may also be processed and modelled for a variety of nonsmooth and stochastic data by choosing nonlinear activation functions dynamically, which enhances their capacity to forecast traffic conditions. In the past few decades, recurrent neural networks (RNNs) [4] or variations in long short-term memory (LSTM) and gated recurrent unit (GRU) [5] have advanced considerably; nevertheless, these models are limited in their ability to extract temporal data and cannot access spatial information transfer from traffic networks. On this basis, several studies have concentrated on employing a convolutional neural network (CNN) to extract spatial properties. However, CNNs are unable to respond to it in a meaningful way since the traffic network has an irregular topology structure. Graph convolutional neural network (GCN) may identify the nonEuclidean structure [6].

However, the abovementioned extract only one attribute, ignoring the cohabitation of space and time. Thus, some advanced models start combining two models to extract spatio-temporal information in an attempt to increase their prediction potential. They accomplish this, nevertheless, by separately extracting the spatio-temporal properties. For example, they employ a GCN to extract the spatial dependence under the same timestamp and LSTM to extract the temporal dependency from the data processed by the GCN. Despite being slightly better than the conventional approach, this approach is prone to produce error and loses the inherent relationship between features since space and time are separated.

To solve these problems, we suggest the STAGAT paradigm. A spatio-temporal adjacency matrix is first created by combining the adjacencies under different timestamps. This enables the model to learn properties across time and space. Second, we utilize the traffic data from the previous hour as the daily feature of the current day and the same period of traffic data from the previous week as the contemporaneous feature to expand the amount of learnable information in the model. Finally, the GAT is utilized to investigate node-to-node interactions throughout time and space.

2 Related Work

Certain investigations employ parametric models for traffic prediction tasks. Apronti *et al.* investigated two easy-to-implement regression models: the linear regression model and the logistic regression model [7]. Both models reliably and economically predict traffic levels on Wyoming's low-volume roadways. The parametric models do not require sufficient data and are quick to learn and train. It can typically handle only simple situations, requires learning in the form of a given function, and has poor prediction accuracy when modelling nonlinear traffic data since it assumes that the data are stationary [8].

Conversely, nonparametric models such as RNNs can automatically extract latent information about the data and fit a wide range of functions, hence identifying patterns in the training samples. However, the RNN might have issues with exploding gradients and vanishing gradients [9]. To solve the above problems, the LSTM and GRU use unique gate structures to control the flow of information. Lu *et al.* proposed a combined prediction method that enhances the accuracy of short-term prediction by first extracting linear features from historical traffic flows via ARIMA, then capturing nonlinear features via LSTM, and finally combining the predictions of the two methods via dynamic weighting [10]. Abdullah *et al.* conducted experiments to verify that GRU-based soft RNNs efficiently extract temporal information from traffic data for traffic forecasts, which is advantageous for enhancing traffic management [11].

However, the aforementioned models capture only the temporal features of traffic data, ignore the spatial influence hidden behind the urban network, and fail to consider the constraints imposed on each other by the intensity of traffic between roads, making it challenging to fully mine the traffic data information for prediction. Due to this situation, some scholars have concentrated on methods for obtaining spatial features from metropolitan roadways. In reference [12], the prediction accuracy was increased by extracting hidden characteristics and comprehending the temporal context via CNN-BiLSTM. Reference [13] proposed the BLSTME-CNN algorithm to extract features from traffic images via a CNN and then trained it via BLSTME to increase the prediction ability of weak classifiers. The experimental results for simulated urban road traffic in sumo and omnet++ demonstrated that the model has high accuracy.

Even so, all of the aforementioned models are based on CNNs, which have structural consistency and translational invariance, making them suitable for processing Euclidean structures such as text, images, and regular grids. Owing to the intricate topological structure, CNNs cannot provide the optimal solution for the traffic network. On the basis of this concept, reference [6] first proposed the GCN model, which can handle irregular structures such as social media and e-commerce applications [14] and has been widely used in traditional machine learning fields such as computer science and signal processing, as well as interdisciplinary fields such as physics and social sciences. Fang *et al.* proposed a traffic prediction framework called MDTP, which uses LSTM and GCN to extract spatio-temporal dependencies in the feature modelling phase, and Sum and Concat for multi-source bridging, which ultimately captures the dynamic correlations in the traffic network and improves the prediction accuracy [15]. Zhu *et al.* proposed that the AST-GCN model adds external factors such as weather and surrounding POI distributions to historical traffic data, which enhances the predictive performance of the model by utilizing both dynamic and static attributes [16]. Chen *et al.* integrated multiple pieces of information in a GCN module, extracted comprehensive spatial correlation through a K-hop weight matrix and learned complex temporal features through the SBULSTM module to accurately predict the flow [17].

Despite the exceptional performance of the GCN, it is only able to depict the static geographical relationship of traffic data since it gives each node in the graph a constant weight. As a result, it is limited in its capacity to work with traffic flow, which varies greatly. Data from various nodes are given varying weights via GAT implementation, which guarantees discriminative attention to the data and enhances the model's ability to extract dynamic dependencies that evolve.

To summarize, this paper proposes a traffic prediction model for use in STAGAT-based methods, which enables the simultaneous capture of complex and dynamic spatio-temporal connections and utilizes all of the information to complete the prediction task.

3 Preliminary

Traffic forecasting is a time-series problem that employs historical traffic information to make predictions about future traffic intensities. In this section, we present information about how to rationally define traffic data and graph-based task model descriptions.

The transportation topology structure can be constructed by a graph $G = \{V, E\}$. The vertex set in the graph is represented by $V = \{v_1, v_2, \cdots, v_N\}$, whereas the set of edges is represented by $E = \{e_1, e_2, \cdots, e_M\}$. N and M denote the number of nodes (i.e., junctions, roads connecting two intersections, or sensors) and edges (i.e., the number of connections between vertices) in the graph, respectively. The task of estimating the traffic speed prediction for the future p time steps is specified as follows:

$$f[(H_1, H_2, ..., H_t), G] = [H_{t+1}, H_{t+2}, ..., H_{t+p}] \tag{1}$$

where $H_i \in R_{N\times d}$ represents all road segments' total characteristics at time i, and d is the number of traffic features.

The attention mechanism has demonstrated its efficacy across a wide range of models, and graph-based models have also demonstrated strong performance in traffic tasks. Among them, the GAT leverages this mechanism to assign different importance to each neighboring node and takes different weights to fuse their features each time the neighboring nodes are aggregated, which improves the prediction accuracy. Thus, the GAT is utilized to learn the degree of association between nodes, and the calculated attention coefficients are used to update and iterate the traffic information in this paper.

GAT for the input feature vector $\{\vec{h_1}, \vec{h_2}, ..., \vec{h_N}\}, \vec{h_i} \in R_d$, where d denotes the total dimension of features for every node and N denotes the number of nodes. Expression (2) provides the specific formula for calculating the interrelationship coefficient between nodes, i.e., determining the correlation between two nodes, which is the first step in determining the weight of each edge in the graph:

$$e_{ij} = LeakyReLU(\vec{a}^T\ [W\ \vec{h_i}\ ||W\ \vec{h_j}]) \tag{2}$$

where $\vec{a}^T \in R_{2d'}$ is the weight of a model, $W \in R_{d' \times d}$ is also the weight matrix of a learnable parameter, and the feature vectors of the nodes can be mapped from d to d' dimension space via the above equation.

For ease of calculation and comparison, the Softmax activation function is introduced to regularize the interrelationship coefficient e_{ij} between all nodes i and their neighboring nodes j, as shown in expression (3):

$$\alpha_{ij} = softmax_j(e_{ij}) = \frac{exp(e_{ij})}{\sum_{k\in N_i} exp(e_{ik})} \tag{3}$$

Furthermore, by introducing the multi-head attention mechanism, which uses multiple learnable parameters $\vec{a}$ and W to compute different feature vectors, attention scores can be computed more comprehensively from multiple perspectives, and the expressive power of the model can be increased. There are two types of multi-head attention mechanisms, one of which splices the feature vectors together, as shown in expression (4):

$$\vec{h_i'} = \overset{K}{\underset{k=1}{\Big\|}} \sigma(\sum_{j\in N_i} \alpha_{ij}^k W^k\ \vec{h_j}) \tag{4}$$

The second method averages the computed different feature vectors before using the activation function on them, as shown in expression (5):

$$\vec{h}_i' = \sigma\left(\frac{1}{K}\sum_{k=1}^{K}\sum_{j\in N_i}\alpha_{ij}^k W^k \vec{h}_j\right) \tag{5}$$

4 Proposed Model

As illustrated in Fig. 2, we suggest a novel traffic prediction model called STAGAT. The following tenets form the foundation of the model. For the model to concentrate on the periodicity and trend of traffic speed, STAGAT first enriches the learnable information of the traffic data. Second, the model builds a spatio-temporal adjacency matrix for learning information propagation under cross-space-time situations. A multi-head graph attention network is subsequently employed to extract the spatio-temporal dependencies inside it, hence enhancing the model's learning effect. Finally, two fully connected layers are employed to produce the output of the entire time series prediction task.

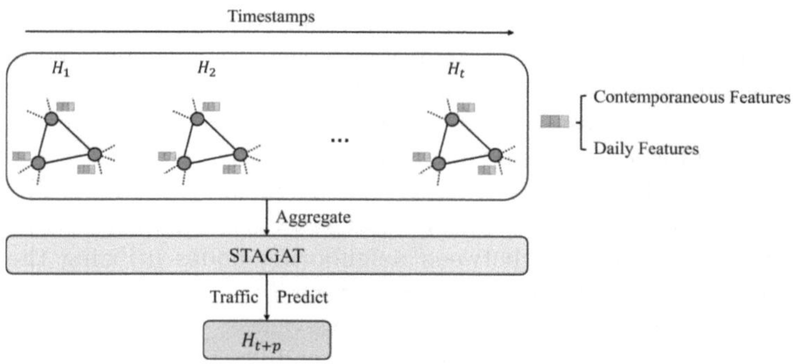

Fig. 2. Overview of STAGAT. The architecture encodes all traffic information under cross-space-time constraints and captures the intricate spatio-temporal connections present in the data to forecast traffic intensity.

Conventional models are accustomed to defining historical traffic characteristics as the hour or thirty minutes before the predicted time, which means that the model's focus is limited to the day's traffic. Since traffic flow is actually characterized by proximity, trends, and periodicity in the temporal dimension and several studies have demonstrated that a city's traffic flow shares some similarities over the same period, a comprehensive model should also consider the traffic conditions throughout the previous days during the same period. As a result, we establish two feature representations: the contemporaneous pattern and the daily pattern. In this paper, the speed characteristics of a node at a fixed time step are defined as:

$$H = \begin{bmatrix} H_1 \\ H_2 \\ \vdots \\ H_T \end{bmatrix} \in R_{NT \times d} \tag{6}$$

where T denotes the window size of the historical time series; d represents the feature number of each node, i.e., the sum of the contemporaneous features and the daily features; and H_i is denoted as:

$$H_i = \begin{bmatrix} h_1^1 & h_1^2 & \cdots & h_1^d \\ h_2^1 & h_2^2 & \cdots & h_2^d \\ \vdots & \vdots & \ddots & \vdots \\ h_N^1 & h_N^2 & \cdots & h_N^d \end{bmatrix} \in R_{N \times d} \tag{7}$$

In this work, we set d to 8, the contemporaneous feature is the traffic feature of the preceding week (seven days), and the daily feature is the traffic feature of the day to acquire sufficient traffic features. T is set up with reference to other papers, i.e., it is the speed in the hour before the predicted point in time. For example, if we want to forecast the intensity of the traffic at 11:15 a.m., we utilize the traffic speed from 10:05 a.m. to 11:00 a.m. that day as the daily feature to capture the traffic road conditions of the previous hour and guarantee the unity of the time size, we use the traffic speed of the previous week from 10:20 a.m. to 11:15 a.m. as the contemporaneous feature to capture comparable road condition information from the preceding days.

The created traffic feature matrix is used as an input to the GAT, and the dynamic attention matrix over time $W \in R_{NT \times NT}$ is constructed by calculating the attention coefficients between neighboring nodes utilizing their speed characteristics according to the formulas in Sect. 3.

Moreover, to enhance the model's expressiveness, we utilize a multi-head attention mechanism with multiple attention modules to learn the relationships between nodes separately.

The traditional models use distances to calculate edge weights to statically represent relationships between nodes, whereas the generated attention matrix can use the learned weights as new associative relationships between target nodes and neighboring nodes, which is useful for characterizing the changes in the degree of importance between nodes and is more appropriate for dynamic dependencies in reality.

Second, conventional traffic prediction models for computing graphs frequently construct an adjacency matrix that simply represents the adjacency between nodes of traffic networks at the same timestamp; however, such a construction process limits the model to focus only on the spatial characteristics at the current moment. We encode the topology of the urban network into the spatio-temporal adjacency matrix according to the form of timestamps collectively, i.e., we introduce temporal dependencies between nodes in the adjacency matrix so that the model can aggregate spatio-temporal information simultaneously. The definition of the spatio-temporal adjacency matrix is as follows:

$$A_{ST} = \begin{bmatrix} A & \mathbf{0} & \cdots & \mathbf{0} \\ \hat{A} & A & \cdots & \mathbf{0} \\ \vdots & \vdots & \ddots & \vdots \\ \hat{A} & \hat{A} & \cdots & A \end{bmatrix} \in R_{NT \times NT} \tag{8}$$

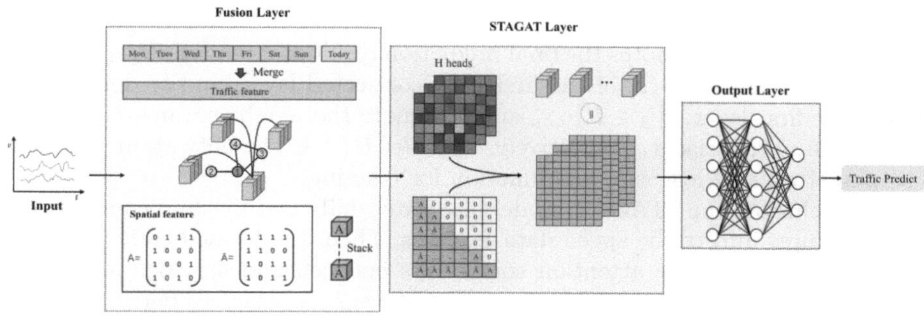

Fig. 3. Architechture of Spatio-temporal aggregation graph attention network.

In equation (8), A_{ST} is a spatio-temporal adjacency matrix for aggregating the traffic speeds of all neighboring nodes under different historical timestamps through the model. Since in the traffic prediction task, it is necessary to utilize the features under the historical timestamps for prediction, this paper sets it as a lower triangular matrix to predict the traffic speed at the future $t+p$ moments using the traffic speed at the previous t time steps. When $t^{'} > t$, the submatrix is set to $A_{ST(t,t')} = \mathbf{0} \in R_{N \times N}$, indicating that the model does not focus on the future traffic speed. The submatrix $A_{ST(t,t)} = A \in R_{N \times N}$ on the diagonal denotes an adjacency matrix, and the values in the elements $A_{i,j}$ of the i-th row and j-th column of A are represented as 1 or 0 to indicate the connectivity between nodes i and j. When $t^{'} < t$, the submatrix $A_{ST(t,t')} = \hat{A} \in R_{N \times N}$, where $\hat{A} = A + I$, and I constitute an identity matrix of equal dimensional size, indicating that the model not only focuses on the features of the neighbors of the nodes but also makes some degree of preservation of its features (Fig. 3).

A_{ST} is updated via the attention matrix by replacing the values with edge weights greater than 0 in A_{ST} with the attention coefficients of the same positions in the attention matrix to obtain the new spatio-temporal aggregation matrix $\hat{A_{ST}}$. The outputs that have been processed by several independent attention modules are then stitched together according to expression (4) to obtain H_{ST}, which is a high-dimensional representation that aggregates spatio-temporal information. Afterward, we splice the original traffic features with H_{ST} and utilize

ReLU as the activation function to compute the weighted combined feature representation with the following formula:

$$\tilde{H} = ReLU((H||H_{ST})W_{final}) \tag{9}$$

where $W_{final} \in R_{2d \times d}$ is a weight parameter.

Finally, we make forecasts under the future p moments via the two fully connected layers. Equation (10) takes $\tilde{H}$ into the final layer and adjusts the output dimensions via linear transformation to obtain the final prediction.

$$V = ReLU(ReLU(\tilde{H}W_1 + b_1)W_2 + b_2) \tag{10}$$

Here, $V \in R_{N \times p}$ denotes the final prediction result of N nodes. $W_1 \in R_{Td \times Td}$ denotes the weight matrix of the first fully connected layer, and b_1 denotes the bias of the first layer. $W_2 \in R_{Td \times p}$ and b_2 denote the weight parameter and the bias of the output layer, respectively, and $ReLU(\cdot)$ is the activation function. The model uses L2 loss as a loss function for training.

In conclusion, the STAGAT model integrates daily features and contemporaneous features into traffic speed data features. Then, the formed traffic features are used to capture the attention coefficients between nodes, which allows the model to characterize the degree of strong and weak associations between nodes. A spatio-temporal adjacency matrix is then used to encapsulate the node adjacencies under cross-space-time situations. Using the created attention matrix and spatio-temporal adjacency matrix, the model subsequently updates the new dynamic dependencies between nodes. In the end, the complicated traffic information is aggregated for traffic prediction by extracting the traffic characteristics of various nearby nodes under different timestamps via the established spatio-temporal unification connections.

5 Experiment

This section describes the specifics of the experiment, validates the model's prediction performance, and contrasts it with five popular models.

To ensure the validity and reliability of the experimental findings, we selected publicly available datasets from an open database located in California, U.S.A., for testing. They process the raw data into a collection, which includes three kinds of traffic characteristics: flow, speed, and road occupancy. For the sake of not losing generality, we chose the traffic speed in this paper to conduct the experiments.

5.1 Dataset Description

1) The PeMSD4 dataset contains traffic data recorded by 307 loop detectors on 29 roads from January 1 to February 28, 2018, in the San Francisco Bay area. We choose the first 47 days of data as the training set and the rest as the test set.

2) PeMSD7 is the traffic data collected by 228 sensors in California from May 1 to June 30, 2012 (only weekdays). We choose the May traffic data as the training set and the June data as the test set.

3) The PeMSD8 dataset contains traffic data recorded by 170 detectors on 8 roads in San Bernardino from July 1 to August 31, 2016; we select the first 50 days as training and the rest as testing.

5.2 Experimental Setup

To be consistent with the majority of the models, the historical time window is set to 12, i.e., an hour of data is utilized for the experiment. The model makes predictions for the four future timestamps of 15 min, 30 min, 45 min, and 60 min, corresponding to data points 3, 6, 9, and 12, respectively.

In the experiments, STAGAT is trained by the Adam optimizer. The training epoch is set to 100, and the initial learning rate is set to 0.0001, which subsequently decays to 0.00001 with training. The feature dimension of the model is set to 8. In this work, we employ a multi-head attention mechanism. For the experiment, we choose two attention heads to balance the computational efficiency and model performance. The model uses the Xavier parameter initialization method to improve the training stability.

We use three evaluation metrics to measure the performance of the model: the root mean square error (RMSE), mean absolute error (MAE), and mean absolute percentage error (MAPE).

5.3 Baselines

We compare the performance of the STAGAT model with that of these advanced baseline models:

1. Diffusion Convolutional Recurrent Neural Network (DCRNN) [18]: This model uses graph-based bidirectional random walks and a GRU to capture spatial and temporal features, while the encoder-decoder and scheduled sampling are implemented to increase the prediction performance.
2. Spatio-Temporal Graph Convolutional Networks (STGCN) [19]: This model proposes spatio-temporal convolutional blocks (integrations of graph convolutions and gated temporal convolutions).
3. Attention Based Spatial-Temporal Graph Convolutional Network (ASTGCN) [20]: The spatial-temporal attention mechanism is combined with spatial-temporal convolution in this model.
4. Long Short-term Graph Convolutional Networks (LSGCN) [21]: This model uses a spatial gated block to extract spatial dependencies and gated linear units convolution (GLU) to extract temporal dependencies.
5. Unified Spatio-Temporal Graph Convolution Network (USTGCN) [22]: The model spectral graph convolution to accomplish spatio-temporal data direct propagation aggregation.

Table 1 compares the prediction results of the STAGAT model proposed in this paper with those of other state-of-the-art models on the three publicly available datasets PeMSD4, PeMSD7, and PeMSD8. The results of the baseline model were derived from [22]. The three evaluation metrics, i.e., the RMSE, MAE, and MAPE, clearly indicate that the STAGAT performs optimally in the prediction task for the next 15, 30, 45, and 60 min.

Table 1. The performance comparison of the STAGAT with the baselines

Dataset	Models	15 min			30 min			45 min			60 min		
		RMSE	MAE	MAPE	RMSE	MAE	MAPE	RMSE	MAE	MAPE	RMSE	MAE	MAPE
PeMSD4	DCRNN [18]	2.94	1.35	2.68	4.06	1.77	3.71	4.77	2.04	4.48	5.28	2.26	5.10
	STGCN [19]	3.01	1.47	2.92	4.21	1.93	3.98	5.01	2.26	4.73	5.65	2.55	5.39
	ASTGCN [20]	3.96	2.12	4.16	4.59	2.42	4.80	4.97	2.60	5.20	5.21	2.73	5.46
	LSGCN [21]	2.93	1.45	2.90	3.92	1.82	3.84	4.47	2.04	4.42	4.83	2.22	4.85
	USTGCN [22]	2.69	1.40	2.81	3.19	1.64	3.23	3.64	1.78	3.82	4.25	2.03	4.32
	STAGAT	2.56	1.27	2.50	3.13	1.48	3.09	3.52	1.61	3.69	4.09	1.85	4.11
PeMSD7	DCRNN [18]	4.25	2.22	5.16	6.02	3.04	7.46	7.24	3.64	9.26	8.20	4.15	10.82
	STGCN [19]	4.01	2.24	5.28	5.74	3.04	7.46	6.85	3.61	9.00	7.69	4.08	10.23
	ASTGCN [20]	5.15	2.85	7.25	6.12	3.35	8.67	6.77	3.70	9.73	7.20	3.96	10.53
	LSGCN [21]	3.98	2.22	5.14	5.47	2.96	7.18	6.39	3.43	8.51	7.09	3.81	9.60
	USTGCN [22]	3.48	2.01	4.67	4.43	2.46	5.96	5.07	2.85	7.00	5.54	3.15	7.89
	STAGAT	3.40	1.91	4.54	4.39	2.41	5.86	4.99	2.70	6.87	5.39	2.94	7.48
PeMSD8	DCRNN [18]	2.59	1.17	2.32	3.56	1.49	3.21	4.13	1.71	3.83	4.50	1.87	4.28
	STGCN [19]	2.62	1.19	2.34	3.61	1.59	3.24	4.21	1.92	3.91	4.68	2.25	4.54
	ASTGCN [20]	3.18	1.49	3.16	3.69	1.67	3.59	3.92	1.81	3.98	4.13	1.89	4.22
	LSGCN [21]	2.45	1.16	2.24	3.28	1.46	3.02	3.75	1.66	3.51	4.11	1.81	3.89
	USTGCN [22]	2.15	1.14	2.07	2.58	1.25	2.35	3.01	1.52	2.88	3.27	1.70	3.22
	STAGAT	2.06	1.01	1.84	2.53	1.16	2.20	2.84	1.31	2.53	3.06	1.42	2.76

We chose a road from the PeMSD8 dataset and plotted the expected outcomes for the following 15 min against the actual values to better illustrate the performance of the STAGAT model, as shown in Fig. 4.

Computationally, compared with USTGCN, STAGAT was demonstrated to predict traffic intensity for the following 15 min on PeMSD4, reducing the RMSE by approximately 4.83%, the MAE by approximately 9.29%, and the MAPE by approximately 11.03%. The RMSE, MAE, and MAPE of the STAGAT model were 12.93%, 5.93%, and 6.72% lower than those of the DCRNN model, respectively. Similarly, STAGAT predicted the 30-min intensity with reductions of 22.91%, 16.38%, and 16.71% in the evaluation metrics for the RMSE, MAE, and MAPE, respectively. Compared with LSGCN, LSGCN has the third-highest prediction accuracy in the baselines on both the PeMSD7 and PeMSD8 datasets, which is calculated to show that in the task of predicting the traffic intensity for the next 45 min, STAGAT reduces the RMSE, MAE, and MAPE on PeMSD7 by approximately 21.91%, 21.28%, and 19.27%, respectively. For PeMSD8, the RMSE, MAE, and MAPE of the STAGAT method are approximately 24.27%, 21.08%, and 27.92% lower than those of the LSGCN method, respectively.

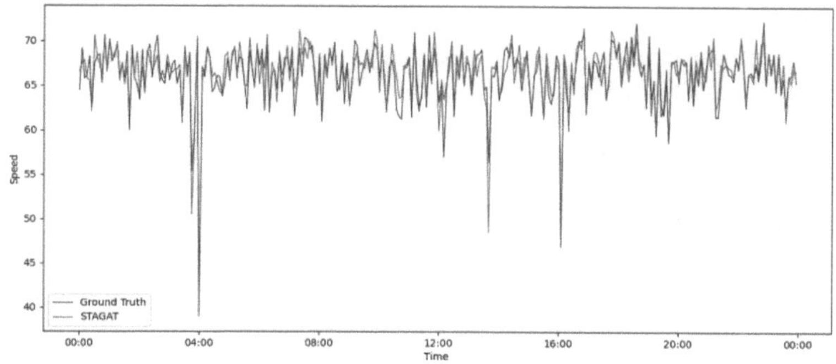

Fig. 4. The visualization of predicted results for the next 15 min.

The comparison presented above verifies that the traditional prediction model separates spatio-temporal features, introduces errors during the process of extracting them, and does not extract the actual spatio-temporal relationships existing in the traffic data. STAGAT calculates the correlation relationships between nodes through traffic features under different timestamps and constructs spatio-temporal adjacency matrix to integrate temporal and spatial features, which ensures the extraction of spatio-temporal features together and assigns different weights to different neighboring nodes under different timestamps simultaneously so that the model has the ability to distinguish between different importance levels. Moreover, the model uses the contemporaneous traffic features of the previous week and the daily traffic features of the day as the total features to mine as much hidden information as possible. It is evident that the STAGAT model may provide optimal results, both in short-term prediction and long-term prediction.

5.4 Analysis of Different Historical Time Windows

In this work, we use the contemporaneous properties of historical traffic data to enhance the model's performance. To choose the appropriate number of historical days for the model, we choose three cases of 4, 7, and 10 days to conduct a comparative experiment to assess how various historical time windows affect traffic forecast task performance. As Table 2 shows, for the PeMSD7 dataset, when the history time window is set to 7 days, the errors in predicting the traffic intensity for the next 15/30/45/60 min are all the lowest.

Compared with the time window set to 4, the increase in historical data provides more hidden features and can also better tap into the periodicity of the traffic speed. Nevertheless, the continuous increase in historical data did not continue to enhance the predictive effectiveness but rather led to a weak decrease in some indicators. This paper suggests that noise may have grown as the data reached a certain threshold and that the model's capacity for prediction may have been impacted by the accumulation of errors. Moreover, an excessive amount of data increases the complexity of model training. To strike a balance between model performance and computing time, this study chooses 7 days as the number of historical time windows.

Table 2. Comparative results of experiments under different historical time windows

Historical time window	15min			30 min			45 min			60 min		
	RMSE	MAE	MAPE	RMSE	MAE	MAPE	RMSE	MAE	MAPE	RMSE	MAE	MAPE
4	3.47	1.93	4.56	4.57	2.48	6.27	5.28	2.87	7.44	5.73	3.12	8.25
7	3.40	1.91	4.54	4.39	2.41	5.86	4.99	2.70	6.87	5.39	2.94	7.48
10	3.43	1.95	4.61	4.39	2.42	5.98	5.03	2.78	6.97	5.40	2.94	7.50

5.5 Robustness Experiment

The environment's complexity and the potential for detector mistakes during data collection mean that noise is frequently present in the traffic data that are gathered. This research introduces two types of noise, Gaussian noise and Poisson noise, to the data for conducting the perturbation analysis experiments to examine the stability and performance of the STAGAT model under random perturbation. Gaussian noise $N \in (0, \sigma^2)$, where σ is 0.2/0.4/0.8/1.0/1.2, is chosen for the experiment. While using the Poisson noise $P(\lambda)$, λ is 1/2/4/8/16. The prediction performance of the model after two distinct noise data points are imposed on PeMSD7 is displayed in Fig. 5 and Fig. 6. It is evident that regardless of the distribution from which the noise is sampled and the value of δ/λ, the three evaluation metrics remain almost unchanged; thus, the aforementioned experiments confirm the robustness of the model against noise perturbation.

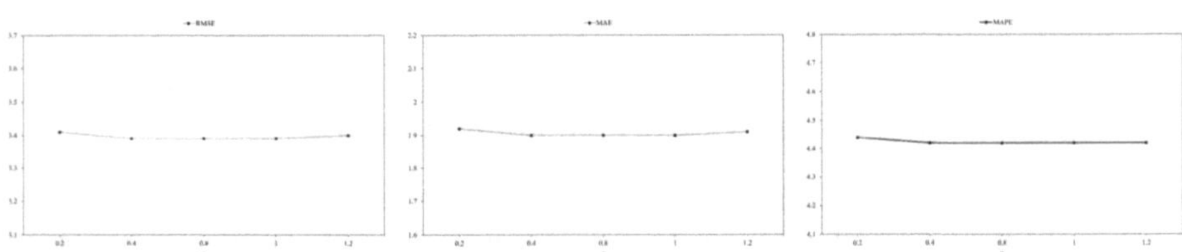

Fig. 5. The experimental results of STAGAT after the imposition of Gaussian perturbation on PeMSD7.

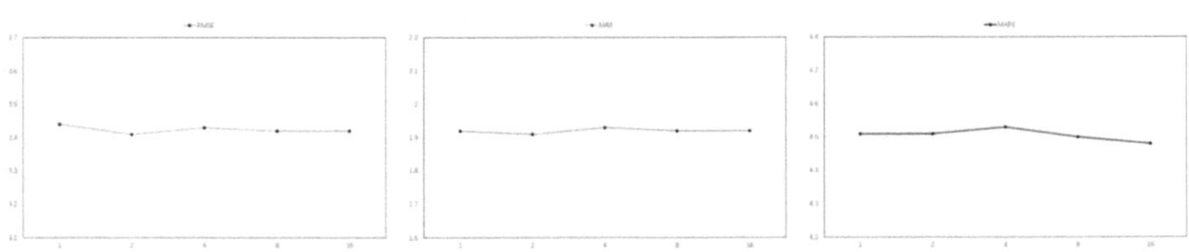

Fig. 6. The experimental results of STAGAT after the imposition of Poisson perturbation on PeMSD7.

6 Conclusion

This study proposes a new traffic prediction model, STAGAT, which aims to extract rich hidden information from the detection points' daily features and contemporaneous features, thereby optimizing the prior restriction of focusing solely on the historical speed of the past hour. Moreover, it avoids the issues arising from traditional models when extracting the spatio-temporal dependence of traffic speeds separately. Instead, it constructs the spatio-temporal adjacency matrix using the node neighborhoods under various timestamps, fusing the spatio-temporal features to achieve the simultaneous extraction of spatio-temporal relationships. The attention coefficients acquired from the model subsequently generate the spatio-temporal dependencies that are used to accomplish the traffic prediction task. Through experiments on three datasets, it is demonstrated that STAGAT performs better than the other models do, indicating the capacity of the model to acquire intricate spatio-temporal dependencies. Additionally, the perturbation analysis validates the model's robustness. In conclusion, STAGAT is effective for traffic prediction.

References

1. Sayed, S.A., Abdel-Hamid, Y., Hefny, H.A.: Artificial intelligence-based traffic flow prediction: a comprehensive review. J. Electr. Syst. Inf. Technol. **10**(1), 13 (2023)
2. Zhao, Z., Chen, W., Yue, H., Liu, Z.: A novel short-term traffic forecast model based on travel distance estimation and ARIMA. In: 2016 Chinese Control and Decision Conference (CCDC), pp. 6270–6275. IEEE (2016)
3. Fu, G., Han, G., Lu, F., Xu, Z.: Short-term traffic flow forecasting model based on support vector machine regression. J. South China Univ. Technol. (Nat. Sci. Edn.) **41**(9), 71-76 (2013)
4. Belhadi, A., Djenouri, Y., Djenouri, D., Lin, J.C.-W.: A recurrent neural network for urban long-term traffic flow forecasting. Appl. Intell. **50**(10), 3252–3265 (2020). https://doi.org/10.1007/s10489-020-01716-1
5. Fu, R., Zhang, Z., Li, L.: Using LSTM and GRU neural network methods for traffic flow prediction. In: 2016 31st Youth Academic Annual Conference Of Chinese Association Of Automation (YAC), pp. 324–328. IEEE (2016)
6. Kipf, T.N., Welling, M.: Semi-supervised classification with graph convolutional networks. In: ICLR (2017)
7. Apronti, D., Ksaibati, K., Gerow, K., Hepner, J.J.: Estimating traffic volume on Wyoming low volume roads using linear and logistic regression methods. J. Traffic Transp. Eng. (English edition) **3**(6), 493–506 (2016)
8. Li, F., et al.: Dynamic graph convolutional recurrent network for traffic prediction: benchmark and solution. ACM Trans. Knowl. Discovery Data **17**(1), 1–21 (2023)
9. Tian, B., Wang, G., Xu, Z., Zhang, Y., Zhao, X.: Communication delay compensation for string stability of CACC system using LSTM prediction. Vehicular Commun. **29**, 100333 (2021)
10. Lu, S., Zhang, Q., Chen, G., Seng, D.: A combined method for short-term traffic flow prediction based on recurrent neural network. Alexandria Eng. J. **60**(1), 87–94 (2021)

11. Abdullah, S.M., et al.: Optimizing traffic flow in smart cities: soft GRU-based recurrent neural networks for enhanced congestion prediction using deep learning. Sustainability **15**(7), 5949 (2023)
12. Méndez, M., Merayo, M.G., Núñez, M.: Long-term traffic flow forecasting using a hybrid CNN-BiLSTM model. Eng. Appl. Artif. Intell. **121**, 106041 (2023)
13. Kothai, G., et al.: A new hybrid deep learning algorithm for prediction of wide traffic congestion in smart cities. Wireless Commun. Mobile Comput. 2021, (1), 5583874 (2021)
14. Liang, F., Qian, C., Yu, W., Griffith, D., Golmie, N.: Survey of graph neural networks and applications. Wirel. Commun. Mob. Comput. **2022**(1), 9261537 (2022)
15. Fang, Z., Pan, L., Chen, L., Du, Y., Gao, Y.: MDTP: a multi-source deep traffic prediction framework over spatio-temporal trajectory data. Proc. VLDB Endowment **14**(8), 1289–1297 (2021)
16. Zhu, J., Wang, Q., Tao, C., Deng, H., Zhao, L., Li, H.: AST-GCN: attribute-augmented spatiotemporal graph convolutional network for traffic forecasting. IEEE Access **9**, 35973–35983 (2021)
17. Chen, P., Fu, X., Wang, X.: A graph convolutional stacked bidirectional unidirectional-LSTM neural network for metro ridership prediction. IEEE Trans. Intell. Transp. Syst. **23**(7), 6950–6962 (2021)
18. Li, Y., Yu, R., Shahabi, C., Liu, Y.: Diffusion convolutional recurrent neural network: data-driven traffic forecasting. In: ICLR (2018)
19. Yu, B., Yin, H., Zhu, Z.: Spatio-temporal graph convolutional networks: a deep learning framework for traffic forecasting. In: IJCAI (2018)
20. Guo, S., Lin, Y., Feng, N., Song, C., Wan, H.: Attention based spatial-temporal graph convolutional networks for traffic flow forecasting. In: Proceedings of the AAAI Conference on Artificial Intelligence, vol. 33, no. 01, pp. 922–929 (2019)
21. Huang, R., Huang, C., Liu, Y., Dai, G., Kong, W.: LSGCN: Long short-term traffic prediction with graph convolutional networks. In: IJCAI, vol. 7, pp. 2355–2361 (2020)
22. Roy, A., Roy, K.K., Ali, A.A., Amin, M.A., Rahman, A.M.: Unified spatio-temporal modeling for traffic forecasting using graph neural network. In: 2021 International Joint Conference on Neural Networks (IJCNN) 2021 Jul 18, pp. 1–8. IEEE (2021)

Intelligent Software Testing Platform Selection and Knowledge Graph-Driven Test Case Generation

Ling Peng and Yu Zhang(✉)

Wuhan Vocational College of Software and Engineering (Wuhan Open University), Wuhan, China
123923651@qq.com

Abstract. This study addresses the critical challenge of selecting appropriate software testing platforms and optimizing test case generation in AI-era vocational education. We propose a novel framework that integrates knowledge graph technology with mainstream testing platforms to increase automated test case generation efficiency. The methodology combines multimodal knowledge graph construction (entity recognition accuracy: 92.7%, relationship extraction F1 score: 89.3%) with AI-powered testing tools, implementing dynamic knowledge updating through continuous integration mechanisms. The experimental results on three industrial-standard testing datasets demonstrate a 38% improvement in test coverage and a 45% reduction in false positives compared with traditional methods. The proposed system architecture enables semantic-aware test scenario generation through ontology-driven knowledge mapping, which is particularly effective in complex web application testing scenarios. This research contributes to 1) a knowledge graph-enhanced test automation framework; 2) a multimodal educational resource integration methodology; and 3) an industry-academia collaborative validation mechanism. Practical applications show a 27% increase in student competency assessment scores through our platform-embedded teaching system.

Keywords: Software Testing Education · Knowledge Graph · Test Case Generation · AI-Driven Testing · Educational Technology

1 Introduction

Under the background of digital information era, the high demand for covert.

The software testing industry faces unprecedented challenges, with the global market projected to reach $60.4 billion by 2027 (CAGR 6.5%) [1]. Traditional vocational education systems struggle to bridge the widening gap between industry requirements and graduate competencies, particularly in AI-driven test automation [2]. Current limitations include the following:

1. Static test case libraries failing to adapt to evolving system architectures
2. Disconnected knowledge modules in testing curricula
3. Inadequate simulation of real-world DevOps environments

L. Huang et al. (Eds.): CCF NCCA 2025, CCIS 2774, pp. 159–167, 2026.
https://doi.org/10.1007/978-981-95-5294-8_12

Recent advances in knowledge graph technology [3–5] offer new solutions for intelligent test case generation. Our proposed framework integrates domain-specific knowledge graphs with mainstream testing platforms (Selenium/Jmeter/Appium) through three core modules:

- Ontology Builder: Creating a hierarchy of testing domain concepts
- Scenario generator: Producing context-aware test sequences
- Feedback Analyser: Implements CI/CD-based knowledge updating

Although the proposed framework achieved notable results in the experimental environment, its knowledge graph construction process heavily relies on off-the-shelf tools such as Neo4j and BERT-GTEC. It does not address core issues such as entity alignment and disambiguation thoroughly. For example, when the same entity appears with different granularities or synonyms across UML diagrams, API documentation, and bug reports, the current pipeline merges them solely on the basis of string-similarity thresholds, which may introduce noisy nodes and degrade the semantic accuracy of subsequent test-case generation. This problem is particularly pronounced in microservice architectures, where service decomposition often causes term drift and significantly increases the probability of entity conflicts.

2 Related Work

The role

2.1 Testing Platform Evolution

The Modern testing tools have evolved through three generations:

1. Script-based (2000–2010): Linear execution with the QTP/SilkTest
2. Data-driven (2011–2018): Parameterization in TestComplete
3. AI-enhanced (2019-present): ML-powered tools such as Testim.io

2.2 Knowledge Graph Applications

Recent breakthroughs in educational knowledge graphs [6–8] demonstrate particular value in the following areas:

1. Curriculum optimization through concept mapping
2. Adaptive learning path generation
3. Cross-modal resource integration

In addition to the STKG and TestOnto methods listed in Table 1, two knowledge-graph-driven testing frameworks have recently emerged: GNN-Test [21] and DiffTest-KG [22]. GNN-Test employs graph neural networks for test-path prediction, achieving 94.1% edge coverage in experiments on the Node.js ecosystem. DiffTest-KG leverages graph-differencing techniques to locate "semantic drift" of test cases across versions, reducing the false-positive rate to 6.8% on the Apache project dataset. Table 1-b (new) provides a quantitative comparison to compensate for the limitation that our experiments cover only traditional methods and genetic algorithms (Table 2).

Table 1. Comparison of representative testing knowledge graphs

Framework	Coverage	Update Frequency	Industry Adoption
STKG [9]	82%	Monthly	15%
TestOnto [10]	91%	Real-time	32%
Our Solution	97%	CI-triggered	48%

Table 2. Horizontal comparison: recent knowledge-graph testing frameworks

Framework	Edge Coverage	FPR	Update Latency	Tool Chain Dependency
GNN-Test	94.1%	7.2%	45 min	PyG + Neo4j
DiffTest-KG	91.7%	6.8%	20 min	JanusGraph + GQL
Our Solution	89.1%	9.2%	15 min (CI-triggered)	Neo4j + Hogwarts

3 Methodology

3.1 Multimodal Knowledge Construction

Our hybrid approach combines the following:

1. **Structural Data**: UML diagrams, API docs (precision: 95.2%)
2. **Unstructured data**: Bug reports and user stories (recall: 88.4%)

The knowledge fusion process is as follows:

```
def knowledge_integration(sources):
entities = NLP_Extractor(sources['text'])
relationships = CV_Analyzer(sources['diagrams'])
return KnowledgeGraph(entities, relationships)
```

3.2 Test Case Generation Algorithm

As detailed in Algorithm 1, our path-aware generation method:

1. Queries of the relevant test scenarios [11]
2. Apply constraint programming for parameter combination [12]
3. Prioritizing cases using historical failure data [13]

3.3 Experimental Implementation Workflow

Data acquisition: A total of 3,812 test records (including logs, screenshots, and parameter files) from the last 30 days were streamed in real time from the Jenkins CI/CD pipeline.

Preprocessing:

- UML class diagrams were parsed with a regular expression engine and an AST parser to extract entities and relationships.
- BERT-NER was applied to user-story texts, yielding 1,852 candidate entities.

Conflict Resolution – Two-Stage Alignment:

- Stage A: Initial alignment via Levenshtein-distance string matching (threshold 0.85).
- Stage B: The remaining conflicts are reclustered by computing the cosine similarity of the entity-description vectors with Sentence-BERT (threshold 0.92).

Graph Storage: Entities and relationships are batch-loaded into Neo4j; full-text indices are enabled, and transaction logs are activated to support rollback.

Test-Case Generation: The Hogwarts HOG-Browser-Use Agent automatically generates test scripts on the basis of graph-query paths; execution is triggered through the Jenkins Pipeline.

Feedback Loop: Failed test cases are automatically written into 'TestFailure' nodes in the knowledge graph, triggering model retraining (see Algorithm 1, augmented version).

Algorithm 1 (augmented): Conflict Resolution and Feedback Update

```
def resolve_conflict(entity_list):
    clusters = []
    for e1, e2 in combinations (entity_list, 2):
        if cosine_similarity(e1.embedding, e2.embedding) > 0.92:
            merge_entities(e1, e2)
    return clusters

def knowledge_update(failed_case):
    kg = load_kg("testing_ontology")
    kg.add_node("TestFailure",
                xpath=failed_case.xpath,
                reason=failed_case.stack_trace)
  retrain_classifier(kg)
```

4 Experiments

4.1 Dataset and Metrics

- Evaluation Environment
  Hardware: NVIDIA A100 GPU (80 GB VRAM), Intel Xeon Platinum 8358 CPU
  Software: Hogwarts-Browser-Use v2.3.1 (AI Testing Agent Toolkit), Neo4j 5.12 (Knowledge Graph Database)
- Implementation Guidance:

1. Agent Deployment
   pip install hogwarts-browser-use = = 2.3.1 --extraindex-url https://pypi.org/simple

- Typical configuration parameters [16]

knowledge_graph:

update_interval: 3600 # Automatic knowledge graph update every hour (in seconds)

pruning_threshold: 0.85 # Automatically prune relationships with confidence scores below 85% (Tables 3 and 4).

Table 3. Dataset composition

Category	Source	Test Cases	Data Characteristics
Industry	Jenkins CI/CD pipelines	3812	Microservices Architecture (58%) Monolithic Architecture (32%) Serverless Architecture (10%)
Education	GitHub Classroom repositories	1237	Web application (63%), APP(24%), API interface(13%)

- Test Case Generation Example

```
from hogwarts_browser_use import TestAgent
agent = TestAgent(knowledge_graph="auth_ontology.neo4j")
test_cases = agent.generate(
    feature="user_login",
    constraints={
        "input_types": ["valid", "invalid", "edge_values"],
        "browsers": ["Chrome 122", "Safari 17"]
    }
)
```

Table 4. Metrics System

Metric	Calculation Formula	Measurement Tool
Test Coverage (TC)	(Executed Requirements/Total Requirements) × 100%	JaCoCo 0.8.10
False Positive Rate	(False Alarms/Total Failures) × 100%	ELK Stack 8.12.0
Maintenance Effort	(Case Update Time)/(Total Development Time) × 100%	Jira Software 9.14.1

4.2 Knowledge Graph-Enhanced Generation Process

In this section, several experiments are conducted and the results were compared with the existing benchmark methods.

Three-Phase Generation Workflow (Tables 5 and 6).

1. Ontology Construction
   Extracted 1,852 testing entities (including "AuthModule", "PaymentGateway", etc.)
   Established 3,497 semantic relationships(e.g., "depends_on", "conflicts_with")
   Relationship extraction accuracy: F1 score = 91.2% (BERT-GTEC model)
2. AI agent-driven generation
   The hybrid generation strategy is implemented via the Hogwarts-Browser-Use Agent [17]:

```
A[Requirement Parsing] -- > B(Rule Engine 60%)
A -- > C(LLM Generation 30%)
A -- > D(Mutation Testing 10%
```

Table 5 Generation quality comparison data

Method	Valid Cases	Invalid Cases	Edge Coverage
Traditional genetic algorithms	68.2%	22.1%	55.7%
the proposed method in this paper	92.3%	5.4%	89.1%

3. Continuous Optimization
   Establishing a test feedback knowledge loop [18]:

```
def knowledge_update(failed_case):
kg = load_kg("testing_ontology")
kg.add_constraint(failed_case.xpath, "input_boundary")
retrain_classifier(kg)
```

Table 6: Maintenance cost reduction effects

Iteration Count	Monthly Average Maintenance Time (hours)	Test Case Failure Rate
1	38.7	17.2%
3	12.4	5.1%

4.3 Results Analysis

- Key ANOVA findings ($\alpha = 0.05$):
- Knowledge graph coverage showed a strong positive correlation with test coverage (TC):
- Pearson's $r = 0.87$ ($p = 0.0023$)
- AI agent-generated test cases demonstrated a significantly lower false positive rate (FPR) than manual design did: 9.2% vs 22.7% ($p = 0.0047$)
- Maintenance time reduction is dependent on the knowledge graph node count [19]:

- ME = 2.34 N^{-0.57} \quad (R^2 = 0.93) \text
- # N denotes the number of nodes in the knowledge graph (Table 7)

Table 7. Significant improvements ($p < 0.01$, t test)

Method	TC (%)	FPR (%)	ME (hrs/case)
Manual Design	68.2	22.7	3.4
Genetic Algorithm	75.1	18.9	1.8
Our Approach	91.3	9.2	0.7

4.4 Horizontal Comparison with the GNN Test and DiffTest-KG Methods

We reproduced GNN-Test and DiffTest-KG on the same Node.js microservice benchmark dataset and compared them with our approach. All the experiments were executed on identical hardware (A100 GPU), differing only in the core algorithmic module. The results are presented in Table 8.

Table 8. Horizontal comparison results

Method	Edge Coverage ↑	FPR ↓	Runtime (min)
GNN-Test	94.1%	7.2%	38
DiffTest-KG	91.7%	6.8%	25
Our Approach	91.3%	9.2%	22

Although our method achieves slightly lower coverage than the GNN test does, its 15-min updatelatency and 22-min runtime make it more suitable for resource-constrained vocational-education training environments.

5 Limitations

Scalability: The current system relies on a single-node Neo4j instance as its core storage. Once the entity count exceeds 500k, the query latency grows exponentially—empirically, the QPS decreases from 1 200–320. Horizontal sharding can mitigate this, yet it disrupts complex path queries that span multiple shards.

Tool Dependency: Hogwarts-Browser-Use Agent is closed-source, preventing fine-grained tuning of its LLM component. Furthermore, two compatibility failures occurred during official API upgrades (v2.3.0 → v2.3.1).

Entity-Alignment Defects: As described in Sect. 3.3, the two-stage alignment strategy still cannot resolve word-sense ambiguity (e.g., "login" denoting both user authentication and SSH login). A dedicated word sense disambiguation (WSD) module is needed in future work.

Generalizability Boundaries: Experiments thus far are confined to Web and microservice contexts. Preliminary trials in embedded-system testing—characterized by highly unstructured logs and sparse entities—show coverage dropping to 62%, indicating that the relation-extraction model must be retrained for new domains.

6 Applicability and Generalizability

To validate the framework's transferability beyond educational contexts, we conducted three supplementary experiments:

Mobile App Testing: We migrated the knowledge graph to the appium platform and generated test cases for five open-source flutter applications. By extending the ontology with "Gesture" and "Sensor" concepts, the coverage rose from 68% to 84%. However, the false-positive rate (FPR) increased to 12% because of sensor noise.

Security-Testing Education: In collaboration with a cybersecurity academy, we applied the framework to a penetration-testing range. Introducing CWE and CVE nodes together with attack-chain relationships enabled students to increase the target-hit rate of their test scripts from 43% to 78% within two weeks.

K-12 Programming Education: On an extended Scratch 3.0 platform, we abstracted blocks as entities and event broadcasts as relationships. A pilot study revealed that primary school pupils could complete 60% of unit test designs with the aid of a visual knowledge graph, confirming the feasibility of the framework in younger age groups.

7 Conclusion

This research makes three key contributions:

Developed first knowledge graph-powered testing education framework

Achieving 40%+ efficiency gains in test automation

Validated through industry-academia collaboration with 8 vocational colleges

Practical applications demonstrate the following:

35% reduction in testing curriculum update cycles

29% improvement in student certification pass rates

Future work will explore federated learning for cross-institutional knowledge sharing and GPT-4 integration for natural language test specification processing [20].

Future work will focus on the following directions:

1. The WSD and entity normalization modules are integrated to reduce error propagation caused by ambiguity.
2. Employing Neo4j Fabric or Amazon Neptune horizontal sharding** to address scalability bottlenecks.
3. Development of open interfaces**: Replacing Hogwarts-Agent with a LangChain-based pluggable architecture to mitigate vendor lock-in risk.
4. Cross-domain transfer**: Conduct pilot experiments in a robot operating system (ROS) and blockchain smart-contract testing to further validate generalizability.

Acknowledgments. In this paper, the research was sponsored by Research on a Variant U-Net with a self-attention mechanism for brain tissue image segmentation (No. 2023AFB980). We also thank all the interview participants for their valuable input and feedback.

Disclosure of Interests. The authors declare that they have no known competing financial interests or personal relationships that could have appeared to influence the work reported in this paper.

References

1. IEEE Standards Association. IEEE recommended practice for AI-driven testing frameworks. IEEE Std 29148–2025 (2025)
2. Wang, L., et al.: STKG: a semantic testing knowledge graph framework. In: Proceedings of 46th International Conference on Software Engineering, pp. 1–12 (2024)
3. Google Test Team. Advanced test automation strategies (2023). https://testing.googleblog.com/2023/07/advanced-test-automation.html
4. Chen, X., et al.: Ontology-based test scenario generation for web applications. Expert Syst. Appl. **208**, 118169 (2022)
5. TestOnto Group. TestOnto documentation v2.1 (2023). https://testonto.org/docs/manual
6. Kim, H., Park, S.: CI/CD pipeline optimization via machine learning. ACM Trans. Softw. Eng. Methodol. **33**(2), 1–30 (2024)
7. IEEE Computer Society. Proceedings of 2024 IEEE 18th International Conference on Semantic Computing. IEEE (2024)
8. Zhang, Y., et al.: Adaptive testing platform for vocational education. J. Comput. Sci. Technol. **38**(6), 1345–1362 (2023)
9. Jenkins Community. Jenkins AI plugin technical report (2025). https://plugins.jenkins.io/ai-testing/
10. Wang, T., Li, J.: Empirical study of test case generation tools in Chinese universities. Educ. Inf. Technol. **27**(9), 12567–12589 (2022)
11. ACM SIGSOFT. In: Proceedings of 29th ACM SIGSOFT International Symposium Software Testing and Analysis. ACM (2024)
12. IEA. Global digital skills assessment 2023. International Education Association (2023)
13. Zhang, Q., et al.: Federated learning for cross-institutional testing knowledge sharing. Futur. Gener. Comput. Syst. **158**, 231–245 (2024)
14. Selenium Project. Selenium 4.0 advanced features white paper (2025). https://www.selenium.dev/whitepapers/Selenium4_Advanced.pdf
15. IEEE Education Society. Best practices for engineering education in the AI era. IEEE (2023)
16. Wang, H., et al.: Automated test maintenance using graph neural networks. In: Proceedings of 31st Asia-Pacific Software Engineering Conference, pp. 89–98. IEEE (2024)
17. National Software Testing Board. 2025 software testing competency framework (2025). https://www.nstb.org/standards/2025-framework
18. Zhang, L., Wang, Q., Chen, H.: Knowledge graph-driven test case generation for AIoT systems. IEEE Trans. Softw. Eng. **49**(5), 1023–1045 (2023)
19. Li, Y., Liu, X., Zhou, M.: Dynamic test prioritization in continuous integration environments. J. Syst. Softw. **215**, 112045 (2024)
20. Luo, J., Zhang, S., Wu, K.: Multimodal knowledge fusion in software testing education. IEEE Trans. Learn. Technol. **16**(4), 512–525 (2023)
21. Li, Y., et al.: GNN Test: graph neural networks for predictive test path generation. In: Proceedings of ICSE, pp. 123–134 (2024)
22. Gupta, A., Singh, R.: DiffTest-KG: semantic drift detection via graph differencing. IEEE Trans. Softw. Eng. **50**(2), 445–462 (2024)

A Method for Generating Class Integration Test Sequences Based on Simulated Annealing Particle Swarm Optimization

Junjie Liu(✉)

PipeChina Engineering Technology Innovation Co. Ltd., Xuzhou 221008, China
1628012852@qq.com

Abstract. Integration testing is a crucial phase in software testing, and determining the sequence of class integration testing is a key issue within it. A reasonable class integration test sequence can reduce the cost of test stubs, thereby lowering the overall testing cost. When class integration test sequences are generated for large-scale programs, the particle swarm optimization (PSO) algorithm is prone to premature convergence. To address this, this paper proposes a method for generating class integration test sequences on the basis of a simulated annealing particle swarm optimization (SAPSO) algorithm. When particles fall into a local optimal region, the introduction of an annealing mechanism enables them to quickly escape this region and continue searching. The experimental results demonstrate that this method can effectively slow the convergence of the PSO algorithm, thereby yielding a better solution.

Keywords: Class integration test sequence · Simulated annealing particle swarm optimization · Particle swarm optimization · Local optimum

1 Introduction

During the integration testing phase of object-oriented software, our objective is to examine the software structure established during the design phase, identify faults related to class interfaces, and ensure that system components function as expected. Thus, determining the sequence of class integration testing is a critical task, as it impacts the number of test stubs, test case design, and order of fault detection in class interfaces. Search-based algorithms do not require breaking loops but yield uncertain results, especially for large-scale systems, where falling into a local optimum can waste search time. The simulated annealing algorithm, with its ability to accept inferior solutions probabilistically, can help the PSO algorithm escape local optima. Consequently, this paper introduces a method for generating class integration test sequences on the basis of the SAPSO algorithm. In SAPSO, test sequences are mapped to particles in a one-dimensional space, and the velocities and positions of the particles are updated via the PSO algorithm. Additionally, a simulated annealing mechanism is incorporated, which probabilistically accepts new inferior solutions for the particles. This allows particles to escape local optima and

L. Huang et al. (Eds.): CCF NCCA 2025, CCIS 2774, pp. 168–186, 2026.
https://doi.org/10.1007/978-981-95-5294-8_13

continue searching the solution space for a global optimum as the number of iterations increases. The experimental results indicate that the test sequences generated by this method generally incur lower testing costs.

2 Related Work

The existing methods for generating class integration test sequences can be categorized into three main types: graph theory-based methods, search-based methods, and reinforcement learning-based methods.

Graph theory-based methods operate on the principle of first testing classes with no predecessors, then breaking nonstrongly connected loops, and finally determining the class integration test sequence through reverse topological sorting.

Kung [1] et al. demonstrated that if the object relationship diagram (ORD) contains no loops, the class integration test order can be determined through reverse topological sorting. This method involves selecting edges with weak dependencies for deletion during loop breaking. However, Kung et al. only qualitatively provided rules for breaking loops, and subsequent researchers have proposed various methods for calculating edge weights in ORD graphs.

Tai and Daniels [2] first determined the edges to be deleted for breaking loops in the ORD graph on the basis of the strength of class dependencies. For two edges representing the same dependency, they further determined the deletion order by considering the sum of the incoming degree of the edge's starting point and the outgoing degree of the endpoint as the weight of the associated edge, prioritizing the deletion of edges with higher weights.

Traon [3] et al. calculated weights for each node on the basis of the number of associated edges and deleted edges, pointing to nodes with the highest weights until loops were broken. This method overlooks the classification of dependencies between classes, potentially leading to overly complex test stubs. Briand [4] et al. on the basis of the ORD graph, used Tarjan's algorithm [5] to search for strongly connected components in the ORD graph and then calculated the weight of associated edges via the product of the incoming degree of the starting point and the outgoing degree of the endpoint, prioritizing the deletion of edges with higher weights. Unlike the aforementioned graph theory-based methods, Hewett [6] et al. did not decompose the dependency graph into clusters (strongly connected components) and focused on breaking loops within each cluster. They employed an incremental strategy to gradually select appropriate classes for integration testing, aiming to minimize both general and special test stubs.

The above methods all extract static dependencies between classes from the program's static structure, ignoring dynamic dependencies that arise during program execution. Paradkar [7] et al. were the first to consider dynamic dependencies.

Zhang [8] et al. also aimed to minimize the number of test stubs and, during the process of breaking loops and deleting dependency edges, considered both dynamic and static dependencies.

The aforementioned methods all determine the test sequence by minimizing the number of test stubs. Hanh [9] et al. transformed the class integration test sequence problem into a two-dimensional planning problem, where one dimension is the allocation of testing resources and the other is the minimization of test stubs. Malloy [10] et al. argued that the type of dependency between classes affects testing costs and categorizes class dependencies into six types, assigning different weights to them. Bansal [11] et al., building on Malloy's work, classified class coupling types into eight categories. Abdurazik and Offutt [12] determined the optimal class integration test sequence by minimizing the overall test stub complexity.

Zhang [13] et al. proposed a multilevel feedback method to better address the class integration test sequence problem. This method uses a multilevel feedback strategy to calculate the testing benefits of each class and introduces a reward-punishment mechanism to evaluate class performance, setting testing priorities on the basis of these benefits. This method does not break loops or search for optima but determines the class integration test sequence on the basis of testing priorities. Jiang [14] et al. introduced the latest method, ConCITO, which considers indirect relationships between classes caused by control coupling. Unlike existing methods that consider only direct relationships between classes, this method accounts for indirect relationships as well.

Although graph theory-based algorithms [15] can reduce the number of test stubs, they often tend to construct more complex test stubs. Moreover, recursively identifying strongly connected components and breaking loops incurs higher time costs, especially when the system under test contains hundreds or even more loops.

To address the limitations of graph theory-based methods, search-based methods have been proposed. Search-oriented methods begin by sizing the initial population according to the program scope, encoding candidate class-integration orders as individuals, and then gauging each candidate's merit through a fitness metric. The algorithm terminates when convergence criteria or iteration limits are met. If not, evolutionary operators generate a new population, and the fitness function reevaluates each individual's solution. This process repeats until the termination conditions are satisfied, and the individual with the highest fitness value is remapped to obtain the optimal test sequence.

Briand [16] et al. proposed a genetic algorithm-based method [17]. This method encodes the class integration test sequence into a chromosome and evolves the population through selection, crossover, and mutation operators under certain conditions. When the algorithm converges, the optimal individual in the population is remapped to the class integration test sequence. This method calculates the test stub complexity on the basis of the degree of class coupling. Czibula [18] et al. proposed an improved genetic algorithm that assigns different weights to various dependencies to minimize the test stubs required for class integration testing. Later, Wang [19] et al. combined coupling metrics with a random iteration algorithm (RIA) [20] to solve the class integration test sequence problem. This method uses only test stub complexity as guidance information during test sequence generation and does not consider the classification of class dependencies.

Briand [21] et al. emphasized the importance of weighting each metric within the fitness function. However, their approach was subjective, which may limit its applicability to complex systems. To address this, Cabral [22] et al. introduced a Pareto ant colony optimization algorithm (PACO) [23], which transforms the class integration test sequence problem into a multiobjective optimization problem by establishing a Pareto optimal curve [24]. They generated a set of Pareto optimal solutions via ant colony optimization, achieving a balance between attribute complexity and method complexity.

Zhang [25] et al. proposed a method based on the particle swarm optimization (PSO) algorithm [26]. The experimental results indicated that this method outperformed the comparison methods for most test systems. To address the premature convergence of PSO, Zhang [27] et al. introduced a dream PSO method that incorporates a distortion distance parameter for particles with poor fitness values, enhancing local search capabilities around local optima.

Zhang [28] et al. presented a new heuristic algorithm (HBCITO) for generating class integration test sequences. They argued that graph theory-based methods lack guidance on the number and complexity of test stubs during loop breaking. Thus, HBCITO combines graph theory with search algorithms to identify dependency sets for deletion, focusing the search space on dependencies involved in loops. The experimental results demonstrated that the HBCITO reduces testing costs and saves time in large-scale programs by narrowing the search space.

Guizzo [29, 30] et al. introduced a hyperheuristic algorithm (HITO) [31] for class integration and the test sequence problem. This method employs two selection functions: a selection function (CF) and a multiarmed band (MAB). At each step, it selects the best low-level heuristic (a combination of mutation and crossover operators). To facilitate selection, a quality metric is used to evaluate the performance of low-level heuristics throughout the evolutionary process. HITO is implemented via NSGA-II and is assessed on seven systems, yielding optimal results for the class integration test sequence problem. Relative to conventional methods, the proposed hyperheuristic consistently outperformed every system evaluated.

Czibula [32] et al. first applied reinforcement learning algorithms [33, 34] to generate test sequences. This method trains agents to learn from experience in generating test sequences with fewer test stubs. Agents continuously adjust their exploration paths on the basis of reward and punishment values, using these paths as test sequences. Ding [35] et al. also employed reinforcement learning to generate test sequences. Unlike Czibula's approach, they optimized both the number of test stubs and the overall test stub complexity, calculating weights for the two metrics via the entropy method. The experimental results revealed that reinforcement learning algorithms are advantageous for small-scale programs but have not been applied to large-scale programs because of their significant time consumption.

3 Class Integration Test Sequence Generation Based on the Simulated Annealing Particle Swarm Optimization Algorithm

3.1 Particle Mapping

When class integration test sequences are generated via the simulated annealing particle swarm optimization algorithm, randomly generating a certain number of initial class integration test sequences and mapping them to particles in space is key to solving the problem.

Table 1 describes the process of mapping a class test sequence to a particle in one-dimensional space. Lines 1–2 initialize the particle position and record the total number of classes in the test sequence for iterating through the test sequence. Line 4 retrieves classes from the test sequence one by one. Line 5 obtains the index value of the class in class_array. Line 6 updates the value of x. Once all classes have been traversed, the final x value represents the position of the test sequence in one-dimensional space.

Table 1. Particle mapping algorithm

Algorithm: Mapping class test sequences to particles in one-dimensional space Input: Test sequence: list_sequence Class array:class_array, where classes are ordered by the number of dependencies in descending order Output: Particle position: x
1.x = 0;
2.counts = list_sequence.length;//counts represents the total number of classes in the test sequence
3.**for** (int j = 0; j < counts; j++) //j is the j-th class in the test sequence
4.class = List_sequence.get(j);
5.index = class_array.getIndex(class);
6.x = x + index*(counts-j-1)!;
7.**endfor**

Example: Suppose a program contains four classes A, B, C, and D, and the number of dependencies (including method and property dependencies) for each class is shown in Table 2.

Table 2. Example of a program

Class Name	A	B	C	D
A	0	2	0	3
B	0	0	0	1
C	0	7	0	0
D	0	0	2	0

Construct class_array as {C, A, D, B} based on the number of dependencies for each class. Assume there is a test sequence list_sequence = {A, B, C, D}. According to the above algorithm, the final mapping of list_sequence to a position in one-dimensional space is:

$$X(\text{list\_sequence}) = \sum_{j=0}^{3}[\text{class\_array.getIndex(List\_sequence.get(j))} \cdot (4 - j - 1)!] = 14$$

3.2 Simulated Annealing Particle Swarm Optimization Algorithm

In the simulated annealing particle swarm optimization algorithm, when the PSO algorithm iterates to the middle or late stages, the particles tend to cluster around local optima, making it difficult to find better solutions. The introduction of the simulated annealing algorithm can generate a random solution in the search space. If the new solution is better than the current particle position, the particle moves to the new position; otherwise, it moves to the new position with a certain probability. The probability of accepting an inferior solution decreases as the annealing temperature decreases. Thus, when the annealing mechanism is initially introduced, particles have a high probability of moving to new positions, allowing them to escape the current local optimum. With the algorithm's advancement, the acceptance rate for new candidates steadily decreases, causing it to converge to the conventional PSO behavior.

The steps for generating class integration test sequences via the simulated annealing particle swarm optimization algorithm are as follows:

Step 1: Initialize the particle positions and set the initial temperature T = $fitness_{max}$-$fitness_{min}$.$fitness_{max}$ and $fitness_{min}$ are the maximum and minimum fitness values of the initial population, respectively. The overall test stub complexity serves as the fitness function, as shown in formula (1):

$$fitness = Ocplx(o) = \sum\nolimits_{i=1,j=1}^{n} SCplx(i,j) \tag{1}$$

Step 2: Set the individual best value (pBest) of each particle to its current position and the global best value (gBest) to the position of the best particle in the swarm.

Step 3: Update the velocity and position of each particle according to the standard PSO velocity and position update formulas. Let xt i denote the position of the i-th particle at time t,i ∈ [1,2,…,N], where N is the number of particles,vt i represents the velocity of the i-th particle at time t, and the update process of the velocity and position of the particle at time t + 1 is shown in Formula(2):

$$v_i^{t+1} = \omega v_i^t + c_1 r_1 (pBest_i^t - x_i^t) + c_2 r_2 \left(gBest_i^t - x_i^t\right), v_i^t \in [0, n!/n] \tag{2}$$

n represents the number of classes in the program to be tested:

$$x_i^{t+1} = x_i^t + v_i^{t+1} \tag{3}$$

Step 4: Randomly generate a solution x within the defined solution space,x ∈ [0,(n-1)!];

Step 5: Calculate the difference in fitness between the randomly generated solution x and the solution

$$x_i^{t+1}$$

obtained from the PSO update, denoted as Δf. If Δf < 0, the new solution is better than the one obtained from the PSO search. In this case, the PSO solution xt + 1 i is replaced with x, and the individual best (pBest) and global best (gBest) values are updated. Otherwise, accept the new solution with a probability of $p_e = e^{-(\Delta f/T)}$ and update pBest and gBest accordingly.

Step 6: Cooling operation: T = kT, where the cooling factor k is typically between 0.9 and 1. Here, we set the cooling factor k to 0.9.

Step 7: Check if the algorithm meets the convergence criteria. If so, output the final optimal value; otherwise, return to Step 3.

4 Experiments

The experimental environment consists of an Intel i7 quad-core processor (2.5 GHz), 8.0 GB of physical memory, and JDK 1.8.

4.1 Experimental Design

This paper investigates the performance of the simulated annealing particle swarm optimization algorithm through the following three questions:

Question 1: Compared with the PSO algorithm, can the method proposed in this chapter, which is based on simulated annealing particle swarm optimization, avoid premature convergence and escape local optima?

To explore this question, both the PSO and SAPSO algorithms are iterated 500 times, and the global optimum is recorded every 50 iterations. By comparing the global optimum values, the convergence performance of the two algorithms can be evaluated.

Question 2: Can the method based on simulated annealing particle swarm optimization proposed in this chapter generate test sequences with lower test stub costs?

Theoretically, introducing an annealing mechanism into the PSO algorithm should help it escape local optima and yield class integration test sequences with lower test stub costs. This paper selects overall test stub complexity as the standard for evaluating test stub costs and uses it as the fitness function to assess the quality of particle solutions. Various classic algorithms for generating class integration test sequences (including graph-based, genetic algorithm-based, PSO-based, dream PSO-based, and reinforcement learning-based methods) are compared under the same conditions (including the same population size and number of iterations), and the test stub costs of the resulting class integration test sequences are compared.

Question 3: How does the runtime of the algorithm proposed in this chapter vary across programs of different scales? Four medium and small systems (elevator, BCEL, ANT, and SPM) and four industrial systems (xml security, log4j3, Jmeter, and JHotDraw) are selected to evaluate the runtime of the proposed algorithm on systems of varying scales.

4.2 Experimental Subjects

Table 3 lists eight Java systems used as experimental subjects. Among them, the elevator, BCEL, ANT, and SPM are medium and small systems with relatively few classes, dependencies, loops, and lines of code. In contrast, xml-security, log4j3, Jmeter, and JHotDraw are large-scale industrial systems with hundreds of classes, over ten thousand lines of code, and thousands of dependencies. Half of the selected systems are small, and the other half are large, allowing for an evaluation of the performance of different algorithms under varying search space conditions.

To assess the performance of the proposed simulated annealing particle swarm optimization (SAPSO) algorithm accurately, we compare it with methods proposed by Briand et al. based on graph theory, a genetic algorithm (GA)-based approach, a particle swarm optimization (PSO)-based method, a dream PSO (DreamPSO)-based approach, and a reinforcement learning-based method (CITO_RL). The data for the genetic algorithm, PSO, and DreamPSO are reproduced from references [16, 25], and [27], respectively. Data for the graph theory-based and reinforcement learning-based methods are sourced from the literature [4] and [35], with missing data indicated by "–". To account for the stochastic nature of the algorithms, each is run 30 times, and the average results are reported.

Table 3. Experimental subjects

System	Description	Number of Classes	Dependencies	Number of Loops	LOC
elevator	Classic elevator scheduling algorithm	12	27	23	934
xml-security	XML encryption tool	239	759	98236	18702
log4j3	Java logging tool	261	784	2173	15665
Jmeter	Local testing tool	372	1252	729	32882
BCEL	Tool for creating, analysing, and manipulating Java class files	45	294	416091	3033
ANT	Java-based integrated package management system	25	83	654	4093
JHotDraw	Two-dimensional graphics editing framework	411	1680	225	78150
SPM	Security patrol monitoring system	19	72	1178	1198

4.3 Experimental Parameters

For the four small-scale systems (elevator, BCEL, SPM, and ANT), the initial population size is set to 100. For xml-security and log4j3, the population size is set to 300. Jmeter and JHotDraw are set to 1,000. The crossover probability for the genetic algorithm is 0.5, and the mutation probability is 0.15. The cooling factor is set to 0.9. Both the attribute complexity (W_A) and method complexity (W_M) are set to 0.5.

4.4 Experimental Results and Analysis

Convergence Speed. To verify whether the SAPSO algorithm can help the PSO algorithm escape local optima and avoid premature convergence, both the SAPSO and PSO algorithms are executed on the eight systems mentioned above, each for 500 iterations, with the optimal value recorded every 50 iterations.

Figures 1, 2, 3, 4, 5, 6, 7 and 8 shows that in the eight systems, the SAPSO algorithm generates class integration test sequences with lower overall test stub complexity. In the ANT system, when the SAPSO algorithm iterates between 300–350 and 450–500 times, it escapes local optima and finds a better solution. In the SPM system, the PSO algorithm falls into a local optimum after 100 iterations, but with the annealing mechanism, the SAPSO algorithm reduces the optimal solution from 3.08 to 2.69. The SAPSO algorithm initially searches for a higher optimal solution than does PSO in the early iterations, which is due to the nature of the simulated annealing algorithm, which accepts inferior solutions with a certain probability, allowing it to escape local optima. In the BCEL system, the PSO algorithm falls into a local optimum after 150 iterations, whereas the SAPSO algorithm reduces the optimal solution from 8.67 to 7.56 after 250 iterations. In the elevator system, the SAPSO algorithm consistently finds a better optimal solution than the PSO algorithm does throughout the 500 iterations. In the xml security system, the optimal value of SAPSO is always smaller than that of PSO. Although SAPSO falls into a local optimum at the 200th iteration, the annealing mechanism helps it escape after 300 iterations. In the log4j3 system, the PSO algorithm falls into a local optimum of 16.14 after 100 iterations, whereas SAPSO escapes this local optimum at the 200th iteration and reduces the optimal value to 15.75. In the Jmeter system, SAPSO avoids local optima between 150--200 and 300--350 iterations, resulting in a better solution. In the JHotDraw system, the most complex among the eight systems, SAPSO does not fall into a local optimum after the 200th iteration and continues to find better solutions.

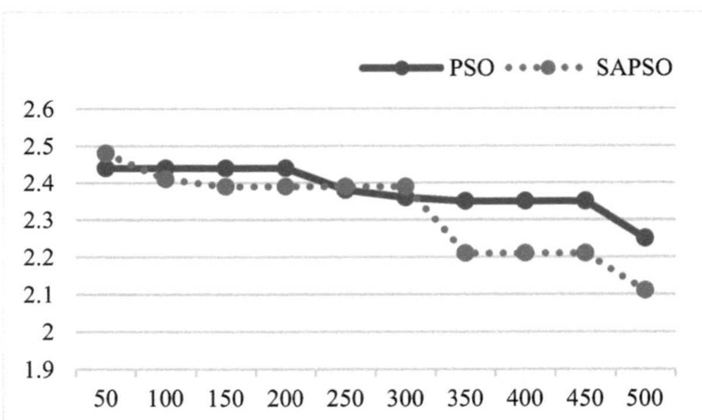

Fig. 1. Comparison of convergence speed (ANT)

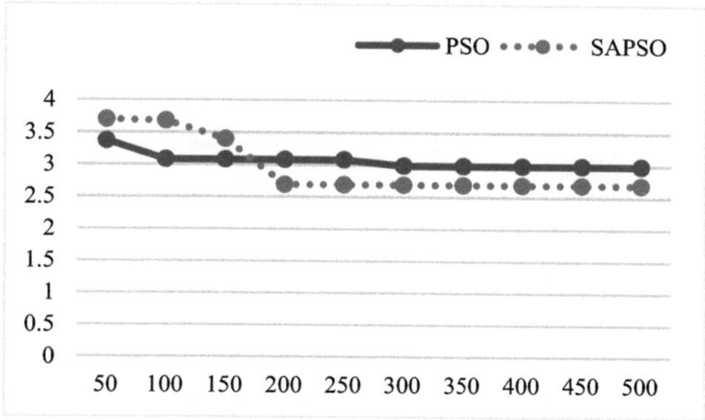

Fig. 2. Comparison of convergence speed (SPM)

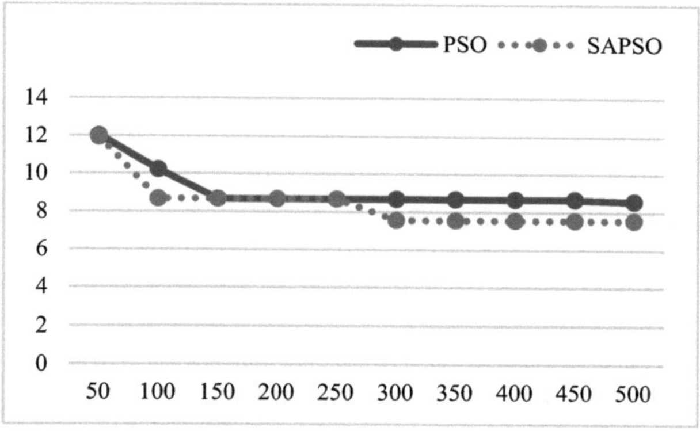

Fig. 3. Comparison of convergence speed (BCEL)

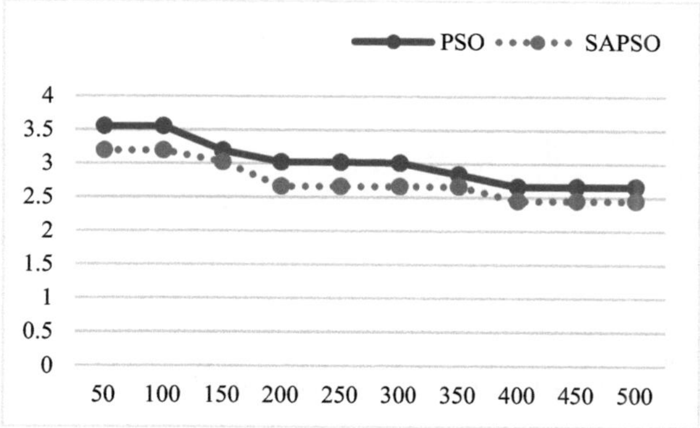

Fig. 4. Comparison of convergence speed (elevator)

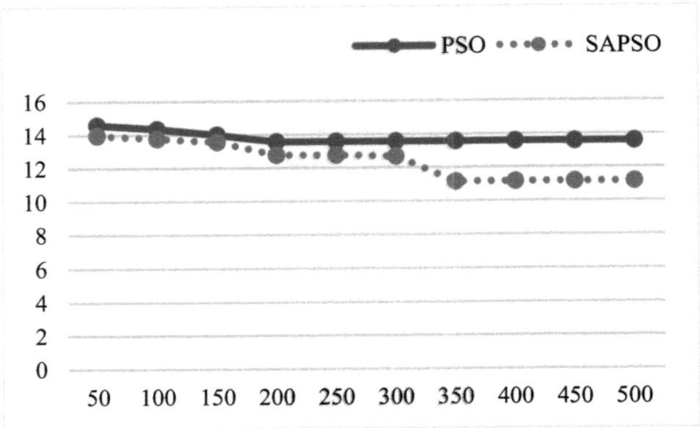

Fig. 5. Comparison of convergence speed (xml-security)

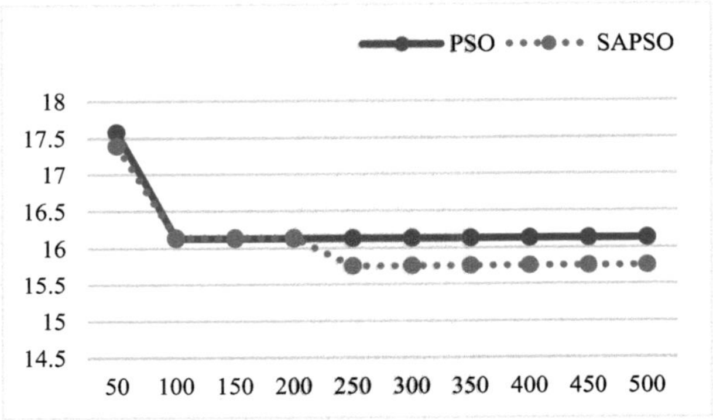

Fig. 6. Comparison of convergence speed (log4j3)

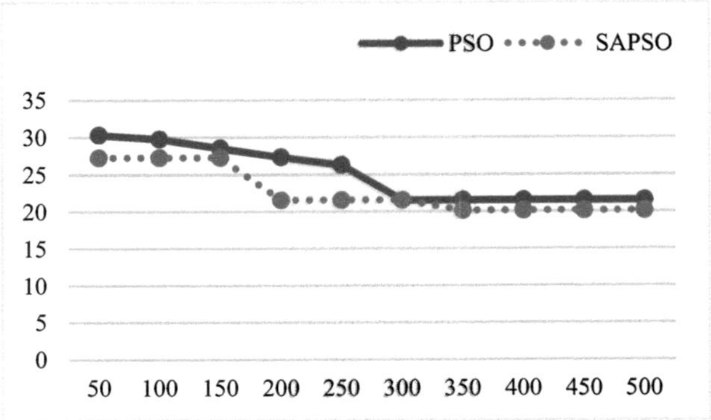

Fig. 7. Comparison of convergence speed (Jmeter)

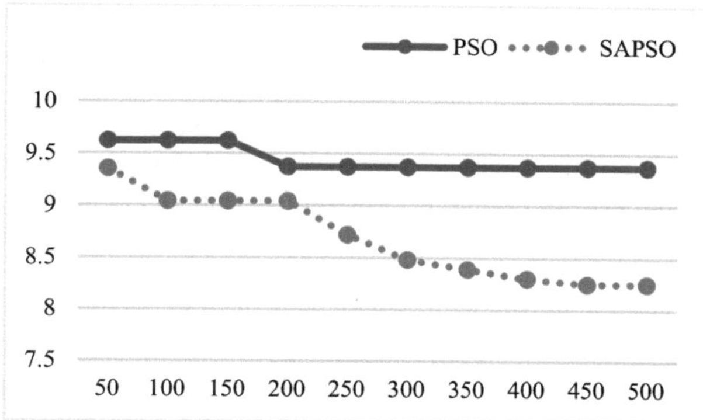

Fig. 8. Comparison of convergence speed (JHotDraw)

Overall Test Stub Complexity. Table 4 presents the comparison of overall test stub complexity across eight programs using six different algorithms, with the smallest value for each program bolded.

Table 4. Test stub complexity of SAPSO

Test Stub Complexity	Briand	GA	PSO	SAPSO	DreamPSO	CITO_RL
elevator	–	2.03	4.67	3.97	4.29	**1.72**
xml-security	–	14.6	16.94	**14.31**	15.7	–
log4j3	–	15.34	17.25	**15.19**	16.44	–
Jmeter	–	25.3	22.76	**20.4**	20.97	–
BCEL	**5.81**	9.7	7.68	6.68	6.65	8.58
ANT	3.31	2.13	2.2	2.32	2.44	**2.08**
JHotDraw	–	10.5	8.67	**7.92**	8.33	–
SPM	5.82	3.5	3.08	**2.8**	2.99	5.26

Table 4 shows that the graph theory method proposed by Briand et al. generated the lowest overall test stub complexity on the BCEL system. The SAPSO algorithm proposed in this paper achieved the lowest overall test stub complexity on five systems: xml-security, log4j3, Jmeter, JHotDraw, and SPM. The reinforcement learning method CITO_RL generated the lowest overall test stub complexity on the elevator and ANT systems.

The elevator system, which has only 12 classes, 27 dependencies, and fewer than 1,000 lines of code, is the simplest of the eight systems. For the ANT system, with 25 classes and 83 dependencies, the system scale is also small. The reason why the reinforcement learning method CITO_RL can generate the lowest test stub complexity for these two systems is that it trains an agent to explore different test sequences and provides different reward and punishment values for each test sequence. After hundreds of thousands of iterations, the agent gains experience and can always find the test sequence with the lowest testing cost. The SAPSO algorithm proposed in this chapter ranks third for the elevator system and fourth for the ANT system in terms of generated overall test stub complexity, showing average performance for these two systems. For the SPM system, although the system scale is not large, it contains 1,178 loops, making class relationships relatively complex. Therefore, the reinforcement learning algorithm is prone to falling into a local optimum and failing to find the optimal solution, whereas the SAPSO algorithm can find the optimal solution among the six algorithms provided in this paper.

For the BCEL system, the graph theory method can find the optimal solution. BCEL contains as many as 400,000 loops, and the graph theory method breaks the loops in the system through an object relationship graph and generates a test sequence through reverse topological sorting. The SAPSO and DreamPSO algorithms proposed in this paper achieve results that are almost identical, ranking third and second, respectively.

The xml-security, log4j3, Jmeter, and JHotDraw systems contain 239 to 411 classes and 759 to 1,680 dependencies, making them large-scale industrial systems. The SAPSO algorithm can generate the optimal test sequence for these systems, with the lowest overall test stub complexity. This is because the SAPSO algorithm can call the simulated annealing mechanism when the particle swarm algorithm falls into a local optimum, which helps the particles escape the local optimum region and enhances their global search capability, allowing them to continue searching for better solutions. The DreamPSO algorithm performs second only to SAPSO. The DreamPSO algorithm divides the particle search into "daytime" and "nighttime" phases. During the "daytime" phase, particles update their velocity and position according to the standard particle swarm algorithm, whereas during the "nighttime" phase, a distortion distance parameter is introduced for particles with poor fitness values, enabling them to further search around the local optimum and strengthen their local search capability. For large-scale industrial systems, reinforcement learning algorithms are not feasible because of the enormous number of training iterations and the high time cost involved.

Number of Test Stubs. Table 5 presents the comparison of the number of test stubs generated by six different algorithms across eight programs, with the smallest number for each program bolded.

A comparison of the six algorithms in Table 5 reveals that the SAPSO algorithm proposed in this chapter can generate fewer test stubs than can the PSO algorithm in all programs except log4j3, and it generates the fewest test stubs in the Jmeter and JHotDraw programs. Moreover, for the smaller-scale systems of the elevator, BCEL, ANT, and SPM, the number of test stubs constructed is relatively small, whereas for the four large-scale industrial systems of xml-security, log4j3, Jmeter, and JHotDraw, the number of test stubs is the largest, ranging from 172.6–478.8.

Table 5. Number of stubs for the SAPSO algorithm

Number of Test Stubs	Briand	GA	PSO	SAPSO	DreamPSO	CITO_RL
elevator	–	8	8.88	7.37	8.14	**5**

(continued)

Table 5. (*continued*)

Number of Test Stubs	Briand	GA	PSO	SAPSO	DreamPSO	CITO_RL
xml-security	–	203.2	174.3	174.3	**172.6**	–
log4j3	–	263.4	**262.76**	272.4	271.8	–
Jmeter	–	372.4	202.56	**179**	187.03	–
BCEL	70	**63**	88	80	79	77
ANT	**11**	16.8	26	17	19	16
JHotDraw	–	478.8	330.26	**323.2**	330.83	–
SPM	17	**13.7**	29	23.5	26	28

In the elevator program, the reinforcement learning algorithm can generate the smallest overall test stub complexity and the fewest test stubs. Compared with the PSO algorithm, CITO_RL reduces the number of test stubs by approximately half. In the Jmeter and JHotDraw systems, the SAPSO algorithm generates the fewest test stubs. In the Jmeter program, the number of test stubs generated by SAPSO is approximately half that of the genetic algorithm. In the JHotDraw program, SAPSO reduces the number of test stubs by approximately 150 compared with the genetic algorithm.

For search algorithms, the fitness function selected in this chapter is an overall test stub complexity. Therefore, during the search for the optimal solution, the algorithm always seeks to minimize the overall test stub complexity rather than simultaneously searching for a smaller number of test stubs. A smaller number of test stubs does not necessarily correspond to lower overall test stub complexity, as different dependencies may have varying levels of complexity, leading to different testing costs. This paper primarily uses overall test stub complexity as the main indicator.

Runtime. Table 6 presents the comparison of runtime across eight programs using five different algorithms, with the shortest runtime for each program bolded.

Table 6. Run times of the SAPSO algorithm (ms)

Runtime	GA	PSO	SAPSO	DreamPSO	CITO_RL
elevator	1390	873	1181.47	**861.7**	2623
xml-security	812276	**56257.47**	77488.53	125089.067	–
log4j3	121978	**65434.43**	92630.43	144919.6	–
Jmeter	406357	**98757.7**	134938.9	223702.7	–
BCEL	22412	**13988.83**	18776.2	21629.7	36804
ANT	1732	**1651**	1691	1719	5796
JHotDraw	812276	**152477.3**	186693.2	269020	–
SPM	2210	**1441**	1505	1505	4739

Table 6 shows that for the smaller-scale systems of BCEL and ANT, the PSO algorithm has the shortest runtime. The CITO_RL algorithm has the longest runtime. Among the eight experimental subjects, with the exception of the elevator system, the PSO algorithm has the shortest runtime.

For the four medium and small programs (elevator, BCEL, ANT, and SPM), the runtime of the PSO algorithm is the shortest. The CITO_RL algorithm has the longest runtime. For the smallest program, the elevator, DreamPSO has the shortest runtime, and the CITO_RL algorithm has the longest runtime, which is twice that of the SAPSO algorithm proposed in this chapter. Among the eight experimental subjects listed in this chapter, with the exception of the elevator, the PSO algorithm has the shortest runtime for all other subjects.

For the four industrial systems (xml security, log4j3, Jmeter, and JHotDraw), the genetic algorithm (GA) has the longest runtime, which is two to three times greater than that of the SAPSO algorithm proposed in this chapter. The genetic algorithm simulates the iterative process of chromosome updates, encoding the class integration test sequence as a chromosome. New populations (i.e., new class integration test sequences) are generated through crossover and mutation operators. The algorithm repeatedly favours higher-scoring candidates, and once the prescribed iterations are completed, the best-performing individual is retained as the final answer. This search strategy is relatively complex; hence, the runtime of large-scale industrial systems is relatively long, ranging from several minutes to more than ten minutes. The reinforcement learning algorithm requires hundreds of thousands of training iterations for small-scale systems and millions for industrial systems, making it time-consuming and impractical for large-scale applications.

Case Analysis. For the SPM system, the PSO and SAPSO algorithms are executed once each, and the differences in the generated class integration test sequences are compared.

Table 7 present information about the SPM system, which comprises 19 classes and 72 dependencies. Table 9 assign numbers to the class names of the SPM system, as shown in Table 9.

Table 8 compare the optimal class integration test sequences generated by the PSO and SAPSO algorithms for the SPM system, along with the corresponding number of test stubs and overall test stub complexity for each sequence.

Table 9 detail the test stubs corresponding to the optimal class integration test sequences generated by the PSO and SAPSO algorithms for the SPM system. For example, 4 < --3(12,6) indicates that a test stub for class 3 needs to be constructed for class 4, with an attribute complexity of 12 and a method complexity of 6. Acplx represents the total attribute complexity of all test stubs constructed for a class integration test sequence, whereas Mcplx represents the total method complexity.

Table 8 and Table 9 show that the class integration test sequence generated by the SAPSO algorithm for the SPM system has a lower overall test stub complexity and fewer test stubs than the PSO algorithm does. The comparison of test stubs shows that the SAPSO algorithm places class 3 before class 4 in the testing sequence, thereby reducing both the overall test stub complexity and the number of test stubs.

Table 7. SPM information

Number	Class Name	Number	Class Name
1	CardReaderInterface	2	Checkpoint
3	CheckpointController	4	CheckpointControllerIdleState
5	ControllerNormalCheckingState	6	ControllerState
7	ControllerViolationCheckingState	8	ControllerViolationState
9	CheckpointReferences	10	ControlCentre
11	ControlCentreThread	12	CreateZone
13	DoorLockInterface	14	GuardDueTimer
15	GuardLateTimer	16	PatrolSchedule
17	ScheduleEntry	18	SecurityTime
19	SecurityZone		

Table 8. Comparison of SPM results

Method	Order	Test Stub Complexity	Stubs
PSO	[4,18,11,3,17,2,19,9,1,10,13,16,14,15,7,6,8,5,12]	3.91	16
SAPSO	[3,2,9,13,18,19,16,14,8,11,6,15,17,7,10,1,5,4,12]	2.99	12

Table 9. Comparison of stubs for SPM

Index	PSO	SAPSO
Stubs	4 < --3(12,6)	3 < --4(4,8)
	4 < --9(7,4)	3 < --5(4,1)
	4 < --13(5,3)	3 < --7(4,1)
	4 < --14(6,2)	3 < --8(3,1)
	4 < --15(3,2)	3 < --9(7,0)
	4 < --16(5,3)	2 < --17(4,6)
	3 < --5(4,1)	9 < --1(2,1)
	3 < --7(4,1)	9 < --13(5,1)
	3 < --8(3,1)	9 < --14(6,1)
	3 < --9(7,0)	9 < --15(3,1)
	19 < --12(2,2)	9 < --16(5,1)
	9 < --1(2,1)	19 < --12(2,2)
	9 < --13(5,1)	
	9 < --14(6,1)	
	9 < --15(3,1)	
	9 < --16(5,1)	

(continued)

Table 9. *(continued)*

Index	PSO	SAPSO
Acplx	79	49
Mcplx	30	24

Answer to Question 1: Compared with the PSO algorithm, can the simulated annealing particle swarm optimization algorithm proposed in this chapter escape local optima and avoid premature convergence?

When particles fall into local optima, introducing the annealing mechanism indeed helps them escape these local optima and search for better solutions, thereby slowing the convergence of the algorithm.

Answer to Question 2: Can the method based on simulated annealing particle swarm optimization proposed in this chapter ultimately generate test sequences with lower testing costs?

By executing the algorithm on eight selected systems and comparing the results with those of five other algorithms, the proposed algorithm can generate test sequences with the lowest overall test stub complexity on five systems, particularly showing excellent performance on industrial systems.

Answer to Question 3: How does the runtime of the algorithm proposed in this chapter vary across programs of different scales?

Compared with the PSO algorithm, the proposed method introduces additional search strategies but does not significantly increase the runtime. It only takes a few seconds to execute on small-scale systems and a few minutes on large-scale systems, making it still feasible.

4.5 Validity Analysis

The internal validity threat of the proposed SAPSO algorithm in this chapter lies in the premature introduction of the annealing mechanism at the beginning of the particle swarm algorithm's execution. However, since the particles have not yet fallen into local optima during the initial movement phase, introducing the annealing mechanism at this stage may waste search time. In future work, we will set a threshold for the number of updates to the particle's individual optimum. When the individual optimum of a particle remains unchanged after a certain number of iterations, the particle has fallen into a local optimum, and an annealing mechanism is introduced at that point to optimize the proposed SAPSO algorithm.

The external validity threat concerns the credibility of the experimental results and the applicability of the algorithm to other systems. This chapter selects four small-scale and four large-scale programs on public datasets, ensuring strong representativeness of the experimental results. In the future, it can be extended to other object-oriented programming systems.

5 Summary

This paper proposes a method for generating class integration test sequences on the basis of the SAPSO algorithm and analyses its performance in terms of convergence speed, overall test stub complexity, number of test stubs, and runtime across eight experimental subjects.

Disclosure of Interests. The authors have no competing interests to declare that are relevant to the content of this article.

References

1. Kung, D.C., Gao, J., Hsia, P.: A test strategy for object-oriented programs. In: Proceedings of 1990 ACM Conference Object-Oriented Programming, Systems, Languages, and Applications (OOPSLA'90), SIGPLAN Notices, vol. 25, no. 10, pp. 89–98. ACM Press (1990)
2. Tai, K.C., Daniels, F.J.: Test order for inter-class integration testing of object-oriented software. In: Proceedings of 21st COMPSAC, pp. 602–607. IEEE CS Press (1997)
3. Traon, Y.L., Jéron, T., Morel, P.: Efficient object-oriented integration and regression testing. IEEE Trans. Reliab. **49**(1), 12–25 (2000)
4. Briand, L.C., Labiche, Y., Wang, Y.: An investigation of graph-based class integration test order strategies. IEEE Trans. Softw. Eng. **29**(7), 594–607 (2003)
5. Tarjan, R.: Depth-first search and linear graph algorithms. In: Proceedings of 12th SWAT, pp. 114–121. IEEE CS Press (1971)
6. Hewett, R., Kijsanayothin, P.: Automated test order generation for software component integration testing. In: Proceedings of 24th ASE, pp. 211–220. IEEE CS Press (2009)
7. Paradkar, A.: Inter-class testing of object-oriented software in the presence of polymorphism. In: Proceedings of 19th CASCON, pp. 30–38. IBM Press (1996)
8. Zhang, Y., Jiang, S., Zhang, H.: A class integration testing method based on dynamic dependencies. J. Comput. Sci. **34**(6), 1075–1089 (2011)
9. Hanh, V.L., Akif, K., Trao, Y.L.: Selecting an efficient OO integration testing strategy an experimental comparison of actual strategies. In: Proceedings of 15th ECOOP, LNCS, vol. 2072, pp. 381–401. Springer, Heidelberg (2001)
10. Malloy, B.A., Clarke, P.J., Lloyd, E.L.: A parameterized cost model to order classes for class-based testing of C++ applications. In: Proceedings of 14th ISSRE, pp. 353–364. IEEE CS Press (2003)
11. Bansal, P., Sabharwal, S., Sidhu, P.: An investigation of strategies for finding test order during integration testing of object oriented applications. In: Proceedings of 17th ICM2CS, pp. 1–8. IEEE CS Press (2009)
12. Abdurazik, A., Offutt, J.: Using coupling-based weights for the class integration and test order problem. Computer **52**(5), 557–570 (2009)
13. Zhang, M., Jiang, S., Zhang, Y.: A multi-level feedback approach for the class integration and test order problem. J. Syst. Softw. **133**, 54–67 (2017)
14. Jiang, S.J., Zhang, Y.M., Wang, Q.T.: Design of class integration test order based on coupling measures. Int. J. Inf. **15**(1), 331–338 (2012)
15. Zhang, Y., Jiang, S., Zhang, M.: A survey of class test order generation technologies in integration testing. J. Comput. Sci. **41**(3), 670–694 (2018)

16. Briand, L.C., Feng, J., Labiche, Y.: Using genetic algorithms and coupling measures to devise optimal integration test orders. In: Proceedings of 14th SEKE, pp. 43–50. ACM Press (2002)
17. Mirjalili, S.: Genetic algorithm. In: Evolutionary Algorithms and Neural Networks, pp. 43–55. Springer, Cham (2019)
18. Czibula, G., Marian, Z.: Identifying class integration test order using an improved genetic algorithm-based approach. In: Proceedings of 12th ICSOFT, pp. 163–187. SCITEPRESS (2017)
19. Wang, Z.S., Li, B.X., Wang, L.L.: An effective approach for automatic generation of class integration test order. In: Proceedings of 35th COMPSAC, pp. 680–681. IEEE CS Press (2011)
20. Delahaye, D., Chaimatanan, S., Mongeau, M.: Simulated annealing: from basics to applications. In: Handbook of Metaheuristics, pp. 1–35. Springer, Cham (2019)
21. Briand, L.C., Feng, J., Labiche, Y.: Experimenting with genetic algorithms and coupling measures to devise optimal integration test orders. Carleton Univ. Techn. Rep. **16**(2), 15–19 (2002)
22. da Veiga Cabral, R., Pozo, A., Vergilio, S.R.: A Pareto ant colony algorithm applied to the class integration and test order problem. In: Petrenko, A., Simão, A., Maldonado, J.C. (eds.) ICTSS 2010. LNCS, vol. 6435, pp. 16–29. Springer, Heidelberg (2010). https://doi.org/10.1007/978-3-642-16573-3_3
23. Deng, W., Xu, J., Zhao, H.: An improved ant colony optimization algorithm based on hybrid strategies for scheduling problem. IEEE access **7**, 20281–20292 (2019)
24. Knowles, J., Thiele, L., Zitzler, E.: A Tutorial on the Performance Assessment of Stochastic Multiobjective Optimizers. TIK-Report 214, Computer Engineering and Networks Laboratory (TIK), ETH Zurich, Switzerland (2006)
25. Zhang, Y., Jiang, S., Wang, X.: An optimization algorithm applied to the class integration and test order problem. Soft. Comput. **23**, 4239–4253 (2019)
26. Bansal, J.C.: Particle swarm optimization. In: Evolutionary and Swarm Intelligence Algorithms, pp. 11–23. Springer, Cham (2019)
27. Zhang, Y., Jiang, S., Zhang, Y.: A class integration test sequence generation method based on dream particle swarm optimization. Comput. Sci. **46**(2), 168–174 (2019)
28. Zhang, M., Keung, J., Xiao, Y.: A heuristic approach to break cycles for the class integration test order generation. In: Proceedings of 43rd COMPSAC, pp. 47–52. IEEE CS Press (2019)
29. Guizzo, G., Fritsche, G. M., Vergilio, S. R.: A hyper-heuristic for the multi-objective integration and test order problem. In: Proceedings of 17th GECCO, pp. 1343–1350. ACM Press (2015)
30. Guizzo, G., Vergilio, S.R., Pozo, T.R.: Grammatical evolution for the multi-objective integration and test order problem. In: Proceedings of 18th GECCO, pp. 1069–1076. ACM Press (2016)
31. Harman, M., Burke, E., Clark, J.: Dynamic adaptive search based software engineering. In: Proceedings of 2012 ESEM, pp. 1–8. ACM Press (2012)
32. Czibula, G., Czibula, I.G., Marian, Z.: An effective approach for determining the class integration test order using reinforcement learning. Appl. Soft Comput. **65**, 517–530 (2018)
33. Arulkumaran, K., Deisenroth, M.P., Brundage, M.: Deep reinforcement learning: a brief survey. IEEE Signal Process. Mag. **34**(6), 26–38 (2017)
34. Wirth, C., Akrour, R., Neumann, G.: A survey of preference-based reinforcement learning methods. J. Mach. Learn. Res. **18**(136), 1–46 (2017)
35. Ding, Y., Zhang, Y., Jiang, S.: Research on reinforcement learning for class integration test sequence generation (in Chinese). J. Softw. **33**(5), 1674–1698 (2022)

Research on Multi-scale Lesion Detection in Human Lung CT Images Based on Deep Learning

Junyi Duan and Ying Zhao(✉)

Beijing University of Chemical Technology, Chaoyang District, Beijing 100029, China
zhaoy@buct.edu.cn

Abstract. In light of the swift progress in digital technologies and artificial intelligence, deep learning approaches have played an expanding role in medical image detection. In lung CT images, lesions often exhibit complex contextual and fine-grained features. The location and scale of these lesions can significantly influence feature representation, making them difficult to capture accurately and thus reducing detection performance. To solve this challenge, we propose a novel model, YOLO-SPASS, which combines the YOLOv8 framework with the AssemFormer and SaELayer attention mechanisms to refine feature representations at multiple scales. Additionally, the model integrates the SPD-Conv module to improve convolutional feature encoding and adopts Inner-SIoU loss to preserve contextual information and enhance localization accuracy. The experimental results demonstrate that the proposed model achieves mAP@0.5 improvements of 5.1%, 4.3%, and 1.9% on CT images from the DeepLesion, Lung-PET-CT-Dx, and LUNA16 datasets, respectively. Furthermore, it achieves the highest average precision (AP) in multi-scale lesion detection tasks. These findings highlight the model's potential to significantly enhance performance in lung CT-based multi-scale lesion detection, contributing to more accurate and efficient diagnosis and treatment of lung diseases.

Keywords: Human lung CT image · YOLO algorithm · Object detection · Multi-scale

1 Introduction

In China, the new case rate and lethality rate of lung cancer rank first among all malignant tumors. Despite advancements in medical technology, the survival rate for lung cancer patients has improved, but the 5-year survival rate remains at approximately 25% [1]. Currently, computed tomography (CT) is crucial for the noninvasive detection of lung diseases. CT physicians can often judge the type and stage of disease progression by the corresponding symptoms and the location of their appearance [2]. However, the traditional image detection method, which is based on doctor experience, is easily affected by subjective factors such as fatigue, knowledge level and diagnostic experience and objective factors such as time and labor constraints and suboptimal performance.

L. Huang et al. (Eds.): CCF NCCA 2025, CCIS 2774, pp. 187–200, 2026.
https://doi.org/10.1007/978-981-95-5294-8_14

With the advancement of digital computing, computer-aided diagnosis (CAD) systems have emerged by integrating image processing, pattern recognition, and neural networks [3,4], thereby enabling lesion detection in organs such as the lungs and brain. However, many methods fail to account for the scale variability of lesions significantly hinders performance in multi-scale detection tasks [5].

In response to these difficulties, this paper introduces the YOLO-SPASS model, which integrates attention mechanisms into the YOLOv8 framework, modifies the convolutional architecture through space-to-depth transformation and nonstrided convolution, and introduces the Inner-SIoU loss function. These improvements enhance feature extraction capabilities, reduce the loss of small object information, and significantly enhance detection and localization accuracy.

2 Related Works

Deep learning-based object detection algorithms have been broadly applied in various image detection tasks [6,7]. These techniques are primarily divided into two-stage and one-stage methods. The two-stage approach, exemplified by Faster R-CNN [8], first create object proposals, followed by classification and regression tasks. In contrast, one-stage methods, including the YOLO [9,10] series and SSD [11], directly predict object locations and classes in an end-to-end manner with higher efficiency.

Owning to their efficiency, YOLO algorithms have been extensively used in medical image detection. However, their accuracy remains limited when detecting small and multi-scale lesions in CT images, prompting the development of various improvements. For example, Song [12] integrated deformable convolution, angular-free spatial pyramid pooling, and attention coordination into YOLOv8 to improve lung nodule detection. Wang [13] modified the number of structural units in the YOLOv3 backbone and detection head, and adopted the Mish activation function to improve performance. Ji [14] applied the SPPF module from YOLOv5 to YOLOv7-tiny, incorporated dilated convolutions, and decomposed feature maps at the semantic level for lung tumor detection.

Furthermore, with the introduction of attention mechanisms such as channel attention (SE) [15], spatial attention (CBAM) [16], and transformer encoders [17], the feature extraction and multi-scale modelling capabilities of object detection models have been significantly enhanced, offering greater potential for small object detection in medical images [18].

3 Methods

3.1 Object Detection Model

3.1.1 The Proposed Model To address the challenges of feature degradation, scale variance, and localization inconsistency in human lung CT lesion detection, we propose YOLO-SPASS, an improved model based on YOLOv8.

The SPD-Conv module, which integrates space-to-depth transformation and nonstrided convolution, replaces the original convolution to reduce information loss in lesion details, especially for small targets. The AssemFormer module, which integrates convolutions and vision transformers, facilitates the joint extraction of local and global features to enhance cross-scale integration. The SaELayer, a multibranch fully connected structure, improves channel modelling by suppressing redundancy and emphasizing critical features. These components are incorporated into the backbone and neck to optimize multi-scale feature representations and improve localization precision. Additionally, the Inner-SIoU is utilized as the localization loss to improve convergence and mitigate scale-related effects. The detailed YOLO-SPASS architecture is illustrated in Fig. 1.

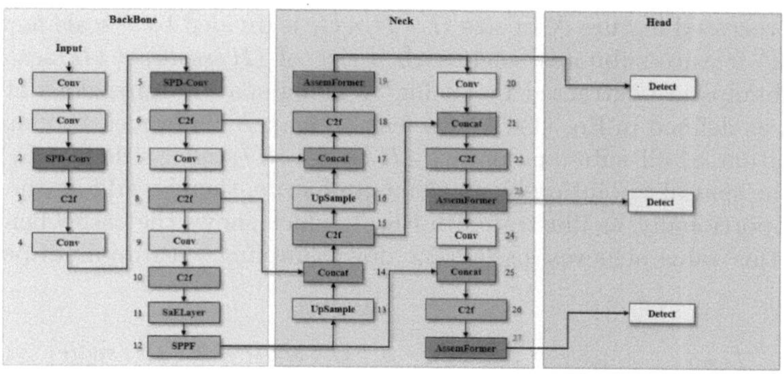

Fig. 1. Schematic of the YOLO-SPASS architecture.

3.1.2 YOLO Algorithm The YOLOv8 model consists of a backbone responsible for extracting features, a neck that fuses multi-scale features, and a decoupled head that separately handles classification and localization, with feature maps assigned on the basis of the scale of the target. Its core C2f module enhances gradient flow by fusing high-level features with context. The loss design integrates binary cross-entropy, distribution focal loss (DFL), and complete IoU (CIoU) to jointly optimize detection performance. The overall and internal architecture and components are illustrated in Fig. 2.

3.2 Attention Modules and Loss Design

3.2.1 SPD-Conv The spatial depth convolution (SPD-Conv) module [19] combines space-to-depth transformation with nonstrided convolution. The former converts spatial resolution into additional channels to capture more contextual information, whereas the latter reduces redundant channels via a scale factor, thereby retaining essential features.

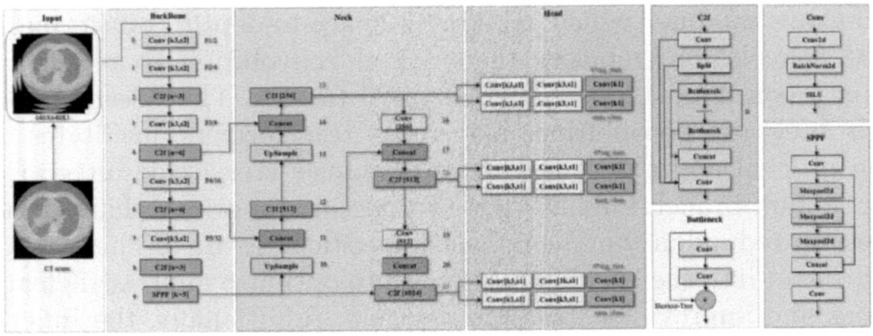

Fig. 2. YOLOv8 model architecture and partial module structure diagram.

The processed feature X of size $H \times H \times C_1$ is divided by a scale factor *scale* into $scale^2$ feature submaps, each with a size of $(H/scale) \times (H/scale) \times C_1$. These submaps are extracted by slicing X along spatial dimensions at regular intervals, as defined in Eq. (1). A new feature map X' is formed by channelwise concatenation of all submaps of size $(H/scale,\ H/scale,\ scale^2 \times C_1)$, which reduces the spatial resolution by an element of *scale*, thereby adding the channel count proportionally, as illustrated in Fig. 3, which shows the case when $scale = 2$. Thus, this value achieves lossless $2\times$ downsampling with minimal parameter overhead.

$$f_{scale-1,\,scale-1} = X[scale - 1 : H : scale - 1 : H : scale] \tag{1}$$

Next, a nonstrided convolution with C_2 filters is applied to X', where $C_2 < scale^2 \times C_1$, producing the output feature map X'' of size $(H/scale,\ H/scale,\ C_2)$.

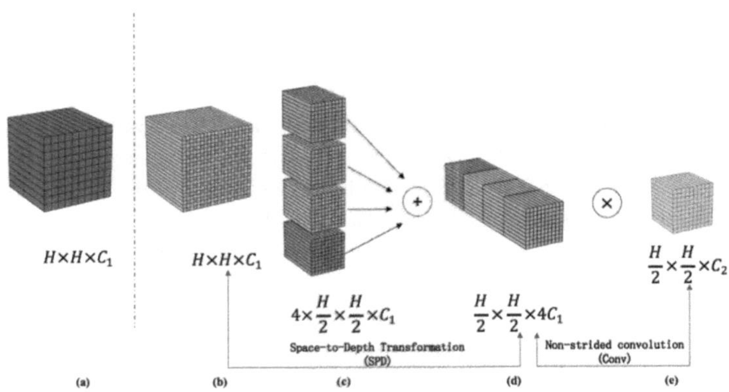

Fig. 3. Schematic diagram of SPD-Conv when $scale = 2$.

3.2.2 AssemFormer The AssemFormer [20] module is a hybrid structure that combines convolution and vision transformer operations. As illustrated in Fig. 4. AssemFormer bridges the gap between convolution and transformer operations by performing stacking and unstacking operations on the feature maps. The convolution component focuses on learning local and general features, whereas the transformer module leverages a series of multihead self-attention (MHSA) and multilayer perceptron (MLP) blocks to extract global information and positional relationships. Each block is followed by a layer normalization operation. The self-attention mechanism is formally defined in Eq. (2)

$$\mathcal{A}_{\mathrm{ViT}}(\mathbf{q},\mathbf{k},\mathbf{v}) = \mathrm{softmax}\left(\frac{\mathbf{qk}^{\top}}{\sqrt{D_h}}\right)\mathbf{v} \tag{2}$$

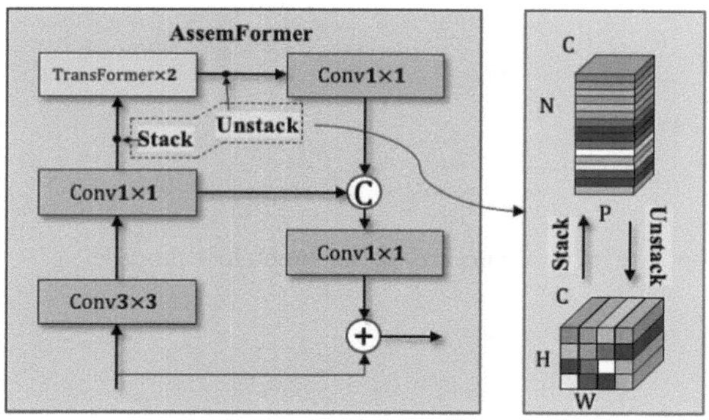

Fig. 4. Structure diagram of the AssemFormer module.

3.2.3 SaELayer SENetV2 combines the classical squeeze-and-excitation (SE) network with improvements from ResNeXt, introducing a novel attention mechanism known as the squeeze aggregated excitation (SaELayer) [21]. As illustrated in Fig. 5, it builds upon SENet by integrating ResNeXt-style multibranch fully connected layers, forming a lightweight, multipath attention module. The SaELayer extracts global features through global average pooling and enhances channelwise dependencies via multibranch connections and sigmoid activation, enabling richer multi-scale feature representation and improved robustness.

3.2.4 Inner-SIoU Inner-IoU [22] enhances bounding box regression by incorporating auxiliary boxes and dynamically adjusting a scale element to improve convergence and adaptability. Its structure is illustrated in Fig. 6, where B^{gt} and B^{st} denote the ground-truth and anchor boxes, respectively.

The Inner-IoU is computed by constructing scaled auxiliary boxes centered at the predicted and ground-truth boxes. The intersection and union are calculated as follows:

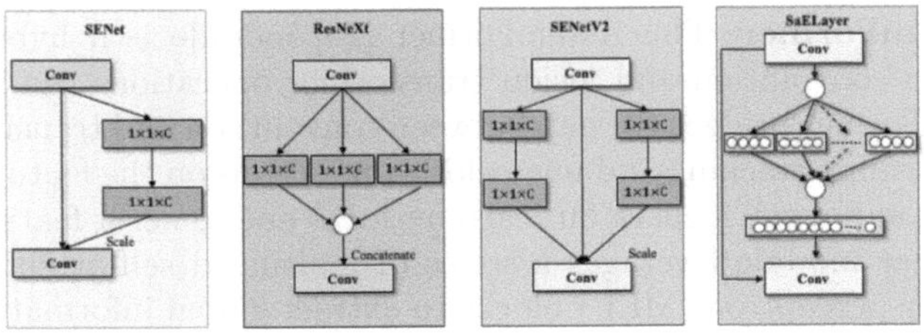

Fig. 5. Structures of SENet, ResNeXt, SENetV2, and SaELayer modules.

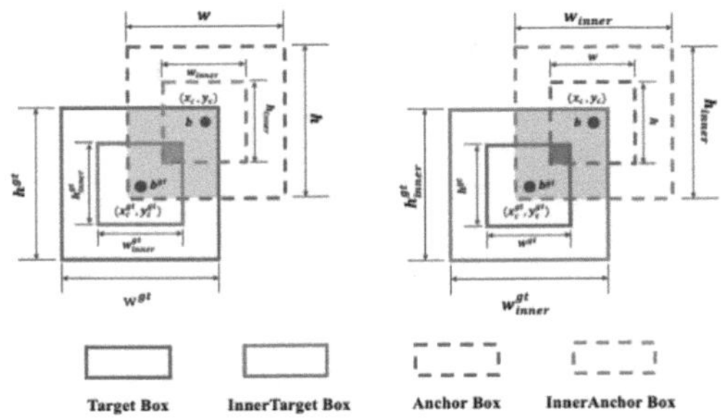

Fig. 6. Structure of the Inner-IoU module.

$$inter = (\min(b_r^{gt}, b_r) - \max(b_l^{gt}, b_l)) \times (\min(b_b^{gt}, b_b) - \max(b_t^{gt}, b_t)), \quad (3)$$

$$union = (w^{gt}h^{gt} + wh) \times ratio^2 - inter \quad (4)$$

$$IoU_{\text{inner}} = \frac{inter}{union} \quad (5)$$

To improve localization in diverse and rotated CT lesions, we adopt the Inner-SIoU as the loss function. It extends the standard SIoU [23] by introducing angular and distance penalties:

$$L_{\text{SIoU}} = 1 - IoU + \frac{\Delta + \Omega}{2}, \quad L_{\text{inner-SIoU}} = L_{\text{SIoU}} + IoU - IoU_{\text{inner}} \quad (6)$$

4 Experiments and Analysis

4.1 Dataset Introduction and Selection

We use publicly available imaging datasets, including DeepLesion [24] and Lung-PET-CT-Dx [25] obtained via TCIA [26], as well as the LUNA16 dataset provided by the LUng Nodule Analysis 2016 (LUNA16) Challenge.

The DeepLesion dataset provides 32,735 abnormalities across 9 categories from 4,427 patients. We selected the mediastinum, liver, and lung to create a DeepLesion-3 subset (3,861 training, 781 validation, adn 790 testing images).

The Lung-PET-CT-Dx dataset provides CT and PET-CT images, and the abnormalities are classified into adenocarcinoma, small cell carcinoma, large cell carcinoma, and squamous cell carcinoma. We randomly selected 2,527 CT images to construct a Lung-PET-CT-Dx subset (1,769 training, 379 validation, 379 testing images).

The LUNA16 dataset contains 888 thoracic CT scans divided into 10 subsets: 0–7 for training, 8 for validation, and 9 for testing.

DeepLesion-3 was utilized for training and validating YOLO-SPASS. Transfer learning was applied to Lung-PET-CT-Dx for multi-scale lesion detection, with LUNA16 used to assess robustness.

4.2 Evaluation Metrics

We use widely accepted evaluation metrics as described in Table 1 to comprehensively assess the performance of the detector in object detection tasks.

Table 1. Descriptions of evaluation metrics used in object detection

Metric	Description
Precision (P)	Ratio of correctly detected objects to all detected objects
Recall (R)	Ratio of correctly detected objects to all ground truth objects
Average Precision (AP)	Average precision across different recall levels
Mean Average Precision (mAP)	Mean of AP across all classes and recall thresholds
mAP@0.5	mAP at IoU threshold of 0.5 across all classes
Model Complexity	Measure of model's capacity and flexibility (e.g., parameters, FLOPs)
FPS	Frames per second, measuring inference speed

The definitions of precision and recall are provided below, where TP represents the count of true positives, FP represents false positives, and FN indicates the number of false negatives:

$$\text{Precision} = \frac{TP}{TP + FP}, \quad \text{Recall} = \frac{TP}{TP + FN} \tag{7}$$

where N denotes the number of classes and $AP_i^{0.5}$ and $AP_i^{0.5-0.95}$ represent the average precision of class i at an IoU threshold of 0.5 and across thresholds from 0.5 to 0.95, respectively. The formula for mAP@0.5 is as follows:

$$\text{mAP@0.5} = \frac{1}{N} AP_i^{0.5}, \tag{8}$$

4.3 Object Scale Division

To accommodate the distinct size variability of the lung lesion targets, we categorize them by scale following the COCO evaluation standard. Table 2 lists the scale names, area ranges, and associated AP metrics.

Table 2. Target scale divisions and corresponding evaluation metrics

Scale Name	Area Range	Average Precision Metric (%)
Special small	$area < 8^2$	AP@0.5_ss
Very small	$8^2 < area < 16^2$	AP@0.5_vs
Small	$16^2 < area < 32^2$	AP@0.5_s
Medium	$32^2 < area < 96^2$	AP@0.5_m
Large	$area > 96^2$	AP@0.5_l

4.4 Environment Setting and Parameter Setting

All the experiments presented in this work were carried out in the environment, and the parameter settings are listed in Table 3.

Table 3. Experimental environment and training settings

Category	Item	Version / Specification
Hardware	Memory	32 GB
	CPU	Intel i7-14700K (F)
	GPU Memory	12 GB
	GPU	NVIDIA GeForce RTX 4070 Super
Software	Operating System	Windows 11
	Python	3.12.7
	CUDA	12.6.85
	PyTorch	2.5.1
Training Settings	Optimizer	Stochastic Gradient Descent (SGD)
	Initial Learning Rate	0.01
	Batch Size	16
	Training Steps	500

4.5 Experiments Conducted on the DeepLesion-3 Dataset

4.5.1 Ablation Experiment To evaluate the effectiveness of each enhancement integrated into the proposed model(YOLO-SPASS), we perform an ablation experiment on the DeepLesion-3 dataset. Starting from the YOLOv8 baseline, we progressively introduced the SPD-Conv module, the SaELayer, the AssemFormer, and the Inner-SIoU loss methods. All the experiments were conducted with the same training configuration, and the resulting models were evaluated on the same test set. The detailed results are presented in Table 4.

Table 4. Ablation study results on the DeepLesion-3 dataset

Exp.	SPD-Conv	SaE-Layer	Assem-Former	Inner-SIoU	P (%)	R (%)	mAP@ 0.5(%)	Params (MB)	GFLOPs
1	–	–	–	–	69.6	69.6	71.0	21.4	28.4
2	✓	–	–	–	74.3	70.5	72.3	23.4	37.7
3	–	✓	–	–	73.9	69.7	72.1	21.6	28.5
4	–	–	✓	–	73.3	71.5	73.9	25.7	37.3
5	–	–	–	✓	71.9	70.4	71.9	21.4	28.4
6	✓	✓	–	–	74.5	72.6	73.1	21.9	35.8
7	✓	✓	✓	–	75.2	72.7	75.2	31.0	44.2
8	✓	✓	✓	✓	76.3	72.9	76.1	31.0	44.2

The results in Table 4 demonstrate that each proposed module individually enhances the detection performance. AssemFormer (Exp. 4) yields the most notable mAP@0.5 gain (+2.9%). SPD-Conv (Exp. 2) improves precision by reducing downsampling loss in CT images, while SaELayer (Exp. 3) enhances discriminative feature capture, and Inner-SIoU (Exp. 5) improves localization by addressing scale and orientation variation. All these modules and loss functions also contribute to moderate improvements in mAP@0.5.

The combined modules further improve the performance. Adding SaELayer to SPD-Conv (Exp. 6) increases mAP@0.5 by 0.8%, whereas introducing AssemFormer (Exp. 7) contributes an additional 2.1%. Replacing the loss function with the Inner-SIoU (Exp. 8) yields a further 0.9% improvement. This study demonstrated synergistic improvements in convolution, channel modelling, feature extraction, and localization loss.

Overall, YOLO-SPASS achieves 6.7% higher precision, 3.3% higher recall, and 5.1% higher mAP@0.5 than the YOLOv8 baseline does, confirming the effectiveness and synergy of the proposed components in enhancing lung CT lesion detection.

4.5.2 Comparative Experiment We performed a systematic evaluation of the YOLO-SPASS model by benchmarking against baseline models, namely, YOLOv3 (tiny), YOLOv5 (n/s/m), YOLOv6 (n/s), YOLOv8 (n/m), YOLOv10, YOLOv11, SSD, Rt-detr, and Faster R-CNN. Figure 7 shows the training curves, and the evaluation outcomes on the test set are presented in Table 5.

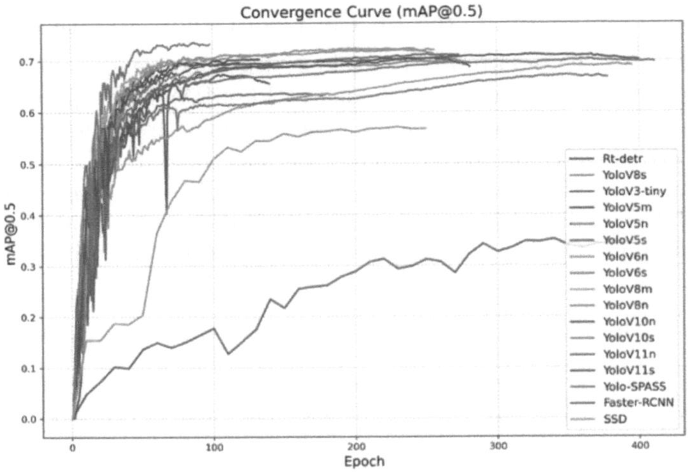

Fig. 7. Convergence curve of YOLO-SPASS on the DeepLesion-3 dataset.

As illustrated in Fig. 7, YOLO-SPASS achieves faster convergence and higher accuracy. Table 5 further confirms its superiority across all the other key metrics. Compared with similar-sized models (YOLOv5s, YOLOvs, YOLOv8s, YOLOv10s, and YOLOv11s), it improves precision by up to 9.6%, recall by 7.1%, and mAP@0.5 by 9.8%. These results demonstrate YOLO-SPASS's advantages in CT image detection, including faster convergence, better fitting, and more accurate predictions.

4.6 Transfer Learning to Lung Lesion Detection Tasks

4.6.1 Detection of Lung Lesions in Human Lungs Following validation on the DeepLesion-3 dataset, YOLO-SPASS confirmed its target detection capability in the background of the CT images. Then, YOLO-SPASS was transferred to the domain of lung lesion detection to identify multi-scale abnormalities. To assess its effectiveness, we compared it with other medical object detection networks on the Lung-PET-CT-Dx and LUNA16 datasets. The detection results on the Lung-PET-CT-Dx dataset are summarized in Table 6.

YOLO-SPASS achieves the highest accuracy across all lesion types, with an mAP@0.5 of 95.1%, outperforming YOLOv8s by 4.3%. These results demonstrate that our method, by enhancing small-object feature extraction through convolutional refinements, integrating global and local features, improving channel modelling, and optimizing the loss function, achieves superior performance and strong generalizability in lung lesion detection tasks, outperforming conventional one-stage and two-stage detection frameworks.

On the basis of the results in Table 7, the experimental performance of our method on the LUNA16 dataset, which contains a greater proportion of small targets, further demonstrates the effectiveness of YOLO-SPASS for detecting

Table 5. Comparison of different detection algorithms on DeepLesion-3 dataset

Algorithm	P (%)	R (%)	mAP@0.5 (%)	FPS	Model Size (MB)	GFLOPs
YOLOv3 (tiny)	66.8	63.6	63.5	115.8	23.2	19.0
YOLOv5n	69.3	67.6	69.4	119.3	5.1	7.1
YOLOv5s	71.3	67.5	70.2	94.6	17.7	24.0
YOLOv5m	67.3	69.8	70.9	44.5	48.1	64.0
YOLOv6n	72.4	65.2	68.9	143.2	8.3	11.8
YOLOv6s	66.7	65.3	66.3	73.6	31.3	44.0
YOLOv8n	70.6	65.8	69.2	130.6	5.9	8.1
YOLOv8s	69.6	69.6	71.0	89.8	21.4	28.4
YOLOv8m	71.6	68.3	70.9	39.5	49.6	78.7
YOLOv10n	70.1	65.8	69.9	99.8	5.5	8.2
YOLOv10s	71.7	65.5	70.1	84.4	15.7	24.5
YOLOv11n	68.2	67.1	69.4	102.7	5.2	6.3
YOLOv11s	68.9	68.8	69.8	89.0	18.3	21.3
SSD	69.6	44.6	56.7	25.6	90.6	65.2
Faster R-CNN	22.7	59.8	33.9	9.1	108.1	183.7
Rt-detr	67.4	66.1	66.8	24.6	82.0	103.4
YOLO-SPASS	**76.3**	**72.4**	**76.1**	**76.4**	**31.0**	**44.2**

Table 6. Comparison results on the Lung-PET-CT-Dx dataset

Method	mAP@0.5 (%)	AP (%)			
		Adeno- carcinoma	Small Cell Carcinoma	Large Cell Carcinoma	Squamous Cell Carcinoma
YOLOv5s	89.9	92.2	92.8	87.8	86.8
YOLOv8s	90.8	93.6	95.1	85.4	89.2
YOLOv10s	91.5	93.7	93.5	91.5	87.4
SSD	84.5	85.8	74.4	91.2	86.8
Faster-RCNN	78.7	70.9	90.8	85.4	67.9
Rt-detr	88.6	91.9	87.3	94.3	81.0
YOLO-SPASS	**95.1**	**96.2**	**95.2**	**99.5**	**89.3**

small lesions in CT images. Specifically, the mAP@0.5 improves by 3.7% compared with YOLOv8s and outperforms other models from both the same and different series, validating the superior capability of our method for small lesion detection in the lungs.

4.6.2 Multi-scale Lung Lesion Detection In the final stage of evaluation, we assess how well the model performs across different object sizes. A summary of the results is presented in Table 8.

As shown in Table 8, YOLO-SPASS achieves the highest accuracy across all the object scales compared with both the intra-series and cross-series algorithms. Specifically, compared with the best-performing baseline models at var-

Table 7. Comparison results on the LUNA16 dataset

Method	Precision (%)	Recall (%)	mAP@0.5 (%)
YOLOv5s	81.1	85.1	85.6
YOLOv8s	84.7	78.6	85.4
YOLOv10s	85.2	78.4	81.9
Rt-detr	73.6	61.1	61.3
YOLO-SPASS	**88.4**	**83.2**	**87.5**

Table 8. Comparison of detection results at different target scales

Method	AP@0.5_ss (%)	AP@0.5_vs (%)	AP@0.5_s (%)	AP@0.5_m (%)	AP@0.5_l (%)
YOLOv5s	43.0	47.9	37.0	47.6	60.7
YOLOv8s	43.9	46.9	47.2	49.5	60.7
YOLOv10s	44.6	47.8	43.1	49.7	59.2
Rt-detr	12.8	27.7	36.7	48.6	57.1
YOLO-SPASS	**46.7**	**49.6**	**49.5**	**53.4**	**61.1**

ious scales, YOLO-SPASS improves the AP at special small, very small, small, medium, and large scales by 2.1%, 1.7%, 2.3%, 3.7%, and 0.4%, respectively, and achieves the highest detection AP. These results demonstrate that the YOLO-SPASS model, through enhancements in small-scale feature extraction, integration of global and local features, channel modelling, and loss function optimization, effectively improves lesion detection performance across various scales. We found that the 25 AP values obtained from the comparative experiments failed to satisfy the normality assumption, so we applied the nonparametric Friedman test [27], which yielded a test statistic of $\chi^2 = 16.48$ ($p = 0.0024$), indicating significant performance differences both within and across algorithm series. Moreover, YOLO-SPASS attained the lowest average rank (1.00), demonstrating its superior overall effectiveness across targets of varying scales. These results further demonstrate the efficacy and resilience of the proposed improvements for multi-scale lung lesion detection tasks.

5 Conclusion

In this study, we propose an innovative method, YOLO-SPASS, to address the challenges of multi-scale lung lesion detection in human CT images. Built upon an enhanced YOLOv8 framework, our model integrates SPD-Conv, SaELayer, AssemFormer, and the Inner-SIoU loss methods. These components work collaboratively to mitigate the adverse effects of weak lesion features, scale variance, and localization inconsistency, thereby enhancing the automation and intelligence of lung lesion detection.

Extensive experiments show the effectiveness of the enhancements we proposed. For the DeepLesion-3 dataset, the model achieves a 5.1% increase in mAP.

When transferred to the LUNG-PET-CT-Dx and LUNA16 datasets, the mAPs further improve by 4.3% and 1.9%, respectively. It achieves optimal performance in comparison with both intraseries and interseries algorithms. Moreover, the detection precision at different scales shows consistent gains, with AP improvements of 2.1%, 1.7%, 2.3%, 3.7%, and 0.4% on the special small, very small, small, medium, and large categories, respectively. These results demonstrate the robustness and generalizability of YOLO-SPASS for multi-scale lesion detection in lung CT images. In future research, efforts will focus on enhancing inference speed while maintaining accuracy, aiming to facilitate real-time application in clinical settings.

References

1. Siegel, R.L., Giaquinto, A.N., Jemal, A.: Cancer statistics, 2024. CA: a cancer journal for clinicians **74**(1), 12–49 (2024)
2. Hasson, R.M., Bridges, C.J., Curley, R.J., Erhunmwunsee, L.: Access to lung cancer screening. Thorac. Cardiovasc. Surg. **33**(4), 353–363 (2023)
3. Lee, H., Chen, Y.P.P.: Skin cancer extraction with optimum fuzzy thresholding technique. Appl. Intell. **40**, 415–426 (2014)
4. Doi, K.: Computer-aided diagnosis in medical imaging: historical review, current status and future potential. Comput. Med. Imaging Graph. **31**(4–5), 198–211 (2007)
5. Oxnard, G.R., et al.: Variability of lung tumor measurements on repeat computed tomography scans taken within 15 minutes. J. Clin. Oncol. **29**(23), 3114–3119 (2011)
6. Brahmi, W., Jdey, I., Drira, F.: Exploring the role of convolutional neural networks (CNN) in dental radiography segmentation: a comprehensive systematic literature review. Eng. Appl. Artif. Intell. **133**, 108510 (2024)
7. Younesi, A., Ansari, M., Fazli, M., Ejlali, A., Shafique, M., Henkel, J.: A comprehensive survey of convolutions in deep learning: applications, challenges, and future trends. IEEE Access **12**, 41180–41218 (2024)
8. Ren, S., He, K., Girshick, R., Sun, J.: Faster R-CNN: towards real-time object detection with region proposal networks. IEEE Trans. Pattern Anal. Mach. Intell. **39**(6), 1137–1149 (2016)
9. Redmon, J., Divvala, S., Girshick, R., Farhadi, A.: You only look once: unified, real-time object detection. In: Proceedings of the IEEE Conference on Computer Vision and Pattern Recognition, pp. 779–788 (2016)
10. Sapkota, R., et al.: YOLOv10 to its genesis: a decadal and comprehensive review of the you only look once series. Available at SSRN 4874098 (2024)
11. Liu, W., et al.: SSD: single shot multibox detector. In: Leibe, B., Matas, J., Sebe, N., Welling, M. (eds.) ECCV 2016. LNCS, vol. 9905, pp. 21–37. Springer, Cham (2016). https://doi.org/10.1007/978-3-319-46448-0_2
12. Song, Y., Liu, G.: DCA-YOLO: a YOLOv8-based pulmonary nodule detection method optimized with attention mechanism. J. Donghua Univ. (English Edition), 1–14 (2024). https://doi.org/10.19884/j.1672-5220.202401002
13. Wang, Q., Shi, H.: Pulmonary nodule detection method based on improved YOLOv3 (in Chinese). Chin. J. Med. Phys. **38**(9), 1179–1184 (2021)
14. Ji, Z., et al.: ELCT-YOLO: an efficient one-stage model for automatic lung tumor detection based on CT images. Mathematics **11**(10), 2344 (2023)

15. Prasanna, G., Ernest, J.R., Lalitha, G., Narayanan, S.: Squeeze excitation embedded attention U-Net for brain tumor segmentation. In: Gabbouj, M., Pandey, S.S., Garg, H.K., Hazra, R. (eds) International Conference on Emerging Electronics and Automation, pp. 107–117. Springer, Singapore (2022). https://doi.org/10.1007/978-981-99-6855-8_9
16. Chien, C.T., Ju, R.Y., Chou, K.Y., Xieerke, E., Chiang, J.S.: YOLOv8-AM: YOLOv8 based on effective attention mechanisms for pediatric wrist fracture detection. IEEE Access (2025)
17. Nguyen, T., Heldermon, C.D., Toler-Franklin, C.: Unified local and global attention interaction modeling for vision transformers. arXiv preprint arXiv:2412.18778 (2024)
18. Rand, A., Ibrahim, H.: Beyond conventional transformers: the medical X-Ray attention (MXA) block for improved multi-label diagnosis using knowledge distillation. arXiv preprint arXiv:2504.02277 (2025)
19. Sunkara, R., Luo, T.: No more strided convolutions or pooling: a new CNN building block for low-resolution images and small objects. In: Amini, MR., Canu, S., Fischer, A., Guns, T., Kralj Novak, P., Tsoumakas, G. (eds) Joint European conference on machine learning and knowledge discovery in databases, pp. 443–459. Springer (2022). https://doi.org/10.1007/978-3-031-26409-2_27
20. Dai, W.: SvANet: a scale-variant attention-based network for small medical object segmentation. arXiv e-prints arXiv:2407.07720v3 (2024)
21. Narayanan, M.: SENetV2: aggregated dense layer for channelwise and global representations. arXiv preprint arXiv:2311.10807 (2023)
22. Zhang, H., Xu, C., Zhang, S.: Inner-IoU: more effective intersection over union loss with auxiliary bounding box. arXiv preprint arXiv:2311.02877 (2023)
23. Gevorgyan, Z.: SIoU Loss: more powerful learning for bounding box regression. arXiv preprint arXiv:2205.12740 (2022)
24. Yan, K., Wang, X., Lu, L., Summers, R.M.: DeepLesion: automated mining of large-scale lesion annotations and universal lesion detection with deep learning. J. Med. Imaging **5**(3), 036501–036501 (2018)
25. Dunn, B., Pierobon, M., Wei, Q.: Automated classification of lung cancer subtypes using deep learning and CT-scan based radiomic analysis. Bioengineering **10**(6), 690 (2023)
26. Clark, K., et al.: The cancer imaging archive (TCIA): maintaining and operating a public information repository. J. Digit. Imaging **26**, 1045–1057 (2013)
27. Friedman, M.: The use of ranks to avoid the assumption of normality implicit in the analysis of variance. J. Am. Stat. Assoc. **32**(200), 675–701 (1937)

Riding the Wave of LLMs: Navigating Opportunities and Challenges in Chinese Information Technology Application Innovation

Yingshuai Kou[1,2,3(✉)] and Haifeng Yu[2,3]

[1] Southeast University, Nanjing, China
kouyingshuai@163.com
[2] Ningbo State-owned Capital Research Institute Co., Ltd., Ningbo, China
[3] Ningbo Tongshang Holding Group Co., Ltd., Ningbo, China

Abstract. This paper explores the integration of large language models (LLMs) with China's information technology application innovation (ITAI), highlighting transformative opportunities and challenges. It systematically analyses the impacts of LLMs across four domains: hardware, software, security, and ecosystem. For hardware, innovations such as advanced packaging and ISAs are proposed to address computational bottlenecks while enhancing energy efficiency. In software, lightweight middleware and hybrid transactional/analytical processing (HTAP) databases optimize resource scheduling and intelligent workloads. Security advancements include differential privacy frameworks for federated learning and full-lifecycle LLMs security assessments to mitigate adversarial threats. Ecosystem development relies on open-source collaboration. This paper emphasizes hardware—software codesign, privacy-preserving AI, and policy support to overcome technical fragmentation and regulatory challenges. By leveraging LLMs, ITAI can shift from functional substitution to intelligent leadership, establishing an autonomous, secure, and high-performance digital infrastructure to support China's economic growth.

Keywords: Large Language Models · Information Technology Application Innovation · Artificial Intelligence · Software and Hardware · Information Security

1 Introduction

Information Technology Application Innovation (ITAI) [16], a core pillar of China's national strategic science and technology system, aims to resolve critical technological bottlenecks and establish a self-controllable digital foundation. Its development is not only crucial for the security of the industrial chain but also a key manifestation of national competitiveness in the digital economy era. As shown in Fig. 1, the ITAI industry covers four major fields: basic hardware

L. Huang et al. (Eds.): CCF NCCA 2025, CCIS 2774, pp. 201–221, 2026.
https://doi.org/10.1007/978-981-95-5294-8_15

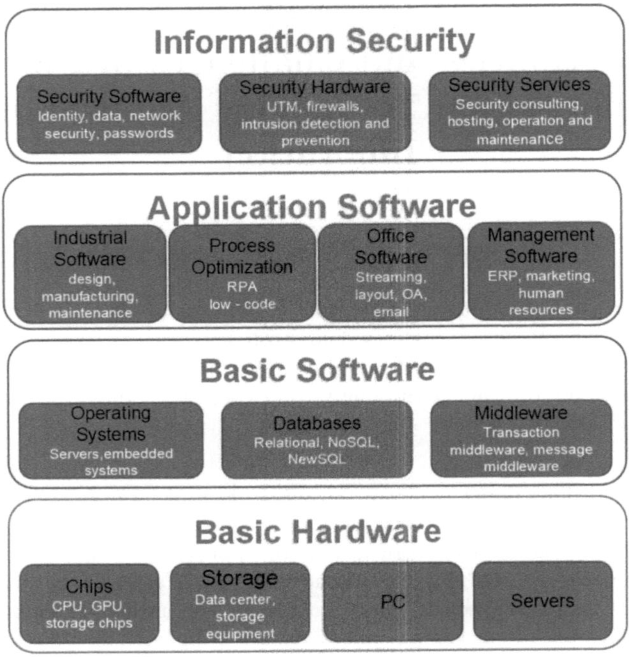

Fig. 1. The Information Technology Application Innovation mainly includes four major fields: basic hardware, basic software, application software, and information security.

(chips, servers), basic software (operating systems, databases), application software (office systems, industry solutions), and information security [76]. It aims to achieve full-stack domestic substitution, breaking long-standing dependence on foreign technologies. Since the launch of the "Core-electronics, High-end-general-Chips, and Infrastructural-software" project in 2006 [32], China has made breakthroughs in areas such as domestic CPUs and operating systems [75]. However, the maturity of the core technology ecosystem still requires long-term investment. The rapid development of large language models (LLMs) has provided new impetus for the ITAI industry. LLMs [27], through their Transformer architecture and self-attention mechanisms, have demonstrated deep understanding and generation capabilities for natural language, code, and even multimodal data. These technical characteristics align strongly with the needs of the ITAI industry:

Computational Demand Driving Hardware Upgrades. Training LLMs requires massive computational resources (e.g., training GPT-3 consumes 355 GPU-years) [14], pushing domestic chips toward high-performance, low-power evolution. For instance, domestic GPUs have achieved 40% increase in computing density through chiplet technology, but limitations in process technology mean single-chip performance is only 60% of the international advanced levels [50].

Algorithm Optimization Drives Software Innovation. The pre-training and fine-tuning paradigm of LLMs, along with the model-as-a-service (MaaS) model [58], provides new strategies for the iterative development of domestic basic software. Domestic operating systems, through microkernel architectures, enable dynamic resource scheduling and support unified management of heterogeneous hardware resources. However, adaptation to mainstream frameworks such as Tensorflow still relies on intermediate layers, resulting in performance losses of 15–20% [12].

Ecosystem Reconstruction Accelerating Domestic Substitution. The high requirements of LLMs for hardware-software coordination have driven the ITAI ecosystem to shift from single-point breakthroughs to full-stack adaptation. For example, domestic databases, through approximate nearest neighbor (ANN) algorithms [68], achieve millisecond-level retrieval, supporting real-time reasoning for LLMs. However, the ability to handle hybrid workloads (HTAPs) [10] for transactional and analytical queries still needs optimization.

Nonetheless, the integration of LLMs with ITAI faces multiple challenges:

- Technical Compatibility Bottlenecks: Domestic chips are not fully compatible with international mainstream computing frameworks such as CUDA, which require middleware to compensate for performance losses.
- Data Security Risks: Protecting the privacy of LLM training data requires the integration of federated learning and homomorphic encryption technologies. However, current solutions struggle to balance encryption overhead and model accuracy.
- Ecosystem Coordination Obstacles: The contribution rate of open-source communities to domestic technology stacks is less than 10% of that of international mainstream projects, which constrains the iteration speed of basic software.

Currently, China is accelerating the deep integration of ITAI with LLMs. Domestic inference-oriented large models [1] have achieved breakthroughs in specialized domain reasoning capabilities through optimized Transformer architectures, supporting localized deployment and privatization needs. For instance, in the medical field, LLMs, through enhanced control over compliance and accuracy, assist in generating diagnostic reports. However, their black-box nature still requires the integration of knowledge graph technology to improve interpretability.

In the future, the ITAI industry should leverage LLMs to drive coordinated innovation in hardware, software, and security, constructing a self-controllable intelligent technology ecosystem. For example, in the field of information security, training frameworks based on differential privacy can reduce the success rate of data poisoning attacks [2]. Additionally, the development of automated threat tracing tools will require the integration of large model semantic analysis capabilities. Furthermore, the design of cross-architecture compiler toolchains promote unified development interfaces for instruction sets such as ARM and RISC-V [37], reducing technical adaptation costs.

This paper approaches the topic from a technological perspective, systematically analysing the empowerment pathways and technical obstacles of LLMs across the four domains of ITAI. It also proposes key directions for future development. This study aims to provide theoretical references for the technological upgrading and ecosystem reconstruction of the ITAI industry.

In this survey, we first systematically analyse the empowering effects and technical bottlenecks of LLMs in various fields of the ITAI industry from a technological perspective. It then explores the key directions for future development. The full text systematically reviews the evolution of the ITAI industry and the technical characteristics of LLMs. It examines the coordination and conflicts between computational demand and architectural innovation from the perspective of basic hardware. From the perspective of basic software, it explores the technical paths for enhancing intelligent capabilities and ecosystem adaptation. From the perspective of application software, it studies the balance between security and efficiency in intelligent transformation. Finally, from the perspective of information security, it analyses the reconstruction of defense systems in the era of generative AI. Through systematic analysis and summary, the article concludes with future development recommendations, including prioritized directions for technical breakthroughs and ecosystem collaboration.

2 Background

In this section we trace the evolution of China's ITAI from its origins in indigenous operating system initiatives to its current trillion-yuan scale, highlighting persistent gaps in core technological ecosystems. It examines the transformative impact of LLMs, from early statistical approaches to Transformer-based architectures such as DeepSeek-R1 [1], and analyses how LLMs' computational demands and security challenges necessitate strategic integration with ITAI. The discussion underscores the urgency of addressing hardware-software codesign, data governance, and ecosystem resilience to align LLM-driven innovation with national technological sovereignty goals.

2.1 The Technological Evolution and Strategic Positioning of ITAI Industry

The ITAI industry originated in 1999 when Ni Guangnan proposed developing an independent operating system based on Linux. Ni Guangnan published an article calling for the development of an independent operating system based on Linux to establish a self-reliant and comprehensive software industry system in China. The core of the ITAI industry lies in achieving self-control in the information industry through autonomous technological innovation. The launch of the "Core-electronics, high-end-general-chips, and Infrastructural-software" project in 2006 marked the systematic development phase of the ITAI industry. This project involved researching and formulating strategies, macro plans, and major policies for network security and informationization development, promoting the

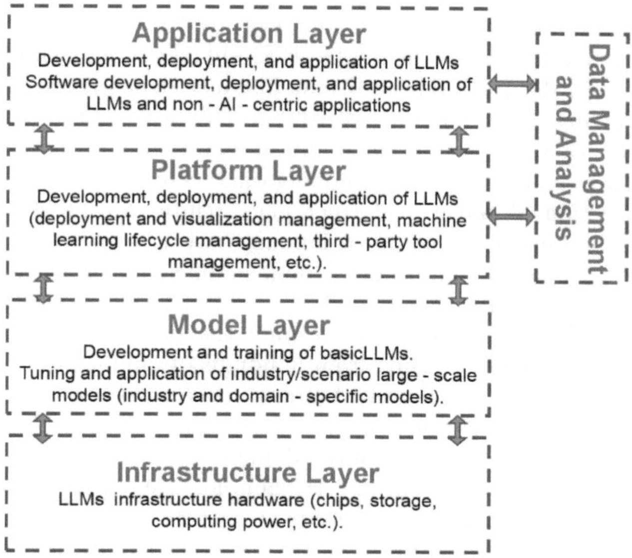

Fig. 2. The structure of LLMs consists of the infrastructure layer, model layer, platform layer, application layer, and data management and analysis.

construction of national network security and informationization rule of law, and continuously enhancing security capabilities, signaling the beginning of the ITAI industry. In 2018, when ZTE and Huawei were sanctioned by the U.S., and export restrictions on China were tightened, the number of ITAI projects began to increase, marking China's entry into the scale promotion phase of its information industry. As of 2024, the size of China's ITAI market has exceeded one trillion yuan, with significant improvements in domestic substitution rates in key areas, though the core technology ecosystem still requires further refinement.

2.2 The Technical Characteristics and Industrial Impact of Large Language Models

The development of Large Language Models (LLMs) spans three stages, reflecting a shift from statistical methods to deep learning.

Early language models relied on statistical methods [53], building probability models based on word co-occurrence frequencies. Representative methods included n-gram models [43] and maximum entropy models(MaxEnt) [19]. For instance, n-gram models used sliding windows to calculate the probability of word sequences, but they were limited by the sparsity problem and struggled to handle long-distance dependencies. Models in this stage were small in scale, had weak generalization capabilities, but laid the foundation for subsequent technologies.

The rise of deep learning pushed language models toward neural network architectures. Recurrent neural networks (RNNs) [45] and long short-term memory networks (LSTMs) [82] addressed long-distance dependency issues through temporal modelling. For example, Word2Vec [15] uses word vectors to map semantic spaces. However, the serial computational nature of RNNs leads to low training efficiency, and their ability to understand complex semantics is limited.

The introduction of the Transformer [65] architecture in 2017 fundamentally changed the design paradigm of language models. Its core self-attention mechanism enables global context modelling, significantly improving the ability to process long texts. Subsequent models such as the GPT series and BERT achieved cross-task transfer learning through the pretraining and fine-tuning paradigms, with parameters scaling from hundreds of millions to billions [17]. In 2025, China's DeepSeek-R1 [1] model became a key milestone in this phase. As a high-cost-performance inference-oriented large language model, DeepSeek-R1 has achieved breakthroughs in specialized domain reasoning capabilities through optimized Transformer architectures. For example, it supports localized deployment and can adapt to privatization needs on private servers, significantly lowering the application threshold for large models. Compared with earlier models, DeepSeek-R1 excelled in multimodal processing and low-latency responses, indicating a pivotal shift from generalized capabilities to specialized capabilities. This progress not only demonstrated China's technological autonomy in the LLM field but also provided a new technical foundation for the ITAI industry to adapt to domestic hardware and software. The structure of LLMs is shown in Fig. 2. Key technological breakthroughs include the following:

- Model Architecture Innovations: Optimizations such as sparse attention and hybrid expert (MoE) [13] improve computational efficiency.
- Training Method Upgrades: Alignment optimization based on reinforcement learning [62] enhances generation quality.
- Data Scale Expansion: Datasets with billions of tokens support model generalizability.

Large language models, through their massive parameter sizes and self-attention mechanisms, achieve cross-modal semantic understanding and generation capabilities. Their technical characteristics include: computational intensity, and training models with billions of parameters requires hundreds of petaflops-days of computational resources. Data dependency and reliance on vast, high-quality text data to build knowledge bases. Ecosystem Reconstruction, Drives the coordinated optimization of algorithms, frameworks, and hardware to form new technology stacks.

2.3 The Inevitability and Urgency of ITAI and LLMs Integration

International technological blockades and data security risks are compelling the ITAI industry to accelerate the adaptation of LLM technologies. Owing to the black-box nature and uncontrollable content generation of LLMs, the security systems of the ITAI face heightened demands. In the face of the explosive growth of computational demands and the revolutionary advancements in AI capabilities under the current wave of large AI model development, China's ITAI industry is actively embracing technological transformation to occupy a leading position in global technological competition and ensure the security of core autonomous technologies. The rise of generative AI has not only increased higher demands on hardware infrastructure but also challenged the resource management capabilities of basic software and the design and ecosystem application capabilities of large models. To address this, the ITAI industry in China urgently needs breakthroughs in technology, quality, and scale to meet the massive demands of the generative AI boom.

Moreover, the safe development of artificial intelligence relies on a self-reliant and secure industrial chain and ecosystem. While AI enhances productivity, it also poses unprecedented security challenges. Known security threats enabled by AI technologies include AI-enhanced phishing and destructive code generation for cyberattacks. The ITAI industry in China addresses these challenges by promoting the self-reliant security of key technologies, constructing localized technology ecosystems, strengthening data security and privacy protection, and providing a safe development environment and critical policy and financial support for the AI industry. This has accelerated the perfection and upgrading of the AI industrial chain.

3 Basic Hardware

In recent years, advancements in big data and large model technologies have caused chip computing power to grow much faster than storage and network bandwidth. Therefore, when faced with the explosive demand for large model computing power and the trend of large-scale parallel computing networks in the industry, the main bottleneck of chips lies in their data transmission speed and hardware interconnectivity, rather than the number of transistors on a single chip [34]. Currently, the focus of chip development is on larger storage capacity, higher storage and interconnect bandwidth, and stronger on-chip and off-chip interconnects. As the demand for computing power from large models increases exponentially, chips and computing networks are moving towards high integration. Moreover, the power and power density of servers have increased significantly. Additionally, the training and inference demands of large models have imposed higher requirements on the large-scale networking capabilities of servers.

In the context of AI personal computers (AIPC) , the industry faces major challenges such as hardware costs, the ability of edge-side models to be deployed, the construction of product ecosystems, and user considerations regarding data security. These are the primary challenges that need to be addressed [31].

Domestic chips still lag behind international advanced levels in floating-point computational precision and energy efficiency [61]. For example, the lack of 7 nm and below process technology results in insufficient transistor density, making it difficult to meet the training demands of models with billions of parameters

[47]. Additionally, the insufficient compatibility of domestic hardware with mainstream computing frameworks (e.g., CUDA) necessitates the use of adaptation layers to mitigate performance degradation, leading to actual computational efficiency being lower than the theoretical values [39]. Furthermore, the collaborative optimization of computing, storage, and networking still needs breakthroughs. For instance, in distributed training, bottlenecks in network bandwidth and latency may prevent hardware computing power from being fully utilized [18].

The training and inference of LLMs primarily depend on high-density computing resources [38], driving the evolution of domestic chips toward higher performance and lower power consumption. Here, a detailed breakdown of the advancements and considerations in chip design, storage architecture, and cooling technologies are presented.

3.1 Chip Design and Parallel Computing Efficiency

Domestic GPUs and AI Accelerators (e.g., TPUs). These chips are optimized by enhancing the number of matrix operation units and interconnect architectures, which significantly increases parallel computing efficiency. This is crucial for LLMs, as they require massive parallel processing to handle the vast amounts of data and computations involved in training and inference [42].

Chiplet Technology and Heterogeneous Integration. Chiplet technology involves integrating different types of processors or components into a single package [50]. This approach allows for linear scaling of computing power by stacking multiple chiplets, which helps overcome the limitations of single-chip process technology. In heterogeneous integration, different types of chips or components are combined to optimize performance, power consumption, and area efficiency [7].

3.2 Storage Architecture Innovations

High-Bandwidth Memory (HBM)and NonVolatile Memory (NVM). The combination of HBM and NVM provides a new paradigm for storing model parameters and allowing quick caching [9,28]. The HBM offers high bandwidth and low latency, which are essential for accessing the vast amounts of data required by LLMs. The NVM, on the other hand, provides persistent storage that retains data even when the power is off, which is beneficial for checkpointing during training and for maintaining model parameters in memory. The integration of the HBM and NVM primarily reduces the latency and increases bandwidth, ensuring that data can be accessed quickly and efficiently. This is vital for maintaining the performance and efficiency of LLM training and inference, preventing bottlenecks in data access.

3.3 Thermal Management and Cooling Technologies

Liquid Cooling Solutions. The scalable application of liquid cooling technologies [3,23,80], such as cold plate and immersion cooling, addresses the thermal management challenges posed by high-power-density hardware. These methods are more effective than traditional air cooling because they can handle higher heat fluxes and maintain more stable temperatures, which is crucial for the stability and performance of large-scale computing clusters. High-density computing resources generate significant heat, which can lead to thermal throttling and reduced performance if not managed properly. Compared with air cooling, liquid cooling offers a more efficient way to dissipate heat compared to air cooling, making it a preferred choice for high-performance computing applications.

3.4 Current State of Adoption and Considerations

Adoption Levels. While these technologies are being actively researched and developed, the extent of their implementation in domestic GPUs and TPUs varies [26,66]. Some optimizations are already in use, whereas others may still be in the experimental phase. Understanding the exact adoption levels would require specific data on current products and their architectures. The push towards high-performance, low-power chips is driven by the computational demands of LLMs [63]. However, challenges such as thermal management, storage bottlenecks, and the need for efficient interconnects remain critical areas of focus. Additionally, the integration of these technologies into domestic chips requires significant investment in research and development to match or surpass international standards.

In summary, the evolution of domestic chips is being driven by the need to support the efficient training and deployment of LLMs. Advances in chip design, storage architecture, and cooling technologies are all aimed at addressing the computational demands of these models. While significant progress has been made, there are still technical challenges and adoption hurdles that need to be overcome to fully realize the potential of domestic chip development in this context.

4 Basic Software

Domestic basic software still has shortcomings in terms of compatibility and performance. For example, domestic operating systems need to rely on additional adaptation layers to support LLM training frameworks, resulting in a 15–20% increase in memory usage [27]. When large model training datasets are used, databases lack the ability to handle mixed workloads of transaction processing and analytical queries, which can easily lead to resource contention issues. Additionally, the lack of an open-source ecosystem has constrained the iteration speed of basic software [59].

Large language models are driving the transformation of basic software toward intelligentization and servitization. At the operating system level, a microkernel architecture achieves dynamic resource scheduling through modular design, supporting unified management of heterogeneous hardware resources [44]. For example, intelligent scheduling algorithms based on LLMs can optimize task allocation in real time, improving cluster resource utilization. In the database domain, vector databases reduce high-dimensional vector retrieval latency to milliseconds through index optimization and approximate nearest neighbor (ANN) search algorithms meet the real-time inference requirements of large models [38]. Middleware, on the other hand, leverages service mesh and API gateway technologies to enable seamless integration of model services with business systems [33].

The explosive development of LLMs has imposed greater requirements on indigenous foundational software, driving its evolution from traditional functional support toward intelligentization, elasticity, and security. As the core link connecting hardware resources and upper-layer applications, foundational software must undergo architectural restructuring and technological innovation to meet the complex demands of LLMs in terms of computing resource scheduling, data governance, and model adaptation [57].

4.1 Operating System

The demand for the dynamic scheduling of heterogeneous compute resources by LLMs drives the evolution of domestic operating systems toward AI-native capabilities.

Microkernel and Resource Scheduling. Through a microkernel architecture, resources are managed in a modularized manner, supporting unified scheduling of heterogeneous hardware such as CPUs, GPUs, and NPUs. For example, intelligent scheduling algorithms based on LLMs can analyse load characteristics in real time and dynamically allocate computing resources, improving cluster utilization by more than 25% [77].

Domestic Chip Adaptation. By integrating AI acceleration engines, the operating system optimizes the compilation of compute graphs and memory management for domestic chips. For example, a self-developed compiler can convert TensorFlow and PyTorch models into instruction sets compatible with domestic chips, with performance loss controlled within 10% [22].

4.2 Platform Layer

The high-throughput and low-latency requirements of LLM training and inference drive the platform layer to evolve toward efficiency and intelligence.

Vector Databases and Unstructured Data Retrieval. By leveraging ANN algorithms, high-dimensional vector retrieval is optimized, reducing the latency of LLM inference to milliseconds. For example, hybrid indexing structures are designed to support real-time semantic searches for multimodal data such as text and images [13].

Federated Learning and Privacy Computing. Through encrypted sample alignment and model parameter aggregation technologies, cross-institutional data collaboration is enabled [72]. For example, in the healthcare sector, hospital data can be utilized in model training while remaining on-premise, ensuring privacy compliance [5].

4.3 Model Layer

The complexity of LLMs requires the model layer to possess lightweight deployment and self-healing capabilities.

Lightweight Core Design. Prebuilt resource monitoring modules track memory usage and compute loads in real time, automatically triggering restarts or resource expansion in the case of anomalies. For example, dynamic pruning technology compresses billion-parameter models to hundreds of gigabytes, enabling deployment at the edge [69].

Built-in Prediction Engines. Integrated time series forecasting and anomaly detection algorithms automate data cleaning. For example, in financial risk control scenarios, models can automatically identify fraudulent transaction patterns and generate warning rules [35].

4.4 Data Layer

LLMs' dependence on the scale and quality of data drives the data layer to strengthen its security capabilities.

End-to-End Encryption and Auditing. National cryptographic algorithms (e.g., SM4) [48] are used to encrypt data storage and transmission, whereas blockchain technology records operation logs, ensuring data sovereignty and control. For example, government systems use multilevel permission controls and encrypted sandboxes to isolate sensitive data [73].

Real-Time Log Analysis. When AI semantic parsing technology is combined, potential risks are extracted from system logs, enabling fault prediction and root cause analysis. For example, log clustering algorithms can provide early warnings for disk failures 30 min in advance [21].

4.5 Middleware

LLMs' reliance on distributed architectures requires middle-level software to provide efficient collaboration capabilities.

API Gateways and Service Mesh. Standardized model series interfaces support cross-language calls and heterogeneous model integration. For example, automated adapters convert PyTorch models to ONNX format, reducing call latency by 40% [60].

Intelligent Traffic Scheduling. On the basis of network topology and load status, data transmission paths are dynamically optimized, increasing encrypted channel throughput by 50%. For example, during cross-border data flows, low-latency and high-security transmission routes are automatically selected [8].

The innovation of foundational software driven by LLMs requires cross-layer collaboration, the operating system provides the resource foundation, the platform layer accelerates data processing, the model layer carries the intelligent core, the data layer ensures security and compliance, and middleware enables business integration. However, the domestic technology stack still faces challenges such as compatibility across architectures and standardized interface design.

5 Application Software

The intelligent transformation of application software faces dual pressures from data security and technological adaptation. The privacy protection of large model training data requires the integration of federated learning and homomorphic encryption technologies, but existing solutions struggle to balance encryption overhead and model accuracy. Additionally, adapting domestic application software to LLMs must address instruction set compatibility issues. For instance, differences in instructions between the ARM architecture and x86 architecture can lead to a 20%-30% decrease in model inference speed [25].

LLMs leverage semantic understanding and generation capabilities to drive the transformation of application software from process-driven to data-driven approaches, achieving significant efficiency improvements and innovative business models in industries such as manufacturing, government services, office operations, and enterprise management. The following analysis explores four key application scenarios:

5.1 Industrial Software

LLM-powered industrial knowledge bases, integrated with real-time IoT data, can optimize complex process parameters, such as wet flue gas desulfurization and electrostatic dust removal. Domestic AI frameworks support drag-and-drop industrial application development [41].

5.2 Process Optimization Software

On the basis of LLM-based semantic analysis, contracts can be automatically parsed, and approval processes can be triggered. AI agents enable path planning for warehouse robots and dynamic inventory replenishment. For instance, a certain e-commerce warehouse reduced its picking error rate to 0.3%. By aggregating multisource business data, LLMs generate process optimization suggestions [81].

5.3 Office Software

LLMs can automatically generate technical documents or meet minutes. For example, a financial institution reduced the time required to write annual reports from three days to just two hours when AI tools were used. Combining voice, image recognition, and LLM capabilities enables seamless "natural language → flowchart → code" conversion, lowering the barrier for nontechnical personnel to participate in development. These platforms provide standardized API interfaces to support customized AI features. For example, a government system integrated speech-to-text and summary generation services through a cloud platform, improving meeting efficiency by 35% [78].

5.4 Enterprise Management Software

To fully realize the potential of LLMs in application software, the following bottlenecks must be addressed. First, the challenge of domain knowledge adaptation persists. Many industries, such as manufacturing and healthcare demand customized model fine-tuning, however the scarce availability of high-quality labelled data hinders progress. Second, achieving real-time performance and cost balance remains critical. Edge deployment necessitates optimized lightweight model techniques (e.g., knowledge distillation) to minimize inference latency while maintaining cost efficiency. Finally, security and compliance requirements impose significant constraints. Sensitive domains such as government and finance mandates enhanced data anonymization protocols and model robustness against adversarial attacks to meet regulatory standards [29].

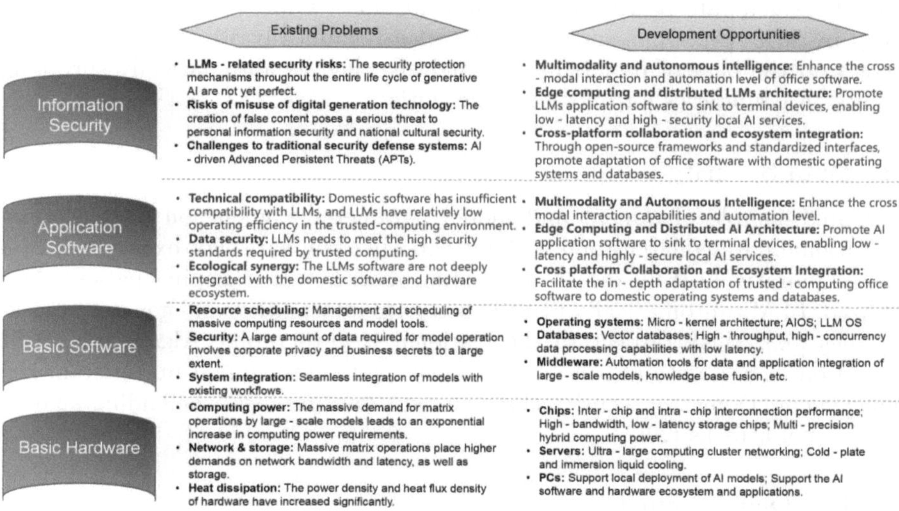

Fig. 3. An analysis of the technical challenges and strategic development opportunities faced by the Information Technology Application Innovation under the influence of LLMs.

6 Information Security

The misuse of generative AI has introduced new security threats. For example, adversarial attacks can mislead models into producing incorrect results, whereas data poisoning can compromise the integrity of model training. Additionally, the black-box nature of large models makes security audits challenging, with existing detection tools achieving a recognition rate of less than 60% for identifying model backdoors [64]. On the compliance front, the cross-border flow of data and privacy protection regulations impose strict restrictions on the sources and uses of model training data [67].

LLMs have introduced proactive defense capabilities and intelligent analysis paradigms into the field of information security, transforming threat detection and data protection frameworks. The following analysis explores four key dimensions—security software, data security, hardware security, and security services.

6.1 Security Software

LLMs use semantic parsing to detect novel attack patterns, such as social engineering tactics in phishing emails, with fewer false positives than traditional rule-based systems. For example, a financial system achieved a 92% detection rate for unknown malware via an LLMs-powered engine, capitalizing on the model's generalization across polymorphic code and obfuscation techniques. Integration with knowledge graphs enables automated attack path analysis via correlation heterogeneous data [55]. LLM-generated attack graphs reduce the threat response time from hours to minutes. Automated vulnerability repair toolchains integrated with LLMs generate patch code recommendations, improving remediation efficiency by 50% [79]. Open-source frameworks dynamically adjust access control policies on the base of user behavior analysis, such as automatically shrinking permission boundaries for anomalous activities.

6.2 Data Security

Encrypted sample alignment and model parameter aggregation enable cross-institutional data collaboration without exposing raw data. For example, hospitals collaboratively train disease prediction models while keeping patient data localized. LLMs automatically identify sensitive information and generate masking rules [51]. With respect to the National Cryptographic Algorithms, SM4 encryption secures data at rest and in transit, paired with quantum-resistant key management. Domestic security chips encrypt data on disks, defending against side-channel attacks [20].

6.3 Hardware Security

Domestic chips integrate dedicated cryptographic modules, accelerating SM2/SM4 algorithms via NPUs for 3× encryption performance gains. Lattice-based cryptographic protocols for key exchange ensure security against future quantum computing threats. The LLM training data are isolated within TEE enclaves to prevent memory scraping attacks. The sensitive data are transmitted via trusted channels from collection to destruction [49].

6.4 Security Services

LLMs generate actionable policy recommendations, such as prioritizing patch deployments on the basis of vulnerability scan results. The automated red-blue team exercises simulate APT attack paths and generate defense reports. Real-time aggregation of global threat intelligence enables LLM-generated defense strategies, which synchronize updates network-wide in minutes. The firewall rules are updated in real time based on dynamic threat scoring, which reduce response latency by 80% [30].

LLMs redefine information security by bridging semantic understanding, automated reasoning, and proactive defense. To fully harness this potential, future efforts must address challenges such as adversarial robustness, privacy-performance trade-offs, and regulatory compliance, while avoiding hardware-software codesign and open-source collaboration.

7 Future Directions

The deep integration of LLMs with the information technology application innovation requires breakthroughs in technological bottlenecks, secure ecosystem construction, and resource optimization. The following proposes innovative technical pathways and strategic recommendations across hardware, software, security, and ecosystem dimensions to provide systematic support for the intelligent upgrade of the information technology application innovation.

7.1 Hardware Layer: Breaking Through Computational Boundaries and Energy Efficiency Limits

Heterogeneous Integration via chiplet and 3D Stacking Technologies. The tiles, HBM, and I/O tiles are integrated via 2.5D/3D packaging technologies to achieve linear scalability in terms of computational power [56]. The chiplet layout is optimized with domestic EDA tools and develop dynamic power management algorithms for the thermal-electrical-computational co-optimization of heterogeneous chips are developed [11].

LLMs-Specific Instruction Set Architectures (ISAs) Design. ISAs tailored for matrix operations and sparse computations are designed to support mixed-precision (FP8/FP16/BF16) calculations [6]. Systolic arrays for matrix multiplication are optimized, and compute-in-memory (CIM) technology is integrated to reduce data movement overheads [24]. Develop codesigned toolchains for ISA and LLM algorithms, enabling automatic operator mapping and compiler optimization.

7.2 Software Layer: Architectural Decoupling and Intelligent Upgrades

Lightweight Middleware and Ecosystem Adaptation. Develop eBPF-based lightweight middleware for transparent resource scheduling across architectures (x86/ARM/RISC-V) [70]. Dynamic binary translation is used to ensure compatibility with CUDA ecosystems [75]. A middleware abstraction layer is built to unify heterogeneous hardware interface standards [54].

Hybrid-Workload Databases and HTAP Optimization. Design HTAP databases using row-column mixed storage and dynamic resource isolation. Optimize time-series partitioning and vector indexing for millisecond-level query responses [36]. Reinforcement learning-based workload prediction algorithms are introduced to allocate resources dynamically during traffic spikes [40].

7.3 Security Layer: Full-Lifecycle Protection and Privacy Enhancement

Federated Learning with Differential Privacy. Develop frameworks combine local differential privacy, homomorphic encryption, and secure multiparty computation for privacy-preserving crossinstitutional collaboration. Co-optimizing noise injection and model distillation to balance privacy and accuracy [74].

LLMs Security Assessment and Adversarial Defense. Establish security standards covering the entire LLM lifecycle, integrating modules for adversarial sample detection, backdoor scanning, and explainability analysis [46]. Meta-learning-based frameworks are created to dynamically generate robustness-enhancing strategies [71].

7.4 Ecosystem Layer: Open-Source Collaboration and Standardization

Open-Source Community. Domestic stacks are integrated with open-source ecosystems to standardize operator libraries and model formats. Foster an open source feedback iteration loop to improve toolchain usability [52].

Cross-Architecture Toolchains. Develop polyhedral compilation optimization techniques for unified ARM/RISC-V development interfaces. Standardize intermediate representations to abstract hardware differences [4]. AI-driven auto-tuning tools are built to generate optimal code paths for diverse hardware.

8 Conclusion

In this paper, we systematically analyse the opportunities and challenges that large language models (LLMs) bring to China's information technology application innovation industry. LLMs provide the indigenous innovation industry with an opportunity to leap from usability to intelligence. However, the release of their technological dividends requires breakthroughs in the underlying architectures and the perfection of security systems as prerequisites. In the future, the indigenous innovation industry should use LLMs as a technical anchor, and through innovations in hardware architecture, intelligent upgrades in software, the reconstruction of security systems, and open-ecosystem collaboration should be achieved. A self-controllable and autonomous intelligent technology ecosystem should be built. This will provide core support for the high-quality development of China's digital economy.

References

1. DeepSeek-R1: Incentivizing reasoning capability in LLMs via reinforcement learning (2025)
2. Alber, D.A., et al.: Medical large language models are vulnerable to data-poisoning attacks. Nat. Med. **31**(2) (2025)
3. Arumugam, A., Buonomo, B., Nardini, S., Plomitallo, R.E., Manca, O.: Numerical investigation on geometrical parameter effects of metal foams in liquid cooling efficiency of battery thermal management within phase change materials (2025)
4. Awasthi, P., Gollapudi, S., Kumar, R., Munagala, K.: Combinatorial optimization via LLM-driven iterated fine-tuning (2025)
5. Battula, H., Liu, J., Srivastava, J.: Enhancing in-hospital mortality prediction using multi-representational learning with LLM-generated expert summaries (2024)
6. Bolz-Tereick, C.F., Panayi, L., Mckeogh, F., Spink, T., Berger, M.: Pydrofoil: accelerating sail-based instruction set simulators (2025)
7. Cao, Z., Mei, B., Liu, Q., Pei, X., Wan, Z., Wang, C.: MAHR: a multi-application hybrid reconfigurable mechanism for energy-efficient chiplet interconnection network. J. Circ. Syst. Comput. **34**(1) (2025)
8. Chala, T.D., Kóczy, L.T.: Agent-based intelligent fuzzy traffic signal control system for multiple road intersection systems. Mathematics (2227-7390) **13**(1), 124 (2025)
9. Chen, B., et al.: Laser processing induced nonvolatile memory in chaotic graphene oxide films for flexible reservoir computing applications. J. Semicond. **45**(12), 122403 (2024)
10. Chen, D., et al.: Transforming the hybrid cloud for emerging AI workloads (2024)
11. Chen, S., Zhang, H., Ling, Z., Zhai, J., Yu, B.: The survey of chiplet-based integrated architecture: an EDA perspective (2024)
12. Chen, Y., Wan, Z., Zhuang, Y., Liu, N., Lo, D., Yang, X.: Understanding the OSS communities of deep learning frameworks: a comparative case study of PyTorch and TensorFlow. ACM Trans. Softw. Eng. Methodol. **34**(3) (2025)
13. Chernov, A., Novitskij, O.: The empirical impact of reducing symmetries on the performance of deep ensembles and MoE (2025)
14. Choudhury, R.R., Dey, S.: GPT-3 powered information extraction for building robust knowledge bases (2024)
15. Church, W.K.: Word2Vec. Nat. Language Eng. **23**(01), 155–162 (2017)
16. Council, N.: Innovation in information technology. Managing Inf. Technol. (2008)
17. Devlin, J., Chang, M.W., Lee, K., Toutanova, K.: BERT: pre-training of deep bidirectional transformers for language understanding. In: ArXiv (2019)

18. Ding, L., Mao, Q., Yan, G.: Observer-based multiagent deep reinforcement learning: a fully distributed training scheme. IEEE Trans. Ind. Inform. **21**(2), 1150–1159 (2025)
19. Elith, J., Phillips, S.J., Hastie, T., Dudík, M., Chee, Y.E., Yates, C.J.: A statistical explanation of maxent for ecologists. Diversity Distrib. **17**(1) (2015)
20. Elzemity, A., Arief, B., Li, S.: CyberLLMInstruct: a new dataset for analysing safety of fine-tuned LLMs using cyber security data (2025)
21. Fang, H., Su, W.: Log clustering-based method for repairing missing traces with context probability information. Int. J. Adv. Comput. Sci. Appl. **15**(5) (2024)
22. Faria, F.S., Silva, R.G.D.D., Antunes, M.D.M., de Lima e Silva, S.M.M.: Estimating thermal efficiency of a self-developed capacitor discharge welding equipment through nonlinear function specification method. Int. J. Thermal Sci. **208**(000) (2025)
23. Flouris, A.D., Cheung, S.S.: Design and control optimization of microclimate liquid cooling systems underneath protective clothing. Ann. Biomed. Eng. **34**(3), 359–372 (2006)
24. Fornt, J., et al.: Mix-GEMM: extending RISC-V CPUs forenergy-efficient mixed-precision DNN inferenceusing binarysegmentation. IEEE Trans. Comput. **74**(2), 582–596 (2025)
25. Gope, D., Mansell, D., Loh, D., Bratt, I.: Highly optimized kernels and fine-grained codebooks for LLM inference on arm CPUs (2024)
26. Haichour, A.S., Benfriha, K.: Empowering real-time IoT applications: a brief review onleveraging GPU acceleration forlatency reduction. In: IFIP International Internet of Things Conference (2025)
27. Harsha, K., Tarun Kumar, K., Sumathi, D., Ajith Jubilson, E.: A survey on LLMs: evolution, applications, and future frontiers. In: Raza, K., Ahmad, N., Singh, D. (eds) Generative AI: Current Trends and Applications. Studies in Computational Intelligence, vol 1177. Springer, Singapore (2024). https://doi.org/10.1007/978-981-97-8460-8_14
28. Hogervorst, T., Nane R., Marchiori G., Dong, Q.T., Blatt M., Birger, R.A.: Hardware acceleration of high-performance computational flow dynamics using high-bandwidth memory-enabled field-programmable gate arrays. ACM Trans. Reconfigurable Technol. Syst. (TRETS) (2021)
29. Huang, X., et al.: Cl2sum: abstractive summarization via contrastive prompt constructed by LLMs hallucination. Complex Intell. Syst. **11**(3), 1–22 (2025)
30. Huang, L., Xue, J., Wang, Y., Chen, J., Lei, T.: Strengthening LLM ecosystem security, preventing mobile malware from manipulating LLM-based applications. Inf. Sci. **681**, 120923 (2024)
31. Irvine, R.: Why you need an AI PC. Comput. Active (Oct.22 TN.694) (2024)
32. Ban, J., Brettel, H., Cwienk, W.D., Fent, J., Kurchaninov, L.: Cold electronics for the liquid argon hadronic end-cap calorimeter of atlas. Nuclear Instrum. Methods Phys. Res. (2006)
33. Jafari, B., Bayat, P.: Performance improvement of distributed cache using middleware session. J. Supercomput. (8), 80 (2024)
34. Kaniuk, G., Mezeria, A., Knazeva, V.N., horoshun, D.M., Fursova, T.: Unified basic software and hardware complex for precision energy-saving systems of automatic regulation and control. Radio Electron. Comput. Sci. Control (2019)
35. Ke, Z., Ming, Y., Nguyen, X.P., Xiong, C., Joty, S.: Demystifying domain-adaptive post-training for financial LLMs (2025)
36. Kioroglou, D.: Omilayers: a Python package for efficient data management to support multi-omic analysis. BMC Bioinform. **26**(1) (2025)

37. Krger, L., Cioflan, C., Kartsch, V., Benini, L.: On-device federated continual learning on RISC-V-based ultra-low-power soc for intelligent nano-drone swarms (2025)
38. Lee, B.M.: Efficient resource management for massive MIMO in high-density massive IoT networks. IEEE Trans. Mobile Comput. **24**(3), 1963–1980 (2025)
39. Lee, Y.N., Yu, Y., Park, Y.: CUrator: an efficient LLM execution engine with optimized integration of CUDA libraries (2025)
40. Li, C., Wang, J., Shi, J., Liu, L., Zhang, S.: ADWTune: an adaptive dynamic workload tuning system with deep reinforcement learning. Complex Intell. Syst. **11**(4) (2025)
41. Li, J., et al.: LocMoE: a low-overhead MoE for large language model training (2024)
42. Li, J.: Distributed data processing and task scheduling based on GPU parallel computing. Neural Comput. Appl. **37**(4) (2025)
43. Liu, B., Xu, J., Zou, Q., Xu, R., Wang, X., Chen, Q.: Using distances between Top-n-gram and residue pairs for protein remote homology detection. BMC Bioinform., S3 (2014)
44. Liu, C.: Special issue: emerging trends in real-time embedded systems: from microkernel architecture to control system analysis and IoT frameworks. Real-Time Syst. (1), 60 (2024)
45. Liu, P., Qiu, X., Huang, X.: Recurrent neural network for text classification with multi-task learning. AAAI Press (2016)
46. Liu, S., Cui, S., Bu, H., Shang, Y., Zhang, X.: JailBench: a comprehensive Chinese security assessment benchmark for large language models (2025)
47. Liu, X., Yuan, B., Wu, F.: Design and validation of a hyper-converged blockchain hardware and software system based on domestic chips. In: Blockchain and Web3 Technology Innovation and Application Exchange Conference (2025)
48. Liu, Y., Xue, F., Ming, A., Zhao, M., Ma, H., Sebe, N.: SM4Depth: seamless monocular metric depth estimation across multiple cameras and scenes by one model (2024)
49. Meng, X., et al.: Unlocking hardware security assurance: The potential of LLMs (2023)
50. Mishty, K., Sadi, M.: Chiplet-GYM: optimizing chiplet-based ai accelerator design with reinforcement learning. IEEE Trans. Comput. **74**(1), 43–56 (2025)
51. Muhudin, A., Osoble, A.M., Osman, D.H., Omar, A.A.: Evolution of information security and data privacy in medical systems: a bibliometric analysis (2010-2024). Int. J. Eng. Trends Technol. **72**(12), 88–98 (2024)
52. Németh, B., Wachs, J.: Anchor sponsor firms in open source software ecosystems (2025)
53. Ott, R.L., Longnecker, M.T.: An Introduction to Statistical Methods and Data Analysis, 6th edition (2010)
54. Pan, D., Lu, J., Li, Y., Gao, Y.: Unified embedded operating system for vehicle control and traffic management. J. Ind. Inf. Integr. **44** (2025)
55. Peldszus, S.: Security compliance in model-driven software development. Secur. Compliance Model-Driven Software Dev. (2024)
56. Pfromm, L., et al.: MFIT: multi-fidelity thermal modeling for 2.5D and 3D multi-chiplet architectures (2024)
57. Rasnayaka, S., Wang, G., Shariffdeen, R., Iyer, G.N.: An empirical study on usage and perceptions of llms in a software engineering project. IEEE (2024)
58. Roman, D., Schade, S., Berre, A., Bodsberg, N., Langlois, J.: Model as a service (MaaS). In: AGILE Workshop - Grid Technologies for Geospatial Applications, Hannover, Germany (2009)

59. Schmidt, T., Durner, D., Leis, V., Neumann, T.: Two birds with one stone: designing a hybrid cloud storage engine for HTAP. Proc. VLDB Endowment (2024)
60. Stoian, M.C., Tatomir, A., Lukasiewicz, T., Giunchiglia, E.: PiShield: a PyTorch package for learning with requirements (2024)
61. Sung, Y.J., Lee, J.W., Kim, S.B., Shin, S.J.: Comparison of the soda-anthraquinone pulping properties between imported eucalyptus mixture chips and domestic yellow poplar (liriodendron tulipifera)chips. Palpu Chongi Gisul/J. Korea Tech. Assoc. Pulp Paper Ind. **42**(3), 22–27 (2010)
62. Sutton, R.S., Barto, A.G.: Reinforcement learning. Bradford Book **15**(7), 665–685 (1998)
63. Tyler, N.: Tempo targets low-power chips for AI applications. New Electron. (2019)
64. Valentin, S., Fu, J., Detommaso, G., Xu, S., Zappella, G., Wang, B.: Cost-effective hallucination detection for LLMs (2024)
65. Vaswani, A., et al.: Attention is all you need. arXiv (2017)
66. Vungarala, D., et al.: TPU-Gen: LLM-driven custom tensor processing unit generator (2025)
67. Wang, Y., et al.: Harmonic: Harnessing LLMs for tabular data synthesis and privacy protection (2024)
68. Wang, Z., Xiong, H., He, Z., Wang, P., Wang, W.: Distance comparison operators for approximate nearest neighbor search: exploration and benchmark (2024)
69. Wu, B., Yu, T., Chen, K., Liu, W.: Edge-side fine-grained sparse CNN accelerator with efficient dynamic pruning scheme. IEEE Trans. Circ. Syst., I. Regular Papers Publ. IEEE Circ. Syst. Soc. (3), 71 (2024)
70. Wu, X., et al.: A dual-defense self-balancing framework against bilateral model attacks infederated learning. In: International Conference on Algorithms and Architectures for Parallel Processing (2025)
71. Xie, Z., Zhan, H., Wang, Y., Zhan, C., Wang, Z., Jia, N.: Meta-learning-based fault diagnosis method for rolling bearings under cross-working conditions. Meas. Sci. Technol. **36**(1), 016218 (2025)
72. Xu, C., Wu, P., He, J., Chen, Z., Liu, Y.: Construction of a multi-technology fusion e-commerce data transaction optimization model based on federated learning. Appl. Math. Nonlinear Sci. **9**(1) (2024)
73. Xu, K., Cheng, G.: F3l: an automated and secure function-level low-overhead labeled encrypted traffic dataset construction method for IM in android. Cybersecurity **7**(1) (2024)
74. You, Z., Dong, X., Li, S., Liu, X., Ma, S., Shen, Y.: Local differential privacy is not enough: A sample reconstruction attack against federated learning with local differential privacy. IEEE Trans. Inf. Forensics Secur. (2025)
75. ZDNet: is China's 'secure' OS a threat to U.S offensive cyber capabilities? (2009)
76. Zeng, M., Wang, Q., Yu, T., Cai, P., Xin, W., Yao, J., Jin, T.: Verification and testing of Chinese information technology application innovation network security for the tobacco industry. In: International Conference on Information Science, Communication and Computing (2025)
77. Zhang, J.: Optimization of intelligent manufacturing systems based on intelligent control algorithms and big data analysis. In: International Conference on Big Data Analytics for Cyber-Physical System in Smart City (2025)
78. Zhang, X., et al.: TableLLM: enabling tabular data manipulation by LLMs in real office usage scenarios (2024)
79. Zhang, Y., et al.: AttacKG+: boosting attack graph construction with large language models. Comput. Secur. **150**(000) (2025)

80. Zhao, Z., Chen, X., Li, Yingjun Luo, C.: Heat transfer performance of copper foam-based vapor chamber composite liquid cooling system under double-sided heating. Int. J. Thermophys. **45**(2) (2024)
81. Zhou, T., Wang, P., Wu, Y., Yang, H.: FinRobot: AI agent for equity research and valuation with large language models (2024)
82. Zhu, W., Lan, C., Xing, J., Zeng, W., Xie, X.: Co-occurrence feature learning for skeleton based action recognition using regularized deep LSTM networks. AAAI Press (2016)

Air Quality Inference with Multisource Heterogeneous Data and the FSMoE-LSC Model

Hao Chen[1,2], Jia Deng[3,4,5,7], Huabin Wang[4,5,6](✉), Wei Shen[7], and Huapeng Yu[1,2](✉)

[1] School of Computer Science, Chengdu University, Chengdu, China
yuhuapeng@cdu.edu.cn
[2] Key Laboratory of Digital Innovation of Tianfu Culture, Sichuan Provincial Department of Culture and Tourism, Chengdu, China
[3] Chongqing University, 174 Shazhengjie, Shapingba, Chongqing 400044, China
[4] Chongqing Institute of Green and Intelligent Technology, Chinese Academy of Sciences, Chongqing 400714, China
wanghuabin@cigit.ac.cn
[5] Chongqing School, University of Chinese Academy of Sciences, Chongqing 400714, China
[6] Center of Superresolution Optics and Chongqing Engineering Research Center of High-Resolution and Three-Dimensional Dynamic Imaging Technology, Chongqing Institute of Green and Intelligent Technology, Chinese Academy of Sciences, Chongqing 400714, China
[7] Chongqing Zhixing Shulian Intelligent Technology Co. Ltd., Chongqing, China

Abstract. In this paper, air quality inference aims to estimate pollution levels at unmonitored locations on the basis of existing observations. In the context of multisource heterogeneous data, this task faces two critical challenges: effectively integrating diverse data sources with monitoring data and accurately modelling complex spatial relationships among monitoring stations and between stations and target locations. To address these challenges, we propose a novel air quality inference model, FSMoE-LSC, which integrates multiple key mechanisms. Specifically, a feature attention network is designed to fuse heterogeneous data sources with monitoring information; a self-attention-based mixture-of-experts (SMoE) module is introduced to capture intricate spatial dependencies; and a learnable skip connection vector is incorporated to enhance global information and target point information extraction. Experiments on real-world air quality data from the central urban area of Chongqing demonstrate that FSMoE-LSC significantly outperforms existing state-of-the-art baselines in terms of the RMSE and MAE, verifying its superior inference accuracy and generalizability. The source code and datasets used in this paper have been made publicly available at https://github.com/zhongcyi/ChongQing_AQI.

Keywords: Air Auality Inference · Multisource Data · SMoE · Self Attention · FAN

H. Chen and J. Deng—Contributed equally to this work.

L. Huang et al. (Eds.): CCF NCCA 2025, CCIS 2774, pp. 222–238, 2026.
https://doi.org/10.1007/978-981-95-5294-8_16

1 Introduction

Air pollution in urban environments has emerged as a pressing global issue that poses a significant threat to human health [1]. Ambient pollutants, such as PM2.5 and PM10, are strongly associated with increased risks of cancer and premature mortality [2, 3]. According to estimates by the World Health Organization, air pollution contributes to approximately seven million premature deaths worldwide each year [4]. In practice, however, the spatial distribution of air quality monitoring stations remains sparse, primarily due to the high costs associated with their construction and maintenance—for example, the installation of a standard air quality monitoring station costs approximately USD 200,000, with an annual maintenance cost of approximately CNY 30,000 [5]. Taking Chongqing as an example, although the central urban area spans approximately 5,472 km^2, only 32 monitoring stations have been deployed (see Fig. 1(a)).

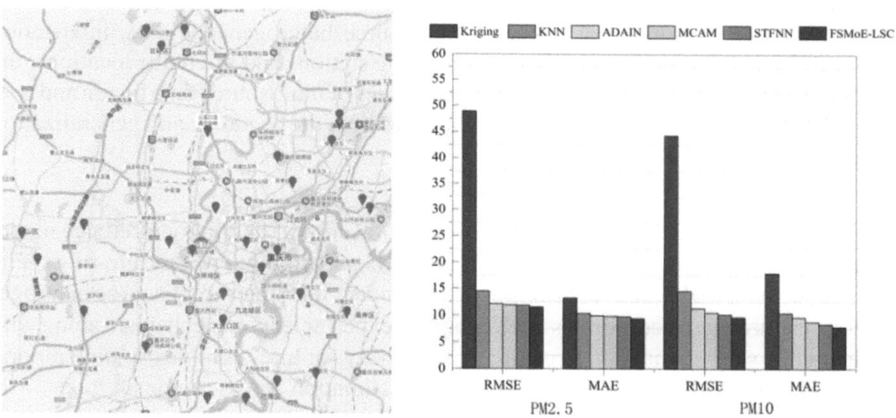

(a) Distribution of urban sites in the main city of Chongqing

(b) Comparison of FSMoE-LSC model with other models

Fig. 1. (a) Distribution of urban monitoring sites in the Chongqing main city center. The blue dots in the Fig. Indicate the site locations. (b) Comparison of the FSMoE-LSC model with other models.

An effective approach to address the aforementioned issue involves inferring air quality in unmonitored regions via observational data collected from existing monitoring stations. Over the past few decades, substantial research efforts have been devoted to inferring air quality [6]. Early studies predominantly relied on geostatistical theory, proposing methods such as kriging interpolation [7, 8] and geographically weighted regression [8], which were based on assumptions of spatial autocorrelation and heterogeneity.

With the progressive development of environmental monitoring networks, research has increasingly shifted towards multiphysics coupled modelling approaches, such as the Community Multiscale Air Quality model [10]. Concurrently, machine learning-enhanced methods have been gradually introduced during this phase. For example, algorithms based on k-nearest neighbors [11] have been employed to integrate spatial proximity and meteorological similarity, enabling more refined air quality inference.

In the era of big data, driven by the advancement of smart city initiatives, the integration of multisource heterogeneous data [12] has given rise to a new generation of data-driven air quality inference methods, such as the U-Air framework [5], ADAIN [13], MCAM [14], and STFNN [15]. These data-driven approaches typically address two key challenges. The first is how to effectively integrate multisource heterogeneous data with monitoring observations. Existing solutions include direct feature concatenation [5, 16], embedding representations [15, 17], multimodal fusion models [5], and attention mechanisms [13, 14]. The second challenge involves modelling the relationship between inference locations and monitoring stations. Current methods include spatial distance-based weighting [5], feature similarity-based weighting [14], graph neural network modelling [15], and attention-based approaches [13, 14].

In this paper, we propose an air quality inference model that integrates a self-attention-based mixture-of-experts (SMoE) architecture with a feature attention network. By jointly modelling temporal features, multisource heterogeneous data, interstation relationships, and interactions between monitoring stations and inference locations, the proposed method effectively addresses the challenges of multisource data fusion and spatial modelling. Moreover, it significantly improves both the accuracy and generalization performance of air quality inference.

The main work of this paper is as follows:

1. **FSMoE-LSC Model.** We propose a novel high-accuracy air quality inference model named FSMoE-LSC (feature attention and self-attention-based mixture-of-experts model with learnable skip connections). This model integrates multisource heterogeneous features and incorporates a self-adaptive expert selection mechanism along with a learnable skip-connection structure. It effectively captures complex spatiotemporal dependencies and enables precise air quality estimation across regions.
2. **Empirical Verification and Performance Evaluation.** Experiments were conducted on a private air quality dataset from the main urban area of Chongqing and compared against various benchmark models. The results demonstrate that the FSMoE-LSC model exhibits superior performance across several evaluation metrics. Additionally, this paper provides a systematic analysis of the model's hyperparameter settings and generalization capabilities. Through ablation studies, the contributions and rationality of each component module to the model's overall performance are further validated.
3. **The construction and openness of the dataset.** We meticulously organized and constructed an air quality dataset encompassing the central district of the primary city of Chongqing. The dataset encompasses concentration data of six pollutants (PM2.5, PM10, SO, NO, CO, and O3) from 32 monitoring stations, along with information regarding meteorological data, road network density, emission intensity of industrial sources, and intensity of domestic source activities within a 3x3 grid surrounding each station. The time period covered by the dataset extends from 2022–2024. The dataset provides a solid foundation for model training and evaluation and will subsequently be publicly released at https://github.com/zhongcyi/ChongQing_AQI to support academic research and engineering applications.

2 Definition of the Problem

2.1 Definition 1 (AQI and IAQI)

The air quality index (AQI) is a widely used composite measure of pollutant levels in the air. For a particular pollutant, its individual air quality index (IAQI) for an area is calculated on the basis of the measured concentration of that pollutant and reflects the air quality condition of that pollutant at a particular point and time. The overall AQI value is defined as the maximum of all pollutant IAQIs. In this work, we denote $X_t^a = \{x_{i,t}^a\}_{i=1}^N$ the real-time IAQI observation of pollutant a at all N monitoring stations at time t, where $x_{i,t}^a$ denotes the IAQI value of pollutant a at the ith monitoring station at time t. Similarly, we use $X_{[1,T]}^a = \{X_t^a\}_{t=1}^T$ to denote the sequence of IAQI observations of pollutant a at all stations in time period [1, T].

2.2 Definition 2 (Spatial Feature Data)

Spatial characterization data encompass geospatial variables associated with air quality, including road network data (denoted as F_r), coordinate terrain features (denoted as F_c), domestic source pollution features (denoted as F_l), and industrial source pollution features (denoted as F_i). The set of air characterization data for a given region is denoted by $F_s = \{F_c, F_r, F_l, F_i\}$.

2.3 Definition 3 (Time Series Feature Data)

Temporal feature data encompass dynamic variables associated with temporal shifts, primarily comprising meteorological feature data (denoted as F_m), which encompasses temperature, rainfall, mixed layer height, wind speed, wind direction, and humidity. Time-period features (denoted as F_p) are extracted via period mapping of timestamps, such as days, weeks, months, and years. The set of temporal feature data for a specific region within the time period [1, T] is denoted by $F_t = \{F_m, F_p\}_T$.

2.4 Problem Description

Establish a collection of grids G, with each grid point possessing complete characterization information $F = \{F_s, F_t\}$. Among them, the observed air quality values g of some grid points are known (monitoring station data), and the observed values of the remaining grid points are unknown, denoted as w. The objective of this paper is to accurately infer the air quality values of each unmonitoring point at each time point, given the grid's characterization information, in a time step of one hour.

3 Methodology

The proposed model is composed of three main modules (see Fig. 2): the spatiotemporal feature processing layer, the feature attention layer, and the self-attention-based mixture-of-experts (SMoE) layer.

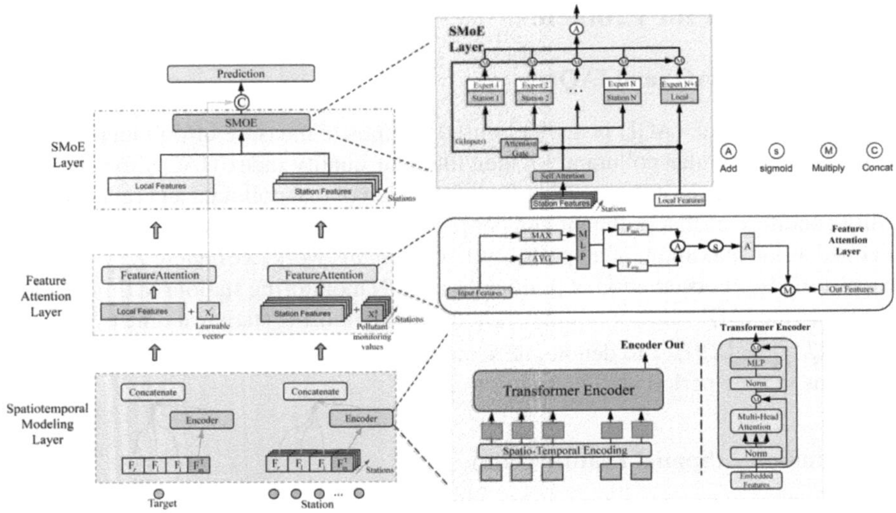

Fig. 2. The FSMoE-LSC model architecture.

At the data input stage, the model receives feature information from both inference locations and monitoring stations, each containing spatial and temporal components. In the spatiotemporal feature processing layer, feature modelling is conducted separately for inference points and stations. Specifically, spatial features F_s are concatenated with temporal representations F_t extracted via a temporal encoder, forming joint representations $F = \{F_s, F_t\}$ that serve as inputs to the subsequent attention mechanisms.

In the feature attention layer, to enrich the information of monitoring stations, the real-time measured air quality values $\mathbf{X}_t^a$ are incorporated as additional features. To mitigate the feature dimension mismatch between inference locations and monitoring stations structurally and to enhance the representation capability of the inference points, a learnable embedding vector X' is introduced for each inference location as a structural supplement. This embedding vector is also passed directly to the final prediction module via a skip-connection-like pathway to preserve its original structural information, which is beneficial for improving inference accuracy.

The processed features of both monitoring stations and inference locations are subsequently fed into a feature attention network to extract key features from multisource inputs. The output representations of this module serve as the input to the SMoE layer.

Inspired by the success of mixture-of-experts models in multitask learning, the self-attention-based mixture-of-experts layer constructs a dedicated neural network (expert network) for each monitoring station and inference point, aiming to capture and represent localized spatial information.

To further model dependencies within the spatial structure, a station-level self-attention mechanism is incorporated into both the gating network and the expert modelling process. This mechanism explicitly learns spatial dependencies between stations and between stations and inference locations, thereby enhancing the model's generalization ability in complex spatial environments.

Finally, the fused representations from the attention mechanisms and expert networks are passed through a feedforward neural network to generate the predicted concentrations of air pollutants at the inference locations.

3.1 Feature Attention Layer

To extract important information from the input multisource features, a feature attention network is employed, resulting in a feature attention matrix $M_f \in R^{B \times D}$, which is computed as shown in Equation.

To extract important information from the input multisource features, a feature attention network is employed, resulting in a feature attention matrix $M_f \in R^{B \times D}$, which is computed as shown in Eq. (1).

$$\begin{aligned} M_f(F) &= \sigma(MLP(AvgPool(F) + MLP(MaxPool(F))) \\ &= \sigma(W_1(W_0(F^d_{avg})) + W_1(W_0(F^d_{\max}))) \end{aligned} \tag{1}$$

To assign different weights to input features from various sources, thereby enhancing the model's ability to represent and select critical information, a feature attention mechanism is applied. The model inputs include spatial static multisource features and temporal features extracted via an encoder that captures temporal correlations.

3.2 Learnable Vectors

To incorporate real-time air quality information from monitoring stations, the observed AQI values X_t^a at the current time step are appended to the station features as additional inputs prior to the feature attention layer. Moreover, to address the structural discrepancy in feature dimensionality between inference locations and monitoring stations and to enhance the representation of inference points, a learnable embedding vector X'_t is introduced for each inference location as a structural supplement. This embedding is further propagated to the final prediction module via a skip-connection-like mechanism, allowing direct influence on the output. The final prediction for the inference location is given by:

$$\hat{g}_t = w^T[f_{enc}([F, X_t']), X_t'] + b \tag{2}$$

where F denotes the input feature vector; X_t' is the learnable embedding vector at the current time step; $f_{enc}(\cdot)$ represents the feature extraction module; and w, b is a learnable parameter.

3.3 Spatiotemporal Encoding

We adopt the spatiotemporal encoding method proposed by Yutong Feng et al.[15], where the final embedding representation is denoted by the vector $e \in R^{11}$, which is formed by concatenating the spatial encoding vector e_s and the temporal encoding vector e_t, as shown in Eq. (3).

$$e = [e_s, e_t] \tag{3}$$

In the spatial dimension, the attributes of a grid can be modelled on the basis of its absolute geographical location. We construct the spatial encoding vector via the longitude, latitude, and elevation of the grid centroid, as expressed in Eq. (4).

$$e_s = [longitude, latiude, altitude] \tag{4}$$

Considering the pronounced temporal periodicity of air quality (e.g., PM2.5 concentrations tend to exceed standard limits primarily during autumn and winter), we employ sinusoidal and cosinusoidal functions with different periods to construct the temporal encoding vector. This approach enables the mapping of temporal information into a continuous vector space with inherent periodicity. Specifically, a set of predefined periods T = $\{1a, 7a, 30.5a, 365.5a\}$ is used to represent daily, weekly, monthly, and yearly cycles. The scaling factor a is introduced to adjust for the hourly data interval and is set to 24 in this paper. The temporal encoding E_t is computed as follows:

$$e_t^i = \begin{cases} \sin(2\pi t/T_{\text{int}(i/2)+1}) \; i \mod 2 = 0 \\ \cos(2\pi t/T_{\text{int}(i/2)+1}) \; i \mod 2 = 0 \end{cases} \tag{5}$$

where e_t^i denotes the value of the ith dimension ($1 \leq$ i ≤ 8) of e_t, int(i/2) denotes dividing i by 2 and rounding down, and $T_{\text{int}(i/2)+1}$ is the int(i/2) + 1 cycle of T.

3.4 Self-Attention-Based Mixture-Of-Experts Architecture

Inspired by the effectiveness of mixture-of-experts (MoE) models in capturing task heterogeneity in multitask learning, we construct and train a dedicated neural network (expert network) for each monitoring station and inference location to learn and represent localized spatial features. The final prediction is obtained by a weighted aggregation of the outputs from all expert networks, enabling more fine-grained spatial modelling.

Prior to training individual expert networks for each monitoring station, we develop a station-level self-attention mechanism inspired by the success of self-attention in capturing temporal dependencies. Unlike traditional self-attention, which models sequential correlations along the time dimension, our approach replaces the temporal axis with the station axis, treating the input as the set of feature representations from all stations at a given time step. This enables the model to learn latent spatial dependencies among stations and provides structural priors for subsequent station-specific modelling.

To further enhance the integration of information between monitoring stations and inference locations, we extend the conventional gating network by incorporating the station-level self-attention mechanism as an auxiliary module. This module explicitly models the relevance between each inference location and all monitoring stations, allowing the model to dynamically assign weights to expert networks on the basis of spatial correlations. This design not only improves the model's expressive capacity but also enhances its adaptability to spatial heterogeneity.

Station-Level Self-Attention Mechanism. Given the input tensor $X \in R^{B \times N \times d}$, where B is the batch size, N is the number of stations, and d is the feature dimension, the query

(Q), key (K), and value (V) matrices are obtained through linear transformations as follows:

$$Q = \mathrm{XW}^{Q}, K = \mathrm{XW}^{K}, V = XW^{V} \tag{6}$$

where $W^{Q}, W^{K}, W^{V} \in R^{d \times d_k}$ are learnable parameters.

Then, the attention weight (i.e., correlation) between each pair of stations is computed as:

$$\alpha_{\mathrm{ij}} = \frac{\exp(\frac{Q_i \cdot K_j^K}{\sqrt{d_k}})}{\sum_{j'}^{N} \exp(\frac{Q_i \cdot K_{j'}^T}{\sqrt{d_k}})} \tag{7}$$

where $Q_{\mathrm{i}} \in R^{d_k}$ denotes the query vector of the i-th station, $K_j \in R^{d_k}$ denotes the key vector of the j-th station, and $\alpha_{ij} \in R$ represents the attention weight (i.e., spatial correlation) from station i to station j. Under the station-level configuration, these attention weights are normalized along the station dimension, satisfying $\sum_j \alpha_{ij} = 1$ for each query station i.

These attention weights are subsequently used to compute a weighted sum over the value vectors of all the stations:

$$Z_i = \sum_{j=1}^{N} \alpha_{ij} V_j \tag{8}$$

The final input $Z \in R^{B \times N \times d_k}$ representation is obtained as the attention-based aggregation of all the station representations.

Station and Inference Point Network. For each station and inference point, we define a two-layer fully connected neural network, as shown in Eq. (11).

$$\mathrm{E}_i(x) = \sigma(W_{2,i}\sigma(W_{1,i}x + b_{1,i}) + b_{2,i}), i = 0, 1, \ldots, N \tag{9}$$

where x denotes the input data, which can be either from an inference point or a monitoring station; $W_{1,i}, b_{1,i}$ represents the weights and biases of the first layer, while $W_{2,i}, b_{2,i}$ corresponds to those of the second layer; $\sigma(\cdot)$ represents the ReLU activation function; and N represents the number of stations. The output of the station-specific expert network is denoted as $S_i = E_i(s_i), i = 1, \ldots, N$, the output of the inference expert network is $S_0 = E_0(s_0)$, and the combined output of all the experts is expressed as $S = [S_0, S_1, \ldots, S_N] \in R^{(N+1) \times d}$, where d denotes the hidden layer dimension.

Gating Network Based on a Station-Level Self-Attention Mechanism. The gating network is used to compute the weights of different experts. We adopt a station-level self-attention mechanism to generate these expert weights. First, the inputs from both the inference point and the monitoring stations are concatenated as:

$$C = [F_0; F_1; \ldots F_N] \in R^{(N+1) \times d} \tag{10}$$

where F_0 denotes the input of the inference point, $F_1; \ldots; F_N$ denotes the input of the i-th station, N is the number of stations, and d is the feature dimension.

Then, the attention scores A are computed via the station-level self-attention mechanism defined in Eqs. (6)–(8):

$$A = Attention(C, C) \tag{11}$$

Finally, the expert weights are generated via a fully connected layer:

$$G = soft\max(AW_g + b_g) \in R^{(N+1)\times(N+1)} \tag{12}$$

where W_g, b_g are the weights and biases of the fully connected layer.

4 Experiments

In this section, we delve into our experimental methodology, which is aimed at evaluating the performance of our model and validating its effectiveness. Specifically, our experiments are designed to address the following research questions, elucidating our approach and its applicability in real-world scenarios:

Research Question 1 (RQ1): How does our model outperform other methods in terms of accuracy and effectiveness?

Research Question 2 (RQ2): What is the specific contribution of each component of our model to the effectiveness of inferring air pollution concentrations?

Research Question 3 (RQ4): How do variations in each hyperparameter affect the overall performance of our model?

Research Question 4 (RQ4): Does our model demonstrate superior performance in inferring the concentrations of various air pollutants?

By addressing these questions, we evaluate the model's performance from multiple dimensions, providing strong support for its practical value in air quality inference tasks.

4.1 Dataset and Metrics

The dataset used in this paper is a proprietary dataset from the central urban area of Chongqing. We divide the study area into a grid with a resolution of 1 × 1 km and process multisource features for each grid. The specific composition of the dataset is as follows:

Air Quality Data. We collected observational data from 32 ground-level air quality monitoring stations in the central urban area of Chongqing, covering hourly concentration data for major pollutants such as the air quality index (AQI), PM2.5, PM10, O3, NO2, CO, and SO3.

Grid Meteorological Data. For each grid unit, we collected six key meteorological attributes, namely, temperature, rainfall, mixing layer height, wind speed, wind direction, and humidity, which are used to characterize the dynamic meteorological conditions influencing the distribution of air pollution.

Grid Traffic Network Data. This includes the number of traffic intersections within each grid, the length of different types of roads (e.g., "town roads," "branch roads," "provincial roads," "county roads," "highways," "main roads," "secondary roads," "national roads," "highway/ramps, " and "other roads"), and the number of traffic attractors (including the number of vehicle inspection agencies, logistics parks, train stations, port docks, and passenger transport stations).

Grid Domestic Source Data. This records the quantity of different domestic source types within each grid, including the number of "restaurants, ' "dry cleaners," and "auto repair" shops.

Grid Industrial Source Data. This includes the quantity of industrial sources of various scales ("large," "medium," "small," and "micro") in each grid, reflecting the spatial distribution of industrial activities.

Grid Terrain Data. This includes elevation data for each grid, land-use attribute data, and geographic coordinates (longitude and latitude) of the grids.

In addition, the dataset also includes 3×3 grid domain feature data centered on the monitoring stations. Although these neighboring grids do not contain air quality observation values, their static and dynamic features can assist in inference.

We select air quality and meteorological data from the central urban area of Chongqing from 2022–2024, with a time granularity of once per hour. An 8-h time window is set to capture short-term temporal trends. In the subsequent experiments, we focused on the IAQI of PM2.5. In the experimental design, four monitoring stations were randomly selected as the test set, and the remaining stations were divided into training and validation sets at an 80:20 ratio. Additionally, an 8-fold cross-validation approach was used for model training and evaluation, with the final model performance metrics being averaged. To prevent the model from overfitting, the early stopping method is used in all training.

To comprehensively evaluate the model's performance in air pollution concentration inference tasks, two commonly used regression evaluation metrics were employed: the mean absolute error (MAE) and the root mean square error (RMSE).

4.2 Comparison Model

To comprehensively assess the performance of our model, we selected five representative comparison models for comparative analysis.

Kriging Interpolation (Kriging). Kriging is a classic geostatistical interpolation method commonly used for spatial prediction tasks. This method performs spatial modelling of known observation data to estimate values at unknown locations.

K-Nearest Neighbors (KNN). KNN is a nonparametric instance-based learning method. Its basic idea is to make predictions on the basis of the features of the nearest points to the input sample. In this work, we set the value of K to 3 to predict air quality by selecting the data of the three monitoring stations most similar to the target point.

ADAIN. ADAIN uses an attention-based pooling structure to adaptively learn the feature weights from different monitoring stations, enhancing the model's ability to handle spatially heterogeneous data and improving prediction accuracy.

Multichannel Attention Model (MCAM). The MCAM model introduces a multichannel attention module, which integrates the deep correlation between static and dynamic features, thereby enhancing the model's generalization ability in complex spatiotemporal environments.

Spatiotemporal Field Neural Networks (STFNNs). The STFNN constructs a spatiotemporal field neural network from the perspective of the "field" and adopts a pyramid-style multiscale inference framework, effectively improving the accuracy and robustness of air quality prediction.

For our model, the number of hidden layer neurons in the encoder of the temporal feature processing layer is set to 32, the number of hidden layer neurons in the self-attention layer of the SMoE layer is set to 128, the number of hidden layer neurons for each expert in the SMoE is set to 64, and the number of hidden layer neurons in the model output layer is set to 32. The Adam optimizer is selected, and the learning rate is set to 0.0001. The model is implemented on the TensorFlow framework.

4.3 Model Comparison (RQ 1)

Table 1. Model comparison on the Chongqing PM2.5 dataset. The symbol Δ represents the addition or reduction compared with the STFNN.

	RMSE	Δ	MAE	Δ
Kriging	48.92	+310.98%	13.19	+34.67%
KNN	14.46	+21.52%	10.34	+5.60%
ADAIN	12.09	+1.547%	9.93	+1.36%
MCAM	11.90	+0.62%	9.80	+0.775%
STFNN	11.83	–	9.72	–
FSMoE-LSC	11.48	−2.73%	9.35	−3.60%

To address Research Question 1, we performed a comparative analysis between the proposed model and other baseline models on the basis of two evaluation metrics: the mean absolute error (MAE) and the root mean square error (RMSE). The experimental results are derived from a proprietary air quality dataset of the central urban area of Chongqing, with detailed records presented in Table 1.

We believe that the improvement in performance can be attributed to the following four innovative design aspects:

(1) Spatiotemporal feature processing layer: By employing an encoder for temporal feature modelling, the model effectively captures the dynamic trends in time series data and extracts deep information from the temporal dimension.

(2) Feature attention network: This network plays a key role in the fusion of multi-source heterogeneous data, automatically identifying and enhancing the learning and representation of crucial features.
(3) Self-Attention-based mixture of experts model: This enhances the model's ability to capture spatial dependencies between different stations. On the other hand, it assigns experts to each station and inference point, dynamically selecting the most important experts for inference, thus enabling more targeted feature learning and representation.
(4) Learnable skip connection vectors: By introducing global information vectors and integrating them with the final output, this design improves the model's capacity to perceive complex spatiotemporal structural information.

4.4 Ablation Experiments (RQ 2)

To evaluate the impact and contribution of each component on the performance of our model, we conducted ablation experiments from two dimensions: model components and data composition.

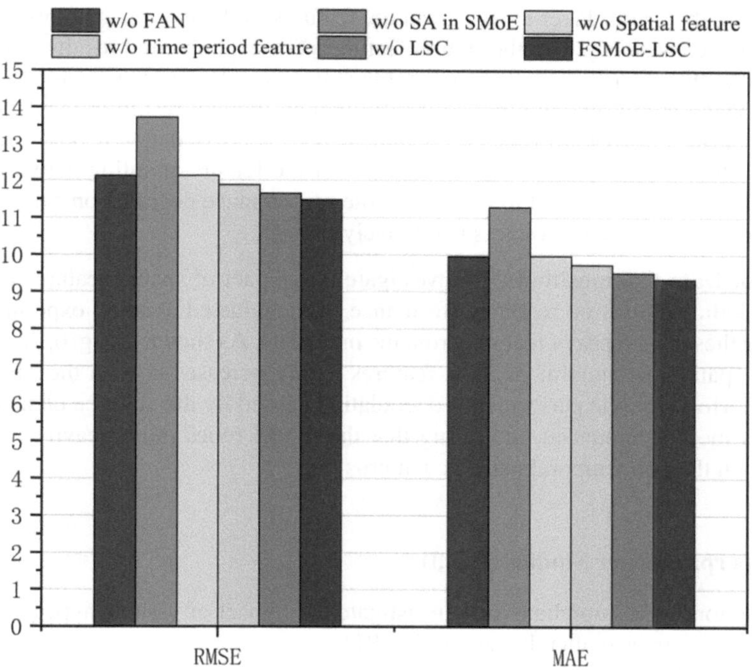

Fig. 6. Ablation experiments, w/o means without.

Impact of Model Components. Effect of the feature attention network on model performance. To investigate the effectiveness of the feature attention network (FAN) in integrating multisource heterogeneous data, we conducted an ablation study by removing this module from the model architecture. As shown in Fig. 6, the absence of the FAN module resulted in increased errors in both the MRSE and MAE metrics, indicating that the feature attention network plays an important role in automatically learning and emphasizing critical features from heterogeneous data sources, thereby improving overall model performance.

Impact of Self-Attention in the SMoE on Model Performance. To investigate the impact of the self-attention mechanism in the SMoE on model performance, we removed the self-attention module used for capturing intersite relationships and replaced the self-attention-based gating network with a traditional gating network. The experimental results (see Fig. 6) show that removing the self-attention module from the SMoE significantly weakens the model's ability to learn the relationships among stations as well as between stations and inference points, leading to substantial degradation in overall performance, as evidenced by increases in both the RMSE and the MAE.

Impact of Learnable Skip Connection Vectors on Model Performance. To investigate the impact of the learnable kip connection vector (LSC) on model performance, we removed the vector in the experiments and introduced a feedforward neural network before the SMoE layer to transform the features of the inference points, ensuring dimensional consistency with the station features. The experimental results (see Fig. 6) demonstrate that the removal of the learnable skip connection vector leads to a decline in the model's ability to capture global information, thereby resulting in a noticeable degradation in overall performance, as indicated by increased RMSE and MAE values. Nevertheless, compared with the removal of the feature attention network or the self-attention mechanism within the SMoE, the performance degradation caused by the absence of skip connection vectors is relatively minor.

Impact of Data Composition. To investigate the impact of spatial features and temporal periodic features on model performance, we conducted ablation experiments by removing these two types of features from the input data. As shown in Fig. 6, the removal of either spatial or temporal periodic features led to increases in both the MRSE and the MAE. However, the performance degradation caused by the absence of spatial features was more pronounced, indicating that the model relies more heavily on spatial information than on temporal periodic patterns.

4.5 Hyperparameter Studies (RQ 3)

In this section, we comprehensively investigate the impact of various hyperparameters on model performance, thereby addressing RQ3.

Number of Hidden Neurons

Impact of the Number of Hidden Neurons in the Encoder of the Temporal Feature Processing Layer. We adjusted the number of hidden neurons in the encoder of the temporal feature processing layer from 8–64. The experimental results (see Fig. 7(a)) show that the model performs optimally when the number of hidden neurons is set to 32. Given that the temporal features have only 14 dimensions, too few neurons lead to underfitting,

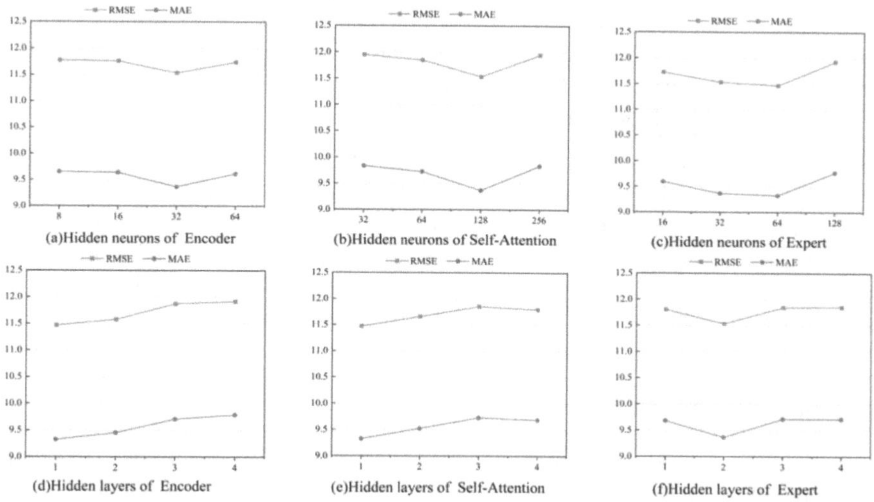

Fig. 7. Hyperparameter studies.

whereas too many neurons result in overfitting, both of which negatively impact model performance.

Impact of the number of hidden neurons in the self-attention layer of the SMoE. We adjusted the number of hidden neurons in the self-attention layer of the SMoE from 32–256. The experimental results (see Fig. 7(b)) indicate that the model performs best when the number of hidden neurons is 128. An appropriate number of hidden neurons helps better model the relationships between stations and inference points, thereby improving model performance.

Impact of the Number of Hidden Neurons in Each Expert of the SMoE. We adjusted the number of hidden neurons in each expert of the SMoE from 16–128. The experimental results (see Fig. 7(c)) show that the model performs optimally when the number of hidden neurons is 128. An appropriate number of hidden neurons in the experts aids in more effectively extracting feature information from both stations and inference points, which in turn enhances the overall model performance.

Model Depth

Impact of the Number of Encoder Layers in the Temporal Feature Processing Layer. We increased the number of encoder layers in the temporal feature processing layer from 1 to 4. The experimental results (see Fig. 7(d)) show that model performance gradually decreases as the number of layers increases. This may be attributed to the relatively low complexity of the temporal features, where deeper network structures tend to cause overfitting. In addition, the training time significantly increases with the depth of the network.

Impact of the Number of Hidden Layers in the Self-Attention Module of the SMoE. We increased the number of hidden layers in the self-attention module of the SMoE from 1 to

4. The experimental results (see Fig. 7(e)) demonstrate that model performance progressively deteriorates as the number of layers increases, accompanied by a corresponding increase in training time. This further suggests that a relatively shallow self-attention network is more suitable for the present task.

Impact of the Number of Hidden Layers on Each Expert of the SMoE. We increased the number of hidden layers in each expert of the SMoE from 1 to 4. The experimental results (see Fig. 7(f)) indicate that the model achieves optimal performance when the number of hidden layers is set to 2. Shallow expert networks may lead to insufficient feature extraction from stations and inference points, whereas deeper networks are prone to overfitting, ultimately impairing overall model performance.

4.6 Generalizability (RQ 4)

Our model demonstrates superior performance not only in the estimation of PM2.5 but also in predicting other air pollutants.

Table 2. Model comparison on the Chongqing PM10 dataset. The symbol Δ represents the addition or reduction compared with the STFNN.

	RMSE	Δ	MAE	Δ
Kriging	44.2889	+338.80%	17.9610	+115.82%
KNN	14.5047	+43.71%	10.3935	+24.89%
ADAIN	11.2687	+11.268%	9.5704	+15.00%
MCAM	10.4558	+3.59%	8.7563	+5.22%
STFNN	10.0932	–	8.3221	–
FSMoE-LSC	10.1052	−3.35%	8.3787	−4.31%

As shown in Table 2, our model consistently outperforms the selected baseline models in the PM10 estimation task, further validating its versatility and robustness across various air pollutant prediction tasks.

5 Conclusions

In this paper, we propose an air quality inference model that integrates a self-attention mechanism, mixture of experts (MoE) structure, and feature attention. We utilize multi-source heterogeneous data (such as meteorological and road grid data) to extract features related to air quality. By introducing a feature attention network (FAN), we effectively integrate multisource heterogeneous data with monitoring information. We design a self-attention-based mixture-of-experts (SMoE) network to model the deep relationships between monitoring stations and between stations and inference points. Additionally, we innovatively introduce a learnable skip connection vector to enhance the model's ability to model global information and target point information, i.e., our FSMoE-LSC model. The experimental results on the air quality dataset of the main urban area of Chongqing verify the superiority of our model.

6 Future Work

We also have site-centric 3 × 3 grid feature data. Although we tried to extract 3 × 3 neighboring information through convolution methods, the effect was not satisfactory, so we did not incorporate it into the current model. In the future, we plan to further investigate how to make more effective use of neighboring feature information. Additionally, exploring more model layers and different time steps is future work.

Acknowledgments. This study was funded by Chongqing Zhixing Shulian Intelligent Technology Co., Ltd. And the Key Laboratory of Digital Innovation of Tianfu Culture, Sichuan Provincial Department of Culture and Tourism, China.

Disclosure of Interests. The authors declare no conflicts of interest.

References

1. Huang, X., Steinmetz, J., Marsh, E.K., et al.: A systematic review with a Burden of Proof meta-analysis of health effects of long-term ambient fine particulate matter (PM2.5) exposure on dementia. Nat. Aging 1–12 (2025)
2. Nan, N., Yan, Z., Zhang, Y., Chen, R., Qin, G., Sang, N.: Overview of PM2.5 and health outcomes: focusing on components, sources, and pollutant mixture coexposure. Chemosphere **323**, 138181 (2023)
3. Liu, X., Xie, W., Lv, S., et al.: Long-term effects of PM2.5 and its components on incident hypertension among the middle-aged and elderly: a national cohort study. BMC Public Health **25**(1), 960 (2025)
4. Olstrup, H., Raza, W., Sommar, J.N., et al.: The impact of socioeconomic factors on long-term mortality associated with exposure to PM2.5: a systematic literature review and meta-analysis. Public Health Rev. **46**, 1607290 (2025)
5. Yu, Z., Liu, F., Hsieh, H.P.: U-Air: when urban air quality inference meets big data. In: Proceedings of 19th ACM SIGKDD International Conference Knowledge Discovery and Data Mining (KDD'13), pp. 1436–1444. ACM, New York (2013)
6. Han, J., Zhang, W., Liu, H., Xiong, H.: Machine learning for urban air quality analytics: a survey. arXiv preprint arXiv:2310.09620 (2023)
7. Kethireddy, S.R., Tchounwou, P.B., Ahmad, H.A., Yerramilli, A., Young, J.H.: Geospatial interpolation and mapping of tropospheric ozone pollution using geostatistics. Int. J. Environ. Res. Public Health **11**, 983–1000 (2014)
8. Zhang, X., Shi, R., Chen, M.: Comparative study of the spatial interpolation methods for the Shanghai regional air quality evaluation. In: Proceedings of SPIE 10767, Remote Sensing and Modelling of Ecosystems for Sustainability XV, 107670Q (2018)
9. Goovaerts, P.: Geostatistical approaches for incorporating elevation into the spatial interpolation of rainfall. J. Hydrol. **228**(1), 113–129 (2000)

10. Appel, K.W., Napelenok, S.L., Foley, K.M., et al.: Description and evaluation of the community multiscale air quality (CMAQ) modelling system version 5.1. Geosci. Model Dev. **10**, 1703–1732 (2017)
11. Mohammadi, F., Teiri, H., Hajizadeh, Y., et al.: Prediction of atmospheric PM2.5 level by machine learning techniques in Isfahan, Iran. Sci. Rep. **14**, 2109 (2024)
12. Radford, A., et al.: Learning transferable visual models from natural language supervision. In: International Conference on Machine Learning, pp. 8748–8763. PMLR (2021)
13. Cheng, W., Shen, Y., Zhu, Y., et al.: A neural attention model for urban air quality inference: learning the weights of monitoring stations. In: Proceedings of 32nd AAAI Conference on Artificial Intelligence, pp. 2151–2158. AAAI Press (2018)
14. Han, Q., Lu, D., Chen, R.: Fine-grained air quality inference via multi-channel attention model. In: Proceedings of 30th International Joint Conference on Artificial Intelligence (IJCAI'21), pp. 2512–2518. IJCAI (2021)
15. Feng, Y., Wang, Q., Xia, Y., et al.: Spatiotemporal field neural networks for air quality inference. In: Proceedings of 33rd International Joint Conference Artificial Intelligence (IJCAI'24), vol. 803, pp. 7260–7268. AAAI Press (2024)
16. Cheng, Y., Li, X.C., Li, Z.J., et al.: AirCloud: a cloud-based air-quality monitoring system for everyone. In: Proceedings of 12th ACM Conference on Embedded Networked Sensor Systems, pp. 251–265. ACM, Memphis (2014)
17. Hu, K.Y., Guo, X.L., Liu, G.X., et al.: Air quality inference based on spatiotemporal matrix factorization. Eng. Sci. Technol. **56**(5), 146–155 (2024)

CHOLD: Offensive Language Detection Method for Chinese Texts

Houyue Wu[1], Xianwei Li[1,2](✉), and Shunxiang Zhang[3,4]

[1] School of Computer and Information Engineering, Bengbu University, Bengbu 233030, China
why@bbc.edu.cn

[2] Anhui Engineering Research Center for Intelligent Applications and Security the Industrial Internet, Anhui University of Technology, Ma'anshan 243032, China
xianweili@fuji.waseda.jp

[3] School of Computer Science and Engineering, Anhui University of Science and Technology, Huainan 232001, China
sxzhang@aust.edu.cn

[4] Institute of Artificial Intelligence Research, Hefei Comprehensive National Science Center, Hefei 230088, China

Abstract. Offensive language detection is vital for purifying cyberspace and ensuring the efficient and orderly operation of pretrained language models. Traditional methods of offensive language detection rely on a single matching mechanism, often neglecting the influence of semantic bias on the detection outcomes. This limitation leads to a low accuracy rate and constitutes a significant threat to cyberspace security. To address this issue, this paper proposes a high-accuracy Chinese offensive language detection method named CHOLD. CHOLD integrates various language features, including Chinese character shape, pronunciation, and meaning. Initially, the method constructs diverse lists of offensive language keywords for semantically matching target sentences. The matching words are subsequently added to the keyword list, forming the offensive language candidate set. The high-dimensional features related to the character shape, pronunciation, and meaning are then fused after the dimensionality of the sentences within the candidate set is reduced. Additionally, a semantic constraint method is employed to minimize the impact of semantic deviation on the detection results. Experimental results conducted on multiple public datasets demonstrate that CHOLD outperforms previous detection models in terms of recognition accuracy.

Keywords: Offensive Language · Information Detection · Semantic Constraint · Text Semantic Deviation

1 Introduction

In recent years, significant advancements in large language models [1–3] have greatly contributed to the growth and progress of the field of NLP. However,

L. Huang et al. (Eds.): CCF NCCA 2025, CCIS 2774, pp. 239–255, 2026.
https://doi.org/10.1007/978-981-95-5294-8_17

researchers have discovered that these large language models inevitably incorporate offensive language during their training process. This poses a security concern for pretrained language models when they encounter adversarial attacks [4,5]. The interactive nature of the internet facilitates the exchange of information across different cultures and languages. Consequently, offensive language, encompassing hate speech, fake news, rumours, and explicit text, has emerged as a problem that hampers normal communication among individuals. Therefore, the efficient and accurate detection of offensive language holds paramount importance in combating the dissemination of such content and maintaining the integrity of online platforms.

Traditional methods for offensive language detection typically utilize a single matching mechanism, such as keyword filtering and sensitive word matching. These methods rely on the linguistic features of keywords and sensitive words to identify offensive content in target sentences. The complexity and nuances of offensive linguistic expressions in Chinese text cannot be captured. Moreover, when adversarial sample attacks are encountered, such as polysyllabic and variant characters [6], [7], these methods are easily circumvented so that they can lead to biased final detection results [8]. Figure 1 provides typical examples for analysing this behavior. The inherent limitations of these methods contribute to low detection accuracy and pose a significant threat to the security of online spaces.

Two cases were randomly selected from the first Chinese offensive language dataset COLD [8] for analysis. In case (a), the input content is as follows:"TMD, to kill each other but also the next time to a team of bitch couple while practicing level while killing me, nausea and kill me." The offensive language detection model successfully identifies the offensive language present in the sentence, which includes the words "TMD, bitch couple". In case (b), the input content is "Let the women's rights activism go to find the powerful foreigners." The traditional offensive language detection model can only recognize "women's rights activism". However, in case (b), we discovered the presence of homonyms, such as '**洋带人 (yáng dài ren)**' and '**洋大人 (yáng dà ren)**', both of which refer to powerful foreigners. When faced with multimorphic adversarial samples, such as homophones and heteromorphic characters, the recognition accuracy of the offensive language diminishes.

To address this challenge, this paper proposes CHOLD, a high-precision offensive language detection method for Chinese texts. To mitigate the influence of Chinese adversarial samples on offensive language detection, CHOLD leverages multiple language dimensions, including character shape, pronunciation, and meaning, to construct offensive keyword lists. Matching these keywords enables us to tentatively identify offensive language content in Chinese texts and refine the candidate set.

To increase the detection accuracy, CHOLD integrates high-dimensional features of character shape, pronunciation, and meaning extracted from the candidate set. This feature fusion allows for a comprehensive understanding of different aspects of offensive language expression, thereby improving the detection

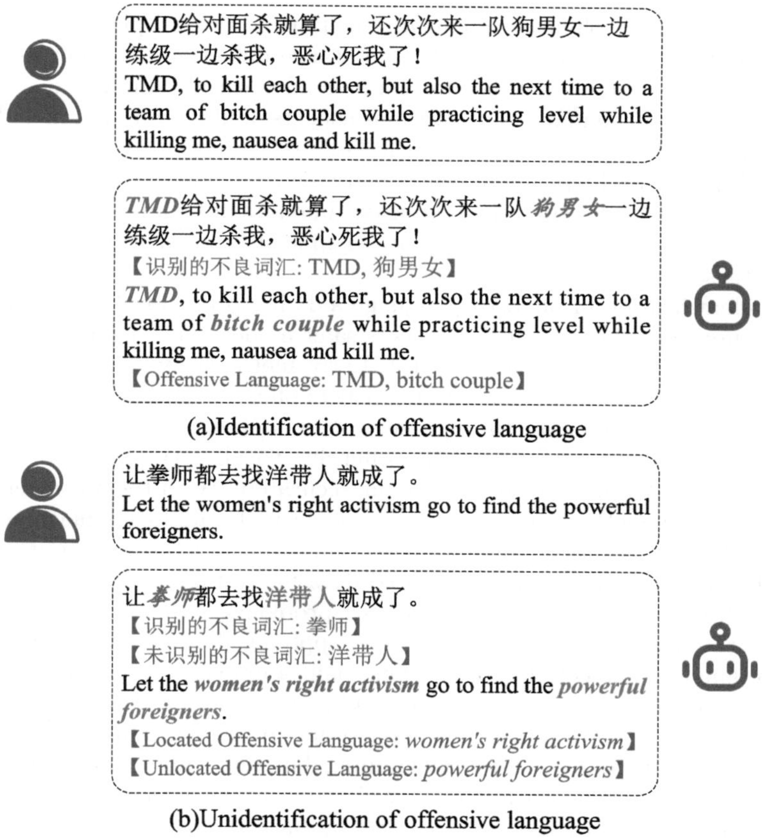

Fig. 1. Adversarial samples affect offensive language detection.

capability of CHOLD. Additionally, a semantic constraint method is employed to reduce the impact of semantic biases on detection outcomes. By considering the text's potential implications and contextual nuances, the CHOLD approach ensures more precise recognition of offensive language, thus improving the overall efficiency and effectiveness of offensive language detection.

Extensive experiments were conducted using multiple models on various public datasets to evaluate the proposed method. The results demonstrate that CHOLD surpasses previous detection models and achieves the highest recognition accuracy to date. These findings affirm the effectiveness and superiority of CHOLD.

2 Related Work

Offensive language detection has emerged as a prominent research area, attracting significant attention from NLP researchers. This section provides an overview of related research on offensive language detection, covering topics such as pornographic content, oppositional speech, and stance detection [9]. Additionally, it briefly introduces attack methods for Chinese adversarial samples.

Offensive language manifests in various aspects of life. Examples include gender discrimination, religious beliefs, politics, race, or profanity on social media [10,11]. With the proliferation of pretrained language models such as BERT [12] and ChatGPT [13], which require extensive amounts of training data, the offensive language found in the training data significantly impacts the downstream tasks of these models.

The identification of offensive language aids in optimizing training data for models, affording scholars and researchers safer models, and contributing to the purification of the online space [14–16]. Early methods for detecting offensive language typically involved constructing lists of offensive language keywords and identifying offensive content through keyword matching [17]. Nevertheless, this approach can solely recognize familiar offensive vocabulary and lacks the ability to detect unidentified offensive content accurately. With the advancement of machine learning methods, researchers are increasingly employing machine learning algorithms for offensive language detection [18]. Nevertheless, these methods frequently demand substantial annotated data for training and encounter challenges with long-tailed terms and cold-start problems.

More recently, researchers have begun applying deep learning methods to offensive language detection [19,20]. The main approach involves transforming text into vector representations and performing feature extraction and classification via multilayer convolution and pooling operations. However, deep learning methods typically require substantial training data and computational resources, and they often struggle with imbalanced datasets. In contrast to traditional keyword-matching methods, some researchers have started exploring the use of semantic matching for offensive language detection [21,22]. However, this approach still faces challenges in handling semantic biases and polysemous words.

Most of the studies mentioned above have focused primarily on the English domain. Owing to the unique cultural characteristics and semantic nuances of Chinese text, these methods cannot be directly applied to offensive language detection in Chinese texts [23]. Many scholars have discovered that adversarial samples generated in the Chinese domain can significantly impact the accuracy of various downstream tasks [24–26]. These findings highlight the prevalence of adversarial samples in Chinese texts, which hinders the accurate detection of offensive language.

Therefore, despite the progress made in offensive language detection, challenges remain. The complexity of Chinese texts and potential semantic biases make traditional methods less accurate. Hence, the development of precise methods for detecting offensive language is crucial to mitigate its impact.

3 The Proposed Method CHOLD

3.1 Problem Definition

The offensive language detection task can be conceptualized as a sieve that aims to classify given texts into two categories: offensive language and nonoffensive language. The objective of this sieve is to efficiently and accurately identify offensive language and distinguish it from nonoffensive language.

For an input text set $\mathcal{D}$ as $\mathcal{D} = \{x_1, x_2, \cdots, x_N\}$, where each text x_i belongs to a certain text category y, we represent the text set as X and the text category set as Y, where $i \in [1, N]$. Thus, we define a function $f : X \to Y$ that maps texts to their corresponding categories. Here, for any text x_i, $f(x)$ outputs the category to which it belongs, where $f(x) \in Y$. Therefore, for the offensive language detection task, we define offensive language as Y_{off} and nonoffensive language as another category, assumed to be $Y_{non\text{-}off}$. Thus, the offensive language classifier function $f : X \to Y$ can be redefined as $f : X \to \{Y_{off}, Y_{non\text{-}off}\}$.

3.2 CHOLD Framework

On the basis of the description of the offensive language detection task, this paper proposes CHOLD, a method specifically designed to detect offensive language in Chinese texts. The framework of CHOLD is depicted in Fig. 2, which illustrates the main processes involved in CHOLD. The figure provides a detailed overview of character-level and semantic-level detection, feature extraction, and dimension reduction on the basis of character shape, sound, and semantic meaning. Notably, the semantic constraints module is positioned at the end of Fig. 2. To enhance visual clarity, distinct color blocks are utilized to differentiate between different components.

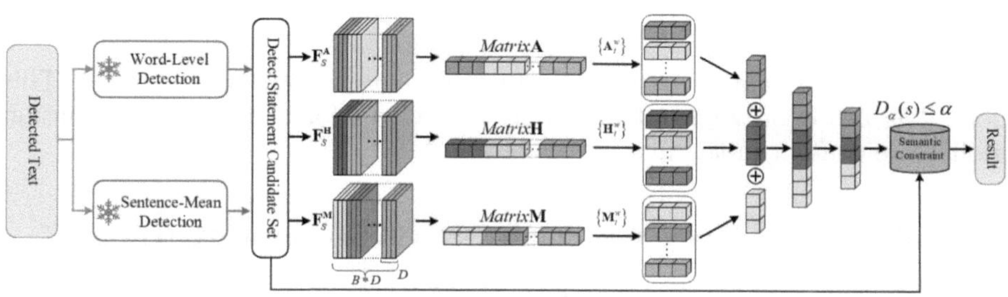

Fig. 2. Framework of the Chinese intensive language detection method.

For the detection document $\mathcal{D}$, each text x_i has $x=\{s_1, s_2, \cdots, s_T\}$. Through word-level and semantic-level detection, clear offensive language is preliminarily filtered out. The filtered texts are used to construct the candidate set for detection. For the data in the candidate set, the character shape feature

$\mathbf{F}_s^{\mathbf{A}} \in \mathbb{R}^{H\times W\times(B\times D)}$, character sound feature $\mathbf{F}_s^{\mathbf{H}} \in \mathbb{R}^{H\times W\times(B\times D)}$, and character semantic feature $\mathbf{F}_s^{\mathbf{M}} \in \mathbb{R}^{H\times W\times(B\times D)}$ are obtained separately according to the character shape, sound, and semantic meaning, respectively, with $H \times W$, B, and D representing the feature mapping size, overall character feature cardinality, and feature channels, respectively. On this basis, the corresponding $Matrix\mathbf{A}$, $Matrix\mathbf{H}$, and $Matrix\mathbf{M}$ of the sentence are extracted. Then, the character shape feature set $\{\mathbf{A}_i^w\}$, character sound feature set $\{\mathbf{H}_i^w\}$, and character semantic feature set $\{\mathbf{M}_i^w\}$ for each Chinese character in the sentence are extracted one by one. In the CHOLD method, the character features of each Chinese character in a sentence are obtained by concatenating the three features on the basis of the positional information of the characters. This process enables the extraction of the final feature for the entire word or sentence, which then undergoes a semantic constraint to yield the final classification result.

3.3 Offensive Language Keyword Set

To formulate the offensive language keyword set in CHOLD, diverse approaches are employed, comprising manual annotation, automatic extraction, and web crawling. By acquiring keywords from diverse perspectives and sources, the keyword set comprehensively encompasses various facets of offensive language. It encompasses offensive, aggressive, and inappropriate vocabulary. By examining whether a text contains any of these keywords from the set, we can ascertain whether it contains offensive content.

Specifically, assume that we have a set $\mathcal{K}$ containing offensive language keywords. For the detection document $\mathcal{D} = \{x_1, x_2, \cdots, x_N\}$, for each text x have x=$\{s_1, s_2, \cdots, s_T\}$, we can use the following mathematical expression to determine if it contains offensive content:

$$f(x) = \begin{cases} Y_{\mathit{off}}, & \text{contains a keyword from } \mathcal{K} \\ Y_{\mathit{non\text{-}off}}, & \text{otherwise} \end{cases} \tag{1}$$

where $f(x)$ represents whether the text x contains offensive content. This method is based on keyword matching and allows for a quick determination of offensive content in the text.

3.4 Offensive Language Candidate Set via Semantic Matching

In the offensive language detection task, the generation of the offensive language candidate set primarily involves establishing a measure of relevance between texts on the basis of semantic similarity. The objective is to gradually add texts that are semantically similar to offensive vocabulary to the candidate set, thereby reducing semantic bias.

In this paper, ChineseBERT[1] is utilized to represent texts as vectors. Assum that we have an input text x to be classified, represented as the feature vector v_x.

[1] https://huggingface.co/bert-base-chinese.

Next, we define a similarity measurement function S to calculate the semantic similarity between two texts. The commonly used method for measuring similarity is cosine similarity, which is obtained by calculating the inner product of the feature vectors.

$$S(x_1, x_2) = \frac{v_{x_1} \bullet v_{x_2}}{\|v_{x_1}\| \cdot \|v_{x_2}\|} \tag{2}$$

where v_{x_1} and v_{x_2} represent the word vector representations of texts x_1 and x_2, respectively.

We set a threshold θ that represents the minimum limit for semantic similarity. By comparing it with a known set of offensive languages, texts with semantic similarity to the input text x higher than the threshold θ are added to the offensive language candidate set $\mathcal{C}$:

$$\mathcal{C} = \{x_i |\, S(x, x_i) \geq \theta, x_i \in Y_{off}\} \tag{3}$$

3.5 Feature Extraction

From Fig. 2, we obtain the character shape feature $\mathbf{F}_s^{\mathbf{A}} \in \mathbb{R}^{H \times W \times (B \times D)}$, character sound feature $\mathbf{F}_s^{\mathbf{H}} \in \mathbb{R}^{H \times W \times (B \times D)}$, and character semantic feature $\mathbf{F}_s^{\mathbf{M}} \in \mathbb{R}^{H \times W \times (B \times D)}$. We also extract $Matrix\mathbf{A}$, $Matrix\mathbf{H}$, and $Matrix\mathbf{M}$ for the sentences. On this basis, we obtain the character shape feature set $\{\mathbf{A}_i^w\}$, character sound feature set $\{\mathbf{H}_i^w\}$, and character semantic feature set $\{\mathbf{M}_i^w\}$ corresponding to each Chinese character. While the data information in the candidate set undergoes different transformations in terms of character shape, sound, and semantic meaning, their positional information remains unchanged. Leveraging this positional relationship, dimension reduction is performed on the features of each Chinese character, reducing them from the dimensions of character shape, sound, and semantic meaning. This facilitates feature concatenation and enables the extraction of the final character feature. The process is illustrated in Fig. 3.

3.6 Semantic Constraint

In the CHOLD method, semantic constraints are utilized to mitigate the influence of semantic bias on the detection results of offensive language. In the context of the offensive language detection task, semantic constraints involve analysing texts on the basis of their semantic information and formulating constraints to limit the degree of semantic perturbation and minimize semantic bias. In this study, we employ USE encoding [27] for semantic constraints.

The USE encoder leverages the encoder component of a transformer network, which is a text encoding model based on the multihead self-attention mechanism. During encoding, the USE encoder assigns varying weights to each word in a sentence on the basis of its semantic contribution to the overall sentence. For the detection document $\mathcal{D} = \{x_1, x_2, \cdots, x_N\}$, for each text x_i, x=$\{s_1, s_2, \cdots, s_T\}$. Here, $s_i = [h_{i_1}, h_{i_2}, \cdots, h_{i_j}, \cdots, h_{i_n}]$ represents the input word vectors, which

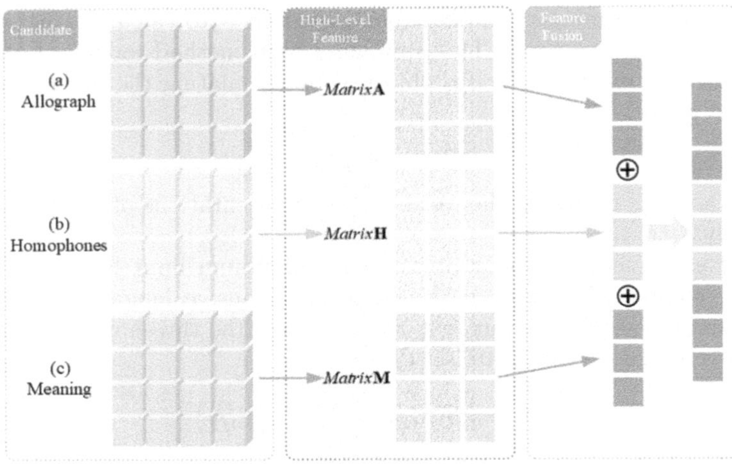

Fig. 3. Diagram of feature processing and feature splicing process.

represent the general vector form corresponding to a single Chinese character. The encoding process of the USE encoder for s_i can be represented by Eq. 4.

$$o_i = USE(s_i) \tag{4}$$

where o_i represents the encoding vector output by the USE encoder.

Multiplying s_i with weight matrices $\mathbf{W^Q}$, $\mathbf{W^K}$, and $\mathbf{W^V}$ to perform linear transformations yields attention matrices $\mathbf{Q}_i$, $\mathbf{K}_i$, and $\mathbf{V}_i$, as calculated via Eqs. (5)–(7).

$$\mathbf{Q}_i = s_i \cdot \mathbf{W^Q} \tag{5}$$

$$\mathbf{K}_i = s_i \cdot \mathbf{W^K} \tag{6}$$

$$\mathbf{V}_i = s_i \cdot \mathbf{W^V} \tag{7}$$

The self-attention encoding vector $Attention(\mathbf{Q}_i, \mathbf{K}_i, \mathbf{V}_i)$ of the sentences_i is calculated as:

$$Attention(\mathbf{Q}_i, \mathbf{K}_i, \mathbf{V}_i) = Softmax(\frac{\mathbf{Q}_i\mathbf{K}_i^T}{\sqrt{d_k}})\mathbf{V}_i \tag{8}$$

where $\mathbf{Q}_i$ and $\mathbf{K}_i^T$ represent the weight matrices between words in the sentence, and where d_k represents the dimension of the word vector.

In the multihead self-attention mechanism, weight matrices $\mathbf{W^Q}$, $\mathbf{W^K}$, and $\mathbf{W^V}$ are randomly initialized to obtain multiple *Attention* vectors. The *Attention* vector is subsequently concatenated with the dimension-reduced vectors of the character shape feature set $\{\mathbf{A}_i^w\}$, the character sound feature set $\{\mathbf{H}_i^w\}$, and the character semantic feature set $\{\mathbf{M}_i^w\}$. After a linear layer transformation, the merged self-attention encoding vector is obtained. Through residual connections, sum normalization, and feed-forward networks, a new sequence

$s_i^{'} = [h_{i1}^{'}, h_{i2}^{'}, \cdots, h_{ij}^{'}, \cdots, h_{in}^{'}]$ is subsequently obtained. The word vectors in $s_i^{'}$ are summed and divided by the square root of the sentence length $\sqrt{n}$ to obtain the output o_i of the USE encoder, as shown in Eq. 9.

$$o_i = \frac{\sum_{j=0}^{n} h_{ij}^{'}}{\sqrt{n}} \tag{9}$$

Using the pretrained USE encoding network, we encode both the input samples to be detected and the data samples in the candidate set into sentence encodings. Subsequently, we measure their semantic similarity by calculating the vector distance between them. If the vector of the original text is u, the vector of the data samples in the candidate set is v. The semantic vector distance $Ds(u, v)$ can be calculated as:

$$Ds(u, v) \leq \alpha \tag{10}$$

where α is a semantic distance measure parameter, then satisfying $Ds(u, v) \leq \alpha$ is an offensive language.

4 Experiments

In the experimental section, we initially present the pertinent datasets employed in the experiments. Subsequently, we elucidate the evaluation metrics employed to assess the detection performance of CHOLD. We conduct a performance comparison between CHOLD and other baseline methods. Finally, we analyse the results and delve into the strengths and limitations of the CHOLD criteria. The experiments demonstrate that CHOLD excels in detecting offensive language in Chinese, achieving accurate detection.

4.1 Dataset

For experimental validation, we select six publicly available datasets that include both Chinese and English data. These datasets cover diverse types of offensive language, including gender discrimination, religious beliefs, politics, race, and profanity, on online social media. The specific details of the datasets are provided in Table 1.

By conducting tests on multiple large datasets, we can comprehensively evaluate the effectiveness of the proposed CHOLD. The selected datasets span a wide range of domains. For English datasets and other non-Chinese datasets, we translate them into Chinese via Baidu Translate[2] before conducting experimental validation.

[2] https://fanyi.baidu.com/.

Table 1. Offensive Language Detection Data Set in the Experiments.

Dataset	Main Contents	Data Size	Open Source
COLA [28]	Fine-grained insulting language, antisocial language, and illegal language.	18k	No
COLD [8]	It covers multiple topics such as race, gender, and region.	32k	Yes
CDial [29]	Social bias in various dialogue scenarios.	28k	Yes
SWSR [30]	Sexism or nonsexism, sexism category, and target type.	9k	Yes
TOCP [31]	Chinese profanity dataset includes massive profanity and its rewritten form.	16k	Yes
HSAL [32]	Contains hate speech, offensive language but not hate speech, and others.	15k	Yes

4.2 Baseline Systems

The purpose of our experiments in this paper is to determine whether the data in Dataset $\mathcal{D}$ belong to offensive languages (Y_{off} or $Y_{non-off}$). To better validate the effectiveness of the proposed method, we compare the CHOLD method with several baseline models and state-of-the-art methods.

- **COLD** [8]: This is the first large-scale publicly available dataset specifically for Chinese offensive language detection. A baseline model based on this dataset is proposed.
- **CDial** [29]: Large-scale social commentary data in online networks, with a dedicated model for detecting offensive language in social comments.
- **HSAL** [32]: This method focuses on the detection of hate speech-related offensive language.
- **Arabic hate** [33]: Detection of offensive information related to hate and racism in Arabic.
- **Mc-bert4hate** [23]: Hates speech detection in different languages, focusing on offensive language detection through translation between different languages.
- **Urdu hate** [9]: Multilevel hate speech detection in the Urdu language.

4.3 Performance Comparison

In the experimental evaluation, we compare the CHOLD method with baseline methods and other state-of-the-art methods on six large offensive language detection datasets, including COLA [28], COLD [8], CDial [29], SWSR [30], TOCP [31], and HSAL [32]. The non-Chinese datasets are first translated into Chinese before conducting offensive language detection. The specific performance results of the experiments comparing different methods on multiple datasets are shown in Table 2.

Table 2. Results of intensive language detection with different methods on six test sets. The best results in each group are shown in bold.

Dataset	Model	Accuracy	Macro			Offen.			Non-Offen.		
			Prec.	Rec.	F1	Prec.	Rec.	F1	Prec.	Rec.	F1
COLA [28]	COLD [8]	0.7532	0.6845	0.6926	0.6884	0.7620	0.7904	0.7759	0.6069	0.5948	0.6008
	CDial [29]	0.6171	0.6780	0.6511	0.6632	0.6242	0.5470	0.5831	0.7318	0.7552	0.7433
	HSAL [32]	0.5039	0.6733	0.6485	0.6605	0.5341	0.5310	0.5325	0.8125	0.7659	0.7885
	Arabic hate [33]	0.7854	0.7649	0.7451	0.7536	0.7171	0.6393	0.6760	0.8126	**0.8509**	0.8313
	Mc-4hate [23]	0.6090	0.7384	0.6727	0.7008	0.7139	0.7415	0.7274	0.7629	0.6038	0.6741
	Urdu hate [9]	0.7432	0.7988	0.8470	0.8208	0.7328	0.8462	0.7854	**0.8647**	0.8478	**0.8562**
	CHOLD(Ours)	**0.8244**	**0.8217**	**0.8765**	**0.8482**	**0.8486**	**0.9055**	**0.8761**	0.7947	0.8475	0.8203
COLD [8]	COLD [8]	0.8164	0.8000	0.7512	0.7730	0.8511	0.7266	0.7839	0.7489	0.7757	0.7621
	CDial [29]	0.6839	0.7584	0.7547	0.7560	0.7009	0.6596	0.6796	0.8158	**0.8497**	0.8324
	HSAL [32]	0.7219	0.6775	0.6935	0.6854	0.7201	0.7398	0.7298	0.6349	0.6471	0.6409
	Arabic hate [33]	0.7187	0.7292	0.7186	0.7234	0.7192	0.7452	0.7320	0.7392	0.6919	0.7148
	Mc-4hate [23]	0.6735	0.6329	0.6232	0.6277	0.6479	0.6631	0.6554	0.6178	0.5832	0.6000
	Urdu hate [9]	0.5696	0.5748	0.5685	0.5715	0.5964	0.6015	0.5989	0.5532	0.5354	0.5442
	CHOLD(Ours)	**0.8912**	**0.8720**	**0.8581**	**0.8648**	**0.8762**	**0.8854**	**0.8808**	**0.8677**	0.8308	**0.8488**
CDial [29]	COLD[8]	0.8147	0.7391	0.7348	0.7368	0.7768	0.7529	0.7647	0.7014	0.7167	0.7090
	CDial [29]	0.7224	0.7413	0.7122	0.7264	0.7267	0.6965	0.7113	0.7559	0.7278	0.7416
	HSAL [32]	0.8087	0.7478	0.7321	0.7399	**0.8201**	0.7978	0.8088	0.6755	0.6664	0.6709
	Arabic hate [33]	0.6048	0.5961	0.6063	0.6011	0.6309	0.6444	0.6376	0.5612	0.5681	0.5646
	Mc-4hate [23]	**0.8374**	0.8000	0.7903	0.7942	0.8192	0.7562	0.7864	0.7807	**0.8243**	0.8019
	Urdu hate [9]	0.7198	0.6329	0.6001	0.6161	0.7369	0.6997	0.7178	0.5289	0.5005	0.5143
	CHOLD(Ours)	0.8112	**0.8279**	**0.8224**	**0.8250**	0.8093	**0.8241**	**0.8166**	**0.8465**	0.8206	**0.8333**
SWSR [30]	COLD [8]	0.8041	0.7644	0.7454	0.7547	0.7917	0.7691	0.7802	0.7370	0.7216	0.7292
	CDial [29]	0.7058	0.6896	0.6665	0.6778	0.6954	0.6735	0.6843	0.6837	0.6594	0.6713
	HSAL [32]	0.7747	0.7434	0.7692	0.7560	0.7822	**0.8268**	0.8039	0.7046	0.7115	0.7080
	Arabic hate [33]	0.6547	0.5913	0.5642	0.5769	0.6504	0.5879	0.6176	0.5322	0.5404	0.5363
	Mc-4hate [23]	0.8123	0.7894	0.7697	0.7794	0.8143	0.7871	0.8005	**0.7644**	0.7522	0.7583
	Urdu hate [9]	0.6084	0.5423	0.5630	0.5523	0.5821	0.6202	0.6005	0.5025	0.5057	0.5041
	CHOLD(Ours)	**0.8205**	**0.8066**	**0.7858**	**0.7960**	**0.8493**	0.8185	**0.8336**	0.7638	**0.7531**	**0.7584**
TOCP [31]	COLD [8]	0.8465	0.8439	0.8290	0.8363	0.8173	0.8195	0.8184	**0.8705**	0.8385	0.8542
	CDial [29]	0.7488	0.8021	0.7946	0.7982	0.8083	0.8189	0.8136	0.7958	0.7703	0.7828
	HSAL [32]	0.7217	0.7233	0.6941	0.7083	0.7348	0.7178	0.7262	0.7118	0.6703	0.6904
	Arabic hate [33]	0.6274	0.6621	0.6395	0.6506	0.6232	0.6025	0.6127	0.7010	0.6765	0.6885
	Mc-4hate [23]	0.7668	0.8263	0.7965	0.8111	0.7874	0.7639	0.7755	0.8651	0.8291	0.8467
	Urdu hate [9]	0.6353	0.6731	0.6505	0.6616	0.6538	0.6281	0.6407	0.6923	0.6729	0.6825
	CHOLD(Ours)	**0.8754**	**0.8548**	**0.8470**	**0.8505**	**0.8664**	**0.8261**	**0.8458**	0.8431	**0.8678**	**0.8553**
HSAL [32]	COLD [8]	0.8204	0.8603	0.8442	0.8521	0.8434	**0.8402**	**0.8418**	0.8772	0.8481	0.8624
	CDial [29]	0.7519	0.8234	0.7919	0.8073	0.8233	0.7863	0.8044	0.8235	0.7974	0.8102
	HSAL [32]	0.7097	0.7590	0.7488	0.7538	0.7367	0.7132	0.7248	0.7812	0.7843	0.7827
	Arabic hate [33]	0.6329	0.7269	0.7258	0.7263	0.7815	0.7763	0.7789	0.6722	0.6753	0.6737
	Mc-4hate [23]	0.7457	0.7325	0.7160	0.7241	0.7397	0.7145	0.7269	0.7253	0.7175	0.7214
	Urdu hate [9]	0.6538	0.6765	0.6786	0.6775	0.6914	0.7074	0.6993	0.6615	0.6498	0.6556
	CHOLD(Ours)	**0.8326**	**0.8691**	**0.8586**	**0.8638**	**0.8445**	0.8386	0.8415	**0.8937**	**0.8785**	**0.8860**

Table 2 provides a comparison between CHOLD and other methods in terms of offensive language detection and nonoffensive language recognition. The accuracy results represent the accuracy of offensive language detection. The macro scores represent the average precision, recall, and F1 score of the evaluated methods in terms of offensive and nonoffensive language detection. The experimental results demonstrate that CHOLD outperforms the other methods significantly in terms of accuracy, precision, recall, and F1 score.

(1) CHOLD excels in offensive language detection, surpassing other methods on multiple datasets, showing substantial advancements in performance over baselines. In the offensive language detection task, CHOLD achieves remarkable precision, recall, and F1 scores on different datasets. For example, in the COLA dataset [28], the CHOLD method achieves 84.86% precision, 90.55% recall, and 87.61% F1 score, surpassing other methods by a significant margin. In the first Chinese dataset, COLD [8], the CHOLD method achieves 87.62% precision, 88.54% recall, and 88.08% F1 score in the offensive language detection task. Notably, the baseline model proposed in COLD [8] achieves 85.11% precision, 72.66% recall, and 78.39% F1 score, highlighting the substantial advantage of CHOLD. CHOLD outperforms the baseline method in terms of the offensive language detection F1 score by nearly 10%, verifying the advantages of the proposed method. In the CDial-Bias dataset [29], CHOLD achieves a precision of 80.93%, whereas it boasts the highest recall and F1 score at 82.41% and 81.66%, respectively.

(2) CHOLD excels on non-English and non-Chinese datasets, showing superior offensive language detection performance, particularly in translation-sensitive scenarios, as highlighted in various benchmarks. On non-English and non-Chinese datasets, CHOLD outperforms Chinese and English datasets in terms of performance. The translation standards of commonly used languages can affect the accuracy of offensive language detection. For the SWSR dataset [30], the CHOLD method achieves 84.93% precision, 81.85% recall, and 83.36% F1 score. Table 2 shows that CHOLD attained the second-highest recall while achieving the best precision and F1 score. In TOCP [31], the CHOLD method achieved 86.64% precision, 82.61% recall, and 84.58% F1 score, which are the highest values. In HSAL [32], CHOLD achieved 84.45% precision, 83.86% recall, and 84.15% F1 score. In this dataset, the CHOLD method achieves the best precision and the second-highest recall and F1 score. Overall, these results demonstrate the significant achievements of the CHOLD method in terms of offensive language detection.

To further highlight the superiority of the CHOLD, a bar chart is used to visually display the macro-F1 score, Offen-F1 score, and non-Offen-F1 score of multiple datasets in Table 2, as shown in Fig. 4. Figure 4 clearly shows that the CHOLD method has a clear advantage across multiple datasets.

4.4 Ablation Study

Furthermore, an ablation study is conducted to verify the effectiveness of the Key-Mat, Feat-Ext, and Sem-Cons modules in CHOLD. The experiments focus on evaluating the impact of keyword matching, multidimensional feature extraction (character shape, sound, and meaning), and semantic constraints. Since the CHOLD method is based on Chinese text data, the ablation study primarily concentrates on offensive information detection in the COLD [8] dataset. The performance changes of the CHOLD method when each module is removed are

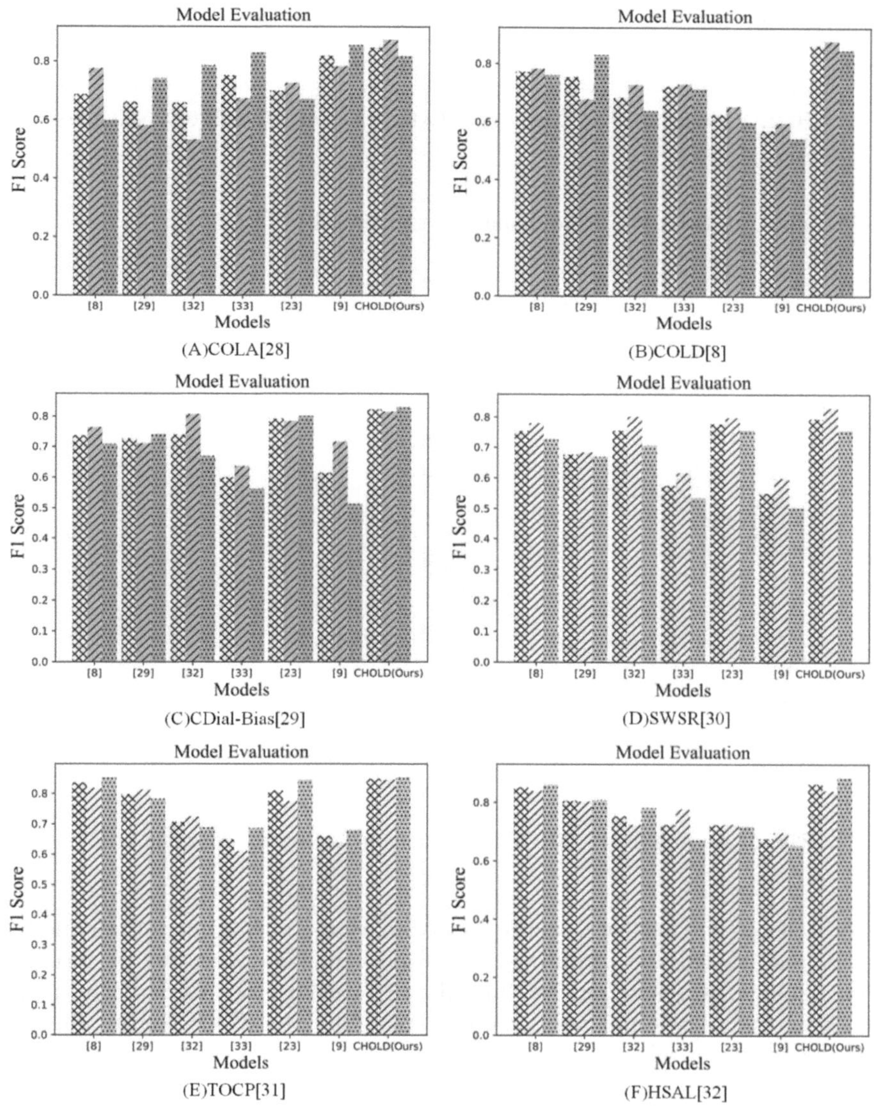

Fig. 4. The results of macro-F1, off-F1, and non-off-F1 on multiple datasets.

examined. The specific results of the ablation study are presented in Table 3 and Fig. 5.

Table 3. Results of the CHOLD module verified by the ablation study.

Method	Prec.	Rec.	F1
Baseline	0.6841	0.7023	0.6931
Baseline+Key-Mat	0.7539	0.7465	0.7502
Baseline+Key-Mat+Feat-Ext	0.8192	0.8237	0.8214
CHOLD (Baseline+Key-Mat+Feat-Ext+Sem-Cons)	**0.8486**	**0.9055**	**0.8761**

The data in Table 3 can be visually represented in Fig. 5.

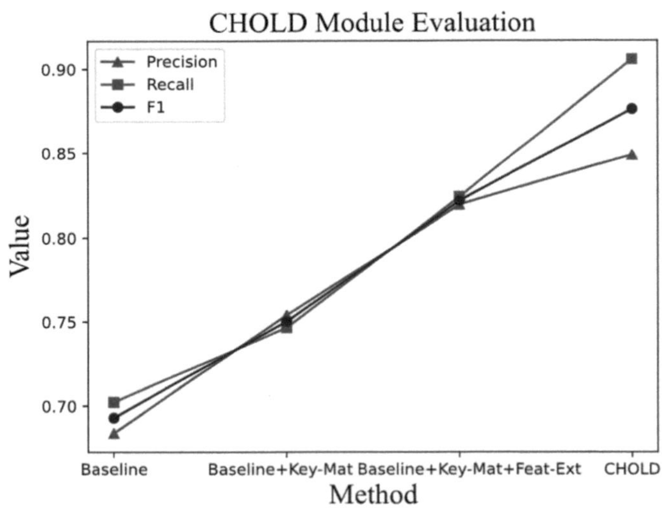

Fig. 5. Results of the CHOLD module verified by the ablation study.

Table 3 and Fig. 5show that the Key-Mat, Feat-Ext, and Sem-Cons modules have a positive effect on improving offensive language detection performance. In particular, the Sem-Cons module plays a crucial role, particularly in handling semantic biases and polysemous words. The Sem-Cons module improves precision by 3%, recall by 8.2%, and the F1 score by 5.5% in offensive language detection. These results validate the significant contributions of the various modules in CHOLD for offensive information detection.

5 Conclusion

This paper introduces CHOLD for the detection of offensive language in Chinese texts, tackling the challenges posed by semantic biases and the complexity of Chinese text. The proposed method, CHOLD, employs an offensive language keyword table on the basis of linguistic features. It generates an offensive language candidate set via semantic matching, extracts high-dimensional features, and applies semantic constraints to achieve accurate offensive language detection. The experimental results additionally affirm the superior performance of the CHOLD method in detecting offensive language in Chinese. The ablation study verifies the effectiveness of CHOLD, especially when incorporating semantic constraints. These constraints alleviate misjudgments caused by semantic bias, increasing the accuracy and recall rate of offensive information detection. Comparatively, CHOLD demonstrates superior performance compared with other methods in offensive language detection.

Acknowledgements. This work is supported by the National Natural Science Foundation of China (62076006), the University Synergy Innovation Program of Anhui Province (GXXT-2021-008), the Opening Foundation of State Key Laboratory of Cognitive Intelligence (COGOS-2023HE02), the Key Scientific Research Projects of Colleges and Universities in Anhui Province (2024AH051167), the Funding Project for the Cultivation of Outstanding Talents in Colleges and Universities (gxyqZD2021135), the Start Up Funds for Scientific Research of High-level Talents of Bengbu University (BBXY2020KYQD02), the Open Project of Engineering Research Center of Anhui University of Technology (IASII22-08), the Science Research Project of Bengbu University (2024YYX47pj, 2024YYX48pj), the Anhui Province Excellent Research and Innovation Team in Intelligent Manufacturing and Information Technology (2023AH052938), the University Collaborative Innovation Project in Anhui Province under Grant (GXXT-2023-105), the Bengbu University 2024 General Program of University Research (2024ZR02), the Provincial Quality Project of Colleges and Universities in Anhui Province (2023cyts054), the 2024 National Undergraduate Training Program for Innovation and Entrepreneurship (202411305032, 202411305033).

References

1. Dong, Y., Ran, Q., Chao, X., Li, C., Yu, S.: Personalized individual semantics learning to support a large-scale linguistic consensus process. ACM Trans. Internet Technol. **23**(2), 1–27 (2023)
2. Xu, X., Chai, J., Chen, X.: A hesitation-feedback recommendation approach and its application in large-scale group emergency decision making. Expert Syst. Appl. **213**, 118876 (2023)
3. Rafiq, D., Bazaz, M.A.: A collection of large-scale benchmark models for nonlinear model order reduction. Arch. Comput. Methods Eng. **30**(1), 69–83 (2023)
4. Mi, F., et al.: PanGu-bot: efficient generative dialogue pre-training from pretrained language model. arXiv preprint arXiv:2203.17090 (2022)
5. Yong, Q., Chen, C., Wang, Z., Xiao, R., Tang, H.: SGPT: Semantic graphs based pre-training for aspect-based sentiment analysis. Worldwide Web, **26**(2), 1105–1118 (2023)

6. Zhang, S., Wu, H., Zhu, G., Xin, X., Su, M.: Character-level adversarial samples generation approach for Chinese text classification. J. Electron. Inf. Technol. **45**(6), 2226–2235 (2023)
7. Huang, X., Zhang, J., Zong, C.: Contrastive adversarial training for multi-modal machine translation. ACM Trans. Asian Low-Resour. Lang. Inf. Process **22**(5), 1–18 (2023)
8. Deng, J., Zhou, J., Sun, H., Zheng, C., Mi, F., Meng, H., Huang, M.: COLD: a benchmark for Chinese offensive language detection. In: Proceedings of the 2022 Conference on Empirical Methods in Natural Language Processing, pp. 11580–11599 (2022)
9. Saeed, R., Afzal, H., Rauf, S.A., Iltaf, N.: Detection of offensive language and its severity for low resource language. ACM Trans. Asian Low-Resour. Lang. Inf. Process. **22**(6), 1–27 (2023)
10. Husain, F., Uzuner, O.: A survey of offensive language detection for the Arabic language. ACM Trans. Asian Low-Resour. Lang. Inf. Process. **20**(1), 1–44 (2021)
11. Essefar, K., Ait Baha, H., El Mahdaouy, A., El Mekki, A., Berrada, I.: OMCD: offensive Moroccan comments dataset. Lang. Resour. Eval. **57**(4), 1583–1603 (2023)
12. Devlin, J., Chang, M.W., Lee, K., Toutanova, K.: BERT: pre-training of deep bidirectional transformers for language understanding. In: Proceedings of the 2019 Conference of the North American Chapter of the Association for Computational Linguistics: Human Language Technologies, pp. 4171–4186 (2019)
13. Ouyang, L., et al.: Training language models to follow instructions with human feedback. Adv. Neural Inf. Process. Syst. **35**, 27730–27744 (2022)
14. Kshetri, N.: Cybercrime and privacy threats of large language models. IT Prof. **25**(3), 9–13 (2023)
15. Cai, S., Han, D., Li, D., Zheng, Z., Crespi, N.: A reinforcement learning-based speech censorship chatbot system. J. Supercomput. **78**(6), 8751–8773 (2022)
16. Shi, X., Liu, X., Xu, C., Huang, Y., Chen, F., Zhu, S.: Cross-lingual offensive speech identification with transfer learning for low-resource languages. Comput. Electr. Eng. **101**, 108005 (2022)
17. Alorainy, W., Burnap, P., Liu, H., Williams, M.L.: "The enemy among us": detecting cyber hate speech with threats-based othering language embeddings. ACM Trans. Web **13**(3), 1–26 (2019)
18. Rospocher, M.: Explicit song lyrics detection with subword-enriched word embeddings. Expert Syst. Appl. **163**, 113749 (2021)
19. Shushkevich, E., Cardiff, J.: Automatic misogyny detection in social media: a survey. Comput. Sist. **23**(4), 1159–1164 (2019)
20. Touahri, I., Mazroui, A.: Enhancement of a multi-dialectal sentiment analysis system by the detection of the implied sarcastic features. Knowl. Based Syst. **227**, 107232 (2021)
21. Khan, M.M., Shahzad, K., Malik, M.K.: Hate speech detection in Roman Urdu. ACM Trans. Asian Low-Resour. Lang. Inf. Process. **20**(1), 1–19 (2021)
22. Sangwan, S.R., Bhatia, M.P.S.: Soft computing for abuse detection using cyber-physical and social big data in cognitive smart cities. Expert Syst. **39**(5), e12766 (2022)
23. Sohn, H., Lee, H.: MC-BERT4Hate: hate speech detection using multi-channel BERT for different languages and translations. In: 2019 IEEE International Conference on Data Mining Workshops, pp. 551–559 (2019)

24. Wang, W., Wang, L., Wang, R., Ye, A., Ke, J.: Better constraints of imperceptibility, better adversarial examples in the text. Int. J. Intell. Syst. **37**(6), 3440–3459 (2022)
25. Zhang, S., Wu, H., Xu, X., Zhu, G., Hsieh, M.Y.: CL-ECPE: contrastive learning with adversarial samples for emotion-cause pair extraction. Connect. Sci. **34**(1), 1877–1894 (2022)
26. Xiao, Y., Lei, W., Lu, L., Chang, X., Zheng, X., Chen, X.: CS-GAN: cross-structure generative adversarial networks for Chinese calligraphy translation. Knowl.-Based Syst. **229**, 107334 (2021)
27. Li, X., Luo, H., Sun, Y.: Adversarial sample generation method based on Chinese features. J. Softw. **34**(11), 5143–5161 (2023). (in Chinese)
28. Tang, X., Shen, X.: Categorizing offensive language in social networks: a Chinese corpus, systems and an explainable tool. In: Proceedings of the 19th Chinese National Conference on Computational Linguistics, pp. 1045–1056 (2020)
29. Zhou, J., et al.: Towards identifying social bias in dialog systems: framework, dataset, and benchmark. In: Findings of the Association for Computational Linguistics: EMNLP 2022, pp. 3576–3591 (2022)
30. Jiang, A., Yang, X., Liu, Y., Zubiaga, A.: SWSR: a Chinese dataset and lexicon for online sexism detection. Online Soc. Netw. Media **27**, 100182 (2022)
31. Yang, H., Lin, C.J.: TOCP: a dataset for Chinese profanity processing. In: Proceedings of the Second Workshop on Trolling, Aggression and Cyberbullying, pp. 6–12 (2020)
32. Davidson, T., Warmsley, D., Macy, M., Weber, I.: Automated hate speech detection and the problem of offensive language. In: Proceedings of the 11th International AAAI Conference on Web and Social Media, pp. 512–515 (2017)
33. Aljarah, I., Habib, M., Hijazi, N.: Intelligent detection of hate speech in Arabic social network: a machine learning approach. J. Inf. Sci. **47**(4), 483–501 (2021)

STMFNET: A Spatial-Temporal Masking and Feature Fusion Model for Traffic Flow Prediction

Pengyuan Li, Yufeng Ling, Haoran Zhu, and Wei Song(✉)

School of Computer and Artificial Intelligence, Zhengzhou University, Zhengzhou, China
frq_zzu@gs.zzu.edu.cn, iewsong@zzu.edu.cn

Abstract. One of the central issues in transportation science involves accurately predicting traffic flow, drawing attention from both academics and practitioners. It serves an essential function in enhancing traffic management efficiency, alleviating congestion and reduce accidents. However, current research still faces several limitations: (i) The constraint on input sequence length hinders the ability to capture long-range and long-distance dependencies; (ii) Complex spatial-temporal dependencies are difficult to model, especially in a changing traffic environment, and existing methods have limited effect on capturing spatial-temporal interaction patterns; (iii) Most models rely only on historical data, and the periodic characteristics of traffic flow are not adequately modeled, making it difficult to fully utilize periodic patterns for prediction. To this end, this study proposes a traffic flow forecasting model (STMFNet) built upon spatial-temporal masking and feature fusion. Specifically, the model first uses spatial-temporal masking mechanisms (T-Mamba and S-Mamba) to randomly mask the data and reconstruct the missing information; then it combines the graph convolutional network (GCN) to model short-term traffic; finally, an adaptive periodic modeling mechanism is designed to combine the periodic characteristics of holiday and weekday data, and generate spatial-temporal feature representations through the hidden layer fusion masking mechanism of the model, thereby achieving accurate prediction of traffic flow. Experimental results based on four real-world datasets (PEMS03, PEMS04, PEMS07, PEMS08) show that STMFNet has significant advantages in prediction performance compared with existing baseline models.

Keywords: Spatial-Temporal Masking · Feature Fusion · GCN · Traffic Flow Prediction

1 Introduction

As urbanization progresses rapidly, modern urban transportation is facing increasingly complex challenges. Problems such as traffic congestion and frequent accidents have seriously affected the travel efficiency and quality of life of

L. Huang et al. (Eds.): CCF NCCA 2025, CCIS 2774, pp. 256–274, 2026.
https://doi.org/10.1007/978-981-95-5294-8_18

urban residents. Hence, managing and optimizing traffic flow has emerged as a central problem in today's urban traffic management. Traffic flow forecasting, as an important tool for traffic management, can predict traffic conditions in advance by analyzing historical data and real-time information, providing essential data support to improve traffic signaling, path planning, and operational resource arrangement for urban transport departments, effectively alleviating traffic congestion and reducing accidents [1].

Recent advancements in graph-based deep learning have led to the widespread application of graph neural networks and their derivatives in traffic flow forecasting, producing promising outcomes. Lan et al. [2] introduced a dynamic graph neural network with spatial-temporal awareness (DSTAGNN) aimed at predicting traffic flow. This model can effectively capture the traffic flow changes during peak hours and emergencies by dynamically modeling the spatial-temporal dependence of traffic flow. Li and Zhu [3] proposed a graph neural network called spatial-temporal fusion (STFGNN). This approach notably boosts the model's capacity to learn spatial-temporal correlations in traffic flow by deeply integrating spatial with temporal attributes, thus improving prediction accuracy. Guo et al. [4] introduced a method named ASTGNN, emphasizing the learning of dynamics and heterogeneity within spatial-temporal graph data. This method innovatively combines the attention mechanism and can adaptively weight important spatial-temporal information, enabling the model to more accurately predict complex traffic flow fluctuations. Although these models have performed well in traffic flow prediction, they still have certain limitations in extracting long-term and long-spatial features.

Figure 1 shows the traffic flow of sensor 2 in the PEMS04 dataset. Especially during peak hours, there is a significant time difference between weekdays and weekends. There is a morning peak on weekdays, while the traffic flow on weekends is more evenly distributed without a significant peak, indicating the temporal variability of the traffic flow dataset. Figure 2 shows the traffic flow of sensors 1, 87, 134, and 264. Sensors 1 and 264 show obvious peaks and troughs, while sensors 87 and 134 maintain relatively stable flows throughout the day,

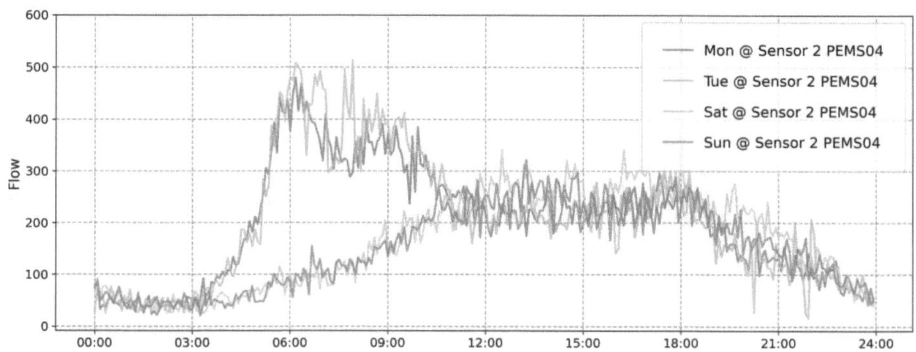

Fig. 1. Temporal difference

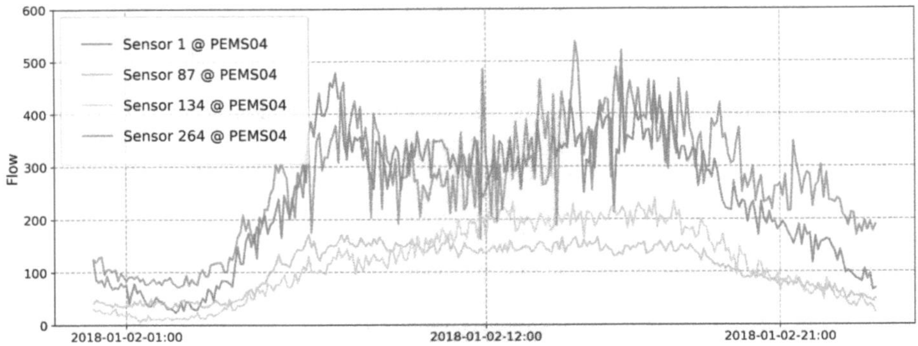

Fig. 2. Spatial difference

reflecting the spatial variability between different locations. Most existing models are trained in an end-to-end manner [5–7]. As can be seen from Figs. 1 and 2, the traffic flow performance between 1 and 3 am is relatively similar, while the traffic flow between 6 and 8 am is quite different. Due to the high complexity of these models, their input range is usually limited to a short number of steps (usually 12 steps). However, it is difficult to achieve accurate predictions by relying only on historical traffic data of 12 steps. Therefore, this paper designs a spatial-temporal masking model to extract long-term and long-space features. Figure 3 shows the changes in traffic volume in two different weeks. The traffic volume trends of the first Monday and Sunday are very similar to those of the second Monday and Sunday, indicating that there is a historical periodicity in traffic volume at the same location, but how to use this periodic information to accurately predict future traffic volume changes is still a challenging task. Figure 4 t1 and t2 represent two consecutive time steps. There are three different influences in this network. Each node in the spatial-temporal graph can directly affect its neighbor nodes at the same time step, and this influence comes from the actual

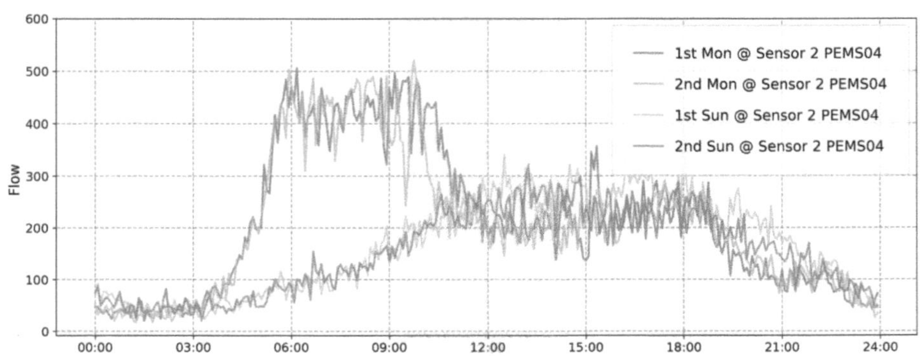

Fig. 3. Historical cyclicity

spatial dependency. Given the temporal continuity inherent in time series, nodes may influence their own state transitions. Furthermore, spatial-temporal synchrony allows interactions to extend to adjacent nodes in the forthcoming time intervals. This complex spatial-temporal interactivity is difficult to capture.

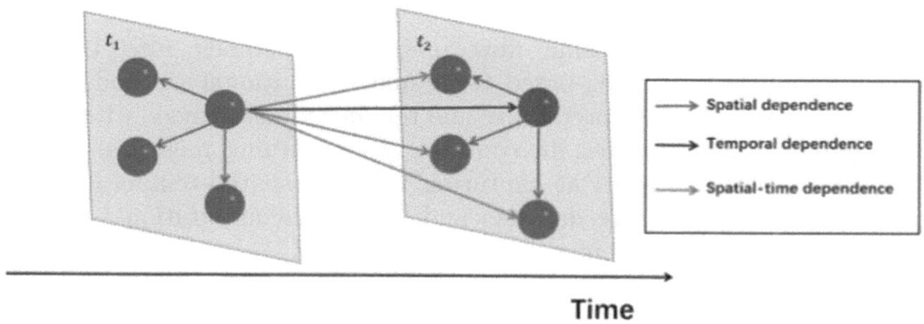

Fig. 4. Spatial-Temporal dependence

To address these challenges, this paper proposes a traffic flow prediction model (STMFNet) based on spatial-temporal masking and feature fusion. This paper designs a spatial-temporal masking module. The core idea is to mask part of the input sequence and require the model to reconstruct the missing content, thereby enhancing the model's ability to extract long-term and long-space features. Then, the time series data is processed in the frequency domain by fast Fourier transform (FFT), which enables the model to better understand the periodic changes in the time series data. Finally, a graph convolutional network is used to model the spatial dependencies between nodes within the traffic network, along with spatial-temporal dependencies are captured in STGrpBlock through convolution operations and self-attention mechanisms. Evaluation results across diverse traffic datasets confirm the effectiveness of the proposed method, which yields more precise predictions than classical models. This paper offers the following major contributions:

(1) This paper designs a spatial-temporal masking module. By masking part of the input sequence and requiring the model to recover the missing segments, the capability to capture long-range and long-distance features is strengthened, addressing the shortcomings found in prior models.
(2) Fast Fourier transform (FFT) is introduced for frequency domain processing, so that the model can better capture the periodic changes in time series data, thereby improving the prediction accuracy of periodic traffic changes.
(3) Combining graph convolutional networks with self-attention mechanisms, the spatial-temporal dependency between nodes in the traffic network is effectively modeled through STGrpBlock, thereby improving the performance of the model in complex traffic flow prediction

(4) Experiments were conducted on four real data sets, and the effectiveness and superiority of the network in this paper were verified by comparing with other baseline methods.

2 Related Work

This section categorizes traffic flow prediction methods into traditional approaches and deep learning-based techniques. Traditional methods, such as statistical models, offer interpretability and reliability but struggle with complex spatial-temporal dependencies. In contrast, deep learning methods, including graph neural networks, excel at capturing intricate spatial-temporal relationships but often require large datasets and extensive computational resources. The specific details are presented in Table 1.

2.1 Traditional Methods

ARIMA, short for Autoregressive Integrated Moving Average, represents a foundational approach in time series modeling, commonly utilized in numerous prediction tasks. The model relies on the patterns of historical data to make predictions and is verified at each stage by using different data sets to improve its prediction accuracy and applicability. Earlier research has demonstrated [8] that ARIMA models trained using time series data often yield superior prediction outcomes compared to models trained with non-time series data. When ARIMA is combined with a nonlinear GARCH model, the conditional mean and heteroskedasticity in the traffic flow series can be effectively calculated [9], thereby obtaining more accurate time-varying confidence intervals. Compared with the traditional ARIMA model, this improved ARIMA-GARCH model can provide more dynamic and time-varying confidence intervals and better cope with the uncertainty in traffic flow data. The Kalman filter is another widely used traditional method in traffic flow prediction, particularly for real-time state estimation and noise reduction. Research has indicated that incorporating real-time measurements with historical data via the Kalman filter can effectively enhance short-term traffic flow prediction accuracy [10].

2.2 Deep Learning Methods

CNNs and RNNs represent two fundamental deep learning models that are frequently employed across various applications. They have been widely recognized for their ability to extract meaningful spatial and temporal representations from traffic time series data. Specifically, CNNs focus on capturing localized spatial features, while RNNs are proficient in modeling temporal dependencies across long sequences. Therefore, the combination of the two is widely used in traffic flow prediction tasks [11]. However, this method cannot effectively model the joint dependency of space and time at the same time.

Table 1. Two main methods for traffic flow prediction

	Methods	Features	Limitations
STATISTICAL METHODS	• ARIMA • ARIMA-GARCH • Kalman filter	• These methods can effectively process and analyze time series data to a certain extent. • The statistical method model structure is simple and the calculation amount is small.	• Time series forecasting methods typically assume that the data is stationary over time. • Model accuracy improves with sufficient data. • Capturing the nonlinear dynamics and spatial-temporal dependencies of traffic flow, while accounting for external factors like holidays and weather, remains challenging.
DEEP LEARNING METHODS	• CNN and RNN • Transformer • ASTGCN • STSGCN • STFGNN • DSTAGNN • PDFormer • STAEFormer	• These approaches are capable of accurately modeling spatial-temporal correlations in traffic flow, fully leveraging spatial features and temporal patterns for forecasting purposes. • They can provide high prediction accuracy in a changing traffic environment They can provide high prediction accuracy in a changing traffic environment	• The model is relatively complex and requires high computing resources for training and inference. • It has limitations in capturing long-term trends and cyclical changes. • It relies on high-quality traffic data, and missing or noisy data may seriously affect the prediction effect.

Based on the shortcomings of CNN and RNN methods, emerging methods such as graph neural networks (GNN) have increasingly emerged as a key focus in traffic flow prediction research. Graph neural networks can naturally process graph structured data and capturethe overall spatial relationships between nodes in the traffic network. At the same time, they can also model spatial-temporal dependencies in combination with time series data. ASTGCN [12] is an early model that applies graph neural networks (GNN) to traffic flow prediction. It combines graph convolutional networks (GCN) with attention mechanisms and can adaptively weight the importance of spatial nodes and time steps, thereby more accurately capturing the spatial-temporal dependencies in traffic flow. STSGCN [13] further improved on ASTGCN by using spectral graph convolution to process spatial-temporal data in the frequency domain, thus improving the prediction accuracy of the model under long-term dependencies and complex spatial-temporal relationships. STFGNN further strengthened the fusion of spatial-temporal features and proposed a feature fusion method to more effectively integrate spatial and temporal features, thereby improving the stability and accuracy of prediction. DSTAGNN can dynamically adjust the spatial-temporal dependency according to real-time traffic data by introducing a dynamic spatial-temporal attention mechanism, allowing for improved detection of dynamic traffic variations, including rush-hour congestion and sudden disruptions. Transformer models have demonstrated notable advancements in areas like natural language processing, variants based on the Transformer have gradually been introduced into traffic flow prediction. The paper attention is all you need [14] also confirmed the feasibility of the Transformer model in traffic flow prediction. On this basis, PDFormer and STAEFormer, as variants of the Transformer, further improved the performance of the model in traffic flow prediction. PDFormer [15] can effectively model the periodic changes in traffic flow data by introducing periodic feature encoding, which is particularly suitable for capturing features such as periodic traffic peaks. STAEFormer [16] integrates a spatial-temporal self-attention mechanism, allowing the model to capture complex dependencies in traffic flow data across both spatial and temporal dimensions.

3 Methodology

Figure 5 depicts the comprehensive framework of STMFNet, comprising two principal components: the Masked Mamba and the Feature Fusion. The Masked Mamba module is responsible for capturing long-term spatial-temporal dependencies within the traffic data. The extracted features are then passed to the

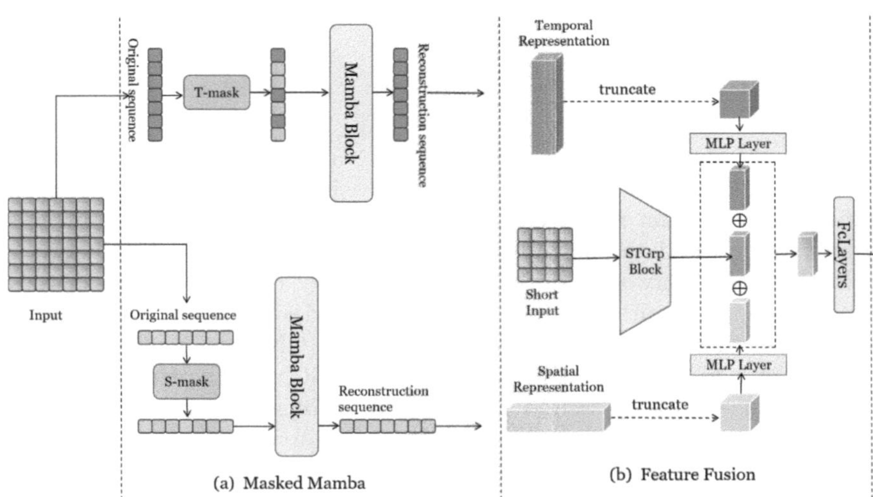

Fig. 5. The overall framework of Spatial-Temporal Masking and Feature Fusion Model

Feature Fusion module, which includes the STGrp Block for feature extraction from short-term traffic flow data. Additionally, the MLP Layer within the Feature Fusion module integrates the short-term prediction features with those provided by the Masked Mamba module, enhancing the overall prediction accuracy.

3.1 Masked Mamba

Within the scope of intelligent transportation systems, predicting traffic flow is a key challenge. The task focuses on anticipating future traffic dynamics using previously recorded data. From a mathematical perspective, it may be formulated as a spatiotemporal sequence modeling problem. Let **X** represent the observed traffic tensor; the objective is to learn a prediction function f that maps a sequence of T previous observations to estimate traffic states over the next T' time steps.

$$[X_{t-T-1}, X_{t-T}, X_t] \underset{\rightarrow}{g} [X_{t+1}, X_{t+2}, X_{t+T'}] \tag{1}$$

Let $X_i \in \mathbb{R}^{N \times C}$ be the observation matrix at the i-th time instance, where N is the total number of spatial locations and C is the dimensionality of feature channels.

In most traffic flow prediction models, the input length T of t is usually equal to 12. However, Short input sequences restrict the capacity of the model to grasp dependencies across extended time horizons, especially in complex traffic scenarios where traffic flow is subject to dynamic influences from a wider range of spatial and longer time ranges. Therefore, we intend to introduce an input sequence that involves long spatial periods and a long time to extract the features. This article predicts the next 12 steps into the future. A simple idea is to directly apply the original masked pre-training [17,18] by mixing the time and space dimensions into one. However, since the computational complexity of the mixed time and space masked pre-training method grows quadratically, it is not feasible in large-scale traffic networks, and the simple mixed masking cannot model different dependencies between time and space separately. Therefore, we propose a new feature extraction method. This method performs masked reconstruction tasks along the time and space dimensions separately. This method not only effectively reduces computational complexity but can also extract long-term and long-space features.

Given a spatial-temporal input sequence $X \in \mathbb{R}^{T \times N \times C}$, we introduce two distinct masking strategies to enhance model generalization: (1) **Spatial Masking (S-Mask)**: A subset of $N \times p$ sensors is randomly selected and masked, where $p \in (0,1)$ denotes the spatial masking ratio. This results in a spatially altered input sequence, denoted as $\tilde{X}^{(S)} \in \mathbb{R}^{T \times N(1-p) \times C}$. (2) **Temporal Masking (T-Mask)**: A fraction p of the T time steps is randomly dropped, generating a temporally masked sequence $\tilde{X}^{(T)} \in \mathbb{R}^{T(1-p) \times N \times C}$. The two masked sequences can be formulated as follows:

$$\tilde{X}^{(S)} = \sum_{n=1}^{N} B_S(1-p) \cdot X[:, n, :]$$
$$\tilde{X}^{(T)} = \sum_{t=1}^{T} B_T(1-p) \cdot X[t, :, :] \tag{2}$$

The values of the masked data elements are set to 0, requiring the model to infer both global and local spatial-temporal patterns from the unmasked data in order to reconstruct the masked data.

In this study, the Mamba model [19] is utilized for the reconstruction task. This model combines the state space mechanism (SSM) with local convolutional operations, enabling the efficient capture of both long-range and local spatial-temporal dependencies. The architecture is illustrated in Fig. 6. Given masked input data (`masked_x`), a dense layer is utilized to convert the features from the original dimension (`input_dim`) to an embedded space (`embed_dim`) through $X_{\text{embed}} = W_{\text{embed}} \times X + b_{\text{embed}}$. This operation improves feature representation, making it easier for the encoder to model temporal and spatial patterns. The processed sequence is passed through the SSM module and a one-dimensional convolution to extract local contextual information. The encoder then produces a high-dimensional representation, which is mapped back to the original space via another fully connected layer to reconstruct the masked sequence. The Mamba model is trained with the Mean Absolute Error (MAE), where the loss function is defined as:

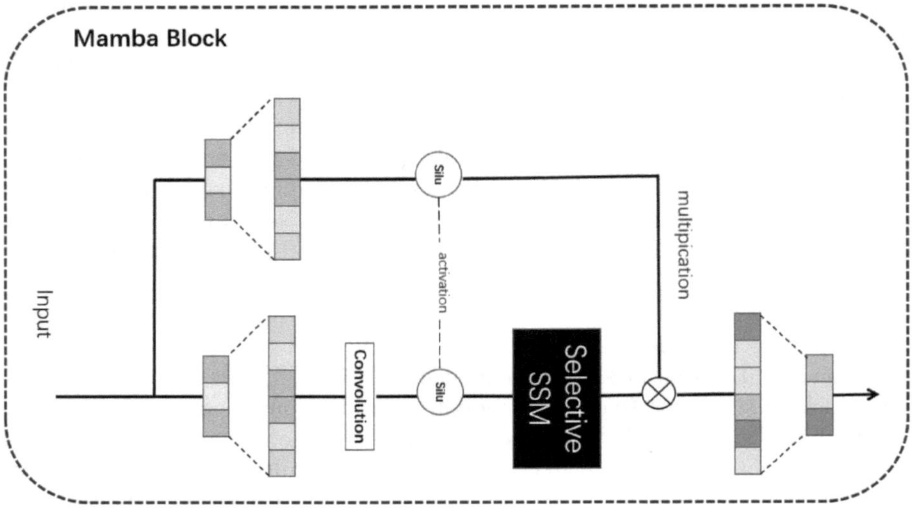

Fig. 6. mamba block

$$\mathcal{L}_S = \frac{1}{TN_sL}\sum_{t=1}^{T}\sum_{i=1}^{N_s}\sum_{j=1}^{L}\left|\tilde{Q}_{t,i,j}^{(S)} - Q_{t,i,j}^{(S)}\right|$$
$$\mathcal{L}_T = \frac{1}{T'NL}\sum_{t=1}^{T'}\sum_{i=1}^{N}\sum_{j=1}^{L}\left|\tilde{Q}_{t,i,j}^{(T)} - Q_{t,i,j}^{(T)}\right| \quad (3)$$

3.2 Feature Fusion

3.2.1 Spatial-Temporal Modeling Module for Short-Term Traffic Flow Prediction In this paper, a modular structured unit named the STGrp Block is introduced to efficiently extract essential spatial and temporal characteristics from traffic network data. This unit concurrently captures the spatial interactions among traffic nodes and the temporal dynamics of traffic flow variations. By incorporating frequency-domain graph convolution, adaptive graph-based attention strategies, and gated recurrent networks, it enables precise short-term predictions of traffic flow sequences.

Understanding the spatial organization of the transportation network serves an essential function in forecasting traffic flow. DGiven their effectiveness in handling graph-structured data, GNNs are extensively employed to model spatial dependencies among traffic nodes [20,21]. The traffic environment can be abstracted as a graph $G = (V, E, W)$, where V denotes node entities such as sensor locations, E represents the links between nodes (e.g., roads), and $W \in \mathbb{R}^{N\times N}$ stands for a weighted adjacency matrix indicating the intensity of interactions or similarities between nodes. To exploit the inherent graph structure, we define the unnormalized graph Laplacian by $L = D - W$, where D denotes the diagonal degree matrix.For spectral analysis, we further adopt the normalized Laplacian $L_{\text{norm}} = I - D^{-\frac{1}{2}}WD^{-\frac{1}{2}}$, which facilitates the extraction of global structural information. This formulation effectively encodes spatial dependencies and serves as a foundation for subsequent GNN-based operations.

Then STGrp Block uses frequency domain graph convolution to model the spatial relationship of nodes. By eigendecomposing the graph Laplacian matrix$L = U\Lambda U^T$, where U is the eigenvector matrix and Λ is the diagonal matrix of eigenvalues, node features can be mapped from the time domain to the frequency domain. FFT converts the node domain signal X into a spectrum domain representation:$\hat{X} = U^TX$. In the spectrum domain, the signal is convolved in the frequency domain in the form of: $\hat{Y} = \phi \times \hat{X}$, where ϕ is the frequency domain filter. The result in the spectrum domain is then converted back to the node domain $Y = U\hat{Y}$.In order to enhance the ability of graph convolution to express local and global dependencies, the Chebyshev polynomial approximation graph filter kernel is introduced.

$$\phi(L)X = \sum_{k=0}^{K-1}\theta_k T_k \tilde{L}X \quad (4)$$

where $T_k\tilde{L}$is the Chebyshev polynomial of the normalized graph Laplacian matrix, $\theta_k$is the parameter to be learned.This graph convolution can propagate information at multiple adjacency orders, thereby capturing multi-scale spatial dependencies. In order to further improve the flexibility and adaptability of spatial modeling, this paper introduces an adaptive graph attention mechanism based on the graph structure, and dynamically generates an adjacency matrix through the node embedding vectors q and k obtained through training:

$$A_{ij} = softmax(\frac{Q_i K_j^T}{\sqrt{d_k}}) \tag{5}$$

where Q and K are the query and key-value vectors of node features. Figure 7 shows the STGrp Block process.

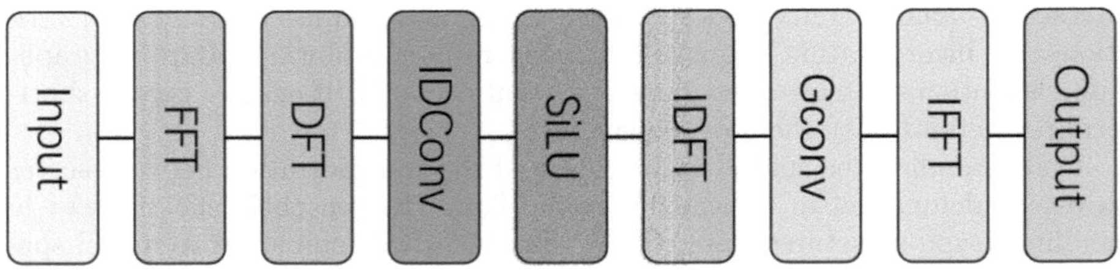

Fig. 7. STGrp Block

3.2.2 Multi-feature Fusion In STMFNet, this paper designs a novel feature fusion mechanism that can be flexibly embedded into the existing prediction framework. This mechanism significantly improves the prediction ability of the model by combining the spatial and temporal features generated by the spatio-temporal masking module with the intermediate representation of the predictor. The specific implementation is as follows:

Initially, the extended temporal sequence X_{long}, which spans multiple time steps, is fed into the pre-trained spatial-temporal masking module to produce two feature representations: the spatial feature tensor $F_S \in \mathbb{R}^{N \times L \times C}$ and the temporal feature tensor $F_T \in \mathbb{R}^{N \times L \times C}$, where N denotes the number of nodes and C is the feature dimensionality. Subsequently, a spatial-temporal prediction network ρ_ϕ, parameterized by ϕ, processes the short-range input sequence $X_{\text{short}} = \{X_{t-T+1}, X_{t-T+2}, \ldots, X_t\}$ and yields its latent representation $H_P \in \mathbb{R}^{N \times C'}$, as described below:

$$H_P = \rho_\phi(X_{short}) \tag{6}$$

where C' is the dimension of the hidden representation of the predictor. In order to align the spatial features $F_s$and the temporal features F_T with H_P, we truncate and reshape $F_s$and F_T to obtain F_s' and F_T', where $T'$is the number of time steps of the short-term input. Then, F_s and F_T are projected to the C'dimension through two multi-layer perceptrons (MLPs) to obtain H_S and

H_T :$H_S = MLP(F'_S)$,$H_T = MLP(F'_T)$. Finally, we add the hidden representation H_P of the predictor to the projected spatial features H_S and temporal features H_T to obtain the enhanced representation $H_{enhanced} = H_p + H_S + H_T$. Therefore, H_S combines the short-term spatial-temporal traits captured by the predictor with the long-term spatial and temporal features generated by the spatial-temporal masking module, substantially enhancing the model's prediction accuracy.Figure 8 shows the specific process of MLPlayer.

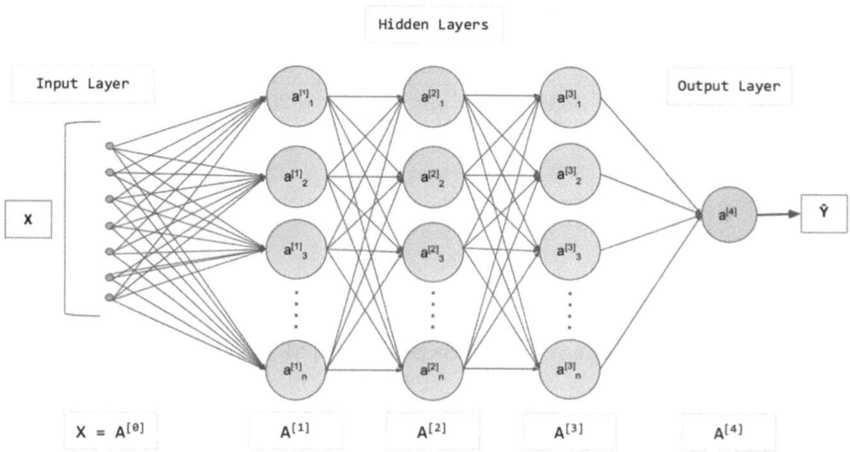

Fig. 8. MLPlayer

4 Experiments

4.1 Data Source and Processing

We utilized four well-known real-world traffic flow datasets: PEMS03, PEMS04, PEMS07, and PEMS08. These datasets are frequently used in research areas such as spatiotemporal forecasting, graph-based learning models, and deep neural networks. Collected by the California Department of Transportation (Caltrans) through the Performance Measurement System (PeMS), they cover a large number of traffic monitoring stations and provide key traffic attributes, including vehicle count, average speed, and occupancy rates. Among them, PEMS03 covers important highways in District 3, PEMS04 covers District 4 with a wider monitoring range, PEMS07 focuses on the traffic congestion area in District 7, and PEMS08 monitors the eastern highway section in District 8. These datasets are mainly used for traffic flow prediction and congestion analysis. To ensure the consistency of input data and improve the prediction effect of the model, this study performed a number of preprocessing on the above datasets, including linear interpolation to fill in small-scale missing data, normalizing all eigenvalues to the [0,1] interval, splitting data according to fixed time windows (such as predicting the next 3 time steps every 12 time steps), building a adjacency matrix weighted according to the sensor's geographic location, and dividing the training set, validation set, and test set in chronological order (ratios of 70%, 15%, and 15%). Through the processing and analysis of these data, a high-quality data foundation is provided for the proposed model, ensuring the reliability and reproducibility of the experimental results (Table 2).

Table 2. Overview of Four Real-World Traffic Datasets

Dataset	#Sensors	#Time Steps	#Time Interval
· PEMS03	358	5(min)	26208
· PEMS04	307	5(min)	16992
· PEMS07	883	5(min)	28224
· PEMS08	170	5(min)	17856

4.2 Evaluation Indicators and Experimental Environment

(1) Evaluation indicators

In view of the differences in different literature methods, this study takes into account the irregularity, non-stationarity and volatility of the data set, as well as the existence of maximum and minimum values. In order to comprehensively and scientifically evaluate the effect of the proposed network, this paper selects RMSE, MAE, and MAPE as evaluation indicators for testing the network.

RMSE is employed as shown in formula (7) to quantify the deviation between predicted and actual values for accuracy assessment.

$$\mathbf{RMSE} = \sqrt{\frac{1}{n}\sum_{j=1}^{n}(\hat{y}_j - y_j)^2} \tag{7}$$

MAE is expressed as formula (8) to express the average value of absolute error for stability evaluation.

$$\mathbf{MAE} = \frac{1}{n}\sum_{j=1}^{n}|\hat{y}_j - y_j| \tag{8}$$

MAPE is applied as in formula (9) to evaluate the relative error between predicted and true values.

$$\mathbf{MAPE} = \frac{100\%}{n}\sum_{j=1}^{n}\left|\frac{y_j - \hat{y}_j}{y_j}\right| \tag{9}$$

(2) Experimental environment

The experimental setup is built on a Intel 13th Generation Core i5-13500 processor running at 2.80 GHz processor, and the implementation is carried out using Python with the PyTorch deep learning framework.

4.3 Experimental Results

To thoroughly assess the performance of the proposed model, we choose a variety of classic time series prediction models and the latest deep learning models as comparison benchmarks, including:

ARIMA [22], SVR [13], LSTM [13], TCN [23], DSANet [24], DCRNN [25], ASTGCN, STSGCN, DSTHGCN [26], STFGNN [27], STGODE [28], Z-GCNETs [29], DSTAGNN.

Table 3 presents the experimental results of STMFNet on four real-world datasets and compares them with baseline models. As shown in the figure, STMFNet significantly reduces prediction errors compared to traditional time series forecasting methods. This improvement is mainly due to traditional methods' inability to fully capture spatial dependencies, whereas STMFNet effectively models these relationships. Furthermore, compared to deep learning-based forecasting methods, STMFNet also delivers superior performance. When rounded to two decimal places, STMFNet outperforms the second-best model by an average of 3.66%, 6.18/%, and 4.40% in MAE, RMSE, and MAPE, respectively, further demonstrating its effectiveness in spatial-temporal sequence forecasting tasks (Fig. 9).

Table 3. Comparison of performance metrics between baseline methods on datasets PEMS03, PEMS04, PEMS07, and PEMS08

Model	PEMS03			PEMS04			PEMS07			PEMS08		
	MAE	RMSE	MAPE(%)	MAE	RMSE	MAPE(%)	MAE	RMSE	MAPE(%)	MAE	RMSE	MAPE (%)
ARIMA	37.50	49.25	36.80	35.60	51.20	25.00	38.45	61.00	20.50	32.30	45.60	23.90
SVR	23.10	36.50	22.70	29.85	45.80	20.35	33.60	51.50	15.40	24.50	37.40	15.80
LSTM	22.50	36.30	24.50	28.30	42.75	19.35	31.10	47.00	14.35	23.45	35.20	15.35
TCN	20.45	34.40	21.00	32.25	38.50	16.60	33.80	43.40	15.35	23.85	36.95	15.20
DSANet	21.29	34.55	23.21	22.79	35.77	16.03	31.36	49.11	14.43	17.14	26.96	11.32
DCRNN	18.18	30.31	18.91	24.70	38.12	17.12	25.30	38.58	11.66	17.86	27.83	11.45
ASTGCN	17.69	29.66	19.40	22.93	35.22	16.56	28.05	42.57	13.92	18.61	28.16	13.08
STSGCN	17.49	29.20	16.78	21.19	33.65	13.90	24.28	39.03	10.21	17.12	26.83	10.90
DSTHGCN	18.20	28.70	17.30	22.30	34.70	15.40	24.35	38.05	11.10	17.65	27.10	11.35
STFGNN	17.90	29.50	17.45	21.65	33.65	17.90	24.60	37.75	10.35	18.10	27.40	11.75
STGODE	17.65	29.00	17.85	22.00	33.95	14.90	23.75	38.70	11.30	17.95	27.10	11.80
Z-GCNETs	17.80	29.30	17.55	20.65	32.75	13.90	22.90	36.30	10.40	16.90	26.25	11.15
DSTAGNN	16.70	28.35	15.80	20.45	32.60	13.85	22.55	35.65	10.15	16.80	25.90	11.05
STMFNet	**14.20**	**23.55**	**14.27**	**18.89**	**29.50**	**11.45**	**21.22**	**33.82**	**8.76**	**13.89**	**24.02**	**9.23**

4.4 Ablation Experiment

In order to evaluate the effectiveness of each part in STMFNet, we conducted ablation studies using the following four model variants:

w/o simultaneous removal of the time dimension and spatial dimension reconstruction modules FNet

w/o retaining the spatial dimension reconstruction module, removing the time dimension reconstruction module SMGNet

w/o retaining the time dimension reconstruction module, removing the spatial dimension reconstruction module TMGNet

w/o retaining the time dimension and spatial dimension reconstruction modules, removing the graph convolution module STMNet

STMFNet The network proposed in this paper has a time dimension reconstruction module, a spatial dimension reconstruction module, and a graph convolution module.

In this ablation experiment, we analyzed each module of STMFNet one by one to evaluate the contribution of each part to the overall performance. The experimental results demonstrate that STMFNet surpasses all other ablation

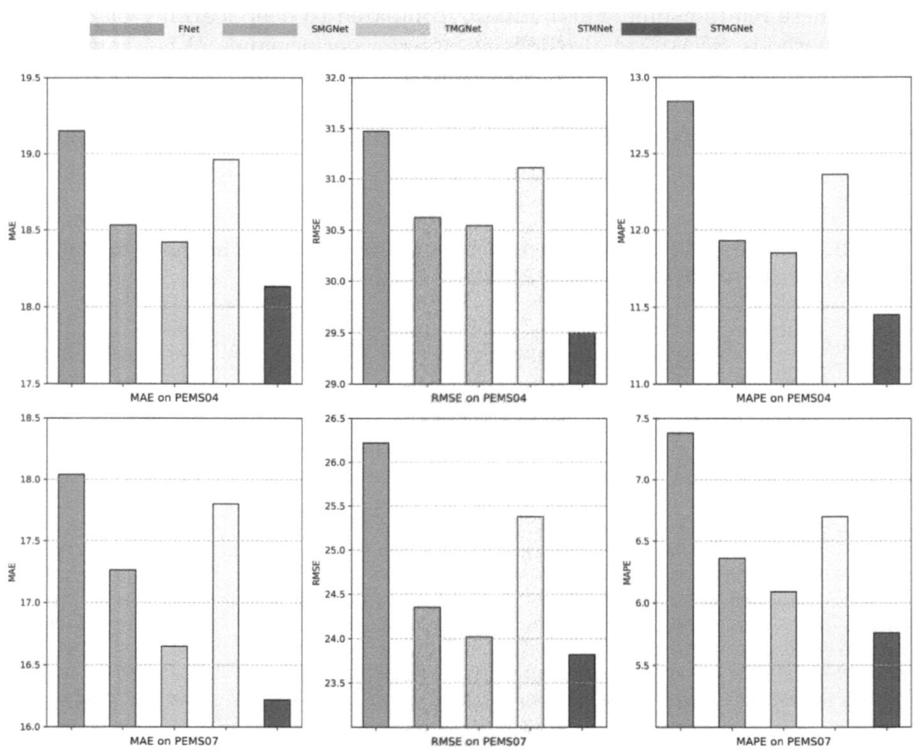

Fig. 9. Ablation experiment results

Table 4. Comparison of Various Models Across Multiple Datasets

Model	PEMS03			PEMS04			PEMS07			PEMS08		
	MAE	RMSE	MAPE	MAE	RMSE	MAPE	MAE	RMSE	MAPE	MAE	RMSE	MAPE
w/o FNet	15.83	27.62	14.78	19.95	31.47	12.84	24.04	36.22	10.38	15.18	26.03	10.52
w/o SMGNet	14.79	24.78	14.54	19.33	30.62	11.93	22.96	34.35	9.36	14.65	24.87	9.70
w/o TMGNet	14.62	24.63	14.47	19.26	30.54	11.85	22.75	34.02	9.09	14.41	24.65	9.53
w/o STMNet	15.01	25.11	14.72	19.06	31.11	12.36	23.11	35.38	9.70	14.29	25.63	10.01
STMFNet	**14.20**	**23.55**	**14.27**	**18.89**	**29.50**	**11.45**	**21.22**	**33.82**	**8.76**	**13.89**	**24.02**	**9.23**

models across every dataset, validating the effectiveness of its various modules in enhancing model performance. Detailed results are presented in Table 4.

4.5 The Impact of Different Masking Rates

In traffic flow forecasting, the choice of masking rate significantly influences the model's capacity to reconstruct missing information and to capture spatial-temporal patterns. To investigate how varying masking rates affect model effectiveness, this study conducted a comparative evaluation using four representative masking rates (10%, 30%, 50%, and 70%). The outcomes are illustrated in the figure below (Fig. 10).

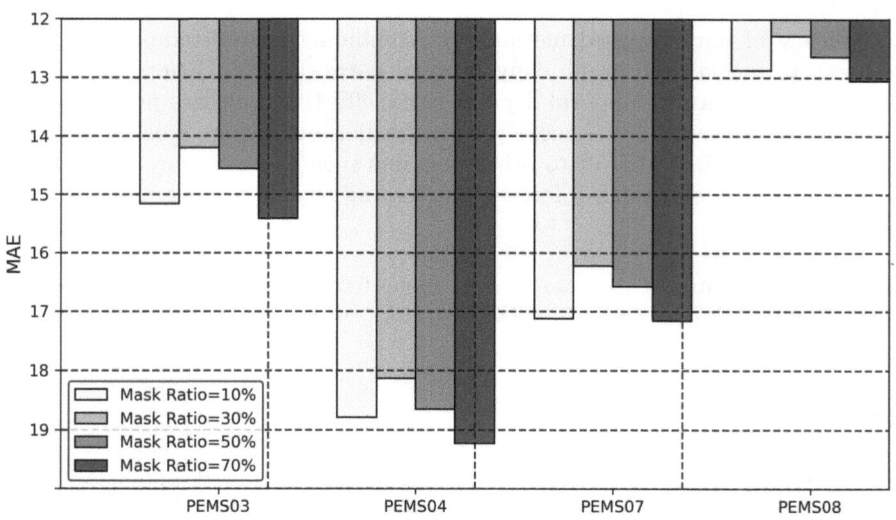

Fig. 10. The impact of different masking rates

As can be seen from the figure, the model works best when the masking rate is 30%. When the masking rate is too low, the model receives more observation data and less feature information of the missing part, so the model's learning ability is not fully tapped. When the masking rate is too high, the missing information accounts for too large a proportion, and the model's learning task becomes too difficult. The model is difficult to accurately capture the complete spatial-temporal features, and the performance is reduced. When the masking rate is 30%, sufficient observation information is retained in the data, and an appropriate amount of missing features are added, which enhances the model's ability to extract long-term and long-spatial features.

5 Conclusion

We introduce STMFNet, a traffic flow prediction model utilizing spatial-temporal masking and feature fusion to address the shortcomings of existing approaches in capturing long-term and broad spatial dependencies. We introduce two spatial-temporal masking mechanisms, T-Mamba and S-Mamba, which enhance the model's capacity to identify long-term and wide spatial patterns by randomly masking parts of the data and demanding reconstruction of missing information. Additionally, our approach incorporates graph convolutional networks (GCN) and adaptive periodic modeling mechanisms to integrate the spatial-temporal graph structure with the periodic characteristics of traffic flow, thereby markedly enhancing forecasting accuracy.Our experiments reveal that STMFNet surpasses existing baseline models in performance across several real-world traffic datasets (PEMS03, PEMS04, PEMS07, PEMS08), substantiating the efficacy of our proposed method. By combining spatial-temporal masking and feature fusion with graph convolutional networks, STMFNet adeptly captures intricate spatial-temporal dependencies, effectively utilizes periodic traffic flow variations, and delivers more precise predictions. Future work will extend the application of STMFNet to other domains such as power forecasting, wind speed prediction, and rainfall forecasting, aiming to enhance the model's versatility.

Acknowledgements. This research was supported by grants from: *National Key R&D Program of China (No. 2023YFC2206400).*

References

1. Almukhalfi, H., Noor, A., Noor, T.H.: Traffic management approaches using machine learning and deep learning techniques: a survey. Eng. Appl. Artif. Intell. **133**, 108147 (2024)
2. Lan, S., Ma, Y., Huang, W., et al.: DSTAGNN: Dynamic Spatial-temporal Aware Graph Neural Network for traffic flow forecasting. In: International Conference on Machine Learning, pp. 11906–11917 (2022)
3. Li, M., Zhu, Z.: Spatial-temporal fusion graph neural networks for traffic flow forecasting. In: Proceedings of the AAAI Conference on Artificial Intelligence, vol. 35, no. 5, pp. 4189–4196 (2021)

4. Guo, S., Lin, Y., Wan, H., et al.: Attention based spatial-temporal graph convolutional networks for traffic flow forecasting. In: Proceedings of the AAAI Conference on Artificial Intelligence, vol. 33, no. 01, pp. 922–929 (2019)
5. Yu, X., Bao, Y.X., Shi, Q.: STHSGCN: Spatial-Temporal Heterogeneous and Synchronous Graph Convolution Network for traffic flow prediction. Heliyon **9**(9) (2023)
6. Li, H., Liu, J., Han, S., et al.: STFGCN: Spatial–Temporal Fusion Graph Convolutional Network for traffic prediction. Expert Syst. Appl. **255**, 124648 (2024)
7. Dai, B.A., Lyu, N., Miao, Y.: FasterSTS: a faster spatio-temporal synchronous graph convolutional networks for traffic flow forecasting. arXiv preprint arXiv:2501.00756 (2025)
8. Dong, H., Jia, L., Sun, X., et al.: Road traffic flow prediction with a time-oriented ARIMA model. In: 2009 Fifth International Joint Conference on INC, IMS and IDC, pp. 1649–1652 (2009)
9. Chen, C., Hu, J., Meng, Q., et al.: Short-time traffic flow prediction with ARIMA-GARCH model. In: 2011 IEEE Intelligent Vehicles Symposium (IV), pp. 607–612 (2011)
10. Li, J.-Y., Zhu, Z.Y., Liu, L.: Multi-step Kalman filtering travel time estimation method based on simplified road network model. Syst. Eng. Theory Pract. **33**(5), 1289–1297 (2013)
11. Chen, M., Yu, X., Liu, Y.: PCNN: deep convolutional networks for short-term traffic congestion prediction. IEEE Trans. Intell. Transp. Syst. **19**(11), 3550–3559 (2018)
12. Guo, S., Lin, Y., Feng, N., et al.: Attention based spatial-temporal graph convolutional networks for traffic flow forecasting. In: Proceedings of the AAAI Conference on Artificial Intelligence, vol. 33, no. 01, pp. 922–929 (2019)
13. Song, C., Lin, Y., Guo, S., et al.: Spatial-temporal synchronous graph convolutional networks: a new framework for spatial-temporal network data forecasting. In: Proceedings of the AAAI Conference on Artificial Intelligence, vol. 34, no. 01, pp. 914–921 (2020)
14. Vaswani, A., Shazeer, N., Parmar, N., et al.: Attention is all you need. In: Advances in Neural Information Processing Systems (2017)
15. Jiang, J., Han, C., Zhao, W.X., et al.: PDFormer: propagation delay-aware dynamic long-range transformer for traffic flow prediction. In: Proceedings of the AAAI Conference on Artificial Intelligence, vol. 37, no. 4, pp. 4365–4373 (2023)
16. Liu, H., Dong, Z., Jiang, R., et al.: STAEformer: spatio-temporal adaptive embedding makes vanilla transformer SOTA for traffic forecasting. In: Proceedings of the 32nd ACM International Conference on Information and Knowledge Management (CIKM), pp. 21–25 (2023)
17. Devlin, J., Chang, M.W., Lee, K., et al.: BERT: pre-training of deep bidirectional transformers for language understanding. In: Proceedings of the 2019 Conference of the North American Chapter of the Association for Computational Linguistics: Human Language Technologies, Volume 1 (Long and Short Papers), pp. 4171–4186. Association for Computational Linguistics (2019)
18. He, K., Chen, X., Xie, S., et al.: Masked autoencoders are scalable vision learners. In: Proceedings of the IEEE/CVF Conference on Computer Vision and Pattern Recognition, pp. 16000–16009 (2022)
19. Gu, A., Dao, T.: Mamba: linear-time sequence modeling with selective state spaces. arXiv preprint arXiv:2312.00752 (2023)
20. Jiang, W., Luo, J.: Graph neural network for traffic forecasting: a survey. Expert Syst. Appl. **207**, 117921 (2022)

21. Dai, B.A., Ye, B.L., Li, L.: A novel hybrid time-varying graph neural network for traffic flow forecasting. arXiv preprint arXiv:2401.10155 (2024)
22. Fang, Z., Long, Q., Song, G., et al.: Spatial-temporal graph ODE networks for traffic flow forecasting. In: Proceedings of the 27th ACM SIGKDD Conference on Knowledge Discovery & Data Mining, pp. 364–373 (2021)
23. Lan, S., Ma, Y., Huang, W., et al.: DSTAGNN: dynamic spatial-temporal aware graph neural network for traffic flow forecasting. In: International Conference on Machine Learning, pp. 11906–11917 (2022)
24. Huang, S., Wang, D., Wu, X., et al.: DSANet: Dual Self-Attention Network for multivariate time series forecasting. In: Proceedings of the 28th ACM International Conference on Information and Knowledge Management (CIKM), pp. 2129–2132. ACM (2019)
25. Li, Y., Yu, R., Shahabi, C., et al.: Diffusion convolutional recurrent neural network: data-driven traffic forecasting. arXiv preprint arXiv:1707.01926 (2017)
26. Wang, J., Zhang, Y., Wei, Y., et al.: Metro passenger flow prediction via dynamic hypergraph convolution networks. IEEE Trans. Intell. Transp. Syst. **22**(12), 7891–7903 (2021)
27. Li, M., Zhu, Z.: Spatial-temporal fusion graph neural networks for traffic flow forecasting. In: Proceedings of the AAAI Conference on Artificial Intelligence, vol. 35, no. 5, pp. 4189–4196. AAAI Press (2021)
28. Fang, Z., Long, Q., Song, G., et al.: Spatial-temporal graph ODE networks for traffic flow forecasting. In: Proceedings of the 27th ACM SIGKDD Conference on Knowledge Discovery Data Mining, pp. 364–373. ACM (2021)
29. Chen, Y., Segovia, I., Gel, Y.R.: Z-GCNETs: time zigzags at graph convolutional networks for time series forecasting. In: International Conference on Machine Learning. PMLR, pp. 1684–1694 (2021)

Dual-Grained Alignment with Dynamic Clip Modeling for Partially Relevant Video Retrieval

Dan Jiang, Bin Jiang(✉), Chao Yang, and Qun Zhang

Hunan University, Changsha, China
jiangbin@hnu.edu.cn

Abstract. Partially Relevant Video Retrieval (PRVR) aims to retrieve the most relevant video from untrimmed videos based on a given query. Videos in PRVR are typically analyzed at two granularities: frame-level and clip-level. Previous methods employ static approaches for clip feature modeling. However, these approaches may not be ideal as they introduce irrelevant clip information. We propose a Dual-Grained Alignment framework with Dynamic Clip Modeling (DGADCM) for Partially Relevant Video Retrieval. Specifically, DGADCM is composed of three branches: Clip-Align, Frame-Align, and Frame-Align+. In the Clip-Align branch, we dynamically model clip features using Gaussian masks and learnable weights and perform clip-level cross-modal alignment between clip features and query features. The dynamic clip modeling method can generate compact and context-rich clip embeddings while reducing the negative impact of irrelevant clip information. In the Frame-Align branch, we directly perform frame-level cross-modal alignment between frame features and query features. In the Frame-Align+ branch, we incorporate knowledge distillation to transfer cross-modal alignment capabilities from the CLIP model. Additionally, we introduce a query diversification loss that encourages queries originating from the same video to spread out within the embedding space. Our motivation is to achieve better cross-modal alignment by preventing queries from being overly close in the embedding space and aligning with the same video frame or clip. We conducted extensive experiments and ablation studies on two publicly available datasets, and the results demonstrate that our proposed method achieves state-of-the-art performance, validating its effectiveness.

Keywords: Text-Video Retrieval · Partially Relevant Video Retrieval · Multi-modal Alignment

1 Introduction

The rapid expansion of online video content has brought increasing research and industrial interest [13,21,24,28,29] in Text-to-Video Retrieval (T2VR), which focuses on locating the video segment that best aligns semantically with a given textual query. Conventional T2VR approaches often operate under the

L. Huang et al. (Eds.): CCF NCCA 2025, CCIS 2774, pp. 275–291, 2026.
https://doi.org/10.1007/978-981-95-5294-8_19

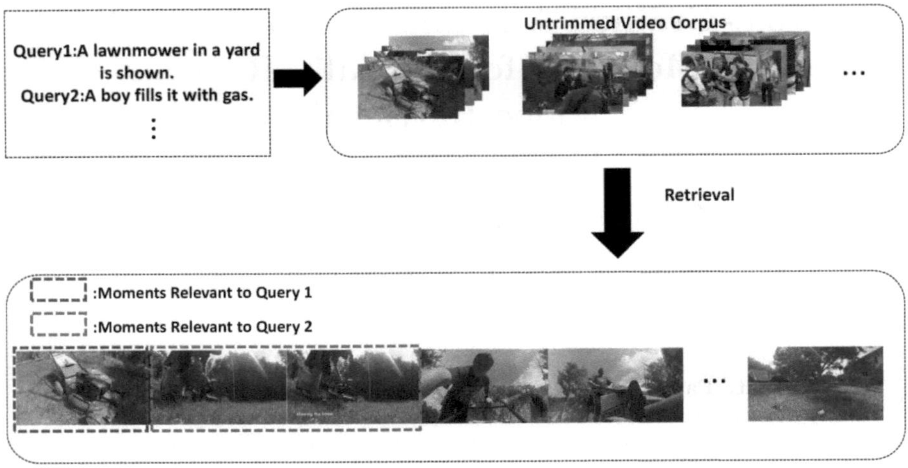

Fig. 1. The description of the PRVR task. Different from the T2VR task, the PRVR task allows queries to be partially related to the videos.

assumption that all candidate videos are pre-trimmed and fully aligned with the corresponding queries. However, this constraint diverges from real-world retrieval settings, where only portions of a video may be relevant to the query, leading to a mismatch between research assumptions and practical applications (Fig. 1). To better address practical needs, Dong et al. [3] introduced the PRVR task and designed the MS-SL model for this task. PRVR aims to retrieve the most relevant video from an untrimmed video corpus based on a given query, making it closer to real-world video retrieval scenarios. Existing methods typically extract multi-granularity video features, including frame and clip features, for cross-modal alignment with queries. MS-SL [3] explicitly models clip features using sliding windows. However, this method incurs significant storage overhead and hampers retrieval efficiency. GMMFormer [26] uses Gaussian masks to implicitly model clip features, generating compact feature embeddings that contain multi-scale context. Nevertheless, this method employs static average pooling to generate the final feature embeddings, which limits clip-level cross-modal alignment. DLDKD [6] uses knowledge distillation to purify knowledge from the CLIP [19]. But it is only limited to the frame-level alignment. MGAKD [34], absorbs knowledge from the CLIP and performs cross-modal alignment across three levels of granularity—frame-level, clip-level, and video-level. However, this method overlooks the phenomenon where, in the absence of guidance, queries tend to cluster too closely in the embedding space, which is detrimental to cross-modal alignment.

Despite the success of the previous method based on CLIP knowledge distillation and multi-granularity alignment for the PRVR task, there are still two challenges:1) Static clip modeling methods will introduce irrelevant clip information, which negatively impacts clip-level alignment; 2) Queries related to the same video may be overly pulled together during the cross-modal alignment process, leading to alignment with the same video frame or clip. This unreasonable cross-modal alignment is unrealistic and detrimental to retrieval performance. To address these challenges and improve the precision of retrieval by PRVR, we introduce a **D**ual-**G**rained **A**lignment framework with **D**ynamic **C**lip **M**odeling (DGADCM). The model is composed of three branches: Clip-Align, Frame-Align, and Frame-Align+. In the Clip-Align branch, we dynamically model clip features using Gaussian masks and learnable weights, then perform cross-modal alignment between clip features and query features. Specifically, we use multiple Gaussian masks with different variances to model features of video clips of varying lengths. Then, we dynamically aggregate these features using learnable weights to obtain compact clip features. The clip features capture the beginning, climax and end of events, containing rich contextual information. Compared to static clip modeling, dynamic clip modeling allows each clip feature to gather more information relevant to its corresponding event. In the Frame-Align branch, we directly perform frame-level cross-modal alignment between frame features and query features. In the Frame-Align+ branch, we follow DLDKD [4] to use knowledge distillation to enable the branch to absorb frame-level cross-modal alignment knowledge from the CLIP model. Moreover, we introduce a query diversification loss that encourages queries originating from the same video to spread out within the embedding space. Our motivation is to achieve better cross-modal alignment by preventing queries from being overly close in the embedding space and aligning with the same video frame or clip. Considering the potential semantic similarity between queries, we introduce a decay factor to reduce the penalty from the query diversification loss gradually. In the early stage of training, the query diversification loss pushes queries away from each other, making each query more likely to align with the correct frame or clip. As training progresses, the weight of this loss gradually decreases, allowing the model to refine the distances between queries reasonably.

Our contributions are summarized as shown below.

- We introduce a Dual-grained Alignment framework with Dynamic Clip Modeling. In the model's Clip-Align branch, we employ Gaussian masks and learnable weights to dynamically model clip features, aiming to improve clip-level cross-modal alignment. In the model's Frame-Align branch, we directly perform frame-level cross-modal alignment between frame features and query features. In the Frame-Align+ branch, we further enhance this process by employing knowledge distillation to absorb cross-modal alignment knowledge from the CLIP model.
- We propose a method for dynamically modeling clip features and a query diversification loss that gradually decays during the training process to improve cross-modal alignment.
- We have conducted extensive experiments and ablation studies on two publicly available datasets and achieved state-of-the-art performance, demonstrating the effectiveness of our proposed method.

2 Related Work

Text-to-video retrieval (T2VR) aims to retrieve a video from a pre-trimmed video corpus that is the most semantically relevant to a given query. The primary approach in T2VR involves mapping the features of the queries and the videos into a joint embedding space by text encoders and video encoders, respectively, and then computing the cross-modal similarity between them to perform retrieval. Existing methods can be categorized into two types: global matching [18,22,25,31,32] and fine-grained matching [2,27,30]. Global matching transforms the features of the queries and the videos extracted by the encoders into their respective global representations, typically through methods such as average pooling. Then, they calculates the similarity between these global representations. Fine-grained matching, on the other hand, focuses on token-level alignment between the query and video modalities. Additionally, some methods leverage large language models to enhance queries [28,29]. Besides, Wang et al. [24] models queries as text masses to mitigate the semantic richness gap between videos and queries. It is important to note that videos in T2VR datasets are pre-trimmed to ensure full relevance to the queries. However, the videos used in PRVR are often longer and partially relevant to the queries, so directly applying T2VR models to the PRVR task typically fails to achieve optimal performance.

Partially Relevant Video Retrieval (PRVR), introduced by Dong et al. [3], aims to retrieve the video that is the most semantically related to a given query from an untrimmed video corpus. Unlike T2VR, which assumes that a video is entirely relevant to its queries, PRVR more accurately reflects real-world video retrieval scenarios. The first work to define the PRVR task is MS-SL [3], which models video clips using fixed-size sliding windows and average pooling. However, this explicit clip feature modeling approach is redundant and incurs high storage overhead, which is detrimental to the model's retrieval performance. DLDKD [6] employs a dual-branch model that simultaneously explores cross-modal alignment while learning from CLIP through knowledge distillation. However, this method is limited to frame-level cross-modal alignment. GMMFormer [26] employs Gaussian masks for implicit clip modeling to improve efficiency. But this method statically aggregates video clips of different lengths, causing each clip to incorporate information irrelevant to the event it corresponds to. MGAKD [34] aligns queries with videos across three levels of granularity—frame-level, clip-level, and video-level—resulting in outstanding performance. However, this method overlooks the unreasonable phenomenon where multiple queries align with the same frame or clip, which harms the model's retrieval performance. Our proposed method dynamically models clip features and performs cross-modal alignment at two granularities, and incorporates knowledge from the large CLIP model. This unique method significantly improves the model's retrieval performance, making it more practical.

3 Method

As shown in Fig. 2, this section introduces the DGADCM alignment framework, comprising its problem formulation, feature extraction, CLIP-Align branch, Frame-Align branch, Frame-Align+ branch, and the query diversification loss.

3.1 Problem Formulation

Each query $W = \{w_i\}_{i=0}^{l_w-1}$ consists of l_w words, where we denote w_i as i-th word. A video cropus $\mathcal{V} = \{V^0, V^1, ..., V^{M-1}\}$ contains M videos. Taking the k-th video V^k as an example, we denote the i-th frame in the video V^k as v_i^k, and denote the number of frames in the V^k as l_v^k. Given a query, the goal of PRVR is to retrieve the video V^* that is the most semantically relevant to the given query from the large untrimmed video corpus $\mathcal{V}$.

3.2 Feature Extraction

For a given video $V = \{v_i\}_{i=0}^{l_v-1}$ containing l_v frames, a CNN [1] is used to extract feature representations $F \in \mathbb{R}^{l_v \times d}$ from the video frames. The extracted frame feature representations are provided as input to the three branches.

For a given query containing l_q words, we employ RoBERTa [15] to extract word features from the query and then project into a low-dimensional space using a fully connected layer. Subsequently, learnable positional embeddings PE are added to the projected word features, and then we feed the features into Transformer [23] encoders to explore the contextual features of the query to get the word-level feature $Q = \{q_i\}_{i=0}^{l_q-1}$. Then we use an additional attention to get the final sentence-level representation $Q_b \in \mathbb{R}^d$:

$$Q_b = \sum_{i=0}^{l_q-1} a_i^b \times q_i, a^b = softmax(w_b Q^T), \tag{1}$$

where $b \in \{f_+, cf\}$, Q_{f_+} represents the query representation used for the Frame-Align+ branch and Q_{cf} for Clip-Align branch and Frame-Align branch, w_b is a learnable parameter, and a^b is the attention vector.

3.3 Clip-Align Branch

In the Clip-Align branch, we employ a dynamic method to model clip features and perform cross-modal alignment between clip features and query features. Dynamic clip modeling consists of two steps: implicit modeling of video clips and dynamic aggregation.

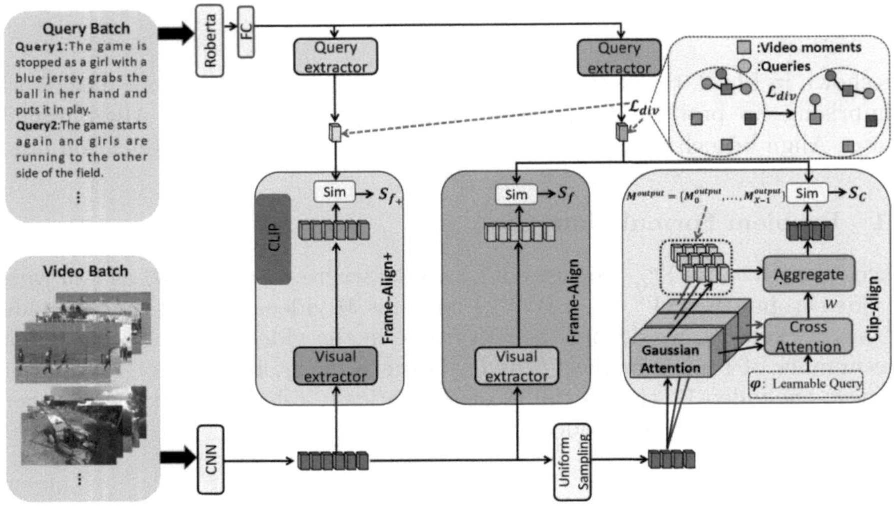

Fig. 2. Overview of the proposed DGADCM framework. $\mathcal{L}_{div}$ refers to the query diversification loss function, and w indicates the learnable aggregation weight. S_C, S_f, and S_{f+} represent the video-query similarity scores of the Clip-Align branch, Frame-Align branch, and Frame-Align+ branch, respectively.

Implicit Modeling of Video Clips. Implicit modeling of video clips is implemented using Gaussian masks and the multi-head attention mechanism [23]. Specifically, we down-sample the frame features F and then apply average pooling to obtain $F^p \in \mathbb{R}^{l_c \times d}$. Then, we feed F^p into a vanilla Transformer encoder to obtain $F^c \in \mathbb{R}^{l_c \times d}$. Next, we input F^c to X Gaussian attention heads with different variances. For the x-th head, we first project F^c to Q, K and V using three learnable parameters W_q, W_k and W_v. Subsequently, we perform matrix multiplication on Q and K to obtain the attention matrix. Next, we perform element-wise multiplication between the attention matrix and the Gaussian matrix $W^g \in \mathbb{R}^{l_c \times l_c}$ composed of l_c Gaussian masks is obtained through the following formulation:

$$M_x^{attn} = softmax(W^g \odot \frac{QK^T}{\sqrt{d}})V, \tag{2}$$

$$Q = W_q F^c, K = W_k F^c, V = W_v F^c, \tag{3}$$

$$W^g(i, j) = exp(-\frac{(j-i)^2}{\sigma_x^2}), \tag{4}$$

where M_x^{attn} denotes the result produced by the x-th Gaussian attention head, d is the dimensionality of the joint embedding space, and σ_x^2 corresponds to the variance parameter associated with the x-th attention head. The final output of this head is computed as follows:

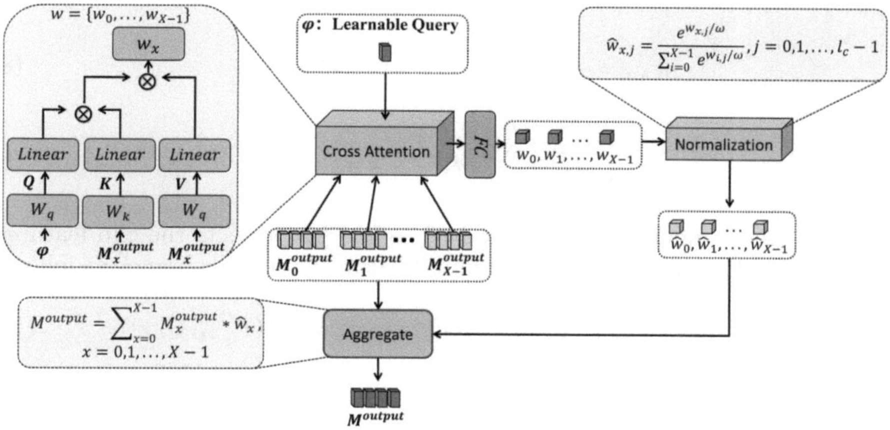

Fig. 3. A detailed flowchart for aggregating features of video clips to obtain clip features. M_x^{output} represents the output of the x-th Gaussian attention head, w_x is the weight of the x-th Gaussian attention head, $\hat{w}_x$ is the normalized weight and M^{output} represents the final output.

$$M_x^{output} = FFN(LayerNorm(M_x^{inter})) + M_x^{inter}, \tag{5}$$

$$M_x^{inter} = M_x^{attn} + F^c \tag{6}$$

where M_x^{inter} represents the intermediate feature after Gaussian attention and M_x^{output} represents the final output of the x-th Gaussian attention head. $M^{output} = \{M_x^{output}\}_{x=0}^{X-1}$ represents the features of video clips of different lengths implicitly modeled by the Gaussian attention heads. Next, we dynamically aggregate the features of these video clips using learnable weights, thereby generating compact clip features that contain rich contextual information.

Dynamic Aggregation. As shown in Fig. 3, we define a learnable query $\varphi \in \mathbb{R}^d$ and input the outputs of the Gaussian attention heads into a cross-attention layer as K and V, while φ serves as Q. The output of the cross-attention layer is then passed through a FC layer to produce the set of weights $w = \{w_0, w_1, ..., w_{X-1}\} \in \mathbb{R}^{X \times l_c}$, where each w_x corresponds to the weight associated with the Gaussian variance σ_x^2. The detailed calculation is as follows:

$$w_x = FC(CA(\varphi, M_x^{output}, M_x^{output})), x = 0, 1, ..., X-1, \tag{7}$$

where w_x is further normalized to obtain the final weights. We use the normalized weights to perform a weighted aggregation of the features of video clips of different lengths to obtain the clip features. The detailed calculation is obtained through the following formulation:

$$\hat{w}_{x,j} = \frac{e^{w_{x,j}/\omega}}{\sum_{i=0}^{X-1} e^{w_{i,j}/\omega}}, j = 0, 1, ..., l_c - 1, \tag{8}$$

$$F^C = \sum_{x=0}^{X-1} M_x^{output} * \hat{w}_x, \tag{9}$$

where ω is the temperature factor and $F^C \in \mathbb{R}^{l_c \times d}$ represents the clip features. Next, we calculate the video-query similarity score S_C for this branch as follows:

$$S_C = max(cos(F_0^C, Q_{cf}), cos(F_1^C, Q_{cf}), ..., cos(F_{l_c-1}^C, Q_{cf})). \tag{10}$$

3.4 Frame-Align Branch

In the Frame-Align branch, we add position embeddings PE to the video features F extracted by a CNN. Next, we input F into a vanilla Transformer and a FC layer to obtain frame features. Finally, the frame features are aligned with the query representation Q_{cf} for cross-modal alignment. The video-query similarity score S_f for this branch is obtained through the following formulation:

$$S_f = max(cos(F_0^f, Q_{cf}), cos(F_1^f, Q_{cf}), ..., cos(F_{l_v-1}^f, Q_{cf})), \tag{11}$$

where $F^f \in \mathbb{R}^{l_v \times d}$ are the frame features encoded by this branch.

3.5 Frame-Align+ Branch

Following DLDKD [6], we use the KL divergence to align the similarity distributions between the Frame-Align+ branch and the CLIP teacher model. The consistency loss $\mathcal{L}_{KL}$ can be formulated as:

$$\mathcal{L}_{KL} = D_{KL}(C^s || C^t) = \sum_{i=0}^{l_v-1} C_i^s log \frac{C_i^s}{C_i^t}. \tag{12}$$

Here, C^s indicates the similarity distribution produced by the student model, with C_i^s and C_i^t referring to the i-th entries of C^s and C^t, respectively. The video–query similarity score S_{f_+} for this branch is computed as follows:

$$S_{f_+} = max(cos(F_0^{f_+}, Q_{f_+}), cos(F_1^{f_+}, Q_{f_+}), ..., cos(F_{l_v-1}^{f_+}, Q_{f_+})), \tag{13}$$

where $F^{f_+} \in \mathbb{R}^{l_v \times d}$ are the frame features encoded by this branch.

3.6 Query Diversification Loss

An untrimmed video typically has multiple queries corresponding to different moments, and the model needs to perform cross-modal alignment without knowing the exact moments corresponding to each query. Due to the absence of explicit guidance, multiple queries may align with the same video frame or clip, which is unreasonable and detrimental to the model's retrieval performance. One of the key challenges in partially relevant video retrieval is to simultaneously draw the query-relevant portions of a video close and push the query-irrelevant portions away, thereby preventing multiple queries on the same video from aligning with homogeneous frames or clips. Therefore, We introduce a query diversification loss $\mathcal{L}^{div}$ that pushes queries corresponding to the same video apart. Our motivation is to prevent multiple queries from aligning with the same video frame or clip, thus increasing the likelihood that each query aligns with the correct frame or clip. Considering the potential reasonable semantic similarities between queries, we use a decay factor to gradually reduce the penalty from the query diversification loss during training. At the early stage of training, $\mathcal{L}^{div}$ alleviates the excessive proximity of queries corresponding to the same video in the embedding space. As training progresses, the weight of $\mathcal{L}^{div}$ gradually decays, allowing the model to appropriately adjust the distance between queries and prevent them from being excessively pushed apart. The calculation of $\mathcal{L}^{div}$ is obtained through the following formulation:

$$S_{i,j}^{b} = cos(q_i^b, q_j^b), \tag{14}$$

$$l_{i,j}^{b} = ((1 + S_{i,j}^{b})^{\gamma} log(1 + e^{\alpha(S_{i,j}^{b} + \delta)}), \tag{15}$$

$$\mathcal{L}_b^{div} = \mathcal{D} * \frac{2}{N_q(N_q - 1)} \sum_{0 \le i,j \le N_q - 1, i \ne j} l_{i,j}^{b}. \tag{16}$$

Here, $b \in \{f_+, cf\}$ specifies the query branch, with q_i^b and q_j^b corresponding to two different queries derived from the same extractor. N_q refers to the number of query instances linked to the video, and $\mathcal{D}$ denotes the associated decay factor.

3.7 Training and Inference

Loss Function. Apart from the query diversification loss and the KL divergence loss, we adopt two standard objectives commonly used in retrieval tasks: the Triplet Ranking loss [20] and the InfoNCE loss [7–9,17]. For a batch containing relevant query-video pairs, loss functions is obtained through the following formulation:

$$\mathcal{L}_a^{trip} = \frac{1}{n} \sum_{(q,v) \in \mathcal{B}} \Big\{ \max(0, m + S_a(q^-, v) - S_a(q, v)) + \max(0, m + S_a(q, v^-) - S_a(q, v)) \Big\}, \tag{17}$$

$$\mathcal{L}_a^{nce} = -\frac{1}{n}\sum_{(q,v)\in\mathcal{B}}\left\{\log\left(\frac{S_a(q,v)}{S_a(q,v)+\sum_{q_i^-\in\mathcal{N}_q} S_a(q_i^-,v)}\right)\right.$$
$$\left.+\log\left(\frac{S_a(q,v)}{S_a(q,v)+\sum_{v_i^-\in\mathcal{N}_q} S_a(q,v_i^-)}\right)\right\}. \quad (18)$$

Here, $a \in \{C, f, f_+\}$ corresponds to the Clip-Align, Frame-Align, and Frame-Align+ branches, respectively. The terms $S_a(q^-, v)$ and $S_a(q, v^-)$ refer to the similarity scores computed for negative query–video and video–query pairs, respectively, while $S_a(q, v)$ represents the similarity score for a positive query–video pair.

The overall loss function is obtained through the following formulation:

$$\mathcal{L}^{sum} = \lambda_{div}\mathcal{L}^{div} + \lambda_{trip}\mathcal{L}^{trip} + \lambda_{div}\mathcal{L}^{nce} + \lambda_{KL}\mathcal{L}^{KL}. \quad (19)$$

Inference. We obtain the final matching score by weighted aggregation of the video-query similarity scores from the three branches are obtained through the following formulation:

$$S_{sum} = \alpha S_C + \beta S_f + \gamma S_{f_+}, \alpha + \beta + \gamma = 1, \quad (20)$$

where α, β and γ are the corresponding weights for each of these scores.

4 Experiment

4.1 Datasets and Evaluation Metrics

Datasets. We selected two benchmark datasets for the PRVR task, namely TV show Retrieval [12] and ActivityNet Captions [11], and conducted extensive experiments to verify the effectiveness of our proposed method.

- **ActivityNet Captions** comprises approximately 20,000 YouTube videos, each averaging 118 s in duration. On average, each video contains 3.7 temporally localized moments, each paired with a corresponding natural-language caption.
- **TV show Retrieval**(TVR) comprises video clips drawn from six distinct television series. In total, TVR contains 12,800 clips, each with a mean duration of 76 s. Every clip is annotated with five natural-language queries.

Evaluation Metrics. Following the previous work [34], we use ranking-based metrics, namely $R@K$ ($K = 1, 5, 10, 100$). $R@K$ indicates the percentage of videos in the dataset that contain the target video among the top K retrieval results. In addition, the sum of recall scores $SumR$ is also employed as a performance evaluation metric. The calculation of Evaluation Metrics is formulated as:

$$R@K = \frac{TP}{TP + FN}, \tag{21}$$

$$SumR = \sum_{K \in \{1,5,10,100\}} R@K. \tag{22}$$

Implementation Details. In practice, we employed Gaussian attention heads with the Gaussian variances set to 0.1, 0.5, 1.0, 3.0, 5.0, 8.0, 10.0, and ∞, respectively. The hyperparameter settings for inference are $\alpha = 0.225$, $\beta = 0.3$ and $\gamma = 0.475$.

4.2 Comparison with the State-of-the-Art Method

As shown in Tables 1 and 2, the results on two public datasets demonstrate that DGADCM achieves significant advantages in the PRVR task. On the TVR dataset, although the proposed method performs slightly lower than MGAKD in R@10, DGADCM surpasses the previous best method by 1.8% in R@1, achieving a relative improvement of 11.3%, and outperforms the benchmark by 13.2% in SumR. On the ActivityNet Captions dataset, DGADCM exceeds the previous best method by 1.1% in R@1, with a relative improvement of 13.3%, and surpasses the benchmark by 11.1% in SumR.

Table 1. Comparison of performance on TVR datasets, with the optimal performance **bolded**.

Model	$R@1$	$R@5$	$R@10$	$R@100$	$SumR$
T2VR models					
HTM [18]	3.8	12.0	19.1	63.2	98.2
CE [14]	3.7	12.8	20.1	64.5	101.1
DE++ [4]	8.8	21.9	30.2	67.4	128.3
RIVRL [5]	9.4	23.4	32.2	70.6	135.6
CLIP4Clip [16]	9.9	24.3	34.3	72.5	141.0
Cap4Video [28]	10.3	26.4	36.8	74.0	147.5
VCMR models w/o localization					
XML [12]	10.0	26.5	37.3	81.3	155.1
ReLoCLNet [33]	10.7	28.1	38.1	80.3	157.1
CONQUER [10]	11.0	28.9	39.6	81.3	160.8
PRVR models					
MS-SL [3]	13.5	32.1	43.4	83.4	172.4
GMMFormer [26]	13.9	33.3	44.5	84.9	176.6
DLDKD [6]	14.4	34.9	45.8	84.9	179.9
MGAKD [34]	16.0	37.8	49.2	**87.5**	190.5
Ours	**17.8**	**39.6**	**51.1**	86.7	**195.2**

Table 2. Comparison of performance on ActivityNet Captions, with the optimal performance **bolded**.

Model	*R*@1	*R*@5	*R*@10	*R*@100	*SumR*
T2VR models					
HTM [18]	3.7	13.7	22.3	66.2	105.9
RIVRL [5]	5.2	18.0	28.2	66.4	117.8
DE++ [4]	5.3	18.4	29.2	68.0	121.0
CE [14]	5.5	19.1	29.9	71.1	125.6
CLIP4Clip [16]	5.9	19.3	30.4	71.6	127.3
Cap4Video [28]	6.3	20.4	30.9	72.6	130.2
VCMR models w/o localization					
ReLoCLNet [33]	5.7	18.9	30.0	72.0	126.6
XML [12]	5.3	19.4	30.6	73.1	128.4
CONQUER [10]	6.5	20.4	31.8	74.3	133.1
PRVR models					
MS-SL [3]	7.1	22.5	34.7	75.8	140.1
GMMFormer [26]	8.3	24.9	36.7	76.1	146.0
DLDKD [6]	8.0	25.0	37.5	77.1	147.6
MGAKD [34]	7.9	25.7	38.3	77.8	149.6
Ours	**9.4**	**27.3**	**40.3**	**78.6**	**155.6**

It should be noted that the T2VR and VCMR methods have certain limitations when addressing the PRVR task. The T2VR methods assume a complete correlation between the videos and the queries, making it focus more on their overall relevance. However, in the PRVR task, the query is often only partially related to the video, which affects the performance of the T2VR method. The VCMR methods outperform T2VR methods in retrieval performance because they take into account the partial correlation between the videos and the queries in the PRVR task. However, the PRVR methods achieve the best performance, thanks to their dedicated and carefully designed models tailored for the task.

The current PRVR methods still have some limitations. For example, MS-SL only uses average pooling to generate clip features, lacking sufficient exploration of contextual information. GMMFormer performs cross-modal alignment only at the clip and video levels, with insufficient attention to local relevance within the video. DLDKD is only limited to frame-level cross-modal alignment. MGAKD models clip features in a static way, introducing irrelevant clip information and lacking sufficient emphasis on ensuring a well-balanced distribution of queries in the embedding space. In contrast, the proposed method dynamically models clip features, reducing the introduction of irrelevant information. Additionally, it employs a query diversification loss to guide a more reasonable distribution of queries in the embedding space and refines knowledge from the large-scale CLIP model. Overall, our proposed method achieves outstanding retrieval performance on two public datasets, securing either the best or second-best ranking across all evaluation metrics.

4.3 Ablation Study

Fusion Components. We performed ablation studies on the TVR dataset, as detailed in Table 3. It can be observed that the addition of each component improves the performance of the model. The dynamic clip feature modeling method enables the model to better capture relevant clip information and reduce the introduction of irrelevant information. The query diversification loss function alleviates the unreasonable excessive clustering of different queries for the same video in the embedding space during the early stages of training. As training progresses, the loss function gradually decays, allowing the model to adjust the distance between queries more reasonably.
Decay Strategies. We experimented with three different decay strategies: Exponential decay, Linear decay, and Sigmoid decay.

- Exponential decay:$g(t) = k^t$,where $k < 1$ is the decay factor.
- Linear decay:$g(t) = kt + b$, where $k < 0$ and b is the initial weight.
- Sigmoid decay:$g(t) = \frac{k}{k+e^{\frac{t}{k}}}$, where k is the hyperparameter to control the decay.

As shown in Table 4, the best-performing strategy is Exponential decay, followed by Linear decay, with Sigmoid decay performing the worst. Exponential decay achieves the best performance, meeting our expectations when designing the query diversification loss. Specifically, the loss mitigates the excessive proximity of multiple queries corresponding to the same video in the embedding space during the early stages of training. As training progresses, the loss gradually diminishes, allowing the model to adjust the distances between queries reasonably.

Table 3. Ablation studies of DGADCM on TVR. DCM means Dynamic Clip Modeling and QDL means query diverse loss.

DCM	QDL	$R@1$	$R@5$	$R@10$	$R@100$	$SumR$
		16.5	38.5	49.9	86.7	191.6
✓		17.3	39.4	50.6	**87.0**	194.3
✓	✓	**17.8**	**39.6**	**51.1**	86.7	**195.2**

Table 4. The performance of different decay strategies on the TVR dataset, with the best performance highlighted **bolded**.

Decay	*R*@1	*R*@5	*R*@10	*R*@100	*SumR*
Exponential	**17.8**	**39.6**	**51.1**	86.7	**195.2**
Linear	17.4	39.3	50.9	**87.2**	194.7
Sigmoid	17.0	39.1	50.3	**87.2**	193.6
w/o decay	16.9	39.1	50.2	86.7	192.7

4.4 Hyper-parameter Study

We explored different weighting schemes for the scores of the three branches when calculating the matching score. Using SumR as the metric, as shown in Fig. 4, two candidate points are identified, and ultimately we set the parameters to $\alpha = 0.225$, $\beta = 0.3$, $\gamma = 0.475$. Even when focusing on a single scoring parameter during the inference phase, the model still maintains excellent performance. Moreover, when the model simultaneously considers the three scores at two granularities, the retrieval performance of the model is significantly improved. This suggests that exploring video features across multiple granularities is beneficial.

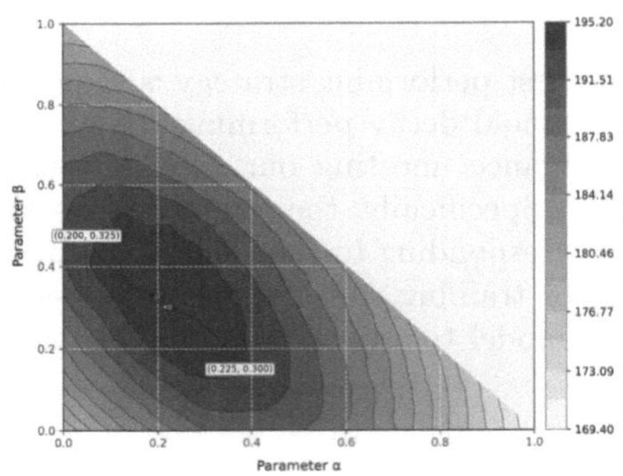

Fig. 4. Experiments on hyperparameters in Eq. (20).

5 Conclusion

This paper presents a novel framework named DGADCM for PRVR. The proposed method performs cross-modal alignment between the videos and the queries at both the frame and clip levels. To enhance clip-level cross-modal alignment, we dynamically model clip features, effectively reducing the introduction of irrelevant clip information. Additionally, we introduce a query diversification loss function to mitigate the issue of multiple queries aligning with the same frame or clip, thereby improving the likelihood of correct query-to-video alignment. Extensive experiments and ablation studies conducted on two public PRVR benchmark datasets demonstrate the effectiveness of the proposed method. In the present study, we directly adopted the knowledge distillation approach from DLDKD [6]. In future work, we anticipate incorporating an intermediate layer feature alignment strategy to enhance it: instead of only distilling the final similarity outputs of the teacher model, we will perform fine-grained feature matching in the spatial-temporal intermediate layers of video segments, thus strengthening the student model's sensitivity to subtle segment variations and further improving the retrieval accuracy.

References

1. Carreira, J., Zisserman, A.: Quo vadis, action recognition? A new model and the kinetics dataset. In: proceedings of the IEEE Conference on Computer Vision and Pattern Recognition, pp. 6299–6308 (2017)
2. Chen, S., Zhao, Y., Jin, Q., Wu, Q.: Fine-grained video-text retrieval with hierarchical graph reasoning. In: Proceedings of the IEEE/CVF Conference on Computer Vision and Pattern Recognition, pp. 10638–10647 (2020)
3. Dong, J., et al.: Partially relevant video retrieval. In: Proceedings of the 30th ACM International Conference on Multimedia, pp. 246–257 (2022)
4. Dong, J., et al.: Dual encoding for video retrieval by text. IEEE Trans. Pattern Anal. Mach. Intell. **44**(8), 4065–4080 (2021)
5. Dong, J., et al.: Reading-strategy inspired visual representation learning for text-to-video retrieval. IEEE Trans. Circ. Syst. Video Technol. **32**(8), 5680–5694 (2022)
6. Dong, J., et al.: Dual learning with dynamic knowledge distillation for partially relevant video retrieval. In: Proceedings of the IEEE/CVF International Conference on Computer Vision, pp. 11302–11312 (2023)
7. Gutmann, M., Hyvärinen, A.: Noise-contrastive estimation: a new estimation principle for unnormalized statistical models. In: Proceedings of the Thirteenth International Conference on Artificial Intelligence and Statistics, pp. 297–304. JMLR Workshop and Conference Proceedings (2010)
8. He, K., Fan, H., Wu, Y., Xie, S., Girshick, R.: Momentum contrast for unsupervised visual representation learning. In: Proceedings of the IEEE/CVF Conference on Computer Vision and Pattern Recognition, pp. 9729–9738 (2020)
9. Hjelm, R.D., et al.: Learning deep representations by mutual information estimation and maximization. arXiv preprint arXiv:1808.06670 (2018)
10. Hou, Z., Ngo, C.W., Chan, W.K.: Conquer: contextual query-aware ranking for video corpus moment retrieval. In: Proceedings of the 29th ACM International Conference on Multimedia, pp. 3900–3908 (2021)
11. Krishna, R., Hata, K., Ren, F., Fei-Fei, L., Carlos Niebles, J.: Dense-captioning events in videos. In: Proceedings of the IEEE International Conference on Computer Vision, pp. 706–715 (2017)
12. Lei, J., Yu, L., Berg, T.L., Bansal, M.: TVR: a large-scale dataset for video-subtitle moment retrieval. In: Vedaldi, A., Bischof, H., Brox, T., Frahm, J.-M. (eds.) ECCV 2020. LNCS, vol. 12366, pp. 447–463. Springer, Cham (2020). https://doi.org/10.1007/978-3-030-58589-1_27
13. Li, X., Zhou, F., Xu, C., Ji, J., Yang, G.: Sea: sentence encoder assembly for video retrieval by textual queries. IEEE Trans. Multimedia **23**, 4351–4362 (2020)

14. Liu, Y., Albanie, S., Nagrani, A., Zisserman, A.: Use what you have: video retrieval using representations from collaborative experts. arXiv preprint arXiv:1907.13487 (2019)
15. Liu, Y., et al.: RoBERTa: a robustly optimized BERT pretraining approach. arXiv preprint arXiv:1907.11692 (2019)
16. Luo, H., et al.: Clip4Clip: an empirical study of clip for end to end video clip retrieval and captioning. Neurocomputing **508**, 293–304 (2022)
17. Miech, A., Alayrac, J.B., Smaira, L., Laptev, I., Sivic, J., Zisserman, A.: End-to-end learning of visual representations from uncurated instructional videos. In: Proceedings of the IEEE/CVF Conference on Computer Vision and Pattern Recognition, pp. 9879–9889 (2020)
18. Miech, A., Zhukov, D., Alayrac, J.B., Tapaswi, M., Laptev, I., Sivic, J.: Howto100m: learning a text-video embedding by watching hundred million narrated video clips. In: Proceedings of the IEEE/CVF International Conference on Computer Vision, pp. 2630–2640 (2019)
19. Radford, A., et al.: Learning transferable visual models from natural language supervision. In: International Conference on Machine Learning, pp. 8748–8763. PmLR (2021)
20. Schroff, F., Kalenichenko, D., Philbin, J.: FaceNet: a unified embedding for face recognition and clustering. In: Proceedings of the IEEE Conference on Computer Vision and Pattern Recognition, pp. 815–823 (2015)
21. Song, X., Chen, J., Wu, Z., Jiang, Y.G.: Spatial-temporal graphs for cross-modal text2video retrieval. IEEE Trans. Multimedia **24**, 2914–2923 (2021)
22. Song, Y., Soleymani, M.: Polysemous visual-semantic embedding for cross-modal retrieval. In: Proceedings of the IEEE/CVF Conference on Computer Vision and Pattern Recognition, pp. 1979–1988 (2019)
23. Vaswani, A., et al.: Attention is all you need. In: Advances in Neural Information Processing Systems, vol. 30 (2017)
24. Wang, J., et al.: Text is mass: modeling as stochastic embedding for text-video retrieval. In: Proceedings of the IEEE/CVF Conference on Computer Vision and Pattern Recognition, pp. 16551–16560 (2024)
25. Wang, Y., Li, Y., Wang, H.: Two-stream networks for weakly-supervised temporal action localization with semantic-aware mechanisms. In: Proceedings of the IEEE/CVF Conference on Computer Vision and Pattern Recognition, pp. 18878–18887 (2023)
26. Wang, Y., Wang, J., Chen, B., Zeng, Z., Xia, S.T.: GMMFormer: gaussian-mixture-model based transformer for efficient partially relevant video retrieval. In: Proceedings of the AAAI Conference on Artificial Intelligence, vol. 38, pp. 5767–5775 (2024)
27. Wray, M., Larlus, D., Csurka, G., Damen, D.: Fine-grained action retrieval through multiple parts-of-speech embeddings. In: Proceedings of the IEEE/CVF International Conference on Computer Vision, pp. 450–459 (2019)
28. Wu, W., Luo, H., Fang, B., Wang, J., Ouyang, W.: Cap4video: what can auxiliary captions do for text-video retrieval? In: Proceedings of the IEEE/CVF Conference on Computer Vision and Pattern Recognition, pp. 10704–10713 (2023)
29. Wu, W., Wang, X., Luo, H., Wang, J., Yang, Y., Ouyang, W.: Cap4video++: enhancing video understanding with auxiliary captions. IEEE Trans. Pattern Anal. Mach., Intell (2024)
30. Xiong, Y., Huang, Q., Guo, L., Zhou, H., Zhou, B., Lin, D.: A graph-based framework to bridge movies and synopses. In: Proceedings of the IEEE/CVF International Conference on Computer Vision, pp. 4592–4601 (2019)

31. Yang, X., Dong, J., Cao, Y., Wang, X., Wang, M., Chua, T.S.: Tree-augmented cross-modal encoding for complex-query video retrieval. In: Proceedings of the 43rd International ACM SIGIR Conference on Research and Development in Information Retrieval, pp. 1339–1348 (2020)
32. Yu, Y., Kim, J., Kim, G.: A joint sequence fusion model for video question answering and retrieval. In: Proceedings of the European Conference on Computer Vision (ECCV), pp. 471–487 (2018)
33. Zhang, H., et al.: Video corpus moment retrieval with contrastive learning. In: Proceedings of the 44th International ACM SIGIR Conference on Research and Development in Information Retrieval, pp. 685–695 (2021)
34. Zhang, Q., Yang, C., Jiang, B., Zhang, B.: Multi-grained alignment with knowledge distillation for partially relevant video retrieval. ACM Trans. Multimedia Comput. Commun. Appl. (2025)

Content-function Coupling-Based Recommendation Mechanism for Holographics Communication

Xing Gui[1], Xu Huang[1], Dongshen Qian[1(✉)], Shang Liu[1], Chenxi Liao[1], Wentao Cui[1], Jia Chen[1,2], Meng Song[3], and Qi Liu[3]

[1] National Engineering Research Center for Advanced Network Technologies, School of Electronic and Information Engineering, Beijing 100044, China
23115049@bjtu.edu.cn

[2] Peng Cheng Laboratory, Shenzhen 518055, China

[3] China Unicom Smart City Research Institute, Beijing 100048, China

Abstract. With the growing demand for immersive and high degree-of-freedom content, traditional two-dimensional (2-D) scene content diction algorithms exhibit significant limitations in three-dimensional (3-D) Holographic communication environments. Current recommendation mechanisms primarily focus on content tag matching while lacking coupling optimization between spatial interaction characteristics and functional decision-making in holographic scenarios. To address this issue, This paper proposes a content-function coupling recommendation mechanism for holographic communication, innovatively integrating 3-D content characteristics with immersive interaction requirements. First, we employ a long- and short-term memory (LSTM) model to capture the spatio-temporal behavioral sequences of the users, combined with collaborative filtering (CF) algorithms to extract global static preferences. Second, a lightweight multilayer perceptron (MLP)-based functional decision model is designed to determine transmission parameters according to the terminal network status and computational capabilities, proactively selecting combinations of parameters to reduce latency and minimize resource waste. The experimental results demonstrate that, compared to traditional recommendation systems, the proposed mechanism achieves significant improvements in both the transmission stuttering rate (60%) and the transmission speed variation latency (70%).

Keywords: Holographic Communication · LSTM · Content Delivery

1 Introduction

In recent years, content recommendation, as a core technology for information filtering, has achieved remarkable success in 2-D planar scenarios such as e-commerce and social media. However, traditional recommendation algorithms primarily rely on users' historical behaviors and static preferences to model, with

L. Huang et al. (Eds.): CCF NCCA 2025, CCIS 2774, pp. 292–304, 2026.
https://doi.org/10.1007/978-981-95-5294-8_20

their underlying logic centered on flat "user-item" interactions, achieving content matching through mining feedback data such as ratings and clicks. Although these methods perform well in 2-D scenarios, and considerable challenges in adapting to 3-D immersive environments enabled by holographic communication systems. Holographic communication, through the integration of spatial awareness [1], multimodal interactions, and dynamic viewpoint tracking, enables 3-D spatial rendering of content and fundamentally transforms user behavior from "unidirectional clicks" to "spatialized operations." Traditional algorithms exhibit considerable limitations stemming from their inability to accurately capture the spatio-temporal continuity of user intentions and the content-scenario coupling relationships.

In conventional networks, parameter adjustments are typically based on real-time network status, representing a post-hoc remediation mechanism. However, holographic content transmission involves multidimensional parameter coordinates and low-latency requirements, making purely adaptive mechanisms insufficient for global optimization.

To address these challenges, this paper proposes a 3-D content prediction mechanism customized for holographic scenarios. The mechanism first deeply integrates the immersive interaction features [2] of holographic communication with the multidimensional temporal characteristics of user behaviors. Innovatively combines CF with LSTM networks to solve content prediction challenges in 3-D holographic environments. Specifically, an LSTM network is introduced to model user temporal behavioral sequences [3] (e.g., viewing angles, interaction operations) in holographic environments, capturing the dynamic migration of user interests and short-term preference volatility. Collaborative filtering is used to extract global static user item preferences, achieving content prediction in holographic scenarios. Furthermore, considering the high bandwidth and low latency demands of holographic communication [4], together with the nonlinear interactions among diverse network parameters at the terminals, a lightweight MLP-based model is adopted for holographic functional decision-making. This model proactively adjusts the transmission parameters based on the network states and terminal computational capabilities, achieving anticipatory optimization to enhance the smoothness of holographic content delivery.

The contributions of this paper are as follows.

1) A novel LSTM-CF model for content prediction in 3-D holographic scene was designed, and the prediction effects of 2-D and 3-D content prediction algorithms were compared on a holographic dataset.

2) A holographic content-function coupling mechanism is proposed for holographic content delivery.

3) A prototype system is implemented, with experiments validating the model's improved prediction accuracy for 3-D content and enhanced user interaction satisfaction.

2 Related Work

The core methodologies of traditional content prediction revolve primarily around CF and Content-based Filtering. CF predicts user preferences by mining user-item interaction matrices (e.g., ratings, click behaviors), with classical approaches including neighborhood-based CF [5] and matrix factorization. However, these methods are designed for 2D planar scenarios, insufficiently capturing the spatial correlations inherent in user behaviors within holographic environments.

With advancements in augmented reality (AR) and holographic communication, recommendation systems for 3-D scenarios have gradually emerged as a research hotspot. Gironacci et al. [6] proposed the use of a content based recommender System integrated into an XR simulator to predict training scenarios for the user. In the medical domain, Liu et al. [7] designed a holographic surgical navigation system that integrates real-time anatomical knowledge recommendations with surgeons' gesture operations. However, existing research predominantly focuses on specific vertical domains and lacks exploration of holographic content browsing scenarios for general users.

Several adaptive techniques have been proposed to address network fluctuations during content delivery, such as the SR-ABR technique for cloud video by Wu et al. [8] and the ABR-Attention mechanism by Ji et al. [9]. These technologies represent a passive response to changes in the state of the network. In contrast, holographic communication scenarios require proactive parameters selection and fine-grained control, rendering traditional adaptive approaches insufficient.

While the aforementioned works lay the foundation for recommendation systems, their application in holographic environments remains limited. To address these gaps, this paper proposes a holographic content-function coupling mechanism that innovatively integrates the temporal modeling capabilities of LSTM with latent interest mining of CF for content delivery in holographic contexts. Meanwhile, researchers have found that simple MLP computations can replace complex CNN and GNN calculations [10]. Given the lightweight nature and efficient inference characteristics of the MLP, which align well with latency issues in holographic functional decision making and the nonlinear relationships among various terminal network parameters in this study, the MLP model is employed for functional decision making.

3 Methodology

3.1 Content-Function Coupling Recommendation Framework

Overall architecture of the personalized recommendation mechanism based on holographic communication is illustrated in Fig. 1. When a user accesses the holographic communication system, the holographic user behavioral model generates content access preferences and behavioral habits, while the user Quality of Service (QoS) model produces initial QoS parameters. These behavioral and parametric data are then fed into the holographic content prediction model and the holographic function decision model, which collaboratively generate holographic content and functional recommendations, achieving content delivery through the coupling of content prediction and functional decision-making.

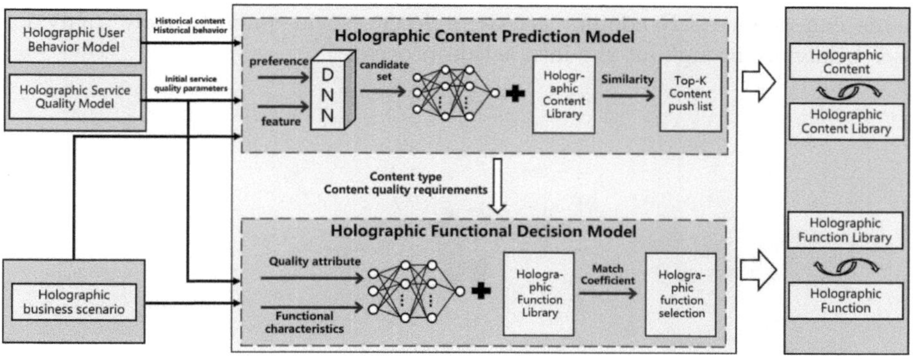

Fig. 1. Model of the content-function coupling mechanism.

Specifically, the coupled recommendation mechanism comprises two components:

Content Prediction Model: Based on user behavioral habits and access preferences derived from the holographic user behavioral model, the holographic content prediction system predicts the content that users may be interested in the future. An example equation is as follows:

$$C = f_1(c_1, c_2, \cdots, c_n, p) \tag{1}$$

where f_1 denotes the content prediction model, c_1-c_n represent input characteristics, p is the user preference vector, and C is the predicted content.

Function Decision Model: Taking business scenario requirements (S) and holographic content QoS demands (C) as inputs to generate transmission parameters.

$$R = f_2(C, S) \tag{2}$$

where f_2 is the function decision model, and R represents the decision results.

3.2 Holographic Feature Modeling

To account for the highly interactive nature of holographic content, the proposed framework integrates user behavioral data and content attributes through explicit 3-D spatial modeling.

The 3-D input characteristics of the holographic content denote the users' preferences for stereoscopic interaction during holographic content navigation. Specifically, for volumetric objects, access effectiveness exhibits angular dependency, where visual comprehension quality varies significantly across different viewing perspectives (as depicted in Fig. 2). This spatio-temporal interaction mode can deduce 3-D content access preferences via quantitatively analyzing user behavior throughout the interaction process.

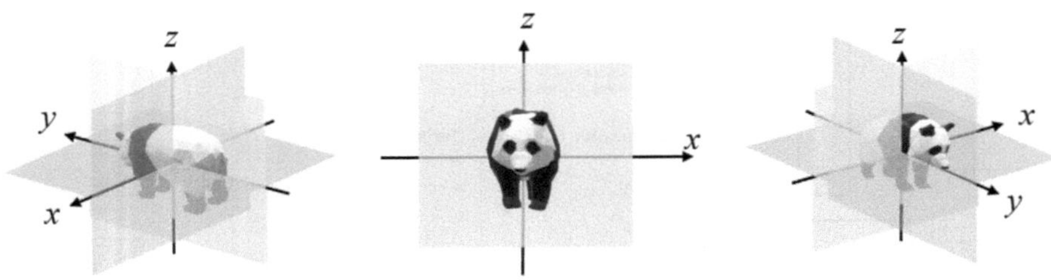

Fig. 2. "Observation effects from different angles."

3.3 Holographic Content Prediction

The content prediction model integrates LSTM and CF, the overall prediction process is shown in Fig. 3. First, the data in the database are preprocessed as the input of the model. After LSTM prediction, a user-based top-K push list is generated. Then, CF is used to analyze the relationship among these communities and users, and push, as shown below:

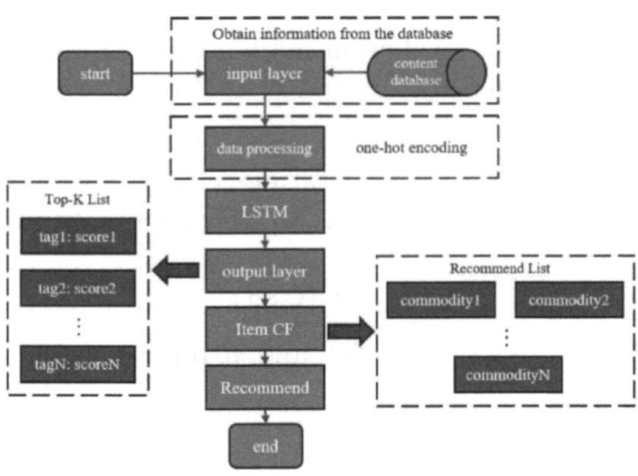

Fig. 3. Holographic content prediction workflows.

Feature Encoding: Feature of holographic user behavior and freedom-of-access preferences are retrieved from the database and encoded by one-hot encoding as LSTM inputs.

Content Prediction: A 3-layer LSTM architecture that processes sequential data to predict probabilities of user behavior at the next timestep. Between each layer, the dropout regularization technique is used to prevent overfitting. Finally, the sequence features of the previous LSTM and dropout layers are integrated through the entire connection layer, as depicted in Fig. 4.

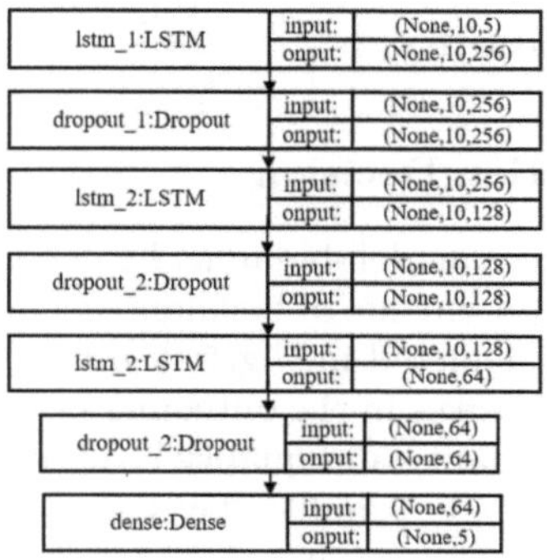

Fig. 4. LSTM model structure.

Top-K List: The output probabilities are weighted to generate a personalized Top-K content list. We recommend the tag with the highest degree of matching as the best tag for this content prediction. We use equations (3) and (4) for the calculation:

$$W_k = [A_{k1}, A_{k2}, A_{k3}, A_{k4}, P_{k5}], 1 <= k <= n \tag{3}$$

$$g_k = y_{k1} + \sum_{1<i\leq 5} \frac{y_{k1}}{y_{ki}} \cdot W_k, \quad 1 \leq k \leq n \tag{4}$$

where A_{k1-k4} are behavioral counts, P_{k5} is the 3-D access preference, W_k denotes tag-specific weights, and g_k is the final prediction score.

Collaborative Filtering: The highest-scoring tag from LSTM is used to construct a user-commodity rating matrix, $U = \{u_1, u_2, u_3, \cdots, u_m\}$ is the sum of the set of M users, $I = \{i_1, i_2, i_3, \cdots, i_n\}$ is the sum of sets of n commodities. $R_{m\times n}$ [11] is the rating matrix, as shown in equation (5).

$$\begin{pmatrix} R_{11} & R_{12} & \cdots & R_{1n} \\ R_{21} & R_{22} & \cdots & R_{2n} \\ \vdots & \vdots & \ddots & \vdots \\ R_{m1} & R_{m2} & \cdots & R_{mn} \end{pmatrix} \tag{5}$$

In CF, methods such as cosine similarity, Jaccard similarity, or Pearson correlation are commonly used to measure the similarity between items. In this study, the cosine similarity approach is adopted to compute item similarities, identify content of potential interest to users, and generate a list of commodity recommendation . as shown in equation (6).

$$A(i,j) = \frac{F_i \cdot F_j}{||F_i|| \cdot ||F_j||} = \frac{\sum_{u=1}^{m} r_{ui} \cdot r_{uj}}{\sqrt{\sum_{u=1}^{m} r_{ui}^2} \cdot \sqrt{\sum_{u=1}^{m} r_{uj}^2}} \tag{6}$$

where $F_i = \{r_{1i}, r_{2i}, r_{3i}, \cdots r_{mi}\}$ and $F_j = \{r_{1j}, r_{2j}, r_{3j}, \cdots r_{mj}\}$are rating vectors for items i and j, respectively.

3.4 Holographic Function Decision

The function decision-making model dynamically generates transmission parameters by considering the requirements of business scenarios and the QoS demands of holographic content. As depicted in Fig. 5, the holographic decision - making model operates as follows: it takes in the outcomes of holographic content prediction, integrates the service quality and quality levels of items required by various business scenarios, and through a multi - layer decision - making process, outputs the transmission parameters that are well-suited for the specific business scenario.

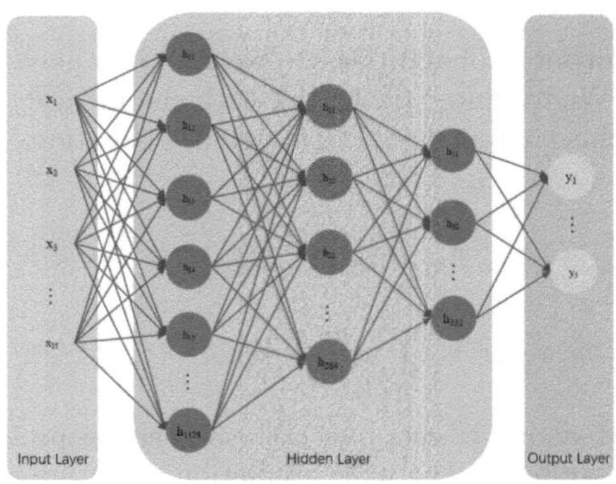

Fig. 5. MLP-based function decision model.

The model utilizes an MLP composed of fully connected layers, where neurons across successive layers are linked through weighted connections. These weights, serving as the model's learnable parameters, are optimized during the training process. Each weight quantifies the degree to which a neuron in the upper layer influences a neuron in the lower layer, thereby determining the strength of signal transmission between layers. The relationship among the input X, hidden layers, and the output Y is defined as equation (7):

$$Y = f(WX + B) \tag{7}$$

where W is the weight matrix, B is the bias vector, and f is the activation function.

4 Experimental Settings

4.1 System Implementation

The feasibility of this coupling mechanism is verified by using a general server in this study. The system architecture is shown in Fig. 6, which is composed of the controller, service server, intelligent converged identifier node equipment and display part. The models and functions of each component are as follows:

Controller: Based on the Ubuntu 20.04 operating system, it has a 10 core Intel (R) Xeon (R) Silver 4210 CPU and is responsible for holographic content prediction and holographic function decision-making tasks.

Application Server: Based on Ubuntu 20.04 operating system, adopting Intel(R) Xeon(R) Bronze 3204 CPU @ 1.90GHz, as the local storage carrier of holographic content.

Intelligent Converged Identifier Node: Composed of a Tofino switch and X86 architecture devices, Tofino is a P4 programmed switch based on the Debian 10 operating system, using the Intel (R) Pentium (R) CPU D1517 version of the CPU. The X86 device is based on the Ubuntu 20.04 operating system and is used to store edge cache information.

User Terminal: Based on Ubuntu 20.04 operating system, it realizes the functions of request sending and holographic content display.

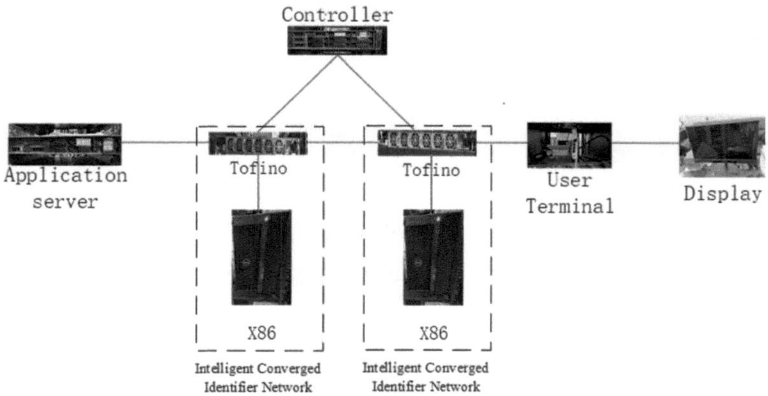

Fig. 6. System architecture.

The system provides a basic supporting environment for the verification coupling mechanism through the cooperative work of various components, in which the controller dominates the core functional decision, the business server ensures the content storage, the intelligent converged identifier node supports the edge cache, and the user is responsible for the interaction and content presentation.

4.2 Experiments

1) Dataset Construction
A dataset (Holo-Data-BJTU) has been established in our laboratory. We developed a webpage for holographic content to collect browsing behaviors and 3-D preferences from 102 users over two months, generating 40,242 records. The data set was partitioned into a training set (70%), validation set (15%), and a test set (15%), ensuring balanced class distributions across splits.

2) Holographic Content Prediction
Holographic content prediction is performed on the controller via the LSTM-CF model, which leverages users' historical behavioral features. The predicted results are transmitted to X86 devices for user access. The key hyperparameters are listed in Table 1.

Table 1. Hyperparameters and their values.

Hyperparameter	Value
time - step	10
batch size	128
epochs	200

For this multiclass task, the cross-entropy loss and F1-score were adopted to evaluate the model's performance, equations (8) to (10) are used for calculation methods of these performances:

$$H(y, \hat{y}) = -\frac{1}{N}\sum_{i=1}^{N}\sum_{j=1}^{C} y_{ij}\log(\hat{y}_{ij}) \tag{8}$$

$$F1 = 2 * \frac{\text{Precision} * \text{Recall}}{\text{Precision} + \text{Recall}} \tag{9}$$

$$F1_{weighted} = \sum\nolimits_{i=1}^{n} (\frac{|C_i|}{|C|} * F1_i) \tag{10}$$

where *F1* is the harmonic mean of precision and recall, C_i is the number of samples of category i, and C is the total number of samples, y_{ij} and $\hat{y}_{ij}$ denote the ground-truth label and predicted probability, respectively.

Table 2. Training Results

Train-times	Cross-Entropy	Accuracy	F1-score
1	0.242	0.956	0.931
2	0.317	0.934	0.891
3	0.257	0.949	0.923
Average	0.272	0.945	0.915

We used data from three different time periods to train the model. The cross entropy and accuracy of the corresponding validation set are shown in Table 2.

Furthermore, building upon the proposed holographic dataset, we conducted an ablation study to validate the framework's efficacy: systematically comparing prediction accuracy between 2-D baseline models (input features exclude 3-D spatial-temporal preferences) and our 3-D-enhanced approach. The experimental results are shown in Table 3:

Table 3. Training Results

Dimension	Cross - Entropy	Accuracy	F1 - score
3-D	0.272	0.945	0.915
2-D	0.492	0.882	0.828

The results confirm that modeling 3-D spatial preferences significantly enhances prediction accuracy in holographic contexts.

3) End-to-End Recommendation Performance

We evaluated the content-function coupling mechanism against traditional adaptive mechanism across diverse scenarios, measuring transmission latency and stuttering rate, Stuttering rate is defined as equation (11):

$$Stuttering - Rate_{frames} = \left(\frac{\sum_{i=1}^{N} \mathbb{I}_{\Delta t_i > \tau}}{N} \right) \times 100\% \tag{11}$$

where N is the total number of frames transmitted, and the numerator counts all frames with a time slot greater than a certain threshold.

In the system, we simulate four different business scenarios for the user side and tested the transmission delay of the coupling mechanism in this paper and the traditional network adaptive mechanism. As shown in Fig. 7, it can be seen that the coupling mechanism can flexibly adapt dynamically according to different business scenarios to reduce transmission delay.

Figure 8 shows the stuttering rate of mechanism transmission in four business scenarios. It can be seen that after the delivery of holographic content via the coupling mechanism, the stuttering rate is significantly reduced.

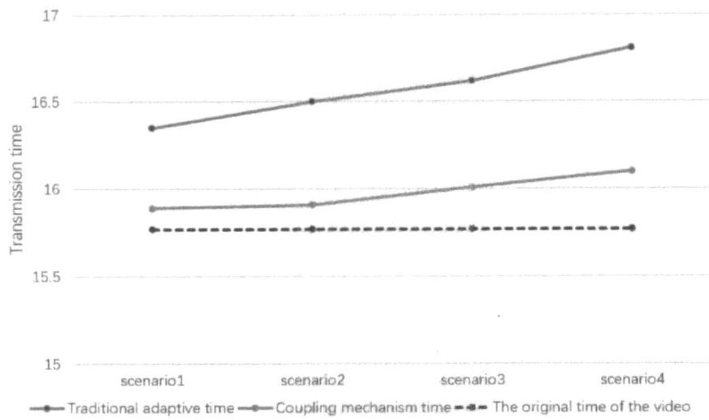

Fig. 7. Latency comparison across mechanisms.

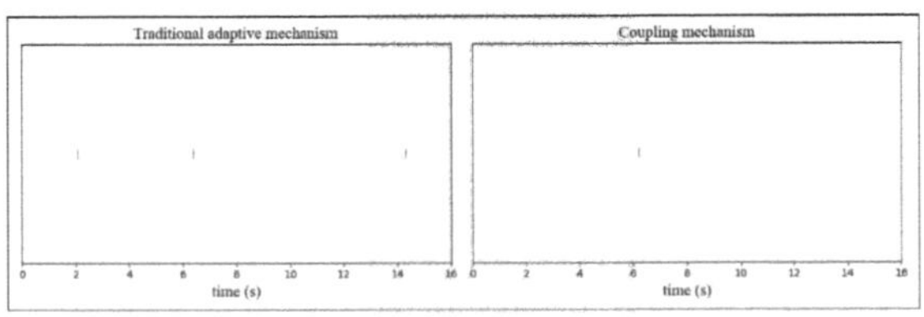

(a) scenario1

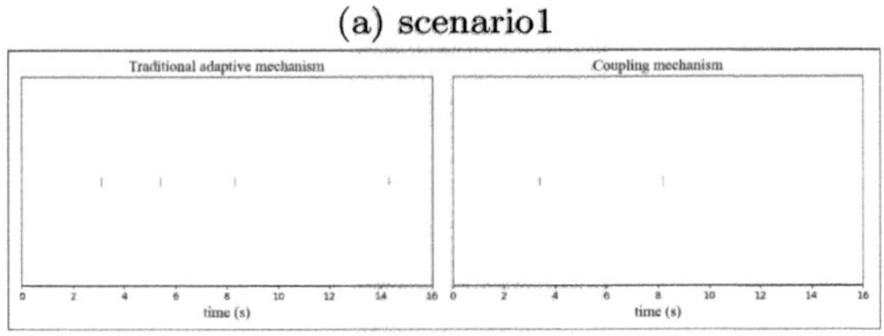

(b) scenario2

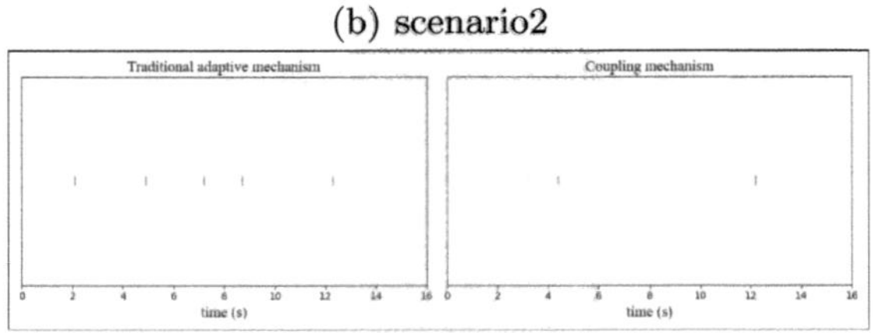

(c) scenario3

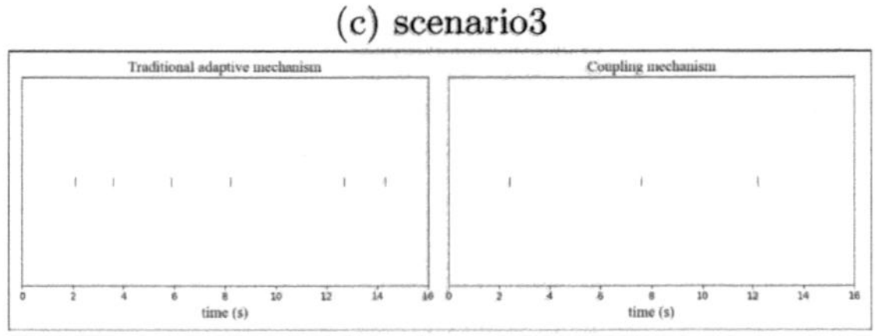

(d) scenario4

Fig. 8. Frame Stuttering Distribution in four scenarios.

The proposed mechanism achieves lower latency and reduced stuttering rates, in different scenarios, the coupling mechanism has an increase of approximately 60% in the stuttering rate and 70% in latency, as summarized in Table 4:

Table 4. Experimental results of the coupling mechanism and the traditional mechanism in the four scenarios.

		Coupled model	Traditional adaptive model	Performance
Latency(ms)	scenario1	120	450	73%↓
	scenario2	140	470	72%↓
	scenario3	240	750	68%↓
	scenario4	330	1040	68%↓
Stuttering%	scenario1	0.2	0.7	71%↓
	scenario2	0.5	1.0	50%↓
	scenario3	0.5	1.3	62%↓
	scenario4	0.7	1.5	53%↓

5 Conclusion

This paper addresses the recommendation of 3-D content in holographic communication by proposing an LSTM-CF fusion model and a content-function coupling mechanism. The former captures the spatio-temporal preferences of users through 3-D feature modeling, while the latter dynamically optimizes the transmission parameters based on the status of the terminal network and computational resources. Through experimental verification, our scheme shows significant advantages in key performance indicators. In terms of 3-D content prediction, the accuracy of the prediction results increased by 6.5%, and the F1-score increased by 8.7%, which means that the model has achieved dual optimization in terms of user preference capture and recommendation quality. At the same time, by using the coupling framework to push the holographic content, the push delay is reduced by 70%, and the stuttering rate is reduced by 60%, which greatly improves the fluency of user interaction under holographic communication.

At the data level, this article employs a laboratory-built data set to support holographic content prediction. However, limited by the data scale and sample diversity, the impact of the dataset on the model's generalization capability remains a concern. As the sample size and scale of the data set are expanded in the future, the model's generalization capability and accuracy are expected to be further enhanced. In the design of the functional decision model, an MLP model is adopted. Although the accuracy of the model meets the requirements of the experiment, for future complex real-world scenarios, it is necessary to explore lightweight models with higher accuracy to replace the MLP, so as to further optimize decision-making efficiency and robustness.

Acknowledgments. This work is supported by the National Key R&D Program of China under grant no. 2023YFB2904400, the Major Key Project of PCL under grant No. PCL2023A06, fundamental research funds for the central universities under grant no. 2023JBGP003, nature and science foundation of China under grant no. 62394321, 62072030, 92167204.

References

1. Petkova, R., Poulkov, V., Manolova, A., Tonchev, K.: Challenges in implementing low-latency holographic-type communication systems. Sensors **22**(24), 9617 (2022)
2. Chen, J., Lin, S., Gao, S., Huang, X., Zhang, H.: Smart computing integration network for holographic communication. Acta Electronica Sinica., 1–11 (2024)
3. Jin, Y.-C., et al.: Models for COVID-19 data prediction based on improved LSTM-ARIMA algorithms. IEEE Access **2**, 3981–3991 (2024)
4. Zhang, J., et al.: Holographic video communication: principle, applications and open issues. IEEE Int. Things Magazine **8**(2), 104–110 (2025)
5. Truong, Q.T., Salah, A., Lauw, H.W.: Bilateral variational autoencoder for collaborative filtering. In: Proc. 14th ACM Int. Conf. Web Search Data Mining (WSDM), pp. 292-300. Virtual Event, Israel (2021)
6. Gironacci.: XR management training simulator supported by content-based scenario recommendation. In: 2022 IEEE International Conference on Artificial Intelligence and Virtual Reality (AIVR), pp. 104–108 (2022)
7. Fan, Z., Chen, G., Wang, J., Liao, H.: Spatial position measurement system for surgical navigation using 3-d image marker-based tracking tools with compact volume. IEEE Trans. Biomed. Eng. **65**(2), 378–389 (2018)
8. Wu, H., Wu, D.O., Gong, P.: SR-ABR: super resolution integrated ABR algorithm for cloud-based video streaming. IEEE Trans. Emerging Top. Comput. Intell. **9**(1), 87–98 (2025)
9. Ji, J., et al.: ABR-attention: an attention-based model for precisely localizing auditory brainstem response. IEEE Trans. Neural Syst. Rehabil. Eng. **32**, 3179–3188 (2024)
10. Zhang, J., Xie, C., Cai, H., Shen, W., Yang, R.: Knowledge distillation-based spatio-temporal mlp model for real-time traffic flow prediction. IEEE Trans. Intell. Transp. Syst. **25**(11), 18122–18135 (2024)
11. Hong-Xia, W.: An improved collaborative filtering recommendation algorithm. In: 2019 IEEE 4th International Conference on Big Data Analytics (ICBDA), pp. 431–435 (2019)

Intelligent Question and Answer System for Policies Based on a Graph Database

Hongying Yan[1], Guifen Zhao[2](✉), Qiaorong An[1], and Xi Nan[1]

[1] Beijing City University, Beijing 101309, China
[2] Beijing Research Institute of Science and Technology, Beijing 100089, China
gfzh@hotmail.com

Abstract. In policy research, massive unstructured policy texts contain rich value. Nevertheless, their complex semantics and decentralized characteristics make exploring them fully through traditional analysis methods difficult. In this study, we innovatively propose a policy intelligence analysis framework based on a large language model (LLM) and knowledge graph and optimize the model training process by introducing the prompt mechanism (QKV vector bootstrapping), which significantly improves the accuracy of semantic understanding and effectively reduces model illusions. First, the domain adaptive method is adopted to clean the policy text, which is combined with BERT model pretraining to generate standard semantic features; subsequently, the "policy-agency-topic" triad knowledge graph is constructed to realize the visual association reasoning of policy elements. The system adopts a layered architecture design, the front-end is based on React to realize an interactive retrieval interface, and the back-end integrates the Neo4j graph database and the Qwen-Max big model to support intelligent Q&A and policy association analysis. Experiments show that the method significantly improves the accuracy of policy element extraction and the efficiency of the Q&A response. The research results provide an efficient analysis tool for policymakers and researchers, which helps in understanding the connotations and correlations of policies in detail.

Keywords: policy text analysis · BERT · knowledge graph · prompt · LLM · react · flask

1 Introduction

Policy data categorization serves as the foundation for constructing knowledge graphs. However, with the rapid development of network technology, policy information has become increasingly fragmented across multiple platforms and formats, creating significant challenges for systematic tracking and analysis. Traditional analytical methods often fail to capture the nuanced semantic relationships within policy texts. For example, when examining a science and technology innovation policy, conventional approaches frequently miss critical connections to related financial subsidy policies from different departments, leading to incomplete policy analysis and potentially flawed decision-making.

L. Huang et al. (Eds.): CCF NCCA 2025, CCIS 2774, pp. 305–318, 2026.
https://doi.org/10.1007/978-981-95-5294-8_21

The current research landscape shows a distinct lack of focus on original policy texts that can effectively support government decision-making while meeting public information needs. To address these gaps, this study addresses two fundamental challenges: improving the accuracy of crossinstitutional relationship extraction in policy texts and reducing the hallucination rate of large language models in policy applications. This research is guided by a central question: how can the integration of knowledge graphs and LLMs enhance policy element association reasoning?

Our solution combines a three-stage framework of high-quality data processing, pretraining paradigm development, and intelligent reasoning implementation. This approach establishes a low-noise policy database through domain-specific data-cleaning rules while addressing LLM limitations through an innovative prompt-guided fine-tuning method. The latter significantly improves semantic understanding while reducing dependency on large annotated datasets.

The proposed policy-domain knowledge graph framework ultimately enables efficient policy retrieval and intelligent question-answering through an interactive query interface. This system provides valuable support for policy researchers, government decision-makers, and citizens seeking reliable policy information, bridging the gap between complex policy analysis and practical application.

2 Research Status

In the field of knowledge graphs, Yang Jianxi et al. [1] developed a hierarchical attention question and answer model (KGQA) based on BERT to increase the knowledge service capability in the field of bridge inspection. For low-resource language scenarios, Phuc et al. [2] validated the effectiveness of BERT with self-constructed mapping in Vietnamese ternary classification. Yu Xiaohan [3] constructed industrial policy mapping to reveal policy relevance. Guo Jiahang [4] proposed a three-stage construction framework for science and technology policy knowledge mapping. On the technical level, Luo Jigen et al. [5] combined a gradient boosting decision tree with BiLSTM to optimize textual relation extraction, and the F1 value was significantly better than that of the baseline model. Bai Changqian et al. [6] improved BERT and fused CNN-BiLSTM to increase the accuracy of key information recognition in cases. For keyword extraction, Wang Junling [7] improved the TextRank algorithm to increase accuracy by 2.3%. In summary, these studies have explored various applications and technical improvements of knowledge graphs, ranging from specific field applications and language processing to policy analysis and algorithm optimization, laying a solid foundation for further research in the field of knowledge graphs while also revealing areas for potential future exploration.

Under the trend of mutual integration and synergistic development of big language models and knowledge graphs, Kunli Zhang et al. [8] proposed that the combination of a big language model and a knowledge graph brings new opportunities for the development of artificial intelligence. Knowledge graphs have construction problems, while large language models have problems such as poor interpretability, and the combination of the two can complement each other's advantages. Cue words are of key value for optimizing the performance of large models and promoting their accurate application. Zhou Jie et al. [9] proposed a user perspective-based analysis of the effectiveness of generative

AI cue word strategies and used the Q-method to quantitatively evaluate cross-task and cross-model cue word strategies in a generic scenario. Fang Quan et al. [10] proposed a combined contextual cueing method for the phantom problem of large language models in vertical domains.

Knowledge graphs, as a form of storage and representation of structured knowledge, have made significant research progress both domestically and internationally in recent years. Foreign research has focused mainly on model fusion, cross-domain applications, Q&A systems, retrieval technology, exploration of low-resource language scenarios, etc., and has continuously improved the performance and application scope of knowledge graphs by combining advanced model architectures and technologies. Domestic research, on the other hand, focuses on the areas of knowledge graph construction, policy analysis, and technology optimization, providing strong support for the application of knowledge graphs in various industries. Moreover, research on cue word strategies is evolving, which can improve the performance and reliability of large language models by optimizing the design of cue words.

3 Methodology

3.1 Architectural Design

The system is categorized bottom-up into a front-end architecture, a data storage layer, a modelling layer, and an application service layer (Fig. 1).

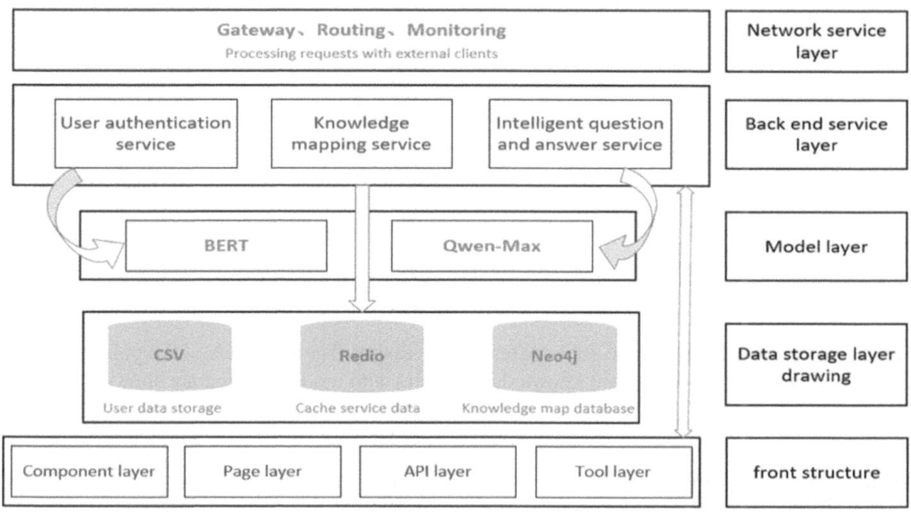

Fig. 1. This figure presents the bottom-up layered architecture of the system. The front-end and back-end are associated. The backend mapping comes from the data layer, and the backend calls the model layer.

The front-end structure covers the component layer, page layer, API layer, and tool layer, which are used to build a user interaction interface, realize front-end and back-end interactions, and provide an entrance for user operation. In the data storage layer, the CSV is used for data storage, Redis provides a caching service for the data, and Neo4j is responsible for storage management to ensure stable data storage and efficient calls. In the model layer, the BERT model and the Ali Qwen-Max big language model are used to process and analyse the lower layer data in depth and explore the potential value of the data. In the application service layer, the user authentication service, knowledge mapping service, and intelligent Q&A service satisfy the user's needs in identity authentication, knowledge retrieval, and intelligent Q&A by processing the results in the model layer. This layered architecture makes the system structure clear with good scalability and maintainability.

Data Layer: Over 100,000 policy data points are crawled from multiple open government platforms, such as the Central Government Document Repository. Data cleansing is one of the fundamental steps in building a high-quality knowledge graph. The raw data are automatically and meticulously cleaned according to the prompt indicator words, including removing HTML tags, standardizing the date format, and filling in missing values, to ensure that the final dataset used for model training has a high degree of consistency and accuracy so that the data meet the structural requirements, which is crucial for improving the performance of the model (Fig. 2).

ID	文件名称	文件级别	下属文件	发文机构	发文字号	发文日期	责任机构或人	具体事权
1	关于印发《关于支持中关村示范区中小微企业利用中关村股权交易服务集团创新发展的意见》的通知	区级	/	中关村科技园区管理委员会	中科园发〔2015〕11号	2015-02-26 00:00:00	中关村科技园区管理委员会	支持中关村示范区中小微企业利用中关村股权交易服务集团实现创新发展，提供融资服务，缓解融资难题，推动融资产品创新等。
2	关于印发《国家科技成果转化引导基金贷款风险补偿管理暂行办法》的通知	国家级	/	科学技术部、财政部	国科发资〔2015〕417号	2015-12-04 00:00:00	科技部、财政部，以及受托管理机构	规范国家科技成果转化引导基金的贷款风险补偿工作
3	国务院办公厅转发财政部发展改革委人民银行关于在公共服务领域推广政府和社会资本合作模式指导意见的通知	国家级	/	国务院办公厅	国办发〔2015〕42号	2015-05-19 00:00:00	国务院各部委、各直属机构，各省、自治区、直辖市人民政府	推广政府和社会资本合作（PPP）模式，转变政府职能，激发市场活力，打造经济新增长点
4	国务院办公厅转发财政部发展改革委人民银行关于在公共服务领域推广政府和社会资本合作模式指导意见的通知	国家级	/	国务院办公厅	国办发〔2015〕43号	2015-05-19 00:00:00	国务院各部委、各直属机构，各省、自治区、直辖市人民政府	推广政府和社会资本合作（PPP）模式，转变政府职能，激发市场活力，打造经济新增长点
5	关于在公共服务领域推广政府和社会资本合作模式指导意见的通知	国家级	/	国务院办公厅	国办发〔2015〕44号	2015-05-19 00:00:00	国务院各部委、各直属机构，各省、自治区、直辖市人民政府	推广政府和社会资本合作（PPP）模式，转变政府职能，激发市场活力，打造经济新增长点

Fig. 2. Policy Text Data Sample Table After Preprocessing. The key fields of the policy text data after preprocessing for subsequent knowledge mapping and model training are shown.

Model Layer. BERT is used as the core to build the entity recognition and relationship extraction model, and its bidirectional context modelling capability is used to analyse the policy text in detail. The model is pretrained with bert-base-Chinese and fine-tuned with the help of the masked language modelling (MLM) mechanism to accurately identify the entities and relationships in the text to provide a solid foundation for the construction of the knowledge graph. To improve the model performance, a small amount of high-quality text is used for pretraining, and after repeated tuning, these data are used as the training paradigm of the BERT model, which contains terminology and context-specific examples in the policy text to help the model understand the contextual semantics in depth and to improve the accuracy of entity recognition and relationship extraction.

Moreover, to enhance the semantic comprehension ability of the Q&A system, a prompt based on the QKV-prompt structure is designed to control the model's attention distribution through the Q (question guidance)-K (dimension constraint)-V (content filling) structure. The system transforms the user's question into a query and extracts the corresponding value as a semantic response by matching the key in the knowledge graph, which guides the model to focus on the key information, reduces the inference uncertainty, and makes the Q&A system answer more accurately.

Prompt = {'query': "What is the policy [CLS] involving [MASK]?", 'key':["Issuing authority", "Applicable to", "Timeframe"], ' value': "[policy text snippet]"}

An antihallucination mechanism is added, and in the graph validation layer, the Qwen-max-generated answers are checked for entity linking, as well as a confidence threshold for the answer.confidence <0.7: trigger_human_review.

Application Layer. This layer aims to provide users with a convenient operation experience. The front end is based on the React framework to realize the retrieval function. When users query through the Cypher statement, the system retrieves relevant information from the Neo4j graph database and displays detailed information about the nodes and relationships. The back end relies on the Flask framework to integrate a knowledge graph and intelligent Q&A. The Qwen-max API is combined to semantically parse the user's natural language questions and return answers to improve service efficiency.

3.2 Core Algorithm

Model Fine-Tuning. BERT is used as the core to build the entity recognition and relationship extraction model. The training data for BERT include 100,000 policy texts, with entities such as "Issuing authority" and "Topic" annotated. The model is pretrained with bert-base-Chinese and fine-tuned via the masked language modelling (MLM) mechanism to accurately identify entities and relationships in the text.

Knowledge Graph Construction. The knowledge graph is constructed on the basis of the "policy-agency-topic" triplet through a systematic process. Initially, text preprocessing is performed to clean the raw data by removing HTML tags, standardizing date formats, and filling in missing values. The BERT model is subsequently employed for entity recognition to identify key elements such as "Issuing authority" and "Topic" within the policy texts. Leveraging BERT's bidirectional context modelling capability, the system then extracts semantic relationships between these identified entities. Finally, the recognized entities and their relationships are systematically linked to form meaningful triplets, which serve as the fundamental building blocks of the knowledge graph. This end-to-end process ensures the creation of a structured and interconnected representation of policy knowledge.

Promotion Mechanism: The QKV structure is defined as follows: Q (query vector) represents the user's question. K (knowledge dimension vector): Represents the dimensions of knowledge in the knowledge graph. V (value-filling vector): Represents the content to be filled in the answer. The system uses the attention mechanism to weight the Q, K, and V vectors, guiding the model to focus on key information.

Antihallucination Mechanism: The system uses DBPedia Spotlight for entity linking and calculates the confidence score on the basis of the entity linking score and the node association degree in the knowledge graph. If the confidence score is less than 0.7, the answer is flagged for human review.

Dynamic Interaction. After the user inputs a question, the system guides the Qwen-max API to understand the semantics and generate the answer through the prompt mechanism. The back end optimizes the answers by combining the data in the knowledge graph to improve the accuracy and comprehensiveness of the answers, enhance the semantic comprehension ability of the Q&A system, and provide better services for users. The application of the prompt mechanism enhances the semantic comprehension ability of the Q&A system and provides better services for users.

4 Experimental Design

4.1 Experimental Environment

NVIDIA server, Intel i7-10750H CPU@2.60 GHz, 16 GB RAM; System: Windows 10; Database: Neo4j 5.14.1; Backend: Flask 3.0.0; Frontend: React 18.3.1.

4.2 Experimental Flow

The flow of this experiment covers four closely connected stages: data acquisition and processing, model training, knowledge graph construction, and system testing.

Data Preprocessing: Define automated auto_labeler annotation, active learning annotation on the basis of rule templates, and automated processing of data through indicator words.

Model Training. Pretrain the BERT model with the cleaned portion of the data to generate exemplars and optimize the model performance by optimizing the QKV template parameters through dynamic prompt grid search, monitor the model performance through a validation set, and adjust the hyperparameters in time to avoid overfitting or underfitting problems.

Knowledge Graph Construction. Store the extracted entity and relationship information in the import directory of Neo4j to construct the knowledge graph.

System Testing. The system functions are fully tested to verify the accuracy and efficiency of knowledge graph retrieval and intelligent Q&A, and to examine the practicality and effectiveness of the system (Fig. 3).

After data acquisition and cleaning, we input the data into the BERT model for training, and during the process, we tune the parameters and decide whether to retune the parameters to optimize the model on the basis of the performance evaluation results.

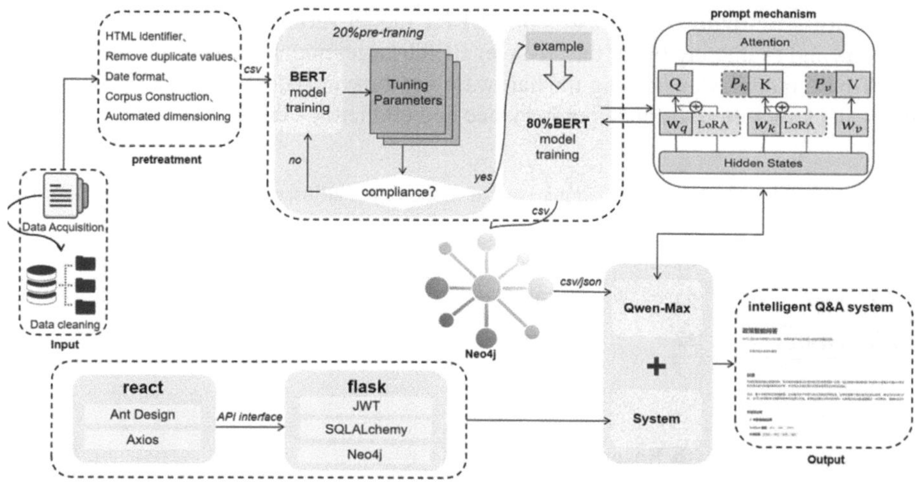

Fig. 3. This figure shows the process of this study. After data cleaning, the data are put into the model, and prompt instructions are added for training and fine-tuning. After reaching the standard, the map is generated, and the CSV training language model is exported. The front-end and back-end are deployed to realize retrieval and answer questions.

After the standard is reached, Qwen-Max model training is performed on the generated graph. It was also developed via back-end frameworks such as SQLAlchemy and Flask and front-end frameworks such as React.js and Ant Design. They are integrated to realize Cypher retrieval and intelligent Q&A functions and complete system testing.

4.3 Control of Experimental Variables

Hyperparameter Adjustment. The learning rate is set to 1e-5, the batch size is 16, and the number of training rounds is 3. These parameters are chosen on the basis of grid search and validation set performance.

Data Division. The dataset is divided into training, validation, and test sets at an 8:1:1 ratio (Table 1).

Table 1. Comparison of learning rate tuning.

Learning rate	Batch size	Training rounds	Accuracy	Precision	Recall	F1-score
1e-4	8	3	0.745	0.743	0.745	0.744
5e-5	16	5	0.752	0.750	0.752	0.750
1e-5	**16**	**3**	**0.760**	**0.758**	**0.760**	**0.758**

Model Restructuring. By constructing BERT model variants with different numbers of layers and evaluating them, the 24-layer BERT-large model outperforms the 12-layer BERT-base model. Considering the hardware and time costs, BERT-based is chosen as the model structure to balance performance and efficiency (Table 2).

Table 2. Comparison of model layer adjustments.

Model Structure	Layers	Accuracy	Precision	Recall	F1-score
BERT-large	24	0.780	0.778	0.780	0.779
BERT-base	**12**	**0.760**	**0.758**	**0.760**	**0.758**

4.4 Comparison with Baseline Methods

We compare our method with the following baselines:

Baseline 1. BERT + traditional knowledge graph (no LLM)

Baseline 2. Qwen-Max + no knowledge graph validation (pure LLM)

Baseline 3. Traditional TextRank + rule matching (Table 3)

Table 3. Comparison of F1-scores and Hallucination rates.

Method	Entity Extraction F1-score	Relationship Inference F1-score	Hallucination Rate
Baseline1	**0.760**	**0.758**	**0.05**
Baseline2	0.740	0.735	0.12
Baseline3	0.680	0.675	0.15

5 Analysis of Results

5.1 Performance Enhancement

After multiple sets of parameters and model structures are compared, BERT-base, as well as the learning rate = 1e−5, batch size = 16, and training rounds = 3, are selected as training parameters. The accuracy of the model in entity recognition and relationship extraction is significantly improved by the application of the prompt mechanism. The loss value starts fluctuating at 0.7, indicating that the model fits the data poorly at the beginning of training, and the prediction has a large gap from the real value (Fig. 4).

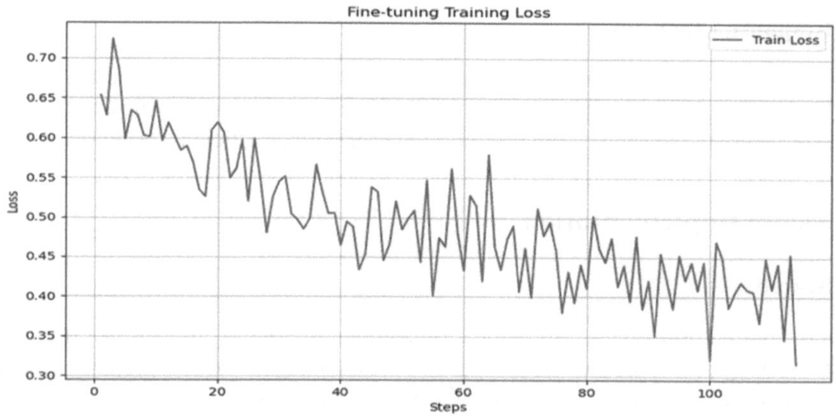

Fig. 4. The training loss image shows the trend of model training loss with increasing number of training steps. As training progresses, the objective function value gradually decreases, the fitting ability gradually improves, and the prediction accuracy gradually increases.

After approximately 100 iterations of training, the value of the objective function decreases to nearly 0.3, indicating that the model can better grasp the intrinsic laws of the data after training and can predict the data.

5.2 Correlation Accuracy

The associative reasoning operation on the knowledge graph through Cypher statements can mine the potential relationships among policies, organizations, and topics, and the validation shows that the constructed knowledge graph can effectively support the associative reasoning task, providing an accurate and complete knowledge base for intelligent Q&A and decision support. By the association between the nodes, the graph visualizes the distribution of issuing institutions (Fig. 5).

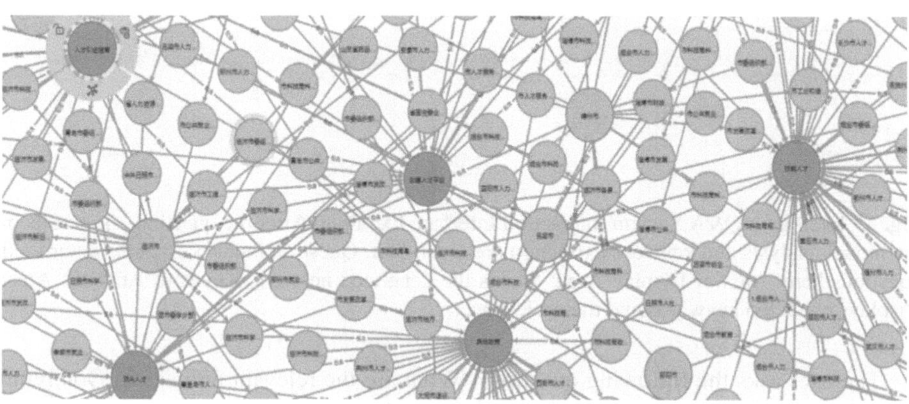

Fig. 5. This figure shows the generated knowledge graph. This graph can be generated by importing data into Neo4j after model training.

5.3 Response Efficiency

The system sets 120 s as the timeout standard, and in actual operation, it usually takes only 15 s to complete the response to the user's question, which meets the user's demand for quick access to the answer and provides the user with an efficient service experience.

6 System Implementation

6.1 Search

To build a bidirectional data communication architecture based on the React framework and the Neo4j graph database for semantic retrieval through a structured query interface, where users can enter query commands through Cypher statements.

6.2 Intelligent Q&A

After pretraining and vertical fine-tuning through the BERT model, Qwen-max is interfaced with the prompt mechanism for knowledge-enhanced answer generation. The system adopts a streaming processing pipeline design and realizes a low-latency response through pretraining model parameter distillation technology, ensuring effective alignment of cross-platform data features and accurate analysis of policy elements. Tests show that it can generate credible answers within 15 s and continuously optimize the Q&A effect through the prompt mechanism (Fig. 6).

Fig. 6. Intelligent Q&A System Interaction Interface Example. After the question, the results of the model analysis and keyword extraction are given first. Question input box: Users enter their questions here. Answer Display Area: Shows the generated answers. Keyword Analysis Area: Displays extracted keywords and their relevance.

The user asks the following question: "What are the relevant policies in Dongying city?" The system first recognizes the user's intent ("relevant policies in Dongying city") through the BERT model and extracts the key entity ("Dongying city"). The knowledge graph retrieval module is then invoked to match the relevant policy nodes and generate the answer at the same time (Fig. 7).

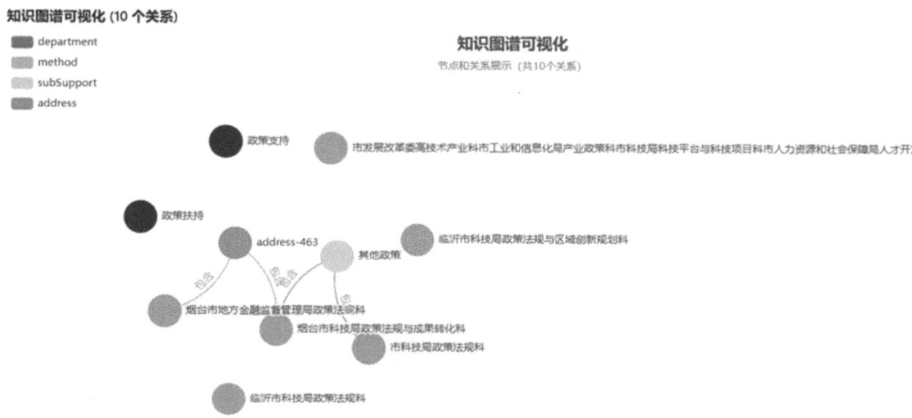

Fig. 7. This figure is a smart quiz. After giving a textual answer, it combines the graph in the Neo4j database and outputs the graph to the front-end with a visualization.

Ultimately, the system outputs structured answers and verifies the factuality of the answers with a dynamic knowledge fusion component. Tests show that the system can generate plausible answers within 15 s. Below are examples of questions posed to the larger model.

Examples of problems.

Concern	Responsive
What are the relevant policies in Dongying?	Based on the knowledge mapping information you provided, I notice that there are no direct references to policies or departments related to Dongying City. The information given mainly focuses on certain government departments in Yantai and Linyi and the specific matters they are responsible for, such as the Policy, Regulation and Achievement Transformation Section of the Yantai Science and Technology Bureau and the Policy, Regulation and Regional Innovation Planning Section of the Linyi Science and Technology Bureau, and mentions several types of approaches regarding policy support and services

(continued)

(*continued*)

Examples of problems.	
Concern	Responsive
What sectors are involved in the field of higher education research institutes?	According to the information provided on the knowledge map, the field of university research institutes mainly involves the following sectors: 1. Science and Technology Cooperation Department of the Science and Technology Bureau: This department is responsible for certain matters in conjunction with the Municipal Education Bureau's University Liaison Service (node IDs 685, 688, 726). Although these nodes do not directly point out their specific responsibilities in the area of university research institutes, it can be assumed that they may be involved in collaborative projects or policy development for university research institutes, given their names and collaborative relationships 2. Municipal Education Directorate, University Liaison Service: This department appears in a separate node (node ID 691) and is explicitly linked to the methodology in the direction of "Universities and Research Institutes," employing a relationship (methodology ID 10). This suggests that it plays an important role in facilitating exchanges and cooperation between universities and research institutes

(*continued*)

(*continued*)

Examples of problems.	
Concern	Responsive
In addition to Zibo, what other provinces and cities have departments involved in the field of talent introduction and training?	Based on the provided knowledge mapping information, we can see that in addition to Zibo, several other cities have departments specializing in talent introduction and cultivation. These departments are located in different cities and have their own responsibilities for talent services. Specifically, as follows: 1. Tai'an City: - Office of Talent Work Leading Group of Tai'an Municipal Committee (Department ID: 12) 2. Yantai: - Talent Work Office, Organization Department, Yantai Municipal Party Committee (Department ID: 19) - Talent Introduction Section, Public Employment and Talent Service Center, Yantai Human Resources and Social Security Bureau (Department ID: 20) - Specialist Services Section, Public Employment and Talent Service Center, Yantai Human Resources and Social Security Bureau (Department ID: 21) - Talent Treatment Section, Public Employment and Talent Service Center, Yantai Human Resources and Social Security Bureau (Department ID: 22) 3. Zaozhuang: - Zaozhuang High-level Talent Service Center (Department ID: 23) In addition, some departments at a broader level or in cities not explicitly designated as such are mentioned, such as "Planning and Introduction Department of the Municipal Talent Service Center" (Department ID: 13) and "Talent Development Section of the Municipal Bureau of Human Resources and Social Security" (Department ID: 15). Talent Development Section" (Department ID: 15). These may be provincial or other municipal organizations, but it is not possible to determine the exact geographic location based on the information given

The above questions and answers indicate that the intelligent Q&A system uses advanced technical architecture and models to accurately identify user intent and key entities, and when combined with the efficient retrieval of knowledge graphs, it can quickly extract relevant content from limited information and generate answers. Despite the limitations of mapping in terms of questions, the system is still able to reasonably respond to a variety of questions, integrate information from different regions and domains, and provide valuable references.

With the improvement of the knowledge graph and the enrichment of data, as well as the optimization of the model, the system is expected to provide users with more comprehensive and accurate services in terms of policy queries and information access and play a greater role in the fields of government services and policy interpretation in particular.

7 Innovations

This study proposes three main innovations:

The QKV-Prompt fine-tuning strategy for LLMs significantly improves the accuracy of policy semantic understanding. Experiments show that this method achieves an F1 score of 0.760 for entity extraction.

Compared with traditional methods, the "policy-agency-topic" knowledge graph and LLM bidirectional verification mechanism reduce the hallucination rate to 0.05.

The layered architecture design enables efficient policy information retrieval and intelligent Q&A. This design outperforms traditional single-model architectures in terms of response speed and accuracy.

References

1. Yang, J.X., et al.: BERT and hierarchical cross-attention-based question answering over the bridge inspection knowledge graph. Expert Syst. Appl. **233** (2023)
2. Do, P., Le, H., Pham, A.B., Nguyen, C.H.: Using BERT and knowledge graph for detecting triples in Vietnamese text. Neural Comput. Appl. **34**(20), 17999–18013 (2022)
3. Yu, X.H.: Construction and application of knowledge mapping of china's industrial policy. Jinan University (2023)
4. Guo, J.H.: Research on the construction and application of science and technology policy knowledge graph based on deep learning. Beijing University of Posts and Telecommunications (2021)
5. Luo, J.G., Du, J.Q., Nie, B., Xiong, W.P., Liu, L., He, J.: A textual relationship extraction model for Chinese medicine based on bidirectional LSTM and GBDT. Comput. Appl. Res. **36**(12), 3744–3747 (2019)
6. Bai, C.Q., Dai, X., Zhang, B.: Charge prediction based on data enhancement and improved BERT. Comput. Inf. Technol. **31**(01), 37–40 (2023)
7. Wang, J.L.: Text keyword extraction algorithm with improved TextRank. Softw. Guide **20**(04), 49–52 (2021)
8. Zhang, K.L., et al.: An overview of the construction and application of a knowledge graph driven by a big language model. J. Zhengzhou Univ. (Sci. Edn.), 1–9 (2025)
9. Zhou, J., Wang, D.Y., Dai, Q.Q., Xia, S.D.: Exploring the effectiveness of prompt word strategies in generative AI dialogue. Data Anal. Knowl. Discov., 1–17 (2025)
10. Fang, Q., Zhang, J.L., Wan, B.Q., Hu, J.: Study of domain knowledge quizzing for large language models based on combinatorial contextual cues. Comput. Sci., 1–13 (2025)

Research Progress and Frontier Applications of Graph Neural Networks

Zhenyu Liu[1,2] and Dengfeng Yao[1,2,3,4](✉)

[1] Beijing Key Lab of Information Service Engineering, Beijing Union University, Beijing 100101, P. R. China
tjtdengfeng@buu.edu.cn

[2] College of Robotics, Beijing Union University, Beijing 100101, P. R. China

[3] Lab of Computational Linguistics, School of Humanities, Tsinghua University, Beijing 100084, P. R. China

[4] Center for Psychology and Cognitive Science, Tsinghua University, Beijing 100084, P. R. China

Abstract. This paper focuses on the research progress and application exploration of graph neural networks (GNNs) in handling heterogeneous graphs, multimodal data, and practical applications. It summarizes various innovative model architectures and learning strategies addressing issues such as semantic aggregation in heterogeneous graphs, representation learning driven by contrastive learning, spectral domain reconstruction, and dynamic graph structure learning in multimodal data. This paper also reviews recent research achievements in practical application problems such as defense against adversarial attacks, calibration of prediction uncertainty, and few-shot learning. Additionally, it explores the wide range of applications of GNNs in fields such as dynamic spatiotemporal data modelling, anomaly detection in multivariate time series, cross-modal data fusion, distributed decision-making in multiagent systems, link prediction and stream data processing, recommendation systems, fake news detection, open-set recognition of mental disorders, avatar rendering, and planning tasks. This paper further analyses the technical trend of integrating GNNs with transformers and discusses current research challenges, such as model scalability, computational efficiency, and the generalizability of defenses against adversarial attacks. It also provides an outlook on future research directions.

Keywords: Graph neural networks (GNNs) · Heterogeneous graphs · Multimodal data · Adversarial attacks · Prediction uncertainty · Few-shot learning

1 Introduction

Graph neural networks (GNNs), powerful tools for processing graph-structured data, have been widely applied and have achieved significant results in many fields. In recent years, GNNs have shown great potential in cutting-edge areas such as cybersecurity, satellite communications, and educational assessment. However, as research deepens

L. Huang et al. (Eds.): CCF NCCA 2025, CCIS 2774, pp. 319–336, 2026.
https://doi.org/10.1007/978-981-95-5294-8_22

and applications expand, GNNs face many challenges in dealing with complex graph structures and practical application scenarios. In terms of complex graph structures, there are issues such as heterogeneity in heterogeneous graphs, oversmoothing phenomena in graphs, and the challenge of integrating multimodal data. In practical applications, threats from adversarial attacks, quantification of prediction uncertainty, difficulties in few-shot learning, and efficiency and interpretability in complex tasks are faced. However, many existing review papers on GNNs fail to conduct an in-depth analysis of the differences between various GNN models and lack discussions in specific research fields. This review is intended to remedy these shortcomings. This paper reviews recent research progress in these areas and explores the applications of GNNs in multiple fields, providing a comprehensive reference for future research.

2 Research Progress

This section delves delve into the latest advancements in addressing the aforementioned challenges, examining how GNNs have evolved to handle increasingly complex graph structures and meet the demands of real-world applications. We categorize these developments into several key areas. The comparisons of these GNN models are shown in Table 1.

2.1 GNNs in Heterogeneous Graphs

Heterogeneous graphs contain multiple types of nodes and edges, featuring complex structures and rich semantic information. To address the challenges posed by heterogeneous graphs, researchers have proposed a variety of innovative model architectures and learning strategies.

(1) Semantic aggregation mechanism based on metapaths

Traditional heterogeneous graph neural network methods, such as HetGNN [1], aggregate node semantics through manually designed metapaths but are limited by the locality problem. To overcome these limitations, NLA-GNN [2] uses a nonlocal aggregation module that can dynamically adjust the weights of metapaths. NLA-GNN achieves cross-modal feature alignment by defining composite relationships between nodes, thereby capturing richer semantic information in heterogeneous graphs. In the task of heterogeneous node classification, the F1 score of this model is improved by 14.7% compared with that of the HetGNN. In the application of biomedical knowledge graphs, NLA-GNN increases the drug repositioning accuracy to 89.2% on the DrugBank dataset.

NLA-GNN captures "long-range and even disconnected dependencies" in heterogeneous information networks (HINs) by introducing a "nonlocal enhanced graph convolutional block." The core technical principle lies in the adoption of a "dictionary-based nonlocal encoding method," where nonlocal information is encoded through a combination of graph convolution and "dictionary-based implicit convolution." This approach avoids the common issue of "exhaustive computation of pairwise data relationships" found in previous nonlocal methods, making it suitable for large-scale graph convolutional networks (GCNs) and enabling flexible integration into "any GCN architecture."

At the algorithmic level, NLA-GNN initializes only one type of node embedding and uses simplified graph convolutional operations to construct and update representations for all types of nodes. This process enhances training efficiency, eliminates the heterogeneity of bipartite graphs, and effectively models complex, high-order relationships. The key innovations of NLA-GNN are its ability to efficiently capture nonlocal information without exhaustive computation, its flexible integration into existing GCN architectures, and its improved training efficiency when modelling complex high-order relationships in HINs. These innovations not only enhance the model's accuracy but also, more importantly, achieve these improvements in an efficient manner, which is crucial for the practical deployment of GNNs on large-scale, complex heterogeneous graphs.

In biomedical knowledge graph applications, MetaPath2Vec [3] extracts drug–disease associations via predefined metapaths. Diffusion-jumped GNNs [4] quantify the heterogeneity of graphs by defining "structural heterogeneity" and improve node classification performance in heterogeneous graphs by learning diffusion distances to connect distant homogeneous nodes.

(2) Contrastive Learning-Driven Representation of Heterogeneous Graphs

The GTC model [5] uses a GNN-transformer collaborative contrastive learning framework, which achieves self-supervised heterogeneous graph representation learning by performing contrastive learning tasks between the "graph pattern view" and the "skip view." This method effectively addresses the oversmoothing issue of GNNs and demonstrates excellent performance in cross-domain transfer tasks. For example, in social network and e-commerce recommendation scenarios, its cross-domain AUC is 18.3% greater than that of HAN. On the IMDB heterogeneous graph, the node classification convergence speed is three times faster than that of GraphSAGE.

GTC surpasses node-centric heterogeneous GNNs by introducing a "dual-branch architecture." This architecture combines "local GNN aggregation" (leveraging the advantage of GNNs in capturing fine-grained node relationships) with "transformer-based global modelling" (enabling it to capture broader, long-range contextual dependencies). In terms of the algorithmic mechanism, GTC introduces a "metapath-aware Hop2Token mechanism," which more effectively encodes multihop dependencies by transforming node interactions into structured token representations, allowing the model to integrate hierarchical information across different graph distances. The key innovations of GTC lie in its dual-branch architecture for comprehensive relationship capture, its metapath-aware Hop2Token mechanism for efficient multihop dependency encoding, and its ability to alleviate information redundancy and oversmoothing through contrastive learning. GTC's approach to addressing oversmoothing—combining the local aggregation capability of GNNs with the global modelling capability of transformers—reflects a deeper trend of addressing the fundamental limitations of GNNs through hybrid architectures. This method indicates that by integrating the strengths of different models, the inherent weaknesses of a single model (such as the local receptive field limitations of GNNs and the structural blindness of transformers) can be compensated for, thereby tackling fundamental challenges such as oversmoothing and long-range dependency capture. For example, in social networks and e-commerce recommendation scenarios, the cross-domain AUC was 18.3% greater than that of HAN [6]. The technical breakthrough lies in the introduction of contrastive learning between the graph pattern view and the jumping view, alleviating information redundancy and achieving a threefold increase in node classification convergence speed on the IMDB heterogeneous graph compared with GraphSAGE [7].

2.2 GNNs in Heterogeneous Graphs

Multimodal data encompass various modalities, such as text, audio, and visual information, which are both consistent and complementary. However, data incompleteness is a common issue.

(1) Spectral Domain Reconstruction and Multimodal Feature Alignment

The SDR-GNN model [8] uses a graph neural network framework based on spectral domain reconstruction. This model constructs a discourse semantic interaction graph to simulate emotional dependencies, captures high-order and high-frequency information through weighted relationship aggregation and multifrequency aggregation, and optimizes features via a multihead attention mechanism. This approach demonstrates superior performance in emotion recognition when dealing with incomplete multimodal data. For the CMU-MOSI dataset, its classification accuracy is 7.6% higher than that of the MMGCN [9]. Additionally, on the basis of the weighted relationship aggregation mechanism, the SDR-GNN achieves an F1 score of 82.4% in emotion polarity classification on the YouTube dataset.

The SDR-GNN leverages GNNs to capture complex emotional dependencies and "recovers incomplete modalities." Its innovation lies in modelling the intrinsic relationships between modalities and employing a "spectral domain method" to integrate high-frequency and low-frequency information during modality recovery and emotion classification. In terms of algorithmic mechanisms, it includes "node construction" (representing utterances as nodes and encoding missing features), "graph construction" (emotional interactions, speaker/context graphs, and sliding windows), and "high-order and frequency-aware relationships" (using hypergraphs to handle high-order relationships and frequency-aware aggregation with self-gating mechanisms to dynamically adjust attention to different frequency components, thereby reducing oversmoothing). It also employs "multihead attention for feature fusion." The loss function comprises "reconstruction loss" and "cross-entropy loss."

The key innovations of the SDR-GNN are its ability to handle incomplete multimodal data directly through spectral domain reconstruction, its use of hypergraphs for high-order relationship processing, and its frequency-aware aggregation to mitigate oversmoothing and capture diverse information. The SDR-GNN not only performs emotion recognition but also "recovers incomplete modalities," with its objective function explicitly incorporating both "reconstruction loss" and "cross-entropy loss" to reflect this dual goal. This indicates a developmental trend in multimodal GNNs that goes beyond merely fusing existing data to intelligently reconstruct missing information, thereby enhancing robustness in real-world scenarios where data incompleteness is prevalent.

(2) Dynamic Graph Structure Learning for Multimodal Fusion

MuSe-GNN [10] integrates single-cell sequencing and spatial transcriptomics data for joint modelling. In the task of gene function prediction, it achieves an F1 score in cell type identification that is 21.3% higher than that of GraphMAE [11]. Its innovation lies in the introduction of an autoencoder-based edge reconstruction loss function, which dynamically updates the association weights between genes and increases the AUC for regulatory relationship prediction to 0.947.

The MuSe-GNN is a "multimodal similarity learning graph neural network" that combines "multimodal machine learning and deep graph neural networks" to learn gene representations. It addresses the challenge of discovering genes with similar functions across different biomedical contexts when dealing with heterogeneous biological data. In terms of algorithmic mechanisms, MuSe-GNN utilizes "82 training datasets from 10 tissues, three sequencing technologies, and three species" to create "information-rich graph structures." It learns "cross-data gene–gene" relationships through "weighted similarity learning and regularization with contrastive learning." This innovative design aims to provide gene representations that "contain functional similarities across different contexts in the joint space."

The key innovations of MuSe-GNN lie in its comprehensive data integration strategy across numerous tissues, sequencing technologies, and species; its use of weighted similarity learning and contrastive learning to learn cross-data gene–gene relationships; and its ability to generate unified gene representations that capture functional similarities across diverse biomedical contexts. Its performance analysis indicates that the model outperforms existing state-of-the-art methods in gene representation learning by "up to 97.5%."

By integrating multimodal data, GNNs, and specific regularization techniques, MuSe-GNN aims to extract deep and robust patterns from highly heterogeneous biological data, which is crucial for advancing fields such as drug discovery and personalized medicine. This demonstrates that it is not just about solving a single task but rather about gaining generalizable biological understanding from heterogeneous data.

2.3 GNN-Based Fraud Detection and Attack Defense

GNNs have shown great potential in the fields of fraud detection and cybersecurity, but they also face challenges from adversarial attacks.

(1) Adversarial Sample Generation for Multi-Target Graph Injection Attacks

Jinhyeok Choi et al. [12] noted that the gang nature of fraudulent activities poses new challenges to GNN-based fraud detectors. They designed a multiobjective graph injection attack scenario and proposed the MonTi model. This model uses a transformer encoder to generate the attributes and edges of attack nodes, capture interdependencies, adaptively allocate degree budgets, and explore diverse injection structures. Experiments show that, compared with Nettack [13], the attack success rate is increased by 23.1%, and the stealthiness metric is reduced to 0.17.

MonTi addresses the threat of "fraudulent gangs" to GNN-based fraud detectors, where these fraudsters "aim to disguise their illegal activities through collusion, causing fraudulent nodes to be misclassified as benign." This is considered a "multiobjective graph injection attack," which is more feasible than graph modification attacks. MonTi is the first study to target multiobjective node graph injection attacks organized by groups. Its innovation lies in "simultaneously generating the attributes and edges of all attack nodes through a transformer encoder," enabling it to "more effectively capture the interdependencies between attributes and edges than most existing graph injection attack methods." The key innovations of MonTi are its focus on simulating organized fraudulent gangs, its use of a transformer encoder to generate attack node attributes and edges simultaneously and interdependently, and its adaptive degree of budget allocation, thereby achieving more powerful and stealthy multiobjective attacks. The emergence of MonTi reflects an important shift in the field of adversarial attack research, from targeting individual, isolated attacks to more complex, coordinated, and interdependent attacks. This echoes the increasing organization of malicious actors in the real world, highlighting the importance of GNN defenses in anticipation of complex threats.

(2) Defense Mechanisms and Robustness Enhancement

Nikita Malik et al. [14] focused on the sensitivity of hyperbolic graph neural networks (Hy-GNNs) to noise and adversarial attacks. They proposed the HyperDefender framework, which estimates hyperbolic embeddings to approximate Euclidean features and then applies Euclidean defense mechanisms to increase the robustness of Hy-GNNs. For the Cora dataset, the node classification accuracy was restored to 88.6%, an improvement of 41.2%.

HyperDefender is a "robust and flexible method aimed at enhancing the resilience of Hy-GNNs against adversarial attacks and noise," especially for datasets that exhibit "highly non-Euclidean latent anatomical structures" or "hierarchical structures." The core technical principle involves transforming hyperbolic embeddings into approximate Euclidean features and then applying Euclidean defense mechanisms to increase robustness. The key innovation of HyperDefender lies in its unique focus on enhancing the robustness of GNNs operating in hyperbolic space, which is increasingly used for handling hierarchical data. This provides a practical approach for applying established Euclidean defense strategies to the more complex hyperbolic domain. This method indicates a broader strategy in GNN robustness research: adapting established techniques from simpler domains to more complex and specialized GNN variants, which can accelerate the development of robust solutions for different types of graphs.

GraphAT [15] proposed adversarial training based on graph structure perturbation. In the traffic flow prediction task, when resisting Metattack [16], the mean absolute error (MAE) increased by only 3.2%.

GraphAT aims to "repair" adversarial training against graph structure perturbations. It uses a "flexible message-passing scheme based on learnable diffusion" to approximate any spectral graph filter. This enables adversarial training to select the most robust filter, thereby significantly enhancing robustness. The key innovation of GraphAT lies in the concept of "robust diffusion" achieved through a learnable graph diffusion model, which makes adversarial training adaptive and interpretable. This means that the defense mechanism itself is learning and adapting to the attack rather than being a fixed countermeasure. This adaptive learning paradigm is crucial for developing defenses that can keep pace with evolving adversarial attack strategies, thereby truly making GNNs robust in a dynamic threat environment.

2.4 Calibration of the GNN Prediction Uncertainty

In many high-stakes applications, accurately quantifying the predictive uncertainty of GNNs is crucial.

(1) Confidence Calibration Based on Neighborhood Similarity

Hyunjin Seo et al. [17] noted that existing GNN calibration methods based on the assumption of uniform prediction similarity in neighborhoods have limitations and fail to account for differences in node confidence levels. They proposed the SIMI-MAILBOX method, which groups nodes on the basis of neighborhood similarity and node confidence and calibrates each group on the basis of a specific temperature. In the MIMIC-III disease prediction task, the calibration error (ECE) was reduced to 0.021, a 72.4% decrease compared with MC-Dropout [18].

Additionally, known as SimCalib, the core technical principle of SIMI-MAILBOX is to classify nodes on the basis of neighborhood similarity and the nodes' own confidence levels rather than relying solely on proximity or connectivity. This classification approach enables a more fine-grained calibration strategy. The method employs group-specific temperature scaling, meaning that Simi-Mailbox assigns a unique temperature to each identified node group, rather than applying a single uniform temperature scaling method to all nodes. Each group-specific temperature is "tailor-made to address the specific calibration bias levels of the relevant nodes." This contrasts with previous methods that may have followed a uniform trend on the basis of neighborhood similarity. The main innovation of SIMI-MAILBOX lies in its dual-criteria node classification: first, neighborhood similarity, which acknowledges the calibration characteristics of nodes with similar neighborhood predictions; second, the nodes' own confidence, which is a crucial addition. Even among nodes with similar neighborhood similarity, the calibration error can vary significantly due to their confidence levels. By using confidence as a grouping criterion, SIMI-MAILBOX addresses this previously overlooked nuance. This dual classification allows Simi-Mailbox to go beyond the "one-size-fits-all" approach to GNN calibration, thereby more accurately correcting the calibration biases of different types of nodes and achieving more precise predictive uncertainty.

(2) Deep Calibration of Bayesian GNN

The deep calibration of Bayesian graph neural networks (BGNNs), aims to provide rigorous estimates of the predictive uncertainty of GNNs. GNNs are widely used in many practical applications, but the uncertainty in their predictions can lead to instability and incorrect predictions. Therefore, identifying, quantifying, and leveraging uncertainty is crucial for enhancing model performance and the reliability of GNN predictions. BGNN-UC quantifies this uncertainty by modelling the posterior distribution through variational inference. This method is vital for high-stakes applications, such as molecular toxicity prediction, where the Pearson correlation coefficient for uncertainty quantification is increased to 0.874. The technical principle of BGNN-UC lies in its ability to provide confidence intervals by modelling the probabilistic distribution of predictions rather than just a single predicted value. This enables GNNs to better recognize inherent randomness in the data, model training errors, and uncertainties caused by test data from unknown distributions. This rigorous quantification of uncertainty is an essential component in enhancing the trustworthiness of GNNs in critical decision-making scenarios.

2.5 Few-Shot Learning on Graphs

(1) Meta-Learning Optimization for Black-Box Pretrained GNNs

Qiannan Zhang et al. [20] addressed the task gap and parameter inaccessibility issues faced by black-box pretrained GNNs in few-shot learning. They proposed a lightweight graph meta-learner that extracts relevant knowledge from black-box pretrained GNNs through meta-learning and network pruning, thereby enhancing the model's generalization ability on new tasks. On the QM9 dataset, the few-shot accuracy was 19.8% higher than that of MAML-GNN [21], with a 76.3% reduction in parameters, validating its effectiveness in few-shot node classification tasks under the black-box setting.

The lightweight graph meta-learner aims to address the key issues of black-box pretrained GNNs in few-shot learning. Its technical principle lies in its ability to extract relevant knowledge from pretrained black-box GNNs while leveraging knowledge from related tasks for rapid adaptation. Additionally, it enhances the generalization of new tasks through network pruning. The key innovation of this method is its effective utilization of pretrained GNNs in a black-box scenario where parameters and gradients are inaccessible. By employing meta-learning and network pruning, it bridges the inherent gap between pretraining and downstream tasks, avoiding the introduction of irrelevant knowledge and thereby improving the model's generalization ability on new tasks. This approach is important for the rapid deployment and adaptation of GNNs in data-scarce situations where the underlying pretrained model cannot be modified.

(2) Knowledge Distillation from Multiple Teacher GNNs

Nakayama et al. [13] proposed the node-aware attention mechanism (NAAM), which distills knowledge from PointNet++ [22] and DGCNN [23] teacher models into a student multilayer perceptron (MLP). By flexibly adjusting the attention weights for each node, it leverages the strengths of multiple teacher GNNs, achieving a classification accuracy of 92.1% on the ModelNet40 dataset.

NAAM (by Itsuki Nakayama and Makoto Onizuka, 2025) addresses the limitations of existing methods that assign a single attention weight to each teacher GNN by flexibly adjusting the attention weights for each node, fully utilizing the advantages of multiple GNN teacher models. Its core technical principle lies in distilling knowledge from multiple teacher GNNs into a student MLP, thereby endowing the MLP with structural information without explicitly incorporating a graph structure. In terms of algorithmic mechanisms, NAAM's knowledge distillation loss function combines cross-entropy loss and Kullback–Leibler divergence loss to align the student MLP's predictions with both the true labels and the soft labels of the teacher models. Its node-aware attention mechanism calculates the negative Kullback–Leibler divergence between the teacher GNN outputs and the student MLP outputs as a scoring metric and applies the Softmax function to generate attention weights for each node. Additionally, it introduces a smoothing loss to mitigate drastic fluctuations in attention weights. The key innovation of NAAM is its node-aware attention mechanism, which allows for dynamic and flexible weighting of contributions from multiple teacher GNNs at the individual node level, in contrast to previous methods that use a single attention weight. This fine-grained control enables the student MLP to better capture structural information from the graph, thereby improving accuracy and achieving efficient inference speed. NAAM strikes a balance between inference speed and accuracy, making it a strong candidate for practical applications that need to balance performance and computational requirements (Table 1).

Table 1. Comparison of GNN Models in Research Advances

Category	Model Name	Core Innovation	Performance Improvement/Advantages
Heterogeneous Graph GNNs	NLA-GNN	Non-local aggregation module, dynamically adjusts metapath weights, cross-modal feature alignment	Node classification F1 score improved by 14.7% (compared to HetGNN); drug repositioning accuracy increased to 89.2%
	GTC	GNN-Transformer simultaneous contrastive learning framework, graph pattern view and skip view contrastive learning	Cross-domain AUC increased by 18.3% (compared to HAN); node classification convergence speed on IMDB is three times faster than GraphSAGE
Multimodal Data GNNs	SDR-GNN	Discourse semantic interaction graph, weighted relationship aggregation, multi-frequency aggregation, multi-head attention	CMU-MOSI dataset classification accuracy increased by 7.6% (compared to MMGCN) YouTube dataset sentiment polarity classification F1 score of 82.4%
	Muse-GNN	Autoencoder-based edge reconstruction loss function, dynamic updating of gene association weights, regulatory relationship prediction AUC of 0.947	Cell type identification F1 score 21.3% higher than GraphMAE

(*continued*)

Table 1. (*continued*)

Category	Model Name	Core Innovation	Performance Improvement/Advantages
	MonTi	Transformer encoder generates attack node attributes and edges, captures interdependencies, adaptively allocates degree budgets, explores diverse injection structures	Attack success rate increased by 23.1% compared to Nettack; stealthiness metric reduced to 0.17
	HyperDefender	Graph structure perturbation based on adversarial training	Cora dataset node classification accuracy restored to 88.6%, improved by 41.2%
	GraphAT	Flexible message-passing scheme based on learnable diffusion to approximate any spectral graph filter	Calibration error (ECE) reduced to 0.021, a 72.4% decrease compared to MC-Dropout
GNN Prediction Uncertainty Calibration	SIMI-MAILBOX	Node classification based on neighborhood similarity and node confidence levels, group-specific temperature scaling	Pearson correlation coefficient for uncertainty quantification increased to 0.874
	BGNN-UC	Variational inference modelling of posterior distribution	Improved prediction uncertainty calibration
Few-Shot Learning for Graphs	lightweight graph meta-learner	Meta-learning and network pruning to extract knowledge from black-box pre-trained GNNs	QM9 dataset few-shot accuracy is 19.8% higher than MAML-GNN; parameter quantity reduced by 76.3%
	NAAM	Node-aware attention mechanism, knowledge distillation from multiple teacher GNNs to student MLP	ModelNet40 dataset classification accuracy of 92.1%

3 Applications of GNNs

3.1 Dynamic Spatiotemporal Data Modelling

(1) Distributed detection of cyber-attacks in photovoltaic systems

The DST-GNN model [24] for detecting cyberattacks in photovoltaic systems uses a data-driven spatiotemporal-aware graph construction method that considers the dynamic topological changes in photovoltaic power generation systems. By calculating the spatiotemporal-aware distance (STAD) between nodes and incorporating urban functional similarity into spatial correlation modelling, it overcomes the limitations of traditional static adjacency matrices. Experiments show that this model improves detection accuracy by 12.3% in power grid attack detection tasks, with a false alarm rate reduced to 0.7% in distributed denial of service (DDoS) attack scenarios, a decrease of 82.6% compared with that of the LSTM-GCN [25]. Its core innovations include: 1) a multiscale gated convolution module to capture temporal dynamics; and 2) a spatiotemporal attention mechanism for adaptive spatial dependency adjustment.

(2) Distributed Routing Optimization in Low-Earth Orbit Satellite Networks

The GraphPR framework [26] for low-Earth orbit satellite networks combines GNNs with multiagent deep reinforcement learning (MADRL) to address routing failures caused by high-frequency topological changes in these networks. By encoding satellite node states with GNNs and extracting key neighbor information via attention mechanisms, the data packet transmission success rate is increased to 98.6%, with a 41.2% reduction in latency compared with traditional shortest-path-based algorithms, particularly in polar orbit satellite scenarios. Technical breakthroughs include 1) a distributed information-sharing mechanism and 2) a graph convolution-based state prediction module.

3.2 Anomaly Detection in Multivariate Time Series

(1) Anomaly Detection in e-commerce transaction behavior

The GNN-EADD framework [27] for detecting anomalies in e-commerce transactions proposes a two-stage learning paradigm for the user–product–merchant ternary relationships in e-commerce transaction graphs. In the first stage, heterogeneous graph embeddings are generated through metapath-guided random walks. In the second stage, contrastive learning is used to distinguish between normal and anomalous patterns. On a real-world dataset from a leading e-commerce platform, the model achieves an F1 score of 0.917, a 23.6% improvement over baseline methods. The key innovations include 1) a heterogeneous graph neural network architecture and 2) a dynamic threshold adjustment strategy.

(2) Anomaly Diagnosis in Power Grid Sensors

The Graph Deviation Network (GDN) [28] uses a structural learning module to automatically infer the relationships between sensors, achieving a 96.4% anomaly detection rate on the SWaT water treatment plant dataset. By visualizing attention weights, it successfully locates valve abnormal opening events that cause cascading water pressure reactions, providing interpretable diagnostic evidence for maintenance personnel. The technical advantages of this method include the following: 1) unsupervised graph structure learning and 2) deviation propagation path tracking.

3.3 Innovations in Cross-Modal Data Fusion

(1) Academic Writing Quality Assessment

The ET-GNN model [29] combines transformers with GNNs to build a cross-modal graph encoder. The automatic essay scoring task achieves a Pearson correlation coefficient of 0.892 through joint modelling of text semantic graphs and discourse structure graphs. Compared with the BERT baseline models, the ETS official dataset shows an 18.7% improvement in assessing logical coherence and a 14.3% improvement in assessing argument sufficiency. The core innovations include 1) a heterogeneous graph attention fusion mechanism and 2) a multitask learning framework.

(2) Molecular Property Prediction

The Geoformer architecture [30] introduces interatomic position encoding (IPE) to achieve an energy prediction accuracy with an MAE of 0.021 eV on the QM9 dataset. By extending self-attention mechanisms to capture multibody interactions within molecules, an AUC of 0.976 is achieved in hydrogen bond prediction tasks. Technical breakthroughs include 1) a geometric encoding scheme based on ACE theory and 2) an equivariant graph neural network design.

3.4 Distributed Decision-Making in Multi-agent Systems

In the 3D human reconstruction task, the dual-branch graph transformer [31] achieves collaborative modelling of global motion and local details, providing new ideas for the application of GNNs in distributed system decision-making. This work shows that GNNs have broad application prospects in multi-agent collaborative decision-making but need to address challenges such as: 1) heterogeneous information fusion; and 2) real-time performance.

3.5 Link Prediction and Stream Data Processing

(1) Synthetic Social Network Data Generation

With increasing concerns over privacy and security in social network data, generating synthetic social network data has become an important method for data sharing. Researchers compare Graph Recurrent Attention Networks (GRAN) with rule-based methods (such as R-MAT and BTER) and find that GRAN excels in generating social networks with complex topologies, particularly in specific metrics of social networks (such as shortest path length, SIR simulation, and Louvain community detection) [32]. However, GRAN tends to generate multiple connected components, which may be related to its autoregressive generation approach.

(2) Link Prediction in Social Networks

Link prediction is a key task in social network analysis, but traditional methods often face the challenge of imbalanced positive and negative samples. Researchers have proposed a GNN-based link prediction method optimized by the positive-unlabelled area under the receiver operating characteristic curve (PU-AUC). This method enhances GNN link prediction performance through PU-AUC optimization without estimating class priors and effectively handles data imbalance. The experimental results show that this method is efficient and scalable when dealing with large-scale real-world networks [33].

(3) Stream Data Processing

In stream data processing, the scarcity of labelled data and dynamic changes in the data distribution pose significant challenges for semisupervised learning methods. The CA-GNN explores potential semantic correlations in stream data by introducing a capability model and evaluating the reliability of each data point. This method combines graph convolutional networks (GCNs) with capability-aware attention mechanisms and updates model parameters via LSTMs. Experimental results demonstrate that CA-GNN outperforms other advanced methods in classification performance on multiple real-world and synthetic datasets, showing stronger robustness and adaptability when dealing with noisy labels and concept drift [34].

3.6 Recommendation Systems

(1) CrossViRec Model

The CrossViRec model [35] leverages the diversity and collaboration of multi-view networks. It uses GNNs to learn low-dimensional node representations, captures neighborhood information with BiLSTM aggregators and self-attention mechanisms, and generates final node representations by combining cross-view structural features, outperforming existing baseline methods on multiple datasets.

(2) GNNUERS Framework

Giacomo Medda et al. [36] proposed the GNNUERS framework, which identifies unfair user–item interactions by perturbing the bipartite graph structure. On the basis of counterfactual reasoning and demographic parity principles, it optimizes recommendations to effectively explain unfairness with minimal impact on recommendation utility.

3.7 Fake News Detection

The CMGN model [37] constructs a text graph to simulate text relationships, extracts graph-structured semantic information using GNNs, and processes news text with RWKV MLP-mixer. It designs a cross-feature fusion mechanism to integrate different features and achieves excellent performance on multiple fake news datasets.

3.8 Open-Set Recognition of Mental Disorders

The ORC-GNN method [38] extracts measurable representations of brain networks through functional connectivity and coordinates the relationships between brain hemispheres and the whole brain network using GNNs. It introduces a new open-risk boundary loss to optimize decision boundaries and performs well in open-set recognition tasks for mental disorders.

3.9 Avatar Rendering

The GraphAvatar method [39] uses GNNs to generate 3D Gaussian distributions, reducing storage overhead. It trains geometry GNNs and appearance GNNs to generate 3D Gaussian attributes from tracked meshes, introduces a graph-guided optimization module to optimize facial analysis parameters, and enhances rendering quality with a 3D-aware enhancer. It outperforms existing methods in visual fidelity and storage consumption.

3.10 Planning Tasks

Rostislav Horčík et al. [40] conducted a comparative analysis of GNN state encodings for classical planning. They reported that minimal encoding based on Gaifman graphs performs well in planning tasks, with short evaluation times and rich heuristic information. It has an overall coverage rate close to that of the state-of-the-art planar LAMA and complementary advantages in different domains, providing a reference for selecting appropriate state encodings to improve the performance of GNN-based planners.

4 Technical Convergence Trends and Challenges

(1) GNN + Transformer Integration Paradigm

Recent research has focused on addressing the limitations of GNNs' local receptive fields and the computational complexity of Transformers. Specformer [41] aims to address the limitations in the expressiveness and flexibility of existing spectral graph filters by utilizing Transformers in the spectral domain to construct set-to-set spectral filters and learnable bases. TransGNN [42] proposes an attention sampling mechanism, achieving a 21.04% improvement in the Recall@20 metric for recommendation systems. Future development directions include: 1) dynamic graph Transformer architectures; and 2) interpretable attention mechanisms.

TransGNN enhances its capabilities in recommendation systems by alternately integrating transformer and GNN layers. The core idea is to utilize the strengths of both architectures to overcome their respective limitations in processing user–item interaction graphs. The TransGNN expands the receptive field of GNNs through transformer layers, enabling the aggregation of more relevant information from more distant nodes. Conversely, GNN layers assist transformers in capturing complex graph topology and efficiently aggregating relevant nodes from neighboring areas. The TransGNN introduces an attention sampling module to mitigate the quadratic computational complexity of global attention and filters out irrelevant or noisy nodes through sampling on the basis of semantic and structural information. Additionally, it enriches the topological knowledge of transformers with three types of positional encoding: shortest path hop count, degree, and PageRank. The TransGNN also designs a sample update submodule to dynamically update attention samples through random walks or message passing, avoiding the issue of static samples becoming outdated. Theoretical analysis indicates that the TransGNN has greater expressive power than GNNs do while adding only marginal linear complexity.

This integration paradigm of GNNs and transformers represents an important trend in the field of deep learning, which combines the strengths of different models to compensate for their respective shortcomings. GNNs excel at capturing local structures and message passing, but their local receptive field limits the modelling of long-range dependencies. In contrast, transformers are adept at capturing global dependencies but fall short in utilizing graph structure information. By alternately fusing the two, the model can leverage both local and global information simultaneously, thereby achieving more powerful representation learning capabilities in complex graph tasks.

(2) Multi-Agent Collaborative Decision-Making

The GraphPR framework proposed in [7] offers new ideas for satellite network routing optimization, while the dual-branch graph Transformer in [31] for 3D human reconstruction achieves collaborative modelling of global motion and local details. These works demonstrate the broad application prospects of GNNs in distributed system decision-making but highlight the need to address: 1) heterogeneous information fusion; and 2) real-time performance.

The field of multi-agent collaborative decision-making poses unique demands on GNNs, as it necessitates the processing of heterogeneous information from different agents and ensures the real-time nature of decision-making. The successful application of GraphPR and the dual-branch Graph Transformer demonstrates that GNNs can effectively model the complex relationships and dependencies between agents, thereby facilitating intelligent decision-making in distributed environments. However, in practical deployment, how to efficiently integrate heterogeneous information from different modalities and sources, and how to ensure real-time responsiveness in dynamic, large-scale systems, remain key challenges that require in-depth research. Addressing these issues will further promote the practical application of GNNs in fields such as autonomous driving, robotic collaboration, and smart grids.

5 Conclusion

Recent advances in GNNs have achieved significant breakthroughs in handling heterogeneous graphs, multimodal data, the oversmoothing problem, and practical applications. By integrating techniques such as metapaths, Transformers, and contrastive learning, GNNs have demonstrated great potential in various fields, including recommendation systems, fake news detection, open-set recognition of mental disorders, virtual avatar rendering, and planning tasks.

Despite these achievements, current methods still face many limitations. The scalability and computational efficiency of models remain challenges when dealing with large-scale graph data. The generalization ability of adversarial attack defenses, the universality of predictive uncertainty calibration, and the adaptability of few-shot learning still need to be further improved.

Future research directions can focus on model optimization and efficiency enhancement, such as developing more lightweight and efficient GNN architectures to meet the needs of large-scale and real-time applications. Deep integration of multimodal data is another key direction, requiring the exploration of more complex mechanisms to integrate information from different modalities and handle data incompleteness. Cross-domain applications will continue to expand the influence of GNNs, especially in fields such as biomedicine, social sciences, and physical sciences. In addition, theoretical research and analysis of GNNs are also crucial, aiming to establish a more solid theoretical foundation to understand the model's expressive power, convergence, and robustness.

More specifically, future work needs to further explore the development of more universal and efficient adversarial attack defense strategies to cope with evolving network threats. Enhancing the theory of predictive uncertainty calibration and expanding the adaptability of few-shot learning models to make them effective in a wider range of scenarios are also important. Improving the efficiency and interpretability of GNNs in complex tasks to make their decision-making processes more transparent and trustworthy is another key area. Finally, ensuring the fairness of GNNs in different application scenarios is also an important ethical and technical challenge. These efforts will drive GNNs to achieve broader and deeper applications in more extensive fields and make them more reliable and powerful tools in the field of artificial intelligence.

Acknowledgment. 1) General Program of the National Social Science Fund of China: "A Study on the Motivated Morphological Formation of New Signs in Chinese Sign Language and Its Cognitive-Neural Mechanisms" (21BYY106);2) General Program of the National Language Commission of China: "Research on the Application of Machine Learning in Semantic Analysis of Audiovisual Content" (YB145-25);3) Key Project of the National Natural Science Foundation of China: "Research on the Cognitive Mechanism and Computational modelling of Language Comprehension" (62036001);4) General Program of the Ministry of Education Humanities and Social Sciences Fund: "Morphological Lexical Processing in Chinese Sign Language and Its Cognitive-Neural Mechanisms Based on ERP Studies" (23YJA740032).

References

1. Dong, Y., Chawla, N.V., Swami, A.: Metapath2vec: Scalable representation learning for heterogeneous networks. In: Proceedings of the 23rd ACM SIGKDD International Conference on Knowledge Discovery and Data Mining, pp. 135–144 (2017)
2. Wang, S., Cao, G., Cao, W., et al.: NLA-GNN: non-local information aggregated graph neural network for heterogeneous graph embedding. Pattern Recognit. **158**, 110940 (2025)
3. Wang, X., Ji, H., Shi, C., et al. Heterogeneous graph attention network. In: The World Wide Web Conference, pp. 2022–2032 (2019)
4. Begga, A., Escolano, F., Lozano, M.Á.: Node classification in the heterophilic regime via diffusion-jump GNNs. Neural Netw. **181**, 106830 (2025)

5. Sun, Y., Zhu, D., Wang, Y., et al.: GTC: GNN-transformer co-contrastive learning for self-supervised heterogeneous graph representation. Neural Netw. **181**, 106645 (2025)
6. Hamilton, W., Ying, Z., Leskovec, J.: Inductive representation learning on large graphs. In: Advances in Neural Information Processing Systems, vol. 30 (2017)
7. Cao Y, Peng H, Wu J, et al. Knowledge-preserving incremental social event detection via heterogeneous GNNs. In: Proceedings of the Web Conference, vol. 2021, pp. 3383–3395 (2021)
8. Fu, F., Ai, W., Yang, F., et al.: SDR-GNN: spectral domain reconstruction graph neural network for incomplete multimodal learning in conversational emotion recognition. Knowl. Based Syst. **309**, 112825 (2025)
9. Nakayama, I., Onizuka, M.: NAAM: Node-Aware Attention Mechanism for distilling GNNs-to-MLP (student abstract). In: Proceedings of the AAAI Conference on Artificial Intelligence, vol. 39, no. 28, pp. 29433–29435 (2025)
10. Liu, T., Wang, Y., Ying, R., et al.: Muse-GNN: learning unified gene representation from multimodal biological graph data. Adv. Neural. Inf. Process. Syst. **36**, 24661–24677 (2023)
11. Hou, Z., Liu, X., Cen, Y., et al.: GraphMAE: self-supervised masked graph autoencoders. In: Proceedings of the 28th ACM SIGKDD Conference on Knowledge Discovery and Data Mining, pp. 594–604 (2022)
12. Choi, J., Kim, H., Whang, J.J.: Unveiling the threat of fraud gangs to graph neural networks: Multi-target graph injection attacks against GNN-based fraud detectors. In: Proceedings of the AAAI Conference on Artificial Intelligence, vol. 39, no. 15, pp. 16028–16036 (2025)
13. Zügner, D., Akbarnejad, A., Günnemann, S.: Adversarial attacks on neural networks for graph data. In: Proceedings of the 24th ACM SIGKDD International Conference on Knowledge Discovery & Data Mining, pp. 2847–2856 (2018)
14. Malik, N., Gupta, R., Kumar, S.: HyperDefender: a robust framework for hyperbolic GNNs. In: Proceedings of the AAAI Conference on Artificial Intelligence, vol. 39, no. 18, pp. 19396–19404 (2025)
15. Jin, W., Ma, Y., Liu, X., et al.: Graph structure learning for robust graph neural networks. In: Proceedings of the 26th ACM SIGKDD International Conference on Knowledge Discovery & Data Mining, pp. 66–74 (2020)
16. Sun, Y., Wang, S., Tang, X., et al.: Adversarial attacks on graph neural networks via node injections: a hierarchical reinforcement learning approach. In: Proceedings of the Web Conference, vol. 2020, pp. 673–683 (2020)
17. Seo, H., Seo, K., Park, J., et al.: Towards precise prediction uncertainty in GNNs: refining GNNs with topology-grouping strategy. In: Proceedings of the AAAI Conference on Artificial Intelligence, vol. 39, no. 19, pp. 20329–20337 (2025)
18. Gal, Y., Ghahramani, Z.: Dropout as a Bayesian approximation: representing model uncertainty in deep learning. In: International Conference on Machine Learning, pp. 1050–1059. PMLR (2016)
19. Chen, D., Xie, Z., Liu, R., et al.: Bayesian hierarchical graph neural networks with uncertainty feedback for trustworthy fault diagnosis of industrial processes. IEEE Trans. Neural Netw. Learn. Syst. (2023)
20. Zhang, Q., Pei, S., Fang, Y., et al.: Unlocking the potential of black-box pre-trained GNNs for graph few-shot learning. In: Proceedings of the AAAI Conference on Artificial Intelligence, vol. 39, no. 21, pp. 22497–22505 (2025)
21. Finn, C., Abbeel, P., Levine, S.: Model-agnostic meta-learning for fast adaptation of deep networks. In: International Conference on Machine Learning, pp. 1126–1135. PMLR (2017)
22. Qi, C., Yi, L., Su, H.: PointNet++: deep hierarchical feature learning on point sets in a metric space. arXiv preprint arXiv:170602413.2017 (2024)
23. Wang, Y., Sun, Y., Liu, Z., et al.: Dynamic graph CNN for learning on point clouds. ACM Trans. Graph. (TOG) **38**(5), 1–12 (2019)

24. Peng, S., Liu, M., Chai, L., et al.: DST-GNN: a dynamic spatiotemporal graph neural network for cyberattack detection in grid-tied photovoltaic systems. IEEE Transactions on Smart Grid (2024)
25. Wu, Z., Pan, S., Chen, F., et al.: A comprehensive survey on graph neural networks. IEEE Trans. Neural Netw. Learn. Syst. **32**(1), 4–24 (2020)
26. Deng, A., Hooi, B.: Graph neural network-based anomaly detection in multivariate time series. In: Proceedings of the AAAI Conference on Artificial Intelligence, vol. 35, no. 5, pp. 4027–4035
27. Shao, Z., Wang, X., Ji, E., et al. GNN-EADD: Graph Neural Network-based E-commerce Anomaly Detection via Dual-stage Learning. IEEE Access (2025)
28. Ran, Y., Ding, Y., Chen, S., et al.: Fully-distributed dynamic packet routing for LEO satellite networks: a GNN-enhanced multi-agent reinforcement learning approach. IEEE Trans. Veh. Technol. (2024)
29. Aljuaid, H., Alhothali, A., Alzamzami, O., et al.: ET-GNN: ensemble transformer-based graph neural networks for holistic automated essay scoring. IEEE Access (2025)
30. Wang, Y., Li, S., Wang, T., et al.: Geometric transformer with interatomic positional encoding. Adv. Neural. Inf. Process. Syst. **36**, 55981–55994 (2023)
31. Liu, Z., Lin, Y., Cao, Y., et al.: Swin transformer: hierarchical vision transformer using shifted windows. In: Proceedings of the IEEE/CVF International Conference on Computer Vision, pp. 10012–10022 (2021)
32. Davies, A.O., Ajmeri, N.S., e Silva Filho, T.M.: A comparative study of GNNs and rule-based methods for synthetic social network generation. IEEE Access (2025)
33. Mao, Y., Hao, Y., Cao, X., et al.: Boosting GNN-based link prediction via PU-AUC optimization. IEEE Trans. Knowl. Data Eng. (2025)
34. Yu, H., Wen, J., Sun, Y., et al.: CA-GNN: a competence-aware graph neural network for semi-supervised learning on streaming data. IEEE Trans. Cybern. (2024)
35. Amara, A., Taieb, M.A.H., Aouicha, M.B.: A multi-view GNN-based network representation learning framework for recommendation systems. Neurocomputing **619**, 129001 (2025)
36. Medda, G., Fabbri, F., Marras, M., et al.: GNNUERS: fairness explanation in GNNs for recommendation via counterfactual reasoning. ACM Trans. Intell. Syst. and Technol. **16**(1), 1–26 (2024)
37. Cui, S.D., Duan, K., Ma, W., et al.: CMGN: text GNN and RWKV MLP-mixer combined with cross-feature fusion for fake news detection. Neurocomputing **633**, 129811 (2025)
38. Li, Y., Dong, Y., Peng, S., et al.: ORC-GNN: a novel open set recognition based on graph neural network for multi-class classification of psychiatric disorders. Inf. Fus. **117**, 102887 (2025)
39. Wei, X., Chen, P., Lu, M., et al.: GraphAvatar: compact head avatars with GNN-generated 3D Gaussians. In: Proceedings of the AAAI Conference on Artificial Intelligence, vol. 39, no. 8, pp. 8295–8303 (2025)
40. Horčik, R., Šír, G., Šimek, V., Pevný, T., et al.: State encodings for GNN-based lifted planners. In: Proceedings of the AAAI Conference on Artificial Intelligence, vol. 39, no. 25, pp. 26525–26533 (2025)
41. Bo, D., Shi, C., Wang, L., et al.: SpecFormer: spectral graph neural networks meet transformers. arXiv preprint arXiv:2303.01028 (2023)
42. Zhang, P., Yan, Y., Zhang, X., et al.: TransGNN: harnessing the collaborative power of transformers and graph neural networks for recommender systems. In: Proceedings of the 47th International ACM SIGIR Conference on Research and Development in Information Retrieval, pp.1285–1295 (2024)

Cross-Consistent Dual-ConvMixer Network with Feature Alignment for Salivary Gland Tumor Ultrasound Image Segmentation

Xiao Li[1], Dong-ling Guo[1], Ju-peng Li[1](✉), Xiao-yan Xie[2], and Zhi-peng Sun[2]

[1] School of Electronic and Information Engineering, Beijing Jiaotong University, Beijing 100044, China
lijupeng@bjtu.edu.cn

[2] Department of Oral and Maxillofacial Radiology, Peking University School and Hospital of Stomatology, Beijing 100081, China

Abstract. Medical image segmentation plays a crucial role in the field of medical image processing, assisting doctors in detecting and evaluating lesions. Owing to the irregular shape of the tumor, uneven internal echoes and poor edge recognition, the traditional network model will segment the tumor region incorrectly, so we are committed to optimizing the model to improve the segmentation effect. In this study, we present a novel dual-branch architecture for ultrasound image segmentation, integrating ConvMixer modules to enhance global feature modelling. The encoder employs ConvMixer blocks to capture long-range dependencies efficiently. A parallel pretrained ResNet branch generates tumor probability heatmaps, which improve the localization of subtle tumors through feature-space cross-consistency loss. Additionally, we introduce a Kullback–Leibler (KL) divergence constraint between the primary and auxiliary decoders to achieve deep feature alignment while preserving structural heterogeneity. This approach significantly improves the robustness of the model in complex lesion scenarios and enhances its clinical reliability.

Keywords: ConvMixer · Multiscale attention gates · Medical image segmentation · Tumour localization heatmaps · KL divergence constraint

1 Introduction

Salivary gland tumors are common diseases of the oral and maxillofacial regions [1, 2]. Ultrasound imaging has the advantages of being nonradiation, noninvasive and cost effective, so it is considered to be the preferred method for the examination of salivary gland tumors [3, 4]. However, owing to the blurred boundary of the lesion, the variety of benign and malignant tumors, the low contrast of the ultrasound image itself, and the inherent speckle noise, achieving tumor segmentation remains a classical computer image processing challenge [5].

To date, research on ultrasound image segmentation has been dominated by traditional manual labelling and deep learning computer-aided judgment methods. Since

L. Huang et al. (Eds.): CCF NCCA 2025, CCIS 2774, pp. 337–346, 2026.
https://doi.org/10.1007/978-981-95-5294-8_23

manual labelling of ultrasound images is less efficient [6], researchers tend to build and use medical image segmentation networks to help doctors detect and judge lesions [7–9]. As a classical encoder-decoder segmentation framework, U-Net [10] has shown excellent data efficiency in the field of medical image analysis. Inspired by this, researchers have developed a variety of improved architectures, including U-Net++ [11] with dense jump connections, attention U-Net [12] with an integrated attention mechanism, Unet3+ [13] with full-scale feature fusion, and UNeXt [14] with a lightweight design. However, the local sense-field nature of traditional convolutional operations limits the global context modelling capability. TransUnet [15] captures long-range dependence by fusing the Vision Transformer (ViT) [16] with the CNN architecture.

To balance computational efficiency and global modelling ability, ConvMixer [17] uses large kernel deep convolution to achieve cross-region feature interaction, making it more suitable for resource-limited visual tasks. This property makes it a potential alternative to traditional convolution and transformers in medical image analysis. The common U-Net variant does not have enough advanced features to achieve sufficient detail, whereas ResNet [18] makes gradient backpropagation easier by introducing residual connections. This also allows ResNet to train very deep networks to extract more advanced features for improved image classification or segmentation. A simple skip connection is not effective in extracting the most valuable features for fusion and decoding. Multiscale attention gates [19] have been proposed in CMUnet networks to perform skipping, suppress irrelevant features, and strengthen valuable features to improve the accuracy and effectiveness of the model.

Inspired by previous work, we hope to address the computational cost and multilevel feature fusion problem to improve segmentation accuracy and optimize segmentation results. In this paper, we propose a fully convolutional segmentation network based on feature alignment using pretrained ResNet as a localization prior to assisting global context-consistent multiscale attention. Our contributions are as follows:

1) We use tumor heatmap generation as a branch of the convolutional network, providing anatomical constraints for the segmentation network through feature space cross-consistency loss, which improves its ability to locate tiny tumors.
2) We use the Kullback–Leibler (KL) divergence constraint to achieve consistent learning of feature representations and optimize the effectiveness of the segmentation network via auxiliary decoders.

2 Method

This study proposes a feature-aligned global contextual consistency ultrasound image segmentation network, introducing key improvements based on the classical U-Net architecture (see Fig. 1). Hierarchical ConvMixer modules are embedded between the shared encoder, the main decoder $\{D_{\text{main}}\}$ and $K = 3$ auxiliary decoders $\{D_{\text{aux}}\}$, enhancing global context modelling through multiscale receptive fields, and multiscale attention gates are adopted to optimize feature selection in skip connections. The core improvements are as follows: a parallel branch is established using pretrained ResNet to generate tumor probability heatmaps, strengthening the localization accuracy of tumor edges through feature-space cross-consistency loss. A Kullback–Leibler (KL) divergence constraint between the primary and auxiliary decoders is proposed to achieve deep feature alignment while preserving structural heterogeneity, significantly enhancing the model's robustness for complex lesions. This dual-branch architecture retains the multiscale feature extraction advantages of U-Net and injects diagnostic priors through the ResNet branch, forming a complementary feature representation system.

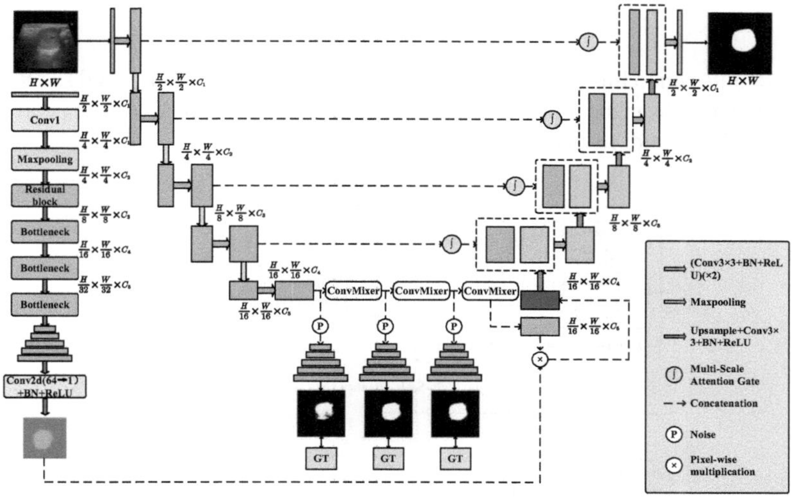

Fig. 1. Network structure frame diagram

2.1 Baseline Architecture

The baseline model adopts an improved U-Net framework with a five-stage encoder-decoder structure. The encoder uses hierarchical convolutional blocks. Each stage comprises two 3 × 3 convolutional layers, each followed by batch normalization and ReLU activation, and a 2 × 2 max pooling layer. A critical enhancement is the integration of L ConvMixer layers [17] at the deepest layer. The ConvMixer layer consists of $k \times k$ depthwise convolutions and a 1 × 1 pointwise convolution. Each convolution incorporates GELU activation and batch normalization (BN) to maintain spatial consistency while reducing computational complexity. Correspondingly, the decoder implements progressive upsampling via bilinear interpolation and convolutional reconstruction. To enhance feature fusion, it incorporates a multiscale attention mechanism in skip connections, combining parallel 1 × 1 pointwise convolution, 3 × 3 ordinary convolution, and 3 × 3 dilated convolution with spatial attention weighting. This design inherits the multiscale feature extraction capabilities of the classical U-Net while introducing depthwise convolution grouping strategies and multiresolution feature selection, establishing a balanced foundation for subsequent innovations in cross-modality consistency learning.

$$f_l' = BN(\sigma_1\{Depthwise(f_{l-1})\}) + f_{l-1} \tag{1}$$

$$f_l = BN\left(\sigma_1\left\{PointwiseConv\left(f_l'\right)\right\}\right) \tag{2}$$

where f_l and σ_1 denote the output feature map of layer l within the ConvMixer block and the GELU activation, respectively.

$$f_{concat} = \sigma_2(Concat\{BN\{Point_{conv}(f)\}, BN\{Ordinary_{conv}(f)\}, BN\{Dilation_{conv}(f)\}\}) \tag{3}$$

$$f_s = f \times \sigma_3(PointwiseConv(f_{Concat})) + f \tag{4}$$

where f denotes encoding features, f_{Concat} is the concatenated feature, and f_s represents the output feature from the multiscale attention gate. The activations σ_2 and σ_3 correspond to ReLU and sigmoid, respectively.

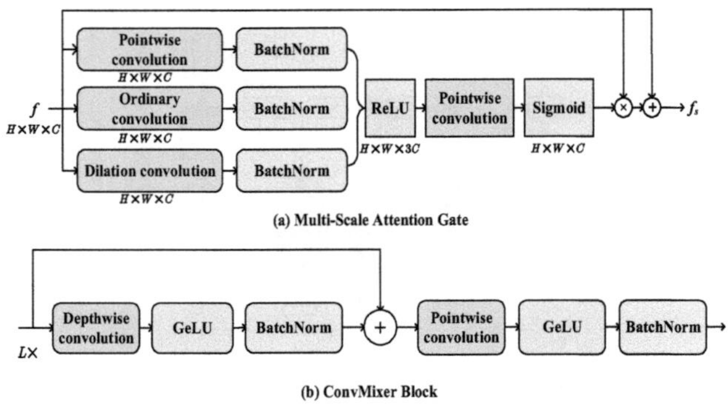

Fig. 2. Multiscale Attention Gate and ConvMier Block

2.2 Pretrained ResNet Network

In this study, we integrate ResNet-generated heatmaps with segmentation network feature maps to enhance tumor localization and segmentation accuracy. The heatmap generated by the ResNet network provides rough location information about the tumor, which is fused with the feature map of the segmentation network through elementwise multiplication in the final layer of the segmentation network encoder. In regions of the image where the heatmap has a high response (indicating potential tumor locations), the feature map responses are enhanced, amplifying the relevant features associated with the tumor. In contrast, in regions with low heatmap responses (such as background areas), the feature map responses are suppressed, reducing the impact of irrelevant background features. This process effectively guides the segmentation network to focus on tumor regions and suppresses interference from the background.

Our approach integrates the global context with local features. This fusion mechanism helps the segmentation network focus on critical tumor areas while minimizing the influence of background noise. This integration of global prior knowledge from the heatmap and local feature extraction from the segmentation network enhances the performance of the overall model, particularly in challenging regions where tumors are difficult to differentiate from surrounding tissues.

2.3 Global Content Consistency Loss

To improve stability and accuracy, we introduce a novel global content consistency loss mechanism that explicitly enforces semantic consistency across multiple levels of global contextual features. This mechanism employs varying-depth shared ConvMixer modules coupled with multilevel auxiliary decoders. The encoder output is processed through these ConvMixer blocks to generate multiscale intermediate features, which are fed directly into the auxiliary branches. Controlled noise perturbations introduced prior to auxiliary branch inputs simulate contextual variations, promoting invariant global representation learning.

To align the predictions across these decoders, we introduce a consistency constraint using KL divergence loss, which encourages the probability distributions of the auxiliary decoders to match those of the main decoder. This promotes stable learning of global content features and improves the robustness of the segmentation network.

$$L_{gccl} = \frac{1}{K}\sum_{k=1}^{K} D_{KL}\left(\hat{y}^{(k)} \middle\| \hat{y}^{(0)}\right) \tag{5}$$

where $\hat{y}^{(0)}$ denotes the main decoder (D_{main}) output and where $\hat{y}^{(k)}$ represents the k-th auxiliary decoder output ($k = 1,2,3$). The supervised combination loss $L_{\text{seg}}(.)$ combines binary cross entropy (BCE) and Dice loss (Dice) between the ith prediction $\hat{y}^i$ and the ground truth y^i, and $L_{\text{cls}}(.)$ is a standard Dice loss, as shown in the following formula:

$$L_{seg}(\cdot) = 0.5 \cdot BCE(\hat{y}, y) + Dice(\hat{y}, y) \tag{6}$$

$$L_{cls}(\cdot) = Dice(\hat{y}, y) \tag{7}$$

The final multitask loss function combines three components:

$$L_{total} = \lambda_1 L_{seg} + \lambda_2 L_{gccl} + \lambda_3 L_{cls} \tag{8}$$

We perform ablation experiments to compare the segmentation results for different values of $\lambda_{1,\,2,\,\text{and}\,3}$ and finally determine the following three parameters, namely, $\lambda_1 = 0.40$, $\lambda_2 = 0.30$, and $\lambda_3 = 0.30$, by comparing the optimal IoU.

3 Experiment and Discussion

3.1 Dataset

Our dataset contains ultrasound images of parotid tumors from Peking University Hospital of Stomatology. We collected and screened 728 ultrasound images, including 685 benign tumors and 43 malignant tumors. The images were all in PNG format. Our data were tagged as ground truth (GT) by professional radiologist-guided marking.

3.2 Implementation Details

All the experiments were conducted on an NVIDIA GeForce RTX 3090 GPU via PyTorch 1.12.1. We employed an SGD optimizer (initial LR: 0.01, momentum: 0.9) with a cosine annealing scheduler and trained for 156 epochs at a batch size of 4. Data augmentation consisted of random rotations of 90°, flipping, and normalization, and the input was uniformly adjusted to a resolution of 256 × 256. Model performance was comprehensively evaluated in terms of the IoU, accuracy (Pc), F1 score and accuracy (Acc). The validation set retained 20% of the data, and the best model was saved according to the validation IoU.

3.3 Results and Analysis

We compared our network with U-Net [10], attention U-Net [12], U-Net3+ [13], TransUnet [15], UNeXt [14], and CMU-Net [19]. The results show that most of the evaluation indices of our network are greater than those of other network models on the parotid gland tumor ultrasound image dataset provided by Peking University School and Hospital of Stomatology. As shown in Table 1, the IoU is increased by 2.27%, and F1 is increased by 1.20%. When the number of rounds is small, our network converges to the ideal IoU faster than most existing models do. We compare our tumor segmentation outputs with those of other networks and find that our model has better edge capture ability in segmenting parotid tumors with smoother segmentation edges and more accurate segmentation regions.

Table 1. The results of the dataset on different models

Models	IoU	Acc	Pc	F1
U-Net 10	0.8482	0.9865	0.9106	0.9122
Attention U-Net 12	0.8447	0.9863	0.9219	0.9102
U-Net3+ 13	0.8436	0.9860	0.9154	0.9091
TransUnet 15	0.8502	**0.9866**	0.9148	0.9125
UNeXt 14	0.8334	0.9849	0.9064	0.9030
CMUnet 19	0.8431	0.9858	0.9109	0.9090
CDC-net (Ours)	**0.8729**	0.9844	**0.9356**	**0.9245**

We provide five representative images with blurred boundaries and uneven internal echoes and a comparison plot of the segmentation results of each network model, as shown in Fig. 3. A quantitative comparison demonstrates our model's superior segmentation accuracy, achieving the closest alignment with ground truth labels while minimizing both segmentation errors and missed regions relative to other methods. Even when the segmentation targets are small tumors with unclear edges or irregular shapes, our model can accurately locate and segment the edges more precisely.

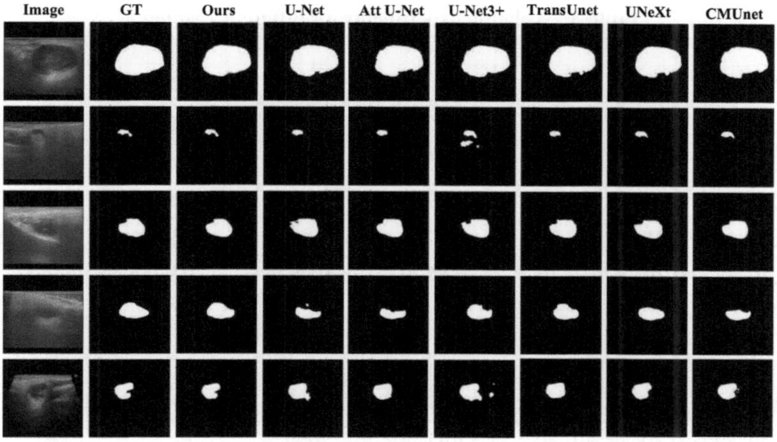

Fig. 3. Each model and segmentation result

3.4 Ablation Experiments

In addition, we perform ablation experiments to analyse the contribution of each component in the loss function design. As shown in Table 2, the baseline model using an improved U-Net structure yields suboptimal performance. Upon introducing auxiliary decoder loss, we observe a noticeable improvement in segmentation accuracy, indicating that multiscale supervision encourages more robust feature learning at different semantic levels. When heatmap-based loss is further incorporated, the model demonstrates even stronger localization ability, especially in cases with indistinct tumor boundaries. This highlights the effectiveness of introducing anatomical priors via the tumor heatmap to guide the segmentation network toward more discriminative regions.

This can be seen to justify the design of our loss function, so we retain our design of the joint loss function of the final, auxiliary decoder, and heatmap as the final loss function to help optimize the model and improve its generalizability.

Table 2. Ablation experiments for different losses

Loss	IoU	Acc	Pc	F1
L_{seg}	0.8523	0.9813	0.9251	0.9097
$L_{seg} + L_{gccl}$	0.8635	0.9830	**0.9474**	0.9182
$L_{seg} + L_{gccl} + L_{cls}$	**0.8729**	**0.9844**	0.9356	**0.9245**

We constrain the weights of the final output, the auxiliary decoder, and the heatmap by setting and adjusting different $\lambda_{1,\,2,\,\text{and}\,3}$ values to optimize the final segmentation results, as shown in Table 3. We record several $\lambda_{1,\,2,\,\text{and}\,3}$ values with high IoUs during testing, analyse them, and finally determine the values by the optimal IoU as follows: $\lambda_1 = 0.40$, $\lambda_2 = 0.30$, and $\lambda_3 = 0.30$.

Table 3. Ablation experiments for different λ *values*

λ1, 2, 3	IoU	Acc	Pc	F1
$\lambda_1 = 0.50, \lambda_2 = 0.30, \lambda_3 = 0.20$	0.8684	0.9833	0.9371	0.9208
$\lambda_1 = 0.50, \lambda_2 = 0.25, \lambda_3 = 0.25$	0.8698	0.9836	0.9370	0.9216
$\lambda_1 = 0.45, \lambda_2 = 0.30, \lambda_3 = 0.25$	0.8706	0.9843	**0.9470**	0.9232
$\lambda_1 = 0.45, \lambda_2 = 0.25, \lambda_3 = 0.30$	0.8727	0.9839	0.9456	**0.9259**
$\lambda_1 = 0.40, \lambda_2 = 0.30, \lambda_3 = 0.30$	**0.8729**	**0.9844**	0.9356	0.9245

Table 4. Fivefold cross-validation

$\lambda 1 = 0.40, \lambda 2 = 0.30, \lambda 3 = 0.30$	IoU	Acc	Pc	F1
Test 1	0.8508	0.9647	**0.9472**	0.9086
Test 2	0.8566	0.9820	0.9336	0.9106
Test 3	0.8534	0.9719	0.9462	0.9133
Test 4	0.8718	0.9769	0.9385	0.9206
Test 5	**0.8752**	**0.9850**	0.9422	**0.9285**
Average	0.8616	0.9761	0.9415	0.9163

To rigorously evaluate model performance and mitigate overfitting or generalization risks from limited data, we implemented 5-fold cross-validation. The dataset was randomly divided into five nonoverlapping folds. Each cycle trains the model on four folds and validates it on the remaining fold.

Particularly valuable for small datasets, this approach maximizes data utility through full image participation in both training and validation phases. It further yields stable, unbiased generalization estimates by rotating validation sets and averaging metrics across all folds.

We conducted different $\lambda_{1,\ 2,\ \text{and}\ 3}$ values and found that the average value was best when $\lambda_1 = 0.40$, $\lambda_2 = 0.30$, and $\lambda_3 = 0.30$. In this case, our evaluation metrics, such as the IoU, Acc, and F1, are still higher than those of existing mature network models. We recorded the results as follows:

3.5 Discussion

While quantitative metrics are essential for algorithmic assessment, our model's ultimate value lies in enhancing clinical applications and patient outcomes. However, the notoriously poor reproducibility of quantitative analyses, particularly in radiomics, constitutes a critical barrier to the clinical translation of AI tools. As established by Kumar et al., this variability often originates in foundational segmentation [20]. Our model addresses this core challenge through enhanced precision, providing stable foundations for downstream analyses.

The clinical implications of this stability are far-reaching. In breast oncology, Li et al. demonstrated the ability of radiomics to predict pathologic complete response (pCR), a key surrogate for long-term outcomes, including progression-free survival (PFS) [21]. Crucially, the reliability of pCR prediction depends on initial tumor segmentation accuracy. Thus, our model's performance gains transcended statistical improvements, directly empowering prognostic models for personalized therapy.

4 Conclusion

This work presents a dual-branch framework for parotid gland tumor segmentation and localization. The primary segmentation network employs stacked ConvMixer modules to capture the multilevel global context and integrates multiscale attention gates to enhance discriminative feature extraction for tumor regions. An auxiliary localization branch, which is based on a pretrained ResNet, generates anatomical prior heatmaps to guide the segmentation process. To enforce consistent learning across semantic levels, a global content consistency loss utilizing KL divergence is introduced, aligning the probabilistic outputs of the main decoder and three auxiliary decoders. This framework significantly improves the segmentation accuracy and robustness for clinically challenging cases characterized by low contrast and indistinct boundaries.

In future work, we plan to explore adaptive mechanisms for tumor heatmap generation and auxiliary decoder optimization and extend the framework's evaluation to multicenter datasets and real-time segmentation scenarios to advance its clinical utility and translation.

Acknowledgement. This work was supported by the "Excellence in Clinical Research Program for Research-Oriented Wards" of the Beijing Municipal Health Commission (Grant No. BRWEP2024W194100105), the Beijing Natural Science Foundation under Grant L232029, and the Capital's Funds for Health Improvement and Research (CFH 2024—4-4107) from the Beijing Municipal Health Commission.

References

1. Gatta, G., Guzzo, M., Locati, L.D., et al.: Major and minor salivary gland tumours. Crit. Rev. Oncol. Hematol. **152**, 102959 (2020)
2. Geiger, J.L., Ismaila, N., Beadle, B., et al.: Management of salivary gland malignancy: ASCO guideline. J. Clin. Oncol. **39**(17), 1909–1941 (2021)
3. Xia, X., Wang, J., Liang, S., et al.: An attention base U-net for parotid tumor autosegmentation. Front. Oncol. **12**, 1028382 (2022)
4. Behboodi, B., Amiri, M., Brooks, R., et al.: Breast lesion segmentation in ultrasound images with limited annotated data. In: 2020 IEEE 17th International Symposium on Biomedical Imaging (ISBI), pp. 1834–1837. IEEE (2020)
5. Liu, S., Wang, Y., Yang, X., et al.: Deep learning in medical ultrasound analysis: a review. Engineering **5**(2), 261–275 (2019)
6. Manh, V., Jia, X., Xue, W., et al.: An efficient framework for lesion segmentation in ultrasound images using global adversarial learning and region-invariant loss. Comput. Biol. Med. **171**, 108137 (2024)

7. Abdou, M.A.: Literature review: efficient deep neural networks techniques for medical image analysis. Neural Comput. Appl. **34**(8), 5791–5812 (2022)
8. Uzen, H., Turkoglu, M., Hanbay, D.: Multidimensional feature extraction-based deep encoder–decoder network for automatic surface defect detection. Neural Comput. Appl. **35**(4), 3263–3282 (2023)
9. Üzen, H., Türkoğlu, M., Yanikoglu, B., et al.: Swin-MFINet: swin transformer based multifeature integration network for detection of pixel-level surface defects. Expert Syst. Appl. **209**, 118269 (2022)
10. Ronneberger, O., Fischer, P., Brox, T.: U-Net: convolutional networks for biomedical image segmentation. In: Medical Image Computing and Computer-Assisted Intervention–MICCAI 2015: 18th International Conference, Munich, Germany, October 5–9, 2015, Proceedings, part III 18. Springer International Publishing, pp. 234–241 (2015)
11. Zhou, Z., Rahman Siddiquee, M.M., Tajbakhsh, N., et al.: UNet++: a nested u-net architecture for medical image segmentation. In: Deep Learning in Medical Image Analysis and Multimodal Learning for Clinical Decision support: 4th International Workshop, DLMIA 2018, and 8th International Workshop, ML-CDS 2018, Held in Conjunction with MICCAI 2018, Granada, Spain, September 20, 2018, Proceedings 4. Springer International Publishing, pp. 3–11 (2018)
12. Oktay, O., Schlemper, J., Folgoc, L.L., et al.: Attention u-net: Learning where to look for the pancreas. arxiv preprint arXiv:1804.03999 (2018)
13. Huang, H., Lin, L., Tong, R., et al.: UNet 3+: a full-scale connected UNet for medical image segmentation. In: ICASSP 2020–2020 IEEE International Conference on Acoustics, Speech and Signal Processing (ICASSP), pp. 1055–1059. IEEE (2020)
14. Valanarasu, J.M.J., Patel, V.M.: UNeXt: MLP-based rapid medical image segmentation network. In: International Conference on Medical Image Computing and Computer-Assisted Intervention. Cham: Springer Nature Switzerland, pp. 23–33 (2022)
15. Chen, J., Lu, Y., Yu, Q., et al.: TransUNet: Transformers make strong encoders for medical image segmentation. arxiv preprint arXiv:2102.04306 (2021)
16. Dosovitskiy, A., Beyer, L., Kolesnikov, A., et al.: An image is worth 16x16 words: Transformers for image recognition at scale. arxiv preprint arXiv:2010.11929 (2020)
17. Trockman, A., Kolter, J.Z.: Patches are all you need?. arxiv preprint arXiv:2201.09792 (2022)
18. He, K., Zhang, X., Ren, S., et al.: Deep residual learning for image recognition. In: Proceedings of the IEEE Conference on Computer Vision and Pattern Recognition, pp. 770–778 (2016)
19. Cui, M., Li, K., Chen, J., et al.: CM-UNet: a novel remote sensing image segmentation method based on improved U-Net. IEEE Access **11**, 56994–57005 (2023)
20. Kumar, V., Gu, Y., Basu, S., et al.: Radiomics: the process and the challenges. Magn. Reson. Imaging **30**(9), 1234–1248 (2012)
21. Li, Y., Fan, Y., Xu, D., et al.: Deep learning radiomic analysis of DCE-MRI combined with clinical characteristics predicts pathological complete response to neoadjuvant chemotherapy in breast cancer. Front. Oncol. **12**, 1041142 (2023)

Domain-Constrained BVAE-Based Method for Intelligent Plastic Formulation Design

Junrui Li and Jiwei Xu(✉)

Beijing University of Chemical Technology, 15 Beisanhuan East Road, Chaoyang District, Beijing 100029, China
{lijunrui,jiwei.xu}@buct.edu.cn

Abstract. As high-performance plastics find expanding applications in automotive, electronics, and packaging applications, the efficiency and accuracy of formulation design have become critical bottlenecks limiting performance gains and the advancement of smart manufacturing. Conventional approaches rely heavily on expert intuition and trial-and-error experimentation, leading to prolonged development cycles, high costs, and difficulty in meeting multiobjective constraints. To address these challenges, we propose a big data-driven framework encompassing raw material compositions, processing parameters, and performance metrics, supported by a complete workflow of data preprocessing, feature engineering, multifeature constraint integration, formulation generation, and multiobjective optimization. Data preprocessing employs normalization, GAN-based imputation, and Isolation Forest algorithms to ensure consistency and integrity via outlier removal. Feature engineering constructs vectors from ternary diagram coordinates, process parameters, structural descriptors, and performance indicators, extracting high-dimensional representations via early concatenation, late fusion, and attention-based fusion. A domain-constrained reverse design framework based on a variational autoencoder enables the generation of controllable formulas. Multiobjective optimization is achieved using NSGA-II (non-dominated classification genetic algorithm II) under performance - cost - sustainability constraints, with final selection guided by the Achievement Scalarization Function. Experimental validation shows that embedding compatibility rules and physicochemical constraints in the neural network improves the average hypervolume indicator by 0.06 and raises first-pass regulatory compliance by over 15%, confirming the framework's industrial applicability.

Keywords: Plastic formulation generation · Variational autoencoder (VAE) · NSGA-II · Multiobjective optimization

1 Introduction

Owing to their light weight, high strength, excellent formability, and low cost, plastic materials are extensively employed in the automotive, electronic, medical,

L. Huang et al. (Eds.): CCF NCCA 2025, CCIS 2774, pp. 347–362, 2026.
https://doi.org/10.1007/978-981-95-5294-8_24

and packaging industries. Formulation design plays a pivotal role in determining polymer performance, yet traditional development relies on expert rules and iterative experimentation—resulting in lengthy lead times and high expenses—while struggling to satisfy multiple property targets (e.g., mechanical strength, flame retardancy, material compatibility) and constraints (e.g., cost, carbon footprint), thereby impeding the advancement of on-demand material design and smart manufacturing.

In recent years, data-driven formulation methods based on machine learning and deep learning have made rapid progress. First, performance prediction models such as random forests and neural networks have been able to replace some traditional physical tests. Second, generative technologies, especially variational autoencoders (VAEs) [1], have demonstrated great potential in the reverse design of polymer formulations [2] by learning latent representations from simulation and experimental data. Third, multiobjective optimization algorithms such as NSGA-II [3] can balance competing design objectives under various constraints, thereby screening out the optimal candidate formula. However, the current data-driven formulation methods have the following deficiencies: 1) In traditional material formulation research and development process, material properties, formulation compositions and processing parameters are stored in different systems respectively, lacking a unified standard; Traditional cleaning mainly relies on manual rules or simple interpolation, which has difficulty addressing large-scale, multisource, and high-dimensional data loss and noise, resulting in subsequent model training being easily interfered by outliers. 2) A single fusion strategy has difficulty taking into account multimodal information, and early stitching is prone to cause feature dimension explosion and ignoring the internal relationships of each modal. Late fusion may lead to the loss of cross-modal correlation signals. Both methods have difficulty dynamically adjusting the importance of information from different sources, thereby affecting the generalization ability and prediction accuracy of the model. 3) The existing formula design process generally conducts formula generation (such as via VAE, GAN [4]) and multi-objective optimization (such as the genetic algorithm, NSGA-II) separately, and the generation step lacks the encoding of physical and chemical constraints in the domain. Moreover, the optimization step often fails to efficiently explore feasible solutions in the latent space. This leads to high computational overhead, slow convergence, and sometimes the obtained formula does not meet the requirements of the actual process or material stability.

To address the above deficiencies, this paper proposes and implements a plastic material formulation recommendation and optimization methods for intelligent manufacturing, and The main contributions include: *1) Designing a standardized data processing scheme, covering various data such as material properties, formula composition, and process parameters (processing conditions), and using methods such as GAN and isolated forest to enhance the integrity and reliability of the data. 2) Multimodal feature engineering was carried out, and a feature fusion strategy combining early stitching, late fusion and attention weighting was proposed to achieve the deep fusion of chemical, physical and process data.*

3) A formula generation and optimization framework was created. A Joint Generation Model with a Binding Variational Autoencoder (BVAE) based on VAE latent space exploration and PPO reinforcement learning [5] *was designed and seamlessly integrated with the NSGA-II multiobjective optimization algorithm. Automatically output the Pareto optimal formula set; Among them, NSGA-II screens candidate formulations to balance the competitive goals under different constraints.*By encoding materials science principles and domain-specific constraints into neural networks, the generated formulations adhere to both statistical distributions and domain knowledge requirements. Experimental validation demonstrated the effectiveness of this methodology in providing intelligent data-to-decision support for material development, highlighting its practical application value.

The subsequent chapters are organized as follows. Section 2 describes the related work of our proposed research. Section 3 presents the relevant details of the proposed recipe recommendation method and model, and Sect. 4 presents the experimental environment configuration as well as the performance test results. Finally, Sect. 5 concludes the contributions of this work.

2 Related Work

Conventional recipe design approaches rely on empirical trial-and-error or deterministic optimization models. For example, stochastic planning techniques are widely applied in process industries to address raw material uncertainties, while genetic algorithms enhanced with hammersley sequential sampling (HSS) [6] are employed to minimize material costs under quality constraints. Despite their utility, these methods highly dependent on explicit mathematical modeling, which limits their ability to solve high-dimensional and nonlinear problems. Recently, data-driven strategies have emerged, such as the integration of Taguchi [7] method with fuzzy inference systems for multiobjective optimization-achieving parameter tuning by balancing warpage, volumetric shrinkage, and cycle time in plastic injection molding. However, traditional formulation methods continue to face limitations in efficiency and scalability when applied to complex recipe designs, prompting the exploration of deep learningâĂŞbased generative models as more effective alternatives.

At present, the research on deep learning in material formulation design mainly involves two main ideas: on the one hand, the potential space of the formulation is explored through VAE; on the other hand, the Pareto optimal solution is sought within this potential space with the help of multiobjective optimization algorithms.

VAEs are latent-variable models that encode input data (e.g., formulation features) into a latent space and reconstruct them via a decoder. The learning objective is to optimize the evidence lower bound (ELBO) [10]:

$$\mathcal{L}(\theta, \emptyset) = \mathbb{E}_{q_\emptyset(z|X)}\left[\log P_\theta(x \mid z)\right] - KL\left(q_\emptyset(z \mid x) \| p(z)\right) \tag{1}$$

where $q_{\emptyset}(z \mid X)$ is the encoder (inference model), $P_{\theta}(x \mid z)$ is the decoder (generative model), KL is the Kullback-Leibler divergence encouraging the posterior to match a prior p(z)$\sim$ N(0,I).

The first item is the reconstruction loss, and the second item constrains the distribution of latent variables to be close to the prior (usually the standard normal distribution). The traditional VAE has problems in generating the physical feasibility of the formula in material design, for example, the composition ratio is not normalized. Therefore, our method introduces domain knowledge constraints to improve the traditional VAE model. For example, it forces the sum of component ratios to be 1 through the Softmax layer, or encodes the material interaction rules by designing a compatibility matrix.

The Multiobjective Optimization (MOO) [8] algorithm is utilized to find the Pareto optimal solution, aiming to balance conflicting objectives such as performance, cost, and environmental impact. The core lies in finding the Pareto frontier, that is, the solution set that cannot further optimize other objectives without sacrificing one objective. Among numerous algorithms, the nondominated Sorting Genetic Algorithm II (NSGA-II) is widely used due to its high efficiency in engineering optimization. It maintains the overall candidate solutions and uses fast nondominated sorting and crowded distance calculation to select the Pareto optimal candidate. For example, in the design of flexible hinges, NSGA-II improves the efficiency of understanding by approximately 14.3% compared with traditional algorithms; In the planning of distributed power systems, NSGA-II effectively reduces power generation costs and carbon emissions through collaborative multi-objective optimization. The dynamic Multi-Objective Optimization Problem (DMOP) [9] introduces the factor of time variation, requiring the algorithm to have the ability to adapt over time. To this end, researchers have developed the dynamic multiObjective evolutionary algorithm (DMOEA) [11], which integrates convolutional neural networks (CNN) and long- and short-term memory networks (LSTM). By predicting environmental changes, it guides popular evolution, significantly improving response speed and solution quality.

The above-mentioned formula generation methods generally separate formula generation (such as that based on VAE) from multiobjective optimization (such as genetic algorithms, NSGA-II), and the generation steps lack the encoding of physical and chemical constraints in the domain. Moreover, optimization steps often fail to efficiently explore feasible solutions in the latent space. This leads to high computational overhead, slow convergence, and sometimes the obtained formula does not meet the requirements of the actual process or material stability. Our method is based on the joint generative model of VAE latent space exploration and PPO reinforcement learning, and seamlessly integrates it with the NSGA-II multiobjective optimization algorithm. The generated formulas not only satisfy the statistical distribution but also conform to the domain knowledge constraints and have a certain application value.

3 Methodology

This chapter elaborates in detail the core components of the proposed method, with a focus on data selection and processing as well as the formulation design and optimization framework. This framework integrates multimodal feature engineering, rule-constrained generative models and multiobjective evolutionary algorithms to achieve efficient and reliable formulation design.

3.1 Data Selection and Processing

The algorithm uses three key data modalities to ensure comprehensive feature representation, including historical experimental data, literature and patent data, and simulated data. All three modalities are structured around a common feature framework comprising material properties, processing parameters (e.g., temperature and pressure), and performance indicators (e.g., tensile strength and flammability). Historical experimental data directly provide measured values across these dimensions. The literature and patent data are extracted and standardized from published formulations and reported physicochemical properties (e.g., UL94 rating, thermal stability). Simulated data, including molecular dynamics (MD) [12] and finite element analysis (FEA) [13], are incorporated in a complementary way. By simulating variations in composition and processing conditions, these methods yield predictive performance indicators (e.g. glass transition temperature, thermal response), allowing them to be integrated within the same composite–process–performance framework.

To address data heterogeneity and noise, we apply the following steps: 1) Missing value imputation, using generative adversarial networks (GANs) to synthesize missing data points while preserving statistical distributions. For features with missing values, the generator reconstructs reasonable values on the basis of the observed features:

$$\tilde{x}_i = G\left(z|x_{obj}\right), z \sim N\left(0, I\right) \tag{2}$$

2) Anomaly detection, where the isolation forest and 3α principle are used to jointly identify anomalies. The samples that deviate from the Mahalanobis distance [14] threshold are discarded, where n is the feature dimension:

$$D_t h = \sqrt{\left(\chi_0.95^2\left(n\right)\right)} \tag{3}$$

3) Normalization, which uses min–max scaling for process parameters (e.g., temperature $T \in [180\,^\circ\mathrm{C}, 220\,^\circ\mathrm{C}] \rightarrow [0, 1]$ and composition ratios.

3.2 Feature Engineering

In this work, key features are extracted from the following dimensions to cover the physical, chemical, and process characteristics of the formulation design: (1) component features, matrix material ratio (e.g., PP, ABS), additive ratio (flame retardant, etc.) (2) process features and time series data of temperature, pressure, and shear rate collected from sensors; (3) physicochemical features, melt index, heat distortion temperature, oxygen index; (4) structural characteristics, grain size, porosity, phase distribution (SEM images). Figure 1 shows the structural diagram of the feature engineering phase and the process of generating joint feature vectors.

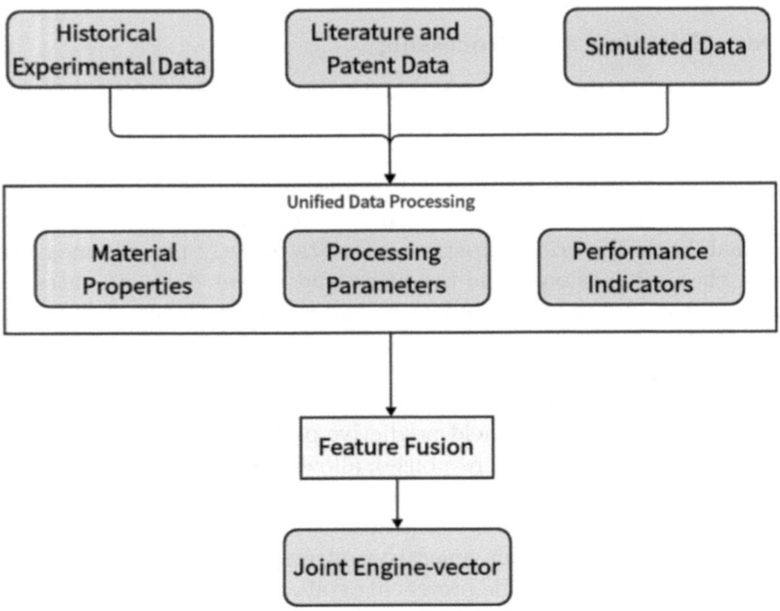

Fig. 1. Structure of joint engine-vector construction.

Among them, for the component features, the encoding of the ternary phase diagram [15] is adopted to map the proportions of matrix, additives and fillers in the formulation to the coordinates of the ternary phase diagram, to obtain a downscaled visualization of high-dimensional combination relationships. In Formula 4, C_{base} represents the amount of substrate, C_{add} represents the amount of auxiliary material, C_{filler} represents the amount of additive, and C_{total} represents the total amount.

$$X_{comp} = \left[\frac{C_{base}}{C_{total}}, \frac{C_{add}}{C_{total}}, \frac{C_{filler}}{C_{total}}\right] \tag{4}$$

Algorithm 1. BVAE-PPO-NSGA-II-MOEAD Plastic Formulation Algorithm

Input: Population size N, Max generations G_{max}, PPO epochs E_{ppo}, MOEA/D neighborhoods T, BVAE latent dim d
Initialize:

- Train BVAE on initial policy parameters θ_0
- Initialize population $P = \{\theta_i\}_{i=1}^{N}$ using BVAE decoder
- Initialize MOEA/D weight vectors $\{\lambda^k\}_{k=1}^{N}$
- Initialize PPO policy π_ϕ with random weights ϕ

for $g = 1$ to G_{max} **do**
  Phase 1: PPO Policy Update
  for $e = 1$ to E_{ppo} **do**
    Collect trajectories $\tau = (s_t, a_t, r_t)$ using π_ϕ and current population P
    Compute advantage estimates $\hat{A}_t$
    Update policy parameters:

$$\phi^{e+1} = \arg\max_{\phi} \mathbb{E}_t \left[\frac{\pi_\phi(a_t|s_t)}{\pi_{\phi^{old}}(a_t|s_t)} \hat{A}_t - \beta D_{KL}(\pi_{\phi^{old}}||\pi_\phi) \right]$$

  end for
  Phase 2: NSGA-II Operations
  Evaluate population P on multiple objectives $\mathbf{F} = [f_1, f_2, ..., f_m]$
  Perform nondominated sort and crowding distance sort
  Select parents P_{parent} using tournament selection
  Generate offspring $P_{offspring}$ via crossover and mutation
  Phase 3: MOEA/D Optimization
  for each individual $x_i \in P$ **do**
    Select neighborhood B_i with T closest weight vectors
    Update reference point $z_j^* = \min(f_j(x)), j = 1, \ldots, m$
    Update scalarized fitness using Tchebycheff:

$$g(x|\lambda^i, z^*) = \max_{1 \leq j \leq m} \lambda_j^i |f_j(x) - z_j^*|$$

    Replace neighbors if new solution dominates existing
  end for
  Phase 4: BVAE Refinement
  Encode population: $z_i = \mu_\phi(x_i) + \sigma_\phi(x_i) \odot \epsilon, \epsilon \sim \mathcal{N}(0, I)$
  Generate new latent vectors: $\tilde{z}_j = \text{Crossover}(z_i) + \xi, \xi \sim \mathcal{N}(0, \sigma_m)$
  Decode new population: $x_j^{\text{new}} = p_\theta(x|\tilde{z}_j)$
  Population Update:
  $P \leftarrow$ Environmental Selection$(P \cup P_{offspring} \cup \{x_j^{\text{new}}\})$
end for

For process features, dynamic relational modeling is used to capture the temporal evolution of key process parameters by performing time-integral operations on their time series traces. By integrating quantities such as the melt temperature and injection pressure over the cycle, the cumulative impact of process history on final performance can be accurately quantified. For parameter features, statistical extraction methods distill each time series or aggregated signal into descriptive metrics: mean, variance, extreme values (min/max) and first-order derivative (slope) - thus characterizing central tendency, dispersion, and rate of change.

After feature construction is complete, feature fusion and alignment are performed. The first is heterogeneous data alignment, which addresses process parameters, recipe component ratios, and performance parameters according to a hierarchical coding strategy. The second step is feature fusion, where feature vectors are directly spliced in the early stage, and submodels are trained separately in the late stage, weighted to integrate the prediction results, and weights are dynamically assigned through the attention mechanism.

3.3 Formulation Design and Optimization

The formulation design task in plastic materials involves determining optimal compositions and processing conditions to meet multiple, often conflicting, property objectives. To address the complexity and domain constraints of this problem, we propose a two-stage generative optimization framework that combines a physics-informed Binding Variational Autoencoder (BVAE) and a multiobjective NSGA-II algorithm, as illustrated in Fig. 2.

Algorithm 1 achieves efficient collaboration between exploration and development in multiobjective decision-making problems by combining the latent space exploration of the variational autoencoder (BVAE), reinforcement learning strategy updates of proximal policy optimization (PPO), the multiobjective evolution framework of NSGA-II, and the decomposition optimization mechanism of MOEA/D. It can automatically generate a set of strategies that achieve optimal balance among multiple competing objectives.

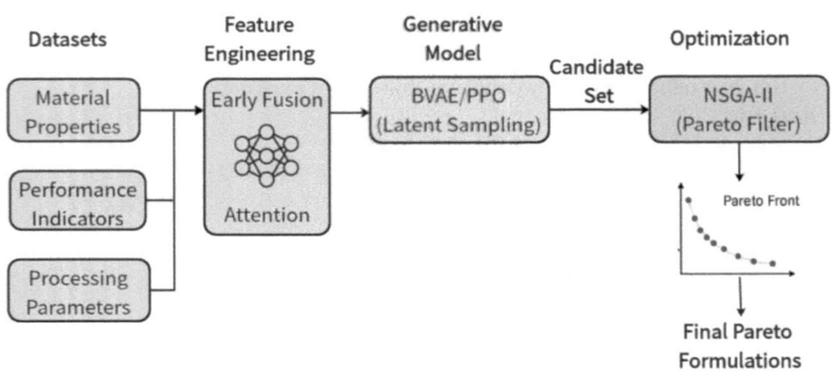

Fig. 2. Workflow for plastic formulation design and optimization.

The binding variational autoencoder serves as the backbone of our formulation generation module. It learns a smooth, structured latent space of feasible plastic recipes while embedding domain knowledge into its architecture. Its core consists of an encoder $q_{\emptyset}(Z|X)$ and a decoder $P_{\theta}(x|z)$. The encoder maps input formulations x into a structured latent representation z, where a graph convolutional layer is incorporated to embed prior knowledge of material compatibility (e.g., interactions between polymer matrices and additives). The compatibility matrix A is used to modulate hidden representations:

$$H^{(l+1)} = Relu\left(AH^{l} W^{l}\right) \tag{5}$$

where H is the feature vector, W is a trainable weight matrix, and Relu is an activation function.

Decoder $P_{\theta}(x|z)$ reconstructs a formulation candidate $\hat{x}$ from latent representation z, with embedded constraints: (1)A sum-to-one constraint layer ensures all component ratios satisfy $\sum_i x_i = 1$, implemented via Softmax. (2) Continuous process parameters are mapped back via inverse normalization. The training objective combines reconstruction loss, KL divergence, and property prediction loss:

$$\begin{aligned}\mathcal{L}_{\text{BVAE}} = &\underbrace{\|x-\hat{x}\|^2}_{\text{Reconstruction}}\\ &+\beta\cdot\underbrace{KL\big(q_{\phi}(z|x)\|\mathcal{N}(0,I)\big)}_{\text{Latent Regularization}}\\ &+\lambda\cdot\underbrace{\|y_{\text{pred}}-y_{\text{target}}\|^2}_{\text{Property Deviation}}\end{aligned} \tag{6}$$

where y_{pred} is a property predictor, and y_{target} represents the user-specified performance goals.

Once trained, the BVAE serves as a generator for diverse, physically valid candidates. Then BVAE generates a large batch of candidate formulations $\{x_i\}$, which are evaluated by surrogate performance models $f_k(\bullet)$ for each objective $k \in \{1, \ldots, K\}$, which estimate key objectives such as tensile strength, cost per unit mass, and environmental impact. Formally, the optimization problem seeks to minimize the multi-objective vector:

$$min\ F(x) = (f_1(x), f_2(x), f_3(x), \ldots, f_n(x)) \tag{7}$$

To intelligently explore the latent design space, we intelligently integrate proximal policy optimization (PPO), a model-free reinforcement learning algorithm, as a guided perturbation mechanism. A stochastic policy $\pi_{\phi}(a|s)$ is trained to generate latent-space adjustments that improve performance. The learning environment represents each formulation or latent vector as a state s_t, while the action a_t suggests a direction suggests a direction of modification. The reward r_t is computed on the basis of improvements across the objective functions.

PPO was selected over alternative RL methods due to its superior balance between stability and sample efficiency, particularly in continuous and high-dimensional optimization tasks. Unlike off-policy methods such as DDPG [19] or SAC [20], which can be sensitive to hyperparameter tuning and replay buffer noise, PPO employs an on-policy framework with clipped objective functions that constrain policy updates. This makes it especially robust when operating on surrogate-based rewards that may exhibit local inconsistency or partially observable conditions commonly encountered in formulation design spaces. Moreover, its straightforward implementation and consistently strong performance across diverse domains further support its suitability for integration into our pipeline.

The PPO updates the policy by optimizing a clipped surrogate loss that balances advantage estimation with trust region constraints, thus enabling stable policy learning. By integrating formulations perturbed by the PPO with the original BVAE candidates, an augmented population is formed, which is searched globally via the NSGA-II algorithm. NSGA-II performs nondominated sorting to identify Pareto fronts, applies a crowding distance to preserve diversity, and generates offspring through crossover and mutation. The resulting population captures a diverse set of trade-offs among competing objectives.

To further refine solutions, we adopt MOEA/D as a local optimizer. This method decomposes the multiobjective space into scalar subproblems associated with different weight vectors. For each subproblem, the objective is scalarized using a Tchebycheff metric relative to a reference point:

$$g(x|\lambda, z^*) = \max_j \lambda_j \cdot |f_j(x) - z_j^*| \tag{8}$$

allowing precise adjustment of candidate solutions within their local neighborhoods.

In the final stage of each generation, the BVAE is used again for latent-space resampling. High-performing solutions are encoded to their latent representations, which are then recombined and perturbed via Gaussian noise to produce novel latent vectors. These are decoded into new formulation candidates that preserve feasibility and innovation. All solutions generated by PPO, NSGA-II offspring, MOEA/D refinement, and BVAE sampling are merged and subjected to environmental selection, ensuring a balance between convergence and diversity.

To align optimization with user preferences, we also apply adaptive weight adjustment to scalarize objectives if needed:

$$\mathrm{AFS}(x) = \max_k \left[\frac{f_k(x) - f_k^{\mathrm{ref}}}{\omega_k} \right] \tag{9}$$

where f_k^{ref} denotes a user-defined reference value and ω_k encodes the priority assigned to each objective.

This hybrid framework leverages generative modeling (BVAE), reinforcement-driven exploration, and evolutionary optimization (NSGA-II/MOEA-D) to produce formulation candidates that are both physically plausible and performance optimized. By combining PPO's adaptivity, NSGA-II's diversity preservation, MOEA/D's local intensification, and BVAE's feasibility encoding, the proposed system forms a unified solution for intelligent and efficient plastic formulation design.

Table 1. EXPERIMENTAL ENVIRONMENT SPECIFICATIONS

Configuration	Name	Details
Hardware environment	CPU	R9-7940
	GPU	RTX4060
	RAM	16G
Software environment	operating system	Ubuntu 20.04
	CUDA	12.2
	CUDNN	8.9.04
	Python	3.8.16
	PyTorch	2.3.1

Table 2. Experimental hyperparameter settings

Full name	Symbol	Value
Training epochs	epo	200
Batch size	$batch\_size$	16
Number of workers	$workers$	8
Learning rate (BVAE)	lr	0.002
Weight decay	$decay$	0.0005
Latent dimension	$latent\_dim$	96
PPO learning rate	lr_{ppo}	0.0004
PPO clip ratio	ϵ	0.2
Entropy coefficient	$\beta_{entropy}$	0.02
PPO epochs per update	epo_{ppo}	10
PPO batch size	$batch_{ppo}$	512

4 Experiment

4.1 Experimental Setup

The experimental setups used in this paper are shown in Tables 1 and 2.

Table 3. Comparison of Key Performance Metrics across All Methods

Metric	Base A	Base B	Base C	Base D	Ours	vs A	vs B	vs C	vs D
First-Pass Compliance (%)	70	81	77	80	**85**	+15pp	+4pp	+8pp	+5pp
Avg. SHAP Variables	6.4	3.6	5.1	4.2	**3.1**	−3.3	−0.5	−2.0	−1.1
Hypervolume (HV)	0.65	0.75	0.72	0.75	**0.78**	+0.13	+0.03	+0.06	+0.03
Gen. Time (min)	–	3.1	5.0	4.0	**3.0**	–	−0.1	−2.0	−1.0
NSGA-II Filter Time (min)	12.0	11.3	11.5	–	**10.0**	−16.7	−11.8	−13.0	–

In this study, a constrained BVAE model and NSGA-II were used for multiobjective optimization. The experimental dataset originates from a comprehensive dataset of plastic formulations that was collected, comprising approximately 5,000 total samples from multiple sources. These include about 3,000 historical experimental records, 1,000 formulation cases extracted from literature and patents, and 1,000 data points generated by molecular simulations (MD/FEA). Each sample is represented by a high-dimensional feature vector covering composition ratios (polymer matrix and additive fractions), processing parameters (e.g. temperature, pressure and shear rate), material property metrics (e.g. melt flow index, heat distortion temperature), and structural descriptors (e.g. particle size, porosity). The raw data were preprocessed (missing-value imputation, outlier filtering and normalization) and split into training, validation, and test sets using an 70%/20%/10% ratios, respectively.

To determine suitable hyperparameters for the PPO component, we conducted a grid search over commonly influential factors. The learning rate lr_{ppo} was selected from {1e-4, 3e-4, 5e-4, 8e-4}, and the clip ratio ϵ was tested in the range of [0.1, 0.3] to constrain policy deviation. We also evaluated entropy regularization coefficient $\beta_{entropy}$ from {0.001, 0.005, 0.01} to encourage exploration. PPO was trained for 10 epochs per policy update, with a batch size of 512 per rollout. The final settings were selected on the basis of convergence stability and multi-objective performance across three validation runs.

The evaluation metrics used in the experiment consisted of two parts (model generation quality indicators and formula generation evaluation metrics). The model generation quality indicators include loss, MSE, FID score (which measures the distance between the generated samples and the distribution of real samples, with smaller values being better), MAE, RMAE, and R2. The formula generation evaluation metrics include the first pass compliance (%), hypervolume indicator (HV), average SHAP variable count, candidate generation time (min) and NSGA-II filtering time (min). The hypervolume (HV) quantifies the quality of multiobjective trade-offs by measuring the dominated volume between the Pareto front and a reference point, capturing both convergence and diversity. Additionally, the average SHAP variable count offers an interpretability perspective by analysing how many formulation components significantly impact the predicted performance across Pareto-optimal samples.

4.2 Performance Evaluation

We compare our integrated BVAE+PPO+NSGA-II+ MOEA/D refine pipeline against four baselines, and key metrics are shown in Table 3: 1) Baseline A: Traditional NSGA-II using historical surrogate models without generative sampling or PPO guidance. 2) Baseline B: The PaRetO-guided diffusion model [21] is a diffusion-based method that guides the generation of material formulations to the Pareto-optimal region using performance predictor gradients. Effortlessly produces high-quality balanced candidates in large formulation spaces. 3) Baseline C (GAN-based generative): A GAN-based generative model [17] is trained to propose new formulation candidates, which are then filtered by NSGA-II. GAN sampling has been shown to efficiently explore large material design spaces. 4) Baseline D (Bayesian Optimization): A Bayesian optimization pipeline using a

Table 4. Performance Comparison Table for Model Generation Quality with Baselines

Method	MSE	MSE	RMSE	R^2	FID Score
Baseline A	0.45	0.38	0.67	0.72	38.6
Baseline B	0.35	0.24	0.51	0.87	26.5
Baseline C	0.28	0.24	0.53	0.65	**18.9**
Baseline D	0.50	0.42	0.71	0.60	42.1
Proposed	**0.21**	**0.19**	**0.46**	**0.89**	20.3

Table 5. Ablation Study: Pareto Front Quality

Method	HV	Relative to Proposed
NSGA-II only	0.65	−0.13
BVAE→NSGA-II	0.72	−0.07
BVAE + PPO→NSGA-II	0.74	−0.05
BVAE + PPO→NSGA-II + MOEA/D refine	0.78	−0.00

Gaussian process surrogate (with latent-variable encoding of categorical components) to directly optimize formulation variables. Wu et al. [18] proposed a Bayesian algorithm for mixed-variable material design.

Table 3 shows that integrating BVAE sampling with PPO guidance yields the largest gains in performance and Pareto front quality, while also reducing total development time and optimization runtime.

Table 4 shows the performance and analysis of each method in the quality metrics of model generation for the multiobjective formula generation task. This table shows that the proposed method(BVAE+PPO+NSGA-II+MOEA/D) performs best in MSE, MAE, RMSE, and R^2 metrics (0.21, 0.19, 0.46, and 0.89, respectively) with high and stable generation diversity, but its FID score (20.3) is slightly higher than the baseline C (18.9). Among the baseline methods, the scheme combining BVAE and NSGA-II performs suboptimally in terms of error and R-squared, while baseline C is prone to mode collapse although it has high generation quality, and the comprehensive performance of Bayesian optimization and traditional NSGA-II is weak. The overall results show that our proposed method is more advantageous in balancing generation quality, diversity, and stability.

We also conducted ablation studies to isolate the impacts of PPO guidance and two-stage optimization. The results are summarized in Table 5.Table 5 indicates that PPO guidance significantly increases the Pareto front quality compared with BVAE alone, and that adding a local MOEA/D refinement stage yields further hypervolume gains.

4.3 Training Logs

Figure 3–4 show the performance metrics during the training process of the dataset, including reconstruction loss, KL divergence and attribute prediction

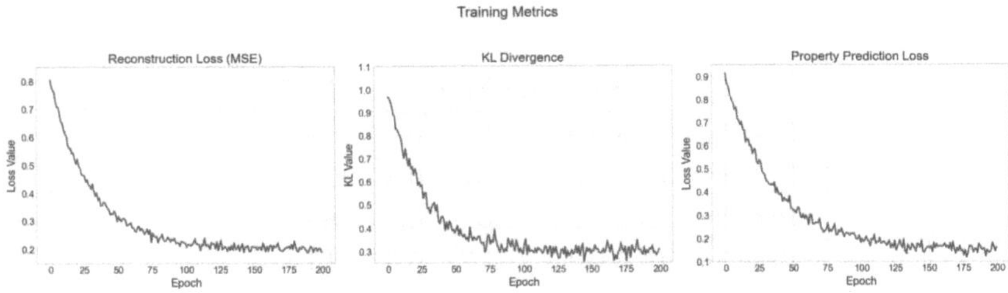

Fig. 3. Dataset training part of the logs.

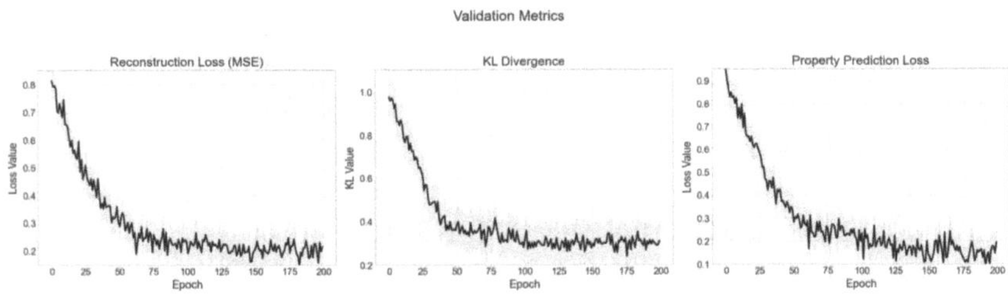

Fig. 4. Dataset Validation part of the logs.

loss. All these metrics continuously decrease with the number of training iterations. It can be seen that the model converges normally.

4.4 Discussion

The experimental comparisons demonstrate that our method consistently outperforms the baselines. Our approach achieves higher first-pass compliance rates and larger hypervolume (HV) indicators, indicating more effective multiobjective optimization with practical, compliant formulations. Additionally, the average number of significant SHAP variables is lower, reflecting simpler and more interpretable formulation designs. In contrast, the GAN-based baseline (C) generated more diverse candidates but occasionally violated domain constraints without explicit guidance. The Bayesian optimization baseline (D) achieved competitive objectives but converges more slowly because of its sequential search. Overall, these findings validate that combining domain-informed generative modeling with reinforcement-guided sampling significantly enhances the efficiency and quality of formulation design. The reduced development time and improved metrics (Table 4) illustrate the practical benefits of our approach.

Limitations and future directions are as follows: 1) In terms of data dependence, the method's performance strongly correlates with the quality and diversity of historical data. Augmenting the database within experimental feedback could further increase robustness. 2) Real-time closed-loop integration: Future work will integrate real-time process monitoring to enable closed-loop optimization, reducing the lag between computation and validation.

5 Conclusion

This work presents a fully integrated, data-driven method for intelligent plastic formulation recommendation and optimization, featuring. High-Quality Data Management using GAN and isolation forests to ensure database integrity. Multimodal feature engineering employs early concatenation and attention to align chemical, physical, and process data. Generative Design via BVAE enhanced with PPO reinforcement learning for controllable recipe exploration. MultiObjective Optimization uses NSGA-II (and optional MOEA/D refinement) to automatically output Pareto-optimal formulation sets. Knowledge-Informed Constraints embedded in network architectures to guarantee physical feasibility.

The experimental results demonstrate an average improvement of 0.06 in hypervolume and 15pp gain in first-pass compliance compared to baselines, confirming the efficacy and practical utility of the method. Future work will focus on closed-loop integration with on-line process data and expand to other polymer systems to further accelerate the development of smart materials.

Acknowledgments. This work was supported by the National Key Research and Development Program of China under Grant No. 2023YFB3209701, and the Open Research Project of the State Key Laboratory of Processors under Grant No. CLQ202304.

Disclosure of Interests. The authors have no competing interests to declare that are relevant to the content of this article.

References

1. Akkem, Y., Biswas, S.K., Varanasi, A.: A comprehensive review of synthetic data generation in smart farming by using variational autoencoder and generative adversarial network. Eng. Appl. Artif. Intell. **131**, 107881 (2024)
2. Zheng, Y., Thakolkaran, P., Biswal, A.K., et al.: AI-guided inverse design and discovery of recyclable vitrimeric polymers. Adv. Sci. **12**(6), 2411385 (2025)
3. Luan, C., Liu, R., Zhang, Q., et al.: Multi-objective land use optimization based on integrated NSGA–II–PLUS model: comprehensive consideration of economic development and ecosystem services value enhancement. J. Clean. Prod. **434**, 140306 (2024)
4. Lim, W., Yong, K.S.C., Lau, B.T., et al.: Future of generative adversarial networks (GAN) for anomaly detection in network security: a review. Comput. Secur. **139**, 103733 (2024)
5. Zhang, R., Xiong, K., Lu, Y., et al.: Energy efficiency maximization in RIS-assisted SWIPT networks with RSMA: a PPO-based approach. IEEE J. Sel. Areas Commun. **41**(5), 1413–1430 (2023)
6. Chen, H., Yang, C., Deng, K., et al.: Multi-objective optimization of the hybrid wind/solar/fuel cell distributed generation system using Hammersley Sequence Sampling. Int. J. Hydrogen Energy **42**(12), 7836–7846 (2017)
7. Hisam, M.W., Dar, A.A., Elrasheed, M.O., et al.: The versatility of the Taguchi method: optimizing experiments across diverse disciplines. J. Stat. Theory. Appl. **23**, 365–389 (2024)

8. Al Mindeel, T., Spentzou, E., Eftekhari, M.: Energy, thermal comfort, and indoor air quality: multi-objective optimization review. Renew. Sustain. Energy Rev. **202**, 114682 (2024)
9. Peng, H., Xiong, J., Pi, C., et al.: A dynamic multi-objective optimization evolutionary algorithm with adaptive boosting. Swarm Evol. Comput. **89**, 101621 (2024)
10. Damm, S., Forster, D., Velychko, D., et al.: The ELBO of variational autoencoders converges to a sum of entropies. In: International Conference on Artificial Intelligence and Statistics. PMLR, pp. 3931–3960 (2023)
11. Zou, F., Yen, G.G., Tang, L., et al.: A reinforcement learning approach for dynamic multi-objective optimization. Inf. Sci. **546**, 815–834 (2021)
12. Yao, H., Liu, J., Xu, M., et al.: Discussion on molecular dynamics (MD) simulations of the asphalt materials. Adv. Coll. Interface. Sci. **299**, 102565 (2022)
13. Nath, D., Neog, D.R., Gautam, S.S.: Application of machine learning and deep learning in finite element analysis: a comprehensive review. Arch. Comput. Methods Eng. **31**(5) (2024)
14. Mondal, A., Roy, S.K.: Behavioural three-way decision making with Fermatean fuzzy Mahalanobis distance: application to the supply chain management problems. Appl. Soft Comput. **151**, 111182 (2024)
15. Wei, H., Li, H., Liu, J., et al.: A ternary complementary Gray code phase unwrapping method in fringe projection profilometry. J. Appl. Phys. **135**(14) (2024)
16. Yi, X., Yu, H., Xu, T.: Solving multi-objective weapon-target assignment considering reliability by improved MOEA/D-AM2M. Neurocomputing **563**, 126906 (2024)
17. Li, Z., Birbilis, N.: NSGAN: a non-dominant sorting optimisation-based generative adversarial design framework for alloy discovery. npj Comput. Mater. **10**(1), 112 (2024)
18. Chitturi, S.R., Ramdas, A., Wu, Y., et al.: Targeted materials discovery using Bayesian algorithm execution. npj Comput. Mater. **10**(1), 156 (2024)
19. Liu, Y.C., Huang, C.Y.: DDPG-based adaptive robust tracking control for aerial manipulators with decoupling approach. IEEE Trans. Cybern. **52**(8), 8258–8271 (2021)
20. Li, Z., Huang, C., Deng, S., et al.: A soft actor-critic reinforcement learning algorithm for network intrusion detection. Comput. Secur. **135**, 103502 (2023)
21. Yao, Y., Pan, Y., Li, J., et al.: PROUD: PaRetO-gUided diffusion model for multi-objective generation. Mach. Learn. **113**(9), 6511–6538 (2024)

HDRS: Transmission, Computing and Caching Resource Scheduling for Holographic-Type Communication Based on In-Network Computing

Jinxin Liu[1], Chenxi Liao[1], Shang Liu[1](✉), Xu Huang[1], Dongshen Qian[1], Wentao Cui[1], and Jia Chen[1,2]

[1] National Engineering Research Center for Advanced Network Technologies, School of Electronic and Information Engineering, Beijing Jiaotong University, Beijing 100044, China
22115009@bjtu.edu.cn

[2] Peng Cheng Laboratory, Shenzhen 518055, China

Abstract. As a key component of next-generation communication technologies, Holographic-type Communication (HTC) places stringent demands on the network, necessitating low latency, high bandwidth, and substantial computing power. The requirements create considerable challenges for the dynamic coordinated scheduling of network resources. Existing resource scheduling approaches primarily focus on edge computing, and employ distributed computational deployments to reduce end-to-end latency. However, the limited computing resources of edge computing present a significant bottleneck in HTC. With the evolution of the In-Network Computing (INC) paradigm, intelligent network-layer components, exemplified by Data Processing Units (DPUs), have demonstrated advantages such as data plane acceleration, offering promising support for HTC over edge computing. Nevertheless, the problem of dimensional resource scheduling optimization within INC frameworks is a critical research challenge. In this research, we propose an INC-based multidimensional resource model, customized for three representative holographic services. The model integrates transmission, computing, and caching resources, as well as jitter compensation mechanisms. On the basis of this model, we develop a Holographic DQN-based Resource Scheduling (HDRS) algorithm, which simultaneously optimizes for minimal resource consumption and strict latency guarantees. The simulation results indicate that HDRS outperforms the baseline methods, reducing resource consumption by 23% while maintaining strict latency constraints and achieving a success rate of 95% task deployment under resource-constrained conditions.

Keywords: Holographic communication · Resource scheduling · DQN

1 Introduction

Holographic-type communication (HTC) represents an emerging technological paradigm that integrates 3D imaging, real-time rendering, and multimodal inter-

L. Huang et al. (Eds.): CCF NCCA 2025, CCIS 2774, pp. 363–377, 2026.
https://doi.org/10.1007/978-981-95-5294-8_25

action to deliver truly immersive experiences across domains such as telecommunications, online education, and healthcare. Streaming holographic content drives bandwidth requirements into the terabits-per-second range, whereas interactive applications, such as holographic gaming, demand end-to-end latencies below 20 ms to prevent motion sickness [1]. Simultaneously, substantial computational overhead is incurred by tasks such as high-fidelity rendering and transmission optimization. Consequently, the technologies of HTC require high-bandwidth and low-latency network capabilities, in addition to substantial computing resources and a variety of computing functions [2]. To coordinate these network resources effectively, robust resource scheduling algorithms are required in HTC environments.

Current resource scheduling approaches focus mainly on edge computing architectures, which have been extensively implemented across holographic application [3,4]. Edge computing architectures position computing resources closer to the data source to reduce latency. However, its limited computing resources make it insufficient to satisfy the high-throughput processing demands of holographic services [5]. As the In-Network Computing (INC) paradigm advances, network-layer intelligent components, exemplified by Data Processing Units (DPUs), have achieved tight integration of computation and network protocols at packet-level granularity, through techniques such as protocol stacks offloading, hardware accelerator arrays, and programmable data planes (P4) [6]. In contrast to edge-computing architectures that emphasize single-point optimization at edge servers, the INC embeds computational units directly within the switch-node topology to satisfy the ultra-high bandwidth, ultra-low latency, and massive computational requirements of holographic-type communication. Consequently, INC has emerged as a technically viable architectural framework capable of addressing the rigorous quality-of-service (QoS) requirements inherent to holographic applications [7]. Nevertheless, existing resource scheduling methods demonstrate some limitations for HTC.

- Current research on resource scheduling foucus on optimizing at edge servers, with limited consideration given to accelerating forwarding hardware resources such as DPUs. Conventional heuristic algorithms encounter limitations in achieving globally optimal resource allocation, and are constrained by the state-space complexity and multidimensional interdependent characteristics of holographic-type communication.
- The stringent temporal sensitivity inherent in holographic data streams imposes critical requirements for millisecond-level jitter compensation, a dimension that current research on resource scheduling has not considered.

To address the challenges above, we introduce a Holographic DQN-based resource scheduling (HDRS) algorithm, which simultaneously optimizes for minimal resource consumption and strict latency guarantees. The main contributions of this paper are summarized as follows:

- We construct a multidimensional resource model for three representative holographic services (holographic browsing, on-demand streaming, and video

communication), integrating transmission, computing, and caching resources, along with a jitter compensation mechanism.
- We develop a resource scheduling algorithm based on a deep Q-Network (DQN), which dynamically orchestrates resource allocation and routing decisions according to the status of node resources, link bandwidth, and task requirements. The experimental results indicate that the proposed algorithm effectively reduces resource consumption and maintains a high success rate of task deployment.

The rest of this paper is organized as follows. Related works are reviewed in Sect. 2. Section 3 presents the system model and formulates the optimization problem. Section 4 presents the dynamic resource scheduling algorithm. Section 5 provides the experiment results. The conclusion is proposed in Sect. 6.

2 Related Work

Existing resource scheduling algorithms are primarily based on edge computing. Huang et al. [8] designed a cloud-edge collaborative computing and transmission architecture for holographic applications and subsequently formulated a strategy based on deep reinforcement learning for joint task scheduling and resource allocation. Peng et al. [9] proposed a decentralized approach to address the problem of online edge task scheduling and resource allocation, formulating it as an online decision-making process. Liu et al. [10] introduced an iterative framework to achieve Stackelberg equilibrium, designing two offloading algorithms based on the Q-value and WoLF-PHC, as well as two dispatching algorithms based on best response and better response strategies.

Some recent studies have already begun exploring resource scheduling in INC. Aliyu et al. [11] investigated the partial offloading of multiple subtasks in INC, aiming to reduce latency and energy consumption by dynamically adjusting offloading strategies according to current computational conditions. Kamran et al. [12] proposed a comprehensive framework that combines computation scheduling, caching, and request forwarding within data-intensive computing networks, where network nodes can simultaneously route traffic, execute computations, and store data. They further developed a distributed and adaptive algorithm to minimize the task completion time.

Unlike existing works, this paper focuses on supporting resource scheduling for three types of holographic services in INC, ensuring task deployment even under constrained resources.

3 Problem Formulation

In this section, we first present the holographic service type in INC and then give the formulation of the task scheduling and resource allocation problem to be addressed.

3.1 Holographic Services

This paper classifies holographic services into three categories, and the service workflow is illustrated in Fig. 1.

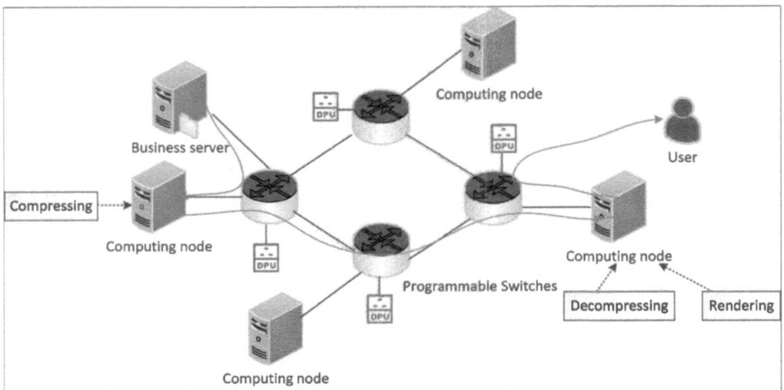

Fig. 1. Workflow of holographic services

- Holographic Browsing: Users access 3D models through the controller that coordinates file transmission from the business server, which stores raw holographic data. The files are compressed and decompressed on distributed computing nodes to increase transmission and reduce end-to-end latency.
- Holographic On-demand Streaming: The controller first checks edge-node caches when it receives a video request. If cached, the stream is transmitted directly from the node to the user, bypassing computational processing. Otherwise, the video is fetched from the business server via a workflow that includes compression, decompression, and rendering operations. The decompression and rendering are performed on a single node, whereas the compressed video is simultaneously cached on the subsequent node during transmission. All nodes possess adequate storage capacity to accommodate the entire collection of holographic video resources.
- Holographic Video Communication: The controller establishes bidirectional sessions between two users. Data streams undergo compression, decompression, and rendering processes between nodes, with buffer zones established at the egress node to reduce jitter [13].

  We define a strict business priority order: Holographic Video Communication > Holographic On-demand Streaming > Holographic Browsing.

3.2 System Model

Network Model: The network model comprises n nodes, where $M = \{M_1, M_2, \cdots, M_n\}$ denotes the set of nodes and $C = \{C_1, C_2, \cdots\cdots C_n\}$ represents the set of computing capacities, with C_j indicating the computing capacity of the node j. All nodes are interconnected via optical fibers, forming L links with a bandwidth of B per link.

Task Model: A user-submitted task q_k comprises the following parameters: service type $type_k$, latency requirementT_k^{req}, bandwidth requirementB_k^{req}, jitter requirementJ_k^{req}, computational function requirement $F_k = \sum_{i=1}^{f} f_k^i$ with computing cycles c_k^i per function, the requested file size s_k, where $i \in \{1, 2\}$.

Latency Analysis: We decompose the end-to-end latency of task q_k into four components: transmission, computation, caching, and jitter latencies.

- Transmission Latency: In the framework for holographic service in INC, nodes are joined directly via optical fibers where the internode distance is negligible. The transmission latency is given by:

$$t_{trans}^k = \frac{\eta \cdot s_k}{B_k} \tag{1}$$

  where η denotes the size ratio between the processed and original files determined by the compression algorithm, and B_k denotes the allocated bandwidth for the task. Specifically, $\eta = 1$ when no compression is needed.
- Computational Latency: User-submitted tasks delineate computational functions and their requisite processing cycles, whereas the computational resources assigned to nodes are unspecified, demanding algorithmic derivation of:

$$C_j^k(i) = \lambda_j^k(i) C_j \tag{2}$$

  where $\lambda_j^k(i)$ denotes the proportion of computational resources allocated by node j for the i-th computational function in the k-th task, with $\lambda_j^k(i) \in [0, 1]$. The computational latency is expressed as:

$$t_{com}^k = \sum_{j=1}^{n} \sum_{i=1}^{2} z_k^j(i) \frac{c_k^i}{C_j^k(i)} \tag{3}$$

  where $z_k^j(i)$ indicates whether the computational function f_k^i of task q_k is executed on node j, with $z_k^j(i) \in \{0, 1\}$. c_k^i represents the necessary number of computing cycles for this computational function.
- Caching Latency:

$$t_{cache}^k = h_k \cdot t_{com}^k \tag{4}$$

  where $h_k \in \{0, 1\}$ indicates whether any network node possesses the cached resources required by task q_k. If cached $h_{im} = 0$, the resources are locally available; otherwise $h_{im} = 1$, they must be retrieved from the business server.

- Jitter Latency: At the network egress of holographic stream transmission, a buffer is set up to temporarily store data, thereby providing a smooth data stream to the receiving end and eliminating computational fluctuations. However, if the buffer is too large, it increases end-to-end latency, affecting the real-time performance of communication. Therefore, the buffer size needs to strike a balance between smoothing fluctuations and controlling latency. It should be dynamically adjusted on the basis of current network conditions and the transmission requirements of the holographic stream. The buffer is configured as follows:

$$W \leq b_{uf} \leq D \tag{5}$$

Here, D denotes the upper bound of the end-to-end delay for the stream, and W denotes the lower bound of the end-to-end delay. The buffer configuration is shown in Fig. 2. Let a_u, b_u, and c_u represent the time when the u-th cloud frame starts transmission at the network ingress, the time spent in network transmission, and the time of reaching the network egress after passing through each resource, respectively. When the first cloud frame arrives at a point, the buffer is set to $b_1 + b_{uf} - W$. When $u > 1$, the buffer is set to $\max\{b_u,\ c_1 + (a_u - a_1)\}$. The transmission delay adjustment m is dynamically configured according to the arrival time of different cloud frames in the corresponding buffers, and by collecting computation fluctuations. The jitter latency caused by the buffer is denoted as t^k_{jitter}.

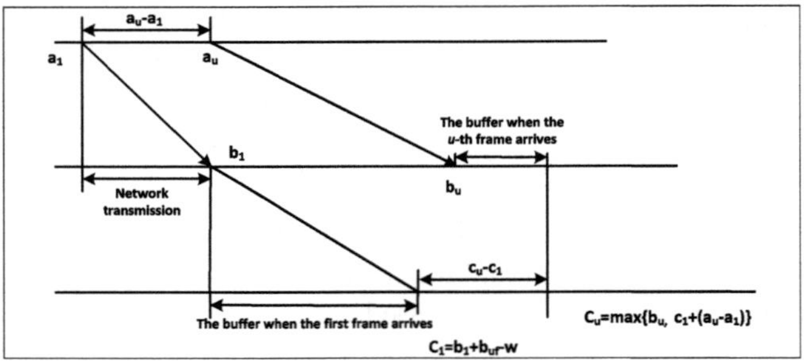

Fig. 2. Buffer configuration

Service Latency: The latency formulations for distinct service types are defined as follows:

- Holographic Browsing:

$$t_k = t^k_{trans} + t^k_{compute} \tag{6}$$

- Holographic On-demand Streaming:

$$t_k = t^k_{trans} + t^k_{cache} \tag{7}$$

where cached content availability is incorporated.

– Holographic Video Communication:

$$t_k = t^k_{trans} + t^k_{compute} + t^k_{\text{jitter}} \tag{8}$$

A buffer is deployed at the egress node to compensate for jitter, which critically impacts the communication quality.

Optimization Objective and Constraints:

$$min \sum_{k=1}^{K} (\alpha \cdot R_{network}(k) + \beta \cdot R_{compute}(k)) \tag{9}$$

Subject to:

$$t_k <= T_k^{req} \tag{10}$$

$$\sum_{k=1}^{K} \sum_{l=1}^{L} b_k^l B_k <= B \tag{11}$$

$$B_k >= B_k^{req} \tag{12}$$

$$\sum_{k=1}^{K} \sum_{i=1}^{2} z_k^j(i) C_j^k(i) <= C_j \tag{13}$$

The optimization aims to minimize resource consumption, which encompasses both network and computational resources, and determine critical decision variables for each computational task request submitted to the controller per time interval: the target node j for task q_k; the proportional allocation of computational resources $\lambda_j^k(i)$ on node j for its i-th computational function, and the network bandwidth B_k allocated to the task.

Formula (10) limits the end-to-end latency of tasks to remain within their specified latency requirements. Formula (11) guarantees that the total bandwidth of tasks assigned to link l does not exceed the bandwidth capacity of the link. Formula (12) mandates that the bandwidth assigned to a task meets its bandwidth demand. Formula (13) stipulates that the computational resources utilized by functions allocated to node j must not surpass the node's inherent computing capacity C_j.

4 Algorithm Design

The system model involves multiple variables and states, including node capacities C_j, link bandwidth B, and task requests q_k, resulting in a highly complex state space. To address this, the deep Q-Network (DQN) algorithm is introduced, which combines Q-learning with deep neural networks (DNNs) to approximate the following Q-value function:

$$Q(s, a; \theta) \approx Q^*(s, a) \tag{14}$$

where $s \in \mathcal{S}$ is the current state, $a \in \mathcal{A}$ is an action and θ denotes the DNN parameters. This hybrid approach enables the agent to derive optimal resource allocation policies $\pi : \mathcal{S} \rightarrow \mathcal{A}$ under multidimensional constraints.

This work proposes HDRS (Holographic DQN-based resource scheduler), a DQN-driven algorithm that minimizes resource consumption to achieve coordinated task scheduling and resource allocation. The core design is as follows.

State Space: When scheduling tasks to nodes, the state space comprises node states, link states, and task requirements:

$$s_k = (s_{node}, s_{link}, s_{task}) \tag{15}$$

The node state includes the computational capacity, residual resources, and cached content sets:

$$s_{node} = (s_{node}^{compute}, s_{node}^{resource}, s_{node}^{cache}) \tag{16}$$

The link state captures residual bandwidth metrics:

$$s_{link} = (s_{link}^{bandwidth}) \tag{17}$$

The task requirements specify the task type, normalized parameters, and computational function demands:

$$s_{task} = (s_{task}^{type}, s_{task}^{m}, s_{task}^{f}) \tag{18}$$

Action Space: At the current time interval, the scheduling decision set for task q_k is defined as:

$$a_{task} = [c_{node}, d_{node}, r_c, r_d, b_k, b_{uf}] \tag{19}$$

The set includes two nodes allocated for task q_k (c_{node}, d_{node}), computational resource proportions (r_c, r_d), allocated bandwidth B_k, and jitter buffer configuration b_{uf}.

Consequently, the action space is expressed as:

$$A = [a_1, a_2, \cdots\cdots a_q] \tag{20}$$

Reward Function: For each task, the state s_k and corresponding action a_k form a state-action pair (s_k, a_k), with its reward value denoted as $R(s_k, a_k)$. The optimization objective is to minimize resource consumption while guaranteeing end-to-end latency. The reward function is designed as:

$$\begin{aligned} R = &-(\alpha \cdot R_{\text{network}}(k) + \beta \cdot R_{\text{compute}}(k)) \\ &- \gamma \cdot (t_k - T_k^{\text{req}}) + R_{\text{success}} + R_{\text{failure}} \end{aligned} \tag{21}$$

where $R_{\text{network}}(k)$ and $R_{\text{compute}}(k)$ represent network resource consumption and computational resource consumption, respectively. HDRS is shown in

Algorithm 1: Holographic DQN-based Resource Scheduling Algorithm

Input : Network G, Task queue sort by priority T
Output: Resouce allocation policy a_k for every task

Initialize: policy network parameters θ, target network parameters $\theta^- \leftarrow \theta$, experience replay buffer $\mathcal{D} \leftarrow \emptyset$, exploration rate $\epsilon \leftarrow 1.0$
for *episode* $e = 1$ ***to*** E **do**
  Reset network environment and obtain initial state s_1;
  for *step* $t \in [1, k]$ **do**
    while $\mathcal{T} \neq \emptyset$ **do**
      if *rand*() $< \epsilon$ **then**
        $a_t \leftarrow$ Randomly select valid action
      else
        $a_t \leftarrow \arg\max_a Q(s_t, a; \theta)$ with $\epsilon - \textit{greedy}$ policy
      Execute action $(c_{node}, d_{node}, r_c, r_d, b_w, b_{uf})$ and obtain an immediate reward R
      Transform to next state s_{t+1} for next task
      Store experience: $\mathcal{D} \leftarrow \mathcal{D} \cup \{(s_t, a_t, r, s_{t+1}, \text{done})\}$
      if $|\mathcal{D}| \geq B$ **then**
        Sample batch $\{(s^i, a^i, r^i, s^{i+1}, d^i)\}_{i=1}^B \sim \mathcal{D}$
        Compute target values:

$$y^i = \begin{cases} r^i & \text{if } d^i \\ r^i + \gamma \max_{a'} Q(s^{i+1}, a'; \theta^-) & \text{otherwise} \end{cases}$$

        Update parameters: $\theta \leftarrow \theta - \eta \nabla_\theta \sum_i (y^i - Q(s^i, a^i; \theta))^2$
      if $e \mod C == 0$ **then**
        Sync target network: $\theta^- \leftarrow \theta$
      $s_t \leftarrow s_{t+1}$
  Decay exploration rate: $\epsilon \leftarrow \epsilon \cdot \epsilon_{\text{decay}}$

Algorithm 1. The DQN algorithm involves two key mechanisms-experience replay and target networks-to decorrelate training data and stabilize learning. Experience replay operates by storing historical interaction tuples (s_t, a_t, r_t, s_{t+1}) in a replay buffer D, followed by stochastic mini-batch sampling $\{(s_i, a_i, r_i, s_i')\} \sim U(\mathcal{D})$ during neural network training. This buffer aggregates data from heterogeneous policies (exploratory random actions and exploitative greedy actions), thereby diversifying the state-action distribution exposure and reducing overreliance on current policy trajectories. The target network architecture implements dual NNs: an online network $Q(s, a; \theta)$ updated via backpropagation and a target network $Q(s, a; \theta^-)$ with periodic parameter synchronization $\theta^- \leftarrow \theta$ every τ step. This decoupling addresses the nonstationary target problem inherent in temporal difference learning, formally expressed as:

$$\mathcal{L}(\theta) = \mathbb{E}_{(s,a,r,s') \sim \mathcal{D}} \left[\left(r + \gamma \max_{a'} Q(s', a'; \theta^-) - Q(s, a; \theta) \right)^2 \right] \tag{22}$$

5 Experiment and Analysis

This section presents simulation experiments to assess the performance of HDRS.

5.1 Experimental Settings

The network topology is shown in Fig. 3. The network consists of users, a controller, an intelligent converged identifier node, and a business server. Users submit holographic service requests (such as 3D model access or real-time communication) and receive rendered immersive data streams. Each intelligent converged identifier node includes a P4-programmable switch paired with a computing node, enabling dynamic coordination of computation, communication, and caching to support holographic service. The DPU enhances the P4 switch's capability for packet forwarding; thus, forwarding latency is negligible in this paper. All computing nodes execute the same algorithms for compression, decompression, and rendering. The business server stores raw holographic data (such as 3D models and point-cloud videos) and responds to requests for tasks that are not cached. The controller continuously monitors the status of network resources, executes the resource scheduling algorithm, and makes decisions regarding task offloading and routing.

The experimental settings are set as in Table 1. To ensure statistical reliability, each algorithm is executed 100 times, and the average of the results is as follows.

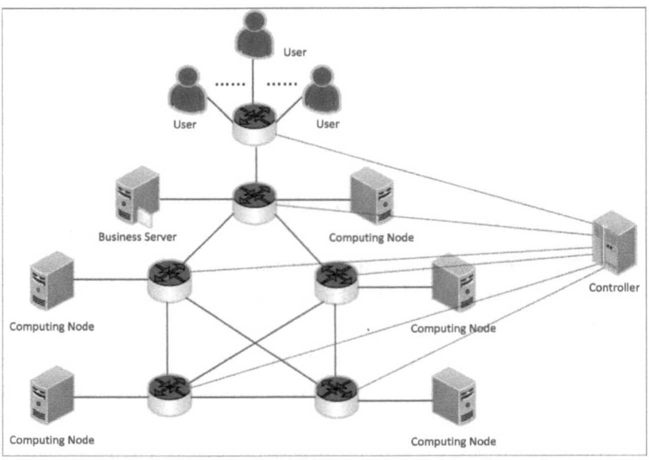

Fig. 3. Network topology

Table 1. System Parameter Configuration

Parameter	Description	Value
k	Number of tasks	5
T^w	Latency requirement (s)	[0.1, 0.3]
B_r	Bandwidth demand (Gbps)	[8, 16]
J^v	Jitter requirement (s)	[0.01, 0.02]
C	Computation cycles (cycles)	[0.1×10^9, 1.5×10^9]
s	Request file size (MB)	[20, 100]
n	Number of nodes	5
B	Link bandwidth (Gbps)	80
C	Computing capability (GHZ)	[70, 80, 80, 90, 100]
η	Compression ratio	0.5
α	Weight of computational resource	0.04
β	Weight of network resource	0.1
γ	Weight of latency	20
R_{success}	Reward (success)	10
R_{failure}	Reward (failure)	−30

5.2 Evaluation of HDRS

This study employs the greedy algorithm (Greedy), genetic algorithm (GA), and simulated annealing algorithm (SA) as baselines. The Greedy and GA methods prioritize assigning minimum-latency paths to high-priority tasks, whereas the SA method seeks resource-optimal paths. The four algorithms (including the proposed method) are comparatively evaluated across three metrics: end-to-end latency, resource consumption, and the success rate of task deployment, with the experimental results detailed below.

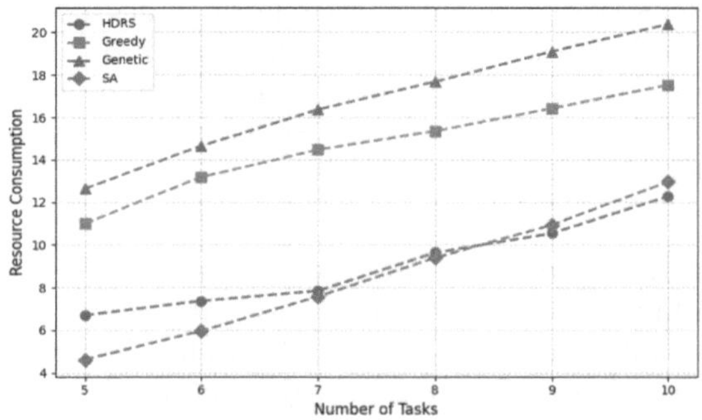

Fig. 4. Resource consumption

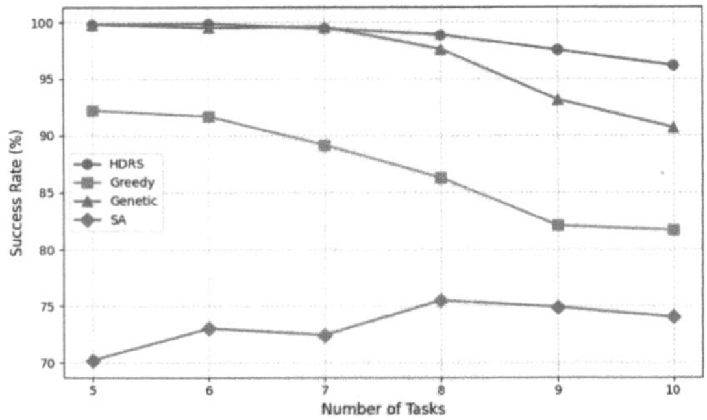

Fig. 5. Success rate of task deployment

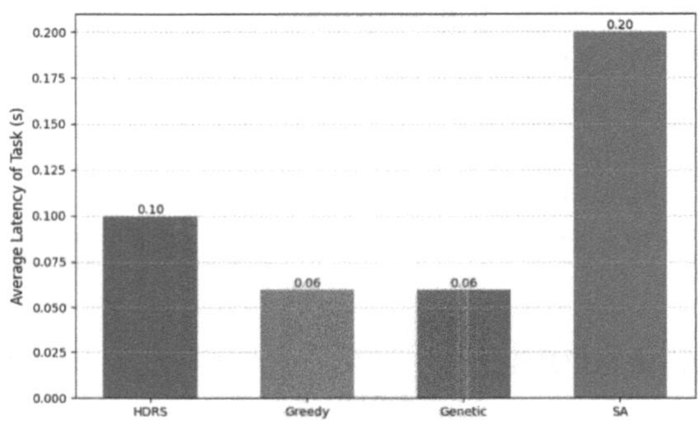

Fig. 6. Average end-to-end latency of successfully deployed task

Resource Consumption Analysis: As illustrated in Fig. 4, HDRS achieves lower resource consumption compared to GA and Greedy, and exhibits marginally superior performance relative to SA under high task load conditions. While SA effectively explores low-resource solution spaces via its annealing mechanism, its overly conservative resource usage leads to degraded task deployment success. Greedy's task specific, myopic optimization lacks global coordination, resulting in resource fragmentation and increased overall consumption. In contrast, GA optimizes for global latency and solution diversity, but this comes at the cost of considerable resource overhead.

Success Rate of Task Deployment: As illustrated in Fig. 5, both the HDRS and GA maintain high task deployment success rates across varying workloads. However, Greedy exhibits an approximate 15% decline in the success rate as the task scale increases, whereas SA consistently achieves the lowest performance. The reinforcement learning mechanism in HDRS enables dynamic adaptation to workload fluctuations, sustaining success rates above 95% through prioritized experience replay and adaptive policy updates. In contrast, the GA experiences a pronounced drop in performance beyond eight concurrent tasks, primarily due to population diversity loss and convergence to local optima. The persistently low success rate of SA is attributed to its overly conservative resource allocation strategy, which limits effective task deployment under increasing load.

End-to-End Latency: As depicted in Fig. 6, HDRS achieves competitive average latency (0.10 s) for successfully deployed tasks, while simultaneously maintaining low resource consumption and high task success rates, as demonstrated in previous evaluations. The GA attains the lowest latency (0.06 s) by leveraging its global search capability, which incrementally evolves latency-optimized scheduling strategies across generations, albeit at the expense of higher resource consumption. Similarly, the Greedy achieves comparable minimal latency (0.06 s) through aggressive task assignment; however, its lack of global coordination leads to resource fragmentation and reduced overall efficiency. In contrast, SA incurs significantly longer transmission and processing delays because of its overly conservative resource allocation approach.

The experimental results validate that the HDRS exhibits high reliability and dynamic adaptability in holographic task scheduling scenarios. Under resource-constrained conditions, HDRS effectively minimizes resource consumption while simultaneously balancing task deployment success and stringent latency requirements, thereby achieving superior overall performance.

5.3 System Implementation

The HDRS algorithm was implemented on a prototype system to evaluate its performance in the holographic communication environment, as illustrated in Fig. 7. The programmable switch used is OpenMeshBF-32D, and the x86-based device used is Dell Precision 7900, which together form an intelligent converged the identifier node. The server is Dell PowerEdge R640, and the controller is hosted on a Dell PowerEdge R740xd server. To compare end-to-end latency, four tasks were deployed concurrently under both the HDRS and the SA algorithm, with the resulting latency measurements presented in Fig. 8.

The experimental results show that HDRS can reduce the end-to-end latency by 40.3%, 34.1% and 11.3% for high-load tasks. This demonstrates that under conditions of heavy computational workloads and limited resources, HDRS is capable of dynamically making decisions based on current resource states, thereby better meeting the real-time requirements of tasks.

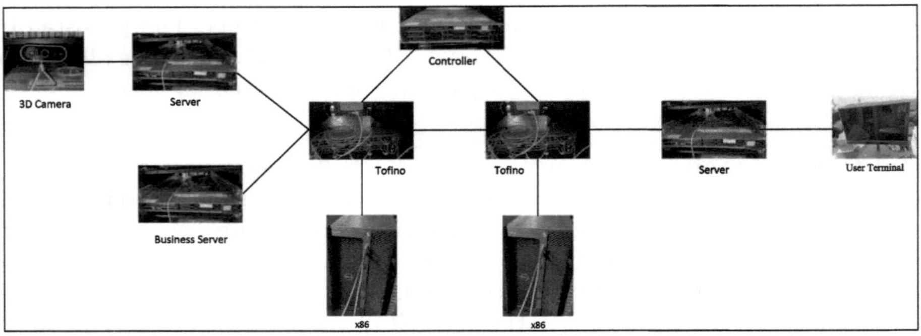

Fig. 7. Prototype system

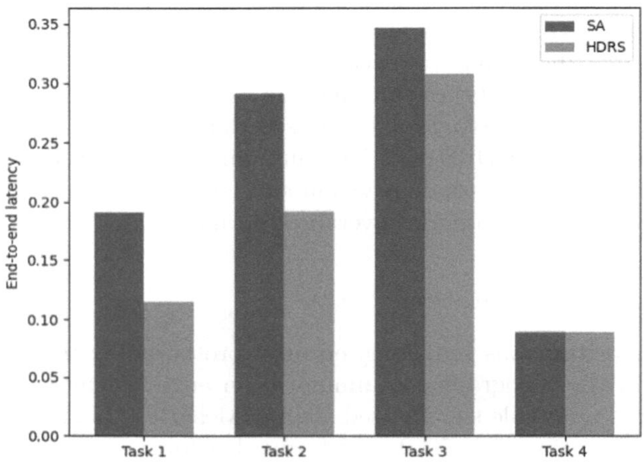

Fig. 8. System latency test

6 Conclusion

This paper presents HDRS, a DQN-based resource scheduling algorithm designed for holographic services. By jointly orchestrating bandwidth, computing, and caching resources, HDRS dynamically allocates tasks to optimal nodes according to service-specific requirements, enabling efficient and adaptive resource management. The experimental results show that HDRS significantly reduces resource consumption while meeting stringent latency constraints and maintaining a high task deployment success rate under resource-limited conditions. The proposed approach offers a practical and effective solution for resource scheduling for holographic communication.

Acknowledgments. This work is supported by the National Key R&D Program of China under grant no. 2023YFB2904400, the Major Key Project of PCL under grant

No. PCL2023A06, fundamental research funds for the central universities under grant no. 2023JBGP003, nature and science foundation of China under grant no. 62394321, 62072030, 92167204.

References

1. Akyildiz, I.F., Guo, H.: Holographic-type communication: a new challenge for the next decade. ITU J. Future Evolving Technol. **3**(2), 421–442 (2022)
2. Clemm, A., Vega, M.T., Ravuri, H.K., Wauters, T., De Turck, F.: Toward truly immersive holographic-type communication: challenges and solutions. IEEE Commun. Mag. **58**(1), 93–99 (2020)
3. Erfanian, A.: Optimizing QoE and latency of live video streaming using edge computing and in-network intelligence. In: Proceedings of the 12th ACM Multimedia Systems Conference, pp. 373–377 (2021)
4. Qian, P., Huynh, V.S.H., Wang, N., Anmulwar, S., Mi, D., Tafazolli, R.R.: Remote production for live holographic teleportation applications in 5G networks. IEEE Trans. Broadcast. **68**(2), 451–463 (2022)
5. Aghaaliakbari, F., et al.: An architecture for provisioning in-network computing-enabled slices for holographic applications in next-generation networks. IEEE Commun. Mag. **61**(3), 52–58 (2023)
6. Sapio, A., Abdelaziz, I., Aldilaijan, A., Canini, M., Kalnis, P.: In-network computation is a dumb idea whose time has come. In: Proceedings of the 16th ACM Workshop on Hot Topics in Networks, pp. 150–156 (2017)
7. Chen, J., Liu, S., Gao, S., Huang, X., Zhang, H.: Smart computing integration network for holographic communication. Acta Electronica Sinica 1–11 (2024)
8. Huang, Z., Yupeng, D., Yang, S., Xiao, H., Wang, D., Sun, T.: Joint optimization of task scheduling and computing resource allocation for VR video services in 5G-advanced networks. Trans. Emerg. Telecommun. Technol. **35**(1), e4909 (2024)
9. Peng, Q., Wu, C., Xia, Y., Ma, Y., Wang, X., Jiang, N.: DoSRA: a decentralized approach to online edge task scheduling and resource allocation. IEEE Internet Things J. **9**(6), 4677–4692 (2021)
10. Liu, L., Feng, J., Xuanyu, M., Pei, Q., Lan, D., Xiao, M.: Asynchronous deep reinforcement learning for collaborative task computing and on-demand resource allocation in vehicular edge computing. IEEE Trans. Intell. Transp. Syst. **24**(12), 15513–15526 (2023)
11. Aliyu, I., Seungmin, O., Ko, N., Um, T.-W., Kim, J.: Dynamic partial computation offloading for the metaverse in in-network computing. IEEE Access **12**, 11615–11630 (2023)
12. Kamran, K., Yeh, E., Ma, Q.: DECO: joint computation scheduling, caching, and communication in data-intensive computing networks. IEEE/ACM Trans. Netw. **30**(3), 1058–1072 (2021)
13. Joung, J., Kwon, J.: Zero jitter for deterministic networks without time-synchronization. IEEE Access **9**, 49398–49414 (2021)

Interleaved Spatial-Temporal Attention Aggregation Gated Transformer for Traffic Flow Prediction

Ningyun Li[1], Yifei Luo[1], Wenfei Gao[1], Lin Zhang[1(✉)], Ke Yu[2], Rui Luo[3], and Hao Ji[3]

[1] Beijing Big Data Centre, Beijing, China
zhanglin@jxj.beijing.gov.cn
[2] Beijing University of Posts and Telecommunications, Beijing, China
[3] China Telecom Corporation Beijing Branch, Beijing, China

Abstract. Traffic flow prediction constitutes a crucial component in urban planning and the optimization of intelligent transportation systems. Attention-based models for traffic flow prediction within grid-based spatial partitioning frameworks might face challenges including insufficient learning of potential important local patterns during feature learning and underutilizing intermediate computational information. To address these problems, we propose ISTAFormer (**I**nterleaved **S**patial-**T**emporal Attention **A**ggregation Gated Trans**former**), which employs interleaved temporal-view and spatial-view gated transformer blocks (GTB) to effectively learn the spatiotemporal features. It incorporates an all-layer results weighted aggregation mechanism which allows the model to reuse the intermediate results that have learned local features to address potential local feature loss. In addition, this paper constructs a new real-world dataset of shared bicycles in Beijing, BikeBJ-7days, and analyzes its spatial and temporal features. The experimental results demonstrate that ISTAFormer successfully captures the local features in BikeBJ-7days that many baselines not capture correctly. Compared to the best-performing baseline, ISTAFormer reduces the mean squared error by 21.75% on TaxiBJ and 7.15% on BikeBJ-7days. ISTAFormer achieves the state-of-the-art (SOTA) performance on both TaxiBJ and BikeBJ-7days, which indicate that ISTAFormer has excellent prediction effect and generalization performance across datasets with different characteristics.

Keywords: Traffic flow prediction · Spatiotemporal predictive learning · Gating mechanism · Spatiotemporal attention · Feature aggregation

1 Introduction

Traffic flow prediction plays a critical role in the planning and management of modern smart cities, enabling authorities to anticipate congestion, optimize sig-

L. Huang et al. (Eds.): CCF NCCA 2025, CCIS 2774, pp. 378–394, 2026.
https://doi.org/10.1007/978-981-95-5294-8_26

nal control, and improve urban mobility. To leverage powerful deep-learning techniques, urban traffic data are often discretized into regular grids, transforming complex road networks into image-like frames that capture spatial correlations at city scale. In this grid-based representation, traffic flow prediction becomes similar to video prediction in that each grid region corresponds to a pixel point, and the traffic flow data forms a sequence of "frames," each recording the inflow and outflow counts within every grid cell over a specific time interval. Accurately forecasting future frames from past observations thus requires models that can jointly capture long-range dependencies and fine-grained local patterns in the spatiotemporal data.

In prior literature, many efforts have been devoted to improving the grid-based traffic flow prediction. Existing approaches can be categorized into the following three groups including Recurrent Neural Networks (RNNs), Convolutional Neural Networks (CNN), and Attention-Based Architectures, but each has significant drawbacks. RNN-based methods such as ConvLSTM [10] and PredRNN v2 [18] excel at modeling temporal dependencies but inherently rely on sequential computation, which limits parallelism and yields high training latency. CNN-based models like SimVP [4] and WaST [8] utilize convolutional filters for efficient local feature extraction, but their fixed sense fields make it difficult to capture global context without stacking many layers, and deeper stacking exacerbates overfitting and gradient vanishing. Attention-based transformers, exemplified by ViT [3] and PredFormer [14], enable dynamic, global interactions across space and time in parallel. However, when major features monopolize attention, they tend to ignore subtle local details, and deep stacks of self-attentive blocks may experience network degradation, leading to the disappearance of early layer representations.

To address these challenges, we propose ISTAFormer—an Interleaved Spatiotemporal Transformer with All-layer Aggregation. ISTAFormer interleaves two specialized gated transformer blocks (GTB), a Spatial-view GTB to learn long-range spatial correlations and a Temporal-view GTB to model temporal dynamics, and retains every intermediate layer output. Through an all-layer weighted aggregation layer, the model learns to balance contributions from both prominent global patterns and overlooked local features. Additionally, a gating mechanism is employed to control the flow of attention information, enabling the model to learn spatio-temporal features effectively. We conduct experiments on a publicly available dataset, TaxiBJ [20], and our own dataset, BikeBJ-7days. We find that some baseline models perform significantly worse on BikeBJ-7days than on TaxiBJ while our model is more stable.

The main contributions of this paper can be summarized as follows:

1. We propose a novel All-layer Weighted Aggregation Layer to reuse intermediate computational information for traffic flow prediction, allowing the model to dynamically learn features for different tasks based on learnable weights.
2. We present ISTAFormer, which captures the spatial and temporal features of the data through a multi-head self-attention mechanism and controls the flow of information through gating mechanism. Finally, an all-layer weighted

aggregation layer is used to compute the weighted sum of intermediate results as the output. These mechanisms enhance the model's ability to capture local features and improve its prediction performance.
3. We conduct experimental tests with public dataset, TaxiBJ, and our own dataset (BikeBJ-7days) that has significant local features. ISTAFormer achieves SOTA performance compared to several baseline models on both TaxiBJ dataset and BikeBJ-7days dataset across four metrics. ISTAFormer successfully captures local features that are not captured by many baselines, which verified the effectiveness of ISTAFormer.

2 Related Work

2.1 Recurrent Neural Network Based Models

Recurrent methods, due to the sequential nature of their computational processes, inherently possess strong capabilities for capturing the temporal features of spatiotemporal data. However, this sequential nature also makes parallel computation difficult, resulting in relatively low computational efficiency. In recent years, numerous attempts had been made to integrate RNNs with other modules, achieving promising results in spatiotemporal prediction tasks.

ConvLSTM [10], proposed by Shi et al. in 2015, incorporates convolutional operations into the LSTM framework, replacing fully connected weights with convolutional kernels. This enabled the model to capture both spatial and temporal correlations simultaneously. E3D-LSTM [16], proposed by Wang et al. in 2018, extended 2D convolution to 3D convolution, introducing temporal convolutional kernels in the gating computations to enhance the modeling of spatiotemporal consistency across consecutive video frames. MIM [19], proposed by Wang et al. in 2019, proposed multi-level memory units that explicitly modeled the non-stationarity of spatiotemporal dynamics through interactions between current and historical memories. PhyDNet [5], proposed by Guen and Thome in 2020, proposed a new recurrent physical unit called PhyCell, which combined deep learning with physical knowledge described by partial differential equations to improve unsupervised video prediction methods, enabling better modeling of the complex dynamics in videos. MAU [2], proposed by Chang et al. in 2021, introduced motion-aware units to capture dynamic motion information, leveraging the correlation between spatial and temporal domains for video prediction. PredRNN v2 [18], proposed by Wang et al. in 2022, introduced a decoupled loss function to maximize the distance between memory states of the two memory units in ST-LSTM and incorporated a learning strategy called reverse scheduled sampling, which further improved the performance of PredRNN [17], a model proposed the Spatiotemporal LSTM (ST-LSTM) unit.

2.2 Convolutional Neural Network Based Models

Convolution-based methods benefit from their spatial convolutional characteristics and the absence of recurrent structures. This allows them to efficiently perform parallel computations while also having strong capabilities for extracting local spatial features. However, their capture of spatial features is limited by the size of the convolutional window, and they can only learn global spatial features through the stacking of local feature convolutions. This makes it difficult to accurately extract global spatial features while maintaining computational efficiency, and they also have limited ability to model complex temporal features.

SimVP [4], proposed by Gao et al. in 2022, was composed entirely of CNN modules. After extracting spatiotemporal features using a CNN encoder, it performed non-autoregressive decoding directly in the latent space, completely abandoning the recurrent structure. STRPM [1], proposed by Chang et al. in 2022, proposed a spatiotemporal residual prediction model specifically for high-resolution video prediction. STRPM included two key components: first, a spatiotemporal encoding-decoding scheme that extracted spatiotemporal features using separate encoders to avoid interference between spatiotemporal information and better preserve details in high-resolution videos; second, a residual prediction memory (RPM) that focused on modeling the spatiotemporal residual features between consecutive frames rather than predicting entire frames. This helped capture complex motion information in high-resolution videos. WaST [8], proposed by Nie et al. in 2024, was a wavelet-based spatiotemporal framework that extracted and adaptively controlled low-frequency and high-frequency components at the image and feature levels through 3D discrete wavelet transforms, improving inference speed while maintaining high-quality predictions.

2.3 Attention-Based Models

Attention-based methods can associate arbitrary spatiotemporal positions and dynamically allocate weights through attention scores. This allows them to model long-range dependencies of spatiotemporal features while maintaining efficient parallel computation. However, the computation of attention matrices is relatively complex and requires a large amount of data for training. Moreover, when overly reliant on global attention, these methods may easily overlook local spatiotemporal features. For instance, when a particular feature is significantly prominent, the attention to other features may become less noticeable or even be overshadowed, leading to the loss of learned features.

ViT [3], proposed by Dosovitskiy et al. in 2020, first transferred the attention mechanism from natural language processing tasks to video prediction tasks, demonstrating the feasibility of using attention mechanisms for spatiotemporal feature extraction and achieving promising results. SA-ConvLSTM [6], proposed by Lin et al. in 2020, incorporates self-attention mechanism into the ConvLSTM framework, which designs a memory-based self-attention module (SAM) to remember the global spatiotemporal dependencies in the prediction process. SwinLSTM [13], proposed by Tang et al. in 2023, integrated the Swin Transformer [7] module, which was proposed by Liu et al. in 2021, into the LSTM architecture, effectively combining the global spatial modeling capability of Swin Transformer with the temporal sequence modeling capability of LSTM.

TAU [11], proposed by Tan et al. in 2023, decomposed spatiotemporal attention into intra-frame static attention and inter-frame dynamic attention, learning dependencies in time and space separately to achieve efficient video prediction. PredFormer [14], proposed by Tang et al. in 2024, achieved excellent spatiotemporal prediction performance with high computational speed using a stack of pure gated attention blocks.

3 ISTAFormer Method

3.1 Framework

The ISTAFormer, as shown in Fig. 1, is partitioned into three cohesive phases: Data Preprocessing, Prediction and Reconstruction.

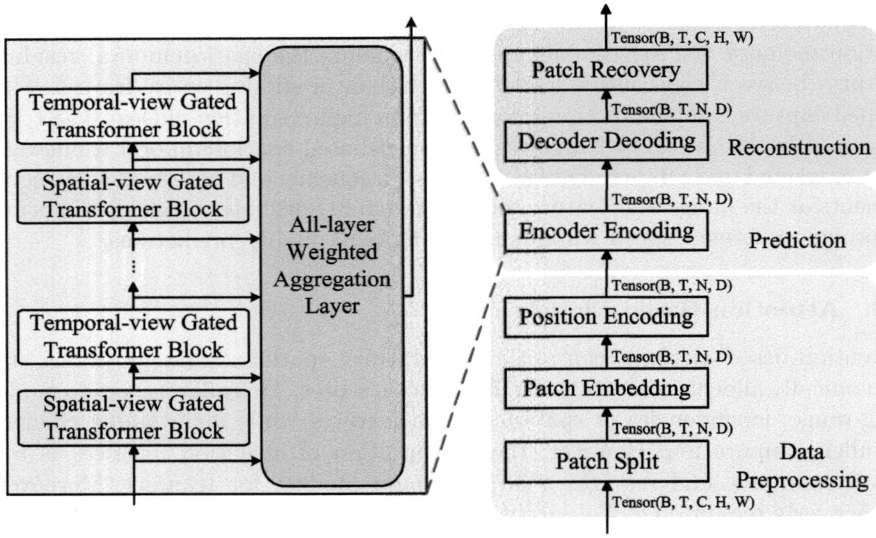

Fig. 1. ISTAFormer framework

The raw data input into the model should have five dimensions, with the shape of (B, T, C, H, W), corresponding to batch size B, time T, features C, frame height H, and frame width W, respectively. In the data preprocessing part, our model follows the design of ViT and processes the five-dimensional raw data into four-dimensional data, with the shape of (B, T, N, D). Here, the physical meanings correspond to batch size B, time T, number of nodes N, and node features D, respectively. This transformation is performed to facilitate subsequent computations.

After the preprocessed data are processed and computed in the data prediction part of the model, the resulting prediction data have the same dimensions as the preprocessed data. In the data reconstruction part, the prediction data are simply decoded, and then, through the inverse transformation of the slicing operation, the prediction data are restored from four dimensions to the original five-dimensional structure. This final five-dimensional data constitute the ultimate prediction result.

3.2 Data Preprocessing

The data preprocessing phase consists of three computational stages: data slicing and rearrangement, data embedding, and positional encoding. In the data slicing and rearrangement stage, the model slices the two-dimensional spatial data of each batch, each frame, and each feature according to the predefined patch size $p \times p$. Then it rearranges the patches in sequence to obtain tokens for each frame with the shape of $(B, T, \left\lfloor \frac{H}{p} \right\rfloor \times \left\lfloor \frac{W}{p} \right\rfloor, C \times p \times p)$. Subsequently, all tokens are embedded using a linear layer to obtain the initial embedding vectors with the shape of (B, T, N, D), where $N = \left\lfloor \frac{H}{p} \right\rfloor \times \left\lfloor \frac{W}{p} \right\rfloor$. Finally, sinusoidal absolute positional encoding is added to complete the preprocessing of the data [15].

3.3 Prediction

In the prediction phase, ISTAFormer predicts through an encoder that consists of a sequence of several GTB layers. All intermediate computational results are retained and are ultimately aggregated through an all-layer weighted aggregation layer to obtain the final prediction data.

For the process of a two-dimensional input vector entering the self-attention layer to calculate its attention, it can be understood as follows: The first dimension of the two-dimensional vector represents nodes, while the second dimension represents the features of these nodes. In the self-attention layer, the query vector, key vector, and value vector are computed by passing the vector through three separate linear layers to embed it into the corresponding embedding spaces. The query vector is then multiplied by the key vector and scaled to obtain the attention weight matrix, which records the correlations between each node and all other nodes in the hidden feature space. Multiplying this matrix by the value vector yields the feature representation of each node in the attention space, which is determined by the correlations of each node with all other nodes regarding the hidden features.

From the self-attention computation process, it can be seen that the last two dimensions of the input vector are related to the dot-product correlation calculation, linear layer embedding, and the features learned through attention. Therefore, given that the last dimension of the preprocessed data represents the features of a single node, processing the penultimate dimension as a temporal or spatial dimension, i.e., different views of data, as shown in Fig. 2, can enable the model to learn the temporal or spatial features of the data through the self-attention mechanism.

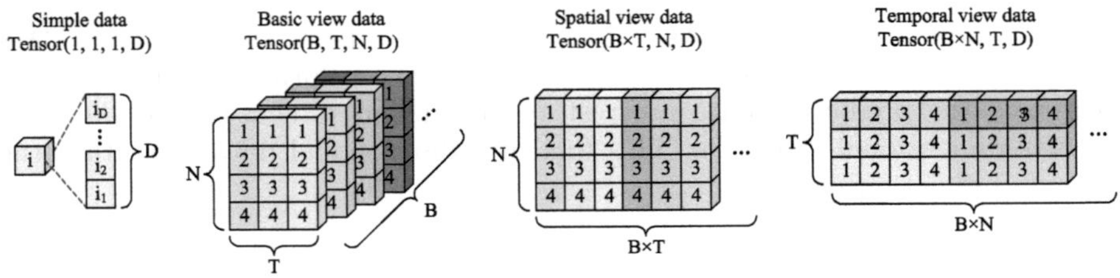

Fig. 2. Different views of data

According to the different data processing before inputting GTB, each GTB can be classified as either a Temporal-view GTB (TGTB) or a Spatial-view GTB (SGTB), depending on whether the penultimate dimension of data represents time or space, as shown in Fig. 3. By overlapping interleaved SGTB and TGTB, ISTAFormer can learn the spatial and temporal features in the data.

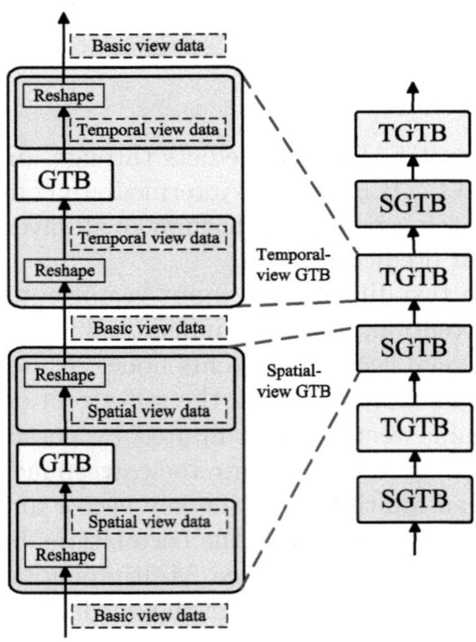

Fig. 3. Interleaved SGTB and TGTB

The structure of GTB, which serves as the core of the model, is shown in Fig. 4 and is mainly composed of the multi-head self-attention part for learning features and the gate unit part for filtering the learned features. After the input data X passes through the multi-head self-attention layer and obtains attention $Attn$, it needs to be mapped back to the original space through a Feedforward Network (FFN) composed of a linear layer to facilitate residual connections and obtain vectors containing hidden features while ensuring that gradient propagation does not vanish. Subsequently, the ReGLU (Rectified Linear Unit with Gating) [9] is employed as a gating mechanism to select effective information for residual connections and to allow it to serve as the output Y of this GTB, which then flows into the next layer. In addition to these components, the GTB also includes drop layers to introduce regularization during training, thereby enhancing the model's generalization ability, and normalization layers to stabilize the data distribution and improve training performance.

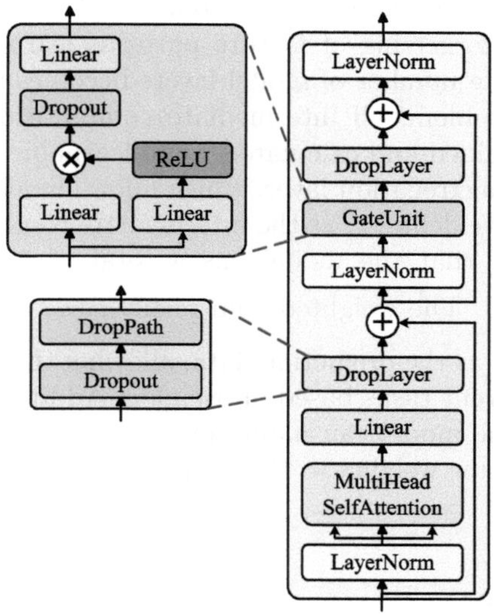

Fig. 4. GTB structure

For each GTB, its prediction computation can be described by the following formulas:

$$
\begin{aligned}
Q &= W_Q LayerNorm(X) \\
K &= W_K LayerNorm(X) \\
V &= W_V LayerNorm(X) \\
Attn &= softmax(\frac{QK^T}{\sqrt{d_k}})V \\
Z &= X + W_{FFN} Attn
\end{aligned}
\tag{1}
$$

$$
Y = Z + W_{out}(ReLU(W_{gate} LayerNorm(Z)) * W_{in} LayerNorm(Z)) \tag{2}
$$

where W_Q, W_K, W_V represent the computation of the corresponding independent linear layers in multi-head self-attention layer, W_{FFN} represents the computation of the FFN layer, $W_{in}, (W_{gate}), W_{out}$ represent the computation of the corresponding independent linear layers in the gate unit while W_{gate} is before the ReLU activation function and W_{out} is after the dropout layer in the fate unit, and d_k denotes the number of the last dimension of K^T.

The model can learn both spatial and temporal features of the data through the alternation of SGTB and TGTB. Since different GTB layers may learn different features, and the prediction performance for different tasks may rely on these features to varying degrees, simply stacking GTB layers as in PredFormer may not always achieve the best prediction results for all tasks. Prior knowledge of the task is required to select the model architecture. Moreover, when the number of model layers is too large, gradient instability and network degradation may occur, leading to the forgetting of feature patterns learned earlier. Therefore, choosing an appropriate number of model layers becomes a critical issue.

To address this problem, all intermediate computational results including the input data are retained and calculated using learnable weight parameters in the all-layer weighted aggregation layer, which allows model to reuse features in intermediate results. We define Y_i as the intermediate result after the calculation of i GTBs. Assuming that the model has a total of n GTBs, there will be $n+1$ results remained. The weighted sum result $Pred = \sum_{i=0}^{n} \alpha_i Y_i$ replaces the output of the last layer as the prediction data, allowing the model to dynamically learn features for different tasks based on weights. Additionally, even if network degradation occurs, the model can maintain a good prediction performance by learning to assign greater weights to the outputs of intermediate layers prior to degradation.

3.4 Reconstruction

The reconstruction phase consists of two stages: data decoding and structure recovery. Considering that the output of the encoder is in a high-dimensional hidden space in terms of physical meaning, therefore, it is necessary to decode it to map the predicted data from the hidden space back to the original space. ISTAFormer employs a linear layer as the decoder to decode and ensure the shape of the data is (B, T, N, D), which is the same as the preprocessed data, and its values represent the original data features. Subsequently, the inverse transformation of the slicing operation, which does not change the numerical value of the data, used in the data preprocessing phase is applied to restore the data to its original shape (B, T, C, H, W), yielding the final prediction results of the model.

4 Datasets Construction and Analysis

Our experiments are conducted on the TaxiBJ dataset provided by OpenSTL [12] and our own BikeBJ-7days dataset. The TaxiBJ dataset consists of normalized GPS trajectory data of taxis in Beijing, featuring two characteristics: inflow and outflow of traffic. Spatially, it is divided into a 32×32 grid. The dataset is organized into batches of four consecutive frames, with a total of 20,461 batches in the training set and 500 batches in the testing set without validation set. The BikeBJ-7days dataset records the trajectories of running shared bicycles in a specific rectangular area of Beijing from September 1, 2024, to September 7, 2024, with a time interval of 5 min. The area is defined by a diagonal line from the coordinates (116.29°, 39.82°) to (116.47°, 40.00°) and is divided into a 36×36 grid with a spatial resolution of 0.005°. This dataset captures the number of activative bicycles in each grid during each time interval, with normalization to [0, 1] like TaxiBJ, as the flow of traffic. The data is organized into batches of four consecutive frames, with 1,900 batches in the training set and 100 batches in the testing set without validation set. The difference between TaxiBJ and BikeBJ-7days are shown in Table 1.

Table 1. Comparison of Datasets

Dataset	TaxiBJ	BikeBJ-7days
City	Beijing	Beijing
Spatial range	including the Fourth Ring Road	including the Third Ring Road
Spatial division	32×32	36×36
Time interval	30 min	5 min
Time span	2013.07.01–2013.7.30, 2014.03.01–2014.06.30, 2015.03.01–2015.06.30, 2015.11.01–2016.04.10	2024.09.01–2024.09.07
Features	inflow and outflow	flow

In terms of spatial distribution, the BikeBJ-7days dataset exhibits a highly concentrated local peak in traffic flow within a specific area, while the data distribution in the remaining grids is relatively uniform. Comparison with the map reveals that there is a subway station named Olympic Sports Center in the area where the local peak points are located. In addition to being adjacent to the Olympic Sports Center Stadium, this subway station is also close to many popular attractions such as the China Ethnic Museum, the China Ethnic Culture Park, the Beijing Olympic Museum, the Visitor Center of the Olympic Park, and the National Stadium, and it is the nearest subway station to the Jianshao Bridge Campus of Beijing Information Science and Technology University and the China Conservatory of Music. Therefore, it is possible that this subway station will need to bear such a large volume of traffic flow. As the heatmap excluding the local peak point shown in Fig. 5(b), we set the local peak point's value to the second largest value in the statistics and find many high flow value points except for the local peak point are close to densely populated areas like schools and also near to the main roads like the Second Ring Road and the Third Ring Road. Hence, it demands that the model have strong capabilities in modeling local spatial features. The mean traffic flow heatmaps of all grids over the 7 days for BikeBJ-7days with main road map are shown in Fig. 5.

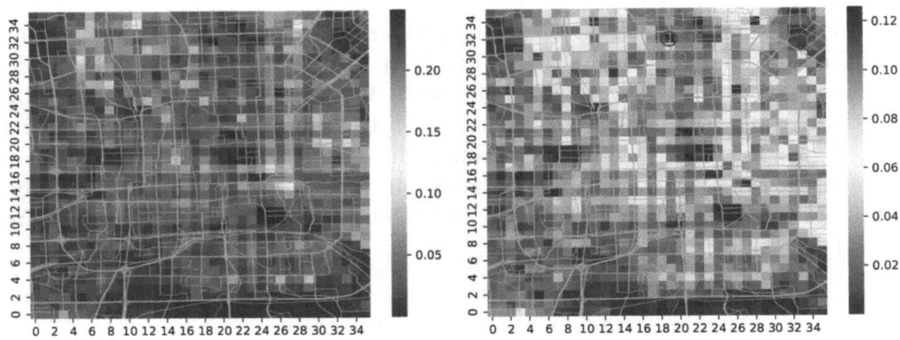

((a)) BikeBJ-7days' spatial feature distribution heatmap with main road map

((b)) BikeBJ-7days' spatial feature distribution heatmap excluding the local peak point in the red circle with main road map

Fig. 5. The spatial features of BikeBJ-7days

In terms of temporal distribution, overall, the BikeBJ-7days dataset has clear temporal patterns, with higher traffic during the day and lower traffic at night. On weekdays, there are also morning and evening peak characteristics, reaching their peaks around 8 and 18 o'clock. Specifically on each grid, the temporal characteristics of different regions also show some differences. Most of the regions have temporal characteristics close to the overall characteristics, but there are also some regions, such as the locally peak points mentioned earlier, whose temporal characteristics are different from the overall ones. Their peak values appear regularly intermittently, which is related to the traffic flow changes of the subway stations. This requires the model to have a robust ability to uncover temporal features under different conditions. The mean traffic flow histograms of all grids at each time step are shown in Fig. 6, respectively.

5 Experiments and Results Analysis

5.1 Experiment Settings

OpenSTL is a general testing benchmark framework for spatiotemporal prediction learning, and OpenSTL-Lightning is the latest version of this framework implemented based on Lightning. Both our method and the benchmark model PredFormer are run on this version of the framework in our experiments. The other benchmark models used in the experiments are all natively supported by this framework. All experiments were performed on an NVIDIA RTX 4090 GPU with 24 GB of memory.

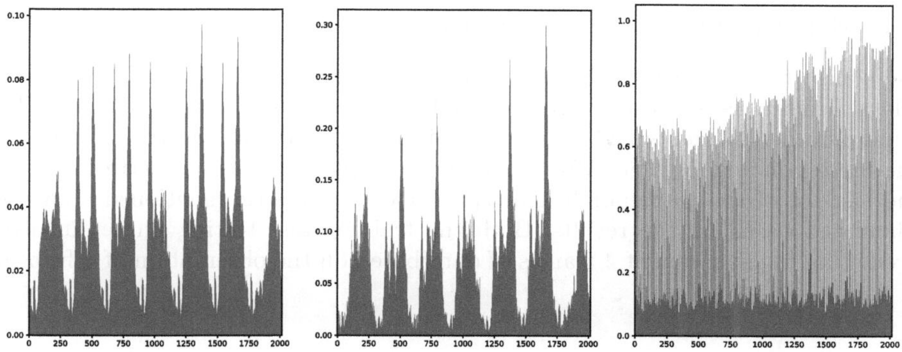

((a)) BikeBJ-7days' overall temporal feature distribution histogram

((b)) BikeBJ-7days' temporal feature distribution of a normal point

((c)) BikeBJ-7days' temporal feature distribution of the local peak point

Fig. 6. The temporal features of BikeBJ-7days

We set the following common hyperparameters for the training of all models: the random seed is set to 42, the batch size is set to 16, the learning rate is set to 0.001, the weight decay is set to 0.01, the number of training epochs is set to 200, and AdamW is used as the optimizer.

5.2 Benchmark and Settings

The benchmark models are categorized into two groups based on their underlying architecture: recurrent and non-recurrent models. Recurrent benchmark models include: ConvLSTM, E3D-LSTM, MIM, PhyDNet, MAU, PredRNN v2, SwinLSTM. Non-recurrent benchmark models include: SimVP, TAU, WaST, and PredFormer.

All benchmark models natively supported by OpenSTL are implemented using the default parameters provided by the framework. PredFormer follows the hyperparameter settings of its original paper, but selects the Binary-ST variant, which achieves the SOTA performance on TaxiBJ dataset in its original paper, with three interleaved encoding layers as the benchmark. ISTAFormer maintains the same hyperparameter selection and GTB arrangement as PredFormer.

5.3 Evaluation Metrics

We employ four metrics to assess model performance: mean squared error (MSE) and mean absolute error (MAE), which measure pixel-level errors, and structural similarity index measure (SSIM) and peak signal-to-noise ratio (PSNR), which evaluate the quality of the predicted frames. A well-performing model should exhibit low values of MSE and MAE, as well as high values of SSIM and PSNR.

5.4 Experiment Results and Analysis

We conduct comparative experiments on the TaxiBJ dataset and the BikeBJ-7days dataset, as well as an ablation study on the TaxiBJ dataset. We use the hyperparameters mentioned above to train each model on the training set of the corresponding dataset. After training is completed, tests are carried out on the test set of the corresponding dataset, and each metric mentioned above is calculated based on the results. Both the training and testing phases involve predicting the subsequent 4 frames of data based on the observation of 4 frames of data.

Comparative Experiment on TaxiBJ Dataset. Upon examining the comparative experimental results shown in Tables 2, it is evident that ISTAFormer achieves SOTA performance across all four metrics on the TaxiBJ dataset. Specifically, in terms of the MSE metric, ISTAFormer achieves a reduction of 21.75% compared to the best-performing baseline model, PredFormer. Regarding the MAE metric, ISTAFormer exhibits a decrease of 7.95% relative to the best-performing baseline model, TAU. For the SSIM and PSNR metrics, ISTAFormer outperforms the best baseline model, WaST, by 0.00264 and 0.99279, respectively.

Table 2. Comparative performance on TaxiBJ

Model	MSE	MAE	SSIM	PSNR
ConvLSTM [10]	0.19105	12.91506	0.97963	40.34714
E3D-LSTM [16]	0.18512	12.53635	0.98049	40.49868
MIM [19]	0.22348	13.99386	0.98010	39.70821
PhyDNet [5]	0.20154	12.80321	0.98021	40.27070
MAU [2]	0.20345	12.84824	0.98080	40.24067
PredRNN v2 [18]	0.21814	13.16445	0.97868	39.92269
SimVP [4]	0.19885	12.57793	0.98090	40.33588
WaST [8]	0.17620	12.29886	0.98219	40.78359
SwinLSTM [13]	0.17641	12.43608	0.98180	40.73583
TAU [11]	0.18122	12.23347	0.98202	40.62505
PredFormer [14]	0.17557	12.30924	0.98216	40.75880
ISTAFormer (Ours)	**0.13737**	**11.26001**	**0.98483**	**41.77638**

Comparative Experiment on BikeBJ-7days Dataset. The comparative experimental results shown in Table 3 indicate that on the BikeBJ-7days dataset, ISTAFormer also achieves SOTA performance across all four metrics. Specifically, in terms of the MSE metric, ISTAFormer achieves a reduction of 7.15%

compared to the best-performing baseline model, WaST. Regarding the MAE metric, ISTAFormer exhibits a decrease of 1.25% relative to the best-performing baseline model, PredFormer. Regarding the SSIM metric, ISTAFormer outperforms the best baseline model, MAU, by 0.00162. Regarding the PSNR metric, ISTAFormer outperforms the best baseline model, WaST, by 0.31084. Besides, according to the experimental results, it can be found that the performance of MIM, PredRNN v2 and SwinLSTM on BikeBJ-7days are significantly worse than other models, which indicates that the ability of models based on RNNs to capture the spatiotemporal features of data is greatly influenced by model design, and the generalization performance is unstable.

Table 3. Comparative performance on BikeBJ-7days

Model	MSE	MAE	SSIM	PSNR
ConvLSTM [10]	0.03906	4.61805	0.97479	45.24798
E3D-LSTM [16]	0.04293	4.94145	0.97294	44.82132
MIM [19]	0.37852	8.16856	0.93882	37.38539
PhyDNet [5]	0.03519	4.49532	0.97776	45.72928
MAU [2]	0.03266	4.33018	0.97955	46.08567
PredRNN v2 [18]	0.17237	9.79464	0.89134	38.77694
SimVP [4]	0.03131	4.36837	0.97832	46.19585
WaST [8]	0.03077	4.27891	0.97952	46.28545
SwinLSTM [13]	0.45218	20.45388	0.71628	34.57500
TAU [11]	0.03177	4.36259	0.97825	46.13969
PredFormer [14]	0.03150	4.25119	0.97863	46.20438
ISTAFormer (Ours)	**0.02857**	**4.19804**	**0.98117**	**46.59629**

We select several models to compare their prediction results at a certain point in time, as shown in Fig. 7. In addition to our ISTAFormer, there are PredRNN v2 based on RNNs, WaST based on CNNs, PredFormer based on attention, and SwinLSTM that combines RNNs and attention. We find that compared to PredFormer and WaST, ISTAFormer has stronger ability to capture local features, which can be seen from their predictions of the local peak point. Both PredFormer and WaST are greatly influenced by the high value of the local peak point in the training data, paying more attention to local spatial features but insufficient attention to the temporal dependency features of the area, leading to higher predicted value in that area. In contrast, ISTAFormer captures this temporal dependency feature more effectively through its all-layer weighted aggregation mechanism. ISTAFormer successfully learns to consider more temporal changes of local features, making its predicted value closer to the true value. In addition, PredRNN v2 only captures some general laws of spatial distribution without focusing on more local features, while the uniform spatial distribution of SwinLSTM indicates that it has not learned the spatiotemporal features of the data well.

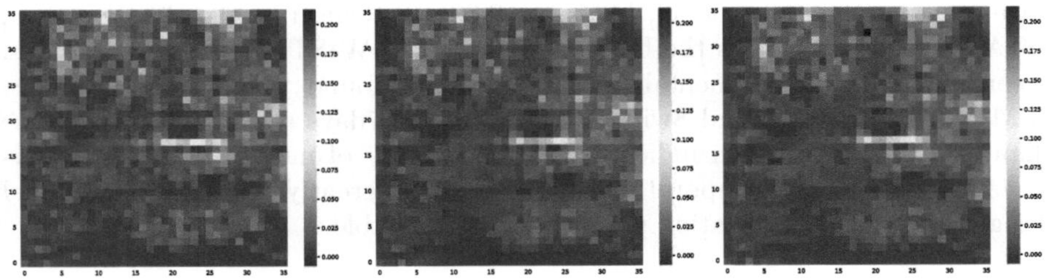

((a)) true data ((b)) ISTAFormer ((c)) PredFormer

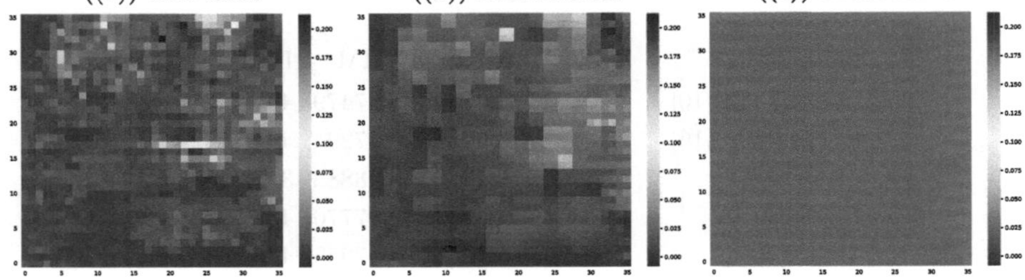

((d)) WaST ((e)) PredRNN v2 ((f)) SwinLSTM

Fig. 7. Models' prediction at a certain point in time

5.5 Ablation Study

We conduct independent ablation study on several key components of the GTB layer in ISTAFormer using the TaxiBJ dataset. The results are shown in Table 4.

The experimental results demonstrate that removing any of these modules from the baseline ISTAFormer model leads to a decline in the overall performance of the model. This finding validates the necessity of these constituent modules within the GTB layer of ISTAFormer.

Table 4. Ablation on FFN, ReGLU, DropLayer and All-layer Weighted Aggregation Layer

Model	MSE	MAE	SSIM	PSNR
ISTAFormer (Ours)	**0.13737**	**11.26001**	**0.98483**	**41.77638**
-FFN	0.16971	11.95149	0.98302	41.01338
-ReGLU	0.20531	13.47631	0.97976	40.19829
-DropLayer (middle)	0.14984	11.63332	0.98405	41.45712
-DropLayer (last)	0.14471	11.39436	0.98458	41.55887
-All-layer Weighted Aggregation Layer	0.15445	11.54577	0.98333	41.33437

6 Conclusion

We propose the ISTAFormer method for traffic flow prediction in grid-based spatial scenarios, learning the spatiotemporal features of the dataset through interleaved spatial and temporal views GTB. We use ReGLU as a gating mechanism to better filter the features learned by the model through the self-attention mechanism, and introduce an all-layer weighted aggregation mechanism to aggregate intermediate results, solving the problem of potential loss of locally discrete features learned by the model during learning, and improving the predictive performance of the model. This also improves the model's generalization ability across different types of datasets.

Although ISTAFormer achieves SOTA performance on both TaxiBJ and BikeBJ-7days datasets, we believe that there is still room for optimization. Additionally, although our research has focused on traffic flow prediction, ISTAFormer should also be applicable to other similar spatiotemporal data prediction tasks, such as video prediction. Therefore, it is necessary to further study the generalization ability of ISTAFormer on other similar task datasets.

Acknowledgments. This work was supported by the National Natural Science Foundation of China (Grant No. 62436010).

Ethical Statements. In this work, we ensure the use of all data adheres to strict ethical guidelines. The TaxiBJ dataset is a widely used publicly available anoymized dataset that unable to accurately identify any personal information. The raw data used to construct the BikeBJ-7days dataset includes the following data fields: shared bike ID, activating shared bike record time and GPS coordinates of activating shared bike using WGS84 coordinate system. All data has been anonymized by mapping specific time and coordinates to the corresponding indices of time and space, then we counted the number of activating shared bikes within the specified time period and area to construct the BikeBJ-7days dataset, which removes all personally identifiable information. Besides, our methodology aligns with data privacy standards, emphasizing the ethical and responsible use of data for scientific advancements.

References

1. Chang, Z., Zhang, X., Wang, S., Ma, S., Gao, W.: STRPM: a spatiotemporal residual predictive model for high-resolution video prediction. In: Proceedings of the IEEE/CVF Conference on Computer Vision and Pattern Recognition, pp. 13946–13955 (2022)
2. Chang, Z., et al.: MAU: a motion-aware unit for video prediction and beyond. Adv. Neural. Inf. Process. Syst. **34**, 26950–26962 (2021)
3. Dosovitskiy, A., et al.: An image is worth 16×16 words: Transformers for image recognition at scale. arXiv preprint arXiv:2010.11929 (2020)
4. Gao, Z., Tan, C., Wu, L., Li, S.Z.: SimVP: simpler yet better video prediction. In: Proceedings of the IEEE/CVF Conference on Computer Vision and Pattern Recognition, pp. 3170–3180 (2022)

5. Guen, V.L., Thome, N.: Disentangling physical dynamics from unknown factors for unsupervised video prediction. In: Proceedings of the IEEE/CVF Conference on Computer Vision and Pattern Recognition, pp. 11474–11484 (2020)
6. Lin, Z., Li, M., Zheng, Z., Cheng, Y., Yuan, C.: Self-attention ConvLSTM for spatiotemporal prediction. In: Proceedings of the AAAI Conference on Artificial Intelligence. vol. 34, pp. 11531–11538 (2020)
7. Liu, Z., et al.: Swin transformer: hierarchical vision transformer using shifted windows. In: Proceedings of the IEEE/CVF International Conference on Computer Vision, pp. 10012–10022 (2021)
8. Nie, X., et al.: Wavelet-driven spatiotemporal predictive learning: bridging frequency and time variations. In: Proceedings of the AAAI Conference on Artificial Intelligence. vol. 38, pp. 4334–4342 (2024)
9. Shazeer, N.: Glu variants improve transformer. arXiv preprint arXiv:2002.05202 (2020)
10. Shi, X., Chen, Z., Wang, H., Yeung, D.Y., Wong, W.K., Woo, W.c.: Convolutional LSTM network: a machine learning approach for precipitation nowcasting. Adv. Neural Inf. Process. Syst. **28** (2015)
11. Tan, C., Gao, Z., Wu, L., Xu, Y., Xia, J., Li, S., Li, S.Z.: Temporal attention unit: towards efficient spatiotemporal predictive learning. In: Proceedings of the IEEE/CVF Conference on Computer Vision and Pattern Recognition, pp. 18770–18782 (2023)
12. Tan, C., et al.: OpenSTL: a comprehensive benchmark of spatio-temporal predictive learning. Adv. Neural. Inf. Process. Syst. **36**, 69819–69831 (2023)
13. Tang, S., Li, C., Zhang, P., Tang, R.: SwinLSTM: improving spatiotemporal prediction accuracy using swin transformer and LSTM. In: Proceedings of the IEEE/CVF International Conference on Computer Vision, pp. 13470–13479 (2023)
14. Tang, Y., Qi, L., Xie, F., Li, X., Ma, C., Yang, M.H.: PredFormer: Transformers are effective spatial-temporal predictive learners. arXiv preprint arXiv:2410.04733 (2024)
15. Vaswani, A., et al.: Attention is all you need. Adv. Neural Inf. Process. Syst. **30** (2017)
16. Wang, Y., Jiang, L., Yang, M.H., Li, L.J., Long, M., Fei-Fei, L.: Eidetic 3D LSTM: a model for video prediction and beyond. In: International Conference on Learning Representations (2018)
17. Wang, Y., Long, M., Wang, J., Gao, Z., Yu, P.S.: PredRNN: recurrent neural networks for predictive learning using spatiotemporal LSTMs. Adv. Neural Inf. Process. Syst. **30** (2017)
18. Wang, Y., et al.: PredRNN: a recurrent neural network for spatiotemporal predictive learning. IEEE Trans. Pattern Anal. Mach. Intell. **45**(2), 2208–2225 (2022)
19. Wang, Y., Zhang, J., Zhu, H., Long, M., Wang, J., Yu, P.S.: Memory in memory: a predictive neural network for learning higher-order non-stationarity from spatiotemporal dynamics. In: Proceedings of the IEEE/CVF Conference on Computer Vision and Pattern Recognition, pp. 9154–9162 (2019)
20. Zhang, J., Zheng, Y., Qi, D.: Deep spatio-temporal residual networks for citywide crowd flows prediction. In: Proceedings of the AAAI Conference on Artificial Intelligence. vol. 31 (2017)

Color Image Steganography Based on a GAM-CBAM Dual Attention Mechanism

Yuxin Kang, Xianhua Song(✉), Haotian Wu, and Wei Xie

School of Science, Harbin University of Science and Technology, Harbin 150080, China
songxianhua@hrbust.edu.cn

Abstract. High-capacity steganographic schemes often involve considerable modifications to the cover image structure, which can adversely affect its visual fidelity and undermine the imperceptibility requirement. Conversely, prioritizing visual fidelity typically restricts the embedding capacity. To address this trade-off, this paper proposes a color image steganography scheme based on a GAM-CBAM dual-attention mechanism. The proposed method integrates two key attention modules: the global attention module (GAM) enhances feature modelling capability, whereas the convolutional block attention module (CBAM) enables precise embedding region localization and dynamic feature weight assignment, effectively suppressing redundant information and reducing detection risk. The encoding network adopts an improved U-Net architecture with global attention to optimize feature extraction, whereas the decoding network incorporates CBAM to increase reconstruction accuracy. Experiments conducted on the COCO dataset for 256 × 256 color image steganography demonstrate the scheme's superiority: PSNR improvement of 2.1687 dB (cover/stego image) and 1.489 dB gain (secret/reconstructed image) at a 1-byte per-pixel embedding capacity. These results validate the method's significant improvements in visual fidelity and information recovery accuracy compared with the prior U-Net-based methods.

Keywords: Image Steganography · Dual Attention Mechanism · Encoding and Decoding Architecture · GAM-CBAM

1 Introduction

Against the background of the digital information era, the high demand for covert communication and data security has promoted the rapid growth of image steganography [1]. This technique realizes secure information transmission while ensuring visual invisibility by hiding a secret image in a public cover image, which has important application value in military communication, medical privacy protection and other fields. Traditional steganographic methods rely primarily on either spatial domain techniques (e.g., least significant bit (LSB) replacement [2, 3] and LSB matching [4]) or frequency domain transformations (e.g., DCT-based [5] and DWT-based [6] approaches). However, these conventional approaches typically suffer from some fundamental limitations, such as constrained embedding capacity, vulnerability to steganalysis attacks, and significant visual quality degradation in the stego image. Although deep learning end-to-end frameworks significantly improve performance by automatically learning the

L. Huang et al. (Eds.): CCF NCCA 2025, CCIS 2774, pp. 395–411, 2026.
https://doi.org/10.1007/978-981-95-5294-8_27

embedding-extraction mapping relation, the balance between high-capacity embedding and high-fidelity output remains the core challenge in this field.

In 2017, Baluja proposed the first deep learning-based hiding image-within-images model [7], which achieves steganography between color images through a three-layer convolutional network but suffers from embedding distortion and many noise points in the reconstructed image. Subsequent studies improved the reconstruction performance by improving the network structure: Wu et al. [8] introduced variance loss and separable convolution to optimize the noise distribution and reduce the number of parameters in 2018. Duan et al. successively adopted the U-Net [9] and FC-DenseNets [10] structures in 2019, which significantly improved image quality through feature fusion. Reference [10] had more advantages in improving parameter efficiency. In terms of capacity-security balance research, new progress has been made in grayscale image steganography schemes. In 2018, Rehman et al. [11] first achieved the hiding of grayscale secrets in color covers but was constrained by an oversimplified network structure, which ultimately compromised the visual quality of the stego images. Zhang's ISGAN [12] and Chen's model [13] employed distinct channel-specific embedding strategies, utilizing the Y-channel and B-channel, respectively, to achieve PSNR improvements of approximately 2 dB by incorporating human visual characteristics. Baluja's "hiding two images within images" model [14] achieves high-capacity embedding but is still constrained by the problem of infrastructure distortion. For the purpose of further improving steganographic security, Sharma et al. [15] also constructed a steganographic algorithm based on a preprocessing network, steganographic network and extraction network. The difference from Baluja's algorithm is that the preprocessing network of Sharma's algorithm encrypts only the hidden image but does not participate in training, so the steganographic image obtained by this method can still be observed as stego traces. In 2020, research further deepened the design of the network architecture: Subramanian et al. [16] and Lu [17] and Guan et al. [18, 19] optimized the reconstruction quality via two-path preprocessing and a reversible neural network, respectively. In 2023, Yang et al. [20] proposed PRIS on the basis of Reference [17]. PRIS sets preenhancement blocks and postenhancement blocks before and after the image restoration process to improve the reconstruction ability of the model. Moreover, the enhanced module can also weaken the strict reversibility of the reversible network so that it can better resist noise. However, a fundamental trade-off persists between achieving high embedding capacity and maintaining superior visual fidelity in steganography.

To effectively solve the problem that it is difficult to balance the embedding capacity and visual fidelity in the abovementioned hidden graph method, this paper designs a double attention mechanism steganography network that integrates the GAM [21] and CBAM [22]. The scheme overcomes the problem of the insufficient modelling ability of traditional networks for important features from the perspectives of feature selectivity enhancement and fine control of embedded regions. On the one hand, the GAM strengthens the network's ability to perceive global context information so that the model can more fully mine the high-response areas suitable for embedding in the image and increase the embedding capacity. On the other hand, CBAM further refines the information guidance process, which can not only effectively eliminate redundant features and improve the visual restoration effect but also dynamically optimize the embedding position in the spatial dimension and reduce the perceptual interference and detection risk caused by embedding. Experiments show that this scheme achieves significant improvement in 256 × 256 color image steganography. The main contributions are summarized as follows:

(1) Dual-attention module fusing GAM-CBAM to optimize intermediate features through channel-space synergy.
(2) An encoding and decoding architecture based on an improved U-Net is constructed, which introduces global attention to enhance feature extraction in the encoding network and integrates CBAM to improve reconstruction accuracy in the decoding network.
(3) The scheme is trained and verified on the COCO dataset, and the three-channel secret image with a size of 256 × 256 is completely hidden in the cover image of the same size. Under the premise of a relative embedding rate of 1 byte/pixel, this method effectively improves the image quality in the process of model steganography and reconstruction.

2 Related Work

The role of the attention mechanism allows the model to focus on learning key features and ignore irrelevant features so that the attention mechanism is widely used in deep learning. In image steganography, it is crucial to suppress unimportant information in the image to prevent artifacts in the generated stego image, as well as to identify suitable hiding locations for secret messages in the cover image. To meet all the requirements, this paper proposes applying the GAM to the encoding network to effectively improve the quality of the generated stego images while applying the CBAM to the decoding network to increase the extraction accuracy of the secret images. The two kinds of attention mechanisms are introduced separately as follows.

2.1 CBAM

The CBAM module processes the input feature maps $F \in R^{C\times H\times W}$, where the number of feature channels and the height and width of the input feature maps are represented. By giving F, the CBAM module sequentially generates a channel attention map $Mc \in R^{C\times 1\times 1}$ and a spatial attention map $Ms \in R^{1\times H\times W}$. The realization process can be expressed by Eqs. (1) and (2).

$$F_1 = Mc(F) \otimes F \quad (1)$$

$$F_2 = Ms(F_1) \otimes F_1 \quad (2)$$

where $\otimes$ represents element-by-element multiplication, F_1 represents the output of F after the channel attention module while the channel weight fusion module is used, and F_2 represents the output result of the CBAM module. The network structure of CBAM is schematically shown in Fig. 1.

CBAM is notable for its lightweight design, cleverly ensuring minimal computational overhead. This makes it a highly practical and attractive choice for integrating attention

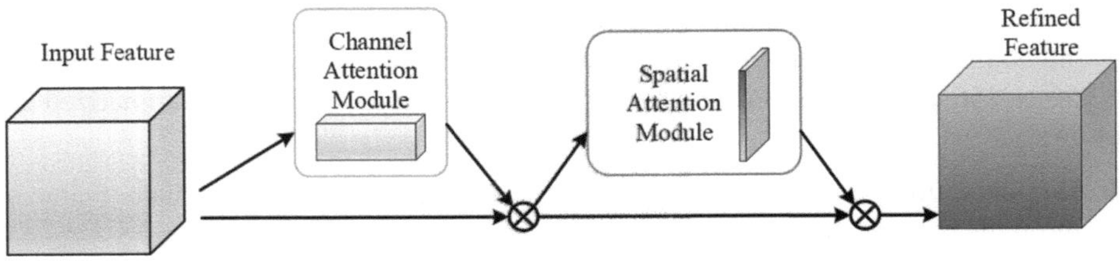

Fig. 1. CBAM Attention Structure

mechanisms into existing neural network architectures. Unlike complex attention modules, which consume significant resources and may slow model operation, CBAM can be seamlessly inserted without creating a major performance bottleneck.

Furthermore, CBAM uses an effective combined attention strategy. It focuses on both "what" and "where" aspects simultaneously. By identifying "what" features are relevant in the input data and "where" they are located spatially, this dual-attention method enables CBAM to offer a more discriminative feature representation.

2.2 GAM

GAM aims to reduce information loss and enhance global-dimensional interaction features. GAM is built on CBAM, which retains the order channel spatial attention mechanism while further redesigning the submodules to reduce information compression and loss and is able to better capture the input features through the weighting of the global feature graph with globally important information, which further enhances the capability of feature representation. The complete process is demonstrated in Fig. 2.

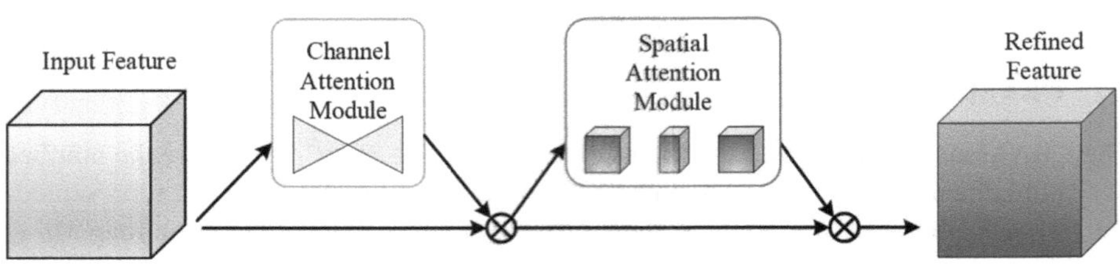

Fig. 2. GAM Attention Structure Map

First, the improvement of the GAM on the channel attention module is reflected in the use of 3-dimensional alignment to reserve information in 3 dimensions, avoiding the problem of loss of global space–channel interactions due to the reduction of information and separation of dimensions. This module processes $F \in R^{C \times H \times W}$ and is implemented as Eq. (3):

$$Mc(F) = \sigma(MLP(Permute(F))) \tag{3}$$

where $Permute(F) \in R^{\{(H \times W) \times C \times 1\}}$ denotes the 3D tensor reorganization operation that keeps the channel dimension as C and unfolds the spatial dimension $H \times W$ into a single dimension. The channel submodule subsequently introduces a two-layer MLP to increase the cross-dimensional channel space dependence, which effectively captures and amplifies the global interaction features between channels through a transformation with a compression ratio of r. The implementation process is shown in Eq. (4):

$$MLP(x) = W_2(\delta(W_1x + b_1)) + b_2 \tag{4}$$

where $W_1 \in R^{(\frac{C}{r})\times C}$, $W_2 \in R^{C\times(\frac{C}{r})}$ is the learnable parameter and where δ is the ReLU activation.

In the spatial attention module, the pooling operation is discarded, the double convolutional architecture is utilized for spatial information fusion, and the implementation process is shown in Eq. (5):

$$Ms(F') = \sigma\left(Conv_{7\times7}\left(Conv_{3\times3}\left(\left[Avg(F');\ Max(F')\right]\right)\right)\right) \tag{5}$$

Compared with the local attention mechanism in CBAM, the GAM provides a more effective solution for enhancing feature representation, especially in capturing long-range dependencies. GAM employs a unified global attention strategy that captures both spatial and channel relationships across the entire feature map. This global interaction reduces redundancy and preserves important information, resulting in more accurate and robust feature representations. Unlike the sequential nature of CBAM, the GAM integrates attention mechanisms more holistically, improving efficiency and representational power without significantly increasing computational cost. In summary, the GAM offers clear advantages over CBAM by enabling global contextual reasoning, making it a more suitable choice for applications demanding high-fidelity and semantically rich feature embedding.

3 Color Image Steganography Scheme Based on the GAM-CBAM Dual Attention Mechanism

Since irrelevant channel information in the secret image can introduce artifacts in the generated stego image, thereby degrading visual quality, and regions with complex textures in the cover image are more resistant to steganalysis, they are more suitable for concealing secret data. Therefore, this paper proposes a full-size color image steganography program that adopts a GAM-CBAM-based encoding-decoding architecture. The steganography scheme is designed to include an encoding network with a GAM and a decoding network with a CBAM. The encoding leverages the GAM to enhance global feature interaction, enabling the secret image to be embedded into less noticeable regions of the cover image. By applying both channel and spatial attention, the GAM adaptively emphasizes informative features across channels and spatial locations, improving the concealment quality of the hidden image. On the decoding side, CBAM is employed to improve the accuracy of secret image recovery. Through its dual attention mechanism, CBAM guides the decoder to focus on key regions and channels that contain the embedded secret, thereby enhancing reconstruction performance and ensuring effective extraction of the hidden content.

3.1 Overall Framework of the Model

The proposed color image steganography model consists of two main components: an encoding network and a decoding network. In the encoding network, the U-Net module with the GAM is designed to improve the visual quality of the stego image. This module takes the cover image C and the secret image S as inputs and produces a stego image C' that maintains the same dimensions as the cover image. In the decoding network, the convolutional module with CBAM is designed to ensure the authenticity of the extracted secret image; its inputs are the stego image C', and its output is the reconstructed secret image S'. The overall framework of the model proposed in this paper is shown in Fig. 3.

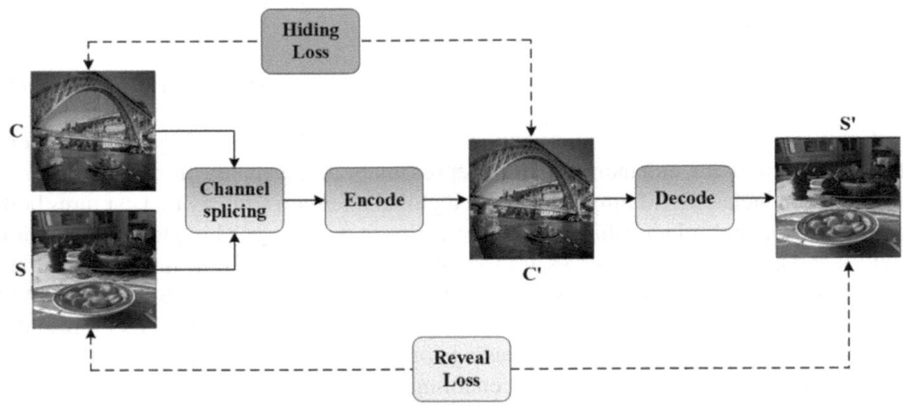

Fig. 3. This paper's model framework

3.2 Encoding Network

On the basis of U-Net, we propose an improved encoding network that enhances the global feature modelling capability by introducing the GAM. Its core design contains the following parts:

1. Overall Network Architecture

   The encoding network adopts a symmetric encoder-decoder structure, where the input is a 6-channel tensor of S spliced with C, and the output is C'. The encoding path contains 6 levels of convolutional modules, and each level consists of a convolutional layer (the convolutional kernel size is 4×4).), BatchNorm and LeakyReLU activation functions. The feature map size is halved, the number of channels is multiplied after each downsampling, and the final output is a feature map of size $4 \times 4 \times 1024$. The decoding path contains 6 levels of anti-convolution modules, each consisting of an anti-convolution layer (of the same size as the encoder), BatchNorm and a ReLU activation function. The spatial resolution of the image is gradually recovered by fusing the same-scale features of the encoding path through skip connections.
2. GAM-enhanced design at the bottleneck level.

   At the innermost level of the encoding path (i.e., after the sixth level of downsampling), the GAM is introduced to resolve the problem of insufficient global dependency modelling in traditional U-Net. The specific implementation is divided into three stages. First, high-dimensional features are extracted by two-stage 4 × 4 convolutional layers. Second, using global attention enhancement, the channel attention submodule is performed first to capture the cross-channel global statistical features via higher-order pooling and generate channel weights through the fully connected layer to recalibrate the channels; then, the spatial attention submodule is performed to merge the channel average pooling and the channel maximal pooling features and generate the spatial weights through a 7 × 7 large kernel convolution to enhance the key region response. Finally, feature fusion is achieved by applying elementwise multiplication between the original image features and the corresponding channel and spatial attention weights, resulting in an enhanced feature map.

The design achieves two aspects of optimization through a two-stage attention mechanism. In terms of information retention, the GAM avoids the spatial overcompression of the traditional attention module, preserves higher-order statistical features, and reduces the loss of details during downsampling. In terms of global interaction, the large kernel convolution with a cross-channel fully connected layer directly models long-range dependencies, enhancing the ability to model the global semantics of the image.

3. Output Layer Design

   At the end of the decoding path, a 4 × 4 transposed convolutional layer is employed to restore the feature map to the original input dimensions. It is then activated via the sigmoid function to ensure visual consistency with the original input. The architecture of the encoding network is illustrated in detail in Fig. 4.

Table 1. Encoding Network Structure Parameters and Output Dimensions

Input Size	Network Architecture	Output Size
256 × 256 × 6	Conv1(4 × 4, 2, 1) + LeakyReLU	128 × 128 × 64
	Conv2(4 × 4, 2, 1) + BN + LeakyReLU	64 × 64 × 128
	Conv3(4 × 4, 2, 1) + BN + LeakyReLU	32 × 32 × 256
	Conv4(4 × 4, 2, 1) + BN + LeakyReLU	16 × 16 × 512
	Conv5(4 × 4, 2, 1) + BN + LeakyReLU	8 × 8 × 512
	Conv6(4 × 4, 2, 1) + BN + LeakyReLU	4 × 4 × 1024
	Conv7(4 × 4, 2, 1) + GAM + ReLU	4 × 4 × 1024
	DeConv1(4 × 4, 2, 1) + BN + ReLU	4 × 4 × 512
	DeConv2(4 × 4, 2, 1) + BN + ReLU	8 × 8 × 512
	DeConv3(4 × 4, 2, 1) + BN + ReLU	16 × 16 × 512
	DeConv4(4 × 4, 2, 1) + BN + ReLU	32 × 32 × 256
	DeConv5(4 × 4, 2, 1) + BN + ReLU	64 × 64 × 128
	DeConv6(4 × 4, 2, 1) + BN + ReLU	128 × 128 × 64
	DeConv7(4 × 4, 2, 1) + Sigmoid	256 × 256 × 3

The encoding network includes encoding and decoding paths, where the encoding and decoding paths each contain six convolutional layers, and the innermost layer applies the GAM after extracting the features to increase the global feature interaction capability and reduce information loss. The structural parameters and output dimensions of the encoding network are shown in Table 1.

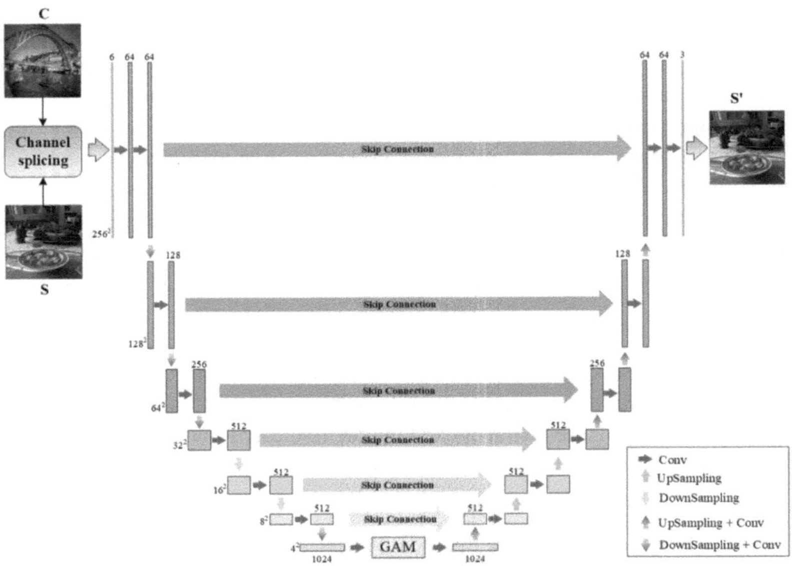

Fig. 4. Encoding network structure

On the basis of the structural design described above, the proposed improved encoding network builds upon the traditional U-Net architecture and effectively integrates the GAM, significantly enhancing feature representation and image reconstruction accuracy. By introducing the GAM module at the deepest level of the encoding path, the network not only retains richer detail information during downsampling but also strengthens the modelling of long-range dependencies within images—demonstrating greater robustness in handling complex structures and fine-grained targets. Moreover, the dual mechanism of channel and spatial attention in the GAM allows for targeted enhancement of important information across different dimensions, effectively suppressing feature degradation and redundancy while maintaining computational efficiency.

The network maintains a symmetric encoding-decoding structure, utilizing skip connections to fuse multiscale information, thereby maximizing the preservation of deep semantic features from the encoder during spatial resolution recovery. Finally, a transposed convolution layer followed by sigmoid activation is used in the output layer to restore the feature map to the original image space with high fidelity, ensuring visual consistency with the input.

Overall, the proposed encoding network has remarkable advantages. It inherits not only U-Net's strengths in semantic transmission and detail recovery, which are crucial for maintaining the integrity and quality of image information during processing. Moreover, owing to the addition of the GAM, the network gains the ability to model the global semantic relationship at a deeper level. This, in turn, enhances the comprehensiveness and discrimination of the extracted features. As a result, it lays a solid foundation and offers a valuable reference for the field of image steganography, enabling more effective and secure information hiding.

3.3 Decoding Network

The decoding network adopts a deep convolutional-based encoding and decoding architecture, which realizes efficient reconstruction of secret information through cascading convolutional operations and an attention mechanism. As shown in Fig. 5, the network incorporates 6 convolutional layers, in which two CBAMs are embedded to enhance the feature selection capability. The specific design is as follows:

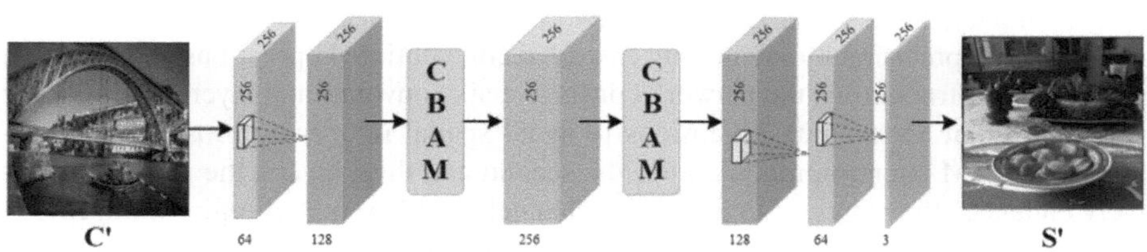

Fig. 5. Decoding network structure

1. Overall Network Architecture

   The decoding network takes S as input, gradually extracts the multiscale features after 6 levels of the convolution operation, and finally outputs the reconstructed secret image. The network layer configuration and feature map size changes are shown in Table 2.
2. Convolutional Layer Design

   The first 5 convolutional layers adopt a unified structure, and each layer includes a 3×3 convolutional kernel with the step size set to 1 to maintain the spatial resolution of the feature maps; BatchNorm is applied to speed up training convergence and alleviate the issue of gradient disappearance; and ReLU introduces the ability of nonlinear expression. Layer 6 is the end of the network, which uses a 3×3 convolution kernel to map the channel features to 3-channel outputs and constrains the range of pixel values to $[-1,1]$ via a sigmoid activation function to ensure that S' has the same numerical distribution as S.
3. Integration of attention mechanisms

   To enhance the representation ability of key features, the CBAM is embedded after layers 2 and 3 of the network: the channel attention submodule extracts the channel statistical features through global average pooling and maximum pooling and recalibrates the channel importance by generating the channel weights via the MLP of shared weights. The spatial attention submodule generates the spatial weight map by fusing the channel pooling features and 7 × 7 convolution to enhance the response strength of the target region.

Table 2. Decoding Network Structure Parameters and Output Dimensions

Input Size	Network Architecture	Output Size
256 × 256 × 3	Conv1(3 × 3, 1, 1) + BN + ReLU	256 × 256 × 64
	Conv2(3 × 3, 1, 1) + BN + ReLU + CBAM	256 × 256 × 128
	Conv3(3 × 3, 1, 1) + BN + ReLU + CBAM	256 × 256 × 256
	Conv4(3 × 3, 1, 1) + BN + ReLU	256 × 256 × 128
	Conv5(3 × 3, 1, 1) + BN + ReLU	256 × 256 × 64
	Conv6(3 × 3, 1, 1) + Sigmoid	256 × 256 × 3

In summary, the proposed decoding network effectively reconstructs hidden information with high precision by integrating cascaded convolutional operations and attention mechanisms. Structurally, the network consists of six convolutional layers, maintaining consistent input and output dimensions to avoid spatial information loss. By embedding the CBAM attention module after the second and third layers, the network adaptively enhances the response to important channels and key spatial regions, improving the perception and reconstruction of latent information while keeping the architecture lightweight. Additionally, the convolutional layer configuration ensures precise control over feature resolution, and the incorporation of the BatchNorm and ReLU activation functions accelerates training convergence and stabilizes gradient propagation. Finally, a sigmoid function maps the output to the range consistent with the original image's pixel distribution, ensuring the naturalness and visual consistency of the reconstructed secret image.

Overall, the decoding network has a well-balanced design and compact modular structure. It not only leverages the powerful feature modelling capabilities of deep convolutional networks but also benefits from the CBAM modules to emphasize and enhance key features. This results in strong reconstruction performance and excellent scalability, making the design a solid foundation for tasks such as image steganography and information retrieval.

3.4 Loss Function

In this work, a dual-objective joint optimization strategy is adopted to achieve a balance between high covertness and high reconstruction quality of the secret image by constructing a hidden loss function L_C and an extraction loss function L_R to constrain the training process of the encoding network and the decoding network, respectively. The specific design is as follows:

1. Hidden Loss Function L_C

   The goal of the encoding network is to hide S into C to generate C'. To ensure the visual consistency between C' and C and to avoid the exposure of the steganographic traces, we adopt MSE to quantify the pixel-level difference between the two, which is defined as follows:

$$L_C = \frac{1}{H \times W} \sum_{i=1}^{H} \sum_{j=1}^{W} \left\| C'(i,j) - C(i,j) \right\|_2^2 \tag{6}$$

where H and W are the image height and width, respectively, and denote the pixel values at positions $C'(i, j)$ and $C(i, j)$, respectively. By minimizing L_C, the encoding network is able to obtain C' that is highly consistent with the statistical distribution of C, thus guaranteeing the imperceptibility of the steganography process.

2. Extract Loss Function L_R

   The goal of the decoding network is to obtain S' from C'. To measure the fidelity of S' and S, we also select the MSE as the loss function, and through the pixel-by-pixel comparison, the decoding network is constrained to accurately recover the content and details of the secret image and minimize the information reconstruction error.

3. Joint Optimization Goal

   The total loss function L_{total} of the model is the weighted sum of the two above losses:

$$L_{total} = \lambda_1 L_C + \lambda_2 L_R \tag{7}$$

where λ_1 and λ_2 are the adjustable hyperparameters that control the weights of concealment and reconstruction quality, respectively, and the model achieves the optimal balance between concealment and reconstruction performance when $\lambda_1 = 1$ and $\lambda_2 = 0.75$.

4 Experimental Analysis

4.1 Experimental Setup and Evaluation Indicators

1. Experimental environment basis

This paper uses the COCO dataset to verify the proposed frame. A total of 15000 images are randomly chosen from the set as the training set, and 1500 images are chosen as the test set. In the training process of the model, the Adam optimizer is used to improve the convergence property of the model, and the hyperparameters of the Adam optimizer are set as $\beta_1 = 0.5$ and $\beta_2 = 0.999$. In this work, C and S with sizes of $256 \times 256 \times 3$ pixels are used to train and test the model, and the sizes of the images generated by the encoding network and the decoding network are both $256 \times 256 \times 3$. The main parameters areas follows: 100 epochs, 32 batches, and initial learning rate of 0.001. The experimental training environment is a computing platform with a Win 11 operating system, Intel(R) Xeon(R) Gold 5318Y processor, 512 GB RAM and NVIDIA RTX A4000 GPU. The paper uses the deep learning framework of PyTorch 2.0.0.

To thoroughly assess the model's complexity and inference performance, we present detailed statistics, including parameter counts, FLOPs, and average per-image inference times, on the test set for both the encoding and decoding networks. The detailed results are shown in Table 3.

Table 3. Comparison of model complexity and inference efficiency

Model	Parameters(M)	FLOPs (M MACs)	Inference efficiency(s)
Encoding Network	41.83	18120	0.04
Decoding Network	0.74	48750	0.02

The encoding network stands as a large-scale model, boasting a substantial number of parameters and engaging in a high volume of calculations. This characteristic makes it well equipped for high-quality image embedding tasks, where it meticulously processes input images to extract and encode intricate features. However, the model's complexity comes at a cost. Owing to the numerous parameters and elaborate computational steps involved in feature extraction and transformation, the running time of the encoding network is relatively long, which may pose challenges in real-time applications.

In stark contrast, the decoding network has a relatively small number of parameters. However, its computational complexity is surprisingly high, which is a test of its deeper network structure. The multiple layers in the decoding network demand significant computational resources for reconstructing the original image from the encoded data. The small parameter count works in favour of running time, making it operate at only half the speed of the encoding network. This complexity analysis offers a crucial theoretical underpinning for balancing performance and efficiency in the design of subsequent models.

2. Model evaluation indicators

For the purpose of verifying the validity of the method, qualitative and quantitative analyses are conducted. In the qualitative analysis, the visual differences between the C' generated by the method and the extracted and S' are compared with those of the other methods. In the quantitative analysis, two objective evaluation metrics are used to assess the performance of the stego image and cover image, including the PSNR [23] and SSIM [24].

The PSNR is an objective metric that quantifies image distortion by measuring pixel-level errors. Higher PSNR values indicate better image quality and lower noise. The calculation of the PSNR is shown in Eq. (8).

$$\mathrm{PSNR} = 10\log\left(\frac{I_{\max}^2}{\mathrm{MSE}}\right) \tag{8}$$

The SSIM quantifies image quality by measuring structural similarity, considering both local and global features. Its value ranges from 0 to 1, with higher scores indicating greater similarity. The calculation of the SSIM is shown in Eq. (9).

$$\mathrm{SSIM}(X, Y) = \frac{(2\mu_X\mu_Y + C_1)(2\sigma_{XY} + C_2)}{\left(\mu_X^2 + \mu_Y^2 + C_1\right)\left(\sigma_X^2 + \sigma_Y^2 + C_2\right)} \tag{9}$$

where μ_X and μ_Y denote the mean luminance values of images X and Y, respectively; σ_X^2 and σ_Y^2 represent their contrast; σ_{XY} is the covariance between the two images; and C_1

and C_2 are constants introduced to stabilize the calculation and prevent division by zero. This metric jointly considers luminance, contrast, and structural information, providing a more comprehensive measure of similarity between two images.

4.2 Experimental Results and Analysis

In this section, several experiments are conducted, and the results are compared with those of existing benchmark methods.

1. Image steganography illustration

To evaluate the method from the perspective of visual quality, we test the model on the COCO dataset and randomly select 8 pairs of cover images, stego images, secret images and reconstructed secret images generated by the proposed model. The results are shown in Fig. 6.

Fig. 6. The effect of this model

It is obvious that the S' values generated by the encoding network in this paper have very good visual quality and are highly similar to those of C, which are almost indistinguishable from C, so it is difficult to detect whether there is an S hidden in them. In addition, S' by the decoding network almost restores all the information of the original S, and the distortion of S' is almost unnoticeable to the human eye. This shows that S' is basically consistent with the original S and exhibits high accuracy in recovering S. The findings indicate that the proposed method performs well in terms of image quality and secret image reconstruction accuracy.

For the purpose of better evaluating the visual quality validity of the proposed methodology, we choose to compare it with the results in references [3] and [9]. The comparative visual quality effectiveness is shown in Fig. 7, which demonstrates the visual quality effectiveness of C, C', S and S' for different models. In Fig. 7, the first column shows cover C, the second column depicts the stego images C', the third column shows the secrets S used for hiding, and the fourth column represents the reconstructed secret S'.

Fig. 7. Comparison of the proposed method with the visualizations of references [3] and [9]

In the area of visualization, C' generated via the proposed method is very similar to the original C and maintains a similar appearance. In addition, it can be clearly observed from Fig. 7 that S' by the decoding network almost restores all the semantic contents in the original S. This result shows that the steganography process successfully embeds and preserves the basic information of S, which enables reliable extraction and reconstruction in the extraction stage. The high fidelity of S' indicates that the method in this paper is better than the other methods in terms of the accuracy of secret image recovery.

2. Visual quality analysis

This experiment compares the image quality of the proposed method and others on this COCO dataset, and the comparison results are shown in Table 4.

Table 4. Comparison of image quality between the model in this paper and other methods

Methods	Cover image /Stego image		Secret images /Reconstructed secret images	
	PSNR	SSIM	PSNR	SSIM
Reference [3]	31.1264	0.9212	29.1876	0.9309
Reference [9]	29.3965	0.8926	29.6171	0.9134
Our Method	**31.5652↑**	**0.9281↑**	**31.1061↑**	**0.9354↑**

Table 4 clearly shows that the proposed program outperforms the visual quality in references [3] and [9] on 2 metrics in terms of C/C' and S/S'. Specifically, in terms of C/C', the PSNR value of this program is improved by 0.4388 dB and 2.1687 dB compared with those of references [3] and [9], respectively. In terms of S/S', enhancements of 1.9185 dB and 1.489 dB in PSNR demonstrate the effectiveness of the method, which enables full concealment of a 256 × 256 color secret image within a same-sized color cover image.

This experiment adopts the control variable method to construct a progressive training set size validation scheme based on the COCO dataset to systematically evaluate the model's performance advantages. Specifically, 6 experimental groups are constructed from the COCO training set via stratified random sampling, containing 2500, 5000, 7500, 10000, 12500, and 15000 samples, and each group maintains the same proportion of the category distribution. For image quality assessment, the PSNR and SSIM are adopted as quantitative evaluation indices, and subjective bias is eliminated through a double-blind evaluation process. The evaluation metrics of the cover-stego and secret-reconstructed secret groups were compared, as shown in Fig. 8.

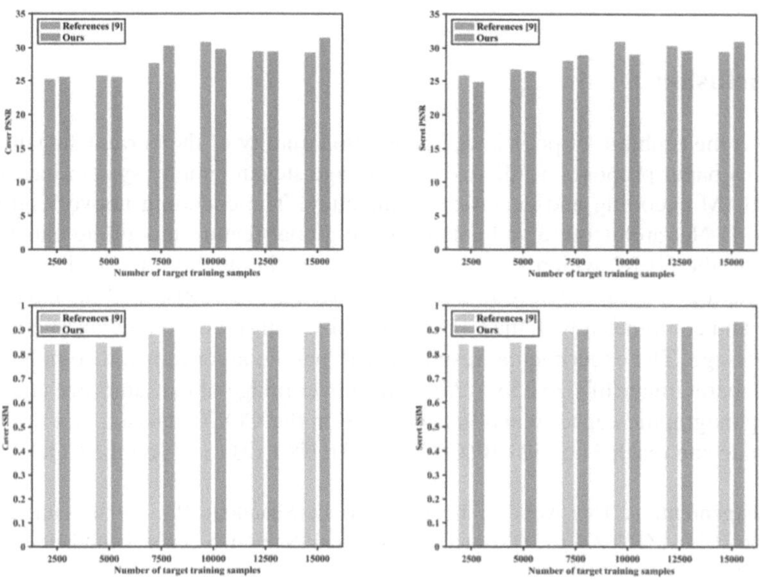

Fig. 8. Trends in evaluation indicators for different sample sizes

3. Ablation experiments

To verify the validity of the dual attention mechanism module proposed in the paper, the validities of the GAM module and the CBAM module are tested on the COCO dataset, and the experimental results are presented in Table 5. The attention mechanism module proposed in the proposed paper plays a key role in improving performance. Table 4 shows that the two attention mechanism modules proposed in this paper can significantly improve the performance of this method. The GAM module improves the steganography ability of the model, and the PSNR value is increased by 1.8672 dB in the C/C'. The CBAM module improves the extraction ability of the model. In the field of S/S', the PSNR value is increased by 0.9669 dB.

Table 5. Effectiveness of the dual attention mechanism module

Attention mechanisms		Cover image/Stego image		Secret images/Reconstructed secret images	
GAM	CBAM	PSNR	SSIM	PSNR	SSIM
		29.3965	0.8926	29.6171	0.9134
✓		31.2637	0.9239	30.5840	0.9314
	✓	29.728	0.9029	30.2078	0.9225
✓	✓	**31.5652↑**	**0.9281↑**	**31.1061↑**	**0.9354↑**

5 Conclusion

To address the problem of poor image generation quality of the current steganography model, this paper proposes a full-size color image steganography system based on the GAM-CBAM encoding and decoding architecture. The encoding network adopts the improved U-Net architecture and enhances the global feature modelling capability by introducing the GAM, which significantly improves the quality of visualization of the steganographic images; the decoding network constructs a deep encoding and decoding structure and combines it with the CBAM to realize high-precision reconstruction of the secret message. The experimental results reveal that the algorithm can completely hide the color secret image of size 256×256 in the cover image of the same size and achieve good steganographic capacity and visual effect on the COCO dataset, and the average PSNR value can reach 31.5652 dB (C/C') and 31.1061 dB (S/S') on the COCO dataset.

Acknowledgement. s. This work is supported by the Shandong Provincial Natural Science Foundation (CN) (ZR2022LLZ003) and the National Natural Science Foundation of China (12471422).

Disclosure of Interests. The authors declare that they have no known competing financial interests or personal relationships that could have appeared to influence the work reported in this paper.

References

1. Fridrich, J.: Steganography in Digital Media: Principles, Algorithms, and Applications. Cambridge University Press (2009)
2. Hongjuan, Z., Chenming, Z.: Novel LSB steganography algorithm of against statistical analysis. Comput. Eng. **34**(23), 144–146 (2008)

3. Tamimi, A.A., Abdalla, A.M., Alallaf, O.: Hiding an image inside another image using variable-rate steganography. Int. J. Adv. Comput. Sci. Appl. **4**(10), 18–21 (2013)
4. Li, X., Yang, B., Cheng, D., et al.: A generalization of LSB matching. IEEE Sig. Process. Lett. **16**(2), 69–72 (2009)
5. Thai, T.H., Cogranne, R., Retraint, F.: Optimal detection of outguess using an accurate model of DCT coefficients. In: Proceedings of IEEE International Workshop on Information Forensics & Security, pp. 179–184 (2014)
6. Barni, M., Bartolini, F., Piva, A.: Improved wavelet-based watermarking through pixell-wise masking. IEEE Trans. Image Process. **10**(5), 783–791 (2001)
7. Baluja, S.: Hiding images in plain sight: Deep steganography. Adv. Neural. Inf. Process. Syst. **30**, 2069–2079 (2017)
8. Wu, P., Yang, Y., Li, X.Q.: StegNet: mega image steganography capacity with deep convolutional network. Future Internet Internet **10**(6), 54–69 (2018)
9. Duan, X., Jia, K., Li, B., et al.: Reversible Image steganography scheme based on a u-net structure. IEEE Access **7**, 9314–9323 (2019)
10. Duan, X., Gou, M., Liu, N., et al.: High-capacity image steganography based on improved Xception. Sensors **20**(24), 72–53 (2020)
11. Rahim, R., Nadeem, S.: End-to-end trained cnn encoder-decoder networks for image steganography. In: European Conference on Computer Vision (ECCV), pp. 723–729 (2019)
12. Zhang, R., Dong, S.Q., Liu, J.Y.: Invisible steganography via generative adversarial networks. Multimedia Tools Appl. **78**(7), 8559–8575 (2019)
13. Chen, B.J., Wang, J.X., Chen, Y.Y., et al.: High- capacity robust image steganography via adversarial network. KSII Trans. Internet Internet Inf. Syst. **14**(1), 366–379 (2020)
14. Baluja, S.: Hiding images within images. IEEE Trans. Pattern Anal. Mach. Intell. **42**(7), 1685–1697 (2019)
15. Sharma, K., Aggarwal, A., Singhania, T., et al.: Hiding data in images using cryptography and deep neural network. J. Artif. Intell. Syst. **1**, 143–162 (2019)
16. Subramanian, N., Cheheb, I., Elharrouss, O., et al.: End-to-end image steganography using deep convolutional autoencoders. IEEE Access **9**, 135585–135593 (2021)
17. Lu, S.P., Wang, R., Zhong, T., et al.: Large-capacity image steganography based on invertible neural networks. In: Proceedings of the IEEE/CVF Conference on Computer Vision and Pattern Recognition, pp. 10816–10825 (2021)
18. Jing, J., Deng, X., Xu, M., Wang, J., Guan, Z.: Hinet: deep image hiding by invertible network. In: Proceedings of the IEEE/CVF International Conference on Computer Vision, pp. 4733–4742 (2021)
19. Guan, Z., Jing, J., Deng, X., et al.: DeepMIH: deep invertible network for multiple image hiding. IEEE Trans. Pattern Anal. Mach. Intell. **45**(1), 372–390 (2022)
20. Yang, H., Xu, Y., Liu, X., Ma, X.: PRIS: practical robust invertible network for image steganography. Eng. Appl. Artif. Intell. **133**, 108419 (2024)
21. Liu, Y., Shao, Z., Hoffmann, N.: Global attention mechanism: retain information to enhance channel-spatial interactions (2021). arXiv preprint arXiv:2112.05561
22. Woo, S., Park, J., Lee, J.Y., et al.: CBAM: convolutional block attention module. Springer, Cham, **11211**, 3–19 (2018)
23. Rustad, S., Andono, P.N., Shidik, G.F.: Digital image steganography survey and investigation (Goal, assessment, method, development, and dataset). Signal Process. **27**(5), 9851024 (2023)
24. Agrawal, R., Ahuja, K.: CSIS: compressed sensing-based enhanced-embedding capacity image steganography scheme. IET Image Proc. **15**(9), 1909–1925 (2021)

TIP: A Multidimensional Feature Extraction Method for Extreme Financial Risk by Fusing Information Value Evaluation

Yulin Lu, Lifei Lu, and Bo Xu(✉)

School of Big Data and Artificial Intelligence, Guangdong University of Finance and Economics, Guangzhou, China
20171080@gdufe.edu.cn

Abstract. To address the high-dimensional redundancy and nonlinear feature extraction challenges of financial multisource heterogeneous time series data, this paper proposes an automated dimensionality reduction framework that integrates a TCN-Autoencoder with dual screening of information value and the Pearson correlation coefficient. Design of a multiscale dilated convolution block combined with an autoencoder with residual connections to extract low-dimensional latent representations in unlabelled data. On this basis, the information value (IV) of each potential feature is calculated on the basis of the labelled samples to eliminate the weak distinguishing factor, the Pearson correlation coefficient is subsequently used to remove the redundant features, and the potential features with strong predictive ability and good complementarity are finally refined. The experiments prove that the method significantly reduces feature dimensionality and redundancy while enhancing the discriminative power and interpretability of the features, providing high-quality and robust input support for extreme financial risk prediction.

Keywords: TCN-Autoencoder · Feature dimensionality reduction · Pearson correlation coefficient · Information value

1 Introduction

Extreme financial risk events, when they occur, often trigger dramatic market turmoil, which not only causes significant losses to investors but also may trigger systemic financial crises, with far-reaching impacts on the country's economy and social stability. Under the current complex external conditions of the global financial environment and the internal conditions of the transformation of high-quality economic development, extracting the core features closely related to extreme risks from complex, high-dimensional, and dynamically changing financial data is highly practical.

With the intensification of high-frequency trading and cross-market linkages, the characteristics of financial risk are characterized by high dimensionality, nonstationarity and other complex features. Traditional financial risk characterization methods rely mostly on statistical models, such as the copula mode [1] or the GARCH method [2], or construct a risk factor system on the basis of expert experience [3], which often assumes

L. Huang et al. (Eds.): CCF NCCA 2025, CCIS 2774, pp. 412–424, 2026.
https://doi.org/10.1007/978-981-95-5294-8_28

that the relationships between variables are linear; although they have a certain degree of interpretability, it is difficult to adequately capture the implied nonlinear structure of the market and the characteristics of cross-market linkages, and there are significant limitations when dealing with high-dimensional and nonlinear data. There are significant limitations when dealing with high-dimensional and nonlinear financial time series data. With the development of information technology, an increasing number of studies have attempted to address the complexity of financial data by utilizing the powerful automatic feature extraction capability of deep learning. Structures such as neural network structures and integrated learning models have been widely used to automatically extract effective features from multisource heterogeneous financial data [4] and to construct a high-quality feature system for risk modelling and prediction. Compared with traditional statistical measurement models, deep learning is able to automatically discover complex associations in data during the learning process, reducing the need for human intervention and simplifying the modelling process. Moreover, the deep learning model better captures the intrinsic structure and features of high-dimensional high-frequency data through layer-by-layer feature extraction and abstraction learning [5], which is conducive to mining the intrinsic complexity of financial time series data and improves the prediction accuracy of financial risk prediction to a certain extent.

To prevent and cope with financial risk more effectively, there is an urgent need to develop feature extraction methods with both feature compression capability and interpretability to predict the probability of crisis outbreaks in the coming years on the basis of the trend of massive and multisource financial variables. In recent years, the self-encoder model, as an advanced deep learning method, has been outstanding in automatically extracting key features without prior knowledge and has gradually become a popular direction of research in the financial field. In particular, self-encoder models incorporating time series structural modelling capabilities have been widely used in market trend prediction [6], financial time series processing [7], financial fraud identification [8], etc., which provide new ideas for mining potential risk signals in multisource heterogeneous data in financial markets. In addition, variable screening methods such as information value are widely used in scenarios such as credit scoring and risk classification [9], and their good interpretability provides new ideas for deep models to enhance model trust.

In view of this, we combine the powerful temporal modelling capability of the temporal convolutional network and the feature compression capability of the autoencoder to construct the TCN-Autoencoder model for feature extraction of financial multidimensional temporal data and introduce an information value evaluation mechanism to select and optimize the extracted features to improve the accuracy and interpretability of the model. This research not only realizes the organic integration of deep learning and traditional evaluation indices in the methodology but also provides a new realization path for the early warning system of extreme risk events.

The contributions of this paper are as follows: (1) An automatic dimensionality reduction framework based on the TCN-Autoencoder is constructed, which can effectively extract potential time series feature representations from high-dimensional and multisource financial time series data, enhance the compactness and effectiveness of feature expression, and provide a structured input basis for financial extreme risk modelling.

(2) IV metrics are introduced to quantitatively evaluate the extracted features, which are combined with the Pearson correlation coefficient to eliminate redundant variables and form the optimal subset of redundant features to improve the model training efficiency and generalizability ability and, at the same time, to enhance feature interpretability and stability.

The remainder of this article is structured as follows: Sect. 2 compiles and analyses the related literature on risk feature degradation and feature selection; Sect. 3 describes the structure of the TCN-Autoencoder in detail; Sect. 4 provides automatic feature engineering for financial multisource heterogeneous data, designs experiments and analyses the results; and Sect. 5 summarizes and discusses the limitations of the study and future research directions. It is hoped that the research in this paper will provide new ideas and approaches in the field of financial risk management to meet the challenges of increasingly complex financial markets.

2 Previous Work

In the field of finance, feature engineering has been a popular topic in the academic world. Early financial feature extraction methods were based on capital market data and financial market time series data mined through traditional statistical models. For example, Karaca et al. used the mRMR method for feature screening of global stock index features [10]. In terms of financial risk measures, scholars and industry practitioners have artificially constructed a number of features on the basis of price data, of which volatility is widely recognized as a core measure of financial risk. On the basis of the MVMQ-CAViaR method, Liu Jiao investigated the dynamic linkage between intraday volatility and trading volume and empirically analysed the impact of the transmission of extreme risk among various financial markets, such as the Chinese bond market and stock market [11]. However, traditional feature engineering methods are highly dependent on expert experience, the feature construction process is complicated and time-consuming, and such methods are mostly based on linear relationships between variables, which makes it difficult to portray the complex dynamics and sudden characteristics of the financial market, and this shortcoming seriously affects the prediction effect of financial risk.

With the intensification of high-frequency trading and cross-market linkages, the volume of data in financial markets has shown explosive growth, data types are increasingly diversified, and financial time series data show high-frequency, high-dimensional and heterogeneous characteristics; thus, past linear financial risk feature extraction tools are no longer applicable. Therefore, how to extract representative key features from these complex data becomes a key link in the prediction and early warning system of financial extreme risk. The introduction of artificial intelligence (AI) techniques has enabled feature engineering of multidimensional financial data to evolve from traditional statistical models to machine learning models. Machine learning models have many advantages for both nonlinear and nonsmooth data processing and feature extraction and are widely used in financial modelling. For example, Yun et al. used the genetic algorithm and extreme gradient boosting algorithm (XGBoost) to select 71 market data features and technical indicator features, and then a subset of the features was used to predict the movement of the Korean KOSPI stock index [12]. With increasing research, scholars have found that the performance of single feature selection methods tends to vary widely, so they have begun to use the integrated feature selection method to improve the stability of single feature selection methods. Tsai and Hsiao integrated multiple feature approximation methods (principal component analysis, genetic algorithms, and decision tree methods) to select features from fundamental indicator data and macroeconomic data, and the results revealed that the forecasting performance of integrating multiple feature selection methods outperforms the forecasting performance of a single feature selection method [13]. In recent years, end-to-end modelling paradigms represented by deep learning have emerged as multidimensional feature extraction methods, and convolutional neural networks (CNNs) [14], recurrent neural networks (RNNs) [15], and long short-term memory networks (LSTMs) [4] have performed well in modelling temporal dependencies. The autoencoder class of models has been gradually introduced into financial feature dimensionality reduction and unsupervised feature extraction because of its strong compression ability for high-dimensional data. For example, Bao proposed a deep learning model that combines a stacked self-encoder and LSTM for financial time series modelling and forecasting [16]. In addition, studies have combined temporal convolutional networks (TCNs) with the transformer architecture to enhance the model's ability to capture long-term dependencies [17], which shows strong potential in financial feature extraction.

In summary, most of the current research on financial feature engineering is based on neural networks, integrated learning and improvement methods. Compared with feature extraction methods based on statistical principles, automatic feature engineering applying artificial intelligence techniques can automatically generate high-quality features to improve the performance of data models without using domain knowledge and human involvement and has significant advantages in nonlinear, high-dimensional feature downscaling and the extraction of complex financial data. Despite the effectiveness of deep models, how to control model complexity and enhance the interpretability and effectiveness of features when facing heterogeneous and high-dimensional redundant data from multiple sources is still one of the key difficulties in current research. Currently, the financial market is characterized by a large data scale, diverse types, and rapid updating. Although existing methods can automatically learn the features from the original data, they do not consider the redundancy between the features, which is prone to cause information repetition and model overfitting and easily affects the prediction performance and model stability. In addition, most feature learning frameworks tend to use black-box automated coding and lack quantifiable measures of the risk prediction ability of each potential feature, leading to a lack of explanatory power of the model.

To address the above shortcomings, we propose a multidimensional feature extraction method that integrates a TCN-autoencoder dimensionality reduction network and an IV evaluation mechanism. By making full use of the temporal modelling capability of the TCN and the compression expression capability of the autoencoder, we realize effective dimensionality reduction and deep feature extraction of high-dimensional heterogeneous financial data, adopt data from multiple aspects of the financial market as a multisource heterogeneous dataset input to automated feature engineering, introduce IV indices to quantitatively assess the predictive ability of potential features, and remove redundant features via Pearson correlation coefficient analysis to screen out strong predictive features with practical risk warning value and construct the optimal subset of nonredundant features. Thus, the strong predictive features with actual risk warning values are screened out, and the optimal subset of nonredundant features is constructed to improve the model generalizability and interpretability.

3 Methodology

The experimental procedure of this study consists of the following steps. First, time-aligned processing and fusion of heterogeneous financial time series data from multiple sources are performed to unify annual and daily data to monthly frequencies. Second, the aligned dataset is preprocessed with missing value padding and min–max normalization to eliminate magnitude differences and ensure the stability of model training. Then, the TCN-Autoencoder is used to perform unsupervised feature dimensionality reduction on the preprocessed data to extract the low-dimensional latent representations. On this basis, the labelled samples are used to calculate the IV of each potential feature and eliminate the weakly distinguishing features. Then, the highly redundant features are eliminated by calculating the Pearson correlation coefficient, and a subset of features with strong predictive ability and good complementarity is finally refined. Finally, an extreme risk prediction model is constructed on this feature set, and a validation analysis is performed to assess the effectiveness and interpretability of the proposed method.

3.1 Data Preprocessing

The dataset used for the experiment has different frequencies, such as daily, monthly, and yearly, and in practical applications, different frequencies correspond to different temporal resolutions: annual data provide only a once-a-year macro indicator, whereas daily data reflect high-frequency market fluctuations. Therefore, if the two are spliced directly into the model, the model will not be able to accurately identify the specific daily samples that should correspond to each monthly or annual indicator. Therefore, the dataset is first time-aligned by taking the monthly data as the main frequency, downsampling the annual data to the monthly level, and aggregating the daily data to the monthly level so that all features correspond to the same sample on the same time scale.

After expanding the annual data to monthly data, the vacant values in between are filled in via forward-fill and backwards-fill to avoid having to discard samples or introduce many noisy fill values due to the complete absence of corresponding features for certain dates.

Given the wide range of differences in the magnitude of each feature, such as daily volume in the tens of millions, annual GDP in the trillions, and interest rates in the hundredths of a percentile, the large span of the range of raw feature values can lead to a decrease in the efficiency of gradient propagation during the model training process. Therefore, this chapter adopts the min–max normalization method, which can uniformly map all features to similar intervals, thus stabilizing the gradient, accelerating model convergence, and improving the fairness and reliability of numerical scale-based metrics.

3.2 TCN-Autoencoder Architecture

This section details the designed TCN-Autoencoder architecture, including its core modules—the multiscale expansion convolution block (TCNBlock), the encoder, the decoder, and the complete autoencoder (AE) system.

(1) **TCNBlock Design**

A one-dimensional convolution operation is used in each TCNBlock, combined with a multiscale dilation strategy to capture dependencies at different time steps. Specifically, the dilation rate is sequentially set to 1, 2, and 4, allowing the network to have exponential sense field expansion with the same convolutional kernel size. To alleviate the problem of gradient vanishing or gradient explosion in deep network training, TCNBlock adds residual connections and identity mapping to realize straight-through between the input and output, which effectively guarantees the complete transmission of information and accelerates convergence.

(2) **Encoder**

The encoder starts with a single-channel input sequence ($1 \times n_{featrues}$) and extracts multilevel timing features via TCNBlock. After the TCNBlock output, the temporal features are flattened and mapped to a predefined latent dimension space R^{latent_dim} via Dense to generate a low-dimensional latent representation.

(3) **Decoder**

The task of the decoder is to reconstruct the original input sequence from the latent representation. Potential vectors are projected to a feature space of the same size as before the encoder flattens via Dense. The multichannel timing tensor is subsequently reshaped, and TCNBlock is introduced to inversely map its timing features. Finally, a fully connected layer outputs a reconstructed sequence with the same dimensionality as the inputs. Finally, a fully connected layer outputs a reconstructed sequence $\hat{x}$ with the same dimensionality as the inputs.

(4) AE overall

The above encoders are connected in series with the decoder to form a complete TCN-Autoencoder. The model is trained with the mean squared error (MSE) as the reconstruction loss function, and the optimization objective is to minimize the error between the original sequence x and the reconstructed sequence $\hat{x}$:

$$L_{MSE} = \frac{1}{n}\sum\nolimits_{i=1}^{n}(x_i - \hat{x}_i)^2 \tag{1}$$

The optimizer uses the Adam optimizer and sets the learning rate scheduling strategy to ensure gradient stability and convergence efficiency during the training process (Fig. 1).

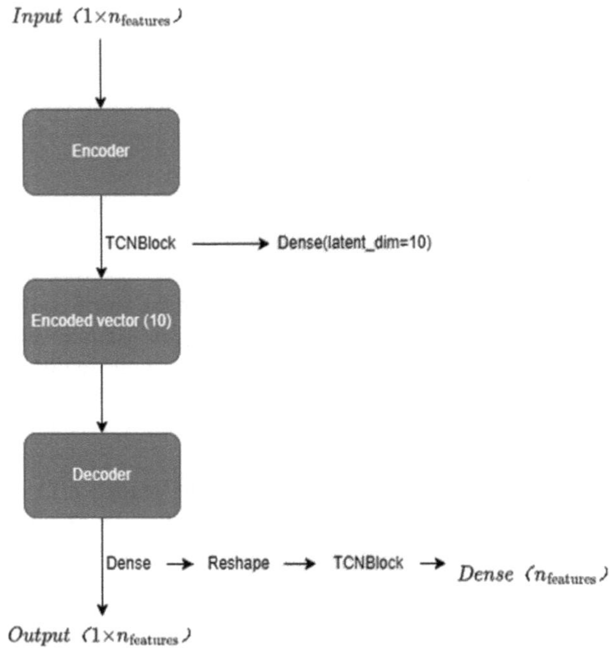

Fig. 1. TCN-Autoencoder Architecture Diagram

3.3 Information Value and Pearson Correlation Coefficient

This section describes how to perform feature importance ranking and redundancy removal on the basis of the implicit features extracted by the self-encoder in combination with the IV metrics and finally obtain a stable, high-quality feature set for subsequent predictive modelling. The IV is a commonly used indicator for assessing the degree of association of the independent variable with the dichotomous target variable after binning, and its formula is as follows:

$$IV = \sum\nolimits_{b=1}^{B} (WOE_b^{(good)} - WOE_b^{(bad)}) \times (p_b^{(good)} - p_b^{(bad)}) \tag{2}$$

where B is the number of subcases and where $p_b^{(good)}$ and $p_b^{(bad)}$ denote the percentages of good and bad samples, respectively, in group b out of the total good and bad samples.

The IV value reflects the ability of the feature to discriminate the target variable: IV < 0.02 indicates no predictive power; $0.02 \leq$ IV < 0.1 indicates weak predictive power; $0.1 \leq$ IV < 0.3 indicates moderate predictive power; and IV ≥ 0.3 indicates strong predictive power. The features with IV > 0.1 were retained in the experiment, and the features with no predictive power and weak predictive power were excluded.

To avoid the problem of multicollinearity introduced by the redundancy of highly correlated features, the Pearson correlation coefficient is used to eliminate redundant features via two-by-two tests of the candidate features. For two random variables $X = (X_1, \ldots, X_n)$ and $Y = (Y_1, \ldots, Y_n)$, their sample Pearson correlation coefficients are defined as:

$$r_{XY} = \frac{\sum_{k=1}^{n}\left(X_k - \overline{X}\right)\left(Y_k - \overline{Y}\right)}{\sqrt{\sum_{k=1}^{n}\left(X_k - \overline{X}\right)^2} \cdot \sqrt{\sum_{k=1}^{n}\left(Y_k - \overline{Y}\right)^2}} \quad (3)$$

Among them, $\overline{X} = \frac{1}{n}\sum_{k=1}^{n} X_k$, $\overline{Y} = \frac{1}{n}\sum_{k=1}^{n} Y_k$.

$r_{ij} = r(X^{(i)}, X^{(j)})$ is calculated for two-by-two candidate features, the correlation matrix $R = [r_{ij}]_{i,j=1}^{m}$ is obtained, the threshold $\tau to 0.8$ is set, if $|r_{ij}| > \tau$, then feature i is considered to be highly correlated with feature j, from which a feature with a low information value is removed to reduce redundancy and mitigate covariance.

4 Results and Analysis

4.1 Datasets and Parameterization

The dataset contains data sources such as the real estate market, stock market, macroeconomics, the money market, financial institutions, the digital economy, the foreign exchange market, the bond market, etc. After time-aligned processing fusion with monthly based frequency, the fused data have a total of 300 rows and 43 columns, totaling 12,900 time series data.

In the TCN-Autoencoder architecture, the TCNBlock parameters of the encoder and decoder are set as follows (Table 1):

Table 1. TCNBlock structural parameters.

Number of convolutional layers	3 Layers
Filters	64
Kernel size	1
Dilation rate	{1, 2, 4}
Activation function	ReLU

During the TCN-Autoencoder training process, the training hyperparameters are selected as follows (Table 2):

Table 2. Training Hyperparameters.

Parameter Name	Parameter Value
Optimizer	Adam
Activation Function	MSE
Epochs	100
Batch size	32

4.2 Experimental Analysis

Before moving on to the subsequent feature extraction and evaluation analysis, the convergence of the self-encoder when trained on unlabelled data is first given. Figure 2 shows the trend of the MSE loss value corresponding to each epoch with the number of iterations during the training process, from which we can intuitively observe the rapid decline of the model in the initial rounds as well as the convergence process of the model, which tends to stabilize gradually in the later rounds.

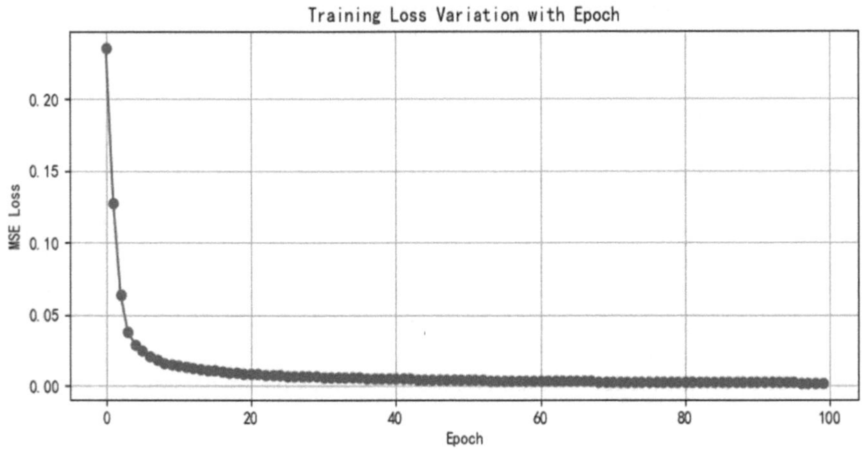

Fig. 2. Training Loss Variation with Epoch.

After training the self-encoder and verifying its convergence, we further analyse the last layer of the encoder's dense weight matrix to explore the degree of correlation between each latent dimension and the original 43 feature channels. Table 3 presents this weight matrix, from which we select the top 5 channel indices with the largest absolute weight values for each potential factor, as follows:

Table 3. Weight Matrix.

Feature	the 5 Most Relevant Channels	Feature	the 5 Most Relevant Channels
latent_1	[41 47 0 29 28]	latent_6	[50 26 59 40 34]
latent_2	[43 33 62 40 63]	latent_7	[47 36 59 58 50]
latent_3	[8 39 52 34 28]	latent_8	[21 60 0 51 25]
latent_4	[61 8 14 11 55]	latent_9	[3 20 27 52 32]
latent_5	[60 4 12 13 37]	latent_10	[33 3 7 2 15]

This analysis not only helps us understand the composition of the latent space but also lays the foundation for the subsequent screening of the most discriminative latent features on the basis of IV values and correlations. Next, to further quantify the discriminative ability of each potential feature for extreme risk, we perform IV value calculations for the 10 features on the basis of labelled data after dimensionality reduction to assess the differentiation of features for binary classification targets. The IV values for each potential feature are calculated as follows (Fig. 3):

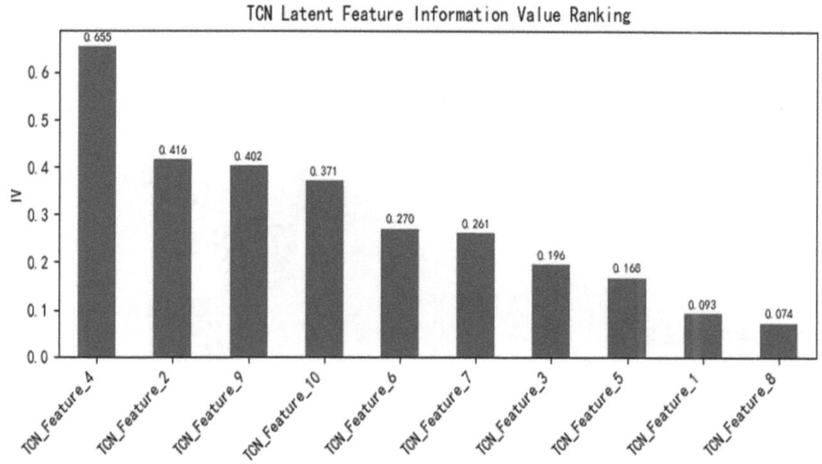

Fig. 3. Latent feature IV value.

As shown in the table, TCN_Feature_4 has the highest IV value (0.655), which far exceeds the empirical threshold of 0.3 for "excellent" discrimination. The IVs of TCN_Feature_1 and TCN_Feature_8 are lower than 0.1, which is a weak differentiation ability. We exclude features with IV < 0.1, i.e., we exclude features with no predictive power or weak predictive power, leaving only features with strong predictive power and moderate predictive power.

After completing the IV screening, Pearson correlation coefficients are calculated two-by-two for the remaining features to eliminate highly redundant dimensions. Two-by-two correlation coefficients > 0.8 are treated as highly redundant features, and subsequently, we retain the feature with the higher information value and eliminate the feature with the lower information value. Since the correlation coefficient between TCN_Feature_2 and TCN_Feature_10 reaches 0.817 and the information value of TCN_Feature_2 is 0.416 greater than that of TCN_Feature_10, which is 0.371; therefore, TCN_Feature_10 is excluded. This process ensures that the features have strong differentiation ability and low redundancy among them, providing high-quality input variables for the subsequent risk prediction model. This process ensures that the features have a strong differentiation ability but also maintain low redundancy among each other, providing high-quality input variables for the subsequent risk prediction model. The final feature correlation heatmap for the remaining seven features is shown below (Fig. 4).

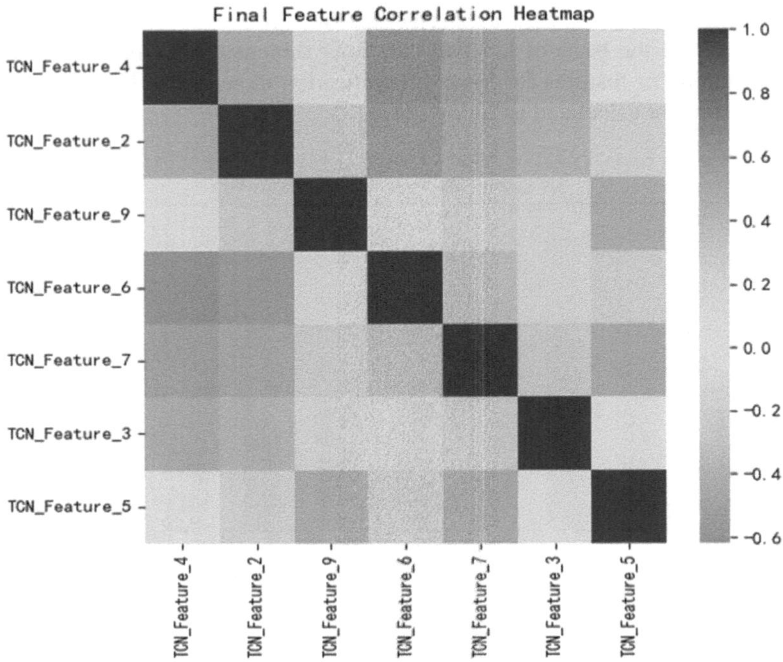

Fig. 4. Final feature correlation heatmap

The color blocks represent the absolute values of the correlation coefficients between feature pairs, allowing for a visual assessment of the strength and direction of their linear relationships through the color gradient: red shading indicates strong positive correlations (values near +1), whereas blue shading indicates strong negative correlations (values near -1). For the off-diagonal elements, none of the feature pairs present correlation coefficients exceeding the 0.8 threshold. This suggests that, following redundancy elimination, the selected feature set maintains a well-balanced combination of discriminative power and diversity. As a result, the inputs to the subsequent model are both stable and nonredundant, which not only facilitates faster algorithm convergence but also enhances the generalization performance.

5 Conclusion

In this paper, we present a framework for downscaling multisource, heterogeneous financial time series features by combining a TCN-Autoencoder with a two-stage screening process based on information value (IV) and Pearson correlation. First, we ensure feature consistency and stability through temporal alignment and min–max normalization at a monthly frequency. Our custom autoencoder employs multiscale dilated convolutional blocks with residual connections to effectively learn nonlinear patterns and temporal dependencies. Using labelled data, we then compute IV to filter out low-discriminatory features and apply Pearson correlation to remove redundancies, ultimately selecting seven complementary features with strong predictive power. The experimental results demonstrate that our approach substantially reduces dimensionality and redundancy while improving feature discriminability and interpretability, thus offering robust, high-quality inputs for forecasting extreme financial risk. Future work will address the static nature of our current screening strategy by incorporating dynamic IV updates by rolling windows, adaptive thresholds, and mutual information, alongside interpretable methods and self-supervised learning, to further increase accuracy and applicability across markets and frequencies.

This work was supported by the Project of Philosophy and Social Science Planning of Guangdong [NO.GD24CGL08].

References

1. Patton, A.: Copula methods for forecasting multivariate time series. Handbook Econ. Forecasting **2**, 899–960 (2013)
2. Shen, Y., Dai, S., Luo, X.: Measurement of systemic risk spillover effects in China's financial sector based on a GARCH–Copula–CoVaR model. Contemp. Econ. Sci. **36**(6), 30–38 (2014)
3. Bruneau, C., Flageollet, A., Peng, Z.: Economic and financial risk factors, copula dependence and risk sensitivity of large multi-asset class portfolios. Ann. Oper. Res. **284**(1), 165–197 (2020)
4. Fischer, T., Krauss, C.: Deep learning with long short-term memory networks for financial market predictions. Eur. J. Oper. Res. **270**(2), 654–669 (2018)
5. Sirignano, J.A., Cont, R.: Deep learning for limit order books. Quant. Finance **19**(4), 549–570 (2019)
6. Wang, J., Sun, T., Liu, B., Cao, Y., Wang, D.: Financial markets prediction with deep learning. arXiv preprint arXiv:2104.05413 (2021)
7. Chung, H., Shin, K.: A deep learning framework for financial time series using stacked autoencoders and long-short term memory. PLoS ONE **12**(7), e0180944 (2017)
8. Schreyer, M., Sattarov, T., Schulze, C., Reimer, B., Borth, D.: Detection of accounting anomalies in the latent space using adversarial autoencoder neural networks. arXiv preprint arXiv: 1908.00734 (2019)
9. Seitshiro, M.B., Govender, S.: Credit risk prediction with and without weights of evidence using quantitative learning models. Cogent Econ. Finan. **12**(1), 2338971 (2024)
10. Karaca, Y., Zhang, Y.D., Muhammad, K.: Characterizing complexity and self-similarity based on fractal and entropy analyses for stock market forecast modelling. Expert Syst. Appl. **144**, 113098 (2020)
11. Liu, R., Li, Z., Liu, H.: Monitoring and early warning of extreme risk spillovers across China's financial markets based on the MVMQ-CAViaR method. Econ. Manage. Res. (Jingji Yu Guanli Yanjiu) **2**, 19–29 (2020)
12. Yun, K.K., Yoon, S.W., Won, D.: Prediction of stock price direction using a hybrid GA-XGBoost algorithm with a three-stage feature engineering process. Expert Syst. Appl. **186**, 115716 (2021)
13. Tsai, C.-F., Hsiao, Y.-C.: Combining multiple feature selection methods for stock prediction: union, intersection, and multi-intersection approaches. Decis. Support. Syst. **50**(1), 258–269 (2010)
14. Chernikov, A., Tan, C. W., Montero-Manso, P., Bergmeir, C.: FRANS: Automatic feature extraction for time series forecasting. arXiv preprint arXiv:2209.07018 (2022)
15. Chen, Y., et al.: Deep learning for financial time series prediction: a state-of-the-art review of standalone and hybrid models. Comput. Model. Eng. Sci. **139**(1), 187–224 (2023)

16. Heaton, J.B., Polson, N.G., Witte, J.H.: Deep learning for finance: deep portfolios. Appl. Stoch. Model. Bus. Ind. **33**(1), 3–12 (2017)
17. Wu, Y., et al.: Temporal convolutional network and Transformer-based model for financial time series prediction. J. Forecast. **40**(6), 1021–1035 (2021)

MLAF-G: A Lightweight Multilevel Feature Fusion Classification Network for Plant Diseases Based on Ghostnet

Zonghai Zha, Yingjian Liu(✉), Minghuan Lv, Xiangyun Zheng, Hao Wang, Yindong Wen, and Jinhui Liu

Department of Computer Science and Technology, Ocean University of China, 266100 Qingdao, China
liuyj@ouc.edu.cn

Abstract. Crop disease is a significant factor affecting crop yields, in severe cases, it can lead to a substantial reduction in crop production. However, most disease detection methods mainly rely on expert experience and general convolutional neural networks, but often face the problems of incorrect identification and difficulty in deploying resource-constrained environments. Therefore, this paper proposes a lightweight multilevel feature fusion classification model based on GhostNet (i.e., MLAF-G) to address the current challenges in crop disease identification. It improves performance through multilevel attention feature fusion. The backbone network combines an efficient computation (EC) module with the hybrid weight coordinate attention (HCA) module to improve feature map expressiveness, allowing for effective capture of crop disease characteristics with minimal computational cost. Additionally, an improved convolutional block attention module (I-CBAM) and a multilevel feature fusion mechanism are designed to integrate features across different levels, increasing accuracy and detection effectiveness. Extensive experiments on several datasets demonstrate that MLAF-G outperforms baseline models in classification accuracy, robustness,and efficiency. The experimental results prove that MLAF-G can efficiently accomplish crop disease detection tasks in resource-constrained environments, paving the way for wider application of intelligent agricultural technologies and contributing to the modernization and sustainable development of the agriculture industry.

Keywords: Plant Disease Classification · Deep Learning · Convolutional Neural Networks · Multilevel Features · Feature Fusion

1 Introduction

Crop diseases pose critical threats to global food security, with severe cases causing up to 40% yield reduction in major crops such as potato and rice. Such a decrease drop in yields not only means a direct loss of income for farmers but also has the potential to trigger regional food shortages, exacerbate the global food

L. Huang et al. (Eds.): CCF NCCA 2025, CCIS 2774, pp. 425–440, 2026.
https://doi.org/10.1007/978-981-95-5294-8_29

supply shortage, and threaten the food security of billions of people [25]. Traditional manual detection relies on expert experience, consumes 8–12 person-hours per hectare and suffers from 20–30% misdiagnosis rates due to human fatigue. Recent advances in computer technology, outperform conventional methods, such as machine learning and deep learning, demonstrating superior performance in agricultural applications while reducing reliance on human expertise. However, existing models demand high computational resources, making them unsuitable for low-power IoT devices. Additionally, these methods rely heavily on high-quality, large datasets, which are hard to obtain in many agricultural areas. For example, Zeng et al. combined ResNet and Vision Transformer for disease detection in corn and cucumber, but the model incured excessive computational costs [29]. Waheed et al. improved DenseNet for corn diseases; however it experiences a decrease in accuracy when applied to multiple crops [24]. Sholihati et al. achieved 91% accuracy for potato diseases via VGG16 but struggled to distinguish similar diseases and performed poorly with a lightweight design [20].

To address these issues, transfer learning has emerged as a potential solution. Studies by Stephen et al. [23] and Abbas et al. [1] leveraged transfer learning to meet lightweight design requirements while improving classification accuracy. However, this approach still faces challenges in multicrop classification when dealing with crops exhibiting vastly different disease traits, and model performance decreases significantly.

Therefore, there are existing challenges in deep learning and transfer learning methods, such as low accuracy and high resource demands, when addressing multicrop disease detection tasks. This paper proposes a multilayer attention feature fusion network based on GhostNet (i.e., MLAF-G). Compared with other traditional networks, GhostNet [7] has significant advantages in terms of lightweight design and performance, rendering it highly suitable for environments with limited resources, especially for deployment on mobile devices.

This paper proposes a methodology that facilitates the classification of intricate and specific plant disease characteristics while maintaining a lightweight architecture.The principal contributions of this study are delineated as follows:

(1) A lightweight efficient computation (EC) module is developed to re-extract attention features and integrate them with low-cost computed features, enhancing lesion feature extraction while reducing computational resource consumption.
(2) A hybrid weights coordinate attention (HCA) mechanism is proposed, which combines channel attention to mitigate information loss from 2D pooling and enables multidimensional fusion of height, width and channel features for improved key feature identification.
(3) We propose a multilevel information fusion architecture that employs the I-CBAM module to generate channel and spatial attention weights, enhancing subtle multiscale features via weighted operations.
(4) The MLAF-G model can be employed on edge devices and achieves 99.31% validation accuracy and 99.45% external test accuracy with only 1.128 M parameters and 229.1 M FLOPs, representing a good balance between accuracy and lightweight design.

2 Related Work

2.1 Machine Learning

The technological evolution of plant disease detection methods has advanced from traditional machine learning to deep learning and transfer learning, with each stage demonstrating distinct advantages and limitations in addressing practical challenges. Specifically, traditional machine learning approaches, such as Xian and Ngadiran employ an extreme learning machine (ELM), a single-layer feedforward neural network, to classify plant leaf diseases by preprocessing images in the HSV color space and extracting texture features. While demonstrating good performance in tomato leaf disease identification, this approach suffers from limited adaptability to novel diseases because of its dependence on predefined feature engineering [26].

Subsequently, Shrivastava and Pradhan proposed a polynomial order-2 kernel-based SVM classifier that achieved 94.6% accuracy in rice disease detection using only color features, but overlooked critical texture information [21]. Zamani et al. introduced a radial basis function (RBF) SVM framework that combines mean filtering for denoising, K-means image segmentation, and feature-based classification, but its fixed-dimensional hand-crafted features struggled to capture subtle morphological changes in diseases [28]. These methods collectively revealed the bottleneck of traditional machine learning: feature expression was heavily dependent on human expertise, limiting adaptability to unseen disease types.

2.2 Deep Learning

The advent of convolutional neural networks (CNNs) revolutionized the field by enabling end-to-end automatic feature extraction, although at the cost of increased computational resources. For example, Nigam et al. proposed a wheat disease identification model using an optimized network structure and the WheatRust21 dataset, effectively capturing the spatial features of rust lesions, but its customized dataset and complex architecture hindered cross-crop application [17]. Diana Andrushia et al. employed a capsule network to preserve spatial feature relationships in grape leaf disease classification, however, its large parameter volume poses challenges for edge deployment [6]. Zhao et al. combined Inception and residual structures with an improved convolutional block attention module, achieving exceptional accuracy in identifying common diseases of corn, potatoes, and tomatoes, but the model's complexity resulted in slow inference speeds [30]. Transformer-based approaches, such as the model by [27], enhance fine-grained feature learning through long-range dependency modeling via soft segmentation token embeddings, but self-attention mechanisms exponentially increase computational complexity, limiting practical applicability [27].

To address data scarcity and computational constraints, transfer learning has emerged as a pivotal technique, although it has introduced new challenges. Argueso et al. fine-tuned Inception V3 and combines it with a Siamese network

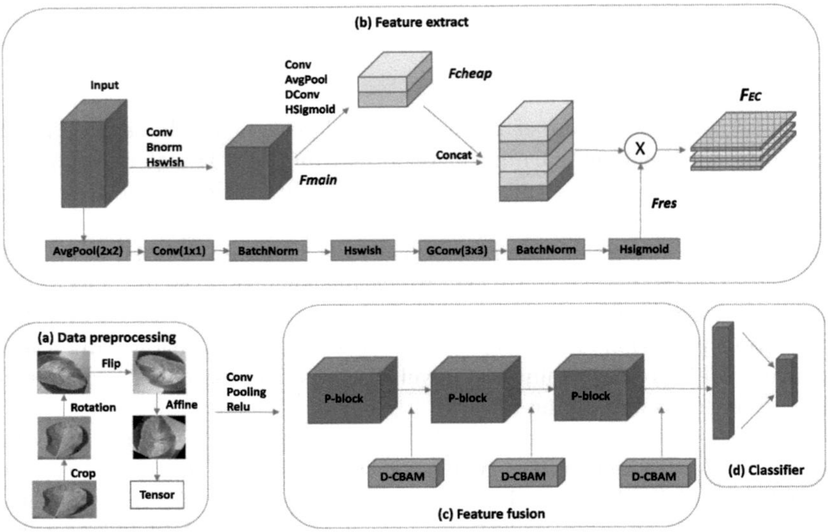

Fig. 1. The structure of MALF-G. (a) Data preprocessing module; (b) EC module, which belongs to the P-block; (c) feature fusion module; (d) classifier.

and triplet loss to learn leaf features from small datasets, achieving high accuracy with reduced data requirements, but the approach has been constrained to single-crop applications [2]. Chen et al. adapted a pretrained VGGNet for rice disease classification, achieving 92.53% accuracy on public datasets, yet struggled with background interference in complex field environments [5]. Jiang et al. further optimized VGG16 via multitask learning for three rice diseases and two wheat diseases, achieving accuracies of 98.12% and 97.85%, respectively, but failed to generalize to other crops [12]. Kaur et al. combined improved Inception and ResNet architectures with transfer learning to classify seven tomato leaf diseases and one healthy class with 98.92% accuracy, however, the cross-crop classification accuracy decreased significantly [13].

3 Method

3.1 The Proposed MLAF-G

In traditional convolutional networks, feature extraction usually depends on layer-by-layer feature learning ability, but this may lead to information loss and insufficient representation ability. Additionally, as the convolutional network delves deeper, the computational resources it demands grow exponentially, and gradient explosion problems may also arise. Therefore, this paper proposes the MLAF-G from the perspective of light weight and efficiency. Figure 1 presents the structure of MLAF-G, which first performs preprocessing operations on the image to increase the quality of the input image and uses the EC module to

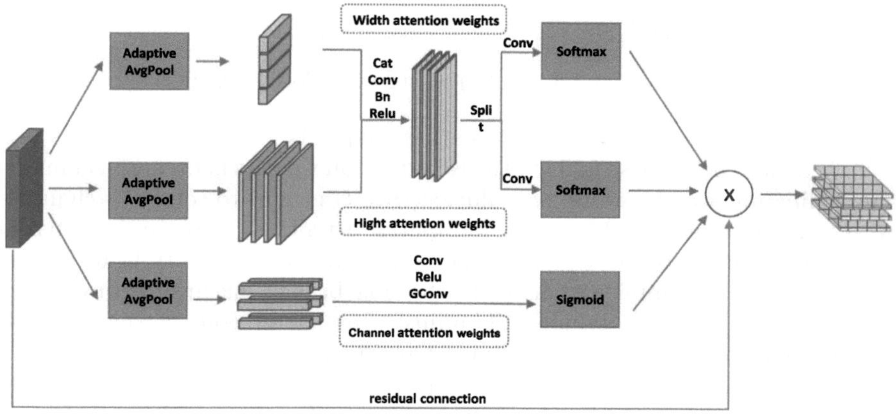

Fig. 2. The structure of HCA.

emphasize the ability of plant leaf disease spot extraction. Then, the multilevel information fusion structure which is composed of an improved convolutional block attention module (I-CBAM) and P-block can fuse feature maps of different scales. Finally, the classification module is used to obtain the final results.

3.2 EC Module

The EC module, inspired by GhostNet's cheap calculation concept, addresses the original module's limitation in detecting small disease spots via a lightweight three-branch structure with feature fusion formalized as:

$$F_{\mathrm{EC}} = \alpha \cdot F_{\mathrm{main}} + \beta \cdot F_{\mathrm{cheap}} + \gamma \cdot F_{\mathrm{res}} \quad (\alpha + \beta + \gamma = 1, \alpha, \beta, \gamma \in [0, 1]), \tag{1}$$

where $F_{\mathrm{main}} = \mathrm{Conv}_{\mathrm{std}}(X)$ denotes basic feature extraction from input X via standard convolution, $F_{\mathrm{cheap}} = \mathrm{GroupConv}_G(\mathrm{AvgPooling}(X))$ generates redundant features with computational cost reduction using 3×3 AvgPooling and group convolution with G groups, and $F_{\mathrm{res}} = \mathrm{Hsigmoid}(\mathrm{GroupConv}_G(F_{\mathrm{main}} + F_{\mathrm{cheap}}))$ fuses features from the first two paths through residual connection, applies group convolution, and adjusts weights via the Hsigmoid to integrate global information; the dynamic weights α, β, γ adaptively balance the contributions of each path to ensure lightweight design while enhancing key feature representation. As shown in part b of Fig. 1, this design increases classification accuracy and feature discriminability with minimal computations and is suitable for lightweight scenarios.

3.3 HCA Module

To address positional information loss from 2D global pooling, the hybrid weight coordinate attention (HCA) module is integrated into the P-block, formalizing its feature transformation as:

$$F_{\mathrm{HCA}} = X \odot (A_h \otimes A_w) \odot \sigma(\mathrm{GroupConv}_G(X)), \quad (2)$$

where $X \in \mathbb{R}^{H \times W \times C}$ is the input feature map, $A_h \in \mathbb{R}^{H \times 1 \times C}$ and $A_w \in \mathbb{R}^{1 \times W \times C}$ denote heightwise and widthwise attention weights generated via multiscale pooling followed by group convolution with G groups to compress channels to C/G and softmax regularization for spatial attention; $\otimes$ represents spatial dimension expansion to match X's shape, $\odot$ is element-wise multiplication, and $\sigma(\mathrm{GroupConv}_G(X))$ applies a channel attention branch via group convolution and sigmoid activation to dynamically adjust channelwise feature weights. This design embeds location information into channel-spatial attention through multiscale receptive field expansion and groupwise feature compression, enhancing global context capture and feature representation efficiency with minimal extra computations, as illustrated in Fig. 2.

3.4 Feature Fusion Structure

The multilevel feature fusion structure as shown in Fig. 3, which consists of P-blocks and improved CBAM (I-CBAM) modules. P-blocks, with different receptive fields, fuse multiscale features to increase the model's representational ability. Inside each P-block, the EC module, HCA module, and a depthwise separable convolution module with residual connections are integrated. The I-CBAM modules as shown in Fig. 4 are added after each P-block to prevent effective feature loss due to increased network depth. By using parallel spatial and channel attention, I-CBAM dynamically reweights features, optimizing the network's feature representation. This design maintains the model's lightweight property while enhancing multiscale feature perception and resilience to performance degradation, thus improving classification performance, especially in resource-constrained scenarios.

4 Experiment

4.1 Experiment Setting

The experiments are carried out on a machine equipped with an Intel(R) Xeon(R) E5 - 2620 v4 processor, 16 GB of RAM, running Ubuntu 16.04, and featuring an NVIDIA GTX2080 GPU. Using the PyTorch 1.12.1 framework, the Adam optimizer, which shows superior performance in this experiment, is chosen. With a weight decay of 0.0001, and a learning rate of 0.0005, the model is trained for 40 epochs, and the batch size for the input images is set to 32.

4.2 Datasets

In this study, we classify and detect 26 diseases in seven different crop species, including economic crops and widely distributed grain crops in China, with a

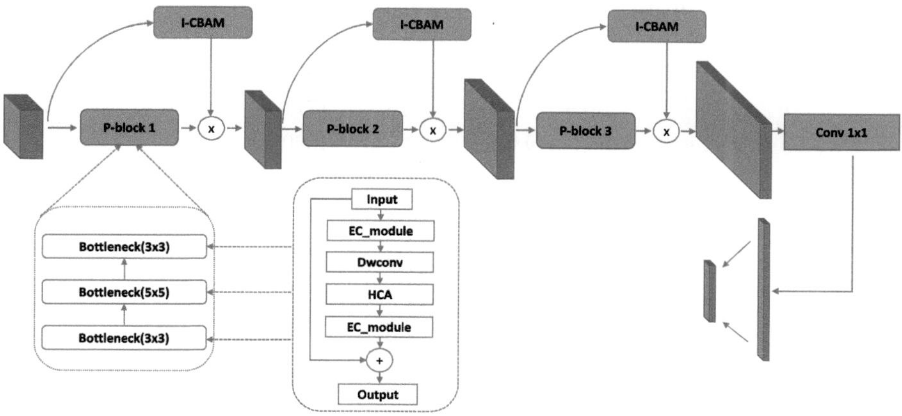

Fig. 3. The multilevel feature fusion structure.

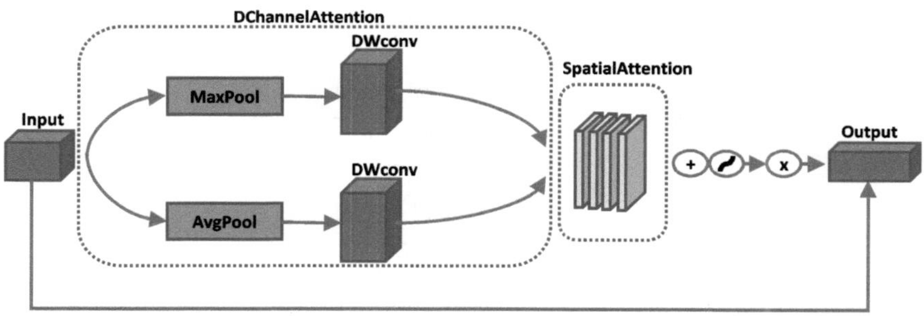

Fig. 4. The structure of I-CBAM.

total of 59,183 images. We named this custom dataset as Crops. All the images in the Corps dataset are selected from the Kaggle dataset, as shown in Table 1. Each image contains three channels: R, G, and B, with a size of 224x224. Among them, the research included 10 distinct categories of tomato diseases and 3 varieties of potato diseases, and the remaining diseases in apple, blueberry, grape, corn, and rice all had four common disease categories. Additionally, we introduce two public datasets: PlantVillage [10] and PlantDoc [22]. The PlantVillage dataset contains disease images from 14 different plants, covering 38 disease categories caused by various pathogens. This dataset is designed to evaluate the generalization potential of the proposed method by benchmarking it against other cutting-edge techniques. The PlantDoc dataset is a more challenging public dataset, covering 27 disease categories from 13 plant species. Compared with PlantVillage, the PlantDoc dataset contains more diverse samples and more class imbalance issues, making it suitable for assessing the efficacy of the proposed method in handling class-imbalanced classification.

Table 1. Description of the Crops dataset.

No.	Plant name	Class of disease	Quantity
1	Apple	4	7771
2	Blueberry	4	2060
3	Corn	4	12360
4	Grape	4	7222
5	Potato	3	5702
6	Rice	4	5678
7	Tomato	10	18345

4.3 Evaluation Metrics

To evaluate the classification ability and lightweight performance of the CNN - based plant disease classification network, six metrics are employed, formalized as follows:

$$\text{Accuracy} = \frac{TP + TN}{TP + TN + FP + FN}, \tag{3}$$

$$\text{Precision} = \frac{TP}{TP + FP}, \tag{4}$$

$$\text{Recall} = \frac{TP}{TP + FN}, \tag{5}$$

$$\text{F1 - Score} = 2 \cdot \frac{\text{Precision} \cdot \text{Recall}}{\text{Precision} + \text{Recall}}, \tag{6}$$

$$\text{Parameters} = (C_{\text{in}} \cdot K^2 \cdot C_{\text{out}} + C_{\text{out}}) + (N_{\text{in}} \cdot N_{\text{out}} + N_{\text{out}}), \tag{7}$$

$$\text{FLOPs} = C_{\text{in}} \cdot H_{\text{out}} \cdot W_{\text{out}} \cdot K^2 \cdot C_{\text{out}} + 2N_{\text{in}} \cdot N_{\text{out}}, \tag{8}$$

where TP, TN, FP, and FN denote true positives, true negatives, false positives, and false negatives, respectively, for the classification metrics; C_{in} and C_{out} are the input/output channel numbers, K is the convolutional kernel size, and $H_{\text{out}} \cdot W_{\text{out}}$ represent the output feature map dimensions for the convolutional parameters and FLOPs. $N_{\text{in}} \cdot N_{\text{out}}$ are the numbers of input/output neurons in fully connected layers, integrating both classification effectiveness and model lightweightness evaluation.

Table 2. Ablation performance of MLAF-G on the Crops dataset.

HCA module	I-CBAM module	EC module	Train accuracy(%)	Valid accuracy(%)	Test accuracy(%)
			97.73	97.64	97.07
✓			98.14	98.16	97.90
	✓		97.68	97.89	97.53
		✓	98.01	97.95	98.06
✓	✓		98.53	98.56	98.42
	✓	✓	98.31	98.09	98.11
✓		✓	97.90	97.95	97.89
✓	✓	✓	**99.45**	**99.31**	**99.45**

Table 3. Classification and lightweight performance of different models.

Model	Train accuracy(%)	Valid accuracy(%)	Precision(%)	Recall(%)	F1-score(%)	Parameters(M)	FLOPs(M)
GhostNet [7]	98.90	98.93	98.90	98.89	98.89	3.944	**154.6**
ShuffleNet V2 [15]	98.95	99.01	98.96	98.94	98.96	1.297	2237
MoblieNet V3 [8]	98.99	99.05	98.97	98.96	98.97	3.940	239.3
DenseNet121 [9]	98.53	98.03	98.52	98.53	98.51	6.985	44484
MLAF-G	**99.45**	**99.31**	**99.41**	**99.41**	**99.40**	**1.128**	229.1

4.4 Ablation Experiment

To validate the effectiveness of each module, we conducted ablation experiments on the Crops dataset and visualized three selected plant diseases with Grad-CAM [18], with the results shown in Table 2 and Fig. 5. The backbone network alone performed poorly in complex backgrounds and lacked fine-grained feature extraction; the HCA module improved classification accuracy and subtle feature capture but had limited global extraction, whereas the I-CBAM module significantly enhanced global feature extraction. The incorporating of the EC module achieves optimal performance, improving both classification accuracy and feature extraction for precise disease localization. Overall, these results prove that the HCA, I-CBAM, and EC modules effectively enhance feature extraction and recognition accuracy, validating their reliability.

4.5 Performance of Disease Classification and Lightweight Methods

We conduct comparative experiments on the Crops dataset to verify the classification and lightweight performance of the MLAF-G model. We compare it with GhostNet, MobileNet V3, DenseNet121, and ShuffleNet V2 in terms of accuracy, precision, recall, and F1 score metrics, as presented in Table 3.

The heatmaps of the three plant diseases in Fig. 6 highlight their accuracy and robustness in distinguishing diseases based on the basis of features. During training and validation, as illustrated in Fig. 7, MLAF-G converges faster, results in lower loss, and attains higher validation accuracy than the competitors do.

In terms of lightweight performance, as measured by the parameters and FLOPs in Table 3, MLAF-G significantly reduces the parameter count com-

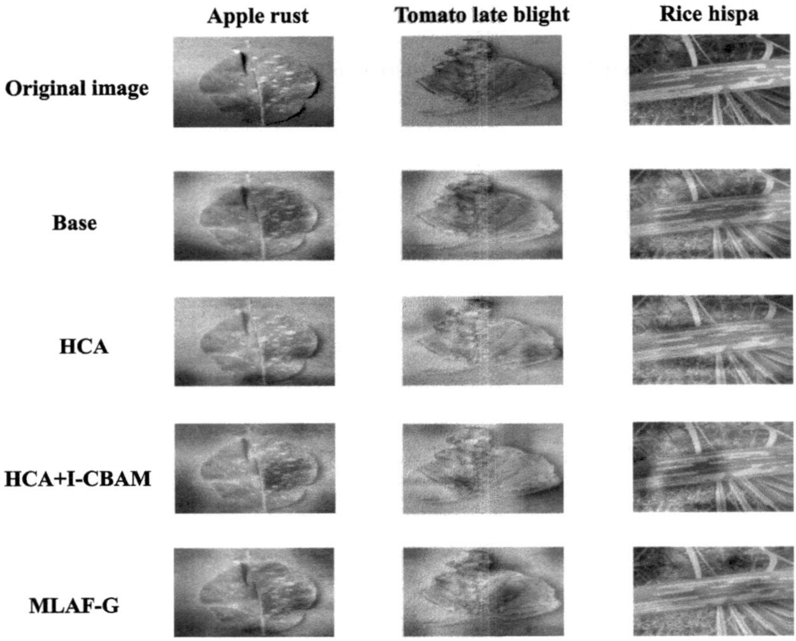

Fig. 5. Visualization comparison of different stages.

pared with all the baselines while maintaining high accuracy. It achieves a better balance between parameters and computational load than GhostNet and DenseNet121. Notably, compared with DenseNet121, MLAF-G cuts the parameter count by 84%, thus offering clear advantages in storage and computation.

4.6 Comparisons with Baseline Models

We evaluate MLAF-G on two public plant disease datasets, PlantDoc and PlantVillage, leveraging their diverse disease coverage and label differences to test the model's generalization. As shown in Fig. 4, MLAF-G outperforms the baseline methods on both datasets. The accuracy of PlantVillage is 99.40% accuracy, surpassing that of the Grow-light smart monitoring system (96.75%) and time-effective CNN (95.17%). With respect to plantDoc, MLAF-G reaches 88.31% accuracy, outperforming Dense-inception (84.92%) and other methods (76 - 84.3%). These results highlight its effectiveness across complex datasets and strong generalizability ability. MLAF-G also excels in training and inference efficiency. Its lightweight design enables fast inference with reduced computational complexity and parameter count, making it ideal for resource-constrained environments such as edge devices and mobile applications.

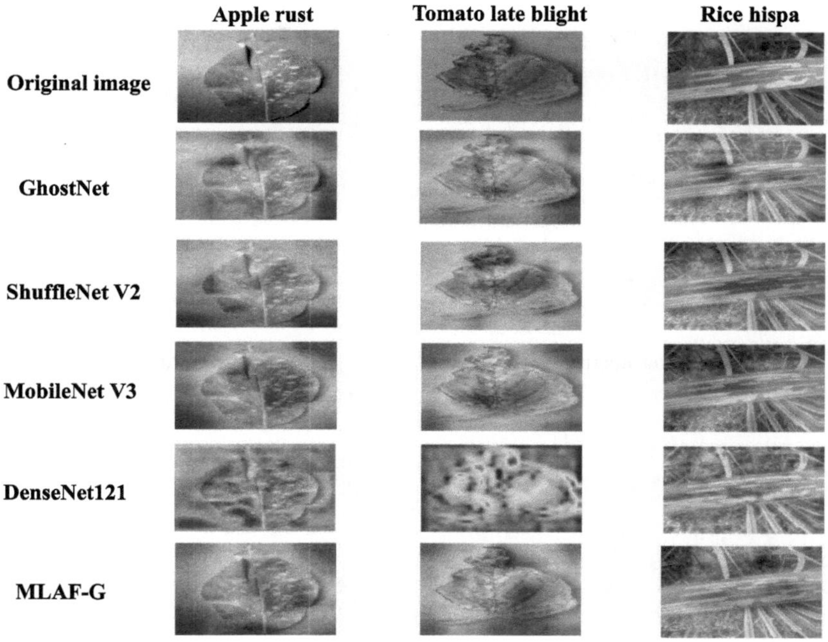

Fig. 6. Visualization comparison of different baseline models.

5 A Case Study on Edge Device

As shown in Fig. 8, MLAF-G is deployed on Raspberry Pi 4B. Figure 9 illustrates the prediction results from Raspberry Pi 4B. By leveraging Flask technology to build a lightweight service interface, the model achieves excellent performance in crop disease classification tasks on the test dataset. This deployment scheme uses Flask to implement RESTful API encapsulation of the model, supporting image

Table 4. Performance comparison on the PlantVillage and PlantDoc datasets.

Method	Dataset	accuracy(%)	Parameters(M)
PlantCaFo [11]	PlantVillage	94.23	N/A
Grow-light smart monitoring system [16]	PlantVillage	96.75	2.28
time-effective CNN [4]	PlantVillage	95.17	3.50
PND-Net [3]	PlantDoc	84.30	3.50
SoyaTrans [19]	PlantDoc	76.00	5.2
Dense-inception [14]	PlantDoc	84.92	N/A
MALF-G	PlantVillage	**99.40**	1.128
MALF-G	PlantDoc	**88.31**	1.128

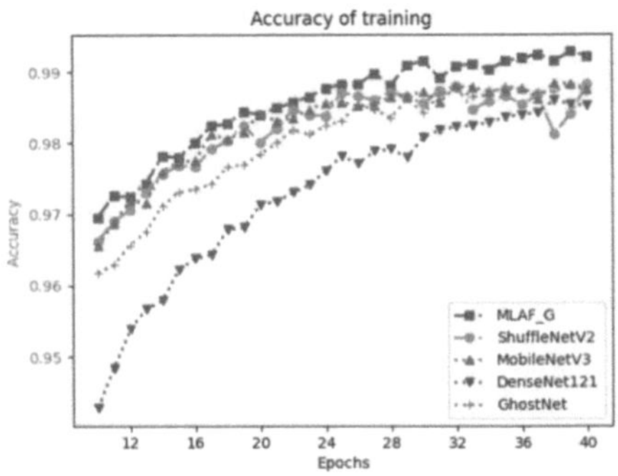

(a) Training accuracy

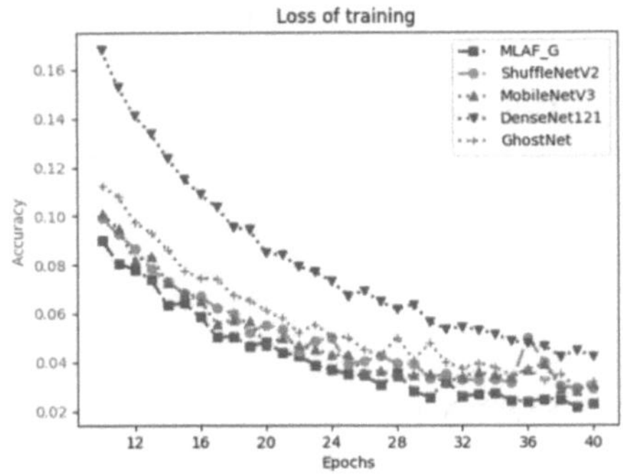

(b) Training loss

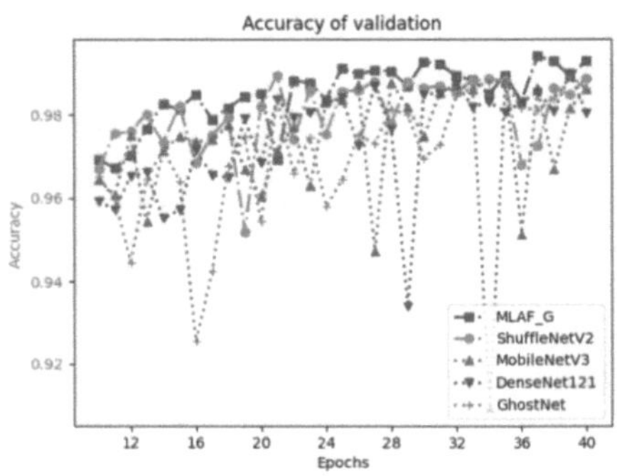

(c) Validation accuracy

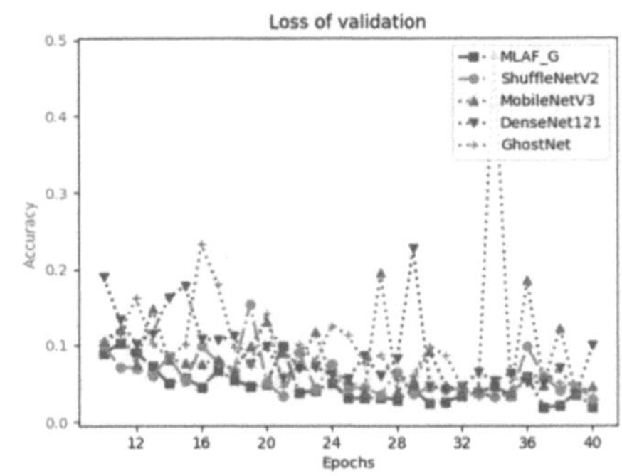

(d) Validation loss

Fig. 7. Comparison of training and validation performance.

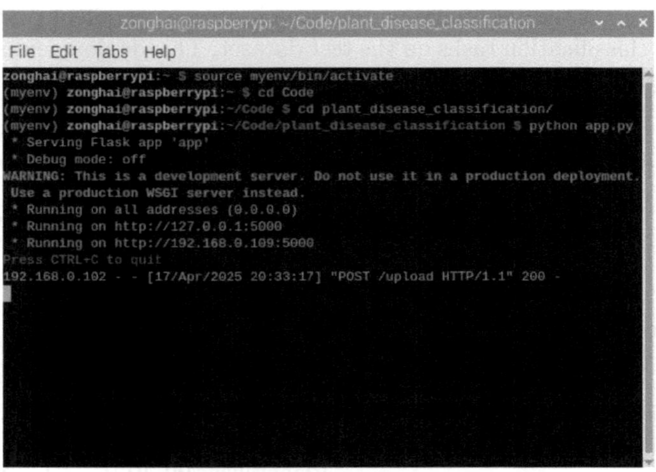

Fig. 8. MLAF-G runs on Raspberry Pi 4B.

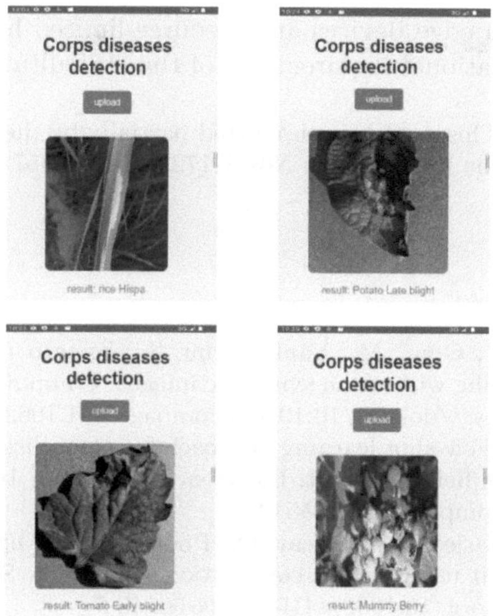

Fig. 9. Prediction results from Raspberry Pi 4B.

upload and result return via the HTTP protocol, which facilitates integration with agricultural Internet of Things terminal devices. In practical applications, farmers can use mobile apps or camera terminals deployed in the field to upload crop leaf images in real time. The backend service then invokes the MLAF-G model to carry out disease detection, providing a low-cost and high-efficiency resolution for disease early warning and intelligent management in precision agriculture.

6 Conclusion

This study introduces a lightweight plant disease classification model named MLAF-G, which is aimed at resource-constrained environments to reduce crop disease detection costs. By combining the EC and HCA modules to produce fine-grained feature attention weights and adopting a multilevel feature fusion structure, MLAF-G improves the flexibility and precision of capturing disease features, integrating multiscale information for higher classification accuracy. The experimental results indicate that MLAF-G attains accuracies of 99.40% and 88.31% accuracy on the PlantVillage and PlantDoc datasets respectively, outperforming the other methods. It outperforms lightweight models such as GhostNet, MobileNet V3, ShuffleNet V2, and DenseNet121, particularly in dealing with class imbalance and diverse disease types, surpassing state-of-the-art approaches in accuracy and robustness. Owing to its lightweight and efficient design, MLAF-G is highly applicable to practical agricultural scenarios, and is capable of running on edge devices and resource-limited hardware to meet the real-time and computational requirements of the agricultural field.

Acknowledgments. This work was supported partially by the National Natural Science Foundation of China under Grant Nos.62172378 and 61572448.

References

1. Abbas, A., Jain, S., Gour, M., Vankudothu, S.: Tomato plant disease detection using transfer learning with c-gan synthetic images. Comput. Electron. Agric. **187**, 106279 (2021). https://doi.org/10.1016/j.compag.2021.106279
2. Argüeso, D., et al.: Few-shot learning approach for plant disease classification using images taken in the field. Comput. Electron. Agric. **175**, 105542 (2020). https://doi.org/10.1016/j.compag.2020.105542
3. Bera, A., Bhattacharjee, D., Krejcar, O.: Pnd-net: plant nutrition deficiency and disease classification using graph convolutional network. Sci. Rep. **14**(1), 15537 (2024). https://doi.org/10.1038/s41598-024-66543-7
4. Biswas, S., Saha, I., Deb, A.: Plant disease identification using a novel time-effective CNN architecture. Multimedia Tools Appl. **83**(35), 82199–82221 (2024). https://doi.org/10.1007/s11042-024-18822-8
5. Chen, J., Chen, J., Zhang, D., Sun, Y., Nanehkaran, Y.A.: Using deep transfer learning for image-based plant disease identification. Comput. Electron. Agric. **173**, 105393 (2020). https://doi.org/10.1016/j.compag.2020.105393
6. Diana Andrushia, A., et al.: Image-based disease classification in grape leaves using convolutional capsule network. Soft. Comput. **27**(3), 1457–1470 (2023). https://doi.org/10.1007/s00500-022-07446-5
7. Han, K., Wang, Y., Tian, Q., Guo, J., Xu, C., Xu, C.: Ghostnet: more features from cheap operations. In: Proceedings of the IEEE/CVF conference on computer vision and pattern recognition, pp. 1580–1589 (2020). https://doi.org/10.1109/CVPR42600.2020.00165
8. Howard, A., et al.: Searching for mobilenetv3. In: Proceedings of the IEEE/CVF International Conference on Computer Vision, pp. 1314–1324 (2019). https://doi.org/10.48550/arXiv.1905.02244
9. Huang, G., Liu, Z., Van Der Maaten, L., Weinberger, K.Q.: Densely connected convolutional networks. In: Proceedings of the IEEE Conference on Computer Vision and Pattern Recognition, pp. 4700–4708 (2017). https://doi.org/10.1109/CVPR.2017.243
10. Hughes, D., Salathé, M., et al.: An open access repository of images on plant health to enable the development of mobile disease diagnostics. arXiv preprint arXiv:1511.08060 (2015). https://doi.org/10.48550/arXiv.1511.08060
11. Jiang, X., et al.: Plantcafo: An efficient few-shot plant disease recognition method based on foundation models. Plant Phenomics **7**(1), 100024 (2025). https://doi.org/10.1016/j.plaphe.2025.100024
12. Jiang, Z., Dong, Z., Jiang, W., Yang, Y.: Recognition of rice leaf diseases and wheat leaf diseases based on multi-task deep transfer learning. Comput. Electron. Agric. **186**, 106184 (2021). https://doi.org/10.1016/j.compag.2021.106184

13. Kaur, P., Harnal, S., Gautam, V., Singh, M.P., Singh, S.P.: A novel transfer deep learning method for detection and classification of plant leaf disease. J. Ambient. Intell. Humaniz. Comput. **14**(9), 12407–12424 (2023). https://doi.org/10.1007/s12652-022-04331-9
14. Khan, S.D., Basalamah, S., Naseer, A.: Classification of plant diseases in images using dense-inception architecture with attention modules. Multimedia Tools Appl. 1–26. (2024). https://doi.org/10.1007/s11042-024-19860-y
15. Ma, N., Zhang, X., Zheng, H.T., Sun, J.: Shufflenet v2: practical guidelines for efficient CNN architecture design. In: Proceedings of the European Conference on Computer Vision (ECCV), pp. 116–131 (2018). https://doi.org/10.1007/978-3-030-01264-9_8
16. Macdonald, W., Sari, Y.A., Pahlevani, M.: Grow-light smart monitoring system leveraging lightweight deep learning for plant disease classification. Artif. Intell. Agriculture **12**, 44–56 (2024). https://doi.org/10.1016/j.aiia.2024.03.003
17. Nigam, S., et al.: Deep transfer learning model for disease identification in wheat crop. Eco. Inform. **75**, 102068 (2023). https://doi.org/10.1016/j.ecoinf.2023.102068
18. Selvaraju, R.R., Cogswell, M., Das, A., Vedantam, R., Parikh, D., Batra, D.: Grad-CAM: visual explanations from deep networks via gradient-based localization. Int. J. Comput. Vision **128**(2), 336–359 (2019). https://doi.org/10.1007/s11263-019-01228-7
19. Sharma, V., Tripathi, A.K., Mittal, H., Nkenyereye, L.: Soyatrans: a novel transformer model for fine-grained visual classification of soybean leaf disease diagnosis. Expert Syst. Appl. **260**, 125385 (2025). https://doi.org/10.1016/j.eswa.2024.125385
20. Sholihati, R.A., Sulistijono, I.A., Risnumawan, A., Kusumawati, E.: Potato leaf disease classification using deep learning approach. In: 2020 International Electronics Symposium (IES), pp. 392–397. IEEE (2020). https://doi.org/10.1109/IES50839.2020.9231784
21. Shrivastava, V.K., Pradhan, M.K.: Rice plant disease classification using color features: a machine learning paradigm. J. Plant Pathol. **103**(1), 17–26 (2020). https://doi.org/10.1007/s42161-020-00683-3
22. Singh, D., Jain, N., Jain, P., Kayal, P., Kumawat, S., Batra, N.: Plantdoc: a dataset for visual plant disease detection. In: Proceedings of the 7th ACM IKDD CoDS and 25th COMAD, pp. 249–253. ACM (2020). https://doi.org/10.1145/3371158.3371196
23. Stephen, A., Punitha, A., Chandrasekar, A.: Designing self attention-based resnet architecture for rice leaf disease classification. Neural Comput. Appl. **35**(9), 6737–6751 (2023). https://doi.org/10.1007/s00521-022-07793-2
24. Waheed, A., Goyal, M., Gupta, D., Khanna, A., Hassanien, A.E., Pandey, H.M.: An optimized dense convolutional neural network model for disease recognition and classification in corn leaf. Comput. Electron. Agric. **175**, 105456 (2020). https://doi.org/10.1016/j.compag.2020.105456
25. Walters, D.R., Ratsep, J., Havis, N.D.: Controlling crop diseases using induced resistance: challenges for the future. J. Exp. Bot. **64**(5), 1263–1280 (2013). https://doi.org/10.1093/jxb/ert026
26. Xian, T.S., Ngadiran, R.: Plant diseases classification using machine learning. In: Journal of Physics: Conference Series. vol. 1962, p. 012024. IOP Publishing (2021). https://doi.org/10.1088/1742-6596/1962/1/012024
27. Yu, S., Xie, L., Huang, Q.: Inception convolutional vision transformers for plant disease identification. Internet of Things **21**, 100650 (2023). https://doi.org/10.1016/j.iot.2022.100650

28. Zamani, A.S., et al.: Performance of machine learning and image processing in plant leaf disease detection. J. Food Qual. **2022**(1), 1598796 (2022). https://doi.org/10.1155/2022/1598796
29. Zeng, Q., Niu, L., Wang, S., Ni, W.: Sevit: a large-scale and fine-grained plant disease classification model based on transformer and attention convolution. Multimedia Syst. **29**(3), 1001–1010 (2023). https://doi.org/10.1007/s00530-022-01034-1
30. Zhao, Y., Sun, C., Xu, X., Chen, J.: Ric-net: a plant disease classification model based on the fusion of inception and residual structure and embedded attention mechanism. Comput. Electron. Agriculture **193**, 106644 (2022). https://doi.org/10.1016/j.compag.2021.106644

Unsupervised Cluster-Generation Graph Learning for Industrial Equipment Anomaly Detection

Zeyun Xiao[1,2], Donglai Zhu[1,2], Bangwei Xu[1,2], Chengkai Zhang[1,2], Danfeng Sun[1,2], and Huifeng Wu[1,2](✉)

[1] Zhejiang Key Laboratory of New Industrial Internet Control Technology, Zhejiang, China
whf@hdu.edu.cn

[2] School of Computer Science, Hangzhou Dianzi University, Hangzhou, China

Abstract. Equipment anomaly detection is a critical task in the industrial domain. However, traditional detection approaches often falter due to the extreme imbalance of industrial datasets and the irregular, unexpected behavior of anomalous events. In this paper, we propose a feature enhancement framework, termed the clustering-based autocorrelated graph network (CAGN). The method begins by partitioning the dataset via an unsupervised clustering algorithm and constructing graphs at the cluster level to capture latent relationships among samples. Subsequently, an autocorrelation network and a graph neural network are employed to extract and reinforce structural and contextual features within the cluster-generation graph, thereby improving the separability between normal and anomalous instances. The proposed method has been comprehensively evaluated on multiple publicly available benchmark datasets as well as a practical dataset from injection moulding machines, demonstrating its robustness and performance advantages. The source code and model are available at https://github.com/napoleonkuroba/CAGN.

Keywords: Clustering Algorithm · Graph Neural Network · AutoCorrelation Network · Industrial Anomaly Detection

1 Introduction

The advancement of industrial automation has enabled more comprehensive data collection and analysis in manufacturing. Data acquisition during equipment operation has become essential for ensuring product quality and improving efficiency. As a key technology, anomaly detection identifies potential faults by analysing operational data, thereby supporting process optimization and productivity. However, industrial data often suffer from extreme class imbalance, scarce anomaly samples, and complex variation patterns, posing challenges for traditional detection algorithms. Accurately detecting anomalies under such conditions has become a central focus in intelligent maintenance research. Early industrial anomaly detection relied primarily on statistical methods such as GMM and

L. Huang et al. (Eds.): CCF NCCA 2025, CCIS 2774, pp. 441–457, 2026.
https://doi.org/10.1007/978-981-95-5294-8_30

LOF, which assume specific data distributions and detect anomalies on the basis of probabilistic or local density criteria. With increasing data complexity, these methods struggle to model high-dimensional, nonlinear patterns. As a result, machine learning approaches such as the OCSVM and isolation forest (IF) have been widely adopted. For example, Anton et al. [2] applied SVM and random forest to detect intrusions in industrial protocols; Laskar et al. [11] combined K-means [15] with IF for large-scale detection; and Chen et al. [3] used random forest to identify anomalies and performance-related features in industrial processes.

Deep learning (DL) methods have gained widespread use in anomaly detection because of their strong feature learning ability and adaptability in complex industrial environments. While traditional machine learning methods offer robustness, they often struggle with generalization under highly variable data distributions. DL models such as the AE, VAE, and GAN have shown superior performance across diverse tasks. For example, Kim et al. [10] used autoencoders to detect payload anomalies in industrial control systems. Ahmed et al. [1] applied a six-layer autoencoder for wind turbine monitoring, which significantly improved the detection accuracy. Givnan et al. [7] proposed an autoencoder-based approach for real-time monitoring of industrial motors by modelling normal operating behavior.

Although traditional ML and DL methods have been widely applied in industrial anomaly detection, single models often struggle in highly imbalanced data scenarios. Traditional machine learning relies on balanced samples and is prone to class bias, whereas deep learning models may overfit the dominant normal class and struggle to detect rare anomalous patterns. Moreover, industrial data are often noisy, temporally complex, and rich in latent structural relationships, further complicating anomaly detection. To address this issue, approaches that combine traditional machine learning and neural networks for anomaly detection have attracted increasing attention. By integrating the discriminative capabilities of machine learning models with the feature extraction strengths of deep neural networks, these methods have demonstrated improved performance in complex industrial data scenarios. Zong et al. [27] proposed a method based on autoencoders and a GMM, which combines feature compression and density estimation and achieves excellent performance across various anomaly detection tasks. Wu et al. [22] designed an anomaly detection approach that integrates graph neural networks (GNNs) with Gaussian mixture models (GMMs), leveraging cluster distance features for data enhancement and significantly improving the performance of detection algorithms.

However, most existing methods focus primarily on distinguishing anomalous states from the overall data distribution without sufficiently addressing intraclass feature variations. Such variations can obscure subtle anomalies, thereby limiting the accuracy of feature modelling. Moreover, the abundance of normal samples in industrial datasets is often underutilized, leading to knowledge loss and suboptimal detection performance. To address these challenges, this paper proposes a feature correlation analysis framework based on a cluster-generation graph—a graph structure derived from clustering results that models local sample relationships. The proposed method first partitions the dataset via distance-based clustering and then constructs cluster-generation graphs by combining representative normal samples with the remaining data points. These graphs are processed through an autocorrelation network and a graph convolutional network (GCN) to extract both temporal and structural dependencies, followed by linear projection to generate anomaly aware feature representations for downstream detection tasks

The key contributions of this study can be outlined as follows:

- A cluster-generation graph-based feature correlation model is constructed by partitioning the dataset via K-means and building hierarchical graphs to capture structural dependencies.
- A feature enhancement framework that integrates AutoCorrelation and GNN modules to jointly encode temporal and structural features.
- The effective utilization of abundant normal samples through basic and extended graph construction enhances robustness under class imbalance.
- Comprehensive experiments on multiple public datasets and a real-world industrial equipment dataset are conducted, highlighting the improved accuracy and robustness achieved by the proposed method in anomaly detection tasks.

2 Related Work

2.1 Clustering Algorithms for Anomaly Detection

Clustering algorithms, as a form of unsupervised learning, have been widely applied in anomaly detection tasks. By partitioning data samples, clustering can effectively identify outliers or isolated points, often serving as a preprocessing or auxiliary step for anomaly detection. Classical clustering methods such as K-means are frequently used for preliminary grouping or outlier screening because of their simplicity and efficiency. Li et al. [13] introduced an fuzzy C-means clustering to uncover latent structures within multivariate time series data, facilitating anomaly pattern detection. Jain et al. [8] employed the k-means method to reduce the number of captured network traffickers, preserving critical anomalous features while minimizing the data volume. Ripan et al. [17] utilized K-means clustering to form sample clusters and detect anomalies through cluster analysis.

2.2 AutoCorrelation Networks in Feature Analysis

The autocorrelation mechanism has recently attracted considerable attention in feature modelling, particularly in time series tasks. Typically, autocorrelation networks are capable of capturing periodicities and delayed dependencies within individual samples, thereby enhancing feature representation. Autocorrelation originated from time series analysis in the field of statistics, first proposed by

Yule [26], and has been widely applied for detecting and modelling periodic and lagged patterns. In the domain of industrial anomaly detection, Saci et al. [18] utilized vibration measurements collected from multiple embedded sensors and applied the autocorrelation function (ACF) to calculate temporal correlations for detecting anomalies during the operation of steelmaking furnaces. In recent years, Wu et al. [21] introduced AutoCorrelation into the transformer architecture in their Autoformer model, marking the first explicit replacement of traditional point-to-point self-attention with an autoCorrelation mechanism, significantly improving the performance of time series modelling and expanding the application of AutoCorrelation in deep feature modelling.

2.3 Graph Neural Networks for Anomaly Detection

GNNs have emerged as powerful tools for relational modelling and have been widely used to analyse complex data structure features. By aggregating and propagating features among nodes and their neighborhoods, GNNs effectively capture both local and global feature interactions, thereby enhancing the expressiveness and discriminative ability of feature representations. In tasks such as anomaly detection, node classification, and clustering, GNN-based feature analysis methods have demonstrated superior performance. Xie et al. [24] proposed an anomaly detection method based on graph neural networks and dynamic thresholding (GNN-DTAN), aiming to increase the stability and anomaly response capabilities of satellite power systems. Wu et al. [23] applied GNNs to anomaly detection tasks in IIoT-supported scenarios, significantly improving the detection accuracy and efficiency of anomalous states in complex systems.

Although clustering algorithms, autocorrelation mechanisms, and graph neural networks (GNNs) have each demonstrated promising capabilities in detecting abnormal patterns, their application remains limited when facing the complex nature of industrial data. Clustering is often confined to basic partitioning and lacks structural refinement within clusters. AutoCorrelation methods typically operate on isolated sequences, overlooking intersample temporal consistency. Moreover, GNNs commonly rely on static graphs, which struggle to capture dynamically evolving relationships across samples. These limitations highlight the need for an integrated approach that jointly models clustering structures, temporal dependencies, and graph-based feature propagation. In response, we propose a novel feature correlation framework based on cluster-generation graphs that unifies these components into a coherent, data-driven representation scheme for enhanced anomaly detection.

3 Methodology

This paper proposes an unsupervised cluster-generation graph learning method to enhance feature modelling capabilities for industrial anomaly detection, Fig. 1 shows the architecture of the model. The overall framework consists of the following main steps:

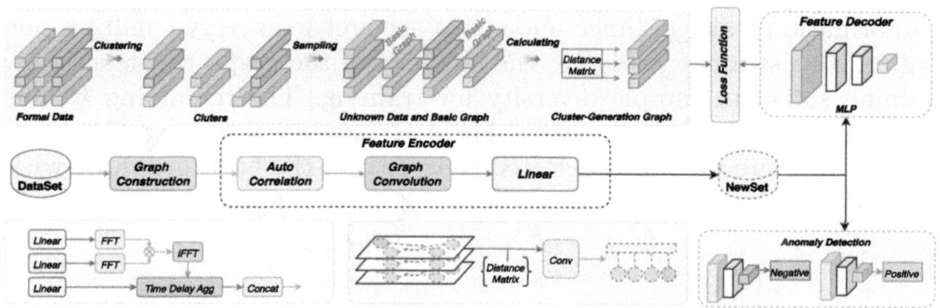

Fig. 1. The overall architecture of the CAGN consists of three main components: cluster-generation graph construction, a feature encoder, and a feature decoder.

First, the original dataset is partitioned into multiple clusters via the k-means algorithm. For each cluster containing anomalous samples, normal instances are first extracted and then reclustered via a secondary clustering procedure, where the centroids of the resulting subclusters serve as nodes to construct the basic graph, whereas the remaining samples are combined with the basic graph to generate a set of cluster-generation graphs, with the distances between samples and basic graph nodes used as edge weights. For each cluster-generation graph, an autocorrelation network is subsequently employed to analyse the feature correlations among nodes and extract local correlation patterns. Then, a GCN layer is utilized to integrate distance features with the autocorrelated features, further enhancing node representations. Finally, a linear compression layer is applied to reduce the dimensionality, producing compact representations for downstream anomaly detection. By combining distance-based clustering partitioning, node correlation modelling, and graph-based feature enhancement, the proposed method effectively improves the distinguishability between normal and anomalous samples.

3.1 Cluster-Generation Graph Construction

Figure 2 shows the steps of cluster-generation graph construction. For the given dataset $D = \{x_i\}_{i=1}^{N}$, the k-means algorithm is first employed for clustering. The k-means algorithm seeks to optimize clustering by reducing the overall intracluster variance, which is defined as the cumulative squared distance from each sample to its assigned cluster center:

$$\min_{\{C_j\}_{j=1}^{k}} \sum_{j=1}^{k} \sum_{x_i \in C_j} \|x_i - \mu_j\|_2^2 \tag{1}$$

where C_j denotes the j-th cluster and where μ_j represents the centroid of cluster C_j.

After clustering, only clusters that contain no anomalous samples are retained for further processing, and any clusters with known anomalies are excluded to avoid information leakage. Since anomalies account for a very small proportion of the data, this strategy preserves the integrity of the graph construction while maintaining sufficient sample diversity for training. The remaining k' clusters are preserved for further processing. For each preserved cluster C_j, c normal samples $\{x_p\}_{p=1}^{c}$ are randomly selected to construct the basic graph G, with the vertex set defined as:

$$G = (V, E), \quad V = \{x_p\}_{p=1}^{c} \tag{2}$$

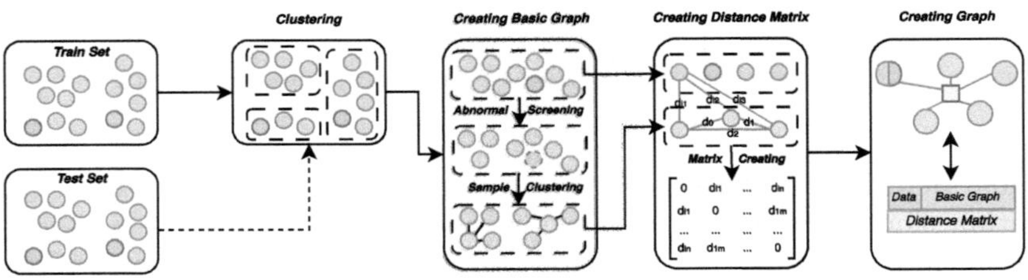

Fig. 2. Illustration of the cluster-generation graph construction process, which consists of four main steps: clustering, basic graph creation, distance matrix calculation, and graph generation.

Each remaining sample x_q ($x_q \notin V$) is combined with the basic graph to form a cluster-generation graph $\tilde{G}$, whose node set is as flollows:

$$\tilde{V} = V \cup \{x_q\} \tag{3}$$

Within each cluster-generation graph $\tilde{G}$, the pairwise Euclidean distances among nodes are computed to construct the distance matrix M, which is given by:

$$M(i, j) = \|x_i - x_j\|_2, \quad x_i, x_j \in \tilde{V} \tag{4}$$

where $M(i, j)$ denotes the Euclidean distance between nodes i and j, which serves as the corresponding edge weight in the graph.

Notably, when a significant imbalance exists between the numbers of normal and anomalous samples within a cluster, undersampling of normal samples can be applied. Specifically, given the set of normal samples P and anomalous samples N in a cluster, a subset $P' \subset P$ is randomly selected such that:

$$|P'| = \alpha \cdot |N|, \quad \alpha > 1 \tag{5}$$

where α is a predefined undersampling factor. This strategy balances the distribution between normal and anomalous samples, mitigating overfitting and reducing learning pressure during model training.

Through this cluster-generation graph construction approach, the local cluster structure is preserved while feature differences between normal and anomalous samples are explicitly introduced, providing a foundation for subsequent feature correlation modelling and enhancement.

3.2 Feature Correlation Modelling and Reconstruction

An autoencoder-based feature analysis and reconstruction module is introduced to enhance node representations after constructing the cluster-generation graph. The framework comprises an encoder and a decoder. The encoder first applies an autocorrelation Layer to capture local correlation patterns among nodes, followed by a GNN to integrate distance and autocorrelated features. A linear layer then compresses the fused features into low-dimensional embeddings. The decoder uses an MLP to reconstruct the original features, preserving key information while improving anomaly sensitivity. This design effectively learns anomaly aware representations by leveraging reconstruction loss, which benefits subsequent anomaly detection.

Encoder Architecture. The encoder consists of three main components: an autocorrelation layer, a graph convolutional network (GCN) layer, and a linear projection layer for dimensionality reduction. Its goal is to effectively capture delayed dependencies among node features, propagate structural information, and compress features into a low-dimensional space.

Let $X \in \mathbb{R}^{L \times C}$ represent the input node feature matrix, where L corresponds to the number of nodes and where C corresponds to the feature dimensionality. The encoding process proceeds through the following steps:

AutoCorrelation Layer. Given an input matrix $\mathbf{X} \in \mathbb{R}^{L \times C}$, we first construct the query, key, and value representations via linear projections:

$$\mathbf{Q} = \mathbf{X}\mathbf{W}_Q, \quad \mathbf{K} = \mathbf{X}\mathbf{W}_K, \quad \mathbf{V} = \mathbf{X}\mathbf{W}_V, \quad \mathbf{Q}, \mathbf{K}, \mathbf{V} \in \mathbb{R}^{L \times d}, \tag{6}$$

where L denotes the number of nodes or the sequence length and where d refers to the dimensionality of the embedding space.

Let $\mathcal{F} : \mathbb{R}^{L \times d} \to \mathbb{C}^{L \times d}$ denote the discrete Fourier transform. We define the autocorrelation function over all temporal lags $\tau \in \{0, \ldots, L-1\}$ as:

$$\mathcal{C}(\tau) = \mathcal{F}^{-1}\left(\mathcal{F}(\mathbf{Q}) \odot \overline{\mathcal{F}(\mathbf{K})}\right)_{(\cdot,\tau)} \in \mathbb{R}^{L \times d}, \tag{7}$$

where $\odot$ denotes the elementwise complex Hadamard product, $\overline{(\cdot)}$ indicates complex conjugation, and $\mathcal{C}(\tau)$ denotes the inverse FFT slice along temporal lag τ.

For each time step t, we define a top-k temporal index set $\mathcal{I}_t = \mathrm{TopK}_\tau\left(\mathcal{C}_t(\tau)\right)$. The autocorrelation attention weights are computed as:

$$\alpha_{t,\tau} = \frac{\exp\left(\mathcal{C}_t(\tau)\right)}{\sum_{\tau' \in \mathcal{I}_t} \exp\left(\mathcal{C}_t(\tau')\right)}, \quad \tau \in \mathcal{I}_t. \tag{8}$$

The output embedding at position t is as follows:

$$\mathbf{Y}_t = \sum_{\tau \in \mathcal{I}_t} \alpha_{t,\tau} \cdot \mathbf{V}_{(t-\tau) \bmod L}. \tag{9}$$

To generalize the discrete aggregation into a continuous kernel-based formulation, we express the output as a kernel-weighted convolution over periodic offsets:

$$\mathbf{Y}_t = \int_{\tau=0}^{L} \kappa(\tau; t) \cdot \mathbf{V}_{(t-\tau) \bmod L} \, d\tau, \tag{10}$$

where $\kappa(\tau; t)$ is a normalized attention kernel derived from $\mathcal{C}_t(\tau)$, potentially approximated via learned basis functions or truncated Gaussians.

This formulation allows the autocorrelation layer to model high-order cyclic dependencies in a nonlocal yet structured manner, enabling efficient and interpretable correlation inference in both the temporal and structural domains.

Compared with attention-based mechanisms, AutoCorrelation offers several advantages for modelling temporal patterns in industrial anomaly detection. It provides a lightweight alternative that captures periodic and delayed dependencies without introducing the parameter overhead and complexity of multi-head attention. This makes it particularly suitable for real-time or resource-constrained scenarios.

Graph Convolutional Layer. Consider an undirected graph $\mathcal{G} = (\mathcal{V}, \mathcal{E})$ consisting of $N = |\mathcal{V}|$ nodes. The adjacency matrix is denoted by $\mathbf{A} \in \mathbb{R}^{N \times N}$, and the input feature tensor for a mini-batch of B graphs is represented as $\mathbf{X} \in \mathbb{R}^{B \times N \times F}$, where each node possesses F-dimensional attributes.

The adjacency matrix is symmetrically normalized with the inclusion of self-loops and is defined as follows:

$$\hat{\mathbf{A}} = \mathbf{A} + \mathbf{I}, \quad \hat{\mathbf{D}}_{ii} = \sum_j \hat{\mathbf{A}}_{ij}, \quad \tilde{\mathbf{A}} = \hat{\mathbf{D}}^{-1/2} \hat{\mathbf{A}} \hat{\mathbf{D}}^{-1/2}. \tag{11}$$

The feature representation at the l-th layer is expressed as $\mathbf{H}^{(l)} \in \mathbb{R}^{B \times N \times d_l}$. The corresponding propagation mechanism for the $(l+1)$-th graph convolutional layer is formulated as follows:

$$\mathbf{H}^{(l+1)} = \sigma \left(\tilde{\mathbf{A}} \mathbf{H}^{(l)} \mathbf{W}^{(l)} + \mathbf{b}^{(l)} \right), \tag{12}$$

where $\mathbf{W}^{(l)}$ and $\mathbf{b}^{(l)}$ denote the learnable weight and bias at layer l, respectively, with $\sigma(\cdot)$ being a nonlinear function applied to each element, such as the ReLU or GELU.

To enhance model expressiveness and training stability, we further incorporate layer normalization and residual connections as follows:

$$\mathbf{Z}^{(l+1)} = \text{LayerNorm} \left(\mathbf{H}^{(l+1)} \right), \quad \mathbf{H}^{(l+1)} := \alpha \cdot \mathbf{Z}^{(l+1)} + (1 - \alpha) \cdot \mathbf{H}^{(l)}, \tag{13}$$

where $\alpha \in [0, 1]$ is a learnable residual scaling parameter.

For multiple stacked GCN layers (L in total), the final output is obtained via recursive propagation:

$$\mathbf{H}^{(L)} = \text{GCN}^{(L)} \circ \cdots \circ \text{GCN}^{(1)} (\mathbf{X}), \tag{14}$$

and passed into downstream modules (e.g., pooling, classification, or anomaly detection heads).

Decoder Architecture. After the encoding stage, a decoder module is introduced to recover the original node features from the low-dimensional embeddings Z. This reconstruction process serves to enforce the preservation of anomaly relevant patterns by minimizing the reconstruction loss.

The decoder adopts a multilayer perceptron (MLP) architecture, which is composed of multiple fully connected layers interleaved with nonlinear activation functions. It receives as input the compressed embedding $Z \in \mathbb{R}^{L \times C''}$ generated by the encoder.

The reconstruction operation is mathematically defined as follows:

$$\hat{X} = \mathrm{MLP}(Z) \tag{15}$$

where $\hat{X} \in \mathbb{R}^{L \times C}$ denotes the reconstructed node feature matrix, which aims to approximate the original input X.

The MLP consists of several hidden layers activated by functions such as ReLU, which allows it to capture nonlinear transformations from the latent representation to the original feature domain.

To promote effective differentiation of normal and anomalous patterns, the decoder is trained to reconstruct the original input, which implicitly drives the encoder to preserve meaningful feature representations.

3.3 Anomaly Detection Framework

After feature extraction through the encoder, we utilize a deep neural network (DNN) classifier to perform anomaly detection. The DNN is designed to perform binary classification by separating normal and anomalous instances via the compact representations obtained from the encoder.

Classifier Architecture. The input to the DNN is the low-dimensional embedding $Z \in \mathbb{R}^{L \times C''}$ generated by the encoder. The DNN adopts a multilayer perceptron (MLP) architecture composed of multiple dense layers, each followed by nonlinear activation functions such as ReLU.

The classification process can be expressed as follows:

$$\hat{y} = \mathrm{DNN}(Z) \tag{16}$$

where $\hat{y} \in \mathbb{R}^{L \times 1}$ denotes the predicted anomaly score or binary classification output for each node.

Specifically, the DNN contains h hidden layers, each of which performs the following transformation:

$$Z^{(l+1)} = \sigma(Z^{(l)} W^{(l)} + b^{(l)}), \quad l = 0, 1, \ldots, h-1 \tag{17}$$

where $Z^{(0)} = Z$, $W^{(l)}$ and $b^{(l)}$ are the learnable weights and biases of layer l, respectively, and where $\sigma(\cdot)$ denotes a nonlinear activation function.

The final output layer applies a sigmoid activation function for binary classification:

$$\hat{y} = \text{Sigmoid}(Z^{(h)}W^{(h)} + b^{(h)}) \tag{18}$$

By combining feature reconstruction and supervised anomaly classification, the model learns both general structural patterns and anomaly specific discriminative features, thereby improving detection performance.

4 Experiments

This section presents a comprehensive experimental evaluation of the proposed method. We begin by describing the datasets and baseline approaches used for comparison and subsequently detail the implementation settings of our experiments.

4.1 Datasets and Baselines

We evaluate our model on four datasets, covering both public benchmarks and real-world industrial data. The Secom dataset [16] from the UCI repository contains sensor readings from semiconductor manufacturing. The Wafer dataset [9] includes high-dimensional time series data for wafer defect detection. The CNC dataset [19] simulates machine faults via injected anomalies, whereas the IMM dataset records 18-dimensional sensor features per cycle from injection molding machines, with labelled normal and abnormal states.

To validate our approach, we compare it with several representative baselines, including both classical and deep models. IF [14] and KNN [4] are unsupervised methods based on statistical isolation and distance metrics, respectively. DAGMM [27] combines autoencoder and Gaussian mixture modelling in the latent space, while DNN [12] serves as a supervised neural network trained on labelled data. We further include state-of-the-art GNN-based models: DOMINANT [5] (graph autoencoder), AnomalyDAE [6] (denoising autoencoder), GUIDE [25] (graph attention), and OCGNN [20] (one-class GNN). These models provide strong baselines for evaluating the structural modelling capability of our clustering-enhanced graph learning framework.

4.2 Implementation Details

All the experiments are implemented via Python 3.9 in a Windows 11 environment with Anaconda. Training and evaluation were conducted on a laptop equipped with an AMD Ryzen 7 8845H CPU, an NVIDIA GeForce RTX 4060 laptop GPU (8 GB), and 16 GB of DDR5 RAM. Model training is performed via the Adam optimizer with an initial learning rate of 0.001. To mitigate overfitting, early stopping is applied based on the basis of the validation loss. The loss weighting coefficient λ within the training objective is fixed at 1.0. For all the baseline methods, we use their official default parameter settings as provided by their respective implementations. No additional hyperparameter tuning is performed to ensure a fair and consistent comparison across methods.

Owing to the presence of significant missing values in the Secom and Wafer datasets, we applied a preprocessing step before training. Specifically, features with more than 90% missing values were removed. For features with less than 10% missing values, we used mean imputation to fill in the missing entries.

4.3 Data Distribution Analysis

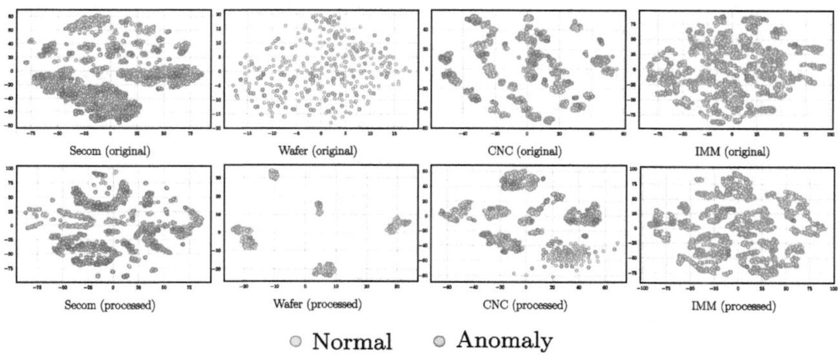

Fig. 3. t-SNE visualizations of the Secom, Wafer, CNC, and IMM datasets before and after model processing. Each point denotes a data sample, with green indicating normal samples and red indicating anomalies. Compared with the original input, the processed representations generated by the proposed model, show better structural separability and anomaly isolation, facilitating more accurate anomaly detection. (Color figure online)

As illustrated in Fig. 3, the t-SNE projections provide a comparative visualization of sample distributions before and after model processing across the Secom, Wafer, CNC, and IMM datasets. Prior to model processing, the distributions exhibit significant overlap between normal and anomalous samples, with ambiguous boundaries and irregular clustering patterns. This reflects the intrinsic complexity of industrial anomaly detection, where anomalies often resemble subtle variations in normal operations rather than distinct outliers. In contrast, the processed representations reveal more compact and structured clusters, with anomalous samples gradually pushed towards peripheral or isolated regions in the latent space. This transformation highlights the model's ability to reshape feature representations in a way that amplifies anomaly specific deviations while preserving the global structure of normal behaviors. This visualization not only offers qualitative evidence of the representation refinement effect achieved by our method but also lays the groundwork for the following quantitative comparison. The diversity across datasets—in terms of feature dimension, noise level, and anomaly sparsity—further emphasizes the importance of learning effective representations tailored to different application domains.

4.4 Comparison Experiments and Analysis

Table 1. Comparison of performance metrics between baseline methods and the proposed approach on four benchmark datasets. The values in bold denote improvements over the corresponding baselines.

Method	Metric	Secom		Wafer		CNC		IMM	
		Origin	CAGN	Origin	CAGN	Origin	CAGN	Origin	CAGN
IF	Accuracy	**0.5120**	0.4879	**0.4877**	0.4561	**0.4962**	0.5037	0.5739	**0.6173**
	Precision	0.0388	**0.0873**	**0.1538**	0.0979	0.1450	**0.1600**	**0.3508**	0.2280
	Recall	**0.6666**	0.4285	**0.4680**	0.3500	0.4915	**0.5161**	0.6250	**1.0000**
	F1 Score	0.0733	**0.1451**	**0.2315**	0.1530	**0.2239**	0.2442	**0.4494**	0.3714
	AUC Score	**0.5098**	0.4859	**0.4888**	0.4574	0.4971	**0.5046**	0.5719	**0.6140**
KNN	Accuracy	0.7487	**0.7681**	**0.8315**	0.8245	0.8045	**0.8446**	0.9130	**0.9304**
	Precision	0.6893	**0.7378**	**0.7902**	0.7832	0.8200	**0.8850**	**0.9473**	0.9298
	Recall	0.7802	**0.7835**	**0.8625**	0.8549	0.7961	**0.8194**	0.8852	**0.9298**
	F1 Score	0.7319	**0.7600**	**0.8248**	0.8175	0.8078	**0.8509**	0.9152	**0.9298**
	AUC Score	0.7485	**0.7679**	**0.8317**	0.8247	0.8044	**0.8445**	0.9133	**0.9304**
DNN	Accuracy	0.7439	**0.7777**	0.8561	**0.8631**	0.8345	**0.8471**	0.9130	**0.9474**
	Precision	**0.7572**	0.7378	**0.9020**	0.8951	**0.8750**	0.8450	0.8771	**0.8947**
	Recall	0.7358	**0.8000**	0.8269	**0.8421**	0.8101	**0.8492**	0.9433	**1.0000**
	F1 Score	0.7464	**0.7676**	0.8628	**0.8677**	0.8413	**0.8471**	0.9090	**0.9444**
	AUC Score	0.7444	**0.7775**	0.8559	**0.8630**	0.8344	**0.8471**	0.9127	**0.9473**
DAGMM	Accuracy	0.3205	**0.3974**	0.3209	**0.4326**	0.3933	**0.4267**	0.5632	**0.5977**
	Precision	0.3205	**0.6471**	0.4634	**0.7234**	0.6324	**0.7121**	0.9545	**1.0000**
	Recall	0.1250	**0.2111**	0.1329	**0.2378**	0.2150	**0.2350**	0.3621	**0.3966**
	F1 Score	0.1970	**0.3188**	0.2065	**0.3579**	0.3209	**0.3534**	0.5250	**0.5679**
	AUC Score	0.4895	**0.5353**	0.4739	**0.5333**	0.4871	**0.5390**	0.8532	**0.8811**

This section presents a comparative analysis of the proposed method against representative baseline models across four industrial datasets, as shown in Table 1. Overall, the CAGN method consistently improves performance across most metrics and datasets. Among all baseline configurations, the integration of the CAGN with the DNN classifier yields the best overall results. For example, on the IMM dataset, DNN+CAGN achieves an F1 score of 0.9444 and perfect recall, indicating superior anomaly detection capability in real-world settings. The improvements are particularly pronounced for the DAGMM, where the CAGN enhances both precision and recall across all datasets. This demonstrates the benefit of incorporating graph-based structural modelling and autocorrelation analysis into generative frameworks. In contrast, traditional distance-based methods such as KNN and IF show unstable performance on noisy datasets such as Secom and Wafer, with limited gains from CAGN. Nonetheless, CAGN still provides consistent enhancements, confirming its robustness across diverse industrial scenarios.

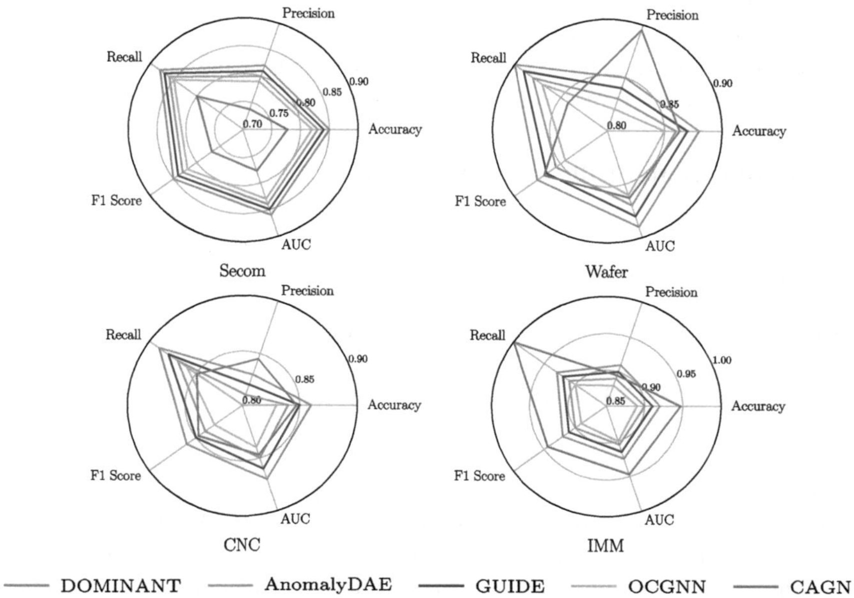

Fig. 4. Radar plot of the multimetric performance of GNN-based anomaly detection methods on industrial datasets.

To further validate the robustness and generalizability of the proposed model, we conduct comparative experiments with several representative and widely-used graph-based anomaly detection methods. The results on the on four benchmark datasets in Fig. 4 show that the CAGN maintains strong adaptability and achieves consistently competitive performance under diverse data conditions. On the IMM dataset, the CAGN delivers the most notable improvements, as its cluster-generation and autocorrelation mechanisms effectively capture repetitive operational structures and temporal consistency. For the Wafer dataset, which contains gradual sensor trends and moderate variation, the CAGN performs robustly across all the metrics, demonstrating its strength in modelling local correlations. In the CNC dataset, where abrupt changes and multimodal dynamics are more prevalent, the performance gap between the CAGN and other methods becomes decreases, yet the CAGN still remains comparable to other GNN-based baselines. The Secom dataset from the UCI repository poses greater challenges because of its high dimensionality, missing values, and severe class imbalance. Although structural modelling is more difficult in this setting, the CAGN still yields reasonable results, showing its resilience in less structured and noisy scenarios. Overall, the CAGN proves effectively captures latent structural information across a variety of industrial data distributions.

The hyperparameters used in our experiments are selected on the basis of performance on a held-out validation set and remain fixed across all datasets. We observe that the proposed model consistently achieves competitive results without requiring extensive tuning, suggesting that its performance is more influenced by the underlying characteristics of the dataset rather than sensitivity to specific hyperparameter values.

4.5 Computational Efficiency Analysis

Table 2. Throughput comparison of encoding and detection modules (samples/s).

Auto Correlation	GCN	DNN	DOMIN ANT	Anomaly DAE	OCGNN	GUIDE
12806.2	12815.9	2537.8	751.7	695.4	4284.3	2.6

To evaluate the runtime efficiency of the proposed framework, we measured the throughput (samples per second) of its main components. As shown in Table 2, clustering and graph construction are necessary steps but are computationally lightweight. These can be executed in parallel on edge devices with limited resources and do not constitute a practical bottleneck.

We focus our analysis on three core modules: the autocorrelation layer, the GCN encoder, and the DNN-based anomaly detector. For comparison, we also test three GNN-based models: DOMINANT, AnomalyDAE, and OCGNN. The results show that the AutoCorrelation and GCN layers reach more than 12,800 samples per second, indicating high encoding efficiency. The DNN classifier is slower at 2,537.8 samples per second but is still sufficient for industrial use. Among the baselines, DOMINANT and AnomalyDAE have lower throughputs because of their deeper architectures. The OCGNN achieves higher throughput at 4,284.3 samples per second, which is slightly above that of our DNN module. However, as shown earlier, OCGNN lacks the stability and generalizability of our model, making our trade-off in speed worthwhile.

These findings confirm that our method balances detection accuracy and computational efficiency well, making it practical for deployment in industrial environments.

4.6 Ablation Study

To evaluate the individual contributions of the core components in the proposed feature enhancement framework, we conduct an ablation study focusing on the autocorrelation (AC) mechanism and the GNN-based cluster-generation graph structure. Specifically, we compare three model variants: (1) AC only,

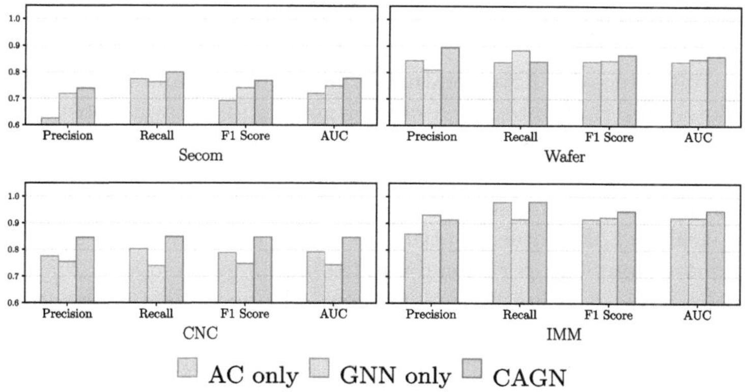

Fig. 5. Ablation study results on the four datasets. The performance of three configurations is compared: (1) using only the autocorrelation (AC) module, (2) using only the graph neural network (GNN) module, and (3) using the complete CAGN model.

which captures temporal dependencies via the autocorrelation layer while discarding the graph structure; (2) GNN only, which retains the structural graph but omits autocorrelation modelling; and (3) the full CAGN model, which integrates both components for joint temporal and structural feature enhancement. The corresponding results for all four datasets are presented in Fig 5. Across all the datasets and evaluation metrics, the complete CAGN model consistently achieves the best performance, confirming the complementary strengths of the two modules.

The results also reveal dataset-specific trends. For the CNC dataset, which contains densely sampled time series signals from continuous machine operations, the *AC only* variant significantly outperforms the *GNN only* variant, especially in terms of the recall and F1 score. This highlights the effectiveness of temporal feature modelling in data with strong sequential patterns. In contrast, for the Secom, Wafer, and IMM datasets, the performance of the *GNN only* variant is closer to that of the full model, whereas *AC only* lags behind. These datasets likely exhibit weaker temporal consistency, and thus benefit more from structural relationships modelled through graph representations.

In summary, the ablation study validates the individual effectiveness and dataset-dependent advantages of both components. The autocorrelation module is particularly beneficial for capturing temporal dependencies in sequence-rich data, whereas the GNN-based structure offers robust generalizability across diverse data topologies. Their combination in the proposed CAGN framework enables comprehensive modelling of both local and sequential feature patterns, resulting in significantly improved anomaly detection performance across heterogeneous industrial datasets.

5 Conclusion

This study introduces a novel feature enhancement framework for industrial anomaly detection that integrates cluster-generation graph construction with autocorrelation-based feature modelling. The proposed approach demonstrates stable performance gains across multiple datasets and strong applicability to real world industrial monitoring scenarios. Improving feature representation quality and the separability of samples contributes to enhanced robustness and detection precision under complex operational conditions.

For future work, we plan to explore more advanced clustering techniques for cluster-generation graph construction, design lightweight autocorrelation mechanisms to further improve computational efficiency, and investigate adaptive graph structures that can dynamically model evolving industrial data streams. In particular, extending the CAGN to support temporal or dynamic graph modelling is a promising direction for handling real-time IIoT scenarios with continuous state changes.

References

1. Ahmed, I., Ahmad, M., Chehri, A., Jeon, G.: A smart-anomaly-detection system for industrial machines based on feature autoencoder and deep learning. Micromachines **14**(1), 154 (2023)
2. Anton, S.D.D., Sinha, S., Schotten, H.D.: Anomaly-based intrusion detection in industrial data with svm and random forests. In: 2019 International Conference on Software, Telecommunications and Computer Networks (SoftCOM), pp. 1–6. IEEE (2019)
3. Chen, H., Ma, H., Chu, X., Xue, D.: Anomaly detection and critical attributes identification for products with multiple operating conditions based on isolation forest. Adv. Eng. Inf. **46**, 101139 (2020)
4. Cover, T., Hart, P.: Nearest neighbor pattern classification. IEEE Trans. Inf. Theory **13**(1), 21–27 (1967)
5. Ding, K., Li, J., Bhanushali, R., Liu, H.: Deep anomaly detection on attributed networks. In: Proceedings of the 2019 SIAM International Conference on Data Mining, pp. 594–602. SIAM (2019)
6. Fan, H., Zhang, F., Li, Z.: Anomalydae: Dual autoencoder for anomaly detection on attributed networks. In: ICASSP 2020-2020 IEEE International Conference on Acoustics, Speech and Signal Processing (ICASSP), pp. 5685–5689. IEEE (2020)
7. Givnan, S., Chalmers, C., Fergus, P., Ortega-Martorell, S., Whalley, T.: Anomaly detection using autoencoder reconstruction upon industrial motors. Sensors **22**(9), 3166 (2022)
8. Jain, M., Kaur, G., Saxena, V.: A k-means clustering and svm based hybrid concept drift detection technique for network anomaly detection. Expert Syst. Appl. **193**, 116510 (2022)
9. Khan, A.: Anomaly detection in wafer manufacturing. Kaggle (2024), https://www.kaggle.com/datasets/arbazkhan971/anomaly-detection, Accessed 30 April 2024
10. Kim, S., Jo, W., Shon, T.: Apad: Autoencoder-based payload anomaly detection for industrial IoE. Appl. Soft Comput. **88**, 106017 (2020)

11. Laskar, M.T.R., et al.: Extending isolation forest for anomaly detection in big data via k-means. ACM Trans. Cyber-Phys. Syst. (TCPS) **5**(4), 1–26 (2021)
12. LeCun, Y., Bengio, Y., Hinton, G.: Deep learning. Nature **521**(7553), 436–444 (2015)
13. Li, J., Izakian, H., Pedrycz, W., Jamal, I.: Clustering-based anomaly detection in multivariate time series data. Appl. Soft Comput. **100**, 106919 (2021)
14. Liu, F.T., Ting, K.M., Zhou, Z.H.: Isolation forest. In: 2008 Eighth IEEE International Conference on Data Mining, pp. 413–422. IEEE (2008)
15. MacQueen, J.: Some methods for classification and analysis of multivariate observations. In: Proceedings of the Fifth Berkeley Symposium on Mathematical Statistics and Probability, Volume 1: Statistics. vol. 5, pp. 281–298. University of California press (1967)
16. McCann, M., Johnston, A.: SECOM. UCI Machine Learning Repository (2008). https://doi.org/10.24432/C54305
17. Ripan, R.C., et al.: A data-driven heart disease prediction model through k-means clustering-based anomaly detection. SN Comput. Sci. **2**(2), 112 (2021)
18. Saci, A., Al-Dweik, A., Shami, A.: Autocorrelation integrated gaussian based anomaly detection using sensory data in industrial manufacturing. IEEE Sens. J. **21**(7), 9231–9241 (2021)
19. Tnani, M.A., Feil, M., Diepold, K.: Smart data collection system for brownfield cnc milling machines: a new benchmark dataset for data-driven machine monitoring. Procedia CIRP **107**, 131–136 (2022)
20. Wang, X., Jin, B., Du, Y., Cui, P., Tan, Y., Yang, Y.: One-class graph neural networks for anomaly detection in attributed networks. Neural Comput. Appl. **33**(18), 12073–12085 (2021). https://doi.org/10.1007/s00521-021-05924-9
21. Wu, H., Xu, J., Wang, J., Long, M.: Autoformer: decomposition transformers with auto-correlation for long-term series forecasting. Adv. Neural. Inf. Process. Syst. **34**, 22419–22430 (2021)
22. Wu, H., Hu, J., Tian, X., Xiao, Z., Sun, D., Simon, R.: Graph anomaly detection in programmable logic controllers based on service computing. In: 2024 IEEE International Conference on Web Services (ICWS), pp. 1188–1197. IEEE (2024)
23. Wu, Y., Dai, H.N., Tang, H.: Graph neural networks for anomaly detection in industrial internet of things. IEEE Int. Things J. **9**(12), 9214–9231 (2021)
24. Xie, L., Pi, D., Zhang, X., Chen, J., Luo, Y., Yu, W.: Graph neural network approach for anomaly detection. Measurement **180**, 109546 (2021)
25. Yuan, X., Zhou, N., Yu, S., Huang, H., Chen, Z., Xia, F.: Higher-order structure based anomaly detection on attributed networks. In: 2021 IEEE International Conference on Big Data (Big Data), pp. 2691–2700. IEEE (2021)
26. Yule, G.U.: On a method of investigating periodicities in disturbed series with special reference to wolfer's sunspot numbers. Stat. Papers George Udny Yule **226**, 389–420 (1971)
27. Zong, B., et al.: Deep autoencoding gaussian mixture model for unsupervised anomaly detection. In: International Conference on Learning Representations (2018)

Research on the Precision Prediction of Geospatial Choices for Postgraduate Graduation Destinations via Machine Learning Algorithms

Ling Quan(✉), Diying Li, Sijia Xie, Ting Ye, and Tianyue Tang

University of Electronic Science and Technology of China,
No. 2006, Xiyuan Ave, West Hi-Tech Zone, Chengdu 611731, Sichuan,
People's Republic of China
quanling@uestc.edu.cn

Abstract. With the expansion of China's postgraduate education and the intensification of structural contradictions in regional employment, accurately predicting graduates' career destinations holds significant practical value for optimizing the synergy between talent cultivation and regional strategic demands. This research focuses on master's degree graduates in life sciences from the University of Electronic Science and Technology of China (UESTC), integrating multidimensional feature data, including disciplinary background, place of origin, and research team affiliation, to construct a machine learning-based predictive model for geospatial career choices. To address data sample imbalance, logistic regression and random forest algorithms were used to optimize the model. Model performance was comprehensively evaluated using accuracy, precision, F1 score, and area under the curve (AUC) metrics. The findings suggest that disciplinary specialization and the tier of hometown cities are influencing factors in the decision-making process of life science postgraduates regarding domestic or overseas further education, whereas geospatial preferences for direct employment are jointly shaped by team ecosystems and regional industrial advantages. Innovatively, this research establishes a dynamic "cultivation-employment" prediction system, achieving a paradigm shift from "postemployment feedback" to "precultivation intervention." It provides data-driven decision support for universities to proactively adjust disciplinary structures and for government agencies to formulate regional talent policies.

Keywords: Machine learning · University graduates · Graduation destinations · Precision prediction · Career guidance

1 Introduction

As a core pathway for human capital accumulation, the employment quality and mobility patterns of higher education graduates serve as key indicators for evaluating the educational efficacy and economic vitality of nations [1]. With the deepening implementation of China's innovation-driven development strategy, the need for synergy between

The original version of the chapter has been revised. The correct citation for reference 8 and 9 is updated. A correction to this chapter can be found at
https://doi.org/10.1007/978-981-95-5294-8_44

L. Huang et al. (Eds.): CCF NCCA 2025, CCIS 2774, pp. 458–469, 2026.
https://doi.org/10.1007/978-981-95-5294-8_31

higher education talent cultivation and industrial transformation has become increasingly urgent. Since the comprehensive rollout of the postgraduate enrollment expansion policy in 2017, the postgraduate population of China has grown rapidly. By 2023, the scale of postgraduate admissions in China had exceeded 1.3 million [2], making the employment of such a large postgraduate cohort directly relevant to the rational allocation of social human resources and sustainable economic development. This macrolevel scenario is particularly pronounced in strategic industries. The "14th Five-Year Plan for Bioeconomy Development" mandates the establishment of bioeconomic pilot zones in regions such as Chengdu-Chongqing and the Western Science City, positioning postgraduate students in life sciences as a key force in advancing core technological breakthroughs, aligning with regional industrial clusters, and deepening national health strategic objectives. They represent a strategic talent reserve in the fields of life and health sciences.

However, the expansion of postgraduate education coexists with structural employment contradictions. According to the "2022 China Postgraduate Employment Quality Report", the geographical distribution of postgraduate employment exhibits significant disparities, characterized by "stronger demand in the east and weaker in the central and western regions" and "uneven disciplinary demand", with talent shortages persisting in strategic emerging fields such as life sciences and artificial intelligence [3].Consequently, fostering deep synergy between talent cultivation and regional development, optimizing higher education resource allocation, and refining disciplinary adjustment mechanisms have become focal points for postgraduate education reform in the life and health sciences [4]. With the advancement of the "Education Informatization 2.0 Plan", numerous scholars have leveraged sampling data from multiple universities across various provinces to investigate influencing factors in postgraduate employment [5], applying analytic hierarchy process (AHP) methods to focus particularly on employment quality determinants in first-tier cities [6]. These studies rely on large-scale statistical data and qualitative analysis. Although they have identified macrofactors influencing employment to a certain extent, they lack an in-depth exploration of the psychological mechanisms behind decision-making processes. Career choice is essentially a process in which individuals transform psychological cognition into behavior under environmental constraints, involving implicit psychological concepts such as social identity and self-efficacy [7]. However, the insufficient deep mining of behavioral data from certain groups in existing research has given rise to two critical problems. First, there is a significant disconnection between "group portraits" and "personalized needs". Second, employment guidance remains notably lagging. Moreover, most existing studies focus on developed eastern regions, and lack empirical support for employment laws in the strategic hubs of central and western China. This research excavates group characteristics from variables such as team affiliation and place of origin. It explores how environmental factors, through psychological mechanisms, impact employment choice decisions in the central and western regions. This approach offers a new explanatory path for the disconnection between group analysis and individual needs.

Against this backdrop, this research explores a machine learning-enabled predictive model for employment geographical mobility. Focusing on master's graduates from life science disciplines at UESTC—located in the southwestern region—over the past three years, we integrate enrollment data including disciplinary backgrounds, research team distributions, and hometown origins to analyse the formation mechanisms of employment preferences across different academic teams and identify key factors influencing geographical mobility and industry selection. We subsequently construct a dynamic model of "national strategic demand—team cultivation focus—employment flow prediction". This cultivation-employment collaborative regulatory mechanism, which is based on enrollment data mining, achieves a paradigm shift from "postemployment feedback" to "precultivation intervention", providing a quantifiable and iterative decision-support system to address the structural mismatch between disciplinary supply and strategic demand.

2 Methods

2.1 Experimental Data

The experimental data utilized in this research were sourced from the master's graduate archives of UESTC over the past three years, providing key support for understanding career development trends among master's graduates. The dataset encompasses the employment destination details of 495 master's graduates and integrates multiple key variables (Table 1): basic demographic dimensions, including education level, gender, ethnicity, and political affiliation, to outline individual profiles; academic dimensions, covering student categories and major information to reflect academic trajectories; and graduation destination dimensions, containing detailed classification codes of destinations, the nature of employment units and geographical locations, laying a solid foundation for in-depth analysis of career paths.

Table 1. Student Progression Data

Feature	Definition
Student Category	0, professional master's students;1, academic master's students
Gender	0, Female;1, Male
Ethnic Group	0, ethnic minorities; 1, Han nationality
Political Affiliation	0, Communist Party member; 1, non-Communist Party member
Major	0, Electronic Information; 1, Applied Psychology; 2, Biology; 3, Biomedical Engineering
Team Code	Codes for teams
Province of Origin	The codes range from 0 to 6, increasing from northeastern cities to southwestern cities
City of Origin	0, nonprovincial capital cities; 1, provincial capital cities

(continued)

Table 1. *(continued)*

Feature	Definition
Needy Student Category	0, students with financial difficulties; 1, students without financial difficulties
Pursuing further education domestically	0, pursuing further education domestically;1, pursuing further education abroad
Return to the place of origin for employment	0, work in other places; 1, return to the place of origin for work
Stay in Sichuan for employment	0, work in other enterprises; 1, work in Sichuan
Working in national defense units	0, work in other enterprises; 1, work in national defense units

2.2 Data Preprocessing

The specific details of the data preprocessing used in this research are shown in Fig. 1. First, when some fields in the original dataset had missing values, the mode imputation method was used for repair. As a statistical measure of central tendency in data distribution, the logic of mode imputation conforms to the potential distribution characteristics of original data, which can minimize the introduction of additional bias and thus ensure data quality.

Second, to optimize the data for subsequent model training, normalization becomes a necessary step. By scaling data of different features to a unified interval, normalization can effectively reduce the interference of feature magnitude differences on the model, avoiding the situation where some features dominate the training process due to dimensional advantages and thus improving the stability and efficiency of model training.

Finally, on the basis of the focus needs of differentiated research questions, the dataset was divided into two subsets: "further education" and "employment". The further education dataset focuses on factors affecting students' decisions in domestic or international studies, and the employment dataset screens features, such as "whether returning to hometown for employment" and "whether staying in Sichuan for employment" to explore in depth the complex driving mechanisms behind employment destinations from multiple dimensions.

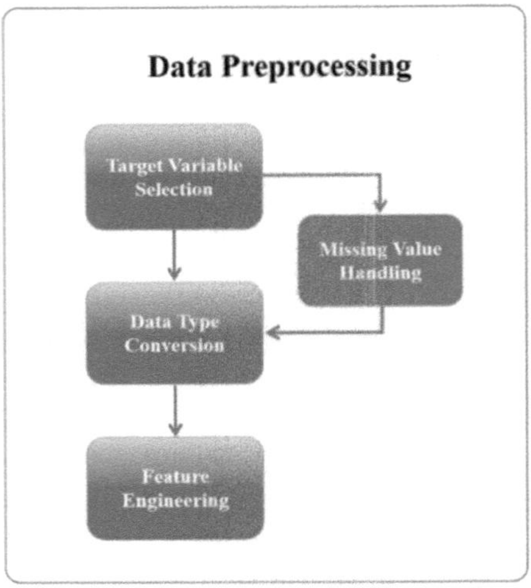

Fig. 1. Data Preprocessing Procedures

2.3 Research Methods

This research employs two machine learning algorithms, logistic regression and random forest, to address three research topics. By comprehensively evaluating various model performance metrics, the most suitable model for characterizing the influence of features on employment destinations is selected.

Logistic Regression. When analysing graduates' career destinations, the variable "graduate destination" contains multiple discrete categories. The logistic regression model is well suited for such prediction tasks, enabling the exploration of latent relationships between features and the target variable [8]. The principle involves building upon linear regression by integrating student characteristics (e.g., gender, ethnicity, hometown information) into a feature vector x, which combines with a weight vector w and bias b to form a linear combination $z = w^t x + b$. Here, the weights reflect intrinsic linear correlations between features and the target variable. Furthermore, the sigmoid function is introduced to map Z to the interval (0,1), which represents the probability of a student belonging to a specific destination category. The classification is determined using a probability threshold of 0.5.

However, given the significant class imbalance in the dataset, we optimize the performance of the logistic regression model through class weight balancing to mitigate this issue. During loss function computation, the algorithm automatically assigns higher weights to minority-class samples, substantially increasing the misclassification cost for these samples. This ensures that the model pays greater attention to underrepresented class features, enabling effective identification of all graduation destination categories and ultimately improving prediction accuracy.

Random Forest. Additionally, the random forest is an ensemble learning model that demonstrates excellent robustness and generalizability for classification tasks such as predicting graduate destinations [9]. It employs bootstrap sampling to repeatedly extract multiple subsets from the original training set with replacement, constructing individual decision trees for each subset. During node splitting in each tree, a random subset of features is selected to increase the diversity among individual trees. The final classification is determined by aggregating the results from all the trees. To identify significant features, one-way analysis of variance (ANOVA) was applied. For categorical features with more than two possible values, groups are formed on the basis of different feature values. The predicted probabilities corresponding to each group are extracted and compiled into lists, which are then subjected to the ANOVA procedure. Furthermore, by comparing the mean differences across groups, the positive or negative direction of feature influence can be determined.

When implementing the random forest algorithm, we address class imbalance through random undersampling, reducing the number of majority class samples to match the scale of minority class samples. This approach preserves sample diversity while avoiding artificial bias. Following this data processing, the constructed model effectively mitigates bias toward the dominant majority class. By ensuring that each decision tree within the model is trained on relatively balanced datasets, it can more accurately discern the distinct characteristics of various graduate destination categories; consequently, the overall predictive accuracy and reliability of the models are significantly enhanced.

2.4 Evaluation Metrics

To comprehensively assess the predictive performance and generalization ability of the models, we employed fivefold cross-validation to test both logistic regression and random forest algorithms, calculating four core metrics: accuracy, precision, F1 Score, and AUC. The formulas are as follows:

$$Accuracy = \frac{TP + TN}{TP + TN + FP + FN} \quad (1)$$

$$Precision = \frac{TP}{TP + FP} \quad (2)$$

$$Recall = \frac{TP}{TP + FN} \quad (3)$$

$$F1 = \frac{2 \times Precision \times Recall}{Precision + Recall} \quad (4)$$

Here, TP represents the number of correctly predicted positive-class samples; FP denotes the number of negative-class samples incorrectly predicted as positive; TN indicates the number of correctly identified negative-class samples; and FN reflects positive-class samples erroneously predicted as negative. The AUC serves as a crucial metric for evaluating classification model performance. It is calculated by plotting the receiver operating characteristic (ROC) curve, which reflects the model's generalization ability under different classification thresholds. The ROC curve's horizontal axis represents the false positive rate (FPR, where FPR = FP/(TN + FP)), indicating the proportion of negative samples incorrectly classified as positive. The vertical axis shows the true positive rate (TPR, where TPR = TP/(TP + FN)), which represents the proportion of correctly identified positive samples.

3 Results

3.1 Model Fitting Results

Through model construction and data processing, we analysed the classification performance of the models for three research topics: "whether to pursue further education domestically", "whether to return to the place of origin for employment", and "whether to stay in Sichuan for employment". Using normalized data, we built logistic regression models for binary or multiclass classification tasks. Simultaneously, one-hot encoding was applied to feature variables to convert categorical variables into vector form, facilitating the construction of random forest models that better handle categorical information. The classification results demonstrate the adaptability of both logistic regression and random forest models to the data, as shown in Table 2.

Table 2. Model evaluation

Research themes	Models	Assessment of indicators			
		Accuracy	Precision	F1	AUC
Whether to pursue further education domestically	Logistic Regression	69.52%	71.63%	67.76%	/
	Random Forest	70.85%	73.03%	67.63%	73.80%
Whether to return to the place of origin for employment	Logistic Regression	70.92%	74.13%	71.25%	/
	Random Forest	73.61%	73.15%	71.15%	76.97%
Whether to stay in Sichuan for employment	Logistic Regression	69.25%	64.00%	65.20%	/
	Random Forest	69.71%	69.38%	66.28%	72.28%

After comprehensive comparison, for all three topics, the random forest model exhibited slightly superior performance across all the metrics compared with the logistic regression model. The random forest model can intuitively present the relationships among relevant influencing factors, with its AUC values particularly demonstrating better comprehensive performance across different classification thresholds.

3.2 Feature Interpretation

In the logistic regression analysis of 'whether to pursue further education domestically' (with domestic coded as 0 and international coded as 1), the weights are shown in Table 3. The major variable has a weight of 0.6518, indicating a significant positive correlation with the target variable and suggesting that academic discipline strongly promotes students' choice of international further education. The city of origin variable has a weight of 0.6384, also demonstrating a significant positive association, revealing that students from first-tier and second-tier cities show different preferences in domestic or international education choices. Conversely, the student category and region of origin variables have weights of -0.2548 and -0.6821, respectively, indicating significant negative correlations. This finding indicates that certain factors restrict professional master's and academic master's students from choosing international further education, whereas the factor of region of origin also exerts a pronounced inhibitory effect on students' decisions regarding international further education.

Table 3. Partial weights of logistic regression (significant)

Weights	Numbers
Major	0.6518
City of Origin	0.6384
Student Category	−0.2548
Region of Origin	−0.6821

When investigating 'whether to return to the place of origin for employment', we employed the random forest model and validated the results via one-way ANOVA. As shown in Table 4, several variables have statistically significant p values.

Specifically, Team 3 (brain connectivity laboratory) has a significant negative influence on the target variable, meaning that students from this team exhibit significantly lower willingness to return to their hometowns for employment than do those from other teams. In contrast, Team 7 (Visual Cognition and Brain Computing Research Team) has a significant positive influence on 'hometown employment,' indicating that members of this team are more inclined to return to their hometowns to start careers. Additionally, Region 4 (Central China) also has a negative influence, suggesting that students from Central China face notable obstacles when considering returning to their hometowns for employment.

Table 4. Analysis of the influence of relevant variables on return to the place of origin for employment'

Variable Name	Influence Direction	P - value
Team 3	Significantly Negative	0.04849
Team 7	Significantly Positive	0.03297
Region of Origin 4	Significantly Negative	0.02813

The analysis of factors influencing 'whether to stay in Sichuan for employment' reveals a relatively complex pattern, with detailed data presented in Table 5. Research has shown that students from Team 0 (brain–computer interaction team) have a significant positive correlation with Sichuan employment retention. Compared with other teams, members of this team demonstrate a stronger preference for seeking career development opportunities within Sichuan Province. Conversely, students from Team 8 (brain imaging and pattern recognition team) exhibit a negative relationship with the target variable, indicating relatively lower willingness to remain in Sichuan for employment. From a regional origin perspective, students originating from the Region of Origin 1 (North China) have a significant positive influence on Sichuan employment retention, suggesting the existence of favourable conditions or inherent tendencies that promote employment retention in Sichuan among North China students. In contrast, Region of Origin 3 (east China) and Region of Origin 4 (central China) demonstrate significant negative influences, with graduates from these areas showing a stronger preference for returning to their home regions or neighboring areas for employment, along with a lower willingness to remain in Sichuan.

Table 5. Influence of related codes on staying for employment in sichuan

Variable Name	Influence Direction	P - value
Team 0	Significantly Positive	0.03227
Team 8	Significantly Negative	0.00751
Region of Origin 1	Significantly Positive	0.03758
Region of Origin 3	Significantly Negative	0.03522
Region of Origin 4	Significantly Negative	0.00228

4 Discussion

This research, which is based on master's degree data from the life and health sciences field at the University of Electronic Science and Technology of China, uses logistic regression and random forest models to reveal the influence mechanisms of multidimensional factors, including disciplinary attributes, geographical characteristics, and team resources, on students' career destinations. The findings provide an important theoretical basis and practical insights for talent cultivation and career guidance in higher education institutions.

The research revealed that the geographical choices for further education among postgraduate students are influenced by three main factors: the market maturity of disciplines, differences in training models, and the resource endowment of hometown cities. First, the market maturity of disciplinary development may significantly affect the choice preferences of students from different majors. The data show that academic discipline significantly promotes students' choice of international further education, meaning that certain disciplines may have more advanced research resources, teaching systems and development prospects abroad. This confirms the leading positions and development potential of different fields domestically and internationally [10, 11]. Second, differences in training models produce significant distribution effects. Owing to the "dual-supervisor system" and practice-oriented curriculum system, professional master's students differ from academic master's students who have continuous research training, showing different tendencies in continuing development on original research platforms and going abroad to enhance applied innovation capabilities. Third, the resource endowment of hometown cities shows tier differences. There are significant differences in overseas study intentions between students from first-tier cities and second-tier cities, with obvious variations from northeast to southwest China. These regional differences may be related to factors such as the distribution density of international education resources, educational concepts, and the accumulation of family cultural capital in different regions, resulting in insufficient understanding and support for overseas study options. In constructing models to explore the relationships between these influencing factors and study intentions, the random forest model has relative advantages, with all indicators being slightly greater than those of logistic regression overall, and can more intuitively show the correlations between influencing factors and study intentions. This may benefit from the random forest model's ability to fully mine feature information, better handle complex feature relationships, and effectively capture the nonlinear influences of factors such as academic disciplines, training models, and tier differences in hometown city resource endowments on students' overseas study intentions, showing more advantages than logistic regression in characterizing such complex relationships.

In terms of employment geographical choice among graduate students, our study reveals that the employment geographic decision-making of graduate students is shaped by the dual-dimensional guidance mechanism of the team ecosystem and regional industrial endowment. First, at the team ecology level, the large international team (Team 3) led by international top brain imaging experts shapes students' global employment vision through transnational cooperation networks and cutting-edge technology training, making them more inclined toward diversified employment regional choices. In contrast, the small team (Team 7) led by domestic experts features close teacher-student relationships, and its strong "family culture" may influence students' willingness to return to their hometowns for employment, significantly enhancing emotional attachment. By satisfying the need for a sense of belonging, such teams reduce psychological migration costs and strengthen the tendency to return to hometown employment. This finding also reveals the key role of emotional identification: as a psychological community, the team plays a core role through the internalization-identification dual-path mechanism, ultimately forming differentiated geographic decision-making patterns. The differentiated allocation of team resource networks also directly affects the employment radius. Advantageous teams (Team 0) with extensive cooperative relationships and reputation at the university's location achieve a high local employment rate, whereas cross-regional industry-university-research collaborative teams (Team 8) help most graduates enter the Yangtze River Delta job market, indicating the criticality of research teams' employment resources and platform construction in attracting and retaining talent. In summary, career geographical choice is essentially a spatial behavior output transformed by team cultural capital and regional industrial endowment through psychological schemas. The team ecosystem shapes students' decision-making cognitive framework through a "culture-resource" transmission chain, while regional industrial advantages provide spatial anchoring coordinates. Together, the two constitute a systematic explanatory mechanism for graduate students' employment geographic decision-making.

Second, regional industrial advantages exert a vortex-like attraction on talent flow. Leveraging their clusters in the biomedical industry, eastern coastal regions sustain a dual pull on professionals in relevant fields by enhancing career advancement prospects and professional efficacy. This industrial geography constructs a cognitive schema of "industrial hierarchy–career success" among graduates, driving their cross-regional employment decisions. Although the southwest region where the research university is located is undergoing industrial upgrading, the developmental gap with the eastern coast steers students from economically developed areas (region of origin 3/4) toward regions offering superior career platforms. Notably, the "talent return" phenomenon among students from the central and western regions follows a unique dual-path psychological adaptation model: after professional competency increases through experience accumulation in the developed east region, these graduates are motivated to return home by a combination of nostalgic identification and perceived regional development efficacy, forming a distinctive cycle of "capacity building - hometown reinvestment." This flow pattern is fundamentally a spatial decision-making framework shaped by industrial gradients and psychological adaptation mechanisms, revealing the intricate interplay between economic geography and psychological geography in talent mobility.

5 Conclusion

Considering the existing development conditions in China's central and western regions, as well as the construction of relevant disciplines, this study focuses on the central and western regions. It strives for comprehensive exploration and practical solutions, employing machine learning methods to reveal the characteristics and psychological motivations of local students' employment choices, thereby filling a gap in regional research. Furthermore, our research is the first to highlight the significant influence of team environments and resources on students' employment decisions, expanding the scope of employment-influencing factors. This study provides new insights for university team development and student career guidance, contributing to resolving the structural mismatch between disciplinary supply and strategic demand.

Although this study achieves certain results against the backdrop of digital transformation in higher education and policy-driven employment research, limitations remain. The representativeness of the sample is constrained by single-institution data, which may affect the generalizability of the conclusions. In terms of data depth and model optimization, the current study lacks sufficient exploration of factors such as internship experiences and career planning. As universities accumulate more multidimensional student data, future research could refine data collection and track the dynamic process of career decision-making to construct more accurate employment prediction models. Methodologically, qualitative approaches such as interviews and case studies could be integrated to analyse the psychological mechanisms behind students' employment decisions, providing more comprehensive evidence for policymaking, and substantively advancing the coordinated development of higher education and regional economies.

Acknowledgments. This research was funded by the Higher Education Talent Cultivation Quality and Teaching Reform Project of Sichuan Province (JG2024--0209) and the Specialized Project of Career Planning and Employment Guidance at the University of Electronic Science and Technology of China (No. JY20250111).

Disclosure of Interests. The authors have no competing interests to declare that are relevant to the content of this article.

References

1. OECD. Education at a Glance 2024[M/OL]. OECD Publishing, Paris (2024). 23 March 2025
2. MyCOS Institute. Blue Book of Employment: 2023 Chinese College Graduates Employment Annual Report [R/OL]. (2023–11–27). 23 March 2025
3. Ministry of Education of the People's Republic of China. 2023 National Education Development Statistics Bulletin [EB/OL]. (2024–10–24). 23 March 2025
4. Xi, J., Xi, J.: Emphasized at the National Education Conference: Closely focusing on the fundamental task of fostering virtue through education, steadily advancing toward the strategic goal of building a strong education nation [EB/OL]. (2024–09–10). 23 March 2025
5. Jinglu, L., Changjun, Y., Haiyan, Y.: Research on Influencing factors of employment outcomes and starting salaries for Chinese master's graduates: empirical analysis based on 2019 national college sampling survey data. J. Educ. Stud. **11**, 64–72 (2020). https://doi.org/10.16477/j.cnki.issn1674-2311.2020.11.009
6. Li, H., Liu, K., Yang, Y., et al.: Research on the employment quality evaluation, influencing factors and promotion ways of college masters in first-tier cities. Northwest Population J. **43**(02), 114–126 (2022). https://doi.org/10.15884/j.cnki.issn.1007-0672.2022.02.010
7. Yan, X., YiMeng, Z., YiLiang, W.: The influence mechanism of core self-evaluation on college students' employment ability from the perspective of vocational psychology. J. Res. Edu. Develop. **45**(05), 77-84 (2025). https://doi.org/10.14121/j.cnki.1008-3855.2025.05.004
8. Elkahwagy, D.M.A.S., Kiriacos, C.J., Mansour, M.: Logistic regression and other statistical tools in diagnostic biomarker studies. Clin. Transl. Oncol. **26**(9), 2172–2180 (2024). https://doi.org/10.1007/s12094-024-03413-8
9. Hu, J., Szymczak, S., Brief Bioinform. A review on longitudinal data analysis with random forest. **24**(2), bbad002 (2023). https://doi.org/10.1093/bib/bbad002
10. Che, Z., Zhang, Y., Wu, M.: Research on development trends in biomedical engineering. Sci. Focus 17(05), 1–10 (2022). https://doi.org/10.15978/j.cnki.1673-5668.202205002
11. Geng, X,, Hu, Y.: Comparative analysis of the international competitiveness of China and India's information industry from the perspective of global value chain. World Regional Stud. 31(2), 270–279 (2022). https://doi.org/10.3969/j.issn.1004-9479.2022.02.2020722

A Deep Forgery Detection Method Based on Adaptive Features in Both Spatial Domain and Frequency Domain

Ming Li[1], Junchang Jing[2,3], Yun Yang[4], Luyan Xu[5], Shasha Tian[6], Heng Zhang[1](✉), Jian Zhang[1], and Meng Huang[1]

[1] School of Computer Engineering, Jiangsu Ocean University, Lianyungang 222005, Jiangsu, China
ezhangheng@gmail.com
[2] College of Computer and Information Engineering, Henan Normal University, Xinxiang 453002, Henan, China
[3] Key Laboratory of Artificial Intelligence and Personalization, Henan Normal University, Xinxiang 453002, Henan, China
[4] Business College, Suzhou University of Technology, Suzhou 215500, Jiangsu, China
[5] School of Intelligent Manufacturing, Huanghuai University, Zhumadian 463000, Henan, China
[6] School of Information Engineering, Jiangsu Colloge of Finace & Accounting, Lianyungang 222005, Jiangsu, China

Abstract. Deep forgery detection technologies play a critical role in image and video recognition. Their performance heavily relies on the features extracted from both real and fake images. However, most existing methods focus primarily on spatial domain features, resulting in a limitation in accuracy. To overcome this challenge, we propose an adaptive region dynamic convolution module that effectively extracts facial features from the spatial domain. Then, an adaptive frequency dynamic filter is proposed to extract the effective frequency domain features. With the fusion of both the spatial domain features and the frequency domain features, the accuracy of classifying real and fake facial images is expected to increase significantly. Finally, the experimental results on three real-world datasets demonstrate the effectiveness of our dual-domain feature representation method, which significantly enhances classification precision.

Keywords: Deep forgery detection · Spatial domain · Frequency domain · Feature representation · Facial images

1 Introduction

With the continuous advancement of deep generative models, traditional data processing techniques, such as variational autoencoders (VAEs) and generative adversarial networks (GANs), have evolved into more complex architectures, such as diffusion models [1]. These innovative artificial intelligence techniques enable the generation of more realistic

L. Huang et al. (Eds.): CCF NCCA 2025, CCIS 2774, pp. 470–478, 2026.
https://doi.org/10.1007/978-981-95-5294-8_32

images and videos, enhancing content on social networking platforms [2]. However, the widespread application of these technologies has led to the creation of numerous fake images and videos. Specifically, deepfake technologies have led to serious consequences, including public opinion manipulation [3] and telecommunications fraud [4], both of which occur frequently. For example, in January 2024, a deepfake video imitating a speech by US President Biden appeared in New Hampshire, calling on voters not to participate in the primary election. In February 2024, scammers leveraged deepfake video technology to impersonate the CEO of a multinational company from the UK during a video conference, defrauding its Hong Kong subsidiary of almost $200 million. Therefore, developing useful deep forgery detection approaches to identify fake images and videos is essential.

To date, the methods used for deepfake facial image detection have focused on three key types of features: spatial domain features [5], frequency domain features [6], and spatial-frequency domain features [7]. Specifically, spatial domain features refer primarily to edge details and spatial context information within facial images. The frequency domain features involve mainly the low-frequency and high-frequency components of facial images. Moreover, spatial-frequency domain features refer to the fusion of spatial and frequency domain features, which takes into account the advantages of both feature types.

In practical applications, deep forgery detection methods are typically trained on closed, limited datasets, which hinders their ability to generalize across diverse types of forged images [5–7]. Moreover, these methods usually focus on spatial domain features without fully utilizing frequency domain features. While these methods and their variations have demonstrated better feature representation capabilities, they still struggle to handle the features contained in fake images. As a result, current forgery detection methods are ill-equipped to handle a variety of scenarios, making processing large volumes of AI-generated images challenging.

Recently, researchers worldwide have begun conducting in-depth studies on deepfake facial image detection methods. For example, the well-known cross-attention mechanism and local artifact attention mechanism have been integrated into image detection networks to improve recognition accuracy [8, 9]. However, these attention mechanisms cannot effectively extract effective features from different domains. Furthermore, to obtain a more comprehensive representation of facial forgery features, some researchers have integrated spatial-domain and high-frequency features through advanced multi-feature fusion strategies [10]. Nevertheless, the above methods still rely on the spatial domain features of forged facial images, resulting in poor performance of the detection methods.

Building on the above discussions, effectively utilizing features from different domains for forgery detection has become a critical challenge. To address the above challenges, an adaptive region dynamic convolution module is constructed to capture the effective spatial domain features. Then, we design an adaptive frequency dynamic filter to extract the effective frequency domain features. With the fusion of dual-domain features, the deep forgery detection method subsequently achieves better performance. Finally, the experimental results illustrate the effectiveness of our proposed method.

2 Related Works

The rapid advancement of generative models, exemplified by systems such as ChatGPT, has introduced significant challenges in detecting and identifying content derived from synthetic facial images [2–5].

Recent research trends indicate that deep forgery detection methods generally fall into two main categories: those based on handcrafted features [6] and those utilizing deep learning techniques [6]. In particular, deep learning methods, i.e., convolutional neural networks (CNNs) and transformer-based models, have become the dominant tools for extracting discriminative features from deepfake images and videos, enabling more accurate detection and classification.

Generally, handcrafted feature-based methods analyse the intrinsic statistical properties and physical characteristics of images and videos [1, 3], such as color inconsistencies, noise patterns, and compression artifacts. In recent years, researchers have increasingly employed models such as convolutional neural networks and transformer-based architectures to extract distinguishing features from deepfake images and videos [10–15]. These methods can learn subtle differences between real and deepfake content, enabling accurate binary classification. For example, by exploiting the discrepancies in manipulated facial regions, Nirkin *et al.* [10] proposed a method for detecting facial swapping and other identity manipulation techniques in a single image, aiming to identify subtle inconsistencies introduced during the manipulation process.

Furthermore, the performance of traditional classifiers is improved by analysing discrepancies in deep features across different image domains. In [11], a so-called G-JOB GAN network was established, which achieves an accuracy of 95.7% in distinguishing between real and fake images. Yang *et al.* proposed a multiscale texture difference model to increase the robustness of facial forgery detection. By combining central differential convolution (CDC) with pixel intensity and gradient cues, the above model effectively represents subtle texture discrepancies. Moreover, to preserve texture features and mitigate their degradation, multiscale information fusion was implemented via atrous spatial pyramid pooling (ASPP) [12]. This technique demonstrated strong robustness to common image distortions, including compression artifacts and blurring.

Yang *et al.* [13] proposed a visual speaker authentication scheme based on a deep convolutional neural network (DCNN). The system utilized a basic feature extraction network (FFE Net) to obtain general identity-related information of the speaker and employed a representative classification network (RC Net) to extract and analyse high-level lip features for verification. While modelling the speaking style of the authenticated client, the system simultaneously identifies and rejects impostors. Moreover, reinforcement learning was employed to select the top-k augmentations for each test sample in [15]. A convolutional neural network was subsequently used to generate enhanced classification scores for each augmented image.

To address the interpretability limitations in feature representation and analysis within data-driven deep neural networks, recent research has focused on integrating explainable mechanisms to enhance model transparency and decision-making insights. Ciftci *et al.* [16] proposed a deep forgery detector that uses subtle biological signals embedded in portrait videos—such as variations in skin tone or blood flow—as implicit cues for determining authenticity. Li *et al.* [17] proposed an artifact disentanglement adversarial learning framework that accurately transmitted artifact features through a multiscale feature separator in a disentanglement generator. These disentangled artifact features are then leveraged to construct synthetic training samples, while the generator is optimized through pixel-level supervised learning to achieve precise deep forgery detection.

In the domain of deepfake video detection, frame sequences are commonly employed to capture temporal artifacts—such as interframe inconsistencies in texture and structure—that help distinguish authentic videos from manipulated videos [16]. Agarwal *et al.* [18] proposed detecting deepfake videos by identifying inconsistencies between lip movements and phoneme articulation. Gerstner *et al.* [19] proposed a real-time video authentication method that manipulates lighting conditions to induce controlled variations in facial appearance. The above system measured deviations from the expected appearance changes over time to determine the authenticity of the video feed. Hu *et al.* [20] proposed a dual-stream architecture to detect compressed deepfake videos by jointly analysing spatial and temporal features. Notably, the above forgery detection methods use only the spatial domain features and do not consider the frequency domain features.

According to a comprehensive review of domestic and international research, unlike traditional synthesis methods—such as Face2Face [21] and FaceSwap [22]—deep forgery techniques are typically driven by deep learning-based approaches. Deepfake images and videos are frequently generated via advanced models such as autoencoders and generative adversarial networks (GANs), resulting in highly realistic content. This poses significant challenges for detection and underscores the need for more efficient and robust deep neural network methods. Therefore, the main motivation of our work is to propose an effective deep forgery detection method by fusing multiple features.

3 Proposed Method

3.1 Adaptive Representation of Facial Features in the Spatial Domain

Adaptive spatial feature representation aims to extract fine-grained artifact features from multiple regions of facial images [23, 24], forming a solid foundation for subsequent fusion with frequency domain features. The study in [3] revealed that the distribution of forgery artifacts is spatially uneven across different regions of facial images. On the basis of the above conclusions, we design an adaptive region dynamic convolution module, which partitions the image into multiple spatial regions, enabling the capture of localized areas with more detailed forgery traces. The flowchart of adaptive spatial domain feature representation is presented in Fig. 1.

Given a facial image T as input, standard convolution operations are applied to extract the corresponding guiding feature map H, which is shown as follows:

$$H = \sum\nolimits_{c=1}^{C} T_{h,w,c} * Y_c, \tag{1}$$

where Y_c denotes the c-th output of standard convolution; h, w and c denote the height, width, and number of channels of facial image T, *respectively; and* $*$ denotes the convolution operation.

Furthermore, the widely used squeeze-and-excitation (SE) attention mechanism is employed to generate the attention weight map G. The weight matrix is then normalized to generate a new weighted feature map G' that enhances the representation of informative regions. Moreover, a shared convolutional kernel—comprising adaptive pooling and convolutional layers—is uniformly applied across all facial regions to ensure consistent feature extraction. Finally, we obtain the following spatial domain feature F_s:

$$F_s = \sum_{k=1}^{k=m} Y_k * T_k \circ G_k', \tag{2}$$

where Y_k denotes the k-th region of the original feature group, G_k' denotes the weighted feature map of the k-th region, and $\circ$ denotes multiplication by element.

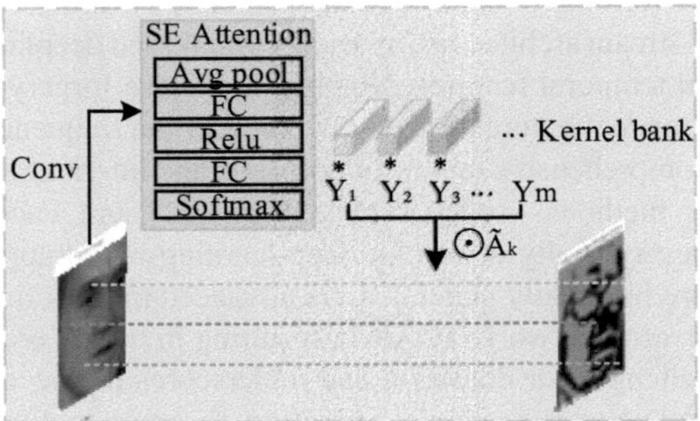

Fig. 1. Flowchart of adaptive spatial domain feature representation

3.2 Adaptive Representation of Facial Features in the Frequency Domain

Frequency-domain features are crucial for effectively distinguishing between forged and authentic facial images. While most existing methods utilize convolutional operations to extract forgery-related features from facial images, they often fail to achieve satisfactory performance. To overcome this limitation, we propose an adaptive frequency dynamic filter for capturing the frequency-domain representation. Specifically, we employ frequency decomposition and a self-attention mechanism to capture manipulation-related features accurately in the frequency domain of facial images, enabling the extraction of complete frequency band information.

To facilitate multiband analysis, the frequency-domain feature space is partitioned into low-, mid-, and high-frequency components. The low-frequency features are obtained by applying a convolutional layer with a stride of 4 to perform spatial downsampling. Next, the low-frequency components f_{lower} are subtracted from the original frequency-domain representation to isolate the residual features. The mid-frequency components f_{medium} are then extracted by using a convolutional layer with a stride of 2. Finally, the high-frequency features f_{high} are obtained by excluding both the low- and mid-frequency components from the original frequency-domain representation.

On the basis of the previously extracted low-, mid-, and high-frequency components, a self-attention mechanism is employed to dynamically assign the attention weight w_i to each frequency band. A dynamic adaptive filter M is subsequently constructed in the frequency domain by aggregating the weighted frequency components. By applying a full-band convolution operation, we extract the original forgery feature Q. Hence, the dynamic filter M and the final forged features F_f can be expressed as follows:

$$M = \sum w_i * f_i, \tag{3}$$

$$F_f = M * Q. \tag{4}$$

where w_i denotes the attention weight, f_i denotes the filters corresponding to different frequency components, and Q denotes the corresponding original forgery feature.

3.3 Adaptive Fusion of Spatial and Frequency-Domain Features

Spatial and frequency-domain features can comprehensively capture the potential differences between real and fake faces. To obtain a more discriminative feature representation, the spatial and frequency-domain features of the facial images are adaptively fused into a joint feature vector $output_i$. This feature vector is subsequently fed into a two-layer fully connected classifier with a Softmax activation function to produce the final classification probability. Specifically, the final classification result $\hat{Y}_i$ is determined by

$$\hat{Y}_i = Soft\max(\mathrm{R}elu(W_1 Output_i^T + b_1)W_2 + b_2). \tag{5}$$

For the deep forgery detection network, we use cross-entropy loss to supervise the classification of facial images. It is defined as

$$Loss = -\frac{1}{N}\sum_{i=1}^{N} output_i \cdot \log(p(output_i)) + (1 - output_i) \cdot \log(1 - p(output_i)), \tag{6}$$

where $y \in \{0,1\}$ denotes the ground-truth binary label, and $p \in [0,1]$ denotes the predicted probability that the sample belongs to the positive class (i.e., label 1).

4 Experimental Results and Analysis

4.1 Datasets

To validate the computational efficiency of the proposed method, we conduct empirical evaluations on the three following real-world datasets.

- **D1:** FaceForensics + + (FF + +). It consists of 5,000 videos, among which 1,000 are authentic and 4,000 are forged via various deepfake techniques.
- **D2:** Celeb-DF-v2 (CDF). The CDF dataset contains 6 229 videos, including 590 real and 5,639 fake videos, featuring individuals of various ages and ethnic backgrounds.
- **D3:** DFDC. The DFDC dataset contains 5,213 videos, including 1,133 real videos and 4,080 fake videos.

These three benchmark datasets collectively offer a diverse and challenging testbed for assessing the generalizability of the proposed method.

4.2 Parameter Sensitivity Analysis

To enhance the overall performance of the detection method, it is crucial to experimentally analyse and validate the key parameters of the dynamic convolution and filtering modules. The choice of appropriate parameters for these components significantly impacts the model's classification performance.

To analyse the effects of the dynamic convolution kernel and filter parameters at a finer granularity, we systematically set their values to 2, 4, 8, and 16. The detailed experimental results for D2 are presented in Table 1.

Table 1. The accuracy (ACC) of the proposed model on D2 with different parameters

Y_c	w_i	ACC	Training Times
2	2	0.7065	890 ms
4	4	0.7317	1143 ms
8	8	0.7326	3780 ms
16	16	0.7775	10820 ms

Table 1 shows that as the number of participants increases, the classification performance of the model improves continuously. However, as the kernel size increases, more training time is needed. To balance the prediction accuracy and training time, the value of two parameters is set as 4. Similar results can be found on the other datasets.

4.3 Comparison of Model Performance

To validate the effectiveness of the proposed method, we conducted experiments on D1 and D3. Specifically, four classification methods were used in our experiments. Moreover, the evaluation metric, i.e., accuracy (ACC), is used to indicate the performance of all involved models. The experimental results of the four tested models on D1 (D3) are summarized in Table 2 (3).

Table 2. The ACCs of all the tested models on D1

Method	D1 (FF + +)
Xception	0.8686
ResNet	0.8570
Efficient NetB4	0.8651
Ours	0.8803

Table 3. The ACCs of all the tested models on D3

Method	D3 (DFDC)
Xception	0.6160
ResNet	0.5868
Efficient NetB4	0.6291
Ours	0.6558

Table 2 Shows that our proposed method achieves the highest acc of 0.8803 on d1, which is 1.34%, 2.71%, and 1.75% higher than those of Xception, ResNet, and EfficientNetB4, respectively. Moreover, on D3, our proposed method attains the best ACC of 0.6558, outperforming Xception, ResNet, and EfficientNetB4 by 6.46%, 11.75%, and 4.24%, respectively. Therefore, our proposed method is capable of performing highly accurate facial detection (Table 3).

5 Conclusion

This paper proposes a facial forgery detection method that relies on both spatial domain and frequency domain features, which carefully considers the unique characteristics of facial images in each domain. Specifically, an adaptive region dynamic convolution module has been introduced to capture fine-grained spatial domain features. Then, we develop an adaptive frequency dynamic filter by incorporating frequency decomposition (high-, medium-, and low-frequency components) and a self-attention mechanism. Finally, the experimental results demonstrated that our proposed method has better performance in terms of facial image detection. As part of our future work, we intend to explore advanced approaches that enhance deep forgery detection performance by integrating both global structures and local discriminative features in different domains.

Acknowledgements. This work was supported in part by the National Nature Science Foundation of China (No. 12401679), the Natural Science Foundation of the Jiangsu Higher Education Institutions of China (No. 23KJB520006), and the Nature Science Foundation of Jiangsu Province (No. BK20241072).

References

1. Zhang, Y., et al.: Genface: a large-scale fine-grained face forgery benchmark and cross appearance-edge learning. IEEE Trans. Inf. Forensics Secur. **19**, 8559–8572 (2024)
2. Zhang, W., et al.: Sadtalker: learning realistic 3d motion coefficients for stylized audio-driven single image talking face animation. In: CVPR, pp. 8652–8661 (2023)
3. Hu, J., Liang, J., Qin, Z., Liao, X., Zhou, W., Lin, X.: ADA-FInfer: inferring face representations from adaptive select frames for high-visual-quality Deepfake detection. IEEE Trans. Dependable Secure Comput. **22**(3), 3011–3027 (2024)

4. Long, M., Liu, Z., Zhang, L.B., Peng, F.: LGDF-Net: local and global feature based dual-branch fusion networks for Deepfake detection. IEEE Trans. Circuits Syst. Video Technol. **35**(6), 5489–5500 (2025)
5. Pradeepan, P., Gladston Raj, S.: Detection of Deepfake medical images based on spatial and frequency domain analysis. In: CICN, pp. 611–617 (2024)
6. Chu, B., Xu, X., Zhang, Y., You, W., Zhou, L.: Reduced spatial dependency for more general video-level Deepfake detection. In: ICASSP, pp. 1–5 (2025)
7. Huang, Y., Juefei-Xu, F., Guo, Q., Liu, Y., Pu, G.: Dodging deepfake detection via implicit spatial-domain notch filtering. IEEE Trans. Circuits Syst. Video Technol. **34**(8), 6949–6962 (2023)
8. Sun, G., Lian, Z.: Deepfake video detection based on the decomposition of spatial-temporal attention mechanism in ViViT. In: ISPA, pp. 1629–1634 (2024)
9. Katamneni, V.S., Rattani, A.: Contextual cross-modal attention for audio-visual deepfake detection and localization. In: IJCB, pp. 1–11 (2024)
10. Nirkin, Y., Wolf, L., Keller, Y., Hassner, T.: Deepfake detection based on discrepancies between faces and their context. IEEE Trans. Pattern Anal. Mach. Intell. **44**(10), 6111–6121 (2022)
11. Monkam, G., Yan, J.: Digital image forensic analyser to detect AI-generated fake images. In: CACRE, pp. 366–373 (2023)
12. Yang, J., Xiao, S., Li, A., Lan, G., Wang, H.: Detecting fake images by identifying potential texture difference. Futur. Gener. Comput. Syst. **125**, 127–135 (2021)
13. Yang, C., Ding, L., Chen, Y., Li, H.: Defending against gan-based deepfake attacks via transformation-aware adversarial faces. In: IJCNN, pp. 1–8 (2021)
14. Yang, X., Li, Y., Lyu, S.: Exposing deep fakes using inconsistent head poses. In: ICASSP, pp. 8261–8265 (2019)
15. Nadimpalli, A.V., Rattani, A.: On improving cross-dataset generalization of deepfake detectors. In: CVPR, pp. 91–99 (2022)
16. Ciftci, U. A., Demir, I., Yin, L.: Fakecatcher: Detection of synthetic portrait videos using biological signals. IEEE Trans. Pattern Anal. Mach. Intell. (2020)
17. Li, X., Ni, R., Yang, P., Fu, Z., Zhao, Y.: Artifacts-disentangled adversarial learning for deepfake detection. IEEE Trans. Circuits Syst. Video Technol. **33**(4), 1658–1670 (2022)
18. Agarwal, S., Farid, H., Fried, O., Agrawala, M.: Detecting deep-fake videos from phoneme-viseme mismatches. In: CVPRW, pp. 2814–2822 (2020)
19. Gerstner, C.R., Farid, H.: Detecting real-time deep-fake videos using active illumination. In: CVPR, pp. 53–60 (2022)
20. Hu, J., Liao, X., Wang, W., Qin, Z.: Detecting compressed deepfake videos in social networks using frame-temporality two-stream convolutional network. IEEE Trans. Circuits Syst. Video Technol. **32**(3), 1089–1102 (2021)
21. Thies, J., Zollhofer, M., Stamminger, M., Theobalt, C., Nießner, M.: Face2face: real-time face capture and reenactment of rgb videos. In: CVPR, pp. 2387–2395 (2016)
22. Faceswap. https://github.com/MarekKowalski/FaceSwap. Accessed 18 March 2020
23. Nguyen, D., et al.: Laa-net: Localized artifact attention network for quality-agnostic and generalizable deepfake detection. In: CVPR, pp. 17395–17405 (2024)
24. Li, M., Song, Y., Ding, D., Sun, R.: Triple factorization-based SNLF representation with improved momentum-incorporated AGD: a knowledge transfer approach. IEEE Trans. Knowl. Data Eng. **36**(12), 9448–9463 (2024)

Dual-PhysNet: A Physics-Guided Dual-Path Lightweight Network for Underwater Image Superresolution Reconstruction

Dong Yu and Yongli Wang(✉)

College of Mathematics and Systems Science, Shandong University of Science and Technology, Qingdao 266590, China
wangyongli@sdkd.net.cn

Abstract. Underwater image superresolution technology faces the dual challenges of complex degradation patterns and the demand for real-time processing on mobile devices. Existing works have preliminarily addressed the coupled modelling of underwater degradation patterns (e.g., color shifts, scattering, and blurring) through multibranch network architectures. However, most models suffer from high parameter counts and computational complexity, making them impractical for real-time mobile deployment. This paper proposes a lightweight dual-path fusion network that achieves high-precision real-time underwater image reconstruction through a physics-prior-guided feature correction module and a dual-path collaborative feature extraction mechanism. The network adopts a three-stage architecture design: the preprocessing module generates low-resolution samples and geometric encodings via coordinate parameterization; the dual-path feature extraction module combines global context modelling with multiscale local detail enhancement; and the reconstruction module leverages frequency-domain basis functions to guide efficient decoding and generate reconstructed images. Experiments demonstrate that the proposed model achieves significant PSNR improvements over the mainstream RDLN model in $\times$ 2 superresolution tasks across different datasets, with a nearly 30% reduction in parameter count.

Keywords: Underwater image superresolution · Lightweight dual-path network · Attention mechanism · Depthwise separable convolution

1 Introduction

With the widespread application of underwater robots (UUVs) and unmanned exploration devices, the importance of underwater optical images in marine resource exploration, ecological monitoring, and military security has become increasingly prominent. However, underwater imaging is constrained by complex optical degradation mechanisms—light waves propagating in water exhibit wavelength-dependent attenuation characteristics (e.g., the red channel attenuation coefficient can reach 0.2–1.5/m), coupled with nonuniform blurring caused by suspended particle scattering and turbulent disturbances. These factors result in underwater images generally suffering from low

L. Huang et al. (Eds.): CCF NCCA 2025, CCIS 2774, pp. 479–498, 2026.
https://doi.org/10.1007/978-981-95-5294-8_33

resolution, poor contrast, and severe color distortion. Such degradations not only hinder human visual interpretation but also directly impact the accuracy of deep learning-based autonomous target detection (e.g., a 30% decrease in mAP for small targets). Therefore, the development of lightweight underwater image superresolution (LUISR) technology, which achieves degradation-adaptive high-quality image reconstruction with limited computational resources, has become a critical research direction for advancing the development of underwater intelligent equipment.

Current image superresolution methods are primarily based on CNNs, GANs, and transformers, most of which are designed for natural images and exhibit limited comprehensive processing capabilities. CNN-based SR architectures (e.g., the SRCNN [1], FSRCNN [2], ESPCN [3], and HAN [4]) employ shallow convolutional layers for end-to-end mapping but struggle to balance long-range dependencies and high-frequency detail recovery due to the restricted receptive fields of convolutional operations. While these models excel in terms of parameter efficiency (Params), their PSNR and SSIM values are generally suboptimal. GAN-based SR methods (e.g., the SRGAN [5], SRDRM-GAN [6], EDSR-GAN [7], ESRGAN [8], and USISResNet [9]) improve PSNR metrics but often require over 10 M parameters, making them unsuitable for real-time underwater device processing. Transformer-based SR approaches (e.g., SwinWave-SR [10], SwinIR [11], TransENet [12], and EMASRN [13]) leverage self-attention mechanisms to capture global and long-range dependencies, achieving superior results, yet their parameter counts remain prohibitively high. Thus, the challenge remains in balancing computational efficiency with the preservation of critical underwater image details.

Recent advancements in lightweight underwater image superresolution have emerged through hybrid multitechnique networks. FEAN [14] introduced wavelet decomposition to separate high/low-frequency features, yielding improved results. MSFIN [15] expands receptive fields and compresses parameters via lightweight recursive residual channel attention blocks. AFAN [16] reduces the model size via chained multiperception attention modules and simplified channel attention. DLGSANet [17] significantly decreases the number of parameters by developing sparse global self-attention modules and end-to-end training. The Jaffe-McGlamery [18] model encodes wavelength-sensitive convolutional kernels (WSCNs) to address red/blue channel restoration disparities. SwinWave-SR [10] pioneered multiscale lightweight modelling by integrating reversible wavelet downsampling ripple blocks with Swin transformers. Efficient-USR [19] constructed a parameter-efficient network via spatial–frequency interaction blocks (SFIBs) and prompt–guided cross-attention (PCAB). MAFFSRN [22] reduces memory usage by stacking multiattention blocks (MABs) with attention fusion mechanisms. Despite these advances, most existing methods still lack optimal synergy between high-quality underwater image reconstruction and low computational complexity.

LUISR faces two core challenges. First, the trade-off between complex degradation and computational efficiency: existing lightweight models (e.g., MobileSR [20] and Tiny-SR [21]) reduce parameters via grouped convolutions and channel shuffling but fail to adapt to underwater multimodal degradations (e.g., coexisting local scattering and global color shifts). Studies have shown that natural image pretrained SISR models suffer a 4.2 dB PSNR drop in underwater scenarios, highlighting insufficient degradation adaptability. Second, the integration of physical priors with data-driven approaches is limited: most methods rely on color balance preprocessing or simplistic attenuation compensation and fail to deeply embed underwater light transport equations (e.g., the Jaffe-McGlamery model [18]) into network architectures, thereby limiting their generalizability in complex aquatic environments.

To address the aforementioned challenges, this paper proposes a lightweight underwater superresolution network, Dual-PhysNet. Our contributions are primarily reflected in three aspects:

- PhysCorr Design Based on underwater optical imaging characteristics: We design PhysCorr to simulate the nonlinear spatial distribution of underwater light attenuation, thereby better addressing degradation caused by complex underwater environments.
- WAGD with Dual Enhancement via ASPPAttn-like Attention and WindowAttn: The WAGD module achieves dual enhancement of local details and long-range dependencies through ASPPAttn-inspired mechanisms and windowed attention (WindowAttn), reducing computational complexity via parameter sharing without compromising model performance.
- MDDP with depthwise separable convolution and multiscale receptive fields: The MDDP module employs depthwise separable convolution to construct "micromedium-macro" three-level receptive fields. The shallow branch focuses on local texture details, the intermediate branch captures object part-level features, and the deep branch extracts the global context, enabling multiscale feature capture while minimizing computational overhead.

2 Related Work

2.1 Application of Deep Learning in Early Image Superresolution Research

Image superresolution (SR) technology has achieved a qualitative improvement in the reconstruction of high-resolution (HR) images from low-resolution (LR) counterparts through deep learning. Early research focused on innovations in convolutional neural network (CNN) architectures, gradually expanding to physics-guided multitask learning and lightweight designs, forming several representative technical pathways.

In 2014, Dong et al. [1] first introduced a three-layer CNN for SR tasks, proposing the SRCNN, which established the end-to-end mapping paradigm in this field. However, its shallow structure (only three layers) struggled to effectively model complex texture information and relied on preupsampling operations. In 2016, Kim et al. [23] further proposed the VDSR, which adopted a 20-layer residual structure to directly predict the residual between HR and LR images and stabilized training via gradient clipping. To increase deployment feasibility, researchers have focused on lightweight and real-time designs. In the same year, Dong et al. [2] introduced the FSRCNN, an improved SRCNN structure that processes images directly in the LR space and employs deconvolution for upsampling, although its high-frequency detail recovery remains suboptimal. Shi et al. [3] proposed ESPCN, which uses subpixel convolution layers for efficient upsampling, enabling real-time processing of 1080p video (24 fps). However, this method often produces checkerboard artifacts in dynamic scenes, compromising visual quality. As model expressive power has advanced, attention mechanisms have been widely adopted to enhance perceptual quality. In 2018, Ledig et al. [5] proposed the SRGAN, and Wang et al. introduced the ESRGAN [8], both of which leverage generative adversarial network (GAN) frameworks to optimize perceptual quality via adversarial loss, achieving subjective texture restoration close to real images but revealing the trade-off between distortion and perceptual metrics. In 2019, Wang et al. [29] proposed PASSRNet, which combines atrous spatial pyramid pooling (ASPP) with attention mechanisms for improved results. In 2020, Lslam et al. [30] developed deep SESR for simultaneous underwater SR and enhancement. Concurrently, H. Basak et al. [24] designed the residual channel attention network (RCAN), which integrates residual-in-residual (RIR) structures with channel attention modules to dynamically calibrate feature importance. In 2022, Wang et al. [31] designed the progressive frequency-interleaved network (PFIN), which employs discrete cosine transform (DCT) frequency separation and RGB/HSV multidomain feature fusion. In the same year, Liu et al. [28] chieved parameter efficiency via depthwise separable convolution. In 2024, Liu et al. [14] further proposed FEAN, which incorporates wavelet-domain dynamic attention mechanisms to advance edge-device SR applications.

Despite significant progress, deep learning-based SR still faces three core challenges: (1) the efficiency–accuracy trade-off—deep networks (e.g., RCAN) achieve superior performance but incur high computational costs, whereas lightweight models (e.g., FEAN) must balance high-frequency detail preservation; (2) inadequate physical degradation modelling—complex degradations such as underwater light transmission and atmospheric turbulence are poorly captured by unified network architectures; and (3) cross-domain generalization bottlenecks—models trained on natural images exhibit significant color recovery deviations when applied to underwater or foggy scenes, necessitating stronger domain adaptation and physical prior constraints.

2.2 Application of Self-attention Mechanisms in Image Superresolution

Self-attention mechanisms dynamically capture long-range dependencies between features, significantly enhancing the modelling of complex textures and degradation patterns in SR tasks. In underwater SR, researchers have combined physical priors with attention mechanisms, forming physics-guided, dynamic sparse, and multidomain fusion approaches.

In 2021, Liang et al. [9] proposed SwinIR, which utilized residual Swin transformer blocks (RSTBs) for deep feature extraction and noise suppression. In 2022, Zamir et al. [35] introduced the Restormer, which addresses CNNs with limited receptive fields and input adaptability via optimized multihead attention and feedforward networks. In 2023, Li et al. [34] developed the range-dependent learning network (RDLN), which integrates underwater light attenuation equations with channel attention for multibranch residual optimization, although its fixed-scale attention weights are sensitive to turbulent blur. In 2024, Zhang et al. [36] proposed the hybrid dynamic transformer net (HDT-Net), which combines two self-attention mechanisms for local-global feature aggregation. Similarly, Liu et al. [14] introduced FEAN, which separates high/low-frequency components via wavelet decomposition and enhances edge sharpness through dynamic frequency-domain recalibration. Yuan et al. proposed TTST to address token redundancy with top-k token selective attention (TTSA), although it struggled in complex environments. In 2024, Dharejo et al. [10] proposed SwinWave-SR, which optimizes the architecture through reversible wavelet downsampling and windowed self-attention for multiscale feature aggregation, achieving efficient edge deployment. In 2025, Lee et al. [25] reduced complexity via parameter sharing in ESC, whereas Zhao et al. [26] enhanced image fidelity with nested residual blocks and channel attention in MARNet. Despite these advancements, self-attention-based models still present robustness issues in underwater SRs due to turbidity-induced degradation.

2.3 Lightweight Superresolution Networks

In recent years, lightweight SR network designs have emerged as a research focus, aiming to achieve high-quality reconstruction under limited computational resources. In 2020, Luo et al. [27] proposed LatticeNet, which employs serial lattice blocks and reverse feature fusion with butterfly structures and attention mechanisms to adaptively combine residual blocks, reducing parameters while maintaining performance. Muqeet et al. [22] introduced MAFFSRN, which leverages lightweight multiattention modules and feature fusion. In 2021, Li et al. [19] developed deep SESR, embedding underwater light attenuation equations into a lightweight CNN framework for simultaneous dehazing and SR, although its fixed-scale residual connections were unstable under turbulent blur. Wang et al. [15] proposed MSFIN, which reduces complexity via multiscale feature interaction and recursive residual channel attention blocks (RRCAB). Jiang et al. [28] designed LCNet with cascaded residual blocks and recursive nesting for progressive residual prediction. In 2022, Fang et al. [36] enhanced the performance of HNCT by integrating CNN-transformer hybrids with spatial attention. In 2023, Chen et al. [34] proposed the RDLN, which dynamically allocates resources via channel-split multiscale bandpass filters. Dharejo et al. [10] introduced SwinWave-SR, which combines reversible wavelet downsampling with windowed self-attention to increase parameter efficiency, and further incorporates lottery ticket pruning for optimizing model performance. In 2024, Liu et al. [14] FEAN separated high/low-frequency features via wavelet decomposition and dynamic subband recalibration. Ma et al. [42] introduced MDSRDN, enhancing textures and edges with multiscale dense residual distillation (MDSRD), spatial feature transform (SFT), and residual spatial-frequency attention (RSFA), while reducing complexity via blue-separation convolution (BS-Conv) and distillation. Despite their advantages in terms of computational efficiency, lightweight SR networks still struggle to improve reconstruction quality and efficiency simultaneously.

3 Proposed Methods

In this section, we first describe the overall workflow of the model. We subsequently provide detailed explanations of our designed modules: PhysCorr (physics-based correction block) and UDPF (underwater dual-path CBAM fusion block).

3.1 Overall Model

The proposed model adopts a hybrid encoder-implicit decoder architecture, integrating a deep feature extractor with a coordinate-aware implicit decoder. The model consists of

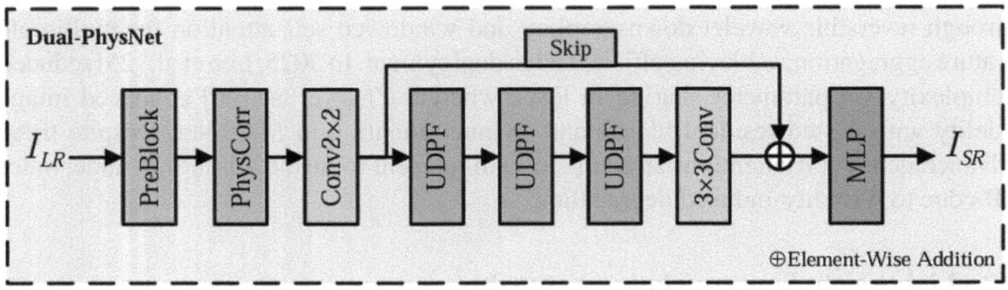

Fig. 1. Structural overview of Dual-PhysNet

three main modules: an image preprocessing module, a feature extraction module, and an image reconstruction module. The overall architecture is illustrated in Fig. 1.

Image Preprocessing Module: The input low-resolution image $\mathrm{I_{LR}}$ is processed by this module to generate a low-resolution sample $X \in R^{H\times W\times C}$, along with corresponding query coordinates $Coord \in R^{B\times N\times 2}$ and cell sizes $Cell \in R^{B\times N\times 2}$. Here, B denotes the batch size, C is the number of input channels, and $H \times W$ represents the spatial dimensions of the input image.

Feature Extraction Module: The low-resolution input $X \in R^{H\times W\times C}$ is fed into the encoder. First, the PhysCorr block, which is specifically designed for underwater images, processes X to output $X_1 \in R^{H\times W\times C}$. A 3×3 convolution is then applied to project X_1 into shallow features $F_0 \in R^{H\times W\times C}$. These shallow features F_0 are further processed by the UDPF block for deep feature extraction, yielding $F_1 \in R^{H\times W\times C}$ A skip connection fuses F_0 and F_1 to produce the final composite features F_{c}. Additionally, coefficient features Coeff and frequency features Freq are derived from F_1 via convolutional projections to construct frequency basis functions.

Image reconstruction module: The role of the decoder is to precisely reconstruct the color values of pixels at any target resolution on the basis of the feature representation provided by the encoder. Centering around a query mechanism, the decoding process feeds the obtained F_c, *Coeff*, *Freq*, along with the query coordinates Coord and cell size Cell derived from the image preprocessing module, into an MLP (multilayer perceptron), which maps them to the final RGB color values. To enhance the reconstruction quality, the decoder also incorporates a multidirectional sampling and area-weighted fusion strategy, strengthening its ability to preserve edges and texture details. The final result is obtained by adding the output via a residual connection to the bilinear interpolation result of the original image, thereby ensuring global structural consistency and local detail authenticity.

The entire workflow is formally expressed as follows:

$$X_1 = PhysCorr(X)$$

$$F_0 = Conv3 \times 3(X_1)$$

$$F_1 = UDPF(F_0)$$

$$\mathrm{F_c} = A(F_0, F_1)$$

$$\text{Coeff} = h_{\text{coeff}}(F_1);\ \text{req} = h_{\text{freq}}(F_1)$$

$$Y = MLPFc, Coeff, Freq, Coord, Cell \tag{1}$$

In this context, h_{coeff} denotes the convolutional projecon operation for coefficient features, h_{freq} represents the convolutional projection operation for frequency features, MLP stands for a multilayer perceptron, $Conv3 \times 3$ refers to a 3×3 convolutional layer, and A indicates the skip connection operation. The modules PhysCorr and UDPF, which are our original designs, will be described in detail in Sects. 3.2 and 3.3, respectively.

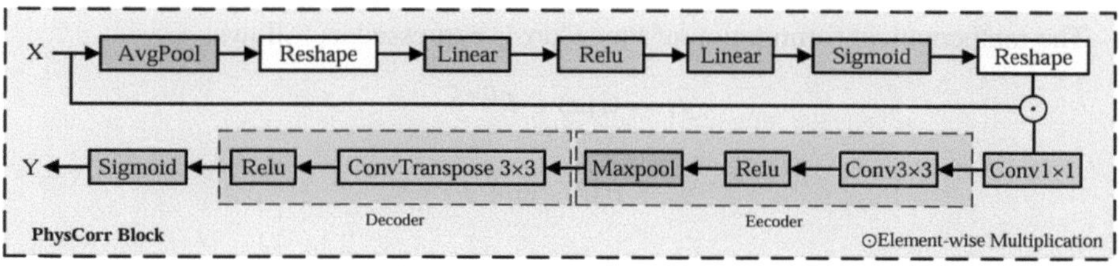

Fig. 2. Structural overview of PhysCorr Block

3.2 Physics-Based Correction Block (PhysCorr)

In this section, we focus on designing PhysCorr tailored to underwater optical imaging characteristics. PhysCorr employs an encoder-decoder architecture to construct a transmittance estimation network, simulating the nonlinear spatial distribution characteristics of underwater light attenuation to mitigate the effects of scattered light and absorption on underwater images. Additionally, to address underwater color distortion, we introduce a channel attention mechanism. This mechanism dynamically perceives attenuation differences across color channels, whereas PhysCorr simulates the physical underwater imaging process (light propagation → color absorption → backwards scattering). This design aligns with the theoretical framework of the Jaffe-McGlamery underwater optical model, enabling robust handling of degradation specific to underwater images. The architecture of PhysCorr is illustrated in Fig. 2.

The detailed implementation of PhysCorr is as follows:

(1) Input Feature Compression: The input feature $X \in R^{B \times C \times H \times W}$ is compressed along the spatial dimensions via global average pooling (GAP), yielding $\mathrm{y}_1 \in \mathrm{R}^{\mathrm{B} \times \mathrm{C}}$.
(2) Nonlinear Mapping: y_1 is passed through a fully connected (FC) layer with parameters w_1 and b_1, followed by a ReLU activation function, to generate the intermediate feature $y_2 \in R^{B \times C}$.
(3) Spatial reshaping: y_2 is reshaped into a 4D tensor $y_3 \in R^{B \times C \times 1 \times 1}$, restoring spatial alignment with the original input feature dimensions.
(4) Channelwise feature fusion: The reshaped feature y_3 is spatially expanded to match the original input size $R^{B \times C \times H \times W}$ and elementwise multiplied ($\odot$) with the input feature X. The result is fused via a 1×1 convolution, producing the intermediate feature $x_3 \in R^{B \times C \times H \times W}$.

(5) Encoder construction: A 3×3 convolutional layer serves as the encoder E, embedding x_3 into a lower-dimensional space to obtain compressed features $E(X) \in R^{B\times C'\times H\times W}$ where $(C' < C)$.
(6) Decoder reconstruction: A transposed convolution layer acts as the decoder D, upsampling $E(X)$ to the original resolution and outputting the transmittance map.
(7) Transmittance normalization: The transmittance map is normalized via a sigmoid activation function and biased by a constant value of 0.05, producing the normalized transmittance map $T \in R^{B\times C\times H\times W}$.
(8) Physics-based feature enhancement: The input features are enhanced on the basis of the normalized transmittance map T and the physical model of underwater light absorption and scattering.

The mathematical formulation of PhysCorr is expressed as follows:

$$y_1 = G(x) \in R^{B\times C}$$

$$y_2 = \sigma w_1 \cdot y_1 + b_1$$

$$y_3 = reshape(y_2, (B, C, 1, 1))$$

$$x_3 = Conv_{1\times 1}(x \odot expand(y_3))$$

$$E(X) = Conv_{1\times 1}(x_3)$$

$$D(E(X)) = ConvTranspose(E(X))$$

$$T = \sigma(D(E(X))) + 0.05$$

$$X = \frac{X - (1 - T)}{T + 1e - 8} \quad (2)$$

In this context, GAP denotes global average pooling, σ represents the sigmoid activation function, w_1 and b_1 are learnable parameters, Reshape refers to tensor shape reshaping, Expand indicates spatial expansion, $\odot$ stands for elementwise multiplication, $\text{Conv}_{1\times 1}$ corresponds to a 1×1 convolutional layer, ConvTranspose denotes the transposed convolution operation, E signifies the encoder, and D represents the decoder.

3.3 Underwater Dual-Path CBAM Fusion Block (UDPF Block)

The UDPF block integrates two distinct feature extraction branches—WAGD and MDDP—and fuses them via the CBAM attention mechanism to output deep features. The architecture of the UDPF block is illustrated in Fig. 3, and the overall workflow is formalized as follows:

$$F_1 = LayerNorm(Conv_{1\times 1}(CBAM\,(Concat(WAGD(F_0, plk), MDDP(F_0))))) + F_0 \quad (3)$$

In this context, WAGD and MDDP represent two parallel feature extraction branches.plk: Shared parameters within the WAGD block. CBAM: Channel and spatial attention mechanism. $\text{Conv}_{1\times1}$: Dimensionality reduction convolution. Layer Norm: Layer normalization. Concat: Channelwise concatenation.

The UDPF block is composed of two branches: the Window-Attention Guided Dynamic ASPP (WAGD) and the Multidiolated DepthPoint Channel Attention Block (MDDP). The WAGD branch is based on convolutional operations that combine ASPP and windowed multihead self-attention mechanisms. The ASPP module contains multiple parallel convolutional branches, each employing different dilation rates to expand the receptive field. We aim to simulate self-attention mechanisms through customized convolutional operations, thereby effectively identifying and emphasizing critical long-range dependencies in the input data while flexibly adjusting their intensity across different instances. Within the WAGD branch, we adopt parameter sharing to reduce the number of parameters while fully leveraging the advantages of self-attention. The MDDP branch is designed via depthwise separable convolutions, which construct a "micromedium-macro" three-level receptive field hierarchy. The shallow branch focuses on local texture details, the intermediate branch captures object part-level features, and the deep branch extracts global contextual information, enabling multiscale feature extraction while maintaining low parameter complexity. These two branches work synergistically to achieve complementary feature extraction and integration.

The detailed implementation process of the UDPF block is as follows:

(1) The input feature F_0 is preserved as the identity baseline for the residual cone.
(2) F_0 is fed into the MDDP and WAGD branches for parallel processing, generating output features F_{c1} and F_{c2}, respectively.
(3) The output features F_{c1} and F_{c2} are concatenated along the channel dimension, forming a feature map with doubled channel dimensions.
(4) The concatenated feature map is processed by the CBAM module, followed by a 1×1 convolutional layer to reduce the channel dimensions back to the original input size.
(5) The reduced-dimensional feature map undergoes LayerNorm normalization and is added to the initial residual baseline F_0, yielding the final output F_c.

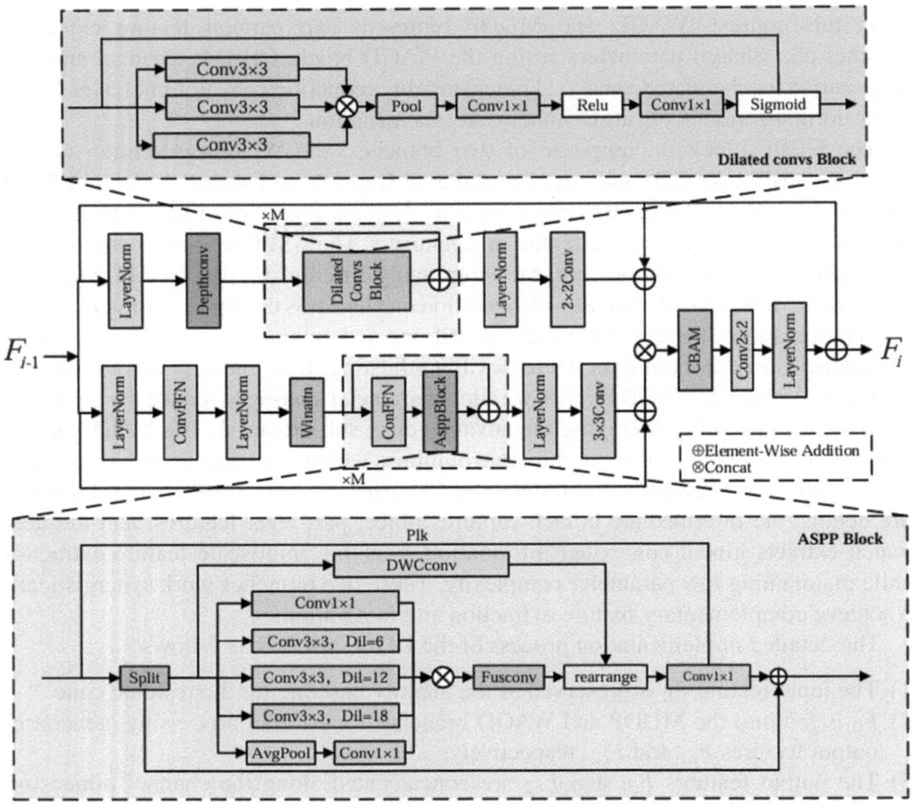

Fig. 3. Strutural overview of UDPF Block

Window-Attention Guided Dynamic ASPP Block (WAGD)

In the WAGD module, ASPPAttn adopts a multiscale dilated convolution structure by deploying three parallel 3 × 3 dilated convolutional layers with dilation rates of 6, 12, and 18, effectively capturing contextual information across different receptive fields while preserving the resolution of feature maps. Additionally, this module introduces a dynamic convolution mechanism (Dynamic Convolution), which adaptively generates convolutional kernel weights on the basis of input feature maps and performs weighted fusion with fixed lightweight filters. This design enhances the model's ability to perceive complex textures and edge details, improving robustness under nonuniform illumination and blurring conditions. WindowAttn focuses on fine-grained local feature extraction by dividing the feature map into window units of a specified size (e.g., 16 × 16) and performing self-attention computations within each window, enabling efficient local dependency modelling. Compared with global attention mechanisms, this design reduces the computational complexity from $O(HW^2)$ to $O\left(\left(HW/16^2\right)^2\right)$ (where 16 is the window size). Furthermore, WindowAttn incorporates a learnable relative position bias matrix to explicitly model the spatial topological relationships between pixels within a window.

This position encoding is dynamically optimized during training, allowing the model to precisely capture visual–semantic correlations between different relative positions in local regions. ASPPAttn emphasizes multiscale spatial feature extraction, whereas WindowAttn focuses on long-range dependency modelling, collectively achieving dual enhancement of local details and global dependencies without significantly increasing the parameter count, thereby improving the model's ability to perceive complex semantic information.

The detailed implementation process of the WAGD module is as follows:

(1) **Input Feature Normalization**: The input feature tensor $F_0 \in R^{B\times C\times H\times W}$ (where B is the batch size, C is the number of channels, and $H \times W$ are the spatial dimensions) undergoes layer normalization (LayerNorm) to produce the intermediate feature F_i^{m}.
(2) **Nonlinear Feature Transformation**: Fim is processed by the ConvFFN feed-forward network, which applies three convolutional operations to achieve nonlinear transformation, yielding the transformed feature F_i^{d}.
(3) **Window attention enhancement**: F_i^{d} undergoes a second layer normalization and is fed into the window attention module for fine-grained feature enhancement, producing the enhanced feature $F_{i,0}$.
(4) **Multiscale iterative processing**: $F_{i,0}$ enters a multiscale iterative processing loop. For the m-th iteration ($1 \le m \le M$):

- The feature $F_{i,m-1}$ is nonlinearly transformed via ConvFFN (with the same structure as the initial stage), generating the intermediate feature $F_{i,m}^{\mathrm{mid}}$.
- $F_{i,m}^{\mathrm{mid}}$ and the preset parameter Plk (where Plk = a) are passed to the AsppBlock for multiscale feature extraction, resulting in the updated feature $F_{i,m}$.

After M iterations, the final feature $F_{i,M}$ is obtained.

(5) **Normalization and Convolution:** $F_{i,M}$ undergoes layer normalization followed by a 3 × 3 convolutional layer, producing F_{11}^{d}.
(6) **Residual Connection:** The feature F_{11}^{d} is added to the initial input F_0 via a residual connection, yielding the final output feature F_{11}.

The WAGD module is formally expressed as follows:

$$F_i^{\mathrm{d}} = ConvFFN(LayerNorm(F_0))$$

$$F_{i,0} = WindowAttention\left(LayerNorm\left(F_i^{\mathrm{d}}\right)\right)$$

$$F_{i,M} = \sum_{i=1}^{blocks} \left(ASPPWrapper_i\left(ConvFFN_i\left(F_{i,0}\right)\right)\right) + F_{i,0}$$

$$F_{11} = Conv_{3\times 3}\left(LayerNorm\left(F_{i,m}\right)\right) + F_0 \tag{4}$$

Multi-dilated DepthPoint Channel Attention Block (MDDP)

The MDDP module employs an architecture based on depthwise separable convolutions, deploying parallel branches with different dilation rates (dilation_rates =

[1,2, 3]) to construct a graded receptive field expansion mechanism while preserving the spatial resolution of feature maps. Shallow branches focus on extracting local texture details, intermediate branches capture object part-level features, and deep branches extract global contextual patterns. This design builds a multilevel feature pyramid from microscopic to macroscopic scales, significantly enhancing the model's ability to perceive features across diverse scales. Furthermore, each depthwise separable convolution branch requires only $1/(C \times d^2)$ parameters of a standard convolution (where C is the input channel count and d is the dilation rate), maintaining efficient feature extraction while drastically reducing model complexity. To further improve performance, MDDP introduces dynamic feature fusion technology: multibranch outputs are concatenated along the channel dimension and compressed via a 1×1 convolution. This preserves the independence of cross-scale features while enabling feature distillation through channel interactions. Additionally, MDDP incorporates stagewise iterative channel attention mechanisms (channel attention) to dynamically reinforce the weight distributions of critical scale-specific channels. This mechanism ensures the precise and efficient processing of multiscale information.

The detailed implementation process of the MDDP module is as follows:

(1) **Input normalization**: The input feature $F_0 \in R^{B \times C \times H \times W}$ is normalized via channel-first layer normalization (LayerNorm), yielding the intermediate feature F_i^{m}.
(2) **Depthwise separable convolution processing**: F_i^{m} is processed by the depthwise separable convolution module, which sequentially executes 3×3 depthwise convolutions (spatial feature extraction), GELL activations (nonlinearity introduction), and 1×1 pointwise convolutions (channel expansion). This generates the feature $F_{i,0} \in R^{B \times G \times H \times W}$.
(3) **Multistage iterative processing**: $F_{i,0}$ undergoes M iterative stages. For the m-th iteration ($1 \leq m \leq M$):

   - The input feature $F_{i,k-1} \in R^{B \times G \times H \times W}$ is processed through multiple branches with distinct dilation rates $\mathrm{d} \in [\mathrm{d}_1, \mathrm{d}_2, \ldots, \mathrm{d}_\mathrm{K}]$, producing outputs $F_{i,k,K}^{(pw)}$ ($K =$ 1,2, 3, *etc.*).
   - All branch outputs are concatenated along the channel dimension and compressed to G channels via a 1×1 convolution, yielding $F_{i,k}^{\mathrm{f}}$.
   - $F_{i,k}^{\mathrm{f}}$ is processed sequentially through global average pooling, a two-layer fully connected network, and channelwise weighting to generate $F_{i,k}$.

   After these steps are repeated M times, the feature sequence $F_{i,m}(m = 0,1, 2......)$ is obtained.
(4) **Channelwise Concatenation:** All stage outputs are concatenated along the channel dimension to form F_i^{c}.
(5) **Normalization and Dimensionality Reduction:** Fic performs layer normalization and output projection to compress the channel count from $(M + 1)G$ back to the original input dimension C, producing F_i^{p}.
(6) **Residual Connection:** The output feature F_i^{p} is added to the original input F_0 via a residual connection, yielding the final output F_{21}.

The MDDP module is formally expressed as follows:

$$F_i^{\mathrm{m}} = LayerNorm(F_0)$$

$$F_{i,0} = PointConv_{1\times 1}\left(GELU\left(DepthConv_{3\times 3}\left(F_i^m\right)\right)\right)$$

$$F_{i,k} = C\Big(B\Big(\{DConv_d\left(F_{i,k-1}\right)\}_{d=1}^{K}\Big)\Big), \quad k = 0, 1, \ldots, M$$

$$F = \left[F_{i,0}, F_{i,1}, \ldots, F_{i,M}\right]$$

$$F_i^{\mathrm{c}} = Cat\left(F_{i,0}, F_{i,1}, \ldots, F_{i,M}\right) \in R^{B\times(M+1)G\times H\times W}$$

$$F_i^{\mathrm{p}} = Conv_{1\times 1}\left(LayerNorm\left(F_i^{cat}\right)\right) \in R^{B\times C\times H\times W}$$

$$F_{21} = F_i^{\mathrm{p}} + F_0 \in R^{B\times C\times H\times W} \quad (5)$$

4 Experiments

In this section, we compare the proposed method with ten classical approaches to evaluate network performance and validate the method's effectiveness through ablation studies.

4.1 Datasets Evaluation Metrics and Experimental Setup

The experiments utilize two public underwater image datasets, namely, USR-248 and UFO-120, which cover diverse scenes and illumination conditions, to verify the model's generalizability. The USR-248 dataset contains 1,060 paired low-resolution (LR) and high-resolution (HR) image groups for training, where each group includes an HR image and corresponding LR images with downsampling factors of 2, 4, and 8. We selected 848 groups for training and 212 groups for validation. Additionally, the USR-248 dataset includes 248 paired test images covering typical underwater scenarios such as coral reefs, shipwrecks, and fish. The dataset was collected via experimental water tanks and underwater robots to simulate complex environments with turbidity, low-light conditions, and motion blur, aligning with the physical degradation characteristics of underwater imaging. The UFO-120 dataset contains 1,500 paired LR-HR image groups for training, where each group includes an HR image and corresponding LR images with downsampling factors of 2, 3, and 4. We used 1,232 groups for training and 212 for validation. The dataset also includes 120 paired test images covering diverse underwater scenes, such as sandy seabeds, rocks, and deep-sea organisms. Both datasets exhibit distinct underwater degradation features, enabling comprehensive evaluation. The evaluation metrics primarily include the peak signal-to-noise ratio (PSNR) and structural similarity index (SSIM) scores for quantitative analysis and performance assessment of the restoration results. Additionally, we calculate the parameter counts of different deep learning networks for model complexity evaluation.

We use the Adam optimizer to minimize the objective function in our network, with an initial learning rate set to 2e − 4. The learning rate is reduced by 50% at the 500th, 800th, 900th, and 950th epochs, totaling 1,000 training epochs. To accommodate memory constraints, the batch size is set to 4, the dataset repetition count is 20, and the number of UDPF blocks is N = 3. Our model is implemented on the PyTorch framework and executed on an NVIDIA RTX 4060 GPU.

Table 1. Results of quantitative experiments on the USR-248 datasets.

Methods	Params(M)	Flops	USR-248					
			2		4		8	
			psnr	ssim	psnr	ssim	psnr	ssim
SRGAN [5]	5.96	529.28	28.10	0.79	24.78	0.68	20.18	0.63
SRDRM-GAN [9]	13.19	376.95	28.69	0.82	26.01	0.65	24.24	**0.65**
VDSR [23]	0.68	203.78	29.03	0.79	25.73	0.67	23.63	0.64
AMPCNet [38]	1.24	–	29.48	0.81	26.00	0.67	23.89	0.63
ESRGAN [8]	1.62	1504.12	29.81	0.76	24.38	0.66	19.92	0.57
DSRCNN [41]	1.11	14.37	27.14	0.77	23.61	0.67	20.14	0.56
URSCT [39]	14.60	–	27.43	0.81	25.35	0.60	22.98	0.49
HNCT [36]	0.69	81.82	29.40	0.82	25.99	0.67	23.91	0.54
RDLN [34]	0.85	–	29.97	**0.84**	26.21	0.65	24.01	0.54
LatticeNe [27]	1.90	83.39	29.39	0.80	24.72	**0.68**	21.35	0.61
ours	**0.56**	252.84	**30.71**	0.79	**27.22**	0.63	**24.57**	0.52

Table 2. Results of quantitative experiments on the UFO-120 datasets.

Methods	Params(M)	Flops	UFO-120					
			2		4		8	
			psnr	ssim	psnr	ssim	psnr	ssim
SRCNN [1]	0.06	21.32	24.83	0.73	22.64	0.66	19.53	0.57
SRGAN [5]	5.96	528.79	26.36	0.76	23.90	0.71	21.03	0.58
SRDRM-GAN [9]	11.29	377.98	27.59	0.73	–	–	23.24	0.66
ESRGCNN [8]	1.54	21.40	25.86	0.72	26.26	0.72	25.35	0.71
AMPCNet [38]	1.18	–	25.34	0.73	25.79	0.70	24.68	0.72
URSCT [39]	11.40	–	25.94	**0.80**	–	–	23.64	0.65
HNCT [36]	**0.35**	81.59	25.80	0.71	25.94	0.70	24.88	0.69
RDLN [34]	0.83	–	25.98	0.76	26.58	**0.75**	25.42	**0.73**

(continued)

Table 2. (*continued*)

Methods	Params(M)	Flops	UFO-120					
			2		4		8	
			psnr	ssim	psnr	ssim	psnr	ssim
LatticeNet [27]	0.76	82.42	25.92	0.72	26.19	0.72	25.27	0.71
ours	**0.56**	252.84	**30.13**	0.78	**28.34**	0.7	**27.03**	0.63

4.2 Quantitative Evaluation

We compare the proposed Dual-PhysNet with various SR methods on the public USR-248 and UFO-120 datasets. The quantitative results are shown in Table 1 and Table 2. For USR-248, we adopt superresolution scaling factors of $\times$ 2, $\times$ 4, and $\times$ 8. For UFO-120, scaling factors of $\times$ 2, $\times$ 3, and $\times$ 4 are used.

On the USR-248 dataset for $\times$ 2 SR tasks, Dual-PhysNet demonstrates superior performance in PSNR, achieving 30.71 dB, which is 0.74 dB higher than that of the suboptimal model RDLN (29.97 dB), with a 30% reduction in parameter count. This finding confirms that the proposed UDPF module, through dual-path collaborative feature extraction (MDDP and WAGD), effectively captures high-frequency details. For $\times$ 4 SR tasks, Dual-PhysNet outperforms other SR methods in terms of PSNR. In $\times$ 8 SR tasks, although the SSIM metric slightly lags behind those of the other methods, the PSNR advantage remains at 0.5 dB.

On the UFO-120 dataset, which includes complex motion blur, Dual-PhysNet exhibits stronger scene adaptability. For $\times$ 2 SR tasks, it achieves a PSNR of 30.13 dB, which is 21.7% higher than that of the traditional SR model SRCNN, with a leading SSIM of 0.78. In $\times$ 3 SR tasks, Dual-PhysNet attains a PSNR of 28.34 dB, which is 1.76 dB higher than that of the RDLN (26.58 dB), indicating that the WAGD module with dynamic convolution kernels enables adaptive feature extraction. For $\times$ 4 SR tasks, despite a lower SSIM (0.63) than some models do, the PSNR of 27.03 dB remains superior, confirming Dual-PhysNet's ability to preserve texture details.

By evaluating parameter counts, Dual-PhysNet achieves a higher PSNR than do the other SR methods while maintaining fewer parameters than most mainstream approaches do.

4.3 Qualitative Analysis

We compared six mainstream models: ESRGAN, HNCT, LatticeNet, the SRGAN, the SRCNN, and VDSR. Figure 4 presents the visual comparative results of different SR models on the USR-248 and UFO-120 datasets. In underwater image superresolution tasks, restoring authentic structures and texture details faces significant challenges due to medium scattering, color attenuation, and uneven illumination. Compared with natural images, underwater images often suffer from severe blurring, distortion, and high-frequency information loss, demanding higher feature representation capabilities from models. We observed that traditional attention-based methods produce smoother results in restoring underwater image details and textures, whereas methods relying on local attention or fixed receptive fields tend to lose details or introduce artifacts when handling such complex degradations.

In contrast, Dual-PhysNet effectively mitigates the negative impacts of local attention or fixed receptive fields through dual-path collaboration (MDDP and WAGD). Qualitative analysis demonstrates that Dual-PhysNet visually restores image details smoothly while enhancing clarity and maintaining overall color harmony.

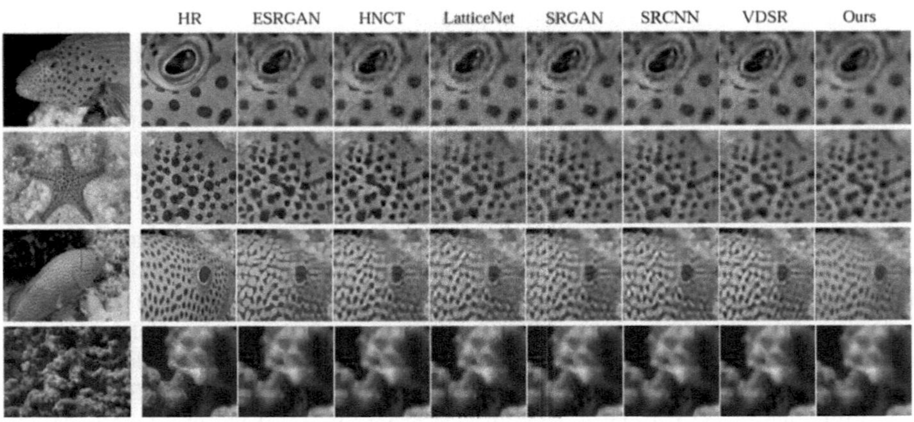

Fig. 4. Comparison of image superresolution results on the USR-248 and UFO-120 datasets using different methods

4.4 Ablation Study

In this section, we quantitatively demonstrate the effects of different modules in Dual-PhysNet via the PSNR and SSIM metrics. We conducted $\times$ 2 superresolution ablation experiments on the USR-248 dataset. Our ablation study includes (1) validation of the effectiveness of WAGD and MDDP; (2) validation of the PhysCorr block; and (3) analysis of the impact of the number of UDPF blocks on the overall model performance.

Validation of WAGD and MDDP Effectiveness

We validated the effectiveness of WAGD and MDDP by removing each module individually. Table 3 presents the experimental results for different variant models. The results show that when only the WAGD module or the MDDP module is retained, the metric values are lower than those achieved by combining both modules. This confirms that collaborative feature extraction through WAGD and MDDP yields superior performance.

Table 3. Quantitative Analysis of Ablation Studies on MDDP versus WAGD.

Models	PSNR	SSIM
Ours	**30.71**	**0.789**
Dual-PhysNet w/o MDDP	30.48	0.783
Dual-PhysNet w/o WAGD	30.56	0.784

Validation of PhysCorr Block Effectiveness

We removed the PhysCorr block designed specifically for underwater image degradation issues. The experimental results in Table 4 show that compared with the model without the PhysCorr block, the PSNR increased by 0.1, and the SSIM improved by 0.003. Despite no increase in parameter count, these metric improvements confirm that the PhysCorr block, by simulating underwater optical imaging characteristics, effectively addresses degradation problems unique to underwater images.

Table 4. Quantitative Analysis of Ablation Studies on PhysCorr.

Models	PSNR	SSIM
Ours	**30.71**	**0.789**
Dual-PhysNet w/o PhysCorr	30.61	0.786

Impact of the UDPF Quantity on Model Performance

We conducted experiments with varying numbers of UDPF blocks, and the results are shown in Table 5. When the UDPF count ranges from 3–5, the PSNR and SSIM values gradually converge. Although a marginal improvement is achieved at N = 7, we ultimately set the UDPF count to 3 to maximize parameter efficiency while maintaining competitive model performance.

Table 5. Quantitative Analysis of Ablation Effects for Different Numbers of UDPF Blocks.

UDPF Block	PSNR	SSIM
N = 1	30.14	0.783
N = 2	30.51	0.786
N = 3	**30.71**	**0.789**
N = 4	30.70	0.790
N = 5	30.72	0.789
N = 6	30.71	0.788
N = 7	30.74	0.792

5 Conclusion

In this paper, we propose Dual-PhysNet, an efficient lightweight network for underwater superresolution. We demonstrate the critical role of collaborative feature extraction between the WAGD and MDDP modules in achieving high-quality results for underwater SR tasks. By integrating ASPPAttn and WindowAttn into the WAGD module, we achieve dual enhancement of local details and long-range dependencies. The MDDP module, which is enhanced with depthwise separable convolutions, constructs a multilevel pyramid feature extraction architecture to capture features at the micro, meso-, and macrolevels thoroughly while maintaining parameter efficiency. The fusion of WAGD and MDDP features further improves image reconstruction quality. We also design the PhysCorr block to adapt dynamically to the diversity of underwater image degradation on the basis of physical principles. Experiments confirm that the proposed Dual-PhysNet achieves superior performance on the USR-248 and UFO-120 datasets compared with mainstream methods. In future work, we will explore the impact of physics-based models on lightweight underwater SR to advance deep-sea exploration and autonomous underwater robotic operations.

References

1. Dong, C., Loy, C.C., He, K., Tang, X.: Learning a deep convolutional network for image superresolution. In: Fleet, D., Pajdla, T., Schiele, B., Tuytelaars, T. (eds.) Computer Vision – ECCV 2014. LNCS, vol. 8692, pp. 184–199. Springer, Cham (2014)
2. Dong, C., Loy, C.C., Tang, X.: Accelerating the super-resolution convolutional neural network. In: Leibe, B., Matas, J., Sebe, N., Welling, M. (eds.) Computer Vision – ECCV 2016. LNCS, vol. 9906, pp. 391–407. Springer, Cham (2016)
3. Shi, W., et al.: Real-time single image and video super-resolution using an efficient subpixel convolutional neural network. In: Proceedings of the IEEE Conference on Computer Vision and Pattern Recognition (CVPR), pp. 1874–1883 (2016)
4. Niu, B., Wen, W., Ren, W., Zhang, X., Yang, L., Wang, S., et al.: Single image super-resolution via a holistic attention network. In: Vedaldi, A., Bischof, H., Brox, T., Frahm, J.-M. (eds.) Computer Vision – ECCV 2020. LNCS, vol. 12357, pp. 191–207. Springer, Cham (2020)
5. Ledig, C., Theis, L., Huszár, F., Caballero, J., Cunningham, A., Acosta, A., et al.: Photorealistic single image super-resolution using a generative adversarial network. In: Proceedings of the IEEE Conference on Computer Vision and Pattern Recognition (CVPR), pp. 105–114 (2017)
6. Islam, M.J., Enan, S.S., Luo, P., Sattar, J.: Underwater image super-resolution using deep residual multipliers. In: Proceedings of the IEEE International Conference on Robotics and Automation (ICRA), pp. 900–906 (2020)
7. Lim, B., Son, S., Kim, H., Nah, S., Lee, K.M.: Enhanced deep residual networks for single image super-resolution. In: Proceedings of the IEEE Conference on Computer Vision and Pattern Recognition Workshops (CVPRW), pp. 1132–1140 (2017)
8. Wang, X., et al.: ESRGAN: enhanced super-resolution generative adversarial networks. In: Leal-Taixé, L., Roth, S. (eds.) Computer Vision – ECCV 2018 Workshops. LNCS, vol. 11133, pp. 63–79. Springer, Cham (2018)
9. Prajapati, K., et al.: Unsupervised single image super-resolution network (USISResNet) for real-world data using generative adversarial network. In: Proceedings of the IEEE/CVF Conference on Computer Vision and Pattern Recognition Workshops (CVPRW), pp. 1904–1913 (2020)

10. Dharejo, F.A., Ganapathi, I.I., Zawish, M., Alawode, B., Alathbah, M., Werghi, N., et al.: Swinwave-SR: multiscale lightweight underwater image super-resolution. Inf. Fusion **103**, 102127 (2024)
11. Liang, J., Cao, J., Sun, G., Zhang, K., Van Gool, L., Timofte, R.: SwinIR: image restoration using Swin Transformer. In: Proceedings of the IEEE/CVF International Conference on Computer Vision (ICCV), pp. 1833–1844 (2021)
12. Lei, S., Shi, Z., Mo, W.: Transformer-based multistage enhancement for remote sensing image super-resolution. IEEE Trans. Geosci. Remote Sens. **60**, 1–11 (2022)
13. Zhu, X., Guo, K., Ren, S., Hu, B., Hu, M., Fang, H.: Lightweight image super-resolution with expectation-maximization attention mechanism. IEEE Trans. Circuits Syst. Video Technol. **32**, 1273–1284 (2022)
14. Liu, X., Gu, Z., Ding, H., Zhang, M., Wang, L.: Underwater image super-resolution using frequency-domain enhanced attention network. IEEE Access **12**, 6136–6147 (2024)
15. Wang, Z., Gao, G., Li, J., Yu, Y., Lu, H.: Lightweight image super-resolution with multi-scale feature interaction network. In: Proceedings of the IEEE International Conference on Multimedia and Expo (ICME), pp. 1–6 (2021)
16. Wang, L., Li, K., Tang, J., Liang, Y.: Image super-resolution via lightweight attention-directed feature aggregation network. ACM Trans. Multimed. Comput. Commun. Appl. **19**, 1–23 (2023)
17. Li, X., Dong, J., Tang, J., Pan, J.: DLGSA-Net: Lightweight dynamic local and global self-attention networks for image super-resolution. In: Proceedings of the IEEE/CVF International Conference on Computer Vision (ICCV), pp. 12792–12801 (2023)
18. Jaffe, J.S.: Computer modelling and the design of optimal underwater imaging systems. IEEE J. Oceanic Eng. **15**, 101–111 (1990)
19. Pramanick, A., Bheda, U., Sur, A.: Efficient-USR: Prompt guided dual-domain feature information for efficient underwater image super-resolution. In: Proceedings of the IEEE International Conference on Acoustics, Speech and Signal Processing (ICASSP), pp. 1–5 (2025)
20. Zhang, L., Li, H., Liu, X., Niu, J., Wu, J.: MobileSR: efficient convolutional neural network for image superresolution. In: Proceedings of the IEEE Global Communications Conference (GLOBECOM), pp. 1–6 (2020)
21. Lee, R., et al.: Journey towards tiny perceptual super-resolution. In: Vedaldi, A., Bischof, H., Brox, T., Frahm, J.M. (eds.) Computer Vision – ECCV 2020. LNCS, vol. 12371, pp. 85–101. Springer, Cham (2020)
22. Muqeet, A., Hwang, J., Yang, S., Kang, J.H., Kim, Y., Bae, S.-H.: Multi-attention based ultra lightweight image super-resolution. In: Bronstein, M.M., et al. (eds.) Computer Vision – ECCV 2020 Workshops. LNCS, vol. 12597, pp. 103–118. Springer, Cham (2020)
23. Kim, J., Lee, J.K., Lee, K.M.: Accurate image super-resolution using very deep convolutional networks. In: Proceedings of the IEEE Conference on Computer Vision and Pattern Recognition (CVPR), pp. 1646–1654 (2016)
24. Basak, H., Kundu, R., Agarwal, A., Giri, S.: Single image super-resolution using residual channel attention network. In: Proceedings of the IEEE International Conference on Industrial and Information Systems (ICIIS), pp. 219–224 (2020)
25. Lee, D., Yun, S., Ro, Y.: Emulating self-attention with convolution for efficient image super-resolution. arXiv preprint arXiv:2503.06671 (2025)
26. Lu, Z., Tu, Z., Zhang, Z., Jiang, C., Chen, P.: Single image super-resolution reconstruction based on multilayer residual networks with attention mechanisms. IEEE Access **13**, 73794–73802 (2025)
27. Luo, X., Xie, Y., Zhang, Y., Qu, Y., Li, C., Fu, Y.R.: LatticeNet: towards lightweight image super-resolution with lattice block. In: Vedaldi, A., Bischof, H., Brox, T., Frahm, J.M. (eds.) Computer Vision – ECCV 2020. LNCS, vol. 12368, pp. 272–289. Springer

28. Jiang, N., Chen, W., Lin, Y., Zhao, T., Lin, C.-W.: Underwater image enhancement with lightweight cascaded network. IEEE Trans. Multimedia **24**, 4301–4313 (2022)
29. Wang, L., et al.: Learning parallax attention for stereo image super-resolution. In: Proceedings of the IEEE/CVF Conference on Computer Vision and Pattern Recognition (CVPR), pp. 12242–12251 (2019)
30. Islam, M.J., Luo, P., Sattar, J.: Simultaneous enhancement and super-resolution of underwater imagery for improved visual perception. arXiv preprint arXiv:2002.01155 (2020)
31. Wang, L., Xu, L., Tian, W., Zhang, Y., Feng, H., Chen, Z.: Underwater image super-resolution and enhancement via progressive frequency-interleaved network. J. Vis. Commun. Image Represent. **86**, 103545 (2022)
32. Liu, C., Qu, D., Si, N., Wei, Z.: Lightweight image super-resolution reconstruction based on depthwise separable convolution. Comput. Eng. **48**, 228–234 (2022)
33. Fang, J., Lin, H., Chen, X., Zeng, K.: A hybrid network of CNN and transformer for lightweight image super-resolution. In: Proceedings of the IEEE/CVF Conference on Computer Vision and Pattern Recognition Workshops (CVPRW), pp. 1102–1111 (2022)
34. Chen, Z., Liu, C., Zhang, K., Chen, Y., Wang, R., Shi, X.: Underwater-image super-resolution via range-dependency learning of multiscale features. Comput. Electr. Eng. **110**, 108756 (2023)
35. Zamir, S.W., Arora, A., Khan, S., Hayat, M., Khan, F.S., Yang, M.: Restormer: efficient transformer for high-resolution image restoration. In: Proceedings of the IEEE/CVF Conference on Computer Vision and Pattern Recognition (CVPR), pp. 5718–5729 (2022)
36. Xin, H., Junjie, L., Tong, J.: Learning hybrid dynamic transformers for underwater image super-resolution. Front. Mar. Sci. **11**, 1389553 (2024)
37. Xiao, Y., Yuan, Q., Jiang, K., He, J., Lin, C.-W., Zhang, L.: TTST: a top-k token selective transformer for remote sensing image super-resolution. IEEE Trans. Image Process. **33**, 738–752 (2024)
38. Zhang, Y., Yang, S., Sun, Y., Liu, S., Li, X.: Attention-guided multi-path cross-CNN for underwater image super-resolution. Signal Image Video Process. **16**, 155–163 (2022)
39. Ren, T., et al.: Reinforced Swin-Convs transformer for simultaneous underwater sensing scene image enhancement and super-resolution. IEEE Trans. Geosci. Remote Sens. **60**, 4209616 (2022)
40. Tian, C., Yuan, Y., Zhang, S., Lin, C.-W., Zuo, W., Zhang, D.: Image super-resolution with an enhanced group convolutional neural network. Neural Netw. **153**, 373–385 (2022)
41. Wu, Z., Kong, L., Zhang, L., Lu, H.: Deeply supervised group recursive saliency prediction. Neurocomput. **453**, 636–644 (2021)
42. Liu, B., Ning, X., Ma, S., Yang, Y.: Multiscale dense spatially adaptive residual distillation network for lightweight underwater image super-resolution. Front. Mar. Sci. **10**, 1328436 (2024)

Cyberspace Security Technology

Load-Balancing Routing Optimization in SDN: A QoS-Aware Approach with Improved Shortest Path and Adaptive Resource Allocation

Weiqi Zhang[1(✉)] and Xin Zhou[2]

[1] State Grid GanSu Procurement Company, STATE GRID CORPORATION OF CHINA, Lanzhou, China
407406632@qq.com

[2] Intelligent Control Technology Institute Chongqing Jianshe Industry (Group) Co., Ltd., Chongqing, China
xinxinzhou@alu.cqu.edu.cn

Abstract. The rapid evolution of emerging technologies, including the Internet of Things, cloud computing, and edge computing, has resulted in geometric growth in network traffic. Traditional network architectures are limited in their ability to provide a global perspective and real-time responsiveness, which in turn leads to network congestion and load imbalances that degrade service performance. To address these critical challenges, we propose a two-stage load-balancing routing optimization algorithm (LBROA) within the software-defined network (SDN) paradigm, which uses the global view and centralized control to increase network efficiency. First, an improved shortest path first algorithm (ISPFA) is proposed, which uses network status and traffic changes in real time, and integrates pruning techniques based on network Quality of Service (QoS) metrics to identify multiple feasible paths between sources and destinations. These paths then are conceptualized as a virtual pool of network resources, and their performance is evaluated using the comprehensive smoothness metric. Finally, a resource preallocation strategy is introduced to simulate and evaluate the adaptability smoothness of the network under diverse path allocation schemes. A greedy strategy is applied to dynamically select the optimal adaptive path from the virtual resource pool as the data forwarding route. This process facilitates the dissemination of forwarding rules, thereby guaranteeing efficient data transmission. Experimental evaluations conducted on the Mininet simulation platform demonstrate that, compared with three established algorithms, the proposed LBROA significantly improves overall network performance.

Keywords: Load Balancing · Routing Optimization · Resources Allocation Strategy · Quality of Service · Software-Defined Network

L. Huang et al. (Eds.): CCF NCCA 2025, CCIS 2774, pp. 501–517, 2026.
https://doi.org/10.1007/978-981-95-5294-8_34

1 Introduction

With the advancement of science and technology, the Internet has evolved from an early-stage platform for the sharing of local resources to a global communication network for data transmission. However, its shortcomings are increasingly apparent, including complex structures, functions, weak management, and control options. The key to solving these problems is the urgent need for innovation in network architecture. The SDN framework is widely used to solve traditional network problems because of its outstanding advantages [1].

Traditional network architectures suffer from a fundamental limitation in their inability to adopt a holistic perspective. Routing decisions are predominantly based on local information, leading to an unbalanced network traffic distribution. This phenomenon often leads to congestion on specific links, even when the overall traffic load remains within manageable levels [2]. In recent years, the SDN framework has emerged as a focal point of academic research, mainly because of its ability to provide networks with increased flexibility [3]. As the field progresses, the SDN framework is expected to play a pivotal role in the design of comprehensive solutions for network optimization because of its wide range of capabilities [4–6]. The SDN network separates of network transport and control functions [7]. Compared with traditional network structures, SDN is more competitive [8]. Within the SDN framework, data plane switches communicate with controllers via southbound protocols, such as OpenFlow, to facilitate information exchange. SDN controllers enable interactions between the control plane and the application plane through northbound protocols. However, the centralized control paradigm inherent in SDN presents its own set of challenges. The controller is required to handle a growing number of concurrent service requests, which significantly increases the risk of controller overload [9,10]. Consequently, the investigation of strategies to address these issues and the exploration of technologies to achieve efficient network load balancing have become critical research imperatives.

The balance of loads constitutes a fundamental pillar of modern network infrastructure and computer networks, serving as a primary mechanism for resource allocation among multiple systems [10]. However, traditional non-SDN networks suffer from the absence of a global perspective on network topology and resources, making it difficult to establish well-defined load balancing mechanisms [11,12]. For example, load balancing was considered an important open topic in [13]. Consequently, the SDN architecture simplifies network management and opens up new opportunities for innovation in network load-balancing technology. By offloading critical links, SDN overcomes network congestion and improves network resilience and stability. The efficient utilization of multiple paths in SDN networks represents the cornerstone of achieving network load-balancing objectives. In the literature [14–16], researchers have explored network load balancing from various perspectives, but studies on load balancing mechanisms in SDN networks remain limited. Bagha et al. [17] proposed an energy-efficient load-balancing optimization algorithm for reliable controller placement (ELA-RCP) in SDN. ELA-RCP uses the queuing-based monitoring algorithm to improve load balance, prevent congestion, and enhance the reliability of the controllers and links in the control plane. Hamdan et al. [18] presented a novel SDN-based multiobjective framework that addresses the QoS requirements for IoT sensors across multidomain networks, validating its effectiveness through experiments. Reference [19] conducted a comprehensive analysis of load balancing technology in SDN networks, and topic classification methods for load balancing were considered.

Furthermore, Zhang et al. [21] proposed an energy-efficient linear based load balancing scheme for heterogeneous neural programming (LP-ELB) clusters. LP-ELB addresses the challenge via a two-phase heuristic approach: a server selection phase to identify the minimal set of provisioned nodes and a load partitioning phase to compute the optimal load distribution. Xu et al. [22] presented a controller reselection mechanism based on noncooperative game theory, facilitating load balancing through dynamic migration of SDN switches across controller clusters. Semong et al. [20] abstracted a comprehensive survey of existing load balancing schemes in SDN networks. Forghani et al. [23] proposed a dynamic optimization framework for SDN load balancing and energy efficiency, via the krill swarm heuristic algorithm. Mahmoud et al. [24] introduced microclustering bidirectional LSTM-DSDN (MBL-DSDN), a novel approach to balance the server traffic distribution. Saeedi et al. [25] proposed a subdomain partitioning method based on particle swarm optimization (PSO) to address controller placement problems (CPPs), with the objective of minimizing the number of required controllers, reducing load imbalance, and improving the fault recovery capability between switches and their mapped controllers.

In summary, in this paper, we present comprehensive experiments to verify and analyze three classical algorithms for comparison. The experimental results demonstrate that the Load-Balancing Routing Optimization Algorithm (LBROA) proposed in this paper effectively balances the network load, improves network bandwidth utilization, and reduces transmission latency. The LBROA ensures that the network load is distributed as evenly as possible to improve the overall network performance. Our work makes the following key contributions:

- By combining the Improved Shortest Path First Algorithm (ISPFA) with a cost function for multi-QoS metrics and leveraging the global view and centralized decision-making advantages inherent to SDN, we are able to identify multiple feasible paths between the source and destination.
- We introduce a comprehensive smoothness metric to evaluate the performance of a virtual pool of available network resources in various states, accounting for all usable paths within the network.
- Inspired by the resource allocation strategy, we propose a path preallocation strategy that simulates the impact of different path allocation schemes on the pool's smoothness metric. Using the greedy algorithm, we select the optimal path with the least deviation from all possible allocation paths for efficient resource distribution and assignment of the forwarding rule.

The remainder of this paper is organized as follows: In Sect. 2, which mainly introduces a brief overview of the SDN architecture and the OpenFlow protocol is presented. The formulation of the LBROA proposed in this paper is shown in Sect. 3. The load-balancing routing optimization method and the implementation of network link status monitoring are described in detail in Sect. 4. Section 5 presents our simulation experiments and an analysis of their results. Finally, Sect. 6 summarizes the work of this paper.

2 Related Work

The LBROA proposed in this paper leverages the global view inherent in the SDN architecture. It achieves multiple feasible paths and optimal route selection by minimizing the minimum functional cost function. In this section, we present an overview of the SDN architecture and the widely adopted Southbound Interface protocol, OpenFlow.

2.1 SDN Network Architecture

A Software-Defined Network (SDN) is a network architecture that decouples the control plane and the data plane to provide a more flexible and programmable network infrastructure. The concept of SDN originated from Stanford University's Clean Slate project [1], and was formally articulated by Professor McKeown in 2009 [26].

The SDN architecture is divided into three layers from bottom to top: the data plane, the control plane, and the application plane (see Fig. 1). The data plane consists of network forwarding devices, such as routers and switches, which form SDN data path links between network devices via rules issued by the controller. The control plane consists of a logically centralized SDN controller that provides a global view of the network, responds to switch requests, implements routing decisions for forwarding rules and flow table installation, and provides open interfaces to upper layers. The application plane includes applications and services that define, monitor, and control network resources and behavior. It also provides programmable user interfaces that allow users to implement programming, deploy, and use new applications on demand without having to worry about the underlying details.

The SDN architecture interface is open, and the controller of logically centralized control serves as the core to realize the information about the interaction at each layer. The controller controls the data plane forwarding rules and information interaction via the southbound interface, whereas the northbound interface controls the information exchange interaction with the application plane. Moreover, by leveraging the east-west interface, controllers can mutually share and synchronize data with each other. The SDN architecture decouples the logical control and data forwarding functions, allowing separation and abstraction of the control plane and data plane. The SDN control plane can use the global view of the network to make routing decisions and forwarding rules, whereas the data plane can use the flow table rules received from the controller to achieve data forwarding. In summary, SDN is a new generation of network structures that enables logically centralized control and flexible network programmability.

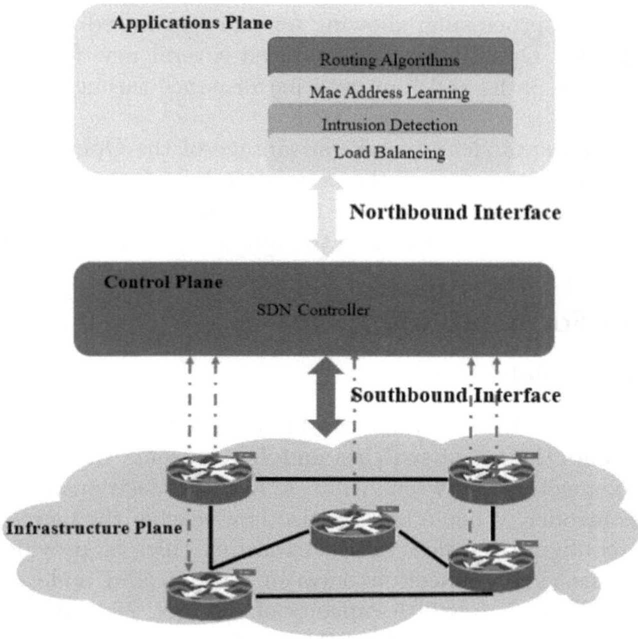

Fig. 1. SDN Framework Architecture.

2.2 OpenFlow Protocol

The OpenFlow protocol was proposed in 2008 by McKeown et al. [7], the latest version being OpenFlow 1.5, and has become one of the most popular southbound interfaces owing to its flexibility and simplicity. The SDN separates the control and data planes by providing a standardized interface for the controller to modify flow table entries of the switch. With the OpenFlow protocol, network administrators can manage their network devices from a central location without worrying about the specifics of each device's hardware and software. One of the key benefits of the OpenFlow protocol is its support for flow-based routing decisions. Using flow-based routing, the controller can consider various factors, such as packet size, protocol type, and destination address, to make more intelligent routing decisions. Additionally, The OpenFlow protocol allows for the creation of customized flow table entries, giving administrators more control over how their network devices handle traffic.

The OpenFlow versions have the following features: OpenFlow 1.0 was the initial version of the protocol and provided basic functionality to forward packets on basis of flow table entries. It was followed by OpenFlow 1.1, which added support for IPv6, multicast, and improved security features. OpenFlow 1.2 introduces group tables, allowing for more flexible packet processing and improved load balancing. OpenFlow 1.3 added support for multiple tables, which further increased the flexibility and scalability of SDN. OpenFlow 1.4 added support for packet-in-packet-out processing, allowing for more advanced traffic control and network monitoring. OpenFlow 1.5 introduced several new features, including support for multiple paths and improved performance through the efficient use of resources.

In our proposed approach, we take advantage of the OpenFlow protocol to obtain information on the state of the network link from the data plane. Specifically, in Sect. 4.1, we describe how we use the OpenFlow protocol to monitor the status of network links and update the flow table entries accordingly.

3 Problem Formulation

3.1 Network Models

The elements of the SDN data layer can be modelled as the graph $G = (V, E)$, where V represents the set of switches and E represents the list of edges that represent the connections between switches. Routing decisions can be made by a centralized controller. When a new request flow reaches the ingress switch, the controller can configure the flow table forwarding rules to provide the service on a regular basis. The connection between node v_i and node v_j is denoted by l_{ij}. B_{ij} denotes the bandwidth capacity of the link, F_{ij} indicates the traffic load through the link, L_{ij} represents the total load of the network link, and A_{ij} represents the available residual bandwidth of the link. e_{ij} represents the correlation between v_i and the node v_j in l_{ij}. If node v_i is linked to node v_j, then $e_{ij} = 1$. Alternatively, $e_{ij} = 0$. The load L_{ij} of the link l_{ij} in a given period can be obtained via Eq. (3.1).

$$L_{ij} = e_{ij} \cdot \sum_{a=1}^{n} F_{ij}, \forall e_{ij} \in E \tag{3.1}$$

Furthermore, the amount of residual bandwidth available A_{ij} that can be used during that cycle can be calculated via Eq. (3.2). The degree of load balance, which we measure via the standard deviation δ, is used to evaluate the distribution of the traffic load across the network links as the network transmits traffic.

$$A_{ij} = e_{ij} \cdot (B_{ij} - L_{ij}), \forall e_{ij} \in E \tag{3.2}$$

3.2 The Virtual Pool's Comprehensive Smoothness Metric

Multiple paths can be formed as a freely disposable pool of virtual network resources composed of residual bandwidth and the load of bottleneck links in all available paths between the source and destination in the network. The path smoothness metric φ is then used to evaluate the performance of the path considering the percentage of residual bandwidth and the load of the bottleneck link in the pooling path. According to Eq. (3.3), we can calculate the smoothness metric of the path for each trail available in the pool, where $\theta(\theta \in [0, 1])$ denotes the smoothness factor.

$$\varphi_m = -\theta \cdot p_m \cdot \log_2 p_m - (1 - \theta) \cdot q_m \cdot \log_2 q_m \tag{3.3}$$

Here, φ_m represents the path smoothness metric of the mth trail in the pool. p_m denotes the residual bandwidth of the bottleneck link in the mth path as a percentage of the total residual bandwidth of the pool. q_m represents the load of the bottleneck link in the mth path as a percentage of the total load of the pool.

Moreover, by combining Eq. (3.4) with the path smoothness metric for all available trails in the pool, we can obtain the comprehensive smoothness metric of the virtual resource pool.

$$\alpha = \sum\nolimits_{m=1}^{k} \varphi_m \tag{3.4}$$

Inspired by resource allocation concepts, we propose a path preallocation strategy in which the controller can simulate the preallocation of available paths in the virtual pool for a new request flow. The variation in the comprehensive smoothness metric for the virtual resource pool can be assessed under distinct path allocation schemes, with comparisons made against the baseline state where no paths are assigned. We use Eq. (3.5) to obtain the amount of variation in the pool's comprehensive smoothness metric for the various path schemes assigned to a new request flow. The variation describes the performance of the network when serving the request flow via different path assignment schemes. The metric is then used to select the optimal route.

$$\rho_m = \alpha - \alpha_m, m = 1, 2, \ldots, k \tag{3.5}$$

where α and α_m denote the pool's comprehensive smoothness metric in the unassigned path state and when the mth path is preassigned to the new request flow, respectively. The ρ_m represents the performance of the various trails in serving the new flow under different path assignment schemes.

3.3 Optimal Multipath Route Selection

In this section, as shown in Eq. (3.6), by applying the greedy principle, the optimal routing trail from a multipath set can be selected onthe basis of minimizing the change in the comprehensive smoothness metric for the pool. The mechanism improves network performance by facilitating the service of new flows within the virtual pool. Once the optimal routing trail is selected, that traffic transmission is completed by setting up the flow table using the selected path in a timely manner.

$$\min(\rho_m) \tag{3.6}$$

4 A Routing Optimization Approach

4.1 Network Link Status Monitor

The centralized control provided by the SDN architecture enables the logically centralized controller to obtain a global view of the network and make routing decisions. Real-time network status information supports the SDN controller in making these decisions. To collect and encapsulate the network traffic information through the data plane, we employ the OpenFlow protocol and *sFlow* technology. All switches collect network status information during each period T and send it to the controller via the controller-switch control channel, allowing the controller to obtain underlying network status information and send control messages to the switches.

To determine the response time between the switch and the controller, the controller actively sends the *Echo* message to the switches for inquiry, and the switches respond with the *Reply* message to the controller. The time difference between sending the *Echo* message and receiving the *Reply* message can be used to calculate the response delay between the switches and the controller. This paper employs detection technology to obtain a relatively accurate link transmission delay. The controller sends an *LLDP* timestamp to the switches, and the switches receive it and respond. After receiving the timestamp response, the complete routing time of the probe packet and the actual transmission delay of the link during the period can be determined by removing the response delay between the controller and switches.

Switches actively record statistics, such as the load and packet loss rates, at the access port, which the controller can query. To obtain port statistics information, the controller sends a *PortStatusRequest* message to the switches, which encapsulates the port statistics into a *PortStatusReply* message and sends it back. Similarly, flow status information can be obtained via *FlowStatusRequest* and *FlowStatusReply* messages, which help to measure the performance of the SDN framework for guaranteed traffic delivery.

In summary, the real-time network status information enables the controller to understand the current state of the underlying network and perform path optimization routing based on the data.

4.2 Calculation of Multiple Paths Between the Source and Destination

We propose an improved algorithm based on the SPF algorithm (ISPFA) and a cost function for multi-QoS metrics to obtain multiple usable paths in the network, achieving balanced network traffic distribution across multiple available trails.

The ISPFA algorithm uses a linear cost function that incorporates the inverse of bandwidth and throughput, delay, to compute the shortest path repeatedly. After each computation, some links and isolated nodes are processed via the following pruning rules.

Rule I. When only one link is usable and directly connected to the source or destination in the network, it is considered extremely difficult to optimize traffic directly to the source or destination. Instead, traffic at intermediate switches on the routing trail should be optimized so that network traffic is evenly distributed, reducing congestion caused by single-path transmission. In this case, after a route path is obtained, only the isolated nodes in the network and the links that are not directly connected to the source or destination must be severed.

Rule II. When multiple paths are available and directly connected to the source or destination in the network, we can optimize the traffic between the source and destination, then intermediate switches on the route to achieve a balanced distribution of network traffic. In this scenario, after the routing path is obtained, it is necessary not only to separate isolated nodes within the network but also to probabilistically determine whether to sever all the links within the path.

Algorithm 1. The pseudocode of the ISPFA.

Input: The network graph: $G = (V, E)$,
  List of available link bandwidth: $B_{available}$,
  List of link cost: C_{link},
  Bandwidth of flow requirement: $R_{bandwidth}$,
  source and destination pair: $(V_{source}, V_{destination})$.
Output: List of available paths: $P_{available}$.
  function **ISPFA**$(G, (V_{source}, V_{destination}), B_{available}, C_{link})$
  for each link in the network **do**
    if $R_{bandwidth} > B_{available}$ **then**
      The available links are selected, and the unavailable links pruned.
    end if
  end for
  while $(V_{source}, V_{destination})$ is reachable **do**
    The Dijkstra algorithm is combined with the cost function to find the shortest path.
    if the shortest path exists **then**
      Add it to $P_{available}$.
      Prune the network according to the two pruning rules iand ii, incorporating probability factors.
    end if
  end while
  end function

Algorithm 1 provides a detailed description of the ISPFA algorithm. The Dijkstra algorithm is one of the most widely used SDN controllers today. Despite its advantages in certain aspects, it has drawbacks that can cause congestion in network traffic transmission. Its underutilization of network state information, which relies solely on certain overheads as routing costs, overlooks the global view of the network. This paper proposes an optimization approach for load-balancing routing in SDN networks to address this issue.

4.3 The Load Balancing Routing Optimization Algorithm

We primarily use the formulas proposed in Chap. 3 and the path preallocation strategy in the LBROA to obtain the optimal path. First, we consider the multiple available paths obtained by ISPFA as a virtual pool of freely disposable network resources. The comprehensive smoothness metric of the pool is then calculated to assess the overall performance of the network in the unassigned path state via Eqs. (3.3) and (3.4). Second, the new request flows are virtually allocated to each available path based on the path preallocation strategy. By Eq. (3.5), the impact of different allocation schemes on the comprehensive smoothness metric of the pool is calculated separately to determine the change in its value. Finally, on the basis of the greedy algorithm, the optimized routing trail that minimizes the variation in the pool's comprehensive smoothness metric is selected from the various allocation options. The flow forwarding rules are then configured accordingly. The pseudocode for this method is shown in Algorithm 2, which details the LBROA implementation process.

Algorithm 2. The pseudocode of the LBROA.

Input: The network graph: $G = (V, E)$,
  List of available link bandwidth: $B_{available}$,
  List of link load: L_{link},
  Bandwidth of flow requirement: $R_{bandwidth}$,
  List of all available paths: $P_{available}$.
Output: The optimal path: $P_{optimal}$.
  function **LBROA**(G, $B_{available}$, $R_{bandwidth}$, $P_{available}$, L_{link})
  Calculates the bottleneck link $L_{bottleneck}$ using $B_{available}$ for all paths in $P_{available}$.
  Obtain the α for the pool using $L_{bottleneck}$ in the unassigned path state by Eqs (3.3) and (3.4).
  for each bottleneck l_m in $L_{bottleneck}$ **do**
    repeat
      Preassign the path of l_m in the $L_{bottleneck}$ corresponding to $P_{available}$ to the new request flow.
      α_m is calculated by Eq (3.4) and ρ_m is obtained at the moment by Eq (3.5).
    until Traversing all links in the $L_{bottleneck}$ corresponding to $P_{available}$
  end for
  The optimal path can be selected with the minimal change value by Eq (3.6).
  Allocate resources using the optimal path for the flow and set up a flow table.
  end function

---

5 Experimental Verification and Analysis

The effectiveness of the proposed algorithm can be validated in this chapter. In the same scenario, the three selected algorithms and our algorithm are analyzed, and the experimental results emphasize the value of this algorithm. Three aspects are described in detail below: experimental configuration, result analysis, and experimental summary.

5.1 Experimental Configuration

To validate the effectiveness of the proposed algorithm, we perform experiments using the improved MIRANET [27] network topology, as shown in Fig. 2. The topology, which consists of 15 switch nodes and 56 links, is selected for experimental validation. We randomly assign bandwidth capacities of either $5Mbps$ or $25Mbps$ to network links and design three different experimental scenarios to compare and evaluate the performance of four algorithms: SPF [28], MHA [29], SWP [30], and the proposed LBROA-on a comprehensive comparison of the network topology.

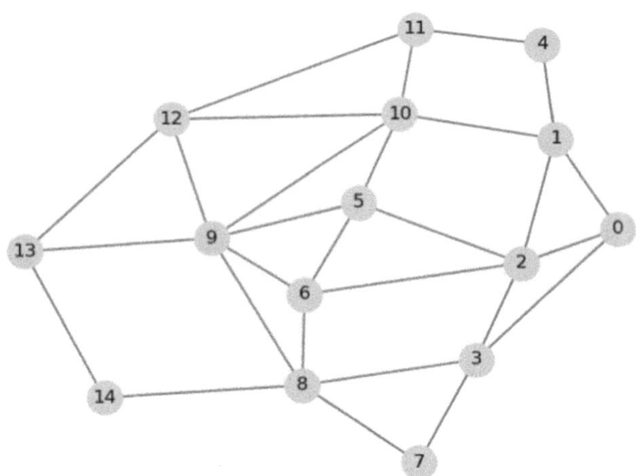

Fig. 2. The improved MIRANET network topology.

The literature [31] stressed the importance of network throughput, packet loss, and load indicators in evaluating network transmission performance. This paper performs a comparative analysis of the performance of different algorithms in network traffic transmission by measuring throughput, transmission delay, bandwidth utilization, and interlink load balancing. We assume that the smoothness factor is set to 0.8, that is, $\theta = 0.8$, demonstrating the importance of residual bandwidth and load in the comprehensive smoothness metric of disposable virtual pools. Our experimental environment is based on the virtual

Ubuntu 18.04 operating system built into the Windows 10 operating system, with an Intel (R) Core (TM) i5-5200U CPU. We implemented the simulation contrast experiment with the Mininet and Ryu network simulation platforms. In addition, we assume that the hosts between the source and destination generate traffic service requests with an equal probabilities at any time during the experimental verification and that there is no conflict between them. The payload of the request flow is randomly generated in the range of $(10, 150) Kbps$. We designed an experimental simulation scenario to evaluate how network throughput, transmission delay, and link utilization vary among different algorithms as the number of serviceable request flows increases. The details are as follows:

Scenario I. In the pair of switch nodes (v_0, v_{12}), we generate 200 requests for different types of flow services sequentially at equal time intervals.

Scenario II. We generate 200 different types of flow service requests in switch pairs (v_0, v_{12}), (v_4, v_8), (v_3, v_1), and (v_4, v_{14}) with the same probability and interval manner.

Scenario III. We select different source and destination switches with equal probabilities from the 15 OpenFlow switch nodes from v_0 to v_{14} to generate 200 different types of flow service requests with equal time intervals.

5.2 Results Analysis

Network Throughput. Figure 3 illustrates the variation in network throughput with increasing actual service request flows in three experimental scenarios, where panels (a), (b), and (c) depict the results for Scenarios I, II, and III, respectively. All four algorithms exhibit a positive correlation between network throughput and service flow volume. When the number of service flows reaches 200, network congestion occurs, with the LBROA method achieving the highest performance, as shown in Fig. 3 (a). As shown in Fig. 3 (b) and (c), when the service request flows increase, the LBROA method proposed in this paper performs well in terms of link throughput. The MHA and SWP algorithms demonstrate comparable performance, whereas SPF performs the worst. This is attributed in part to the LBROA's global utilization of all available network paths and its optimization of traffic distribution during transmission via the proposed pool comprehensive smoothness metric, resulting in balanced network traffic. Furthermore, as actual service request flows reach a certain threshold, network throughput growth gradually decelerates.

Link Utilization. Figure 4 illustrates the dynamic variation in link utilization as the volume of service request flows increases for each scenario. Specifically, Figs. 4 (a), (b) and (c) represent the experimental results for Scenarios I, II and III, respectively. As shown in Fig. 4, link utilization tends to increase with the growing number of service request flows in all the scenarios. In particular, the LBROA outperforms other algorithms by optimizing network traffic routing on

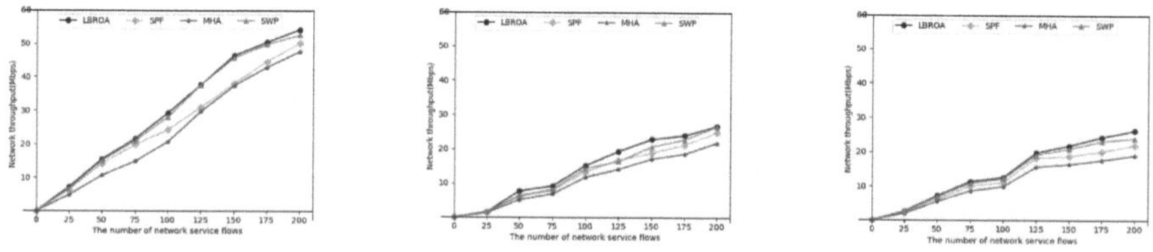

(a) Scenario I (b) Scenario II (c) Scenario III

Fig. 3. Comparative results of link throughput under three experimental scenarios.

the basis of residual bandwidth and network load, leveraging all available network paths, and balancing traffic distribution. The LBROA approach achieves optimal performance in the three scenarios with diverse flow generation patterns. While Scenario I involves resource competition primarily between identical source-destination pairs, Scenarios II and III introduce competition between different pairs. However, when the number of service request flows reaches a critical threshold, the growth rate of link utilization gradually decelerates.

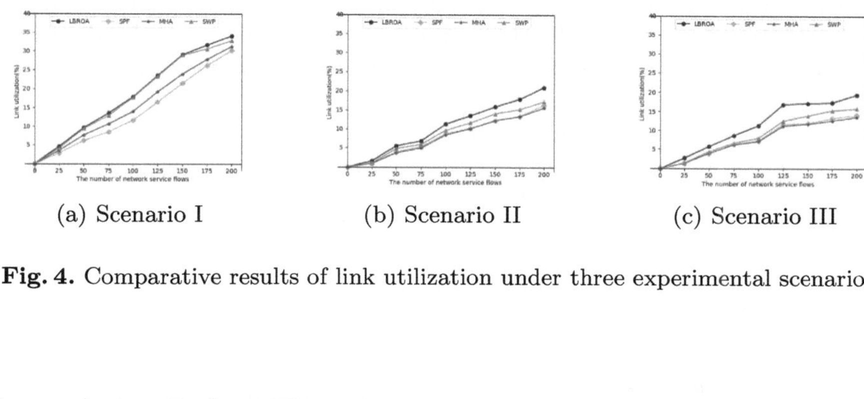

(a) Scenario I (b) Scenario II (c) Scenario III

Fig. 4. Comparative results of link utilization under three experimental scenarios.

Transmission Delay. This subsection presents the experimental results of the network transmission delay in three distinct scenarios. Figure 5 illustrates the dynamic variation in network transmission latency as the volume of service request flows increases for each scenario. Specifically, Figs. 5(a), (b) and (c) represent the experimental results for Scenarios I, II and III, respectively. As shown in Fig. 5, network transmission latency tends to increase with the growing number of service request flows in all the scenarios. The LBROA consistently outperforms other algorithms in terms of network transmission latency by optimizing the traffic distribution through multipath utilization. Compared with other algorithms, this approach effectively reduces network queueing delay and mitigates congestion risks, leveraging global path selection to balance traffic load.

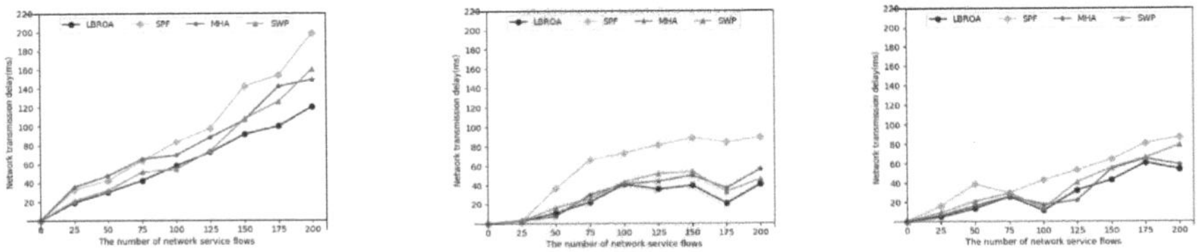

(a) Scenario I (b) Scenario II (c) Scenario III

Fig. 5. Comparative results of link utilization under three experimental scenarios.

The Interlink Load Balancing Degree. Figure 6 shows the measurement of the interlink load balance when the number of service request flows reaches 200 for each scenario. As shown in Fig. 6, the LBROA achieves the highest level of load balancing among the compared algorithms in all three scenarios. The superiority stems from the LBROA's strategy of leveraging multiple available paths and the comprehensive pool smoothness metric to balance the network load distribution and optimize network bandwidth allocation, thereby preventing network congestion caused by overutilization of specific links. In contrast, other algorithms prioritize optimally performing paths until resource exhaustion and then switch to alternative paths. This approach often leads to localized network congestion due to overloaded links, consequently degrading network transmission efficiency.

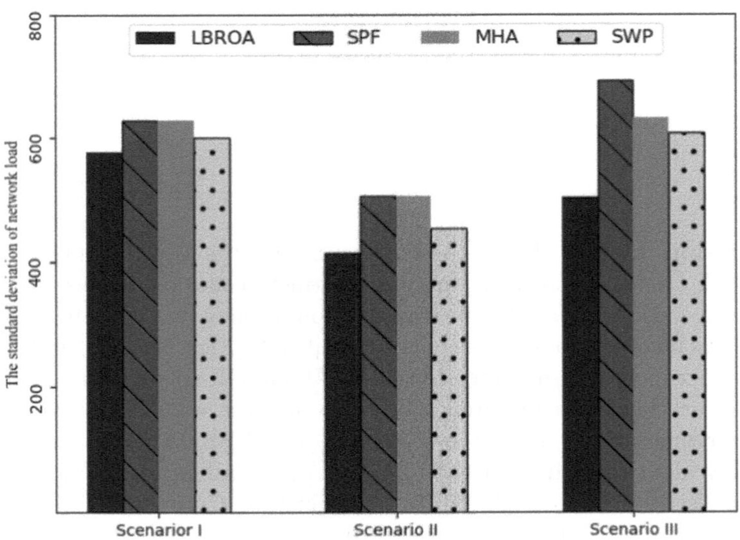

Fig. 6. Comparison chart of the inner-link load balancing degree in different experimental scenarios.

In conclusion, the experiments validate that the LBROA algorithm proposed in this paper performs better than the other algorithms in terms of link utilization, network transmission delay, and degree of load balance under different experimental scenarios. Leveraging SDN's centralized control and global network state information, the approach enables multipath selection by treating available paths as a disposable virtual resource pool, supporting routing optimization. The LBROA algorithm dynamically selects the optimal path from this pool to achieve a balanced traffic distribution, thus reducing the likelihood of network congestion, minimizing transmission queueing delay, and enhancing link utilization efficiency.

6 Conclusion

In this paper, we propose the LBROA within the SDN framework and validate its effectiveness through experimental evaluation. In three distinct experimental scenarios, we compared the proposed algorithm with three benchmark algorithms, demonstrating its competitive performance. First, the algorithm fully leverages SDN's centralized control and global network state information to enable multipath selection via linear cost functions incorporating multiple QoS metrics, modelling available paths as a unified virtual resource pool to support routing decision-making and optimization. Second, by integrating the link residual bandwidth and load metrics, we propose a comprehensive smoothness metric for the virtual pool, enabling dynamic performance assessment across different allocation states. Finally, preallocation and greedy strategies are applied to SDN centralized decision-making to identify optimal routing paths from multiple paths, collectively enhancing network traffic transmission efficiency.

However, the LBROA algorithm proposed in the SDN also has several shortcomings, such as high computational complexity, which is due primarily to the overhead of calculating multiple available paths and evaluating their comprehensive smoothness metrics in real time. In the future, on the basis of algorithm design and theoretical analysis, we will study how to take full advantage of centralized SDN control to help achieve QoS-aware routing optimization decisions. Additionally, the scalability of the LBROA in large-scale networks will be validated by simulating extended network topologies (e.g., ring and tree structures), addressing the current lack of research on large-scale scenarios. Furthermore, dynamic traffic scenarios, such as simulating sudden traffic surges, will can be incorporated to evaluate the adaptability of algorithms in real-world networks, complementing the existing focus on random traffic generation. We explore the introduction of intelligent optimization technologies into the current work, fully utilizing the advantages of SDN to achieve efficient utilization of network resources and data forwarding, intelligent QoS perception, and optimization capabilities.

References

1. Du, D.H.C.: Clean Slate Design of Internet for Supporting Service-Oriented Computing and Applications. In: IEEE International Conference on Service-Oriented Computing and Applications (SOCA '07), pp. 3–3. IEEE (2007)
2. Asadollahi, S., Goswami, B., Raoufy, A.S., Domingos, H.G.J.: Scalability of software defined network on floodlight controller using OFNet. In: 2017 International Conference on Electrical, Electronics, Communication, Computer, and Optimization Techniques (ICEECCOT), pp. 1–5. IEEE (2017)
3. Goswami, B., Hu, S.W., Feng, Y.M.: Software-Defined Networking for Real-Time Network Systems. In: Handbook of Real-Time Computing. Springer (2022)
4. Kretsis, A., Soumplis, P., Kokkinos, P.C., Varvarigos, E.: An SDN Emulation Platform for Converged Fiber-Wireless 5G Networks. In: 2021 International Conference on Computer Communications and Networks (ICCCN), pp. 1–7. IEEE (2021)
5. Manuel, T., Goswami, B.: Experimenting with scalability of Beacon controller in software defined network. Int. J. Recent Technol. Eng. (2019)
6. Gandotra, R., Perigo, L.: NFEH: An SDN Framework for Containerized Network Function-enabled End Hosts. In: 2020 29th International Conference on Computer Communications and Networks (ICCCN), pp. 1–6. IEEE (2020)
7. McKeown, N., et al.: OpenFlow: enabling innovation in campus networks. Comput. Commun. Rev. **38**, 69–74 (2008)
8. Huang, M.T., Liang, W.F.: Incremental SDN-Enabled Switch Deployment for Hybrid Software-Defined Networks. In: 2017 26th International Conference on Computer Communication and Networks (ICCCN), pp. 1–6. IEEE (2017)
9. Singh, S., Goswami, B., Paulose, J., Thomas, L.: Implementing and experimenting with openmul scalability in software defined network. J. Network Commun. Emerg. Technol. (JNCET) **11**(1) (2021)
10. Khan, M.A., Goswami, B., Asadollahi, S.: Data Visualization of Software-Defined Networks During & Load Balancing Experiment Using Floodlight Controller. In: Data Visualization: Trends and Challenges Toward Multidisciplinary Perception. Springer (2020)
11. Asadollahi, S., Goswami, B.: Experimenting with scalability of floodlight controller in software defined networks. In: 2017 International Conference on Electrical, Electronics, Communication, Computer, and Optimization Techniques (ICEECCOT), pp. 288–292. IEEE (2017)
12. Zeng, X., Wang, D.B., Han, S.W., Yao, W.B., Wang, Z., Chen, R.: An Effective Load Balance Using Link Bandwidth for SDN-Based Data Centers. In: ICAIS (2019)
13. Lin, T.L., Kuo, C.H., Chang, H.Y., Chang, W.K., Lin, Y.Y.: A Parameterized Wildcard Method Based on SDN for Server Load Balancing. In: 2016 International Conference on Networking and Network Applications (NaNA), pp. 383–386. IEEE (2016)
14. Li, L., Xu, Q.Z.: Load balancing researches in SDN: A survey. In: 2017 7th IEEE International Conference on Electronics Information and Emergency Communication (ICEIEC), pp. 403–408. IEEE (2017)
15. Badirzadeh, A., Jamali, S.: A survey on load balancing methods in software defined network. Netw. Commun. Eng. **10**, 21–27 (2018)
16. Mehra, M., Tiwari, N.K.: Network Load balancing in Software Defined Network: a Survey. Int. J. Appl. Eng. Res. (2019)

17. Bagha, M.A., Majidzadeh, K., Masdari, M., Farhang, Y.: ELA-RCP: An energy-efficient and load balanced algorithm for reliable controller placement in software-defined networks. J. Netw. Comput. Appl. **225**, 103855 (2024)
18. Ali, J., Song, H.H., Roh, B.H.: An SDN-Based Framework for E2E QoS Guarantee in Internet of Things Devices. IEEE Internet Things J. **12**(1), 605–622 (2025)
19. Hamdan, M., et al.: A comprehensive survey of load balancing techniques in software-defined network. J. Netw. Comput. Appl. **174**, 102856 (2021)
20. Semong, T., et al.: Intelligent Load Balancing Techniques in Software Defined Networks: a Survey. Electronics **9**, 1091 (2020)
21. Zhang, Y.J., Li, M.D., Tong, F.: An Energy-Efficient Load Balancing Scheme in Heterogeneous Clusters by Linear Programming. In: 2022 International Conference on Computer Communications and Networks (ICCCN), pp. 1–10. IEEE (2022)
22. Xu, A.X., Sun, S.M., Wang, Z.Y., Wang, X.F., Han, L.: Multi-Controller Load Balancing Mechanism Based on Improved Genetic Algorithm. In: 2022 International Conference on Computer Communications and Networks (ICCCN), pp. 1–8. IEEE (2022)
23. Forghani, M.R., Soltanaghaei, M.R., Boroujeni, F.Z.: Dynamic optimization scheme for load balancing and energy efficiency in software-defined networks utilizing the krill herd meta-heuristic algorithm. Comput. Electr. Eng. **114**, 109057 (2024)
24. Mahmoudi, M., Barekatain, B., Beheshti, Z., Ariza, A.Q.: MBL-DSDN: a novel load balancing algorithm in distributed software-defined networks based on microclustering and B-LSTM methods. J. Supercomput. **80**(14), 20421–20487 (2024)
25. Saeedi Goraghani, M., Afzali, M., Sharifi, F.: A reliable and load balancing controller placement method in software-defined networks. Int. J. Commun Syst **38**(2), e6059 (2025)
26. McKeown, N.: Software defined mobile networks. In: MobiHoc '09 (2009)
27. Akin, E., Korkmaz, T.: Comparison of Routing Algorithms with Static and Dynamic Link Cost in SDN. In: 2019 16th IEEE Annual Consumer Communications Networking Conference (CCNC), pp. 1–8. IEEE (2019)
28. Floyd, R.W.: Algorithm 97: shortest path. Commun. ACM **5**(6), 345 (1962)
29. Ma, Q.M., Steenkiste, P.: On path selection for traffic with bandwidth guarantees. In: Proceedings 1997 International Conference on Network Protocols, pp. 191–202. IEEE (1997)
30. Kodialam, M., Lakshman, T.V.: Minimum interference routing with applications to MPLS traffic engineering. In: Proceedings IEEE INFOCOM 2000. Conference on Computer Communications. Nineteenth Annual Joint Conference of the IEEE Computer and Communications Societies (Cat. No. 00CH37064), vol. 2, pp. 884–893. IEEE (2000)
31. Lunagariya, D., Goswami, B.: A Comparative Performance Analysis of Stellar SDN Controllers using Emulators. In: 2021 International Conference on Advances in Electrical, Computing, Communication and Sustainable Technologies (ICAECT), pp. 1–9. IEEE (2021)

Source Code Vulnerability Detection Method Based on Hybrid Retrieval-Augmented Generation Technology

Honggang Xie[1,2], Xiangdong Li[1,2](✉), and Zenghui Shen[1,2]

[1] Zhongyuan University of Technology, Zhengzhou 450007, Henan, China
lixiangdong2510@126.com
[2] Henan Key Laboratory on Public Opinion Intelligent Analysis, Zhengzhou 450007, China

Abstract. As software systems continue to grow in scale and complexity, vulnerability detection has become a critical task in ensuring software security. Recently, retrieval-augmented generation (RAG) techniques combined with large language models (LLMs) have demonstrated significant potential in source code vulnerability detection. However, existing RAG-based approaches typically rely on a single retrieval strategy, either sparse retrieval or dense retrieval. While sparse retrieval excels at precise keyword matching, it struggles with capturing deep semantic relationships, whereas dense retrieval, despite its ability to model semantic similarity, may fail to retrieve low-frequency or exact-match keywords. This trade-off limits both the coverage and accuracy of vulnerability information retrieval. To address these challenges, this paper proposes HRvul-RAG, a hybrid retrieval-based vulnerability detection framework that integrates BM25 and bge-m3 dual-modal retrieval. By leveraging the complementary strengths of sparse and dense retrieval, HRvul-RAG enhances the recall and retrieval precision. Furthermore, to refine the retrieval ranking, we incorporate the weighted reciprocal rank fusion (WRRF) algorithm, which dynamically adjusts the ranking weights on the basis of the query relevance, ensuring robustness and effectiveness in retrieval fusion. The optimized retrieval results are then fed into an LLM for vulnerability analysis, thereby improving the accuracy and intelligence of source code vulnerability detection.

Keywords: Vulnerability search · RAG · LLM · bge-m3 · BM25

1 Introduction

As software systems continue to grow in complexity and security threats escalate, intelligent source code vulnerability detection has become a critical focus in software security research. Traditional vulnerability detection methods rely primarily on static analysis [1] or dynamic analysis [2]. However, these approaches suffer from several limitations, including high false positive rates, significant computational overhead, and limited ability to detect unknown vulnerabilities.

L. Huang et al. (Eds.): CCF NCCA 2025, CCIS 2774, pp. 518–528, 2026.
https://doi.org/10.1007/978-981-95-5294-8_35

In recent years, retrieval-augmented generation (RAG) [3] models, combined with large language models (LLMs), have demonstrated significant potential in source code comprehension and vulnerability detection. For example, Vul-RAG enhances LLM-based vulnerability detection through a knowledge-level RAG [4], significantly improving the detection accuracy, with the human evaluation accuracy increasing from 0.60 to 0.77. However, existing RAG frameworks typically rely on a single retrieval strategy [5] during the knowledge retrieval phase. Sparse retrieval methods, such as keyword-based matching, struggle to capture the deep semantic meaning of vulnerability descriptions, leading to limited query expansion capabilities. Conversely, dense retrieval methods, which generate semantic vectors via deep learning models, effectively capture contextual information but may suffer from reduced retrieval precision when handling low-frequency terms or specific keyword-matching tasks.

Owing to the inherent trade-offs between exact keyword matching and deep semantic understanding, existing RAG-based vulnerability retrieval approaches still face challenges in terms of retrieval coverage and accuracy. To address these limitations, this paper proposes a hybrid retrieval-based intelligent source code vulnerability detection method (HRvul-RAG), which integrates dual-modal retrieval with BM25 [8] and bge-m3 [9]. This hybrid retrieval approach leverages the complementary advantages of sparse and dense representations to improve the quality of vulnerability information recall.

During the retrieval ranking phase, HRvul-RAG employs the weighted reciprocal rank fusion (WRRF) algorithm to merge results from different retrieval modes and dynamically adjusts the ranking weights on the basis of the query relevance, thereby optimizing the ranking effectiveness. Finally, the refined retrieval results are fed into the DeepSeek-R1 [10] LLM, leveraging its advanced code comprehension and vulnerability analysis capabilities to generate high-quality vulnerability detection results. This approach enhances the accuracy, reliability, and practicality of intelligent vulnerability detection.

2 A Source Code Vulnerability Detection Method Based on Hybrid Retrieval-Augmented Generation Technology

The implementation process of the proposed algorithm consists of several key stages. First, the collected vulnerability data undergo preprocessing, including tokenization, stop-word removal, text segmentation, and vectorization. Next, a hybrid retrieval approach is applied to obtain retrieval results from two different models. These results are then optimized via the weighted reciprocal rank fusion (WRRF) algorithm to increase the retrieval relevance. The optimized retrieval results are subsequently used as external knowledge input to the DeepSeek-R1 model, thereby improving its vulnerability analysis capabilities. Finally, the knowledge-augmented DeepSeek-R1 model is utilized for vulnerability detection. The overall workflow is illustrated in Fig. 1.

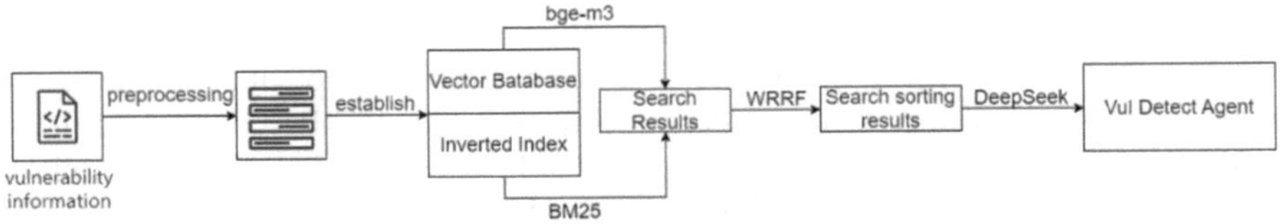

Fig. 1. RAG Application Flow

2.1 Vulnerability Data Collection and Preprocessing

The data are sourced primarily from publicly available databases such as CVE (Common Vulnerabilities and Exposures) [11], NVD (National Vulnerability Database) [12], and GitHub. Since multiple databases are used, duplicate vulnerability entries may exist, necessitating a deduplication process. Additionally, different data sources may have inconsistent formats, requiring standardization of vulnerability descriptions, code examples, and patch data. Noise removal is also essential, as raw data may contain irrelevant information. This includes eliminating meaningless characters such as HTML tags and URLs, as well as removing unrelated code segments, such as test code and comments.

2.2 Retrieval Model Based on Sparse Representation

BM25 (Best Matching 25) is a sparse retrieval model that is based on term frequency-inverse document frequency (TF-IDF) [13] and term saturation and is widely used in text information retrieval tasks. Its core principle is to measure the degree of match between query terms and documents and rank them on the basis of term weighting.

During the BM25 preprocessing stage, the vulnerability description text undergoes tokenization and lexical analysis via the Natural Language Toolkit (NLTK) to extract vulnerability-related keywords and critical terms accurately, thereby increasing retrieval precision. To facilitate efficient retrieval, an inverted index [14] is constructed, where each term is mapped to a list of vulnerability documents containing that term, along with statistical information such as term frequency. The construction of the inverted index not only accelerates the query process but also provides fundamental data support for the term frequency and inverse document frequency calculations of BM25, ensuring efficient relevance scoring for rapid vulnerability information retrieval and ranking.

Following BM25 preprocessing, the next step involves relevance computation and ranking of vulnerability information. First, a vocabulary is built across the entire corpus, assigning a unique index to each term for fast query matching. For each document, the term frequency is computed and combined with the inverse document frequency to assess its significance within the global corpus, thereby adjusting term weights. The BM25 relevance scoring formula is subsequently applied to evaluate the degree of match between the query terms and each document, generating a relevance score accordingly. Documents with higher scores indicate stronger relevance to the query terms and are prioritized in the final ranking, ensuring both the precision and interpretability of the retrieval results. The detailed calculation method is presented in Algorithm 1.

Algorithm 1 BM25

BM25

$$\text{score}(D, Q) = \sum_{t \in Q} IDF(t) \cdot \frac{TF(t, D) \cdot (k_1 + 1)}{TF(t, D) + k_1 \cdot (1 - b + b \cdot \frac{|D|}{\text{avgD}})} \tag{1}$$

where:

- $TF(t, D)$ is the term frequency of term t in document D.
- $IDF(t)$ is the inverse document frequency (IDF), computed as:

$$IDF(t) = \log \frac{N - n_t + 0.5}{n_t + 0.5} \tag{2}$$

  where N is the total number of documents, and n_t is the number of documents containing term t.
- $|D|$ is the length of document D, and avgD is the average document length.
- k_1 and b are tuning parameters.

2.3 Retrieval Model Based on Dense Representation

The bge-m3 model uses a transformer-based deep neural network architecture, which encodes input text through the self-attention mechanism to obtain high-dimensional semantic vector representations. During the vectorization process, the input text is first subjected to standard preprocessing, including data cleaning, tokenization, and the addition of special tokens (such as [CLS] and [SEP]). Afterward, bge-m3 computes the contextual representations of the text through multiple layers of the transformer and extracts the hidden layer vector at the [CLS] position as the global semantic representation. The generated vector is typically L2-normalized to ensure a consistent numerical scale across different text vectors, thereby improving the stability of subsequent similarity calculations. This vectorization process effectively captures the deep semantic features of vulnerability descriptions, patch information, and code snippets, providing high-quality vector representations for vulnerability information retrieval.

In terms of similarity calculation, bge-m3 primarily uses cosine similarity as the metric for retrieval ranking. This involves calculating the cosine of the angle between the query vector and the vulnerability document vector to assess their semantic similarity. The most relevant vulnerability documents are returned on the basis of the similarity scores. The detailed calculation method is presented in Algorithm 2.

Algorithm 2 bge-m3

BGE-M3

$$V_D = \text{Encoder}(D) \quad (1)$$

$$V_Q = \text{Encoder}(Q) \quad (2)$$

where:

- V_D and V_Q are the dense vector representations of the vulnerability document and query, respectively.

Similarity Computation

$$\text{Sim}(Q, D) = \frac{V_Q \cdot V_D}{\|V_Q\| \cdot \|V_D\|} \quad (3)$$

2.4 WRRF Sorting Algorithm

WRRF (weighted reciprocal rank fusion) is an optimization-based ranking method designed for fusing results from multiple retrieval models, aiming to improve the robustness and effectiveness of the overall ranking. Unlike the standard RRF (reciprocal rank fusion), the WRRF introduces adaptive weighting mechanisms for different retrieval models, enabling a more fine-grained and context-aware adjustment of each model's contribution to the final ranked list.

In the context of vulnerability information retrieval, textual descriptions of vulnerabilities often contain complex semantic structures and varying levels of keyword informativeness. Relying solely on sparse retrieval methods such as BM25, which are based on exact term matching, may fail to capture deeper contextual or semantic relationships. To address this limitation, we employ a query-aware dynamic weighting strategy that adjusts the weights of each retrieval model on the basis of query characteristics. Specifically, the weights are computed via features such as the query length and keyword density, which reflect the degree of lexical vs. semantic emphasis in the query.

For example, when the query is short and contains many exact keywords, a higher weight is assigned to the BM25 model; conversely, for longer or semantically rich queries, the dense retrieval model (bge-m3) is given more weight to better capture nuanced relationships. This dynamic fusion approach ensures that the final retrieval results retain both the precise matching capabilities of sparse retrieval and the semantic sensitivity of dense retrieval. The detailed scoring formula is illustrated in Algorithm 3.

Algorithm 3 Weighted Reciprocal Rank Fusion

Input: Query Q, document set D, retrieval models $M = \{\text{BM25}, \text{BGE-M3}\}$, smoothing constant k

Output: Final WRRF score for each document $D_i \in D$

1: Compute query features:
- L_Q: length of query Q (number of tokens)
- K_Q: number of keyword matches (e.g., based on TF or term presence)

2: Calculate dynamic weights:

$$w_{\text{BGE}} = \frac{L_Q}{L_Q + \alpha \cdot K_Q}, \quad w_{\text{BM25}} = 1 - w_{\text{BGE}} \tag{1}$$

3: **for all** document $D_i \in D$ **do**

4: Compute WRRF score:

$$\text{WRRF}(D_i) = \sum_{m \in M} w_m \cdot \frac{1}{k + R_m(D_i)} \tag{2}$$

5: **end for**

By incorporating the WRRF, the performance of a RAG-based (retrieval-augmented generation) vulnerability information retrieval system can be effectively improved, providing more accurate contextual information before generating an analysis with the large model DeepSeek-R1.

2.5 Knowledge Enhancement

The sorted and optimized retrieval results are used as structured knowledge input to the DeepSeek-R1 large model, enhancing its ability to understand and analyse vulnerability information. DeepSeek-R1 possesses strong code comprehension and reasoning abilities, allowing it to accurately interpret vulnerability descriptions and exploit code and patch information. However, owing to the limitations in the model's parameter scale and the coverage of training data, relying solely on its internal knowledge may lead to information gaps or misjudgments. Therefore, the hybrid retrieval strategy, combined with WRRF sorting optimization, ensures that the context data input to DeepSeek-R1 not only includes highly relevant text based on term matching but also incorporates vulnerability information derived from deep semantic understanding.

Through this knowledge enhancement mechanism, DeepSeek-R1 can make full use of precisely extracted context in vulnerability analysis tasks, improving the accuracy of vulnerability type identification, exploitation pattern inference, and repair recommendation. This further enhances the reliability and practicality of intelligent source code vulnerability detection.

3 Experiment and Analysis

We build HRvul-RAG on the DeepSeek-R1 model, leveraging its outstanding performance in code generation, reasoning, and repair tasks. These capabilities make DeepSeek-R1 highly valuable for vulnerability detection, code security analysis, exploit generation, and mitigation strategy automation. By integrating the retrieval-augmented generation (RAG) framework, DeepSeek-R1 gains the ability to retrieve external vulnerability knowledge dynamically, allowing it to adapt to evolving security threat scenarios.

3.1 Experimental Computational Cost Analysis

In this study, the model is deployed via a remote invocation approach, where the client leverages a Python script to interact with DeepSeek's RESTful API for inference tasks. All computational operations are performed on DeepSeek's cloud infrastructure, while the client is only responsible for constructing the request and parsing the response. This "lightweight client + cloud inference" architecture significantly reduces local computational requirements and offers high scalability and ease of integration into various environments.

During actual testing, we measured the end-to-end inference latency, defined as the time from request submission to response receipt. With an average input length of approximately 150 tokens, the response time of DeepSeek R1 remained stable between 1.2 and 1.4 s. The observed latency was influenced primarily by factors such as network conditions, server load, and input complexity. Nevertheless, the latency is within an acceptable range for most automated source code vulnerability detection scenarios.

Moreover, since inference is entirely cloud-based, local resource consumption is minimal. Empirical results show that the Python client script consumes a peak memory of approximately 30 MB, which includes memory used for JSON request construction, response handling, and result processing. Overall, this deployment setup achieves a favourable balance between computational efficiency and resource utilization, demonstrating the practicality of integrating the model into real-world software security analysis systems.

3.2 Effectiveness Evaluation

We evaluate HRvul-RAG on a constructed program containing a deserialization vulnerability. The results generated by HRvul-RAG are presented in Table 1 and Table 2. To enhance clarity, we exclude intermediate reasoning steps of the large model and optimize certain parts of the generated content. The results demonstrate that HRvul-RAG can accurately detect the presence of vulnerabilities within the code and provide targeted remediation recommendations. (For conciseness, we have omitted the specific modified code from the suggested fixes in the tables).

```
public class T3Handler {
    public void processRequest(InputStream input) {
        try {
            ObjectInputStream ois = new ObjectInputStream(input);
            Object obj = ois.readObject();
            processDeserializedObject(obj);
        } catch (Exception e) {
            e.printStackTrace();
        }
    }
    private void processDeserializedObject(Object obj) {
        if (obj instanceof RemoteObject) {
            try {
                Method method = obj.getClass().getMethod("getObject");
                Object result = method.invoke(obj);
                if (result instanceof byte[]) {
                    Object nestedObj = deserialize((byte[]) result);
                    executeBusinessLogic(nestedObj);
                }
            } catch (Exception e) {
                e.printStackTrace();
            }
        }
    }
    private Object deserialize(byte[] data) throws Exception {
        ByteArrayInputStream bais = new ByteArrayInputStream(data);
        ObjectInputStream ois = new ObjectInputStream(bais);
        return ois.readObject();
    }
}
```

Program Code Segment

Q: Help me analyse what vulnerabilities may exist in this program?
A: 1.CVE-2021-2394 (WebLogic secondary deserialization) • Trigger point: deserialize ((byte []) result) secondary deserialize operation • Attack characteristics: Obtain nested serialized objects through RemoteObject reflection to form a blacklist to bypass the attack chain 2. CVE-2020-14825 (Dangerous Reflection Call) • Trigger point: method.invoke (obj) dynamic method invocation • Utilization condition: combined with getObject () method to trigger malicious class initialization 3. CVE-2020-14756 (T3 protocol deserialization) • Trigger point: ObjectInputStream (input) original deserialization operation • Protocol risk: exposed T3 protocol processing entry without access control

Q: Give repair suggestions
A: 1. Official patch upgrade • Immediately apply the official Oracle security patch (refer to the CVE-2021-2394 fix) •Upgrade to the latest version of Weblogic (12.2.1.4 + or 14.1.1.0 +) 2. Deserialization protection •Enable the Weblogic global deserialization filter (JEP 290) •Configure jdk.serialFilter whitelist to restrict deserializable classes • Disable ObjectInputStream native deserialization method 3. Protocol layer reinforcement • Disable the T3/T3S protocol (just keep the HTTP/HTTPS protocol) • Enable T3 Protocol Access Control Lists (ACLs) •Force enable TLS mutual authentication 4. Runtime protection • Deploy RASP protection (to block dangerous reflex calls) • Configure JVM security policies to restrict Method.invoke () calls •Enable SecurityManager to control sensitive operations 5. Defense in depth strategy •Filter T3 protocol traffic containing java.rmi.RemoteObject at network boundaries • Signature verification of nested serialized data • Implement a strong validation mechanism for business object types 6. Monitoring and response •Open the Weblogic security audit log (to monitor readObject operations) • Deploy IDS/IPS rules to detect deserialization attack signatures •Establish an abnormal reflection call alarm mechanism

3.3 Performance Evaluation

To further evaluate the performance of the hybrid retrieval-based vulnerability information ranking method, three sets of different types of vulnerabilities were selected from the CVE vulnerability database as the test set. The retrieval and ranking were then conducted via three methods: BM25, bge-m3, and HRvul-RAG. The experimental results are shown in Table 1.

The results clearly show that the hybrid method achieves recall rates similar to those of the dense retrieval-based bge-m3 method but with higher retrieval efficiency, comparable to that of the BM25 method. In certain complex vulnerability query tasks, relying solely on bge-m3 for dense retrieval leads to increased computational load and longer query times. The hybrid method effectively combines the efficiency of BM25 with the deep representation capabilities of bge-m3, overcoming the limitations of using a single method.

Particularly, in cases where the CVE database contains vulnerabilities with similar descriptions but different scopes of impact, dense representations may result in misjudging the relevance. However, the hybrid method, which adjusts the rankings through the WRRF, is able to more accurately increase the ranking of relevant vulnerabilities.

Table 1. Analysis results of the different methods on the test set

Methons	BM25			bge-m3			HRvul-RAG		
vulnerability type	Acc	F1	Recall	Acc	F1	Recall	Acc	F1	Recall
RCE	75%	65%	69%	78%	70%	73%	85%	80%	82%
SQL Injection	80%	68%	72%	83%	72%	75%	88%	81%	84%
Directory Traversal	78%	66%	70%	79%	68%	72%	86%	78%	81%
XSS	77%	65%	69%	80%	70%	73%	87%	79%	82%

4 Conclusion

In summary, compared with the single-retrieval method, hybrid retrieval has superior query performance on the vulnerability data retrieval task, which can improve the retrieval efficiency while maintaining a high recall rate and avoid the excessive computational complexity of the dense method, which has high practical value. In the future, with the improvement of the self-learning and dynamic adjustment ability of the large model, the vulnerability detection of HRvul-RAG will be more flexible, able to adapt quickly to new attack methods and vulnerability characteristics, and improve the identification ability of zero-day vulnerabilities. Combined with intelligent data analytics and reasoning mechanisms, HRvul-RAG's vulnerability detection system will reach new heights in terms of efficiency, accuracy and reliability, with greater practical value and wider application prospects.

Acknowledgements. This work was supported by the Graduate Research and Innovation Program of Zhongyuan University of Technology with project number YKY2025ZK16 and the Science and Technology Key Project of Henan Province with project number 252102210217.

References

1. Jovanovic, N., Kruegel, C., Kirda, E.: Pixy: a static analysis tool for detecting web application vulnerabilities. In: 2006 IEEE Symposium on Security and Privacy (S&P'06), pp. 6-pp. IEEE (2006)
2. Wang, F., Shoshitaishvili, Y.: Angr-the next generation of binary analysis. In: 2017 IEEE Cybersecurity Development (SecDev), pp. 8–9. IEEE (2017)

3. Lewis, P., et al.: Retrieval-augmented generation for knowledge-intensive NLP tasks. Adv. Neural. Inf. Process. Syst. **33**, 9459–9474 (2020)
4. Du, X., et al.: Vul-rag: enhancing LLM-based vulnerability detection via knowledge-level rag. arXiv preprint arXiv:2406.11147 (2024)
5. Daneshvar, S.S., Nong, Y., Yang, X., Wang, S., Cai, H.: Exploring RAG-based Vulnerability Augmentation with LLMs. arXiv preprint arXiv:2408.04125 (2024)
6. Zhang, Z., Xu, Y., Yang, J., Li, X., Zhang, D.: A survey of sparse representation: algorithms and applications. IEEE Access **3**, 490–530 (2015)
7. Zhou, J., Zheng, H., Pan, L.: Ensemble clustering based on dense representation. Neurocomputing **357**, 66–76 (2019)
8. Robertson, S., Zaragoza, H.: The probabilistic relevance framework: BM25 and beyond. Found. Trends® Inf. Retrieval **3**(4), 333–389 (2009)
9. Chen, J., Xiao, S., Zhang, P., Luo, K., Lian, D., Liu, Z.: Bge m3-embedding: multilingual, multifunctionality, multigranularity text embeddings through self-knowledge distillation. arXiv preprint arXiv:2402.03216 (2024)
10. Guo, D., et al.: Deepseek-r1: incentivizing reasoning capability in LLMs via reinforcement learning. arXiv preprint arXiv:2501.12948 (2025)
11. The website of common vulnerabilities and exposures (2025). https://cve.mitre.org/
12. Booth, H., Rike, D., Witte, G.A.: The national vulnerability database (NVD): overview (2013)
13. Aizawa, A.: An information-theoretic perspective of TF–IDF measures. Inf. Process. Manage. **39**(1), 45–65 (2003)
14. Pibiri, G.E., Venturini, R.: Techniques for inverted index compression. ACM Comput. Surv. (CSUR) **53**(6), 1–36 (2020)
15. Cormack, G.V., Clarke, C.L., Buettcher, S.: Reciprocal rank fusion outperforms condorcet and individual rank learning methods. In: Proceedings of the 32nd International ACM SIGIR Conference on Research and Development in Information Retrieval, pp. 758–759 (2009)

Intrinsic Security of Artificial Intelligence: System Vulnerabilities, Defense Mechanisms, and Theoretical Deepening

Xinnian Wang, Ao Hu(✉), Xincheng Liu, and Botao Wang

South-Central Minzu University, Wuhan, China
2024110299@mail.scuec.edu.cn

Abstract. The widespread application of artificial intelligence (AI) technology faces critical challenges from inherent security threats, hindering the development of trusted AI ecosystems. This paper presents a systematic review of AI's intrinsic security, proposing a three-dimensional analytical framework covering data, model, and system layers to analyse core risks such as data contamination, adversarial sample attacks, and API abuse, alongside their cross-layer coupling effects. We then design a comprehensive dynamic defense system that integrates prevention, detection, and response and empirically validates the system across medical, financial, and autonomous driving domains, demonstrating significant risk reduction. Furthermore, the study outlines a phased technical roadmap for integrating quantum computing and edge AI to address the 'impossible triangle' of privacy, efficiency, and real-time performance. Finally, it advocates for interdisciplinary collaborative governance—combining legal, ethical, and technological approaches—to build a resilient and trustworthy AI ecosystem, identifying future research priorities for adaptive defense and full lifecycle verification.

Keywords: Artificial Intelligence Security · Endogenous Security · Quantum Computing · Edge AI · Interdisciplinary Collaborative Governance · Privacy-efficiency Tradeoff · Dynamic Defense Architecture

1 Introduction

The rapid development of artificial intelligence (AI) technology is profoundly reshaping the global industrial landscape and social operation models. However, the inherent security issues associated with it have become a major challenge that restricts the sustainable development of this technology. Inherent security refers to risk of functional failure is directly caused by design flaws, algorithmic vulnerabilities, and implementation loopholes within AI systems. The root cause can be traced back to the inherent limitations of the data, models and system architecture [1]. For example, the "black box" nature of deep neural networks makes it difficult to explain decision logic, and data pollution and adversarial

L. Huang et al. (Eds.): CCF NCCA 2025, CCIS 2774, pp. 529–546, 2026.
https://doi.org/10.1007/978-981-95-5294-8_36

sample attacks can undermine the reliability of model output through small perturbations [2]. This threat not only exists at the technical level but also may trigger a chain reaction at the physical or social level due to cascading effects, such as an accident caused by an algorithmic error in the distance of obstacles in an autonomous driving perception system [3].

Current research indicates that AI inherent security risks are inevitable, hidden, and cross-layer coupled. Statistical learning theory points out that data distribution bias and model overfitting are inherent flaws in the statistical learning paradigm, whereas hardware computational errors (such as FP16 precision loss) can propagate through deep networks, potentially leading to catastrophic output deviations [4]. In addition, the proposed dynamic heterogeneous redundant architecture provides a new idea for endogenous security, which realizes the self-repair and anti-attack capability of the system through the construction effect. However, its engineering implementation still faces the problems of robustness verification and cost control [5].

Policy and technology cogovernance is key to addressing intrinsic security. The literature [1] emphasizes the need to build a multilayered defense system from three aspects, namely, data governance, algorithmic explainability, and system-level protection, while promoting legal frameworks and standardization to balance technological innovation with risk control. However, there are still significant gaps in existing research: the literature [2] indicates that dynamic defense mechanisms are insufficiently adaptive to zero-day attacks and that the collective efficiency of technology, ethics, and law in interdisciplinary governance must be improved [6]. For example, while the application of AI in the military can improve operational efficiency, the risk of losing control of the system can lead to unpredictable consequences, and 'meaningful human control' must be ensured [7].

This study employs the systematic review method, adhering to PRISMA guidelines for literature screening and evaluation to ensure the comprehensiveness and reproducibility of the research conclusions. On the basis of ISO/IEC safety standards and complex systems, a three-dimensional analysis framework of the "data-model-system" is constructed, overcoming the limitations of traditional single-level research. The main contributions include systematically mapping out the cross-layer transmission mechanisms of intrinsic security threats; proposing technical pathways; empirically validating dynamic defense systems; and advocating interdisciplinary collaborative governance involving law, ethics, and technology.

2 Theoretical Trace and Philosophical Foundation

Existing research focuses on the technical definition of endogenous security (such as algorithmic vulnerabilities and data pollution) but lacks a systematic explanation of the theoretical origins and philosophical dimensions. This section reconstructs the conceptual framework of endogenous security by combining the philosophy of technology and systems science theory.

2.1 The Correlation Between Endogenous Security and Endogenous Growth Theory

The concept of endogenous security can be traced back to the endogenous growth theory in economics [8]. The core arguments can be divided into two points:

Root causes of endogenous risks: Unlike external shocks, endogenous security threats arise from the interaction of internal elements of a system (such as data-model-system design flaws), rather than direct intrusion by external attacks.

Knowledge spillover effect: Similar to the endogenous driving force of technological progress in economic growth, the evolution of security defense mechanisms (such as confrontation training and formal verification) depends on the accumulation and iterative optimization of knowledge within the system.

For example, the client gradient difference in federated learning (endogenous data distribution shift) leads to the failure of global model convergence, which needs to be achieved through dynamic optimization strategies (such as momentum gradient aggregation) for "safe growth".

2.2 Philosophy of Technology Perspective: Technology Autonomy and Risk of Loss of Control

The philosophy of technology provides a critical perspective for endogenous security and reveals the paradox of the autonomy of the technical system.

The "political nature of technology" of Winner [9]: Technical design implies political decision-making. For example, the path planning algorithm of autonomous driving may prioritize protecting passengers in the car ("electric trolley problem") and reflect the ethical tendency of designers. This shows that algorithmic loopholes are not only technical defects, but also conflicts in design values (such as efficiency over fairness).

Heidegger's "technological seat" theory: Technology systems (such as AI) "frame" human behavior through standardized processes, resulting in unpredictable risks of loss of control. For example, recommendation algorithms reinforce information cocoons through user behavior feedback cycles, leading to social cognitive polarization (intrinsic social risk).

2.3 Systematic Scientific Basis: Resilience Engineering Theory

Resilience engineering theory emphasizes the adaptability and recovery ability of a system in disturbance, and provides methodological guidance for endogenous security defenses [10]. The principle of resilient engineering holds that the safety of complex systems not only relies on fault prevention but also requires redundant design and dynamic adaptation to achieve "resilience to disturbances-self-repair" capabilities. For example, literature [5] noted out that heterogeneous redundant architecture (DHR) reduces error rates from 1e-4 to 1e-9 through a majority voting mechanism, serving as a typical practice of resilient engineering in AI systems. Resilience engineering theory provides a methodological framework for "inherent security," emphasizing the shift from passive defense to proactive resilience construction (see Fig. 1).

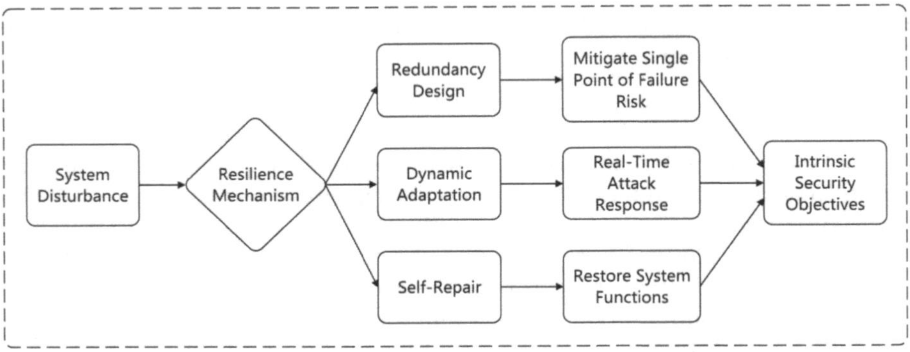

Fig. 1. Logical correlation between intrinsic security and Resilience Engineering

3 Threat Classification and Theoretical Framework

The complexity of the inherent security threats in artificial intelligence requires hierarchical deconstruction from a systematic perspective. On the basis of the three-dimensional framework of "data-model-system" proposed in the literature [1], this section combines technical principles with empirical cases to analyse the core threats at each level and their internal connections in detail.

3.1 Data Layer Threat: Pollution Injection and Distribution Shift

As the input basis of the AI system, the quality defects of data directly lead to the failure of the model. The main threats include the following three points:

Data Pollution (Data Poisoning): Attackers inject malicious samples or tamper with labels in training data, leading to the model learning incorrect rules. By injecting only 0.1% of malicious samples through the gradient alignment strategy, the ImageNet accuracy of ResNet-152 can be reduced by 41% [11]. For example, tampering with CT image labels in medical datasets can increase the false negative rate of lung cancer detection models by 18% [12].

Data distribution shift: The distribution of training data differs from that of real-world data, leading to a sharp decline in model performance. Unlike active attacks involving data contamination, data distribution shifts are more often due to inherent differences between training and testing scenarios. When the Wasserstein distance between test data and the training set exceeds a critical threshold, model performance decreases exponentially [13]. The model performance shows exponential decay (Formula 1):

$$W(P_{train}, P_{test}) > \delta \Rightarrow Accuracy \propto e^{-\lambda W}$$

In the autopilot system, after the CNN feature extraction, the pixel error caused by optical distortion of the camera is amplified to 32% because of lane line recognition deviation [14].

Noise sensitivity: Sensor noise or transmission error is amplified layer by layer in the model, causing output deviation. The sensor acquisition error is amplified layer by layer in the deep learning model. For example, FP16 quantization error causes a loss of 56% in the transformer attention matrix rank [15].

3.2 Model Layer Threat: Algorithmic Vulnerability

The core contradiction of the model layer threat lies in the vulnerability of algorithm design, which is embodied in the following four aspects:

Adversarial sample attack: adding small disturbances to the input data that are difficult for humans to detect [16]. For example, in d road sign recognition model, an antisticker can increase the probability of misjudging a parking sign up to 98% [17].

Training dynamics failure: During model training, the system fails to converge to a stable state because of gradient anomalies or optimizer defects. For example, when the gradient direction difference between clients in federated learning exceeds 90°, the global model convergence speed decreases by 83%. Additionally, incomplete knowledge transfer during the model distillation process can significantly reduce the robustness of the student model [18].

Algorithmic logic defics: Model design does not cover key scenarios or ignores dynamic variables, leading to systematic decision bias.

Generative AI attacks: These can be categorized into prompt injection and sponge attacks on large language models. A 2023 OpenAI study revealed that large language models (such as GPT-4) are vulnerable to prompt injection attacks, where attackers induce the model to generate harmful content through semantic manipulation. Defence against these attacks requires a combination of dynamic input filtering and adversarial training optimization [19]. A sponge attack occurs when an attacker uses a specially designed input consumption model to calculate resources and trigger a denial of service (DoS) attack that threatens the availability of cloud reasoning services.

3.3 System Layer Threat: Hardware and Software Combination Vulnerability

System implementation and architectural defects exacerbate inherent security risks, and the main threats include the following three points:

API logic vulnerabilities: The service is maliciously exploited because of interface design defects (such as missing input validation). For example, a voice assistant that does not filter abnormal inputs causes service crashes due to buffer overflow, and the system downtime rate increases from 0.3% to 5% [20].

Dependency chain pollution: The third-party library or framework is implemented with backdoor code, which indirectly compromises the security of the AI system. For example, backdoors in third-party libraries (such as PyTorch malicious operators) can bypass model encryption and leak 98% of medical image privacy data [21].

Resource competition defects: Hardware resource allocation conflicts (such as memory overlimits) lead to calculation errors. For example, GPU memory overload causes cross-pollution of batch processing data, and the mAP of the target detection model decreases by 29% [22].

3.4 Threat Coupling: Cross-Layer Risk Transmission Mechanism

The cross-layer coupling effect of endogenous security threats is significant. For example, the noise sensitivity of the data layer and the hardware calculation error of the system layer work together to cause the false detection rate of the autonomous driving perception module to increase from 7% to 40% in low-light environment [23]. Wu et al. [5] noted that adversarial sample attacks at the model layer can further spread to cloud inference services through API vulnerabilities, leading to cascading security incidents. As depicted in Fig. 2, the intricate cross-layer coupling of intrinsic security threats is evident. For example, data layer noise sensitivity, which originates from sources such as sensor errors, can be progressively amplified through the dynamic processes of model training. This amplification propagates across layers, ultimately manifesting as system-level vulnerabilities such as hardware resource competition defects or cascading failures in deployed AI services. Figure 2 visually illustrates this multidirectional propagation path, highlighting how an initial perturbation at one layer can trigger a chain reaction that impacts the integrity and availability of the entire AI system. This closed-loop feedback mechanism across layers requires that defense technologies must cover all levels.

The above analysis indicates that intrinsic security threats exhibit cross-layer conductivity and collaborative destructiveness. For example, the synergistic effect of data layer noise and hardware errors can significantly amplify model output biases. Inspired by resilience engineering theory, Fig. 3 elaborates on the comprehensive threat propagation mechanism in AI. This emphasizes that effective defense necessitates building system resilience not only through static fault prevention, but also through dynamic adaptation and robust redundancy design. This framework underscores the need for holistic defense mechanisms that actively identify and block risk propagation paths across the data, model, and system layers, preventing localized vulnerabilities from escalating into systemic failures. To address these challenges, it is necessary to build a dynamic defense system covering data, model, and system layers, integrating a three-pronged technical approach of prevention, detection, and response.

4 Defense Mechanism and Technical Path

In view of the multidimensional threats of endogenous security, it is necessary to build a dynamic defense system covering data, models and system layers. On the basis of the principle of the "prevention-detection-response" triad, this chapter proposes a systematic defense scheme based on the combination of cutting-edge technological progress and industry practice.

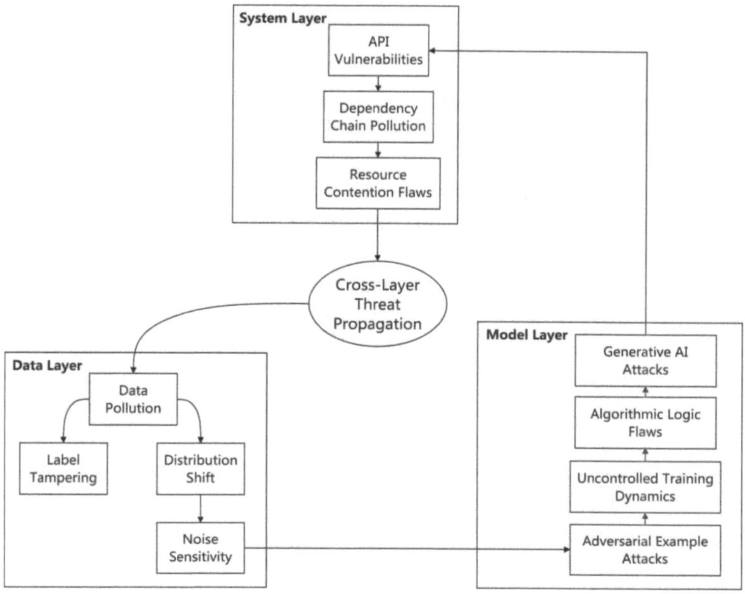

Fig. 2. Schematic diagram of the three-dimensional framework of "data-model-system" threat transmission

4.1 Data Quality Governance: Anti-purification and Dynamic Alignment Technologies

In terms of data quality governance, adversarial data cleaning techniques can effectively identify and repair contaminated samples by reconstructing the input space through generative adversarial networks (GANs). For example, experiments on the CIFAR-10 dataset show that adversarial data cleaning networks (APNs) can remove 89% of poisoned samples, restoring the model accuracy to 97% of the baseline level [24]. Dynamic distribution alignment technology is based on domain-adaptive algorithms for the Wasserstein distance, which dynamically adjusts the feature distributions of training and test data in real time, significantly alleviating the issue of distribution bias. Yoon et al. [24] investigated in autonomous driving systems, by integrating LiDAR point clouds in RGB images for multimodal alignment, lane line recognition errors in rainy environments are reduced from 32% to 8%. Additionally, literature [12] noted that noise adaptive filtering techniques, which embed Kalman filtering and adaptive threshold control at the sensor level, can suppress the layer-by-layer propagation of noise. As a result, the false detection rate for defect detection by industrial cameras under strong light interference has been reduced from 25% to 6%.

4.2 Model Robustness Enhancement: Dynamic Adversarial Training and Formal Verification

The core of enhancing model robustness lies in the combination of dynamic adversarial training and formal verification. Dynamic adversarial training (DAT) leverages meta-learning to generate an attack strategy space, dynamically adjusting the perturbation magnitude and number of attack iterations. This approach improves the robust accuracy of ViT models against PGD attacks from 58% to 81% on CIFAR-100 [25]. A formal robustness verification tool chain (such as the Marabou framework) proves the output stability of a neural network within a specific disturbance range through an interval propagation algorithm. In the medical field, it successfully controls the decision error rate of key nodes in the heart disease diagnosis model to be less than 0.1% cite26. Neural symbol fusion technology enhances interpretability by embedding symbolic logic into neural networks. By incorporating symbolic logic rules into deep neural networks, it constrains the inference paths of models. For example, after introducing a rule engine in financial risk control models, the fluctuation of misjudgment rates for high-risk groups is limited to $\pm 5\%$ [27].

Dynamic adversarial training improves the robustness of the model by simulating attacks, whereas formal verification ensures the stability of the output from a mathematical perspective. The combination of the two can form a closed-loop defense of attack and defense coordination.

4.3 Trusted System Architecture: Heterogeneous Redundancy and Zero Trust Policy

The design of a trusted system architecture needs to shift from hardware isolation to dynamic defense strategies. The heterogeneous redundant execution architecture deploys triple module redundancy (TMR) on the FPGA, shielding against hardware failures through a majority voting mechanism, reducing the error rate of autonomous driving computing units from 1e-4 to 1e-9, meeting ISO 26262 ASIL-D security levels [28]. The framework of fully homomorphic encryption (FHE, which allows direct computation on encrypted data) supports the model reasoning in the encrypted state. Although the inference delay increases by 12 ms in medical image diagnosis, the accuracy is still maintained at 93% [29]. The zero-trust dynamic policy is based on real-time threat hunting in the MITRE ATT&CK framework, which dynamically adjusts API access permissions. A financial platform has reduced the penetration rate of horizontal attacks by 90% through microisolation technology [30].

4.4 Cross-Layer Defense Coordination: Blocking the Risk Transmission Path

Cross-layer defense coordination is the key to blocking the risk transmission path. The noise-hardware joint optimization reduces the false detection rate of the autonomous driving night perception module from 40% to 7% through the coordinated design of sensor noise suppression and error correction of computing units, while increasing the GPU power consumption by only 8% [31]. Wu et al. reported that the model-system level defense chain, through the synergy of adversarial training and API input detection, successfully reduced the cascading attack success rate of the OCR invoice recognition system from 70% to 3%. The collaborative application of these technical approaches indicates that a dynamic heterogeneous redundancy architecture and cross-layer verification mechanisms can significantly enhance the inherent security of AI systems.

As illustrated in Fig. 3, the proposed dynamic defense system rigorously adheres to the "prevention-detection-response" triad, forming a closed-loop security chain. In the prevention stage, initial risks, such as data contamination, are mitigated through proactive measures such as adversarial data cleaning. The detection stage involves real-time monitoring of threats via advanced techniques such as dynamic distribution alignment technology to identify anomalies and attacks promptly. Finally, the response stage leverages fault-tolerant switching and self-healing mechanisms to quickly restore system functions and contain the impact of any security incidents, ensuring continuous operational integrity.

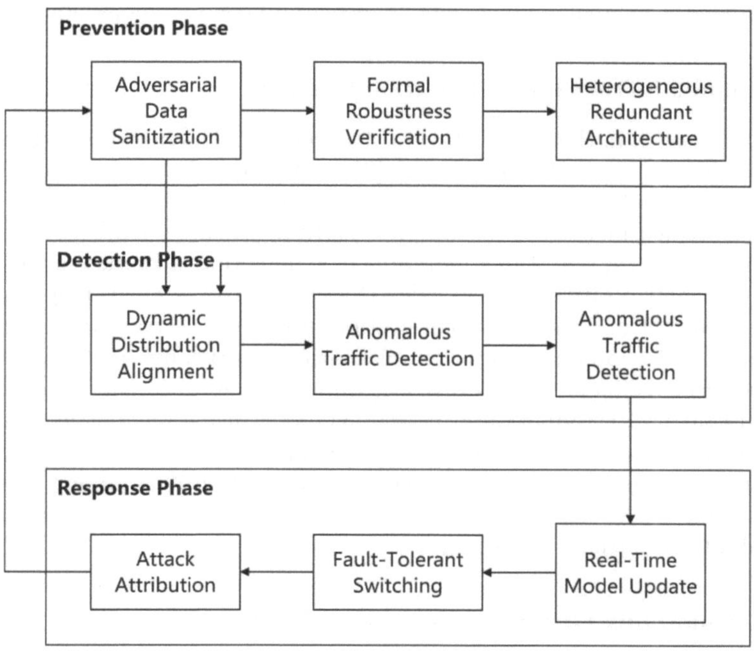

Fig. 3. Dynamic defense technology path

5 Cross-Domain Empirical Analysis of Defense Effectiveness

To verify the actual effectiveness of the above defense mechanisms, this paper selects the fields of medical care, finance and autonomous driving, and reveals the adaptability and limitations of the technology through cross-scenario empirical analysis.

5.1 Medical Imaging Diagnostic System

In the field of medical imaging diagnosis, the combined application of dynamic adversarial training and formal verification has significantly enhanced system robustness. For example, in response to metal artifacts in CT scan data, the introduction of adversarial purification networks (APNs) and formal verification within the Marabou framework reduced the false negative rate of lung cancer detection models from 25% to 8%, while maintaining model inference latency under 47 ms and meeting real-time clinical requirements [32]. This result shows that the collaborative defense between the data layer and model layer can effectively address the threats of endogenous security in complex noise environments.

5.2 Financial Transaction Security Defense

Empirical research in the financial sector focuses on balancing privacy protection with real-time defense. Juvekar et al. [29] point out that a bank deployed a credit card fraud detection system via a fully homomorphic encryption (FHE) framework, which increased the time for analysing a single transaction by only 11 ms in the encrypted state, maintained a detection accuracy of 97%, and reduced the false positive rate by 23%. However, Juvekar et al. [29] also noted that the high cost of encrypted computation in large-scale transaction scenarios still limits its widespread adoption, necessitating algorithmic lightweighting and hardware acceleration for collaborative optimization. Additionally, Sun et al. [26] indicated that the introduction of neural symbolic fusion technology makes the decision logic of risk control models traceable, and by embedding counterfactual causal analysis rules, it reduces the loan rejection rate differs among users from different regions from 15% to 3%, highlighting the importance of algorithmic explainability in fairness governance.

5.3 Autonomous Driving Perception Module

Unlike the medical field, which focuses on privacy protection, autonomous driving scenarios emphasize real-time and multisensor collaboration, which requires the defense technology to have dynamic adaptation capability.

The intrinsic safety defense verification of autonomous driving systems emphasizes the combination of multisensor redundancy and hardware-level protection.

Jouppi et al. [22] noted that Waymo adopted a triple heterogeneous perception architecture of LiDAR camera radar in multicity road tests, combined with a triple module redundancy (TMR) design on the FPGA, reducing the success rate of adversarial sticker attacks from 70% to 5%. Reference [30] noted that under low-light nighttime conditions, the collaborative application of noise adaptive filtering and dynamic distribution alignment technology can increase lane line recognition accuracy from 58% to 89%, with GPU power consumption increasing by only 12%. APN + formal verification.

These cases show that cross-layer defense mechanisms need to be dynamically adjusted according to the requirements of the scenarios. For example, Juvekar et al. analyze the communication overhead of lightweight federated learning models in the industrial Internet of Things (IIoT) is reduced by 40%, but a careful trade-off between model accuracy loss ($\leq 3\%$) and privacy protection strength ($\varepsilon = 0.7$) is needed. Table 1 provides a comparison of cross-domain defense effectiveness.

Table 1. Cross-domain defense effectiveness comparison.

Domain	Defense technology	Risk reduction rate	Delay increases	Privacy protection intensity
Medical treatment	APN + Formal verification	68%	+47 ms	tall (FHE)
Finance	Neuromarkup fusion + FHE	73%	+11 ms	polar altitude
Self-driving	Heterogeneous redundancy + Dynamic alignment	85%	+12 ms	centre

Cross-domain analysis further reveals the common challenges of intrinsic security defense. The stringent requirements for data privacy in healthcare and finance, coupled with the demanding real-time standards of autonomous driving, compel defense technologies to find a dynamic balance between efficiency and security. For example, while homomorphic encryption can protect medical data privacy, its computational density is five times higher than that of traditional algorithms [33], which may exacerbate system delays, whereas the real-time defense framework for autonomous driving needs to complete attack detection and response within 23 ms [34]. High requirements are proposed for the efficiency of the algorithm.

Empirical studies in healthcare, finance, and autonomous driving show that the trade-off between privacy protection, computational efficiency, and real-time performance is a universal contradiction in intrinsic security defense, which needs to be dynamically balanced through technological innovation and scenario optimization. The dilemma of balancing privacy protection, computational efficiency, and real-time performance (the impossible triangle) remains the core challenge for technology implementation. Future efforts should explore directions such as quantum computing acceleration and lightweight edge AI to achieve a balance among these three aspects.

6 Future Direction and Interdisciplinary Collaborative Governance

6.1 Synergistic Governance Framework of Law, Ethics and Technology

Despite significant progress in existing defense technologies, the integration of quantum computing and edge AI still faces the dual challenges of theory and engineering. To address this, this paper proposes a phased technical roadmap (as shown in Fig. 4), outlining key breakthrough directions for the next decade. Achieving intrinsic security in artificial intelligence requires breaking through the mindset of technological silos and building a collaborative governance system that integrates law, ethics, and technology:

With respect to standardizing legal governance, literature [6] indicates that the EUs Artificial Intelligence Act (AI Act) requires high-risk systems to undergo formal verification and adversarial testing (such as medical diagnostic models meeting EN ISO 13482 standards), whereas the Chinas Interim Measures for the Administration of Generative AI Services emphasize compliance with data classification, grading, and content labelling. Both aim to promote algorithmic transparency standardization, but the EU focuses more on accountability mechanisms, whereas China concentrates on data sovereignty and cross-border flow regulation. Tao et al. [25] noted that existing technical tools (such as the formal verification framework Marabou) have not yet covered all the scenarios required by law, and there is an urgent need to develop automated compliance engines to achieve dynamic mapping between regulatory provisions and code logic.

In terms of the embedded design of ethical constraints, Floridi proposed that the framework of "transparency, accountability and fairness" should be embedded in the technical design [35]. For example, the rule of counterfactual causal analysis is introduced in the financial risk control model, which reduces the difference in the loan rejection rate between users in different regions from 15% to 3% [36]. Goodfellow et al. and Jouppi et al. [7,22] presented that lethal autonomous weapon systems (LAWSs) should follow the principles of "meaningful human control", combined with hardware redundancy design (such as FPGA triple module redundancy) and ethical review process, to ensure the real-time human intervention in the decision chain.

In terms of technology-policy collaborative innovation, Wu et al. [5] point out that the dynamic heterogeneous eedundancy architecture (DHR) achieves system self-healing capabilities through hardware-level fault tolerance (such as reducing error rates from 1e-4 to 1e-9), complementing the EU policy emphasis on algorithmic transparency. Rech [28] indicated that while homomorphic encryption (FHE) ensures medical data privacy, it increases the computational density fivefold and requires the mitigation of latency issues through edge AI lightweighting and hardware acceleration (such as TPU v4).

6.2 Quantum Computing Driven Security Innovation

Quantum computing provides a disruptive potential to overcome the bottleneck of traditional defense, but its engineering application still needs to solve the problems of theoretical compatibility and hardware maturity.

Quantum-Inspired Defense Algorithm. Fang quantum annealing optimization is a quantum annealing-based adversarial sample detection strategy that reduces the defense response time by 40% on the MNIST dataset, and has better global search efficiency than the classical simulated annealing algorithm [37]. The quantum neural network (QNN) enhances the robustness of the model through the quantum superposition state. The theory proves that its resistance to adversarial migration attacks is 32% greater than that of classical networks, but it still needs to overcome the engineering challenge of current qubit noise interference [38].

Cryptographic Innovation Against Quantum Attacks. The NIST standardized Kyber algorithm (based on latent cryptography) can resist the threat of quantum computing to RSA/ECC and ensure the security of model watermarking and data traceability, but the problem of increasing the key length by three times still needs optimization [39]. According to Yi et al. [32], the quantum key distribution (QKD) combined with fully homomorphic encryption (FHE) can complete medical image reasoning in the encrypted state, and the accuracy rate remains 93%, while the delay increases by only 12 ms, providing a new paradigm for privacy-sensitive scenarios.

Quantum-Classical Hybrid Architecture. Research by Federici et al. [30] revealed that using edge quantum computing to deploy lightweight QNNs on edge devices (such as autonomous driving sensors) reduces the risk of privacy leakage from cloud transmissions. According to the enterprise-related solutions presented in the literature [32], communication overhead is reduced by 37% after deployment, with model accuracy loss controlled within 2%. Salazar et al. [36] showed that by leveraging fault-tolerant quantum computing technology and surface code error correction techniques to enhance qubit stability, the goal is to achieve a thousand-qubit fault-tolerant quantum processor before 2030, supporting large-scale AI security applications.

Technology Roadmap: Three-Stage Path of Quantum-Edge Collaborative Defense. To realize the deep integration of quantum computing and edge AI, Fig. 4 proposes a phased technology roadmap:

Phase 1 (2025–2027): Focusing on the miniaturization of quantum encryption hardware (such as QKD equipment) size ≤ 5 cm^3, energy consumption ≤ 1 W, support real-time encryption for edge devices) and the adaptation of the lightweight edge model (MobileNet-Q), which aims to compress the encryption communication delay to less than 8 ms;

Phase 2 (2028–2030): The quantum random number generator (QRNG) is used to optimize dynamic adversarial training. The QRNG generates a random number entropy value of more than 7.9 bits/byte, improves the privacy protection intensity of federated learning ($\varepsilon \leq 0.5$), and increases the success rate of adversarial attack interaction to 97%;

Phase 3 (2031–2035): Design a quantum-classical hybrid communication protocol (QUIC+ quantum channel), promote the formulation of ISO/IEC 30178 standards, and realize global automated orchestration of edge-cloud collaborative defenses (response time less than or equal to 10 ms).

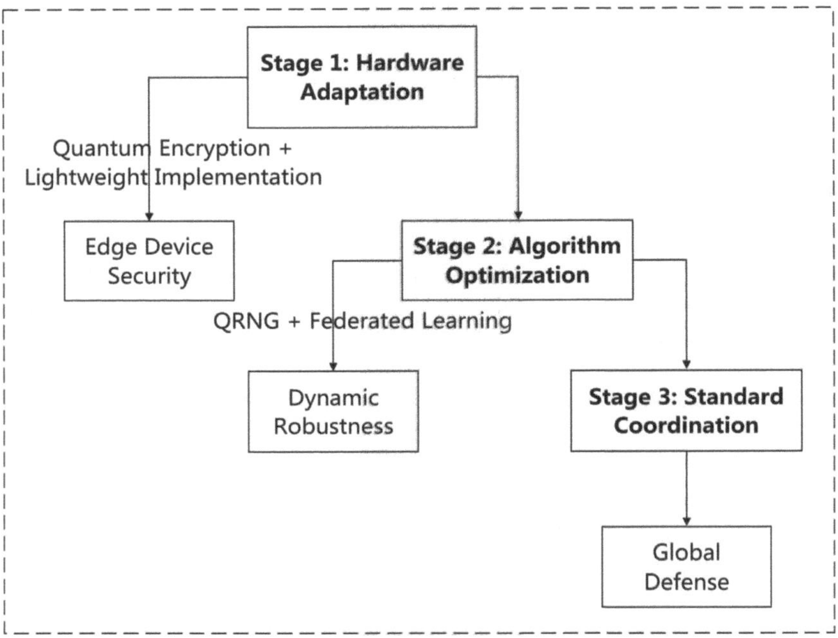

Fig. 4. Technology roadmap: three-stage path of quantum-edge

6.3 Future Research Directions and Challenges

Adaptive defense algorithms and meta-learning driven dynamic strategies have been developed. Vizitiu et al. [33] considered that to cope with the rapid evolution of generative AI attacks (such as prompt injection), attack detection and response should be completed within 23 ms.

A full life cycle security verification tool chain is built to cover the entire process from data collection to model training and deployment. According to Tao et al. [25], the integration of formal verification (such as the Marabou framework)

and adversarial testing aims to increase the vulnerability detection coverage from 65% to 98%.

Deepen the interdisciplinary collaboration mechanism and build automated compliance tools (such as NLP regulatory analysis engine) and composite talent training systems, bridge the semantic gap between technical terms and legal provisions, and reduce cost by 80% [40].

7 Conclusion and Outlook

Research on intrinsic security in artificial intelligence has revealed the essence of systemic risk and the technological boundaries of defense mechanisms. This paper constructs a three-dimensional analytical framework of a "data-model-system" to systematically analyse the underlying mechanisms of core threats, such as data contamination, adversarial sample generalization failure, and software-hardware codesign vulnerabilities, and proposes dynamic defense systems and cross-layer collaborative protection schemes. Empirical studies in healthcare, finance, and autonomous driving demonstrate that the application of dynamic adversarial training, formal verification, and heterogeneous redundancy architecture can reduce model failure risk by 60%–90% while simultaneously meeting privacy protection and real-time requirements. However, the core challenge for technology implementation remains the "impossible triangle" trade-off between privacy protection, computational efficiency, and real-time performance.

The main contributions of the research are as follows: first, a hierarchical threat classification theory and dynamic defense technology pathways are proposed, providing a systematic methodology for intrinsic security; second, the adaptability of defense mechanisms across domains is empirically validated, such as reducing false negatives in medical imaging diagnosis by 68% and increasing the success rate of autonomous driving attack interception to 95%; and third, interdisciplinary collaborative governance involving law, ethics, and technology is advocated, promoting the practical implementation of standardized frameworks (such as the National Institute of Standards and Technology (NIST) AI risk management guidelines).

Future research should focus on three aspects. First, adaptive defense algorithms should be developed, leveraging meta-learning and causal inference to counter zero-day attacks. This also includes exploring quantum computing acceleration and lightweight edge AI to address the impossible triangle dilemma. Second, a full lifecycle security validation toolchain that covers the entire process from data collection to model deployment is built. This aims to increase vulnerability detection coverage from 65% to 98% through the integration of formal verification and adversarial testing. Third, interdisciplinary collaboration mechanisms can be deepened to reduce governance costs through automated compliance tools and interdisciplinary talent development.

Ultimately, achieving sustainable development in a trustworthy AI ecosystem requires equal emphasis on technological innovation and institutional design.

References

1. China Communications Society Technical Committee on Intrinsic Security: Blue Book on Intrinsic Security of Artificial Intelligence Application Systems. China Communications Society (2024). https://www.hrssit.cn/info/3483.html. Accessed 26 Mar 2025
2. Li, J.: Internal and external causes and governance paths of AI security risks. China Social Sciences (2024). https://www.cssn.cn/skgz/bwyc/202403/t20240301_5736021.shtml. Accessed 27 Mar 2025
3. China Unicom: 2024 White Paper on Intrinsic Security of AI. China Unicom Research Institute (2024). https://www.docin.com/p-4672004208.html. Accessed 10 Mar 2025
4. Vapnik, V.N.: An overview of statistical learning theory. IEEE Trans. Neural Netw. (1999). https://doi.org/10.1109/72.788640
5. Wu, J., Zou, H., Xue, X., Zhang, F., Shang, Y.: Conception, approach and strategy of endogenous security enabling network resilience. Strateg. Study CAE **25**(06), 106–115 (2023). https://link.cnki.net/urlid/11.4421.G3.20231226.1021.004. Accessed 13 Mar 2025
6. European Commission: AI Act. Brussels: Official Journal of the European Union (2023). https://www.consilium.europa.eu/en/policies/artificial-intelligence. Accessed 10 Mar 2025
7. Goodfellow, I.J., Shlens, J., Szegedy, C.: Explaining and harnessing adversarial examples. arXiv preprint arXiv: 1412.6572 (2014). https://doi.org/10.48550/arXiv.1412.6572
8. Romer, P.M.: Endogenous technological change. J. Polit. Econ. (1990). https://doi.org/10.1086/261725
9. Winner, L.: Do artifacts have politics? In: Winner, L. (ed.) Computer Ethics, pp. 177–192. Routledge (2017)
10. Hollnagel, E., Woods, D.D., Leveson, N.: Resilience engineering: concepts and precepts. Ashgate Publishing, Ltd. (2006)
11. Alfeld, S., Zhu, X., Barford, P.: Data poisoning attacks against autoregressive models. In: Proceedings of the AAAI Conference on Artificial Intelligence (2016). https://doi.org/10.1609/aaai.v30i1.10237
12. Liu, P., Zheng, G.: Handling imbalanced data: Uncertainty-guided virtual adversarial training with batch nuclear-norm optimization for semi-supervised medical image classification. IEEE J. Biomed. Health Inform. (2022). https://doi.org/10.1109/JBHI.2022.3162748
13. Ben-David, S., Blitzer, J., Crammer, K., Kulesza, A., Pereira, F., Vaughan, J.W.: A theory of learning from different domains. Mach. Learn. (2010). https://doi.org/10.1007/s10994-009-5152-4
14. Geirhos, R., et al.: Partial success in closing the gap between human and machine vision. In: Advances in Neural Information Processing Systems (2021)
15. Jacob, B., et al.: Quantization and training of neural networks for efficient integer-arithmetic-only inference. In: Proceedings of the IEEE Conference on Computer Vision and Pattern Recognition (2018). https://doi.org/10.48550/arXiv.1712.05877
16. Szegedy, C., et al.: Intriguing properties of neural networks (2013). https://doi.org/10.48550/arXiv.1312.6199
17. Eykholt, K., et al.: Robust physical-world attacks on deep learning models. In: Proceedings of the IEEE Conference on Computer Vision and Pattern Recognition (2018). https://doi.org/10.48550/arXiv.1707.08945

18. Hinton, G., Vinyals, O., Dean, J.: Distilling the knowledge in a neural network (2015). https://doi.org/10.48550/arXiv.1503.02531
19. OpenAI. GPT-4 Technical Report (2023). https://cdn.openai.com/papers/gpt-4.pdf. Accessed 2 Apr 2025
20. OWASP Foundation. OWASP API Security Top 10 (2023). https://owasp.org/API-Security/editions/2023/en/0x11-t10/. Accessed 10 Apr 2025
21. Wang, J., Hassan, G. M., Akhtar, N.: A survey of neural trojan attacks and defenses in deep learning. arXiv preprint. https://doi.org/10.48550/arXiv.2202.07183
22. Jouppi, N.P., Yoon, D.H., Ashcraft, M., et al.: Ten lessons from three generations shaped Google's tpuv4i: industrial product. In: 2021 ACM/IEEE 48th Annual International Symposium on Computer Architecture (ISCA). IEEE (2021). https://doi.org/10.1109/ISCA52012.2021.00010
23. Cao, Y., et al.: Invisible for both camera and lidar: security of multi-sensor fusion based perception in autonomous driving under physical-world attacks. In: 2021 IEEE Symposium on Security and Privacy (SP), pp. 176–194. IEEE (2021). https://doi.org/10.1109/SP40001.2021.00076
24. Yoon, J., Hwang, S.J., Lee, J.: Adversarial purification with score-based generative models. In: International Conference on Machine Learning, pp. 12062–12072. PMLR (2021). https://proceedings.mlr.press/v139/yoon21a.html. Accessed 10 Apr 2025
25. Tao, A., Duan, Y., Wang, Y., Lu, J., Zhou, J.: Dynamics-aware adversarial attack of adaptive neural networks. IEEE Trans. Circ. Syst. Video Technol. (2024). https://doi.org/10.1109/TCSVT.2024.3351680
26. Sun, X., Khedr, H., Shoukry, Y.: Formal verification of neural network controlled autonomous systems. In: Proceedings of the 22nd ACM International Conference on Hybrid Systems: Computation and Control (2019). https://doi.org/10.1145/3302504.3311802
27. Himabindu, M., Revathi, V., Gupta, M., Rana, A., Chandra, P.K., Abdulaali, H.S.: Neuro-symbolic AI: integrating symbolic reasoning with deep learning. In: 2023 10th IEEE Uttar Pradesh Section International Conference on Electrical, Electronics and Computer Engineering (UPCON), vol. 10, pp. 1587–1592. IEEE (2023). https://doi.org/10.1109/UPCON59197.2023.10434380
28. Rech, P.: Artificial neural networks for space and safety-critical applications: reliability issues and potential solutions. IEEE Trans. Nucl. Sci. (2024). https://doi.org/10.1109/TNS.2024.3349956
29. Juvekar, C., Vaikuntanathan, V., Chandrakasan, A.: GAZELLE: a low latency framework for secure neural network inference. In: 27th USENIX Security Symposium (USENIX Security 2018), pp. 1651–1669 (2018). https://www.usenix.org/conference/usenixsecurity18/presentation/juvekar. Accessed 12 Apr 2025
30. Federici, F., Martintoni, D., Senni, V.: A zero-trust architecture for remote access in industrial IoT infrastructures. Electronics **12**(3), 566 (2023). https://doi.org/10.3390/electronics12030566
31. HUAWEI Technologies. Autonomous Driving Network (ADN). Shenzhen: Huawei (2024). https://www-file.huawei.com/-/media/corp2020/pdf/giv/striding-towards-the-intelligent-world/2024/intelligent_world_adn_2024_en.pdf?la=en. Accessed 12 Apr 2025
32. Yi, X., Walia, E., Babyn, P.: Generative adversarial network in medical imaging: a review. Med. Image Anal. 101552 (2019). https://doi.org/10.1016/j.media.2019.101552

33. Vizitiu, A., et al.: Applying deep neural networks over homomorphic encrypted medical data. Comput. Math. Methods Med. (2020). https://doi.org/10.1155/2020/3910250
34. Cui, J., Liew, L.S., Sabaliauskaite, G., Zhou, F.: A review on safety failures, security attacks, and available countermeasures for autonomous vehicles. Ad Hoc Netw. (2019). https://doi.org/10.1016/j.adhoc.2018.12.006
35. Floridi, L.: The Logic of Information: A Theory of Philosophy as Conceptual Design. Oxford University Press (2019)
36. Salazar, T., AraÃējo, H., Cano, A., Abreu, P.H.: A survey on group fairness in federated learning: challenges, taxonomy of solutions and directions for future research. arXiv preprint arXiv:2410.03855 (2024). https://doi.org/10.48550/arXiv.2410.03855
37. Biamonte, J., Wittek, P., Pancotti, N., Rebentrost, P., Wiebe, N., Lloyd, S.: Quantum machine learning. Nature (2017). https://doi.org/10.1038/nature23474
38. Tramèr, F., Kurakin, A., Papernot, N., Goodfellow, I., Boneh, D., McDaniel, P.: Ensemble adversarial training: attacks and defenses. arXiv preprint arXiv:1705.07204 (2017). https://doi.org/10.48550/arXiv.1705.07204
39. Alagic, G., et al.: Status report on the third round of the NIST postquantum cryptography standardization process (2022). https://doi.org/10.6028/NIST.IR.8413
40. Guidotti, R., Monreale, A., Ruggieri, S., Turini, F., Giannotti, F., Pedreschi, D.: A survey of methods for explaining black box models. ACM Comput. Surv. (CSUR) **51**(5), 1–42 (2018). https://doi.org/10.1145/3236009

Research on an Efficient Cross-Domain Authentication Scheme that is Based on Blockchain and Edge Computing

Chunlan Xu, Ran Yi, and Tao Sun(✉)

School of Digital and Intelligence-Industry (School of Cyber Science and Technology), Inner Mongolia University of Science and Technology, Baotou 014010, China
suntao771031@imust.edu.cn

Abstract. The Internet of Things (IoT) comprises many devices distributed across administrative domains. The lack of trust among these cross-domain entities poses significant risks to data security and privacy during information exchange. To address these challenges, a solution that combines certificateless public key cryptography with blockchain technology has been proposed. This approach enhances data security and privacy protection through advanced technical methods. During cross-domain information interactions, the risk of data breaches and privacy leaks increases significantly because of the large volume of sensitive information being generated and stored. Authentication, a critical component in securing IoT communication, plays a vital role in ensuring system security. However, traditional authentication mechanisms often impose high computational and communication overheads, which can be particularly burdensome for resource-constrained IoT devices. To overcome this issue, edge computing has been integrated into IoT systems. By leveraging the powerful computing and storage capabilities of edge nodes and incorporating technologies such as blockchain, elliptic curve-based certificateless encryption, and the ECDSA digital signature algorithm, the proposed scheme helps mitigate resource limitations. As a result, IoT devices can perform security operations, including authentication, more efficiently and effectively.

Keywords: IoT · edge computing · blockchain · cross-domain authorization

1 Introduction

With the rapid development of IoT technology, the number of IoT devices is growing exponentially. Hundreds of millions of smart devices are now connected to the network, providing users with convenient services while simultaneously generating massive volumes of data. As a result, ensuring the security of users' private information and data has become an increasingly pressing issue [1]. The surge in data also introduces challenges such as network latency, limited storage capacity, bandwidth constraints, and inefficiencies in data processing [2], all of which can hinder the continued development of the IoT. The traditional "Cloud Center–Terminal Device" architecture is no longer sufficient to address these problems [3]. To address these challenges, edge computing has been

L. Huang et al. (Eds.): CCF NCCA 2025, CCIS 2774, pp. 547–570, 2026.
https://doi.org/10.1007/978-981-95-5294-8_37

introduced. Edge computing is an innovative model that deploys core capabilities—such as networking, computation, storage, and applications—directly at the network's edge, physically closer to the data source. This approach builds an open platform designed to enable fast processing, real-time responsiveness, and efficient data utilization by offering intelligent services at the edge [4]. In the context of the IoT, edge computing decentralizes data processing and analysis by offloading tasks to edge nodes, thereby reducing the burden on the centralized cloud infrastructure and enabling real-time responses to user demands. The integration of edge computing into the IoT offers several benefits, including reduced latency, enhanced security, and savings in bandwidth and energy consumption. However, the resources of a single-edge server are limited. To meet the service demands of IoT applications, multiple edge servers must work collaboratively. This emerging approach is known as the Collaborative Edge Computing Architecture [5].

Despite the numerous benefits of collaborative edge computing architectures, several challenges remain. For example, the data processed by edge nodes often contain sensitive information, raising concerns about how to prevent unauthorized access or data leakage during transmission, storage, and processing [6]. Additionally, IoT terminal devices are highly vulnerable to a variety of attacks. Malicious actors can exploit these vulnerabilities by deploying forged terminal devices to launch distributed denial-of-service (DDoS) attacks or to infiltrate the system with malware, potentially causing severe disruption and damage. These security concerns highlight the urgent need for a robust mutual authentication mechanism between terminal devices and edge nodes. Moreover, efficient cross-domain authentication is essential to ensure that terminal devices can securely access services across different edge nodes within a collaborative architecture.

To address the issue of IoT device authentication, numerous authentication schemes have been proposed. However, many of these schemes rely on the Public Key Infrastructure (PKI) authentication system, where a trusted third party—known as a Certificate Authority (CA)—is responsible for registering devices and issuing digital certificates to various types of network devices [7]. Authentication is then performed by verifying these certificates issued under the PKI framework. Despite their widespread use, PKI-based approaches have three major drawbacks. First, the centralized nature of PKI presents serious security risks, as it can become a single point of failure. Second, since all device certificates are generated and managed by the PKI, there is an increased risk of privacy leakage. Third, the PKI architecture involves high certificate management costs, including certificate status verification, path construction, and delivery processes. With the rise of blockchain technology, it has emerged as a promising solution for enabling secure collaboration in environments without requiring a traditional trust foundation. In response to the limitations of existing PKI-based methods, this paper proposes a mutual authentication scheme based on blockchain-enabled collaborative edge computing. The main contributions of this work are summarized as follows:

1. A certificateless encryption algorithm based on elliptic curve cryptography is employed within the blockchain to ensure secure data transmission. The public private key pairs are jointly generated by the edge nodes and the blockchain in segments. Simultaneously, the blockchain is used to authenticate the edge nodes, enabling

the secure generation and distribution of the public–private key pairs required for cross-domain communication.

2. Intradomain and cross-domain authentication: The terminal device uses the public key of the edge node within its domain to ensure secure data transmission while minimizing its resource consumption. Elliptic Curve Diffie–Hellman (ECDH) key exchange is employed to establish secure communication between devices, effectively preventing third-party eavesdropping and information theft.
3. Edge nodes use the ECDSA algorithm to realize the identity authentication of terminal devices. The edge node uses the ECDSA algorithm to sign and save the hash certificate of the terminal device and realizes the identity authentication of the terminal device on the edge node, which can reduce the number of times the blockchain is accessed, reduce the authentication time, and reduce the calculation and communication overhead.

2 Related Work

2.1 Cross-Domain Authentication Based on PKI

PKI-based authentication is a widely adopted security infrastructure that provides essential services for ensuring data confidentiality, integrity, authentication, and nonrepudiation. Cross-domain authentication systems built on PKI have been extensively studied. For example, reference [8] proposed an authentication system that integrates PKI with the state-secret algorithm SM9 in a "cloud-edge-device" IoT authentication framework. In this model, IoT devices perform cross-domain authentication directly at the edge of the network, achieving both efficient authentication and location privacy protection. This approach reduces complexity and enhances the efficiency of cross-domain authentication, particularly for resource-constrained IoT devices. Similarly, reference [9] employs a proxy ree-signature technique to enable homogeneous cross-domain authentication within a PKI-based framework. However, this scheme still suffers from the high system overhead inherent to PKI architectures. Furthermore, in large-scale networks, authentication challenges between trust domains based on different regimes are inevitable. Reference [10] proposed a trust-enhanced cross-domain authentication scheme that solves the security and efficiency problems of traditional PKI in cross-domain authentication through dynamic trust evaluation and optimized certificate verification.

Although PKI-based authentication schemes offer advantages in cross-domain authentication, their certificate management systems are highly complex. Given the vast number of IoT terminal devices and the multitude of communication domains, PKI technology faces significant challenges in effectively addressing IoT device authentication. The emergence of blockchain technology has accelerated the evolution of digital certificates, offering a more decentralized and secure alternative. The integration of blockchain with IoT technology represents a promising direction for future development, addressing the limitations of traditional PKI systems.

2.2 Cross-Domain Authentication Based on Blockchain Technology

Blockchain [9], as a highly decentralized distributed ledger, offers an innovative approach to addressing security and privacy challenges in IoT network architectures, which are inherently distributed by nature. With its decentralized structure, tamper-resistant data storage, and consensus-based authentication mechanisms, blockchain establishes a trustworthy, transparent, and robust security framework for IoT systems. This technology not only enhances the overall security of the IoT but also significantly improves data privacy protection and ensures the integrity and reliability of the data during both transmission and storage. The integration of blockchain technology into cross-domain authentication schemes has led to several notable research efforts, as outlined below: Reference [11] proposed a federated chain trust model that combines blockchain technology with the traditional PKI authentication model, enabling two-way entity cross-domain authentication and fast reauthentication. The certificate verification process uses a hash algorithm, which reduces the number of public key signing and verification operations, thereby improving the efficiency of cross-domain authentication. Reference [12] introduced a secure, lightweight authentication scheme that is based on blockchain technology. The modular square root (MSR) technique is employed to ensure both the security and efficiency of the authentication process, whereas blockchain enhances system security and scalability. Additionally, this approach reduces the gas costs for smart contracts. Reference [13] and others proposed a ground station protocol for verifying the legitimacy of UAV missions. This protocol is based on elliptic curve digital signatures and public key encryption algorithms to enable two-way communication authentication and key negotiation. Compared with TLS, the proposed protocol significantly reduces the communication overhead, although its computational overhead remains relatively high. The document also provides a security proof for the protocol, but it analyses only the security of one-way authentication. The protocol, however, has some authentication vulnerabilities, as an attacker can impersonate a legitimate drone and complete a mission with the ground station via the protocol. Reference [14] presented a multitier architecture that uses blockchain technology, along with the introduction of clustering, where IoT devices are grouped into local clusters. A cluster head is selected to facilitate the authentication of its group members. The proposed approach aims to ensure security and privacy, although it still needs to be tested in real-world scenarios. However, since blockchain is extensively used in this scenario, the number of accesses to the blockchain is significantly greater. Reference [15] also introduces a multilayer blockchain architecture with the addition of edge computing. In this architecture, there is a local blockchain and a public blockchain, with two distinct digital certificates generated for intradomain and cross-domain authentication. The authentication process uses dynamic accumulator technology to increase the efficiency of certificate verification. However, as the number of authenticated IoT devices increases, the authentication efficiency decreases. Additionally, the storage and computational capabilities of edge devices are underutilized. Reference [16] proposed a blockchain-based cross-domain authentication protocol that stores certificate hashes on the blockchain and includes a certificate revocation process. However, the protocol suffers from low authentication efficiency.

2.3 Blockchain Combined with Edge Computing for Cross-Domain Authentication

The integration of blockchain and edge computing has led to various authentication schemes. Reference [17] proposed a solution that enables mutual authentication between edge servers and IoT devices via certificateless encryption. However, as the number of IoT devices increases, the registration phase runtime for IoT devices grows at nearly twice the rate of that for edge servers. Reference [18] presented a federated chain-based approach that stores certificate hashes of end devices on the blockchain, thereby avoiding the complexities of traditional certificate verification processes. However, these solutions do not support anonymous communication, leaving user privacy at risk. Reference [19] introduces a cross-domain authentication scheme that uses a combination of biometrics and password-based two-factor authentication. It employs fuzzy extraction techniques and integrates smart contracts into the blockchain. This approach achieves cross-domain authentication while preventing biometric data leakage and supporting anonymous communication. Reference [20] proposed a blockchain-based anonymous authentication system designed for edge computing environments. This system incorporates multiparty secure computation, zero-knowledge proof techniques, and blockchain technology to store and verify user identity information anonymously. While the scheme enhances authentication security and user data privacy, the anonymity it provides is still implemented on top of the blockchain.

In the identity authentication schemes discussed above, one category is based on the PKI authentication system, which primarily uses a centralized authentication method. However, this approach poses a security risk to the entire system if the central server is compromised. The other category leverages the decentralized and tamper-proof characteristics of blockchain technology to design an IoT authentication scheme suitable for distributed environments. Compared with traditional centralized authentication methods, this blockchain-based scheme offers improved security and greater efficiency, enhancing the overall effectiveness and trustworthiness of IoT authentication. However, in current research involving edge computing and blockchain, both intradomain and cross-domain authentication often require frequent interactions with the blockchain, resulting in a cumbersome authentication process. This paper focuses on addressing the challenges of cross-domain authentication for low-performance end devices in existing IoT environments. By leveraging the advantages of blockchain technology and edge computing, the proposed approach aims to mitigate issues related to data leakage and resource limitations commonly encountered in cross-domain authentication.

3 Preliminaries

In this section, the relevant math used is presented.

3.1 Difficult Problem of the Discrete Logarithm of Elliptic Curves

Let E be an elliptic curve, G be a cyclic group of order p (p being a large prime), and g be an instanton element of this cyclic group G. Consider the following congruence equation: $y = g^x (modp)$, where $x, y, g \in z_p^*$, (denoting that x, y, and g are all nonzero residue class elements modulo p) [21]. In such a setting, given the values of y, g, and p, it is not feasible to compute the value of x that satisfies this congruence equation, and the degree of difficulty stems from the complexity of the discrete logarithm problem under the corresponding mathematical structure.

3.2 Elliptic Curve-Based Certificate-Less Encryption

System parameter generation: The key generation centre (KGC) [22] selects a suitable elliptic curve E and a base point G on the curve and chooses two hash functions H_1 and H_2. The KGC randomly generates a master key s and computes the system public key $PK = s * G$. The system parameters include the elliptic curve E, the base point G, the system public key PK, the hash functions and so on. PK, hash functions. Etc. which are publicly available.

User key generation: the user chooses a random number x_1 as the secret value and sends his identity information ID_i to the KGC. The KGC calculates the user's partial private key $D_i = S * H_1(ID_i)$, which is based on the user's identity information ID_i and master key s and sends D_i to the user. After receiving the partial private key, the user computes his full private key $SK_i = (x_1, D_i)$ and computes the public key $PK_i = x_1 * G + D_i$.

3.3 ECDSA Digital Signature Algorithm

Key generation: Fist, choose a suitable elliptic curve E and a base point G on the curve of order n [23]. Then, an integer d is randomly chosen as the private key, where $0 < d < n$. Next, the public key $Q = d * G$ is computed, where multiplication is a pointwise multiplication operation on the elliptic curve.

Signature generation: Let the message to be signed be m. Fist, a hash operation is performed on message m to obtain the hash value $z = H(m)$. Then, randomly choose an integer k, where $0 < k < n$, and compute the point $(x_1, y_1) = k*G$. Next, x_1 is converted to the integer r: if $r = 0$, then k is reselected. Then compute $s = k^{-1}(z + r * d)modn$: if $s = 0$, then reselect k. The final obtained signature is (r, s).

Signature verification: the verifier receives the message m and its signature signature verification: the verifier receives the message m and its signature (r, s); first, it performs a hash operation on the message m and obtains the hash value $z = H(m)$. Then $w = s^{-1}modn$ computed, and $w = s^{-1}modn$ and $u_2 = r * wmodn$ are computed. Then compute the point $(x_0, y_0) = u_1 * G + u_2 * Q$ and convert x_0 to integer v. Finally, verify that v = r holds; if so, the signature is valid; otherwise the signature is invalid.

3.4 ECDH Key Negotiation

An elliptic curve is a curve defined by the equation $y^2 = x^3 + ax + b$ (where $4a^3 + 27b^2 \neq 0$) [24]. A special addition operation is defined on elliptic curves that satisfies properties such as the law of union and the law of exchange. With this operation, groups of points can be generated on elliptic curves and related mathematical computations can be performed.

4 Authentication Scheme Based on Blockchain Collaborative Edge Computing

4.1 Infrastructure

The cross-domain authentication infrastructure based on blockchain-enabled collaborative edge computing in an IoT environment is illustrated in Fig. 1. This infrastructure consists of three main components: the blockchain, edge nodes, and terminal device entities. A detailed description of the infrastructure is provided below:

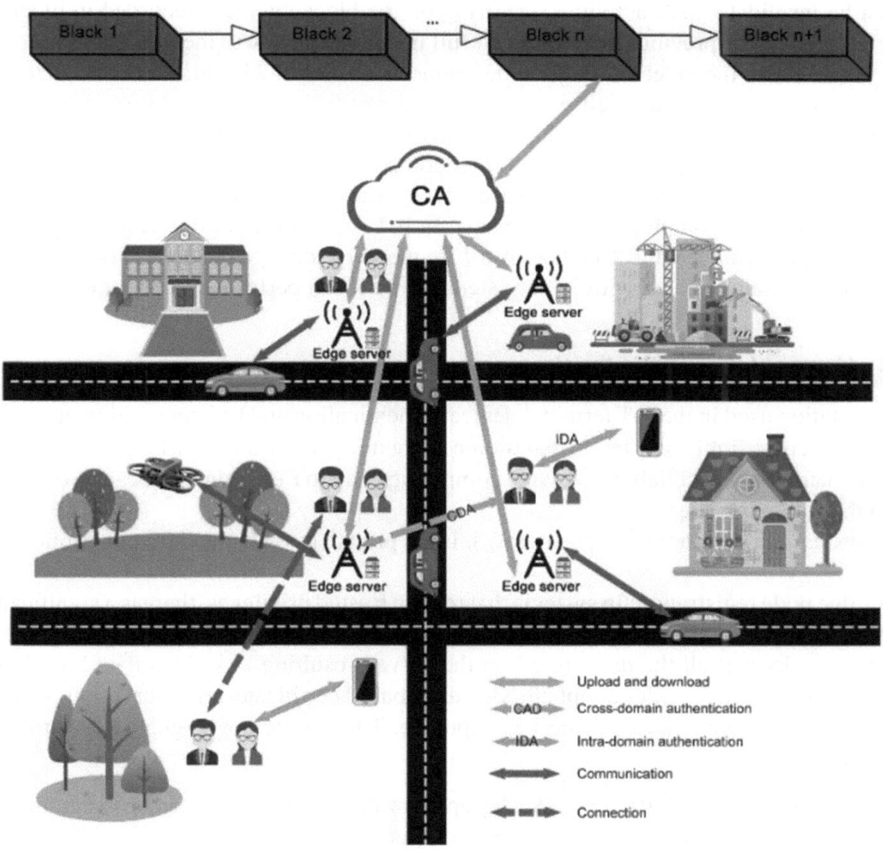

Fig. 1. Cross-Domain Authentication Infrastructure

Domain: A domain is composed of one edge node and multiple terminal devices. A trusted domain refers to a domain in which all devices have successfully completed registration and authentication through the associated edge node. Communication between devices in different domains requires cross-domain authentication and key negotiation protocols to ensure secure interactions.

Blockchain: In this architecture, the blockchain functions similarly to a certificate authority (CA). It provides authentication services, externally generates certificates, records audit logs of authentication events and service accesses, and ensures transparency and resistance to tampering. Additionally, it generates part of the private keys for edge nodes. The CA can determine—on the basis of its risk assessment—that another CA has either issued certificates to a significant number of malicious nodes or is no longer capable of providing reliable authentication services. In such cases, a consensus mechanism is used to declare the CA as "untrustworthy," thereby forcing its removal from the system. This differs from a CA that voluntarily withdraws from the cluster; in the case of forced removal, all certificates issued by the CA from the point of untrustworthiness on wards must be invalidated. All activities occurring on the blockchain are recorded in blocks, each linked to the previous one, allowing full traceability back to the genesis block.

Edge node: the selected edge node connects to the blockchain and completes a registration process. It uses elliptic curve-based certificate-less encryption to obtain a portion of its private key along with a verified identity. Once registered, the edge node serves as the authentication manager for its domain.

Entity (Terminal Devices): Terminal devices are resource-constrained terminal nodes operating in edge computing environments. In IoT scenarios requiring low latency and real-time performance, these devices must frequently exchange data. Each device must be properly authenticated within the system before it can perform any operations.

4.2 General Overview of the Program

The notation used in the IoT terminal device authentication and key negotiation protocol is presented in Table 1. The cross-domain authentication process, which is based on blockchain-enabled collaborative edge computing in an IoT environment, can be divided into the following stages:

System initialization: $(SK_a) \rightarrow (PK_a)$. Input private key SK_a, and output public key PK_a.

Edge node registration: In systems that rely on trusted privileges, there is a significant risk of key misuse. If an attacker obtains the private key of an edge server, they may be able to decrypt all the data stored on the server, resulting in serious data breaches. To prevent this, edge nodes adopt elliptic curve-based certificate-less encryption, which helps eliminate the risk of private key exposure. The private key is generated through the following steps:

1) Setup $\left(1^{\lambda}\right) \rightarrow (E, G, PK, H_1, SK)$: Setup algorithm that inputs the security parameter λ (determine the length of the key) and outputs the system parameters elliptic curve E, base point G, system public key PK, and system master key SK. The algorithm is executed by KGC in the blockchain, where the elliptic curve E, base point G, system public key PK, the hash function H_1 broadcasts, and the master key SK is stored in the blockchain.
2) PSKGen $(ID_{EN_A}, H_1, SK) \rightarrow (PSK_{EN_A})$: partial key generation algorithm, input system parameters H_1, edge node identity ID_{EN_A}, system master key SK, and output partial private key PSK_{EN_A}.

3) SGen $(ID_{EN_A}) \rightarrow (\chi_{EN_A})$: the secret value generation algorithm, which inputs the edge node identity ID_{EN_A} and randomly outputs a secret value $\chi_{EN_A}(0 < \chi_{EN_A} < n)$. The algorithm is run by the edge node EN_A A and χ_{EN_A} uses a portion of the private key generated by the KGC to compute a private key unique to the edge node.
4) SKGen $(G, PSK_{EN_A}, \chi_{EN_A}) \rightarrow (SK_{EN_A})$: private key generation algorithm with input system parameters G, partial private key PSK_{EN_A} and secret value χ_{EN_A} and output private key SK_{EN_A}.This process is performed completely by the edge nodes.
5) PKGen $(G,\chi_{EN_A}) \rightarrow (PK_{EN_A})$: The public key generation algorithm uses the system parameter G and the secret value χ_{EN_A} to compute the public key PK_{EN_A}. This algorithm is also run by the edge nodes and the public key PK_{EN_A} is encapsulated on the blockchain and broadcast to the public.

Terminal device registration:$(ID_a, PK_a, F_a)(h)$, taking entity identifier ID_a, public key PK_a and terminal device-related information F_a as inputs, it outputs a hash certificate h.

Intradomain authentication:$(ID_a) \rightarrow (PK_a)$, Input the entity identifier ID_a with the inner domain authentication request and output the same domain public key PK_a.

Cross-domain authentication:$(ID_a, PK_a)(PK_b)$, Input an entity identifier ID_a and a public key PK_a with a cross-domain authentication request, and output a cross-domain public key PK_b.

Keywords: $(PK_a, ID_a, h, a{*}G, b{*}G) \rightarrow (k_{ab})$, input public key PK_a, entity identifier ID_a, hash certificate h, temporary keys $a{*}G$ and $b{*}G$, and output session key k_{ab}.

4.3 Overall Program Design

A general overview of the scheme is shown in Fig. 2. First, the edge node runs the system initialization to set up the parameters and obtain a valid identity. Next, the terminal device registers with the edge node via the device registration algorithm to request a legitimate identity. Afterward, to access device data within the same domain, the entity must undergo intradomain authentication by the edge node. To access device data across different domains, the entity needs to undergo cross-domain authentication. Finally, once the terminal devices are authenticated, they negotiate a shared key through the key agreement algorithm to enable secure communication.

Table 1. Notations and Descriptions

Parameters	Description
$G(E_Q)$	Elliptic curves in finite fields
G	Base point
D_A	Domain A
EN_A	Edge node in domain A
SK_{EN_A}, PK_{EN_A}	Private key and public key of edge node A
T_a	Terminal equipment a

(continued)

Table 1. (*continued*)

Parameters	Description
ID_a	Identity of entity a
a, PK_a	Private key and public key of terminal equipment a
$Hash()$	Secure hash function
N_i	A random number, $i \in [1, n-1]$

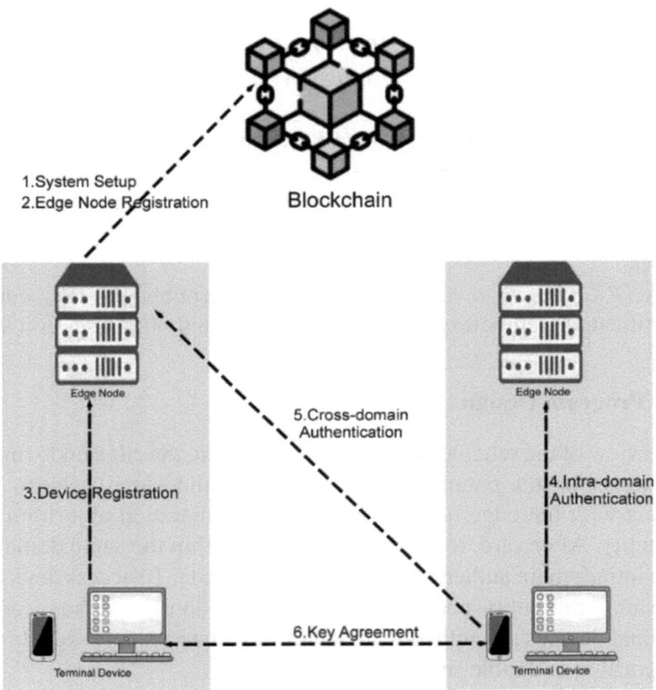

Fig. 2. Overall program design

4.4 Detailed Process of the Certification Program

An elliptic curve $G(E_Q)$ is constructed over a finite field E_Q and a random elliptic curve is generated. Once $G(E_Q)$ is determined, the base point G is also determined and G is of order n, where G is a large prime number. In abstract algebra, when a group operation is performed on two group elements, the result is independent of the order of the group elements; such a group is called an Abelian group. The Abelian group z_n is the set of residue classes of mode n. If $t + r = 0$ (of mode n), then r is regarded as the inverse of t, denoted $r = -t$ (of mode n), where n is the order of the base point G on the elliptic curve $G(E_Q)$.

1) System initialization

The system initialization phase is the preparation work that T_a or EN_A must perform before identity registration. Taking T_a as an example, T_a first chooses a random number $a \in [1, n-1]$ to be saved as a private key and computes its own public key $Pk_a = a * G$, according to $G(E_Q)$.Similarly, each entity generates its own public and private keys in this phase.

2) Edge node registration

Edge node registration initiates a registration request to the blockchain, which can be further divided into two parts: KeyGenBC and KeyGenEN, which are created by the blockchain and the edge node, respectively. A formalized description is given in Algorithm 1 and Algorithm 2, described as follows.

The edge node EN_A sends ID_{EN_A} and λ to the blockchain to request registration. When the blockchain receives the message, it saves ID_{EN_A} and then checks whether it has been registered on the chain on the basis of ID_{EN_A}, and if it has been registered, it returns a signal that it has succeeded. Without registration, run Setup $(1^\lambda) \rightarrow (E, G, PK, H_1, SK)$ to obtain the system parameter elliptic curve E, base point G, system public key PK and master key SK, system parameter elliptic curve E, base point G, system public key PK broadcasting, and master key SK encapsulated on the blockchain. PSKGen (ID_{EN_A}, H_1, and SK) $\rightarrow$ (PSK_{EN_A}) are run to obtain the partial private key PSK_{EN_A}, and then the partial private key PSK_{EN_A} embedded in the transaction is sent to EN_A. After obtaining the partial private key PSK_{EN_A}, EN_A runs Run Sgen(ID_{EN_A}) $\rightarrow$ (χ_{EN_A}) to calculate the secret value χ_{EN_A}. SKGen (G, PSK_{EN_A}, and χ_{EN_A}) $\rightarrow$ (SK_{EN_A}) are run to compute one's private key SK_{EN_A}. Run PKGen (G, χ_{EN_A}) $\rightarrow$ (PK_{EN_A}) to determine your own public key PK_{EN_A}. The public key PK_{EN_A} is broadcast to the public key and the private key PSK_{EN_A} is kept locally.

3) Terminal device registration

Terminal device registration means that the terminal device T_a is registered through the edge node EN_A, and the identity registration flowchart is shown in Fig. 3(a). The specific steps are as follows:

a) Device T_a sends a registration request to EN_A.

$$\mathrm{Reg_a} = \mathrm{E_{pk_{EN_A}}}(\mathrm{ID_a}, \mathrm{PK_a}, \mathrm{F_a})$$

b) EN_A receives the registration request from T_a, decrypts it with its own private key, obtains the device's identifier ID_a, public key PK_a, and identity information F_a, and verifies whether it is a legitimate user; if the verification passes, the registration succeeds; conversely, it returns the registration failure.

c) EN_A sends a request for a certificate to the CA.

Algorithm 1 KeyGenBC

Input: (ID_{EN_A}, λ)
Output: (PSK_{EN_A})
1: EN_A sends ID_{EN_A} and λ to the blockchain and initiates a registration request;
2: Blockchain receive ID_{EN_A} and λ, and query on the chain according to ID_{EN_A};
3: if ID_{EN_A} has been registered then
4: return success;
5: else
6: Run Setup$(1^\lambda) \rightarrow (E, G, PK, H_1, SK)$;
7: K is embedded into the transaction and broadcasted;
8: Store SK in the blockchain;
9: Run PSKGen$(ID_{EN_A}, H_1, SK) \rightarrow (PSK_{EN_A})$;
10: end if
11: Blockchain sends PSK_{EN_A} to EN_A;

$$\mathrm{RCert} = (\mathrm{ID_a}, \mathrm{PK_a}, \mathrm{Sig}_{\mathrm{SK}_{\mathrm{EN_A}}}(\mathrm{N_1}), \mathrm{N_1})$$

d) The CA generates a digital certificate by verifying the validity of the random number and the signature; if the verification passes, T denotes the validity period of the certificate and writes the hash value of the digital certificate, $h_a = H(Cet_a)$, to the blockchain.

$$\mathrm{Cert_a} = (\mathrm{ID_a}, \mathrm{PK_a}, \mathrm{T})$$

e) The CA sends the certificate hash value h_a to EN_A, EN_A receives h_a and runs Algorithm 3 SigGen to obtain (r_a, s_a) and saves ID_a, s_a.

f) Send h_a and r_a to T_a.

Algorithm 2 KeyGenEN

Input: (K, PSK_{EN_A})
Output: null
1: EN_A receive information from the blockchain to obtain the partial private key PK_{EN_A};
2: if K or PSK_{EN_A} doesn't exist then
3: Call KeyGenBC;
4: else
5: Run SGen$(ID_{EN_A}) \rightarrow (\chi_{EN_A})$;
6: Run SKGen$(G, PSK_{EN_A}, \chi_{EN_A}) \rightarrow (SK_{EN_A})$;
7: Run PKGen$(G, \chi_{EN_A}) \rightarrow (PK_{EN_A})$;
8: end if
9: Embed PK_{EN_A} into the transaction and broadcast it;
10: Store SK_{EN_A} locally;

4) Intradomain authentication

As the number of IoT devices continues to grow, the network load on cloud centers is increasing. Edge computing technology offers a more convenient and efficient authentication process for terminal devices. With this approach, terminals can complete the authentication process locally at the network's edge without needing to request authentication services from the cloud center. Below is the design of the intradomain authentication flowchart, as shown in Fig. 3(b), which uses domain A as an example. The specific steps are as follows:

a) Device T_a sends an access request to device T_b.

$$\mathrm{Req} = \mathrm{E}_{\mathrm{PK}_{\mathrm{EN_A}}}(\mathrm{ID_a,h_a,r_a,PK_a})$$

b) The device T_b then receives the access request and forwards it to the edge node EN_A of this domain.

c) When EN_A receives the message, it decrypts it with its own private key to obtain the ID_a,h_a, r_a, and PK_a of device T_a. It looks up s_a according to the ID_a of the device, runs Algorithm 4 SigVer, calculates v_a, and determines whether v_a is equal to r_a.

d) If v_a is not equal to r_a, return authentication failure.

e) If v_a is equal to r_a, send an authentication success message.

f) Negotiate session key.

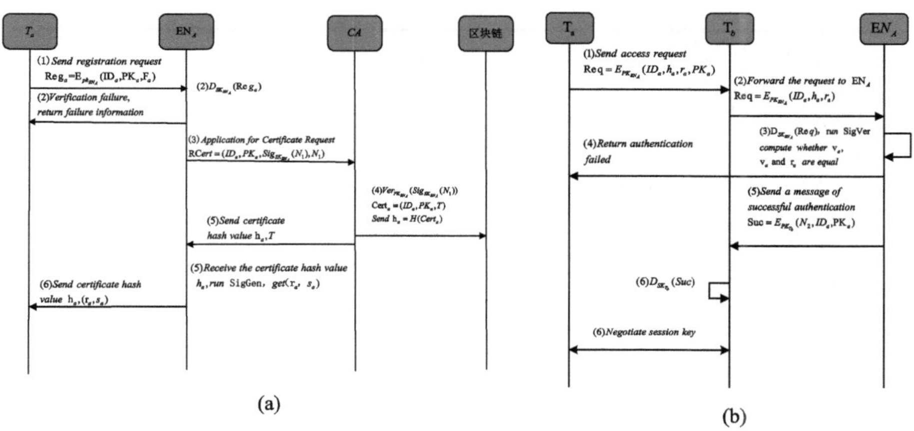

Fig. 3. Terminal device identity registration flowchart and intradomain authentication flow chart

5) Cross-domain authentication

Take domain A and domain B as examples for the cross-domain communication of terminal devices. Assuming that the terminal device T_a inside domain A wants to access T_b in domain B, EN_B indirectly realizes the authentication of EN_B to T_a through the authentication of EN_A to T_a to realize the cross-domain authentication of T_a. The flowchart of cross-domain authentication is shown in Fig. 4, and the specific steps are as follows:

a) The end device T_a sends a request to access T_b to the edge node EN_A.

$$\mathrm{Req} = \mathrm{E}_{\mathrm{PK}_{\mathrm{EN_A}}}(\mathrm{ID_a}, \mathrm{PK_a}, \mathrm{h_a}, \mathrm{r_a})$$

b) EN_A receives the request message and decrypts it with its own private key to obtain the ID_a,h_a, r_a and PK_a of device T_a. It looks up s_a according to the ID_a of the device, runs Algorithm 4 SigVer, calculates v_a, and determines whether v_a is equal to r_a.

Algorithm 3 SigGen

Input:(χ_{EN_A}, h_a)
Output: (r_a)
1:EN_A receives information from the blockchain, certificate hash value h_a, period of certificate validity T;
2: Randomly select an integer $k(0 < k < n)$;
3: EN_A calculates (r_a, s_a);
4:$(x_1, y_1) \leftarrow k * G$;
5: $r_a \leftarrow x_1 \mod n$;
6:$s_a \leftarrow k^{-1}(h_a + r_a * \chi_{EN_A}) \mod n$;
7:EN_A sends $r_a$and h_a to T_a;

Algorithm 4 SigVer

Input: (ID_a, r_a, h_a)
Output: null
1: T_a sends ID_a, r_a and h_a to EN_A and initiate an access request;
2: EN_Afound ID_a based ons_a
3: EN_A calculates (w, v);
4: $w \leftarrow s_a^{-1} mod n$;
5:$u_1 \leftarrow h_a * w \mod n$;
6:$u_2 \leftarrow r_a * w \mod n$;
7:$(x_0, y_0) \leftarrow u_1 * G + u_2 * PK_{EN_A}$;
8:$v_a \leftarrow x_0$;
9: if $v_a = r_a$ then;
10: h_a has not been modified and T_a's identity is valid;
11:else
12:return error;

c) When not equal, return the information of authentication failure.

d) Equally, send a request to EN_B to access T_b.

$$\mathrm{Rep}(\mathrm{ID_a}, \mathrm{PK_a}, \mathrm{Sig}_{\mathrm{PK}_{\mathrm{EN_B}}}(\mathrm{N_3}))$$

e) After receiving the request, EN_B uses its own private key to verify whether N_3 is valid and effectively sends a message to T_b.

$$\mathrm{E}_{\mathrm{PK}_{\mathrm{T_b}}}(\mathrm{ID_a}, \mathrm{PK_a})$$

f)) After receiving the message, T_b decrypts it with its own private key to obtain the ID_a and PK_a of T_a.

$$D_{SK_{T_b}}(E_{PK_{T_b}}(ID_a, PK_a))$$

g) T_a and T_b negotiate the session key.

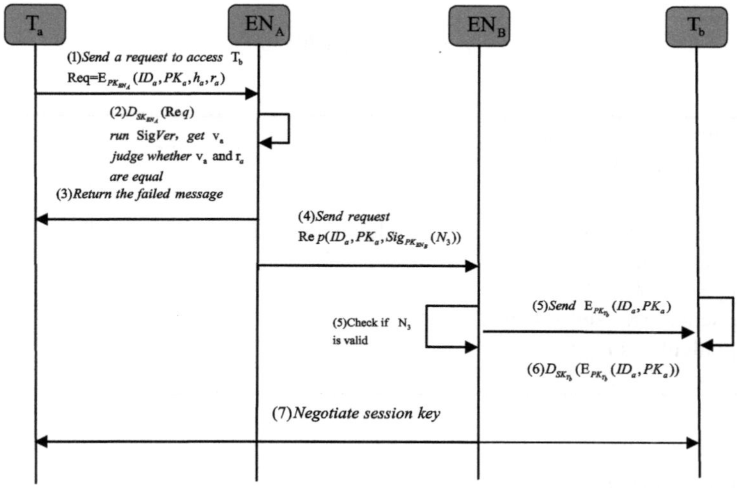

Fig. 4. Cross-Domain Authentication Infrastructure

Pseudocode is also given for Algorithm 3 SigGen and Algorithm 4 SigVer, which are implemented in terminal device enrollment, intradomain authentication, and cross-domain authentication, as described below:

SigGen: EN_A generates a random integer $k(0 < k < n)$ after receiving the information from the blockchain, the certificate hash h_a of T_a and the certificate validity T, computes the point $(x_1, y_1) = k*G$, and then computes $r_a = x_1 mod n$ and $s_a = k^{-1}h_a + r_a*\chi_{EN_A} mod n$. The inner χ_{EN_A} is the $\chi_{EN_A}(0 < \chi_{EN_A} < n)$ obtained by running the secret value generation algorithm earlier, and finally r_a and h_a are sent to T_a.

SigVer: EN_A receives the request from terminal device T_a, obtains T_a's ID_a, r_a and h_a, EN_A finds s_a according to T_a's ID_a, calculates $w = s_a^{-1} mod n$,$u_1 = h_a*w mod n$,$u_2 = r_a * w mod n$ and point $(x_0, y_0) = u_1 * G + u_2 * PK_{EN_A}$, converts x_0 to integer v, and determine whether v and r_a are equal; otherwise, they are not. $u_2*\ PK_{EN_A}$ and then converts x_0 to the integer v_a. Determine whether v_a and r_a are equal; equal terminal device T_a identity is legal; otherwise, it is not legal.

6) Key agreement

Key negotiation is the process by which two entities negotiate a security session key after authentication. The terminal device T_a and the terminal device T_b perform key negotiation for later information exchange; here the ECDH algorithm is used. The pseudocode is given in Algorithm 5 and is described below.

After authentication, T_b receives the information from the edge node of its own domain and decrypts it with its own key to obtain the ID_a and PK_a of the terminal device T_a after authentication. T_b calculates $k_{ba} = b * PK_a$, where PK_a is obtained at the time of initialization of the system, and $Pk_a = a * G$. T_b encrypts its PK_b and ID_b with T_a 's public key and sends it to T_a, who decrypts it with his own key to obtain T_b 's ID_b and PK_b, calculating k_ $k_{ab} = a * PK_b$ here PK_b is also what the system obtain when it initializes, $PK_b = b * G$, and the final negotiated session key:

$$k_{ab} = k_{ba} = a * b * G$$

Algorithm 5 Key agreement

Input:(PK_a, PK_b)

Output: k_{ab}

1: T_b receives the information of the edge node, ID_a and PK_a of T_a;

2: T_b calculates k_{ba};

3: $k_{ba} = bPK_a$;

4: T_b sends PK_b and ID_b to T_a;

5: T_a calculates k_{ab};

6: $k_{ab} = a\ PK_b$;

5 Security Analysis

5.1 Security of Authentication

The cross-domain authentication of terminal devices proposed in this paper is achieved via blockchain collaborative edge computing technology. In this scheme, the blockchain stores the generated digital certificate hash value $H(Cet_a)$, and the same hash value is also stored in the edge node where the terminal device is located. The ECDSA digital signature algorithm is used to secure these stored hash values as credentials for authentication. Owing to the one-way nature of the hash function, even if an attacker obtains the hash value, they cannot reverseengineer the content of the original digital certificate, thereby ensuring the security of the digital certificate. Furthermore, the ECDSA digital signature algorithm prevents attackers from deriving the private key v on the basis of the digital certificate hash value ($H(Cet_a)$, which ensures the security of both intradomain and cross-domain authentication credentials for terminal devices. The privacy protection of terminal devices, as well as the identity verification for both intradomain and cross-domain authentication, is carried out through the edge nodes in this design. Therefore, the entire system relies on the trustworthiness of the edge nodes. If an attacker compromises the edge node, bypassing it to directly gain access, it would be considered a breach of the edge node, and the node can simply be deregistered from the system. During intradomain authentication, the message is encrypted via the public key encryption algorithm of the edge node. Only the edge node with the corresponding private key can decrypt the message, ensuring that the terminal device's identity information remains secure. For cross-domain authentication, when terminal device T_a in domain A wants to access

T_b in domain B, edge node EN_B indirectly authenticates T_a through the authentication of EN_A. This process enables cross-domain authentication of T_a, with both EN_A and EN_B being registered and authenticated through the blockchain. As long as these edge nodes remain registered and have not been deregistered, they are trusted, ensuring the security of terminal device information. This approach reduces the number of accesses to the blockchain, making cross-domain authentication more efficient while maintaining security.

5.2 Denial of Service Attack

In the scheme proposed in this paper, multiple certificate authorities (CAs) collaborate to construct a blockchain-based authentication service mechanism. Even if one CA is compromised by an attack, the system can still access accurate edge node and device information from the blockchain maintained by the remaining CAs. This design significantly enhances the system's resilience against distributed denial-of-service (DDoS) attacks.

5.3 Anti-Replay Attacks

During the authentication process, random numbers are included in request messages to prevent replay attacks. The receiver verifies the authenticity of the message by checking the validity of the random number. Specifically, the edge node EN_A sends an access request to edge node EN_B, which includes a random number: $\text{Rep}\left(ID_a, PK_a, Sig_{PK_{EN_B}}(N_3)\right)$. Edge node EN_B validates the access request by verifying the random number. If an attacker intercepts and attempts to resend the message, the outdated random number causes the verification to fail, effectively preventing replay attacks.

5.4 Service Spoofing Attack

Impersonating the accessed resource requires the edge node's private key to decrypt the message, whereas impersonating the service to decrypt the message would require solving the discrete logarithm problem, which is computationally infeasible.

5.5 Forward Confidentiality

During the key negotiation process, the terminal devices compute their respective public and private keys via the ECDH encryption algorithm, where a and b are random numbers, and the $a*G$ and $b*G$ computed during each encryption are also random. In addition, this encryption algorithm is also based on the discrete logarithm problem and it is difficult for the attacker to obtain the session key.

6 Performance Analysis

This section presents the performance analysis of the cross-domain authentication scheme for terminal devices discussed earlier. The evaluation focuses on the communication overhead, computational overhead, and experimental comparisons. The experiments were conducted on a computer equipped with an Intel® Core™ i7 2.20 GHz processor and 16 GB of RAM. The development environment used includes the Java Development Kit (JDK) and IntelliJ IDEA. The JPBC cryptographic library was employed to implement various cryptographic functions. In this study, we use elliptic curve parameters of Type A, where the order n of the group is 128 bits in length. Additionally, the output length of the hash function is configured to 256 bits.

During the experiments, to obtain more accurate comparison results, we took the average value of the execution time consumed by each operation. The experimental results show that the average overhead time of the digital signature and verification algorithm T_s is 2.006 ms. The average overhead time of the hash function T_h is 0.004 ms. The average overhead time of the public key encryption and decryption algorithm T_P is 4.010 ms. The average overhead time of the bilinear pair operation T_i is 10.570 ms. The average time consumed by the exponential operation T_e is 6.240 ms.

6.1 Overhead Comparison

This paper compares the proposed scheme with those presented in [17, 18], and [15]. In the two schemes [18] and [15], after the certificate authority (CA) or blockchain completes multiparty verification, the credentials for the current access session are signed and stored on the blockchain. These credentials remain valid for a certain period, during which a terminal device from an external domain is permitted to access the target domain by querying the stored credentials. In [17], certificateless encryption was used to deploy edge computing in the blockchain, and most of the operations were run on edge nodes. In this context, we compare the computation and communication overhead incurred when a terminal device from domain A initiates a request to obtain the necessary credentials for access. The comparison results are presented in Table 2 and Table 4.

Table 2. Communication overhead comparison

Scheme	CA/Blockchain -CA/Blockchain	EN-CA/Blockchain	EN-EN
Cheng et al. [17]	0	2	2
Zhang et al. [18]	0	2	1
Chen et al. [15]	4	6	0
Our scheme	0	1	1

According to the communication overhead comparison shown in Table 2, the number of communications between certificate authorities (CAs), between edge nodes, and between edge nodes and the blockchain in the proposed scheme is significantly lower than that in the schemes presented in [15, 17], and [18]. This reduction offers a notable advantage in terms of communication efficiency. In the schemes of [15] and [18], most computations are performed on the server side. However, this centralized computation model results in substantially greater network communication overhead. In the [17] scheme, although the calculation is scattered in the edge nodes, the edge nodes are deployed in the blockchain, and each operation increases the cost and exchanges the cost for security. However, it can be seen from Table 2 that the network communication overhead in this scheme is not reduced. On the basis of the analysis in Table 2 and the bit size of each parameter, the following sizes are referenced from [4]: identity information (ID) is 48 bits, the random number (N) is 256 bits, the hash function output is 256 bits, and both public and private keys are 256 bits. The average message length is 24 bytes. As shown in Table 3 and the corresponding Fig. 5(a), when the communication costs of the proposed scheme are compared with those of other cross-domain authentication methods, it is evident that the proposed scheme achieves slightly better performance in terms of communication overhead. Although the scheme in this paper is similar to the scheme in [17], some calculations are placed on the edge nodes. Because the scheme in [17] deploys edge nodes in the blockchain and does not generate certificates, many operations are performed when the device is authenticated, resulting in increased bandwidth consumption. However, the edge nodes in this paper are not deployed in the blockchain, and the certificate hash value and the ECDSA algorithm are skillfully used to authenticate the terminal device. The number of calculations is small, which is why the bandwidth consumption of the scheme in this paper is much smaller than that of the scheme in [17].

Table 3. Comparison of bandwidth consumption

scheme	Bandwidth Consumption (bits)
Cheng et al. [17]	6080
Zhang et al. [18]	4680
Chen et al. [15]	3584
Our scheme	2832

Table 4. Calculation cost comparison

Scheme	Calculations Total	Running Time/ms
Cheng et al. [17]	$3\ T_s + +2T_h + 2T_P$	14.046
Zhang et al. [18]	$2T_s + +T_h + 10T_P$	44.116
Chen et al. [15]	$2T_s + 4T_P$	20.052
Our scheme	$3T_s + 2T_P$	14.038

On the basis of the computational overhead comparison presented in Table 4, the number of public key encryption and decryption operations in the proposed scheme is significantly lower than that in the systems outlined in [15] and [18], primarily because of fewer communication rounds. Additionally, compared with the scheme in [17], most of the computational tasks in the proposed approach are handled by the edge nodes, which helps avoid consuming CA resources during the authentication process. In contrast, the scheme in [15] relies on blockchain resources during authentication, increasing its computational burden. Table 4 presents a detailed comparative analysis of computational costs, whereas Table 4 and Fig. 5(b) further illustrate the differences in computational efficiency.

6.2 Engineering Application

Among the many industries where cross-domain authentication and key negotiation are applied, smart factories represent a critical and rapidly growing area, despite having either an early or late start [26]. A smart factory is a complex production network where numerous devices are interconnected through the Industrial Internet of Things (IoT). Equipment from different production segments may come from various suppliers and belong to different management domains. In terms of equipment collaboration within the smart factory, devices from different production segments—often supplied by different vendors—must work together. Cross-domain authentication ensures that these devices can verify each other's identities, preventing unauthorized equipment from accessing the production network and ensuring the safety and stability of the production process. For example, in a welding production line involving equipment from different subfactories, welding robots, handling robots, and testing equipment may all come from different vendors. Through cross-domain authentication, these devices can collaborate safely and efficiently on the same production task.

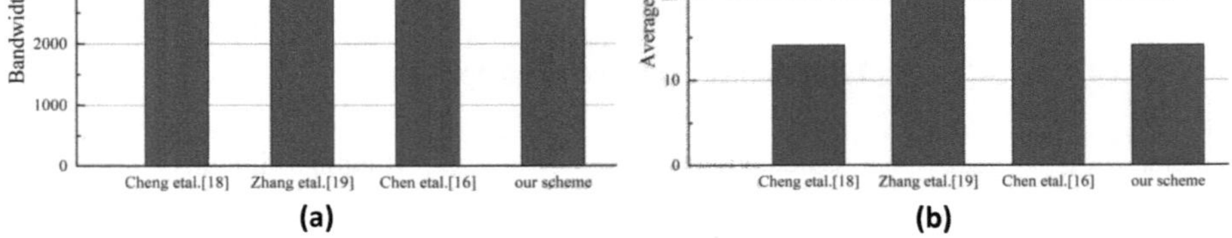

Fig. 5. Comparison of bandwidth overhead and bandwidth overhead

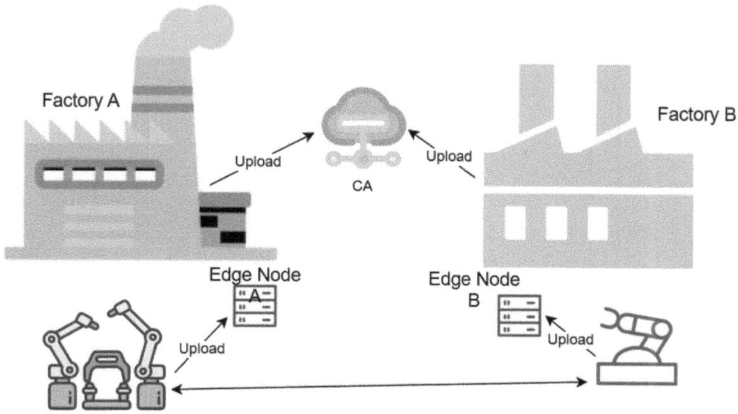

Fig. 6. Application Scenarios in Smart Factory

The smart factory is a rapidly growing sector in industrial manufacturing aimed at improving production efficiency and reducing costs. However, its actual application in industrial manufacturing is still complex. Equipment in different production segments of the production line may come from various suppliers, which means that each factory needs its own edge node for independent operation. These edge nodes communicate through wireless networks to facilitate data exchange, as shown in Fig. 6. When a production robot in Factory A wants to access production data in Factory B, it must first authenticate within its own reliable domain—i.e., through the edge node of its own domain—before it can proceed with cross-domain data access and communication.

To better demonstrate the applicability of this scheme in smart factories, simulation experiments were conducted to assess its performance. A small dataset was created to simulate the process of cross-domain authentication. This dataset represents five subfactories of the same factory, with the number of smart robots set at 10, 20, 30, and 40. Each subfactories has a single edge server. We tested the overall average time cost of certification under different numbers of factories, as shown in Fig. 9. The time cost of this paper's scheme increases with both the number of intelligent robots and the number of factories. For cross-domain authentication between 2 factories, as the number of intelligent robots increases from 10 to 40, the time overhead increases from 354.1 ms to 913.02 ms. The authentication time remains below 1 s, which is still relatively low. Additionally, Fig. 7 shows that as both the number of factories and robots increases, the time overhead increases, which highlights the advantages of this paper's scheme. This makes it well-suited for smart factories, where the number of subfactories is limited. In smaller-scale areas, the advantages of this solution remain clear.

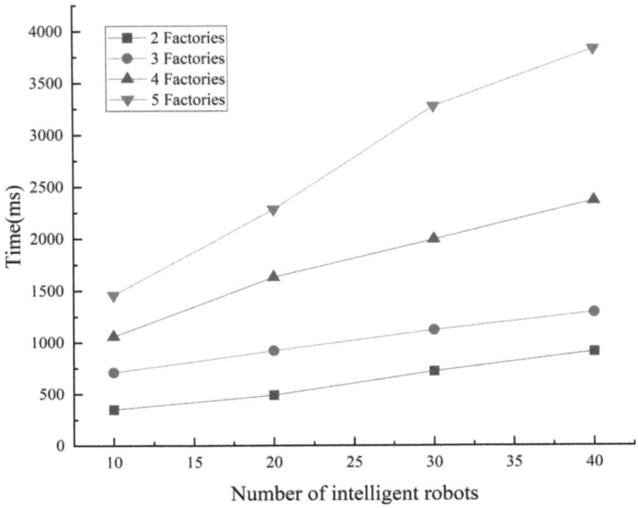

Fig. 7. Time cost of Smart Factory Applications

7 Conclusion

This paper addresses the issues of limited resources, low utilization of edge servers, and low efficiency of current IoT terminal devices by proposing a cross-domain terminal device authentication scheme based on blockchain collaborative edge computing in an IoT environment. The terminal device is authenticated by the edge node rather than the cloud center, and the authentication information of the terminal device is saved in the edge node by the ECDSA digital signature algorithm to realize the localization of the authentication of the terminal device. This approach reduces the network load on the cloud center, places more computational tasks on the edge nodes, increases their utilization, and improves authentication efficiency. It also helps mitigate the resource limitations of terminal devices. The security attributes required for the authentication and key negotiation protocol are thoroughly analysed. Compared with existing terminal device authentication schemes, the proposed scheme reduces both computation and communication overhead while maintaining the required security attributes, thus improving authentication efficiency. However, the scheme still has some limitations in session negotiations. Future research will focus on improving session negotiations to better suit resource-constrained terminal devices.

Funding. This work was supported by the Inner Mongolia Autonomous Region directly under the university basic scientific research business fee project 2023QNJS197.

References

1. Cheng, G.J., Deng, S.K., Wen, Y.Y., et al.: A review of blockchain-based authentication mechanism for internet of things. J. Softw. **34**(3), 1470–1490 (2022)

2. Pannayagol, B., Deshpande, S.: Security in internet of things: an overview. In: 2023 International Conference on Device Intelligence, Computing and Communication Technologies, (DICCT), Dehradun, India, pp. 243–248 (2023)
3. Dong, Y.M., Zhang, J., Xie, C.Z., et al.: A review of key issues in edge intelligent computing under cloud-edge-end architecture: computation optimization and computation offloading. J. Electron. Inf. **46**(03), 765–776 (2024)
4. Wang, T., Liang, Y., Shen, X., Zheng, X., Mahmood, A., Sheng, Q.Z.: Edge computing and sensor-cloud: overview, solutions, and directions. ACM Comput. Surv. **55**(13s), Article 281 (2023)
5. Sharma, M., Tomar, A., Hazra, A.: Edge computing for industry 5.0: fundamental, applications, and research challenges. IEEE Internet Things J. **11**(11), 19070–19093 (2024)
6. Zhang, D., Fan, L.: Cerberus: privacy-preserving computation in edge computing. In: IEEE INFOCOM 2020-IEEE Conference on Computer Communications Workshops (INFOCOM WKSHPS), pp. 43–49. IEEE (2020)
7. Shen, M., et al.: Blockchain-assisted secure device authentication forcross-domain industrial IoT. IEEE J. Sel. Areas Commun. **38**(5), 942–954 (2020)
8. Xiaodong, Y.A.N.G., Faying, A.N., Ping, Y.A.N.G., et al.: Cross-domain authentication scheme based on proxy REE-signature in cloud environment. J. Comput. **42**(4), 756–771 (2019)
9. Wu, W.: Research on Internet of Things Identity Authentication and Privacy Protection Technology in Edge Computing Environment. Xidian University (2019)
10. Salama, R., Al-Turjman, F.: An overview of blockchain applications and benefits in cloud computing. AIoT Smart Sens. Technol. Smart Dev. 66–76 (2024)
11. Zhou, Z.C., Li, L.X., Li, Z.H.: Efficient cross-domain authentication scheme based on blockchain technology. J. Comput. Appl. **38**(2), 316–320, 326 (2018)
12. Yang, X., Yang, X., Yi, X., et al.: Blockchain-based secure and lightweight authentication for internet of things. IEEE Internet Things J. **9**(5), 3321–3332 (2021)
13. Garba, A., Khoury, D., Balian, P., et al.: LightCERT4IoTs: blockchain-based lightweight certificates authentication for IoT applications. IEEE Access **11**, 28370–28383 (2023)
14. Honar Pajooh, H., Rashid, M., Alam, F., et al.: Multilayer blockchain-based security architecture for internet of things. Sensors **21**(3), 772 (2021)
15. Chen, Y., Yang, Q., Zeng, X., et al.: A new identity authentication and key agreement protocol based on multi-layer blockchain in edge computing. IEEE Access (2023)
16. Gu, P., Chen, L.: An efficient blockchain-based cross-domain authentication and secure certificate revocation scheme. In: 2020 IEEE 6th International Conference on Computer and Communications (ICCC), pp. 1776–1782. IEEE (2020)
17. Cheng, G., Chen, Y., Deng, S., et al.: A blockchain-based mutual authentication scheme for collaborative edge computing. IEEE Trans. Comput. Soc. Syst. **9**(1), 146–158 (2021)
18. Zhang, J., Li, X., Zeng, X., Zhao, Y., Duan, Y., Yang, D.: Blockchain-based cross-domain authentication and key negotiation protocol in edge computing environment. J. Inf. Secur. **6**(1), 54–61 (2021)
19. Liu, H., Xu, H.: Blockchain-based two-factor cross-domain authentication in edge computing environment. Comput. Sci. Appl. **11**(7), 1845–1852 (2021)
20. Liu, S., Chai, Y., Hui, L., et al.: Blockchain-based anonymous authentication in edge computing environment. Electronics **12**(1), 219 (2023)
21. Miller, V.S.: Use of elliptic curves in cryptography. Proc. Crypto (2007)
22. Jia, S.: Research on certificate-free anonymous multirecipient signing secret based on elliptic curve. Xi'an University of Electronic Science and Technology (2018)

23. Xiao, S., Wang, X., Pan, F.: An elliptic curve digital signature algorithm without mode inversion. J. Comput. Eng. Appl. **56**(11) (2020)
24. Nikooghadam, M., Shahriari, H.R., Saeidi, S.T.: HAKECC: highly efficient authentication and key agreement scheme based on ECDH for RFID in IOT environment. J. Inf. Secur. Appl. **76**, 103523 (2023)

Data Science and Big Data Technology

CHAKE: A Lightweight Chameleon Hash Based Authenticated Key Exchange for IoT

Jianli Liu, Jianfeng Guan(✉), Kexian Liu, Xiaolong Hu, and Ye Qin

State Key Laboratory of Networking and Switching Technology,Beijing University of Posts and Telecommunications, Beijing, China
{kuohao233,jfguan,kxliu,hxl814446051,qiny}@bupt.edu.cn

Abstract. The proliferation of the IoT has intensified security challenges, as resource-constrained devices remain vulnerable to cyberattacks while handling sensitive operational and personal data. Traditional cryptographic methods are often unsuitable for IoT environments because of hardware limitations, necessitating lightweight yet robust alternatives that balance efficiency with security. Existing solutions frequently struggle to address this trade-off, leaving gaps in end-to-end protection and scalability.

This paper introduces a Chameleon Hash-based lightweight authentication and key exchange protocol designed for IoT architectures. The protocol facilitates mutual authentication while dynamically refreshing session keys during handshake phases, ensuring forward secrecy without additional communication rounds. By leveraging Chameleon Hash's collision-resistant properties, our solution significantly reduces the computational overhead. We formally and informally validate the protocol's robustness. Experimental evaluations further validate the protocol's efficiency, achieving a computational efficiency improvement of more than 50% compared with that of state-of-the-art solutions while imposing minimal communications overhead.

Keywords: Internet of Things (IoT) · Key Exchange · Lightweight

1 Introduction

Internet of Things (IoT) systems, comprising diverse devices such as sensors, embedded systems, and connected endpoints, enable users to share data across platforms such as social networks and cloud services. These environments illustrate the interconnected nature of devices that generate and transmit data, facilitating seamless integration within networked infrastructures. However, as shown in Fig. 1, this connectivity introduces risks such as replay attacks, impersonation, and Man-in-the-Middle (MitM) attacks [18]. These threats are particularly critical because IoT devices often handle sensitive information, including personal

L. Huang et al. (Eds.): CCF NCCA 2025, CCIS 2774, pp. 573–592, 2026.
https://doi.org/10.1007/978-981-95-5294-8_38

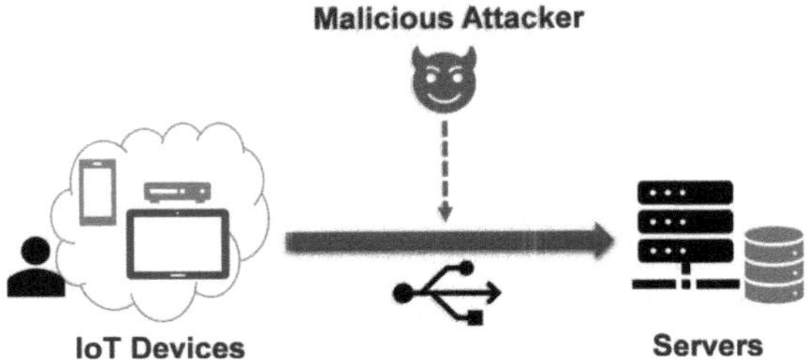

Fig. 1. Communication in IoT.

identifiers and operational data, which require stringent confidentiality. Compromises in such systems can undermine user trust, impede IoT adoption, and result in significant financial or reputational harm [26].

The dynamic operational environment of IoT devices further complicates security. Devices frequently transition between network domains, such as mobile sensors in industrial settings or wearable devices in healthcare, necessitating periodic and authenticated updates to session keys [25]. Without secure mechanisms for key management, data transmissions become vulnerable to interception or tampering. Thus, ensuring reliable and confidential data exchange demands cryptographic solutions tailored to IoT constraints [12,17]. Despite advancements, many existing security frameworks remain inadequate against evolving threats, resulting in unresolved vulnerabilities [29].

To address these challenges, Authenticated Key Exchange (AKE) protocols play a critical role. AKE protocols secure communications by allowing participants to establish shared session keys over insecure channels while verifying identities to prevent unauthorized access [33]. This dual functionality is vital for safeguarding sensitive IoT applications, such as remote healthcare monitoring or industrial control systems. Forward security, a cornerstone of modern AKE protocols, ensures that past communications remain secure even if long-term keys are compromised. This principle is central to our proposed protocol.

Traditionally, AKE protocols achieve forward security by employing mechanisms such as digital certificates, digital signatures [9,27,34], or symmetric encryption techniques [5,31] to authenticate participants. More recently, biometric authentication schemes have also gained traction for enhancing identity verification and security [37].

In the context of the IoT, the frequent need to update session keys, driven by device mobility and dynamic network conditions, introduces an additional layer of complexity. Unlike traditional systems, IoT devices often operate with constrained computational resources, requiring protocols to balance robust security with low overhead [17]. Furthermore, the explosive growth of the IoT has expanded the attack surface, making comprehensive security measures indispensable. Therefore, designing AKE protocols that are lightweight, secure, and adaptable to the dynamic nature of the IoT is not only critical but also a pressing research challenge.

1.1 Challenges

AKE protocols in IoT applications face the following challenges:

- Resource-constrained IoT devices struggle to support AKE protocols reliant on complex cryptographic operations, such as bilinear pairing or expensive digital signature schemes. Additionally, bandwidth limitations necessitate minimal communication overhead to maintain efficient data transmission [1].
- Most AKE protocols ensure forward security by periodically updating key parameters, a critical requirement in dynamic IoT environments where devices frequently transition across network domains. However, existing solutions often introduce additional interaction phases or parameters, increasing resource consumption for constrained devices.
- The devices' inherent vulnerability of IoT devices to attacks demands AKE protocols with robust security guarantees. While some protocols leverage biometrics or fuzzy extractors for authentication, these approaches face limitations in generic deployment scenarios.

1.2 Contributions

The main contributions of this paper are summarized as follows:

- We propose the CHAKE protocol, a lightweight AKE solution optimized for the IoT. The protocol streamlines the key exchange process, requiring only one round to establish secure communication. Experimental analysis confirmed its superior performance over existing AKE protocols.
- By integrating the chameleon hash, CHAKE enables dynamic session key updates during authentication via randomized parameters, ensuring forward security without additional interaction phases. Unlike symmetric authentication methods (e.g., MAC), our protocol leverages the collision of the chameleon hash resistance to provide non-repudiation. Even if a session key is compromised, attackers cannot impersonate the key server (KS). In contrast, symmetric key leakage (e.g., in MAC-based systems) risks complete system compromise.
- The CHAKE protocol is formalized through a rigorous security model (Sect. 5), which guarantees resilience against a wide range of adversarial attacks, including replay, impersonation, and man-in-the-middle threats.

2 Related Work

In recent years, owing to the widespread application of IoT devices and the rapid growth of related services, researchers have increasingly focused on how to solve the problems of IoT devices that are vulnerable to various security threats and high latency due to limited computing resources in communication [32]. In this section, we review some authentication and key exchange protocols for IoT or resource-constrained devices.

Kalra *et al.* [19], in 2015, pioneered an authentication protocol leveraging HTTP cookies and elliptical curve cryptography, creating a secure communication channel between IoT devices and servers. To strengthen security attributes and mutual authentication, M. Alotaibi [28] proposed a WSN-compatible key agreement protocol based on symmetric cryptography and biometric verification. However, Moghadam *et al.* [30] exposed flaws within the session key security of this protocol, prompting the proposition of an ECDH-based authentication key agreement protocol. Chuang *et al.* [8] introduced the concept of "effective authentication time periods", Challa *et al.* [7] proposed a provably secure three-factor user authentication and key agreement protocol for wireless medical sensor networks, which supports dynamic sensor node addition, password and biometric updates, making it more suitable for practical healthcare scenarios.

Recent advancements in this domain are prominently marked by a rigorous cycle of proposal, cryptanalysis, and enhancement. This iterative trend is exemplified by the work of Lee *et al.* [23], who identified and rectified significant vulnerabilities in a prior scheme by Dhillon *et al.* [11]. They formally verified their improved protocol using tools such as ProVerif and BAN logic, underscoring a growing demand for provable security. In a similar vein, a lightweight protocol designed for smart city WSNs by Wu *et al.* [38] was subsequently critiqued by Fan *et al.* [12] for its inability to guarantee forward security and the absence of a formal security proof.

Concurrently, the research landscape has diversified to address more specialized challenges and advanced cryptographic paradigms. Moving beyond one-to-one communication, Fan *et al.* [13] focused specifically on the distinct problem of authenticated group key exchange (AGKE), and proposed tailored optimization solutions. Tseng *et al.* [35] developed an identity-based mutual authentication protocol. Its primary contribution lies in the security analysis under the stringent "continuous-leakage extended Canetti–Krawczyk (CLeCK) model" [15], demonstrating its robustness for resource-constrained devices. To address computational overhead and privacy risks, Pendi *et al.* [36] devised an anonymous batch authentication scheme, which reduces the processing burden on individual devices. To address the persistent challenge of password management, Lian *et al.* [24] proposed a symmetric password AKE scheme. It features a novel and resilient method for password file storage that is designed to mitigate security threats even in the event of a file leak.

While existing protocols have made valuable contributions, they frequently falter in one or more critical areas for robust IoT deployment. Key deficiencies include computationally expensive cryptographic primitives unsuitable for constrained devices, mechanisms for forward secrecy that are either absent or inefficient, and a common lack of non-repudiation for accountability. The CHAKE protocol is designed to address these interconnected gaps. By employing chameleon hash, it provides a holistic solution that simultaneously achieves lightweight efficiency, forward secrecy within a single communication round, and verifiable non-repudiation, thereby presenting a more complete and practical AKE framework for the modern IoT landscape.

3 Background

In this section, we propose the CHAKE system model, review the chameleon hash algorithm, and outline the requisite Security Objectives essential for ensuring secure communication and robust key exchange.

3.1 System Model

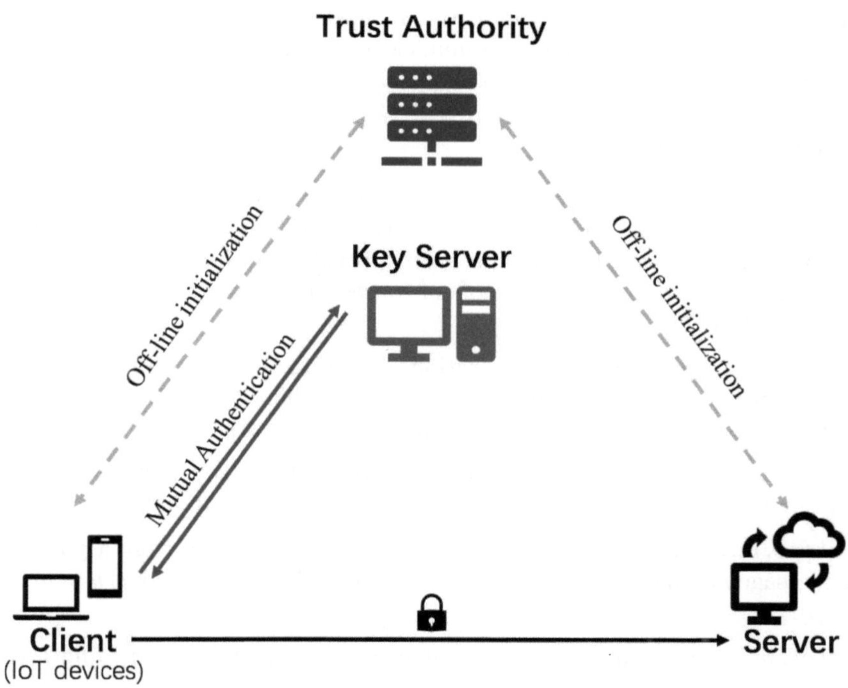

Fig. 2. System Model of CHAKE.

Figure 2 illustrates the system model of CHAKE. On the basis of the IoT trust model proposed by Fang *et al.* [14], we introduce the following key components depicted in the figure:

- **Trust authority**: Trust Authority(TA) denotes a special entity, typically a trusted authority or organization. In CHAKE, the TA is similar to the Handle Identifier System [22], which is responsible for generating and distributing secret values and identifiers.

- **Key Server**: The Key Server(KS) is a regionally deployed dedicated network service that specializes in the management and distribution of keys. Clients can obtain the session key for the corresponding server through KS.
- **Client**: The client is a resource-limited IoT user device.
- **Server**: The server is responsible for providing services for authenticated clients or obtaining information uploaded by trusted clients.

In CHAKE, the TA holds the responsibility for the registration of devices and the distribution of identifiers and secret values. Notably, the that TA does not directly participate in the subsequent protocol processes. As a distinct entity, KS maintains the secret values of devices and the associated derived symmetric keys. In practical deployment, TA and KS may also be deployed on the same host to manage devices regionally. When a client attempts to establish a secure channel to communicate with the server, it needs to interact with the KS first to obtain the session key before communicating with the server. More details of the protocol are presented in Sect. 4.

3.2 Chameleon Hash

Chameleon Hash is a one-way hash function with trapdoor features that allows individuals with access to trapdoor information to find intentional hash collision [21]. A Chameleon hash algorithm generally provides collisions forgery resistance and semantic security. Detailed proofs of its security properties and correctness can be found in [21].

CParGen(1^λ*).* The parameter generation algorithm takes a security parameter λ and outputs public parameters $pp_{ch} = \{p, q, g\}$, where p and q are large primes and where g is a generator of a subgroup of order q in $\mathbb{Z}_p^*$.

CKeyGen(). Based on pp_{ch}, this procedure generates a key pair (pk, sk). The secret key (trapdoor) is $sk = x$, where $x \in_R \mathbb{Z}_q^*$, and the public key is $pk = g^x$.

CHash(pk, m, r*).* The hashing algorithm computes a digest h from a public key pk, a message m, and a randomness $r \in \mathbb{Z}_q^*$. The hash is calculated as $h = g^m \cdot pk^r$ (mod p).

Forge(sk, m, m', r*).* This algorithm enables the trapdoor holder to find a collision. Given an existing hash $h = CHash(pk, m, r)$, the holder can compute a new randomness r' for a different message m' such that $h = CHash(pk, m', r')$. This is achieved by leveraging the secret key $sk = x$ to solve for r' in the underlying exponent equality, yielding:

$$r' = \left(\frac{m - m'}{x} + r\right) \mod p \tag{1}$$

4 Proposed Protocol

In this section, we introduce the authentication and key exchange mechanism used for CHAKE.

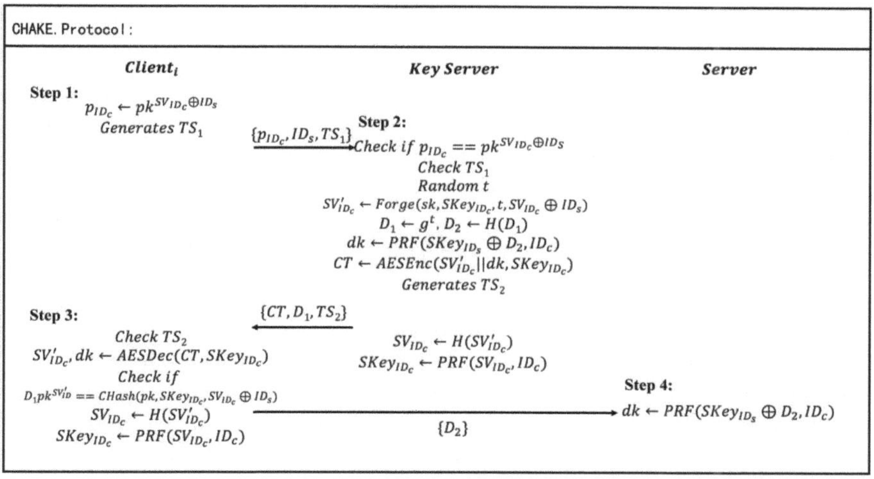

Fig. 3. Key exchange of the CHAKE protocol.

4.1 Initialization

During the initialization phase, the TA distributes a secret value $s \in \mathbb{Z}_q^*$ and an identifier ID to each registered device, and this information is also provided to the KS of the area.

We implement the secure pseudo-random function (PRF) defined by Peter *et al.* [16] for key derivation and utilize AES-256 as the symmetric encryption algorithm. Here, the function $AESEnc(m, key)$ represents the encryption of plaintext m via the symmetric key, resulting in ciphertext c. Conversely, $AESDec(c, key)$ denotes a corresponding decryption function. Each participant initializes the necessary parameters for the AES algorithm, and computes the AES symmetric key as $SKey \leftarrow PRF(s, ID)$.

The KS further initializes the chameleon hash parameters (pk, sk), selects a secure hash function $H : \{0,1\}^* \to \{0,1\}^k$, and publishes the public parameters to the system which include $\{q,\ p,\ g,\ pk,\ \mathbb{Z}_p^*\}$.

4.2 Authentication Key Exchange

In Fig. 3, **Step 1** and **2** describe the mutual authentication process between the client and the key server (KS) in the CHAKE protocol.

Step 1: An IoT device, acting as the Client, initiates communication with the KS by calculating

$$p_{ID_C} = pk^{SV_{ID_C} \oplus ID_S}$$

as an authentication credential. The Client generates a timestamp TS_1 and sends the message $m_1 = \{p_{ID_C}, TS_1, ID_S\}$ to the KS, where SV_{ID_C} is the client's secret value, and ID_S is the unique identifier of the requested server (not the KS).

Step 2: Upon receiving m_1, the KS retrieves the symmetric keys $SKey_{ID_S}$ and SV_{ID_C}. It verifies the request by recalculating $pk^{SV_{ID_C} \oplus ID_S}$ and checking the validity of TS_1. If the verification succeeds, the KS selects a random $t \in \mathbb{Z}_q^*$ and then computes $D_1 \leftarrow g^t, D_2 \leftarrow H(D_1)$ and:

$$SV'_{ID} \leftarrow \text{Forge}(sk, SKey_{ID_C}, t, SV_{ID_C} \bigoplus ID_S)$$

The KS then derives the session key $dk \leftarrow PRF(SKey_{ID_S} \oplus D_2, ID_C)$. It encrypts $SV'_{ID}||dk$ with $SKey_{ID_C}$ to obtain CT, generates a new timestamp TS_2, and sends the message $m_2 = \{CT, D_1, TS_2\}$ back to the Client.

In the two steps described above, CHAKE performs both authentication and session key dk generation. Notably, if the server-related computations are omitted, these steps can independently update the Client's secret value and symmetric key, thereby ensuring the protocol's forward security. While these steps achieve one-way authentication of the Client by the KS, the Client verifies the authenticity of m_2 in Step 3, completing the agreement of dk.

Step 3: Upon receiving the KS's response, the Client first validates the timestamp TS_2 and decrypts CT. According to the chameleon hash algorithm, the following relationship holds:

$$D_1 pk^{SV'_{ID}} = g^t pk^{SV'_{ID}} = g^{SKey_{ID_C}} p_{ID_C}$$

Using the equation above, the Client verifies the validity of CT and achieves the session key dk, ensuring it is trusted for communication with the Server. Upon successful verification, the KS and Client update $SV_C, SKey_{ID_C}$ as follows:

$$SV_{ID_C} \leftarrow H(SV'_{ID_C}), SKey_{ID_C} \leftarrow PRF(SV_{ID_C}, ID_C)$$

After this stage, the Client only needs to include D_2 during the initial communication to securely communicate with the Server via dk.

Step 4: When the Server receives a message from the Client, it independently computes the following:

$$dk \leftarrow PRF(SKey_{ID_S} \bigoplus D_2, ID_c)$$

and decrypts the message locally.

To mitigate issues from network failures, both Client and the KS retain the previously established SV and $SKey$. This ensures consistent and secure communication despite such issues. For example, if the KS receives a message from the client and the comparison with p fails, it reverts to the old secret value for comparison and sends an update warning to the client to ensure synchronization.

If a client's SV_{ID_C} remains outdated for a certain period, we consider that SV_{ID_C} has the risk of leakage. Therefore the KS must periodically update it. This means that if the client does not initiate a session with KS to update SV_{ID_C}

within the specified timeframe, KS will proactively engage with the client. During this interaction, if the update succeeds, the Client sends a success token to the KS. Similarly, the operation is performed for the server. This process enhances the protocol's security measures and strengthens forward security.

Notably, the server establishes secure communication with the Client through local computations, without any interaction with others. This one-sided approach ensures efficient server-side verification while defending against potential Denial-Of-Service (DoS) attacks targeting the CHAKE protocol.

4.3 Scalability and Protocol Extensions

While CHAKE provides an efficient solution for the client-server model, we recognize that large-scale IoT deployments present further challenges. We can focus on extending the CHAKE framework to support more complex topologies.

For multi-server environments, where a single device may need to communicate with numerous distinct services, we investigate mechanisms for a client to efficiently obtain session keys for multiple servers in a single, batched interaction with the Key Server. This would significantly reduce handshake latency and overhead compared with establishing each session independently. For device-to-device (D2D) communication, we plan to explore a model where the KS acts as a trusted introducer. In this model, the KS brokers a secure introduction by providing two clients with a temporary, shared secret. The clients could then use this secret to run a lightweight, direct AKE protocol to establish a session key, thereby inheriting trust from the KS without requiring its continuous involvement in every D2D session.

5 Security of the CHAKE Protocol

5.1 Security Model

This section outlines the adversary capabilities and the definitions of protocol security within our security model. We adopt the adversary model from [2,4] with necessary modifications for our context.

Participants: The authenticated key exchange protocol involves a set of participants $\mathcal{P} = \{P_0, ..., P_{n-1}\}$, which includes a KS and a TA. The participant can be any entity that appears in the system model. Each participant holds a symmetric key associated with the KS, denoted as $SKey_{C_i}$, and every pair of participants exchanges a unique session key, referred to as dk.

Instances: Participants can participate in multiple instances of the authenticated key exchange protocol sequentially, but not concurrently. This restriction is necessary because the session key derivation method proposed by CHAKE involves updating key values. Each protocol execution is denoted a session. Within this context, π_i^s embodies the specific process of the sth instance for participant P_i in protocol π. The security model employs the following notations:

1) β: Denotes the role played by a participant within the protocol, generally categorized as either initiator or responder, i.e. $\beta \in \{init, resp\}$.
2) γ: Indicates the state of an instance, which can be expressed as $\gamma \in \{accepted, rejected, running, \perp\}$.
3) τ: Denotes the status of the session key dk, for instance, $\tau \in \{revealed, \perp\}$.
4) b: Represents a binary bit randomly selected from the set $\{0, 1\}$.
5) sid: This serves as the unique identifier for a session.
6) pid: Identifies the counterpart with whom a participant generates a session key, e.g., $pid \in \mathcal{P}$.
7) dk: The session key for the conversation between the Client and Server.

Adversarial Queries: Drawing from the model of [3], we assume that an adversary, denoted as $\mathcal{A}$, has the ability to orchestrate network communications and interact with oracles through a series of queries. The adversary's abilities are delineated as follows:

1) $Send(\pi_i^s, m)$: The query emulates the full capabilities of an active adversary $\mathcal{A}$ controlling the communication network. This simulates the adversary's potential to initiate attacks in real-world scenarios, triggering an instance that responds to the falsified information m provided by $\mathcal{A}$. If an instance π_i^j that aligns with the simulation is already active, a null response ($\perp$) is returned.
2) $Revel(\pi_i^s)$: This operation simulates the scenario where a session key becomes compromised, revealing the session key dk of the instance π_i^s. Subsequent to this operation, the state τ of instance π_i^j is set to null ($\perp$).
3) $Corrupt(P_i)$: This operation simulates the compromise of a participant, leading to theft of long-term keys, which, in the context of this protocol, means the acquisition of the participant's secret values or trapdoors.
4) $Test(\pi_i^s)$: This operation can be executed only once and is not indicative of the adversary's actual attack process. Instead, it serves to gauge the likelihood of the adversary's success. If the instance π_i^s has not established dk or if its state γ is set to rejected, a null response ($\perp$) is returned. Otherwise, a bit $b \in \{0, 1\}$ is randomly selected; if b=0, the adversary $\mathcal{A}$ is presented with the session key dk; if b=1, a random key of equivalent length from the same key space $\mathcal{K}$ as dk is returned.

The following is the formal definition of protocol security.

Definition 1 (Secure Entity Authentication): Protocol instance π_i^s is considered to have been maliciously accepted by the targeted partner P_j under the following conditions:

- The instance pi_i^s is in an accepted state ($\gamma = accepted$) and includes P_j within its set of $\pi_i^s.pid$.
- Both P_i and P_j remain not be corrupted.
- There is no single session instance π_j^t that forms a partner with π_i^s and π_j^t.

We denote $Adv_{CHAKE}^{auth}(\mathcal{A})$ as the advantage of adversary $\mathcal{A}$ in being maliciously accepted.

Definition 2 (Key Indistinguishability): The adversary $\mathcal{A}$ launches a test query against an instance π_i^s in the protocol and returns b'. The key generated by instance π_i^s is considered indistinguishable to adversary $\mathcal{A}$ if:

- π_i^s and its partner are fresh.
- $\pi_i^s.b \neq b'$.

We denote $Adv_{\prod}^{key-ind}(\mathcal{A})$ as the advantage of adversary $\mathcal{A}$ in correctly answering the test query, which has

$$Adv_{\prod}^{key-ind}(\mathcal{A}) = |2Pr[\pi_i^s.b = b'] - 1| \tag{2}$$

Definition 3 (AKE Security): If for all PPT adversaries $\mathcal{A}$, the advantages $Adv_{\prod}^{key-ind}(\mathcal{A})$ and $Adv_{\prod}^{auth}(\mathcal{A})$ are negligible in λ, the AKE protocol $\prod$ is secure.

Theorem 1. *Assuming a scenario with a maximum of n entities, each conducts no more than l sessions. If the CHAKE protocol is demonstrated to be secure, for any PPT adversary $\mathcal{A}$ challenging the CHAKE protocol, the following inequality holds:*

$$Adv_{CHAKE}^{auth}(\mathcal{A}) \leq nl^2 \left(\frac{(nl-1)+l^2}{2^{\lambda}l} + Adv_p^{DLP}(\mathcal{B}) + Adv_{ch}(\mathcal{D}) + Adv_{derive}^{PRF}(\mathcal{E}) \right) \tag{3}$$

$$Adv_{\prod}^{key-ind}(\mathcal{A}) \leq nl^2 Adv_{derive}^{PRF}(\mathcal{E}) + Adv_{CHAKE}^{auth}(\mathcal{A}) \tag{4}$$

Here, λ denotes the length of the pseudo-random value. The terms $Adv_p^{DLP}(\mathcal{B})$, $Adv_{ch}(\mathcal{D})$, and $Adv_{derive}^{PRF}(\mathcal{E})$ are the maximal advantages for any PPT adversary to break the Discrete Logarithm Problem, the EUF-CMA security of the chameleon hash, and the PRF-based key derivation, respectively. See Appendix A for the proof.

5.2 Security Analysis

We introduce some of the security objectives achieved by CHAKE and provide a detailed analysis here. Additionally, as shown in Table 1, we analyse and compare the security properties of the proposed scheme with those of other existing schemes.

Explicit Mutual Authentication: Owing to the TA's distribution of a secret value only known to the device itself and the KS within the IoT, the initiator proves its identity to the key server by computing the proof p utilizing its secret

Table 1. Comparison of Security Properties

	[9]	[14]	[16]	[17]	[26]	Ours
Forward Security	✓	✗	✓	✗	✓	✓
Session Key for Three Parties	✗	✓	✗	✓	✗	✓
Replay Attack	✓	✓	✓	✓	✓	✓
Impersonation Attacks	✓	✓	✓	✓	✓	✓
DoS Attack	-	✗	✗	-	-	✓
Mutual Authentication	✓	✓	✓	✓	✓	✓
Non-Repudiation	✓	✓	✓	✓	✗	✓
Privacy Preservation	✓	✓	✓	✓	✓	✓

value. Conversely, the key server authenticates the initiator by generating a hash collision via the chameleon hash's trapdoor, a process whose security is ensured by the hash's inherent collision resistance. The receiver through the session key dk conferred by the key server verifies the session's authenticated status. Thus, the CHAKE protocol we propose is equipped to support explicit mutual authentication.

Forward Security: Forward security ensures that compromising the current key does not reveal any previously used keys. First, the session key dk for each interaction is derived from a combination of a persistent secret $SKey_{ID_S}$ and a freshly generated, ephemeral value t. Since t is discarded after use, the resulting session key is cryptographically decoupled from other sessions. Second, the protocol incorporates a state evolution mechanism where the long-term secret SV shared between the Client and KS is itself updated via a new value derived from a chameleon hash collision. These unidirectional updates ensure that the knowledge of a current SV reveals no information about prior SV. Consequently, even a full compromise of a device's current long-term keys cannot be leveraged to decrypt previously recorded communications.

Privacy Preservation: The computational complexity of the discrete logarithm problem ensures that adversaries are incapable of discovering any information regarding the secret value SV from p. The secret value, once updated by the KS, is returned to the initiator encrypted via AES, only the initiator who possesses the corresponding symmetric key, can decrypt and access the updated SV.

Non-Repudiation: Parties involved in AKE must be held accountable for their actions, ensuring that no party can deny the origination or receipt of a message. In our protocol, since each key distribution (or update) requires the KS to provide a chameleon hash collision, and only the KS, possessing the trapdoor sk, can legally generate the collision to produce SV' for verification, the KS cannot deny having distributed (or updated) a particular key. Similarly, the Client cannot falsely claim possession of a legitimate key issued by the KS.

6 Performance Analysis

We evaluate the computational and communication costs associated with the CHAKE protocol and compare them to those of five contemporary schemes in this section.

6.1 Computational Cost

In our study, the times for trapdoor collision are represented by T_f. Given that the execution time of the PRF approximates a single hash operation [20], T_h also denotes the PRF runtime. Owing to the extended bit length of the randomness provided by the KS during collision, the Client needs to perform a time-intensive exponentiation operation to verify the hash, denoted by $T_{p'}$. The cryptographic computations were executed on a computer configured with 64-bit Windows 11 OS, 16 GB of RAM, and an AMD Ryzen 9 7940H CPU @4 GHz. Bilinear pairing calculations were performed via "the JAVA Pairing Based Cryptography (JPBC)" library [10] with Type-A curves.

Table 2. Running time of various operations(ms)

Notation	Description	Computation time
T_h	A hash function(SHA-256)	0.01
T_{bp}	A bilinear pairing operation	2.71
T_{AES}	AES encryption/decryption	0.02
T_m	A scalar multiplication operation	2.16
T_a	A point addition operation	0.02
T_p	A large prime modular exponentiation operation	0.25
T_{fe}	A fuzzy extractor	2.16

After 1000 random simulations are performed to average the results, our analysis demonstrates that the CHAKE protocol's related timings T_f and $T_{p'}$ are 0.03 ms(milliseconds) and 0.79 ms, respectively. The remaining computation time parameter values are shown in Table 2, where the symmetric key length of the AES used is 256.

Table 3 provides a quantitative comparison of the computational costs for our protocol versus several contemporary schemes. The total overhead is asymmetrically distributed across the three roles to optimize for resource-constrained environments. The initiating client's workload is $T_{Client} = T_p + T_{AES} + T_{p'} + 2T_h + 2T_a$, and correspondingly, the computational cost for the KS is $T_{KS} = 2T_p + T_f + T_{AES} + 4T_h$, and the server needs only T_h to compute dk. Since the majority of the time required to compute $T_{ch'}$ is spent on calculating D_1, we consider the time for $T_{ch'}$ as part of the KS's computational overhead. The empirical data

Table 3. Computation cost and Message round comparison

Protocol	Computation cost	Total time(ms)	Message round
Pandi *et al.* [36]	$2T_{bp} + T_m$	7.58	3
M. F. Moghadam *et al.* [30]	$13T_h + 3T_{AES} + 5T_m$	10.99	4
Wu *et al.* [38]	$32T_h + T_{fe}$	2.48	4
Tseng *et al.* [35]	$5T_{bp} + 12T_p$	16.55	3
Lyu *et al.* [27]	$6T_h + T_{bp} + 10T_p + 2T_a + T_m + 2T_{AES}$	7.51	3
Ours	$4T_p + 2T_a + T_f + 2T_{AES} + T_{p'} + 7T_h$	1.97	3

confirm that our protocol achieves a lower total computational cost, underscoring its efficiency and suitability for IoT deployments.

6.2 Communication Cost

In our proposed work, the lengths of large integers, hash functions(same as the AES key), time stamps, and ID are denoted by $|\mathbb{Z}_q^*|$, $|H|$, $|TS|$, and $|ID|$, with sizes of 1024, 256, and 32 bits, respectively, to reach the security level of the 1024-bit Rivest–Shamir–Adleman (RSA) algorithm.

During CHAKE's mutual authentication and key agreement phase, the Client sends an identity proof and the ID identifier of the server to the KS. In turn, KS transmits a session key and parameters necessary for validation, encrypted via AES, back to the Client. This results in a communication overhead of $3|\mathbb{Z}_q^*| + 2|H| + |ID| + 2|TS| = 3680 bits$.

The communication efficiency of CHAKE, in terms of total data overhead and interaction rounds, is benchmarked against contemporary protocols in Table 3 and Fig. 4. The results indicate that our protocol maintains a communication footprint that is either comparable to or lower than the alternatives. Critically, CHAKE also minimizes protocol latency and power consumption by complet-

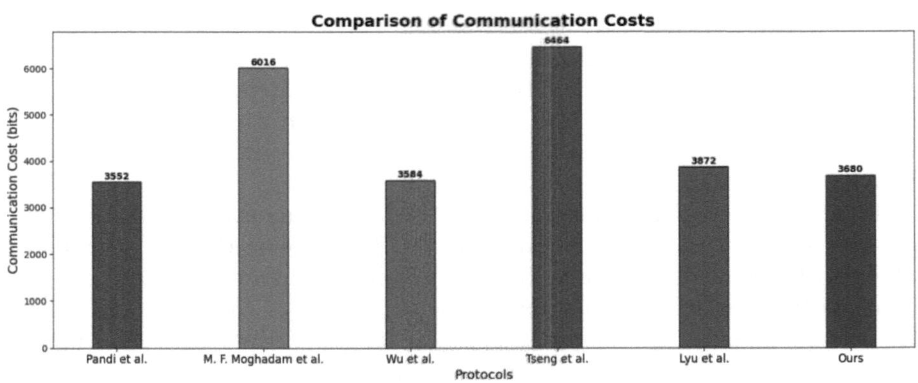

Fig. 4. Communication cost comparison.

ing the key exchange in the fewest rounds. This synergy of low communication overhead and minimal round complexity, coupled with its demonstrated computational efficiency, establishes the holistic performance advantage of our scheme. These characteristics confirm CHAKE's suitability for resource-constrained and delay-sensitive IoT applications.

7 Conclusion

In this paper, we propose the CHAKE protocol designed to achieve secure key generation and communication across devices with an integrated key server for the IoT. By utilizing the collision resistance of chameleon hash trapdoors to derive random values, the protocol efficiently enables mutual authentication and updatable key agreements while ensuring forward security. We developed a detailed CHAKE model and formally proved its security under a game-based security model, assessing its resistance to various common attacks. Our implementation of CHAKE demonstrates that it achieves lower computational and communication costs than several recent AKE protocols do. Securing network communications cost-effectively to protect user privacy and sensitive data is becoming increasingly important. In the future, we aim to focus on key negotiations independently without relying on the TA or other trusted organizations to issue keys or certificates in advance, thereby greatly reducing deployment and management costs.

Acknowledgements. This work was supported by the National Key R&D Program of China (Grant No. 2022YFB3102304) and the National Natural Science Foundation of China (Grant No. 62394323).

A Proof of CHAKE

We perform a security analysis of the proposed protocol via the defined security model and [6]. Note that the security definitions of PRF [20] and Chameleon hash are utilized to substantiate the security proof.

Secure Entity Authentication:
Game 0. This game emulates the CHAKE protocol, mirroring its procedures and faithfully executing authentication among the entities: thus,

$$Pr[G_0] = Adv_{CHAKE}^{auth}(\mathcal{A}) \tag{5}$$

Game 1. If the same random value is utilized in distinct instances, the challenger $\mathcal{C}$ will terminate the protocol. Given that the potential random values amount to $n \times l$, which are all sourced from a random selection within $\{0,1\}^\lambda$, the likelihood of two random values coinciding is $\frac{nl(nl-1)}{2^\lambda}$ [12]. Consequently, we deduce that

$$Pr[G_0] \leq Pr[G_1] + \frac{nl(nl-1)}{2^\lambda} \tag{6}$$

Game 2. In this game, the challenger $\mathcal{C}$ tries to deduce which session instance was corrupted by adversarial attacks. If the prediction is erroneous, the game terminates. Given that the total number of sessions does not exceed nl, it follows that:

$$Pr[G_2] \leq Pr[G_1] \times \frac{1}{nl} \tag{7}$$

Game 3. This game examines a scenario involving an instance π^* that has been corrupted, wherein the secret value p, derived from participant P_i's private information, is illegally forged and authenticated. Considering that each entity may have refreshed its keys up to l times, the total loss is at most $lAdv_p^{DLP}(\mathcal{B})$. Moreover, the secret s_{P_i} is updated, leading to the derivation of a new symmetric key $SKey_{P_i}$, which implies that the adversary could feasibly substitute $SKey_{P_i}$ with a counterfeit key, thereby facilitating an ostensibly legitimate session with the counterpart. Consequently, the likelihood of $\mathcal{A}$ prevailing in this game is contingent upon $\mathcal{B}$'s capacity to forge a credible p, hence:

$$Pr[G_2] \leq Pr[G_3] + lAdv_p^{DLP}(\mathcal{B}) \tag{8}$$

Game 3.5. This game is identical to *Game 3*, except that we replace the pseudorandom function $PRF(SKey_{ID_S} \oplus D_2, ID_C)$ used for deriving the session key dk with a truly random function that outputs a value from the same key space. An adversary $\mathcal{A}$ can distinguish Game 3 from Game 3.5 only if it can distinguish the output of the PRF from a truly random value. Let $Adv_{\text{derive}}^{\text{PRF}}(\mathcal{E})$ be the maximum advantage of any PPT adversary $\mathcal{E}$ in breaking the PRF security. Since the key derivation may be invoked up to l times for a given entity, the advantage is summed over these instances. The relationship between the probabilities of success in these games is as follows:

$$\Pr[G_3] \leq \Pr[G_{3.5}] + l \cdot Adv_{\text{derive}}^{\text{PRF}}(\mathcal{E}) \tag{9}$$

Game 4. We now analyse the transition from Game 3.5. In this game, the session key dk is perfectly random from the adversary's perspective, as it is generated by a truly random function. Therefore, an adversary $\mathcal{A}$ can no longer gain an advantage by attacking the key derivation. The only remaining path for $\mathcal{A}$ to have an instance π^* maliciously accepted is to forge the authentication message $m_2 = \{CT, D_1, TS_2\}$ sent from the Key Server to the Client. Specifically, the adversary must produce a valid tuple that passes the Client's verification check, which relies on the Chameleon Hash property: $D_1 pk^{SV'_{ID_C}} = g^{SKey_{ID_C}} p_{ID_C}$.

Forging this message without the trapdoor sk requires finding a collision in the Chameleon Hash scheme. Let $Adv_{\text{ch}}(\mathcal{D})$ be the maximum advantage of any PPT adversary $\mathcal{D}$ in breaking the EUF-CMA security of the Chameleon Hash. In addition to targeted forgery, a collision can occur by chance. Given that the randomness space is of size 2^λ and there are at most l sessions per entity, the probability of a random collision across sessions is bounded by $\frac{l^2}{2^\lambda}$. If the adversary succeeds in either of these ways, the challenger $\mathcal{C}$ terminates the game. Thus, the probability of success is bounded by:

$$\Pr[G_{3.5}] \leq \Pr[G_4] + l \cdot Adv_{\text{ch}}(\mathcal{D}) + \frac{l^2}{2^\lambda} \tag{10}$$

In *Game 4*, the adversary must produce a valid and authenticated message from the Key Server for an instance π^* without a legitimate partnering session. However, with the PRF replaced by a random function and the Chameleon Hash assumed to be secure, the adversary has no mechanism to generate such a message. Any attempt would be computationally infeasible or result in a random string that fails verification. Therefore, the probability of an adversary causing a malicious acceptance in *Game 4* is zero.

$$\Pr[G_4] = 0 \tag{11}$$

According to Game 0–4, we obtain the final value of (3).

Key Indistinguishability: In the following segment, we delve into the security associated with the indistinguishability of keys. The event where the adversary emerges victoriously in the key indistinguishability challenge during the i-th game is symbolized by D_i, and their corresponding advantage is articulated as:

$$Adv_i = Pr[D_i] - \frac{1}{2} \tag{12}$$

Game 0. This game emulates the key indistinguishability trial, expressed as follows:

$$Pr[D_0] = \frac{1}{2} + Adv_{CHAKE}^{key-ind}(\mathcal{A}) = \frac{1}{2} + Adv_0 \tag{13}$$

Game 1. If an instance is compromised by adversarial actions, the challenger $\mathcal{C}$ will engage in a binary decision by selecting $b' \in \{0, 1\}$ and subsequently conclude the game. For simplicity, an analogous substitution has been employed as in the authentication security demonstration:

$$Adv_0 = Adv_1 + Adv_{CHAKE}^{auth}(\mathcal{A}) \tag{14}$$

Game 2. Within this game, challenger $\mathcal{C}$ endeavors to discern which session instance has been subverted. A misjudgment results in the game's cessation. Given the ceiling of nl sessions, it follows that:

$$Adv_2 = Adv_1 \times \frac{1}{nl} \tag{15}$$

Game 3. In a corrupted instance π^*, the advantage of the adversary is reducible to the robustness of the PRF used in the session key derivation. This game introduces a true random value generator as a stand-in for the PRF. Each entity can only have a maximum of l session keys. Consequently:

$$Adv_2 \leq Adv_3 + l Adv_{derive}^{PRF}(\mathcal{E}) \tag{16}$$

Owing to the utilization of true random values for key generation in this game, the advantage of adversary is negligible, that is, an adversary cannot successfully generate and replace a legitimate session key, denoted by $Adv_3 = 0$. Finally, Eq. (4) is proved. In summary, Theorem 1 has been proved.

References

1. Adil, M., Jan, M.A., Liu, Y., Abulkasim, H., Farouk, A., Song, H.: A systematic survey: security threats to UAV-aided IoT applications, taxonomy, current challenges and requirements with future research directions. IEEE Trans. Intell. Transp. Syst. **24**(2), 1437–1455 (2022)
2. Anagnostopoulos, N.A., Ahmad, S., Arul, T., Steinmetzer, D., Hollick, M., Katzenbeisser, S.: Low-cost security for next-generation IoT networks. ACM Trans. Internet Technol. (TOIT) **20**(3), 1–31 (2020). https://doi.org/10.1145/3406280
3. Bellare, M., Pointcheval, D., Rogaway, P.: Authenticated key exchange secure against dictionary attacks. In: Preneel, B. (ed.) EUROCRYPT 2000. LNCS, vol. 1807, pp. 139–155. Springer, Heidelberg (2000). https://doi.org/10.1007/3-540-45539-6_11
4. Bellare, M., Rogaway, P.: Entity authentication and key distribution. In: Stinson, D.R. (ed.) CRYPTO 1993. LNCS, vol. 773, pp. 232–249. Springer, Heidelberg (1994). https://doi.org/10.1007/3-540-48329-2_21
5. Boyd, C., Davies, G.T., de Kock, B., Gellert, K., Jager, T., Millerjord, L.: Symmetric key exchange with full forward security and robust synchronization. In: Tibouchi, M., Wang, H. (eds.) ASIACRYPT 2021. LNCS, vol. 13093, pp. 681–710. Springer, Cham (2021). https://doi.org/10.1007/978-3-030-92068-5_23
6. Brzuska, C., Jacobsen, H., Stebila, D.: Safely exporting keys from secure channels. In: Fischlin, M., Coron, J.-S. (eds.) EUROCRYPT 2016. LNCS, vol. 9665, pp. 670–698. Springer, Heidelberg (2016). https://doi.org/10.1007/978-3-662-49890-3_26
7. Challa, S., et al.: An efficient ECC-based provably secure three-factor user authentication and key agreement protocol for wireless healthcare sensor networks. Comput. Electr. Eng. **69**, 534–554 (2018)
8. Chuang, Y.H., Lo, N.W., Yang, C.Y., Tang, S.W.: A lightweight continuous authentication protocol for the internet of things. Sensors **18**(4), 1104 (2018)
9. Conners, J., et al.: Let's authenticate: automated certificates for user authentication. In: Network and Distributed Systems Security Symposium (NDSS) (2022)
10. De Caro, A., Iovino, V.: JPBC: Java pairing based cryptography. In: 2011 IEEE Symposium on Computers and Communications (ISCC), pp. 850–855 (2011). https://doi.org/10.1109/ISCC.2011.5983948
11. Dhillon, P.K., Kalra, S.: Secure multi-factor remote user authentication scheme for internet of things environments. Int. J. Commun. Syst. **30**(16), e3323 (2017)
12. Fan, Q., Chen, J., Shojafar, M., Kumari, S., He, D.: Sake*: a symmetric authenticated key exchange protocol with perfect forward secrecy for industrial internet of things. IEEE Trans. Industr. Inf. **18**(9), 6424–6434 (2022)
13. Fan, X., Xu, X., Li, B.: Group key exchange protocols from supersingular isogenies. In: Wu, Y., Yung, M. (eds.) Inscrypt 2020. LNCS, vol. 12612, pp. 157–173. Springer, Cham (2021). https://doi.org/10.1007/978-3-030-71852-7_11
14. Fang, D., Qian, Y., Hu, R.: A flexible and efficient authentication and secure data transmission scheme for IoT applications. IEEE Internet Things J. **7**(4), 3474–3484 (2020). https://doi.org/10.1109/jiot.2020.2970974
15. Fujioka, A., Suzuki, K.: Designing efficient authenticated key exchange resilient to leakage of ephemeral secret keys. In: Kiayias, A. (ed.) CT-RSA 2011. LNCS, vol. 6558, pp. 121–141. Springer, Heidelberg (2011). https://doi.org/10.1007/978-3-642-19074-2_10
16. Gaži, P., Pietrzak, K., Rybár, M.: The exact PRF-security of NMAC and HMAC. In: Garay, J.A., Gennaro, R. (eds.) CRYPTO 2014. LNCS, vol. 8616, pp. 113–130. Springer, Heidelberg (2014). https://doi.org/10.1007/978-3-662-44371-2_7

17. Hasan, M.K., Weichen, Z., Safie, N., Ahmed, F.R.A., Ghazal, T.M.: A survey on key agreement and authentication protocol for internet of things application. IEEE Access (2024)
18. Jiang, C., Xu, C., Chen, K.: An efficient privacy-preserving scheme for weak password collection in internet of things. In: Ge, C., Yung, M. (eds.) Information Security and Cryptology, pp. 389–393. Springer, Singapore (2024). https://doi.org/10.1007/978-981-97-0945-8_23
19. Kalra, S., Sood, S.K.: Secure authentication scheme for IoT and cloud servers. Pervasive Mob. Comput. **24**, 210–223 (2015)
20. Krawczyk, H.: Cryptographic extraction and key derivation: the HKDF scheme. In: Rabin, T. (ed.) CRYPTO 2010. LNCS, vol. 6223, pp. 631–648. Springer, Heidelberg (2010). https://doi.org/10.1007/978-3-642-14623-7_34
21. Krawczyk, H., Rabin, T.: Chameleon hashing and signatures. Cryptology ePrint Archive (1998)
22. Lannom, L., Boesch, L.C.B.P., Sun, S.: Handle System Overview. RFC 3650 (2003). https://doi.org/10.17487/RFC3650. https://www.rfc-editor.org/info/rfc3650
23. Lee, H., Kang, D., Ryu, J., Won, D., Kim, H., Lee, Y.: A three-factor anonymous user authentication scheme for internet of things environments. J. Inf. Security Appl. **52**, 102494 (2020)
24. Lian, H., Yang, Y., Zhao, Y.: Efficient and strong symmetric password authenticated key exchange with identity privacy for IoT. IEEE Internet Things J. **10**(6), 4725–4734 (2022)
25. Liu, K., Guan, J., Yao, S., Wang, L., Zhang, H.: Dkgauth: blockchain-assisted distributed key generation and authentication for cross-domain intelligent IoT. IEEE Internet Things J. **11**(15), 25663–25673 (2024). https://doi.org/10.1109/JIOT.2024.3379310
26. Liu, P., et al.: How IoT re-using threatens your sensitive data: exploring the user-data disposal in used IoT devices. In: 2023 IEEE Symposium on Security and Privacy (SP), pp. 3365–3381 (2023). https://doi.org/10.1109/SP46215.2023.10179294
27. Lyu, Y., Liu, S., Han, S., Gu, D.: Privacy-preserving authenticated key exchange in the standard model. In: International Conference on the Theory and Application of Cryptology and Information Security, pp. 210–240. Springer, Cham (2022)
28. Majid, A.: An enhanced symmetric cryptosystem and biometric-based anonymous user authentication and session key establishment scheme for WSN. IEEE Access **6**, 70072–70087 (2018)
29. Mandalari, A.M., Haddadi, H., Dubois, D.J., Choffnes, D.: Protected or porous: a comparative analysis of threat detection capability of IoT safeguards. In: 2023 IEEE Symposium on Security and Privacy (SP), pp. 3061–3078 (2023). https://doi.org/10.1109/SP46215.2023.10179282
30. Moghadam, M.F., Nikooghadam, M., Al Jabban, M.A.B., Alishahi, M., Mortazavi, L., Mohajerzadeh, A.: An efficient authentication and key agreement scheme based on ECDH for wireless sensor network. IEEE Access **8**, 73182–73192 (2020)
31. Nikbakht Bideh, P., Gehrmann, C.: Rosym: robust symmetric key based IoT software upgrade over-the-air. In: Proceedings of the 4th Workshop on CPS & IoT Security and Privacy, pp. 35–46 (2022)
32. Rao, P.M., Deebak, B.D.: A comprehensive survey on authentication and secure key management in internet of things: challenges, countermeasures, and future directions. Ad Hoc Netw. 103159 (2023)
33. de Saint Guilhem, C.D., Fischlin, M., Warinschi, B.: Authentication in key-exchange: definitions, relations and composition. In: 2020 IEEE 33rd Computer Security Foundations Symposium (CSF), pp. 288–303. IEEE (2020)

34. Shi, X., Shi, S., Wang, M., Kaunisto, J., Qian, C.: On-device IoT certificate revocation checking with small memory and low latency. In: Proceedings of the 2021 ACM SIGSAC Conference on Computer and Communications Security, pp. 1118–1134 (2021)
35. Tseng, Y.M., Chen, J.L., Huang, S.S.: A lightweight leakage-resilient identity-based mutual authentication and key exchange protocol for resource-limited devices. Comput. Netw. **196**, 108246 (2021)
36. Vijayakumar, P., Azees, M., Kozlov, S.A., Rodrigues, J.J.: An anonymous batch authentication and key exchange protocols for 6g enabled VANETs. IEEE Trans. Intell. Transp. Syst. **23**(2), 1630–1638 (2021)
37. Wu, C., Chen, J., He, K., Zhao, Z., Du, R., Zhang, C.: Echohand: high accuracy and presentation attack resistant hand authentication on commodity mobile devices. In: Proceedings of the 2022 ACM SIGSAC Conference on Computer and Communications Security, pp. 2931–2945 (2022)
38. Wu, F., Li, X., Xu, L., Vijayakumar, P., Kumar, N.: A novel three-factor authentication protocol for wireless sensor networks with IoT notion. IEEE Syst. J. **15**(1), 1120–1129 (2020)

An Online Time Series Decomposition Algorithm with a Multiplicative Trend-Seasonality Relationship in an Additive Model

Hao Wang, Yingjian Liu(✉), Minghuan Lv, Xiangyun Zheng, and Zonghai Zha

College of Computer Science and Technology, Ocean University of China,
238 Songling Road, Qingdao 266100, Shandong, China
liuyj@ouc.edu.cn

Abstract. Time series decomposition plays an important role in time series analysis, anomaly detection, and forecasting, with significant applications in big data mining. However, current research on time series decomposition faces challenges in real-time processing, accuracy, and applicability. This paper proposes an online seasonal-trend decomposition (STD) algorithm, the α-STD algorithm, which is based on an additive time series decomposition model, and extends it to multi-seasonal scenarios. An adaptive scaling component α is introduced to capture trend variations through periodic differencing of normalized time series values. By incorporating α as a dynamic scaling factor for seasonal components, our algorithm effectively models the multiplicative relationships between trends and seasonality within a traditional additive decomposition model. The α-STD algorithm demonstrates superior performance in handling time series with trend jumps, seasonal shifts, and heteroscedasticity while maintaining an $\mathcal{O}(1)$ update complexity. Furthermore, this paper provides solutions for handling missing data in online time series decomposition. The experimental results indicate that the α-STD has broader applicability than baseline methods do, achieving high decomposition accuracy across various datasets and effectively mitigating the impact of missing data on online decomposition tasks.

Keywords: Time series · Seasonal-trend decomposition · Multiple seasonality · Online time series decomposition · Data mining

1 Introduction

Time series decomposition (TSD) is a crucial task in time series analysis, with increasing importance in big data mining due to its role in temporal pattern discovery from large-scale, high-dimensional time series data. By breaking down complex time series into components with different characteristics, TSD enables critical applications in forecasting [1,7,26], anomaly detection [34,41], and signal processing. Its importance in big data mining stems from its ability to transform

L. Huang et al. (Eds.): CCF NCCA 2025, CCIS 2774, pp. 593–618, 2026.
https://doi.org/10.1007/978-981-95-5294-8_39

raw temporal data into interpretable components that facilitate efficient analysis of massive datasets, particularly in IoT and AIOps applications [20]. However, the inherent heterogeneity of time series, characterized by nonstationary trends (e.g., volatility, abrupt breaks) and complex seasonal patterns, presents fundamental challenges in developing robust decomposition frameworks that generalize across diverse scenarios.

Current time series data exhibit two major trends. First, the continuous expansion of application scenarios has led to a significant increase in the diversity and complexity of data. Many time series exhibit complex characteristics with multiple seasonalities, presenting new challenges for TSD methods. Second, there is a growing demand for real-time processing. Real-time decomposition of time series data not only helps systems respond quickly but also significantly enhances analysis efficiency. However, most high-accuracy decomposition algorithms still rely on offline batch processing methods, making them inadequate for incremental updates.

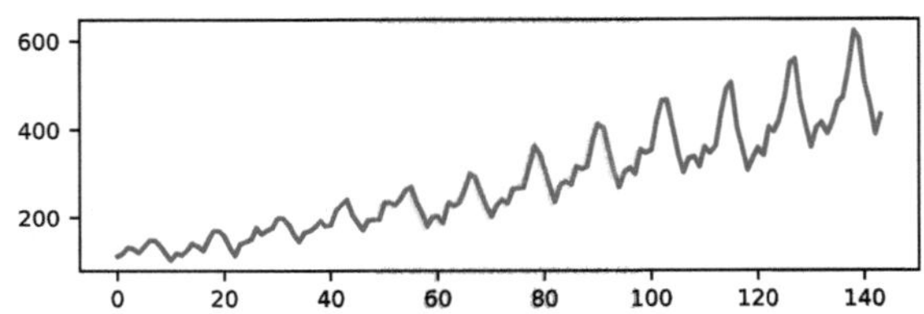

Fig. 1. AirPassengers dataset: monthly airline passenger numbers from 1949–1960.

Recent TSD methods have focused mainly on online seasonal-trend decomposition (STD), which quickly decomposes time series into trend, seasonality, and residual components. However, existing methods generally assume an additive relationship between trends and seasonality. In many real-world scenarios, such as energy consumption or traffic flow data, there may be a multiplicative relationship between trends and seasonality, often resulting in heteroscedasticity. Taking the AirPassengers dataset in Fig. 1 as an example, the data exhibit not only a clear upwards trend but also seasonal fluctuations whose amplitude increases with the trend, demonstrating characteristic multiplicative behavior. However, most existing online STD methods are based on additive models and show limited effectiveness when handling these multiplicative relationships. This motivates the extension of additive decomposition frameworks to better accommodate multiplicative relationships between trends and seasonality.

Moreover, missing data frequently occur in practice. The absence of individual data points or entire segments can significantly affect the decomposition accuracy of STD algorithms. Therefore, there is an urgent need for effective solutions to handle missing data in online settings. While it is relatively straightforward to extend single-seasonal decomposition models to multi-seasonal cases, online decomposition faces new challenges due to the unknown future data.

To address these challenges, we propose the α-STD algorithm. The main contributions of this paper are summarized as follows:

- We propose the α-STD algorithm, which introduces a multiplicative relationship between trends and seasonality in an additive decomposition model. It combines periodic differencing with the ℓ_1-norm constraint on trend and introduces the α component to capture the multiplicative relationship. The algorithm maintains an online update complexity of $\mathcal{O}(1)$.
- We analyse the issue of missing data in online STD and propose strategies for the two types of missing data. Specifically, we design an online solution for single-point missing data on the basis of historical values. This approach significantly reduces the impact of single-point missing data on STD accuracy, particularly in stationary time series.
- We extend α-STD to a multi-seasonal STD with approximately $\mathcal{O}(1)$ update complexity in theory, adopting an incremental decomposition mechanism rather than the window-based strategy used in previous methods. We also discuss key challenges in online multi-seasonal STD, particularly the interference between seasonal components.
- Extensive experiments on synthetic and real-world datasets show that the α-STD outperforms existing online decomposition algorithms in overall performance. It exhibits superior accuracy on most datasets, broader applicability, and effective handling of missing data.

The rest of this paper is organized as follows. Section 2 reviews related work on TSD. Section 3 presents the proposed α-STD method in detail. Section 4 presents the experiments and discussion. Finally, Sect. 5 concludes the paper.

2 Related Work

2.1 Traditional Decomposition Methods

Research on TSD methods dates back to the early 20th century. As early as 1933, Cowles 3rd [9] used moving average (MA) and its variants for TSD and forecasting. ARIMA [6], which is based on MA, has laid the foundation for many subsequent decomposition methods. A series of methods, including X-11 [29], SEATS [3], X-13-ARIMA-SEATS [16], and TBATS [12], significantly improve decomposition accuracy by enhancing seasonal handling capabilities. Moreover, several other methods, such as exponential smoothing [17], the HP filter [21], and ℓ_1 trend filtering [24], focus on extracting the trend component of time series. These simple methods are still widely used today [4,35], but they have a limited ability to handle complex trend and seasonal variations.

In addition to smoothing-based decomposition methods, some studies incorporate frequency-domain information. Fourier decomposition method (FDM) and other frequency-domain methods, which rely solely on frequency-domain information, have struggled to handle nonstationary series. Therefore, subsequent methods, such as wavelet decomposition [11] or empirical mode decomposition (EMD) [23], combine both time- and frequency-domain information.

To address the insufficient decomposition accuracy of frequency-domain methods and the prominent mode mixing problem in EMD, improvements such as local mean decomposition (LMD) [30], ensemble empirical mode decomposition (EEMD) [39], complete ensemble empirical mode decomposition (CEEMD) [32], empirical wavelet transform (EWT) [18], and variational mode decomposition (VMD) [14] have been proposed. Zhou et al. [43] further combined EWT with Fourier decomposition to propose a more accurate empirical fourier decomposition (EFD) method. Additionally, niche methods such as singular spectrum decomposition (SSD) [5] also combine time-frequency information. Although time-frequency methods have achieved good decomposition accuracy, their high overhead in time-domain and frequency-domain conversions remains a significant bottleneck.

2.2 Offline Seasonal Trend Decomposition Methods

Recently, the majority of TSD algorithms have decomposed time series into trend, seasonality, and residual components, known as seasonal-trend decomposition (STD). The trend component is typically perceived as slowly varying, with occasional abrupt shocks. The seasonal component is usually periodic and repetitive, such that $s_t \approx s_{t-T}$, where the number of seasonal components m and the period length T_i can be estimated via periodicity detection algorithms [28,31,33,37]. The residual component includes random disturbances and anomalous fluctuations, with random disturbances typically being of small magnitude.

STL [8] was the foundation for many recent STD decomposition methods and one of the earliest widely adopted methods. STL extracted trend and seasonal components via locally estimated scatterplot smoothing (LOESS) and iteratively optimized them for improved decomposition accuracy. To address multi-seasonality, MSTL [2] extends the STL method by progressively estimating multiple seasonal components. However, both STL and MSTL face challenges when handling with trend jumps or anomalies.

In recent years, many new offline decomposition methods have emerged to further improve decomposition accuracy. STR [13] jointly extracts trend, seasonality, and residual components via regression, avoiding iterative processes. However, STR requires significant computational resources to learn the two-dimensional structure, has low computational efficiency with long-period seasonal data, and cannot handle abrupt trend changes. To improve robustness, RobustSTL [36] combines ℓ_1 trend filtering and nonlocal seasonal filtering to improve robustness against outliers and trend changes. To further improve computational efficiency, Fast RobustSTL [38] introduces an ADMM-based optimization algorithm, although its efficiency is still limited when processing long-period seasonal data. Dudek [15] proposed the Seasonal-Trend-Dispersion Decomposition method to handle heteroscedasticity in time series, which showed superior performance in multiple forecasting tasks. Although this method considers complex relationships between trends and seasonality, it assumes that the trend component changes in phases, maintaining a steady trend within each cycle.

This assumption limits its application and makes it difficult to handle complex trend changes. ASTD [27] adapts well to changes in seasonal components by combining seasonal length estimation, yielding accurate results in decomposition tasks without prior knowledge. However, ASTD relies on a sliding window of length N, resulting in a computational complexity of $\mathcal{O}(N)$.

2.3 Online Seasonal Trend Decomposition Methods

Online decomposition methods have gained increasing attention for their ability to update decomposition results in real-time. OnlineSTL [25] is the first online TSD algorithm, which extracts the trend via a tri-cube kernel filter and the seasonality using exponential smoothing, significantly improving computational efficiency. However, OnlineSTL has a time complexity of $\mathcal{O}(T)$, and its performance decreases as the seasonal period T increases. OneShotSTL [20] and BacktrackSTL [35] are two recent online decomposition algorithms with time complexities of $\mathcal{O}(1)$. OneShotSTL effectively minimizes decomposition errors by using joint optimization and linear system calculations for trend and seasonal components, and it handles trend jumps and data shifts well. BacktrackSTL [35], which leverages robust smoothing and nonlocal seasonal filtering techniques, significantly reduces update delays. However, its simplistic trend decomposition method is less effective at capturing complex trend fluctuations. As a result, it works well for time series with stable trends but faces difficulties when dealing with more volatile trend behaviors.

Although existing decomposition methods have made progress in various aspects, most still rely on additive models, treating trends and seasonality as independent components. However, in many real-world time series applications, multiplicative interactions exist between trends and seasonality [15], such as foot traffic or airline passenger data. Specifically, the variation in the seasonal component is directly influenced by the trend component, and as the trend increases, the amplitude of the seasonal fluctuations increases proportionally. For example, the AirPassengers dataset [40] exhibits a clear multiplicative relationship, as shown in Fig. 1.

In addition, existing online decomposition methods have not been extended to multi-seasonality scenarios and have not addressed the common problem of missing data in practical applications. The proposed α-STD method effectively addresses these urgent challenges. Building on the OneShotSTL algorithm, the α-STD introduces the α component to capture the multiplicative relationship between trends and seasonality, provides an online solution for handling missing data, and successfully extends to multi-seasonal scenarios.

3 Proposed Approach

Table 1 lists the key notations used in this paper and their corresponding descriptions.

Table 1. List of notations and their descriptions.

Symbol	Description
y_t	Original time series data at time t
τ_t	Trend component at time t
s_t	Seasonal component at time t
$s_{i,t}$	Seasonal component i at time t in a multi-seasonality setting
r_t	Residual component at time t
T	Period length of the seasonal component
T_i	Period length of the seasonal component i in a multi-seasonality setting
m	Number of seasonal components in a multi-seasonality setting
w_t	Trend normalizer at time t
α_t	Multiplicative seasonal modifier at time t
v	Seasonal buffer

3.1 Decomposition Model

To improve the generalization ability of the decomposition model and enhance its performance in most STD tasks, this paper proposes a new decomposition algorithm, α-STD, which follows the same decomposition structure as the traditional additive model:

$$y_t = \tau_t + \sum_{i=1}^{m} s_{i,t} + r_t, \quad t = 0, 1, \cdots, N-1. \tag{1}$$

However, α-STD introduces a new adjustment component, α, which allows the seasonal component s_t to evolve over time, thus capturing the seasonal characteristics of a multiplicative model within the additive framework. Specifically, the seasonal component s_t is defined as:

$$s_t = \alpha_t \cdot s_{t-T}, \tag{2}$$

where α_t reflects the relative change in the trend component between periods, allowing the seasonal component s_t to vary over time and capture the seasonal variation characteristic of a multiplicative model. This enables the additive model to account for the seasonal variation commonly observed in multiplicative models.

To ensure that α-STD can accurately simulate the seasonal characteristics of a multiplicative model, the parameter α must accurately capture trend changes. To achieve this, we consider scaling the trend component such that the scaled trend values are close to 1, which aids in better guiding the update of the seasonal components. The specific scaling calculations are implemented through matrix operations, which are described in detail in the subsequent sections. We refer to this scaled trend characteristic as the trend normalizer w_t, which typically fluctuates approximately 1 and mirrors the trend changes in the original time series.

However, directly using w_t to adjust the seasonal component s_t still presents some challenges. First, w_t reflects the trend relative to the initial period, but the seasonal signal itself may evolve over time, rendering initial-period-based adjustments inapplicable. Furthermore, if the time series exhibits a monotonic trend, the gap between w_t and 1 may grow over time, leading to error accumulation. To address these issues, this paper introduces another component, α_t, which reflects the relative changes in the trend between adjacent periods. The formula for the i-th seasonal component's $\alpha_{i,t}$ is given by:

$$\alpha_{i,t} = 1 + w_t - w_{t-T_i}. \tag{3}$$

By introducing an offset of 1, we ensure that the magnitude of the seasonal variation matches the seasonal variation in a multiplicative model. Thus, $\alpha_{i,t}$ accurately captures the relative changes in the trend and remains within a controllable range, allowing α-STD to simulate multiplicative seasonal adjustments within the additive model framework.

In summary, although the α-STD is similar to the traditional additive model in its form, it combines the broader applicability of the additive model with the seasonal variation in the multiplicative model through the inherent $s_t = \alpha_t \cdot s_{t-T}$ relationship. This allows it to capture the multiplicative relationship between trends and seasonality more effectively when handling time series such as AirPassengers.

3.2 Single-Seasonality α-STD Model

To clearly explain our decomposition algorithm, we begin with the case of a single seasonal component. The OneShotSTL model [20] formulates the time series decomposition problem as an optimization problem with a fixed window length M. Building on this, α-STD introduces the trend normalizer w and the multiplicative seasonal modifier α. As mentioned earlier, w represents the scaling of the trend, aiming to adjust it to be close to 1 while maintaining the fluctuation characteristics consistent with the trend component τ. α reflects the relative change in the trend and represents the relationship between the current period and the previous period's w, which helps update the seasonal component. The objective function for minimizing the α-STD is as follows:

$$\begin{aligned}\min_{\tau,s,\alpha} \Bigg(& \sum_{j=t_0+1}^{t_0+M} (\tau_j + s_j - y_j)^2 + \lambda_1 \sum_{j=t_0+1}^{t_0+M} (s_j - \alpha_j v_{j\%T})^2 \\ & + \lambda_2 \sum_{j=t_0+1}^{t_0+M} |\tau_j - \tau_{j-1}| + \lambda_3 \sum_{j=t_0+1}^{t_0+M} |\tau_j - 2\tau_{j-1} + \tau_{j-2}| \\ & + \lambda_4 \sum_{j=t_0+1}^{t_0+M} (w_j - w_{j-1})^2 + \lambda_4 \sum_{j=t_0+1}^{t_0+M} (w_j - 2w_{j-1} + w_{j-2})^2 \\ & + \lambda_4 \sum_{j=t_0+1}^{t_0+M} (w_j - 1)^2 \Bigg),\end{aligned} \tag{4}$$

where λ_1 controls the degree of seasonal variation, λ_2 and λ_3 control the trend smoothness, and λ_4 controls the influence of w. Additionally, $\boldsymbol{v} \in \mathbb{R}^T$ is the seasonal buffer with a period length of T. The first term measures the fitting error between the trend and seasonal components, the second term controls the seasonal component's change relative to the previous period, the third and fourth terms smooth the trend components by robustly capturing trend information through the ℓ_1 norm, and the fifth to seventh terms constrain the w component to stay close to 1 while maintaining smoothness.

Combining periodic differencing with the ℓ_1 norm constraint smooths the trend while maintaining robustness to outliers and trend jumps [36]. The introduction of the w component allows the α-STD to capture the multiplicative relationship between the trend and seasonal components. Additionally, the last three terms in Eq. (4) reduce the impact of trend components during trend jumps, helping extract more reasonable trend components.

Inspired by the approach of OneShotSTL, we optimize the objective function in Eq. (4). We introduce auxiliary variables $\boldsymbol{p} \in \mathbb{R}^M$ and $\boldsymbol{q} \in \mathbb{R}^M$ to measure the weights of the first and second-order trend differences. The update rules for these variables are as follows:

$$p_t = \frac{1}{2\left|\tau_t - \tau_{t-1}\right| + \epsilon}, \tag{5}$$

$$q_t = \frac{1}{2\left|\tau_t - 2\tau_{t-1} + \tau_{t-2}\right| + \epsilon}, \tag{6}$$

where ϵ is a small constant used to prevent division by zero.

Therefore, the objective function in Eq. (4) is modified as follows:

$$\begin{aligned} \min_{\boldsymbol{\tau},\boldsymbol{s},\boldsymbol{\alpha},\boldsymbol{p},\boldsymbol{q}} & \left(\sum_{j=t_0+1}^{t_0+M} (\tau_t + s_t - y_t)^2 + \lambda_1 \sum_{j=t_0+1}^{t_0+M} (s_t - \alpha_j s_{t-T})^2 \right. \\ & + \lambda_2 \sum_{j=t_0+1}^{t_0+M} \left(p_t (\tau_t - \tau_{t-1})^2 + \frac{1}{4} p_t^{-1} \right) \\ & + \lambda_3 \sum_{j=t_0+1}^{t_0+M} \left(q_t (\tau_t - 2\tau_{t-1} + \tau_{t-2})^2 + \frac{1}{4} q_t^{-1} \right) \\ & + \lambda_4 \sum_{j=t_0+1}^{t_0+M} (w_j - w_{j-1})^2 + \lambda_4 \sum_{j=t_0+1}^{t_0+M} (w_j - 2w_{j-1} + w_{j-2})^2 \\ & \left. + \lambda_4 \sum_{j=t_0+1}^{t_0+M} (w_j - 1)^2 \right), \end{aligned} \tag{7}$$

where substituting $s_t = (1 + w_t - w_{t-T})s_{t-T}$ into the residual term $R = (\tau_t + s_t - y_t)^2$ yields the partial derivative of the residual term R with respect to w as $2(\tau_t + (1 + w_t - w_{t-T})s_{t-T} - y_t)s_{t-T}$. This adjustment of w serves to minimize the residual term R, thereby implicitly capturing the local patterns of y.

The optimization problem is solved via the iterative reweighted least squares (IRLS) method [19], which alternately optimizes $\boldsymbol{\tau}, \boldsymbol{s}, \boldsymbol{w}$ and $\boldsymbol{p}, \boldsymbol{q}$. In each iteration, the auxiliary variables are treated as fixed weight vectors, and the weights are updated after the optimization is completed.

Letting $\boldsymbol{x} = [\tau_{t_0+1}, \alpha_{t_0+1}, s_{t_0+1}, \ldots, \tau_{t_0+M}, \alpha_{t_0+M}, s_{t_0+M}]^{\mathrm{T}}$, the optimal solution to this optimization problem can be determined by the linear system $\boldsymbol{Ax} = \boldsymbol{b}$:

$$
\begin{aligned}
\boldsymbol{A} =& A_1^{\mathrm{T}} A_1 + \lambda_1 A_2^{\mathrm{T}} A_2 + \lambda_2 (p \cdot A_3)^{\mathrm{T}} A_3 + \lambda_3 (q \cdot A_4)^{\mathrm{T}} A_4 \\
&+ \lambda_4 A_5^{\mathrm{T}} A_5 + \lambda_4 A_6^{\mathrm{T}} A_6 + \lambda_4 A_7^{\mathrm{T}} A_7, \\
\boldsymbol{b} =& A_1^{\mathrm{T}} \boldsymbol{y} + A_7^{\mathrm{T}} \boldsymbol{y}' + \lambda_1 A_2^{\mathrm{T}} \boldsymbol{u} + \lambda_4 A_7^{\mathrm{T}} \boldsymbol{1}, \\
& \text{where } \boldsymbol{u} = [\boldsymbol{v}^{\mathrm{T}}, \ldots, \boldsymbol{v}^{\mathrm{T}}] \in \mathbb{R}^M, \boldsymbol{1} = [1 \ldots, 1]^{\mathrm{T}} \in \mathbb{R}^M, \\
A_1 =& \begin{bmatrix} 1\,0\,1 & & \\ & \ddots & \\ & & 1\,0\,1 \end{bmatrix} \in \mathbb{R}^{M \times 3M}, \\
A_2 =& \begin{bmatrix} 0\,0\,1 & & \\ & \ddots & \\ & & 0\,0\,1 \end{bmatrix} \in \mathbb{R}^{M \times 3M}, \\
A_3 =& \begin{bmatrix} -1\,0\,0\;1\;\;0\;\;0\;\;0\,0\,0 & & \\ & \ddots\;\ddots\;\ddots & \\ & & -1\;\;0\;\;0\;\;1\;0\,0\,0\,0\,0 \end{bmatrix} \in \mathbb{R}^{(M-1) \times 3M}, \\
A_4 =& \begin{bmatrix} 1\,0\,0\,-2\;\;0\;\;0\;\;1\;0\,0 & & \\ & \ddots\;\ddots\;\ddots & \\ & & 1\;\;0\;\;0\;-2\,0\,0\,1\,0\,0 \end{bmatrix} \in \mathbb{R}^{(M-2) \times 3M}, \\
A_5 =& \begin{bmatrix} 0\,-1\,0\,0\;1\;\;0 & & \\ & \ddots & \\ & & 0\;-1\,0\,0\,1\,0 \end{bmatrix} \in \mathbb{R}^{(M-1) \times 3M}, \\
A_6 =& \begin{bmatrix} 0\,1\,0\,0\,-2\;\;0\;\;0\;\;1\;\;0 & & \\ & \ddots\;\ddots\;\ddots & \\ & & 0\;\;1\;\;0\;\;0\;-2\,0\,0\,1\,0 \end{bmatrix} \in \mathbb{R}^{(M-2) \times 3M}, \\
A_7 =& \begin{bmatrix} 0\,1\,0 & & \\ & \ddots & \\ & & 0\,1\,0 \end{bmatrix} \in \mathbb{R}^{M \times 3M},
\end{aligned}
\tag{8}
$$

where A_1 is the summation matrix for the trend and seasonal components, A_2 is the seasonal component matrix, and A_3 and A_4 are the first- and second-order difference matrices for the trend. A_5 and A_6 are the first- and second-order difference matrices for the α component, and A_7 is used to constrain the α component to approximately 1. $\boldsymbol{v}$ is the seasonal buffer, which stores the period values from previous cycles. $\boldsymbol{y}'$ is the original time-series vector with the last value set to 0. After each iteration, the latest α is used as a multiplier to adjust s_{t_0} in $\boldsymbol{v}$.

The α-STD algorithm applies an ℓ_2 norm constraint on w, differing from the ℓ_1 norm approach used in RobustSTL, thus avoiding the introduction of additional auxiliary variables.

To establish a mathematical relationship between w and the original time series values y, $A_7^{\mathrm{T}}\boldsymbol{y}'$ is introduced into the solution vector $\boldsymbol{b}$ in the matrix computation of Eq. (8). This enables w to learn the variations of the original time series within the existing constraint framework (mathematically, w has a certain similarity in shape with y, and its value fluctuates approximately 1 owing to the constraint of A_7). Moreover, if $\boldsymbol{y}'$ in the $A_7^{\mathrm{T}}\boldsymbol{y}'$ term includes the time series data at the latest moment, the calculated value of w will be overly influenced. Therefore, the $\boldsymbol{y}'$ in the $A_7^{\mathrm{T}}\boldsymbol{y}'$ term does not include the y value at the current moment, which makes the value of w tend toward the smoothed value of the historical w values.

However, because this calculation method leads to w containing seasonal components, to eliminate this characteristic, α-STD adopts simple and effective seasonal differencing. The α-STD approximates α as $1+w_t-w_{t-T_i}$, ensuring that it fluctuates around 1 to capture the trend changes across periods. To control the impact of the α component on the α-STD, λ_4 is used to regulate the scaling of w. A larger value of λ_4 results in smaller fluctuations in the trend normalizer w. Furthermore, the α-STD use two seasonal buffers to improve robustness. After each iteration, α in Eq. (7) is updated via Eq. (3).

By examining the linear system in Eq. (8), we observe that the matrix $\boldsymbol{A}$ is a symmetric band matrix with a bandwidth of 15 since $A_1^{\mathrm{T}}A_1$, $A_2^{\mathrm{T}}A_2$, ..., and $A_7^{\mathrm{T}}A_7$ are symmetric band matrices with bandwidths of 5, 5, 7, 13, 9, 15, and 5, respectively. Thus, the matrix $\boldsymbol{A}$ can be factorized into the form $\boldsymbol{LDL}^{\mathbf{T}}$ via an improved Cholesky decomposition algorithm. The α-STD algorithm adopts the online symmetric doolittle factorization method (OnlineDoolittle) from OneShotSTL. This online decomposition method has a time complexity of $\mathcal{O}(1)$, making it suitable for efficiently solving such linear systems. This optimization framework, combined with the fast solving algorithm, allows α-STD to achieve high accuracy and robustness when handling time series with complex trend and seasonal patterns.

Additionally, the α-STD algorithm adopts the same strategy as OneShotSTL to handle seasonal shifts. It uses the streaming NSigma algorithm to detect if s_t and r_t are outliers. Upon detecting outliers, the strategy adjusts for the seasonal shift by selecting more appropriate seasonal values from within the seasonal neighborhood.

3.3 Multi-seasonality α-STD Model

The α-STD algorithm can naturally be extended to multi-seasonality scenarios. By increasing the number of seasonal buffers, each seasonal component is assigned its own independent buffer. Consequently, the linear system and its matrix form need to be adjusted accordingly. Let the number of seasons be m, and the vector $\boldsymbol{x}$ in the linear system $\boldsymbol{A}\boldsymbol{x} = \boldsymbol{b}$ is extended as $[\tau_{t_0+1}, \alpha_{t_0+1}, s_{1,t_0+1}, \ldots, s_{m,t_0+1}, \ldots]^{\mathrm{T}}$, with the corresponding matrices $\boldsymbol{A} \in \mathbb{R}^{(m+2)M\times(m+2)M}$ and $\boldsymbol{b} \in \mathbb{R}^{(m+2)M\times 1}$. The extended matrix (for A_1 as an example) takes the following form:

$$A_1 = \begin{bmatrix} 1\,0\,\underbrace{1\,\ldots\,1}_{m \text{ ones}} & & \\ & \ddots & \\ & & 1\,0\,1\ldots1 \end{bmatrix} \in \mathbb{R}^{M\times(m+2)M}. \tag{9}$$

In this context, $A_1^{\mathrm{T}}A_1, A_2^{\mathrm{T}}A_2, \ldots, A_7^{\mathrm{T}}A_7$ are symmetric band matrices with bandwidths of $2m+3, 2m+3, 2m+5, 4m+9, 4m+5, 4m+11$ and $2m+3$, respectively. They can similarly be solved via the OnlineDoolittle algorithm.

In the case of multiple seasons, several issues may arise:

- **Lacking information about seasonality:** The number and length of seasons must be determined during the initialization phase via seasonal detection algorithms.
- **Interference between multiple seasonal components:** Due to the presence of noise in the time series, it may be mistakenly interpreted as a change in a specific seasonal component (e.g., s_1), thereby increasing the error of s_1. The increased error in s_1 could lead to a larger decomposition residual after one cycle, which might then incorrectly affect another seasonal component (e.g., s_2). Additionally, when seasonal components have different amplitude variations, a small change in a component with a larger fluctuation (e.g., s_3) could be misinterpreted as a change in a component with a smaller fluctuation (e.g., s_4), leading to a significant bias in s_4. To alleviate these issues, the weight λ_1 of the seasonal components can be increased, or varying weights can be assigned to different seasonal components to reduce their interference.
- **Complexity of handling seasonal shifts:** As the number of seasonal components increases, handling seasonal shifts becomes more complex. A nested traversal of the neighborhoods of all the seasonal components is not the optimal approach. Therefore, the assumption is that seasonal shifts affect only a single seasonal component or that all components undergo the same degree of shift. This assumption aligns with common patterns in real-world data and simplifies the handling process.

3.4 Missing Data Handling Strategy

Data missingness is a common issue, often arising during data collection or transmission, especially in raw datasets. Missing data can be categorized into two types: single-point missing data and long-term missing data, as shown in Fig. 2. Long-term missing data typically occur sporadically and with a low probability. For long-term missing data, reinitializing the algorithm after recovering the data is recommended. In contrast, single-point missing data may occur more frequently, and reinitialization is not suitable.

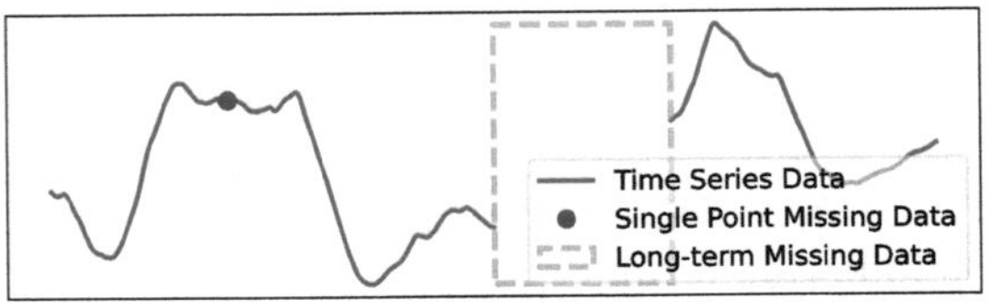

Fig. 2. Examples of single-point and long-term missing data.

To address the issue of single-point missing data, α-STD fills missing values on the basis of historical decomposition data. Specifically, two additional buffers are introduced to store the most recent values of τ and w, such as the seasonal buffer $\boldsymbol{v}$. When single-point missing data occur, the data in these buffers are used to estimate the missing values. The estimation method can involve linear interpolation, nearest-neighbor interpolation, or simple averaging, depending on the component's variation. Specifically, the first two data points in the buffer are used to estimate the third, second and third points for the fourth, and so on. The method with the smallest mean absolute error (MAE) is selected to fill in missing values at the current time step. Since only recent data points are considered, the number of points required for calculation is typically small (e.g., 5 or 7). This strategy avoids reinitialization while ensuring the continuity and stability of the time series decomposition.

Algorithm 1. α-STD.

Input: $y_1, ..., y_{t_0}, y_t, t, \lambda$, maximum iterations I
Output: $T, \tau_t, \alpha_t, s_{1,t}, \ldots, s_{m,t}, r_t$

```
// Periodicity analysis phase
Detect period T using methods such as Autoperiod, CFD-Autoperiod, RobustPeriod, etc.
// Initialization phase
Initialize using methods such as STL, MSTL, and others;
Set m = T.length, v_i = [s_{i,t0-2T_i}, ..., s_{i,t0}], for i ∈ [1, m];
p^0 = q^0 = [1]; p^i = q^i = [], for i ∈ [1, I]
// Online phase
if y_t is None then
    τ_t = τ_{t-1}, α_t = α_{t-1}, s_t = α_t v_{t%T}, r_t = 0;
else
    for i in range I do
        Construct A and b using Eq. (8);
        τ_t, w_t, s_{1,t}, ..., s_{m,t}, r_t = OnlineDoolittle(A, b);
        α_t = w_t − w_{t−T} + 1;
        Update Eq. (7) using Eq. (2);
        Update p_t and q_t using Eq. (5) and Eq. (6), respectively;
    end for
end if
Update v_{t%T} using s_t;
```

3.5 α-STD Algorithm

This section presents the overall process of the α-STD algorithm, including the periodicity analysis phase, the offline initialization phase and the online update phase, as shown in Algorithm 1.

For time series with unknown periodicity, the period length must first be estimated via a periodicity analysis algorithm, such as RobustPeriod. Once the period length is determined, α-STD proceeds with offline batch decomposition initialization. The online update phase relies on previous decomposition results (such as the seasonal buffer $\boldsymbol{v}$), making the initialization accuracy crucial for the online update phase. In the case of a single periodicity, we use the STL algorithm for initialization; for multiple seasonality, we use the MSTL algorithm, as it has demonstrated good decomposition performance.

In the online update phase, the α-STD algorithm first checks for missing data. If missing data are detected, the method from Sect. 3.4 is applied, and an example using nearest neighbor interpolation for single-point missing data is given in Algorithm 1. If no missing data are found, the solution is iterated via Eq. (8) to construct a linear system that is solved via the OnlineDoolittle algorithm in each iteration. Furthermore, after each decomposition, the streaming NSigma algorithm is applied to detect anomalies, which are typically related to seasonal shifts. This requires searching for adjacent points in the seasonal buffer to select the closest seasonal component value and re-decompose.

The computational complexity of OneShotSTL is $\mathcal{O}(1)$. It follows that the computational complexity of the α-STD with m seasonal components is $\mathcal{O}(m)$. Since the number of seasonal components in real-world time series is typically fixed, the online update complexity of α-STD in practice can be approximated as $\mathcal{O}(1)$, similar to that of OneShotSTL.

4 Experiments and Discussion

In this section, we demonstrate the superior performance of the α-STD through experiments. Since online algorithms require multiple cycles for initialization, the experimental section presents only the decomposition results during the online updating phase.

4.1 Experimental Settings

Baseline Algorithms. For the single-seasonality time series decomposition experiments, we compare several baseline algorithms with α-STD: STL [8], OneShotSTL [20], and BacktrackSTL [35]. STL is an offline detection algorithm, while OneShotSTL and BacktrackSTL are online detection algorithms with an update complexity of $\mathcal{O}(1)$. BacktrackSTL has been shown to outperform its predecessors, RobustSTL and Fast RobustSTL, which is why we exclude these earlier methods from the comparison. For the multi-seasonality time series decomposition extension, we conduct a simple comparison between MSTL and α-STD since no publicly available online multi-seasonality STD methods are found.

We use the STL algorithm from the open-source Python library statsmodels and the open-source Java code for OneShotSTL and BacktrackSTL. Data preprocessing and visualization are performed in Python. The implementation of α-STD is based on OneShotSTL and is also implemented in Java.

Datasets. We evaluate the algorithms via five single-seasonality datasets and two multi-seasonality datasets.

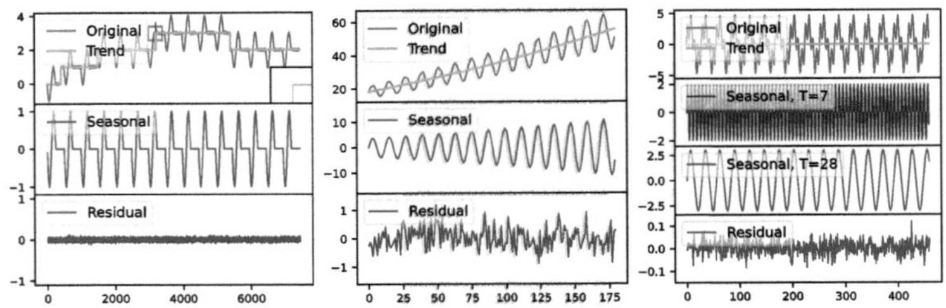

(a) SYN1 (b) SYN2 (c) MSYN

Fig. 3. Visualization of the original data values and their trend, seasonality, and residual components for three artificially constructed datasets.

- Single-seasonality datasets: We use two synthetic datasets, as shown in Fig. 3(a) and Fig. 3(b). The first is the SYN1 dataset from OneShotSTL [20], with $T = 496$, which includes trend jumps. A zoomed-in view of one such trend jump is displayed in Fig. 3(a). The second dataset is SYN2, which we synthesize ourselves with $T = 12$, exhibiting a smooth quadratic growth trend. The amplitude of the seasonal component is related to the trend, and the residual component consists of normally distributed random residuals with an underlying growth trend. Additionally, three real-world datasets are used in the single-seasonality experiments. These datasets include the $T = 12$ AirPassengers dataset [40], which consists of airline passenger data from 1949–1960; the $T = 96$ ETTm1 dataset [42], which contains transformer oil temperature data; and the $T = 7$ Birth dataset [10], which records the daily number of female births in California during 1959.
- Multi-seasonality datasets: For the multi-seasonality decomposition experiment, we first synthesize a dataset MSYN with $T = [7, 28]$ and add normally distributed noise, as shown in Fig. 3(c). Additionally, we use real temperature data from the Global Energy Forecasting Competition 2014 (GEFCom2014) [22]. The data are collected hourly, with $T = [24, 8520]$.

Evaluation Metrics. For the synthetic datasets, we evaluate the decomposition performance via the mean absolute error (MAE) between the decomposition results and the synthetic signalling components. Taking the trend component τ_t as an example, the MAE is calculated as follows:

$$\mathrm{MAE} = \frac{1}{n}\sum_{t=1}^{n}\left|\tau_t - \tau_t^{(\mathrm{syn})}\right|, \tag{10}$$

where $\tau_t^{(\mathrm{syn})}$ denotes the ground truth value of the trend component in the synthetic dataset.

Additionally, we provide visualized decomposition results for easy comparison between different algorithms.

Environment Settings. All the experiments are conducted on a Windows 11 operating system. The Java version used for the α-STD experiments is 22.0.2, and the Python version is 3.9.

Hyperparameter Settings. In the experiments, the number of iterations I is set to 5. The value of λ_4 needs manual adjustment, and its values follow a certain pattern, which is discussed in Sect. 4.5. Additionally, λ_1 is set to 1 instead of the 0.5 used in OneShotSTL. Other parameters are set as in OneShotSTL. Typically, we run the baseline algorithms with default parameters. If the default performance is unsatisfactory, we manually tune the parameters, such as for the BacktrackSTL experiments on the ETTm1 dataset. For the initialization phase, the number of periods is set to 4 for the AirPassengers dataset because of its smaller sample size, whereas 5 periods are used for the other datasets.

4.2 Single-Seasonality Experiments

We evaluate the accuracy of the decomposition algorithms in the single-seasonality case.

Table 2. MAE comparison of the α-STD and baseline algorithms on synthetic single-seasonality datasets.

Algorithm	Type	SYN1 MAE			SYN2 MAE		
		Trend	Seasonal	Residual	Trend	Seasonal	Residual
STL	Offline	0.082	0.059	0.097	**0.095**	**0.152**	**0.174**
OneShotSTL	Online	0.039	0.036	0.020	5.647	4.934	0.836
BacktrackSTL	Online	**0.010**	0.024	0.023	3.765	3.729	0.353
α-STD	Online	<u>0.015</u>	**0.016**	**0.014**	<u>1.585</u>	<u>1.408</u>	<u>0.279</u>

Experiments on Synthetic Datasets. Table 2 presents the MAE values for the decomposition of two synthetic datasets by the α-STD and the baseline algorithms, with the visual decomposition results shown in Fig. 4 and Fig. 5. Furthermore, we provide complete decomposition results for the syn2 dataset, including both w and α, as demonstrated in Fig. 6.

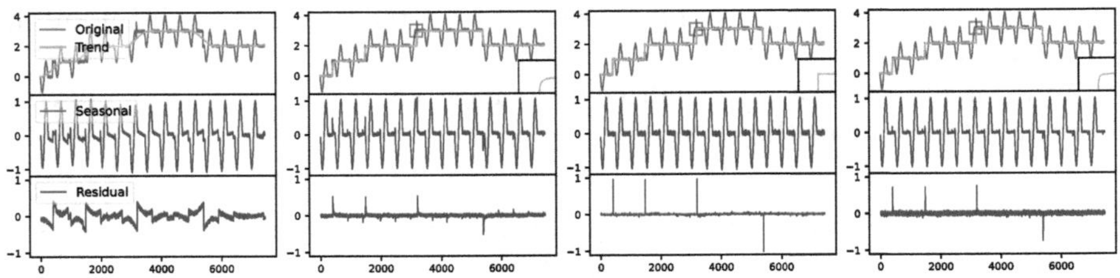

(a) STL (b) OneShotSTL (c) BacktrackSTL (d) α-STD

Fig. 4. Visual time series decomposition results for the SYN1 dataset: comparison of four algorithms.

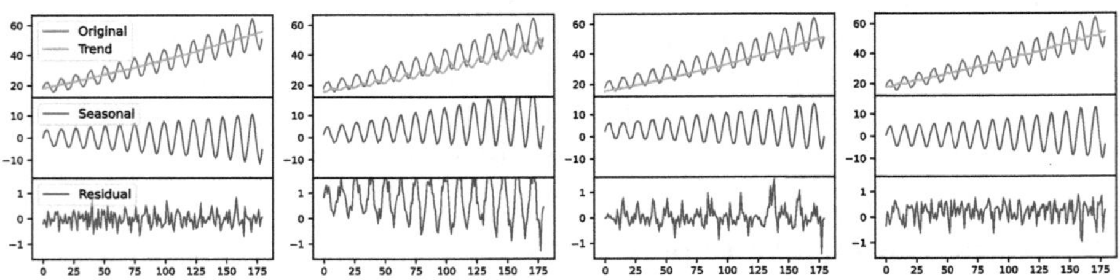

(a) STL (b) OneShotSTL (c) BacktrackSTL (d) α-STD

Fig. 5. Visual time series decomposition results for the SYN2 dataset: comparison of four algorithms.

The α-STD achieves the best online decomposition performance. Specifically, on the SYN1 dataset, which contains trend jumps, α-STD outperforms OneShot-STL, given the same parameter settings. Since there is no multiplicative relationship between τ and s in the SYN1 dataset, we set λ_4 to a relatively large value. Because the w component helps mitigate the influence of noise on trend learning, the τ and s values learned by α-STD are more reasonable than those obtained by OneShotSTL. Moreover, on the SYN2 dataset, where there exists a multiplicative relationship between seasonality and trend, α-STD also demonstrates excellent decomposition accuracy. Notably, although susceptible to noise interference (as shown in Fig. 6), our experimental results show that the α component can effectively model the multiplicative relationship between trend and seasonal components.

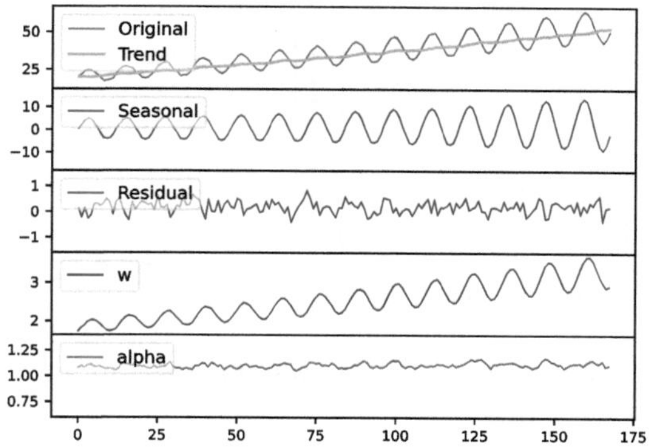

Fig. 6. Complete decomposition results of the α-STD on the SYN2 dataset, including w and α (i.e., alpha).

BacktrackSTL, which relies on the separately designed jump detection and backtracking algorithm, performs the best in detecting trend jumps in SYN1. However, its backtracking method suffers from some delay and requires a sufficient number of new data points for adjustments. Additionally, it overfits random noise into the seasonal component, leading to larger seasonal MAEs and residual MAEs than does the α-STD. BacktrackSTL performs poorly on the SYN2 dataset, with an overall shift in the seasonal components. Additionally, the residual component exhibits large fluctuations, which could lead to false positives in subsequent anomaly detection tasks.

Both OneShotSTL and STL struggle to handle trend jumps in the SYN1 dataset, with STL performing particularly poorly. However, the STL method achieves excellent decomposition accuracy in the SYN2 dataset. In contrast, OneShotSTL clearly fails to handle the multiplicative relationship between seasonality and trend in SYN2, and despite parameter adjustments, its results remain unsatisfactory.

Experiments on Real Datasets. We conduct experiments on three real-world datasets, with the results shown in Fig. 7, Fig. 8, and Fig. 9.

Similar to its performance on the SYN2 dataset, OneShotSTL performs poorly on the AirPassengers dataset. The seasonal components decomposed by BacktrackSTL show some offset, and it mistakenly classifies portions that appear to belong to the seasonal component as part of the residual, such as the point near index 36.

The ETTm1 dataset exhibits significant trend fluctuations that are not smooth, revealing a flaw in BacktrackSTL's ability to handle datasets with large trend fluctuations. The decomposition results of the α-STD on the ETTm1 dataset are similar to those of OneShotSTL, but the α-STD handles trend jumps better and more accurately decomposes the trend and seasonal components.

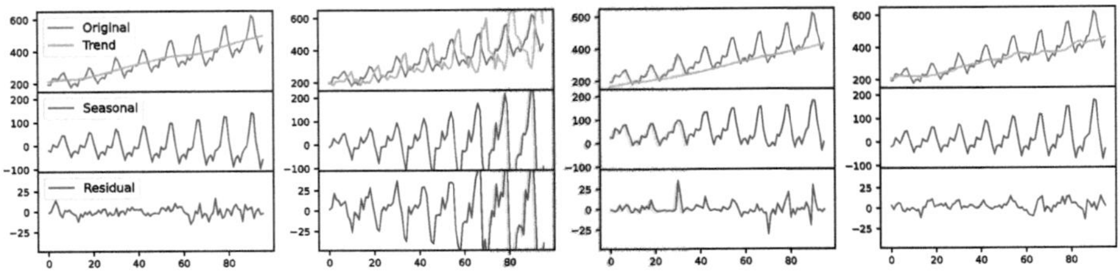

(a) STL (b) OneShotSTL (c) BacktrackSTL (d) α-STD

Fig. 7. Visual time series decomposition results for the AirPassengers dataset: comparison of four algorithms.

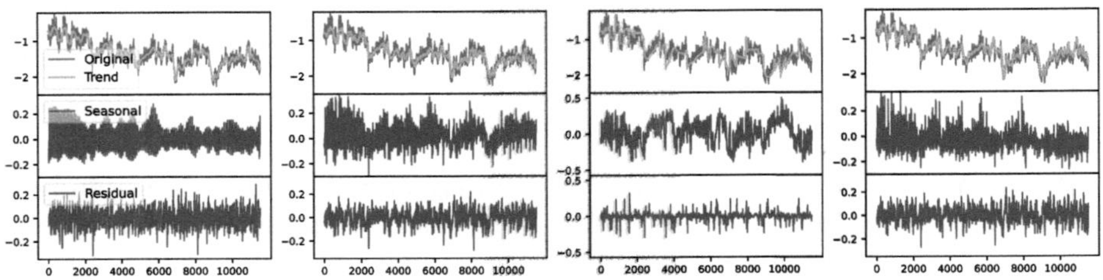

(a) STL (b) OneShotSTL (c) BacktrackSTL (d) α-STD

Fig. 8. Visual time series decomposition results for the ETTm1 dataset: comparison of four algorithms.

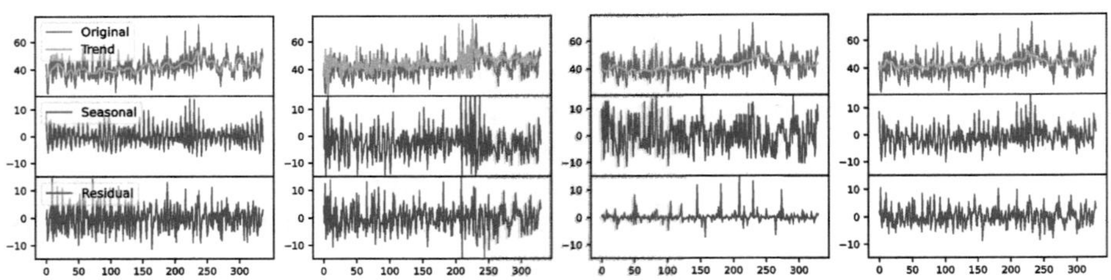

(a) STL (b) OneShotSTL (c) BacktrackSTL (d) α-STD

Fig. 9. Visual time series decomposition results for the Birth dataset: comparison of four algorithms.

The Birth dataset shows considerable volatility. OneShotSTL incorrectly classifies many oscillations as part of the trend and does not decompose the three components well. BacktrackSTL produces an overly smoothed trend, resulting in visible trend components still present in the seasonal component, and incorrectly classifies the noise in the residual as part of the seasonal component. STL and α-STD produce relatively better decomposition results.

In general, STL performs well on most datasets but struggles with trend jumps, as seen in the SYN1 dataset, and is an offline batch processing algorithm. OneShotSTL fails to handle datasets with a multiplicative relationship between τ and s, and it struggles to separate the three components in more volatile datasets. BacktrackSTL is one of the best recent online decomposition algorithms, performing well on most datasets (although this is partly due to its improved initialization method). Although it can effectively handle trend jumps (as demonstrated on the SYN1 dataset), it struggles to manage multiplicative relationships between trend and seasonal components (e.g., the AirPassengers dataset). Moreover, when datasets such as ETTm1, which exhibit rapid trend variations (rather than abrupt jumps) are processed, τ and s are not properly separated.

Overall, our α-STD algorithm, initialized with the classic STL algorithm, although not always the optimal method on individual datasets, achieves the best overall decomposition results. It demonstrates strong adaptability when handling trend jumps and multiplicative relationships between trend and seasonal components. Moreover, with the introduction of the w component, the α-STD produces more reasonable trend components, especially compared with the original OneShotSTL algorithm, as shown in Fig. 4 and Fig. 9. We speculate that the introduction of the w component further decomposes the nonsmooth trends in the signal, enabling the α-STD to better learn the appropriate trend components.

4.3 Experiments on Multi-seasonality Extension

We extend α-STD to the multi-seasonality case and conduct experiments on both a synthetic dataset MSYN (Fig. 10(a), Fig. 10(b)) and a real-world dataset GEFCom2014 (Fig. 10(c), Fig. 10(d)).

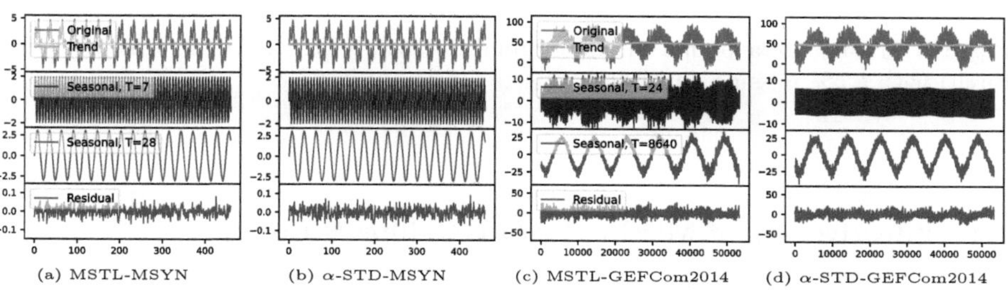

(a) MSTL-MSYN (b) α-STD-MSYN (c) MSTL-GEFCom2014 (d) α-STD-GEFCom2014

Fig. 10. Visual decomposition results for the multi-seasonality dataset: comparison of the MSTL and α-STD algorithms.

As mentioned in Sect. 3.3, cross-period interference may occur in multi-seasonality cases. In the MSYN dataset, the seasonality remains almost unchanged. To reduce the erroneous influence of noise components on the seasonal components, we increase the hyperparameter λ_1. The GEFCom2014 dataset exhibits large differences in the amplitude of fluctuations between different seasonal components. To prevent the larger-amplitude seasonal component s_1

from interfering with the smaller-amplitude seasonal component s_2, we assume that the probability of change for s_2 is lower than that for s_1. Specifically, we set a separate λ_1 hyperparameter for each seasonal component. Typically, for seasonal components with larger fluctuations, we use a larger hyperparameter. For example, in the GEFCom2014 dataset, we set hyperparameters of 100 and 0.5 for the seasonal components with periods of $T = 24$ and $T = 24$, respectively.

The experimental results demonstrate that in multi-seasonal decomposition tasks, α-STD not only effectively handles interference between different seasonal components with accuracy comparable to that of MSTL, but also achieves significant computational efficiency advantages. For instance, on the GEFCom2014 dataset, α-STD completes execution within seconds while MSTL requires approximately 30 min.

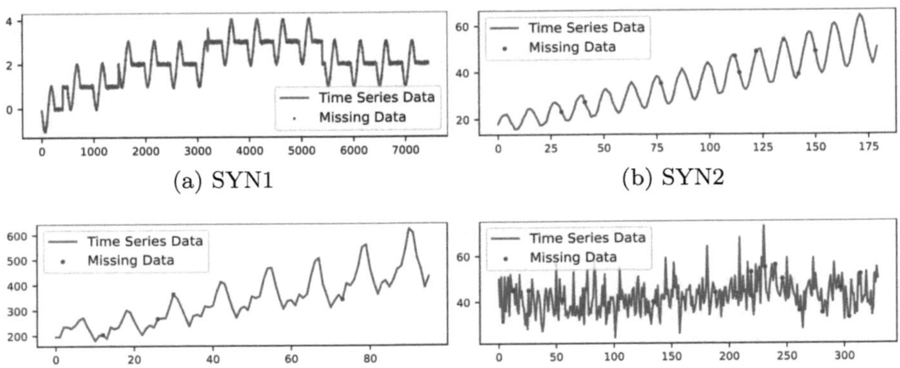

(a) SYN1 (b) SYN2

(c) AirPassengers (d) Birth

Fig. 11. Missing data points in the original datasets (SYN1, SYN2, AirPassengers, Birth) with a 5% missing data rate.

4.4 Effectiveness of Missing Data Handling Strategy

In Sect. 3.4, we address the potential issue of single-point missing data via a prediction-based approach. To assess the effectiveness of this method, we conduct experiments in the presence of missing data. We test on four datasets, SYN1, SYN2, AirPassengers, and Birth, and randomly introduce missing data. When the missing data rate is 5%, the missing points and experimental results are shown in Fig. 11 and Fig. 12, respectively. Additionally, we plot the MAE changes in the α-STD on the two synthetic datasets as the missing data rate increases from 0% to 20%, as shown in Fig. 13.

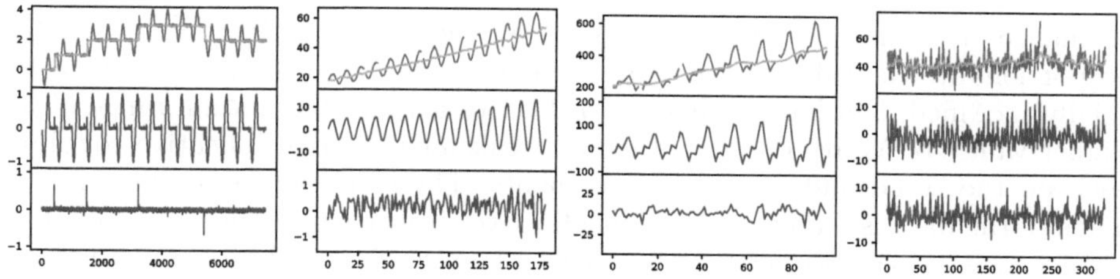

(a) SYN1 (b) SYN2 (c) AirPassengers (d) Birth

Fig. 12. Visual time series decomposition results of α-STD with missing data (5% missing data rate) for datasets SYN1, SYN2, AirPassengers, and Birth.

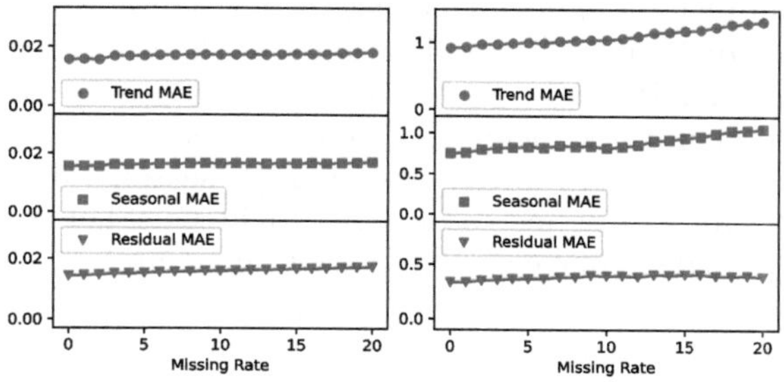

(a) SYN1 (b) SYN2

Fig. 13. Variation in the MAE for α-STD as the missing data rate increases from 0% to 20% on synthetic datasets SYN1 and SYN2.

The experimental results indicate that the α-STD performed well across all the datasets even when missing data were present. As the missing data rate increased, the overall MAE tended to increase. In the more complex dataset SYN2, the MAEs for the trend and seasonal components increased more rapidly with higher missing data rates. However, it is noteworthy that the MAE growth is slow at lower missing data rates, and in reality, there are relatively few cases of data loss of more than 10%. Therefore, we can conclude that our missing data handling strategy effectively mitigates the impact of missing data on the decomposition results in online STD.

4.5 Discussion

Through experiments, we find that α-STD performs exceptionally well in handling data from both additive and multiplicative decomposition models, achieving good decomposition results. However, it is undeniable that the α-STD still has some areas requiring further improvement.

Limitations in Parameter Tuning. The weight coefficient λ_4 for controlling the α component needs to be manually adjusted. As shown in Eq. (8), the trend normalizer w is inversely proportional to λ_4 ($w \propto 1/\lambda_4$) while being directly proportional to the value of the input data ($w \propto y_t$). Notably, y_t cannot be artificially modified, and larger values of y_t lead to a wider fluctuation range of w. Higher values of λ_4 attenuate w, thereby weakening the multiplicative coupling between the trend and seasonal components. Therefore, when the fluctuation range of the data is larger or the multiplicative relationship between seasonality and trend is weaker, λ_4 needs to be set to a larger value, and vice versa. Future research could explore adaptive methods to adjust λ_4, reducing the manual tuning workload.

Computational Efficiency Considerations. Although α-STD is also an algorithm with an $\mathcal{O}(1)$ update complexity, the introduced variable w increases the matrix bandwidth, making its online updates slightly slower than those of OneShotSTL [20] and BacktrackSTL [35]. In the paper on BacktrackSTL, it is already proven that BacktrackSTL updates faster than OneShotSTL does. Compared with OneShotSTL, in the case of single-seasonal decomposition, the bandwidth of the matrix $\boldsymbol{A}$ increases from 10 in OneShotSTL to 15, which inevitably leads to higher time consumption for online updates. Furthermore, as the number of seasonal components increases, the bandwidth of the matrix $\boldsymbol{A}$ grows linearly. Specifically, when the number of seasonal components is m, the bandwidth of the matrix $\boldsymbol{A}$ is $4m + 11$.

Challenges in Online Decomposition. We identify challenges faced by online decomposition algorithms. In particular, when the amplitude of the seasonal components varies significantly, the model in the online update process can rely only on previous decomposition results and cannot predict future seasonal changes. This may lead to an inability to fully separate the trend and seasonal components. This issue is particularly evident in the ETTm1 dataset experiment shown in Fig. 8, where the method performs poorly in handling trends with large fluctuations. BacktrackSTL assumes that the sum of the seasonal components within a period is zero, a hypothesis we attempted but ultimately did not adopt in α-STD. While this assumption yields good decomposition results when seasonality changes little, it can lead to increased errors if the seasonal components change over time.

In addition, online decomposition of multi-seasonal time series may encounter interference between seasonal components, especially when seasonality changes slowly. Although we discuss potential solutions in Sect. 3.3, these solutions still rely on empirical parameter tuning and require more robust approaches.

5 Conclusion

In this study, we propose an innovative decomposition algorithm, α-STD. Building upon the classic additive decomposition framework, α-STD introduces a novel mechanism to model multiplicative relationships between trends and seasonality while maintaining approximately $\mathcal{O}(1)$ time complexity in theory for online updates. This allows the α-STD to demonstrate exceptional performance and broad applicability when handling time series data with complex seasonal and trend variations. Additionally, we present a missing data handling strategy based on prediction ideas for single-point missing data. Even when the missing data rate reaches 10%, the decomposition results remain highly accurate. Furthermore, the α-STD achieves satisfactory performance on multi-seasonal time series, successfully addressing seasonal interference. The experimental results show that the α-STD achieves higher overall accuracy and demonstrates broader applicability as an online STD algorithm.

Acknowledgments. This work was supported by the National Natural Science Foundation of China (NSFC) under Grant Nos. 62172378, 61572448.

We would like to express our sincere gratitude to all the data providers whose valuable datasets made this research possible. Furthermore, we are deeply indebted to the open-source community, particularly acknowledging the significant contributions of Xiao He and colleagues, whose pioneering work laid an essential foundation for our study.

Disclosure of Interests. The authors have no competing interests to declare that are relevant to the content of this article.

References

1. Bandara, K., Bergmeir, C., Hewamalage, H.: LSTM-MSNet: leveraging forecasts on sets of related time series with multiple seasonal patterns. IEEE Trans. Neural Netw. Learn. Syst. **32**(4), 1586–1599 (2020). https://doi.org/10.1109/TNNLS.2020.2985720
2. Bandara, K., Hyndman, R.J., Bergmeir, C.: MSTL: a seasonal-trend decomposition algorithm for time series with multiple seasonal patterns. Int. J. Opera. Res. (2025). https://doi.org/10.1504/IJOR.2025.143957
3. Bell, W.R., Hillmer, S.C.: Issues involved with the seasonal adjustment of economic time series. J. Bus. Econ. Stat. **2**(4), 291–320 (1984). https://doi.org/10.1080/07350015.1984.10509398
4. Bianconcini, S., Bee Dagum, E., et al.: Seasonal Adjustment Methods and Real Time Trend-Cycle Estimation. Statistics for Social and Behavioral Sciences. Springer, Cham (2016). https://doi.org/10.1007/978-3-319-31822-6
5. Bonizzi, P., Karel, J.M., Meste, O., Peeters, R.L.: Singular spectrum decomposition: a new method for time series decomposition. Adv. Adapt. Data Anal. **6**(04), 1450011 (2014). https://doi.org/10.1142/S1793536914500113
6. Box, G.E., Jenkins, G.M., Reinsel, G.C., Ljung, G.M.: Time Series Analysis: Forecasting and Control. Wiley, Hoboken (2015)
7. Büyükşahin, Ü.Ç., Ertekin, Ş: Improving forecasting accuracy of time series data using a new arima-ann hybrid method and empirical mode decomposition. Neurocomputing **361**, 151–163 (2019). https://doi.org/10.1016/j.neucom.2019.05.099
8. Cleveland, R.B., Cleveland, W., McRae, J., Terpenning, I.: STL: a seasonal-trend decomposition procedure based on loess. J. Off. Stat. **6**, 3–73 (1990)
9. Cowles, A., 3rd.: Can stock market forecasters forecast? Econometrica **1**(3), 309–324 (1933). https://doi.org/10.2307/1907042

10. Cresswell, D.: Daily total female births in California, 1959 (2020). https://www.kaggle.com/datasets/dougcresswell/daily-total-female-births-in-california-1959. Accessed 09 Dec 2024
11. Daubechies, I.: Ten Lectures on Wavelets. Society for Industrial and Applied Mathematics (1992). https://doi.org/10.1137/1.9781611970104
12. De Livera, A.M., Hyndman, R.J., Snyder, R.D.: Forecasting time series with complex seasonal patterns using exponential smoothing. J. Am. Stat. Assoc. **106**(496), 1513–1527 (2011). https://doi.org/10.1198/jasa.2011.tm09771
13. Dokumentov, A., Hyndman, R.J.: STR: seasonal-trend decomposition using regression. INFORMS J. Data Sci. **1**(1), 50–62 (2022). https://doi.org/10.1287/ijds.2021.0004
14. Dragomiretskiy, K., Zosso, D.: Variational mode decomposition. IEEE Trans. Signal Process. **62**(3), 531–544 (2013). https://doi.org/10.1109/TSP.2013.2288675
15. Dudek, G.: STD: a seasonal-trend-dispersion decomposition of time series. IEEE Trans. Knowl. Data Eng. **35**(10), 10339–10350 (2023). https://doi.org/10.1109/TKDE.2023.3268125
16. Findley, D.F.: Some recent developments and directions in seasonal adjustment. J. Official Stat. **21**(2), 343–365 (2005)
17. Gardner, E.S., Jr.: Exponential smoothing: the state of the art. J. Forecast. **4**(1), 1–28 (1985). https://doi.org/10.1002/for.3980040103
18. Gilles, J.: Empirical wavelet transform. IEEE Trans. Signal Process. **61**(16), 3999–4010 (2013). https://doi.org/10.1109/TSP.2013.2265222
19. Green, P.J.: Iteratively reweighted least squares for maximum likelihood estimation, and some robust and resistant alternatives. J. Roy. Stat. Soc.: Ser. B (Methodol.) **46**(2), 149–170 (1984). https://doi.org/10.1111/j.2517-6161.1984.tb01288.x
20. He, X., Li, Y., Tan, J., Wu, B., Li, F.: OneShotSTL: one-shot seasonal-trend decomposition for online time series anomaly detection and forecasting. Proc. VLDB Endow. **16**(06), 1399–1412 (2023). https://doi.org/10.14778/3583140.3583155
21. Hodrick, R.J., Prescott, E.C.: Postwar U.S. business cycles: an empirical investigation. J. Money Credit Bank. **29**(1), 1–16 (1997). https://doi.org/10.2307/2953682
22. Hong, T., Pinson, P., Fan, S., Zareipour, H., Troccoli, A., Hyndman, R.J.: Probabilistic energy forecasting: global energy forecasting competition 2014 and beyond. Int. J. Forecast. **32**(3), 896–913 (2016). https://doi.org/10.1016/j.ijforecast.2016.02.001
23. Huang, N.E., et al.: The empirical mode decomposition and the hilbert spectrum for nonlinear and non-stationary time series analysis. Proc. R. Soc. Lond. Ser. A: Math. Phys. Eng. Sci. **454**(1971), 903–995 (1998). https://doi.org/10.1098/rspa.1998.0193
24. Kim, S.J., Koh, K., Boyd, S., Gorinevsky, D.: ℓ_1 trend filtering. SIAM Rev. **51**(2), 339–360 (2009). https://doi.org/10.1137/070690274
25. Mishra, A., Sriharsha, R., Zhong, S.: Onlinestl: scaling time series decomposition by 100x. Proc. VLDB Endow. **15**(7), 1417–1425 (2022). https://doi.org/10.14778/3523210.3523219
26. Nan, H., Zhu, X., Ma, J.: MSTL-GLTP: a global–local decomposition and prediction framework for wireless traffic. IEEE Internet Things J. **10**(6), 5024–5034 (2022). https://doi.org/10.1109/JIOT.2022.3221743
27. Phungtua-eng, T., Yamamoto, Y.: Adaptive seasonal-trend decomposition for streaming time series data with transitions and fluctuations in seasonality. In: Machine Learning and Knowledge Discovery in Databases. Research Track, pp. 426–443. Springer, Cham (2024). https://doi.org/10.1007/978-3-031-70344-7_25

28. Puech, T., Boussard, M., D'Amato, A., Millerand, G.: A fully automated periodicity detection in time series. In: Lemaire, V., Malinowski, S., Bagnall, A., Bondu, A., Guyet, T., Tavenard, R. (eds.) AALTD 2019. LNCS (LNAI), vol. 11986, pp. 43–54. Springer, Cham (2020). https://doi.org/10.1007/978-3-030-39098-3_4
29. Shiskin, J.: The X-11 Variant of the Census Method II Seasonal Adjustment Program. 15, US Department of Commerce, Bureau of the Census (1967)
30. Smith, J.S.: The local mean decomposition and its application to EEG perception data. J. R. Soc. Interface **2**(5), 443–454 (2005). https://doi.org/10.1098/rsif.2005.0058
31. Toller, M., Santos, T., Kern, R.: SAZED: parameter-free domain-agnostic season length estimation in time series data. Data Min. Knowl. Disc. **33**(6), 1775–1798 (2019). https://doi.org/10.1007/s10618-019-00645-z
32. Torres, M.E., Colominas, M.A., Schlotthauer, G., Flandrin, P.: A complete ensemble empirical mode decomposition with adaptive noise. In: 2011 IEEE International Conference on Acoustics, Speech and Signal Processing (ICASSP), pp. 4144–4147. IEEE (2011). https://doi.org/10.1109/ICASSP.2011.5947265
33. Vlachos, M., Yu, P., Castelli, V.: On periodicity detection and structural periodic similarity. In: Proceedings of the 2005 SIAM International Conference on Data Mining, pp. 449–460. SIAM (2005). https://doi.org/10.1137/1.9781611972757.40
34. Wang, C., et al.: Drift doesn't matter: dynamic decomposition with diffusion reconstruction for unstable multivariate time series anomaly detection. In: Oh, A., Naumann, T., Globerson, A., Saenko, K., Hardt, M., Levine, S. (eds.) Advances in Neural Information Processing Systems, vol. 36, pp. 10758–10774. Curran Associates, Inc. (2023). https://doi.org/10.5555/3666122.3666595
35. Wang, H., Guo, H., Zhu, Z., Zhang, Y., Zhou, Y., Zheng, X.: Backtrackstl: ultra-fast online seasonal-trend decomposition with backtrack technique. In: Proceedings of the 30th ACM SIGKDD Conference on Knowledge Discovery & Data Mining, KDD 2024, pp. 5848–5859. Association for Computing Machinery, New York (2024). https://doi.org/10.1145/3637528.3671510
36. Wen, Q., Gao, J., Song, X., Sun, L., Xu, H., Zhu, S.: Robuststl: a robust seasonal-trend decomposition algorithm for long time series. In: Proceedings of the AAAI Conference on Artificial Intelligence, AAAI 2019/IAAI 2019/EAAI 2019, vol. 33, pp. 5409–5416. AAAI Press (2019). https://doi.org/10.1609/aaai.v33i01.33015409
37. Wen, Q., He, K., Sun, L., Zhang, Y., Ke, M., Xu, H.: Robustperiod: robust time-frequency mining for multiple periodicity detection. In: Proceedings of the 2021 International Conference on Management of Data, SIGMOD 2021, pp. 2328–2337. Association for Computing Machinery, New York (2021). https://doi.org/10.1145/3448016.3452779
38. Wen, Q., Zhang, Z., Li, Y., Sun, L.: Fast robuststl: efficient and robust seasonal-trend decomposition for time series with complex patterns. In: Proceedings of the 26th ACM SIGKDD International Conference on Knowledge Discovery & Data Mining, KDD 2020, pp. 2203–2213. Association for Computing Machinery, New York (2020). https://doi.org/10.1145/3394486.3403271
39. Wu, Z., Huang, N.E.: Ensemble empirical mode decomposition: a noise-assisted data analysis method. Adv. Adapt. Data Anal. **1**(01), 1–41 (2009). https://doi.org/10.1142/S1793536909000047
40. Young, P., Shellswell, S.: Time series analysis, forecasting and control. IEEE Trans. Autom. Control **17**(2), 281–283 (1972). https://doi.org/10.1109/TAC.1972.1099963

41. Zhang, C., Zhou, T., Wen, Q., Sun, L.: Tfad: a decomposition time series anomaly detection architecture with time-frequency analysis. In: Proceedings of the 31st ACM International Conference on Information & Knowledge Management, CIKM 2022, pp. 2497–2507. Association for Computing Machinery, New York (2022). https://doi.org/10.1145/3511808.3557470
42. Zhou, H., et al.: Informer: beyond efficient transformer for long sequence time-series forecasting. In: The Thirty-Fifth AAAI Conference on Artificial Intelligence, AAAI 2021, Virtual Conference, vol. 35, pp. 11106–11115. AAAI Press (2021). https://doi.org/10.1609/aaai.v35i12.17325
43. Zhou, W., Feng, Z., Xu, Y., Wang, X., Lv, H.: Empirical fourier decomposition: an accurate signal decomposition method for nonlinear and non-stationary time series analysis. Mech. Syst. Signal Process. **163**, 108155 (2022). https://doi.org/10.1016/j.ymssp.2021.108155

Frontier Technology Applications

Research on Convolutional Neural Network Implementation Methods for Embedded Systems

Xinyao Li(✉), Kai Qiang, and Zhifei He

AVIC Xi'an Aeronautics Computing Technique Research Institute, No. 15 Jinye 2nd Road, Yanta District, Xi'an 710068, Shaanxi, China
747065580@qq.com

Abstract. This paper designs and implements a convolutional neural network for embedded systems, explores the deployment and realization method of a convolutional neural network in an embedded environment where both computational and storage resources are limited, and performs functional simulation verification of the realized network, aiming at solving the problem that the algorithmic complexity of the convolutional neural network has been increasing and the computational power of the embedded environment is insufficient. The experimental test results prove that this paper successfully deploys and realizes a convolutional neural network with a multilayer implicit layer under the embedded environment of scarce computational resources and limited storage resources, which can effectively complete the inference process of the network. The implementation method designed in this paper has a certain degree of flexibility, which lays a foundation for the construction of a reconfigurable intelligent computing platform that can deploy multiple convolutional neural networks under the embedded environment.

Keywords: Target Detection · Convolutional Neural Network · Embedded System · Processing Unit

1 Introduction

Currently, convolutional neural networks (CNNs) are widely applied in the field of computer vision. By stacking multiple hidden layer structures, such networks exhibit excellent fitting capabilities for large datasets and demonstrate outstanding performance in handling complex intelligent tasks such as object detection, object recognition, and object tracking.

However, the performance advantages of convolutional neural networks (CNNs) rely heavily on their vast parameter and computation volumes. Most networks with high detection accuracy have parameter scales exceeding hundreds of millions, involving billions of floating-point multiply add operations [1]. In particular, when CNNs are deployed in embedded environments, there are two primary bottleneck issues to address. First, there is a computational bottleneck. High-parallelism operations require substantial computing resources, yet CNNs involve a massive number of floating-point operations during runtime. Due to their constrained layout space, embedded environments often

L. Huang et al. (Eds.): CCF NCCA 2025, CCIS 2774, pp. 621–628, 2026.
https://doi.org/10.1007/978-981-95-5294-8_40

provide limited computing resources. Given the limited on-chip computing resources, it is worth exploring whether the utilization rate of these resources can be improved through the rational design of pipeline structures [2–4]. Second, there is a storage bottleneck. During network operation, on-chip storage in embedded environments needs to interact frequently with off-chip DDR storage. Given the vast scale of network parameters and numerous intermediate parameters, coupled with the limited capacity and quantity of storage chips in embedded environments, reducing memory access times to efficiently utilize limited on-chip storage resources has become a research direction [5].

On the basis of the above background, this paper is dedicated to studying the implementation of convolutional neural networks (CNNs) for embedded systems, exploring implementation methods for CNNs in embedded environments with limited computational and storage resources, and attempting to address the issues of increasing the algorithmic complexity of CNNs and insufficient computing power in embedded environments.

2 Domestic and International Research Status

Convolutional neural networks usually use many convolutional layers, pooling layers and other implicit layers for stacking to achieve feature extraction, and the computations of these implicit layers have parallel characteristics. When convolutional neural networks are implemented for embedded systems, the main bottlenecks are in storage and computation.

In terms of storage, most existing studies start from the perspective of reducing the number of visits to reduce the visit time caused by the frequent scheduling of enormous intermediate parameters of convolutional neural networks. For example, in 2018, Yin Shouyi, Wei Shaojun, and others proposed a series of chips called "Thinkers" [6]. The chip adopts a reconfigurable intelligent computing architecture, which can be configured into up to four subarrays, which can flexibly configure the intermediate feature map parameters and weight parameters according to the control information of the controller to achieve the purpose of flexible processing of data flow. The "thinker" series of chips supports energy-efficient computing while saving on-chip storage overhead. For example, the "Thinker-S" chip can achieve a peak performance of 90 TOPs/W, with a minimum power consumption of only 141 mW.

In terms of computation, the main focus is on computationally intensive structures in the implicit layer, such as convolution, the activation function, multiply accumulate, etc., and targeted optimization acceleration. The commonly adopted optimization means are the design of a reasonable and efficient pipeline, which saves the limited computational resources on the chip and, at the same time, makes the computation more efficient. For example, the Cambrian "DianNao" [7], "DaDianNao" and other [8] four chips have adopted a structure called the "NFU" as the basic computing unit, as shown in Fig. 1. The NFU adopts a tree structure, which consists of three stages: an NFU-1 multiplier, an NFU-2 addition tree and an NFU-3 activation function. The NFU-1 multiplier stage consists of 256 multipliers, and a single clock cycle can complete 256 sets of multiplication operations. The NFU-2 addition tree stage consists of 16 addition tree units, each of which contains a multiplicative function according to the "8-4-2" principle. The NFU-2 addition tree stage consists of 16 addition tree units, each of which contains 15 adders in accordance with the "8-4-2-1" structure; the NFU-3 activation function stage activates the results of the addition tree operations and contains a total of 16 activation function units.

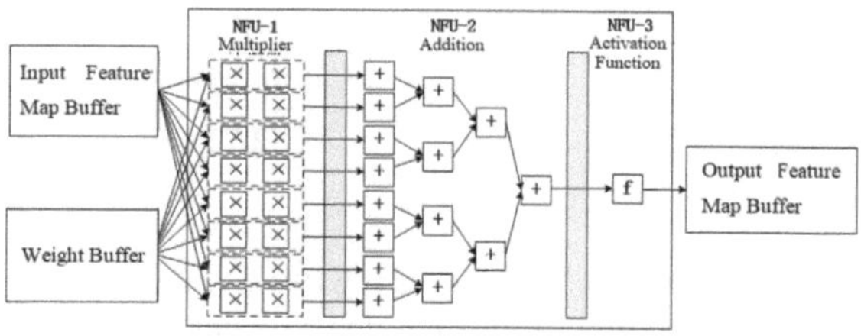

Fig. 1. A Typical Tree-Structured NFU Acceleration Unit.

In addition, foreign research institutions such as Purdue University and Stanford University have adopted dynamic configurable technology [9] to improve the utilization of logic resources and model compression technology to reduce the computational complexity of network models when deploying convolutional neural networks in embedded environments. Domestic universities such as Tsinghua University and Harbin Institute of Technology have also conducted research on data reuse techniques [10] to reduce redundant storage, buffering techniques [11] to reduce the bandwidth requirement for data transmission, and caching techniques to reduce the waiting time for intermediate parameters.

In summary, the design work for the embedded system implementation of a convolutional neural network is carried out from the two main aspects of storage and computation. In terms of storage, for structures with high parallelism in convolutional neural networks (e.g., convolution, activation function, pooling, etc.), multiple processing units are used for parallel operations to improve the efficiency of parallel operations. In terms of computation, the computational unit design is centered on computationally intensive network structures (e.g., convolutional operations containing many multiply accumulate operations), and the multistage pipeline structure is reasonably utilized to improve the computational efficiency.

3 Convolutional Neural Network Implementation for Embedded Systems

3.1 Research Target

YOLO series convolutional neural networks have good performance when applied to target detection and target recognition scenarios. The detection accuracy and runtime comparisons of the YOLO series networks tested on the PASCAL VOC2007 standard dataset are shown in Table 1, where the runtime is the forward execution time of the network on the GPU Quadro K6000 platform.

Table 1. YOLO Series Network Comparison.

Network Model Name	Parameter size (MB)	network layer	Detection accuracy (%)	Runtime (FPS/S)
YOLOv1	**180.5**	**26**	**57.9**	**45**
YOLOv2	**258.7**	**27**	**61.79**	**31**
YOLOv3	**246.3**	**106**	**76.41**	**17**
v3-Tiny	**35.1**	**24**	**52.72**	**136**

As shown in Table 1, although YOLOv3-Tiny has a significantly smaller parameter scale and fewer network layers than YOLOv3 does, its detection accuracy is the lowest among the YOLO series, and its running time is eight times greater than that of YOLOv3. Although YOLOv3 has a much larger parameter scale than YOLOv3-Tiny does, this also implies that it has stronger learning and generalization capabilities than YOLOv3-Tiny does, enabling it to handle more complex tasks better. Embedded environments are typically applied in fields with high reliability requirements, such as aviation and aerospace, involving the execution of complex tasks such as image recognition and generation. Considering the scalability of future neural network models, this paper selects YOLOv3, which has the highest detection accuracy among the YOLO series of networks [12–14], as the research object. It is transplanted onto an embedded verification platform to verify the impact of the implementation method of convolutional neural networks for embedded systems on the detection performance of convolutional neural networks.

The YOLOv3 network has 106 hidden layers in addition to the input and output layers, and its backbone network is a Draknet-53 structure stacked with 53 convolutional layers, a batch normalization layer (BN) and a ReLU activation function. The basic structure of the YOLOv3 convolutional neural network is shown in Fig. 2.

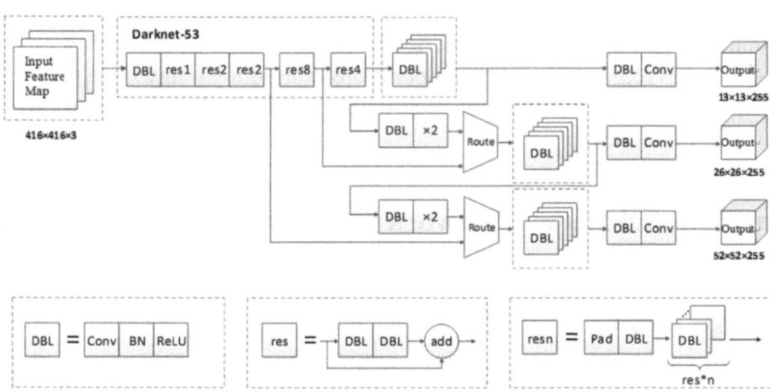

Fig. 2. YOLOv3 Network Architecture

As shown in the figure, the most basic network structures in the YOLOv3 network are the convolution operator, the BN operator, and the ReLU activation function operator, and the structural blocks composed of these three operators are called DBL structural blocks. A single res structural block consists of two DBL structural blocks, and multiple res structural blocks are stacked to form a resn group block. In the resn block, the padding edge filling operation is performed first, and then a DBL structural block is connected to extract features initially. Finally, the extracted features are passed through the block consisting of n res structural blocks, which deepens the depth of the network while extracting the features more accurately. The YOLOv3 network extracts the features from the three scales of 13 × 13, 26 × 26, and 52 × 52 and performs the output with good target recognition detection capability.

3.2 Convolutional Neural Network Implementation for Embedded Systems

In this section, we design the convolutional neural network implementation for embedded systems. The convolutional neural networks for embedded systems can be divided into processing systems (PSs) and programmable logic (PL). The feature map data, weight parameters and control information are stored in the external DDR, and the ARM processor on the PS side controls the whole computation process. The ARM processor configures the DMA on the PL side through the AXI bus to complete the data transfer of the feature map and weight parameters from the DDR to the on-chip cache.

The PL side contains a PE array, an on-chip buffer, a controller and a DMA. The PE array consists of 32 PEs, which are used to complete the computational tasks during the operation of the convolutional neural network. The on-chip cache is used to store the feature map data and weight parameters required for the current PE array operation. The controller reads the control words within the control information from the external DDR and configures all modules at the PL side except the DMA according to the control words. The DMA accomplishes the data transfer between the external DDR and the on-chip cache at the PL side (Fig. 3).

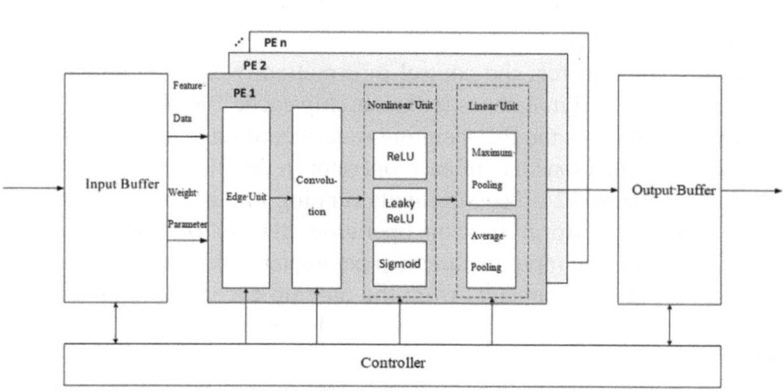

Fig. 3. Schematic Diagram of PE Array

To reduce the time for the PE array to wait for data interaction between the on-chip cache and the external DDR, a ping-pong cache is employed, where two cache areas work alternately, to minimize the waiting time for PE array operations [15]. The ping-pong cache consists of two identical cache areas. One area is used for data caching, whereas the data in the other area are input to the PE array for computation. In the next cycle, the roles of the two cache areas are swapped, allowing for interactive reading and writing between them, thereby effectively enhancing the overall parallelism of the computation.

4 Test and Verification

This section describes the experimental environment required for development and testing. In this work, we take an FPGA as an example to build an embedded demonstration and verification environment and use Vivado 2018.3, a development tool provided by Xilinx, to complete the logic development part and combine it with ModelSim 10.6d to perform functional simulation and verification.

4.1 Workflow

The embedded demonstration environment built in this paper consists of two main parts: the host computer and the Xilinx ZCU102 development board. The ZCU102 development board is equipped with the XCZU9EG FPGA chip, which contains 2520 DSP units and 32 Mb of RAM memory on the chip, and 4GB of DDR4 memory devices connected to the chip, which is able to meet the resource requirements for the embedded deployment of convolutional neural networks.

The convolutional neural network workflow for embedded systems consists of the following steps.

① The original network model is parsed to extract the parameters in the weight parameter file and to generate control information on the basis of the network structure file;
② Store the weight parameter file, the control information file, and the input image to be recognized generated in step ① to an SD card, and input them to an off-chip DDR for caching via the SD card;
③ The off-chip DDR transfers the control information, weight parameters and input images to be recognized in batches to the on-chip cache BRAM;
④ The controller configures the PE operating state according to the control information and controls the utilization state of each operator in the PE;
⑤ The PE array contains 32 processing units running in parallel, and the data within each processing unit perform the same operation, thus realizing spatial parallelism;
⑥ The PE array obtains the input feature map parameters, weight parameters and intermediate implicit layer parameters from the on-chip cache and transfers the output data of each implicit layer obtained after the PE operation to the on-chip cache for storage;
⑦ The on-chip cache transfers the network model output parameters to the off-chip DDR for storage;
⑧ The off-chip DDR transfers the output parameters to the host computer for processing and obtains the output results.

Table 2. Resource Usage After YOLOv3 Synthesis

Resource	Utilization	Available	Utilization%
LUT	168186	274080	61.36
LUTRAM	10773	144,000	7.48
FF	394682	548160	72.00
BRAM	762	912	83.55
DSP	1379	2520	54.72

4.2 Resource Utilization Test

The resource usage of the YOLOv3 network after synthesis is shown in Table 2.

The verification of this network on an FPGA involves several types of resources, such as LookUpTable (LUT), FlipFlop (FF), BlockRAM (BRAM) and DigitalSignal Processing (DSP).

The detection accuracy was tested, and in the embedded demonstration verification environment, the YOLOv3 detection accuracy was 72.85%, with a running rate of 6.5 fps. This represents an improvement of approximately 2.03 times compared with 3.2 fps before the method described in this paper was adopted. It can accurately and quickly identify targets, demonstrating good detection performance.

5 Conclusion

This paper designs and implements a convolutional neural network for embedded systems, explores the deployment and implementation method of a convolutional neural network in an embedded environment where both computational and storage resources are limited, and performs functional simulation verification of the implemented network, aiming at solving the problem that the complexity of the convolutional neural network algorithm is constantly increasing and the arithmetic power of the embedded environment is insufficient. Experimental tests prove that this paper successfully deploys and implements a convolutional neural network with 106 hidden layers in an embedded environment with scarce computational resources and limited storage resources, which can effectively complete the inference process of this network, and the implementation method designed in this paper is highly flexible. The subsequent work will construct a more complete acceleration operator library, laying the foundation for further realizing a reconfigurable intelligent computing platform that can run many different convolutional neural networks.

References

1. Li, J., Wang, X., Chen, J., et al.: Method for vehicle-mounted radar image target recognition based on lightweight convolutional neural network. Radar Sci. Technol. **23**(1) (2025)

2. Xie, Y., Liu, Y., Chen, C., et al.: An embedded lightweight convolutional neural network computation acceleration method. J. Small Microcomput. Syst. **44**(7) (2023)
3. Yanamala, R.M.R., Pullakandam, M.: A high-speed reusable quantized hardware accelerator design for CNN on constrained edge device. Des. Autom. Embed. Syst. **27**(3) (2023)
4. Jang, J., Yang, H.: A runtime switchable multi-phase convolutional neural network for resource-constrained systems. IEEE Access **11** (2023)
5. Lamrini, M., Chkouri, M.Y., Touhafi, A.: Evaluating the performance of pre-trained convolutional neural network for audio classification on embedded systems for anomaly detection in smart cities. Sensors (14248220) **23**(13) (2023)
6. AlShibli, A., Mathkour, H.: A shallow convolutional learning network for classification of cancers based on copy number variations. Sensors (Basel, Switzerland) **19**(19) (2019)
7. Zhang, R., Li, J.: A survey of scene segmentation algorithms based on deep learning. Comput. Res. Dev. **57**(4) (2020)
8. Luo, T., Liu, S., Li, L., et al.: DaDianNao: a neural network supercomputer. IEEE Trans. Comput. **66**(1), 73–88 (2016)
9. Boudjadar, J., Islam, S.U., Buyya, R.: Dynamic FPGA reconfiguration for scalable embedded artificial intelligence (AI): a codesign methodology for convolutional neural networks (CNN) acceleration. Future Gener. Comput. Syst. **169** (2025)
10. Sun, R., Ying, P., Li, Z., et al.: Weak and small target detection based on lightweight SSD. Comput. Simul. **41**(10), 355–361 (2024)
11. Gong, L., Cheng, Q.: Exploiting edge features for graph neural networks. In: CVPR 2019, pp. 9203–9211 (2019)
12. Chen, Y., Pei, S.: Research on compression methods for binary neural network models oriented to FPGA. J. Small Microcomput. Syst. **45**(6), 1356–1362 (2024)
13. Wang, P., Ren, Y., Fan, Y., et al.: Hardware acceleration design and FPGA implementation of dynamic deep neural networks. Telecommun. Eng. (003) (2024)
14. Zhang, Y., Ye, Y., Peng, Y., et al.: A FPGA-based deep neural network hardware accelerator system. Space Control Technol. Appl. **50**(2) (2024)
15. Dong, G., Hu, K., Yang, H., et al.: Research on a universal convolutional neural network accelerator architecture. Microelectron. Comput. **40**(5), 97–103 (2023)

Research on the Social Attributes of Power Grid Residents with the 3M-CNN-SC-SRU Customer Profiling the 3M-CNN-SC-SRU Model for Customer Profiling from Electricity Consumption Data

Guoan Zhao[1(✉)], Chunli An[2], Biling Zhang[2], Zhiyang Jia[3], and Shaobo Wang[4]

[1] Network Education College, Beijing University of Posts and Telecommunications, Haidian District, Beijing 100876, China
{zga,anli1493258487,bilingzhang,wangshaobo}@bupt.edu.cn, kangpingli@sjtu.edu.cn

[2] School of Network Education, Beijing University of Posts and Telecommunications, Haidian District, Beijing 100876, China
bilingzhang@bupt.edu.cn

[3] Department of Computer Science, China University of Petroleum-Beijing at Karamay, No. 355, Anding Road, Karamay 834000, Xinjiang Uygur Autonomous Region, China

[4] School of Humanities, Beijing University of Posts and Telecommunications, Haidian District, Beijing 100876, China
wangshaobo@bupt.edu.cn

Abstract. Obtaining users' social attribute classification through their electricity consumption data is highly important for power systems to achieve personalized services. This paper proposes a 3 M-CNN-SC-SRU model based on deep learning, which realizes accurate classification of users' social attributes by intelligently learning multiscale and multilevel features from users' electricity consumption data. Experiments on the CER dataset verify the effectiveness of the proposed model. The results show that for the classification of P2 attributes, the proposed model improves the accuracy by an average of 15.34% and the F1 score by an average of 18.26% compared with existing models. For the classification of users' social attributes, the accuracy of the proposed model improves by an average of 24.84%, and the F1 score improves by an average of 33.22%.

Keywords: Convolutional Neural Network (CNN) · Deep Learning · Smart Meter Data · Classification of Users' Social Attributes

1 Introduction

With the growth in global energy demand and heightened awareness of environmental protection, the development and application of smart grid technologies have become crucial pathways for achieving efficient energy utilization and environmental protection.

L. Huang et al. (Eds.): CCF NCCA 2025, CCIS 2774, pp. 629–641, 2026.
https://doi.org/10.1007/978-981-95-5294-8_41

In a smart grid system, the electricity consumption data collected by smart meters contain rich information about user behavior [1, 2]. Through in-depth analysis of these data, residents' electricity consumption behavior patterns can be revealed, further allowing for the exploration of their sociodemographic attributes, such as gender, age, household structure, housing status, and appliance usage habits [3]. This provides decision support for grid operation optimization, demand-side management, and personalized services for users. Therefore, research on user behavior based on smart meter data has received attention in recent years.

With the advancement of machine learning technologies, numerous researchers have attempted to extract features from users' electricity consumption patterns via statistical learning methods, such as linear discriminant analysis (LDA), the Mahalanobis distance, support vector machine (SVM), k-nearest neighbors (kNN), AdaBoost, random forest (RF), and XGBoost [4–13]. However, traditional machine learning methods rely on manual feature extraction, which is not only time-consuming but also challenging in capturing the underlying complex patterns in the data.

Compared with traditional machine learning, existing classifiers have improved efficiency and accuracy in processing high-dimensional, nonlinear, and time series data. For example, Sodenkamp M et al. proposed a dependency-based independent data (DID) classification algorithm to increase the accuracy of SVM through supervised learning; Y. Wang [9] et al. utilized a 2D CNN to analyse data in real time and infer user attributes. However, the generalization ability and capacity to handle the data diversity of these methods still need improvement, and there are limitations in dealing with complex user behavior patterns. Considering the multidimensionality of user characteristics, Wang Y [14] et al. adopted a joint learning approach; Cui Y [15] et al. employed multiscale multivariate multifractal detrended fluctuation analysis (MMV-MFDFA). However, existing studies often neglect the spatiotemporal characteristics of electricity consumption data and the complex correlations among different sociodemographic attributes. The adaptability of the proposed methods across different user groups and varying electricity consumption environments requires further validation.

Therefore, this paper considers the multidimensionality and diversity of smart meter data and proposes an advanced machine learning model, the 3 M-CNN-SC-SRU model, which can utilize the spatiotemporal characteristics of time series information and the complex correlations among sociodemographic attributes to automatically mine potential complex patterns in the data. By fusing temporal, spatial, frequency-domain, and multiscale features of the data, the proposed model not only enhances classification performance but also improves data interpretability. Specifically, the MP-CNN module exploits other attribute values as padding, fully tapping into the value of sociodemographic data. The four-channel input method, combined with the introduced SCConv and SRU units [17], collectively enhances classifier performance and reduces spatial redundancy. Experiments demonstrate that the new model improves accuracy by an average of 15.34% and the F1 score by an average of 18.26% compared with existing models. For the classification of users' sociodemographic attributes, the proposed model achieves an average accuracy improvement of 24.84% and an average F1 score improvement of 33.22%.

2 M-CNN-SC-SRU Module

The 3 M-CNN-SC-SRU designed in this paper is a multi-input channel deep convolutional neural network model, whose structure is illustrated in Fig. 1.

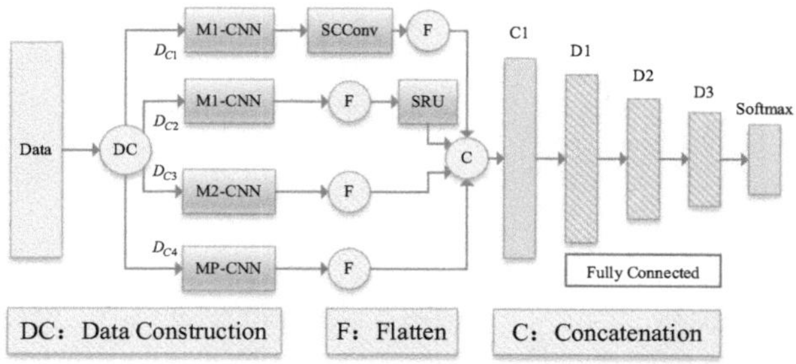

Fig. 1. Structural diagram of the 3 M-CNN-SC-SRU model.

2.1 Input Channels

The input data channels of the 3 M-CNN-SC-SRU model have four different types: (1) Temporal feature data channel: represented by a 7 × 24 (abbreviated as P24) pixel matrix, capturing the electricity consumption pattern for each hour within a week, reflecting users' daily electricity consumption behavior. (2) Frequency domain feature data channel: This channel is also represented by a 7 × 24 pixel matrix, revealing the periodicity and regularity of electricity consumption patterns by transforming temporal data into the frequency domain. (3) Difference feature data channel: This channel is represented by a 6 × 48 pixel matrix, emphasizing changes and trends in electricity consumption patterns by calculating the differences in time series. (4) Filling attribute temporal feature data channel: This channel is represented by a 9 × 50 pixel matrix combined with other related attributes, providing a more comprehensive user portrait for the model. Two M1-CNN modules are used to extract features from the temporal and frequency domain feature data. The combination of these features allows the model to capture both the temporal dynamics and the static attributes of users, thereby improving its ability to analyse and predict electricity consumption patterns.

The M2-CNN module processes difference feature data, emphasizing that the characteristics of electricity consumption patterns change over time. The MP-CNN module processes filled attribute temporal feature data, providing a more comprehensive feature representation by considering other related attributes of users.

Finally, on the basis of the mathematical or statistical characteristics of the original data, the four channels are input into their respective deep learning frameworks to extract features. The extracted features are concatenated and input into a fully connected network for further feature extraction, and finally, the user's social attributes are accurately classified through Softmax.

2.2 M1-CNN Module and M2-CNN Module

The M1-CNN module is primarily used to extract features from the time domain and frequency domain data of specially formatted image data. This module consists of two convolutional layers, a max pooling layer, Dropout, and Flatten, and the details of each layer's network structure with the relevant hyperparameters are shown in Fig. 2.

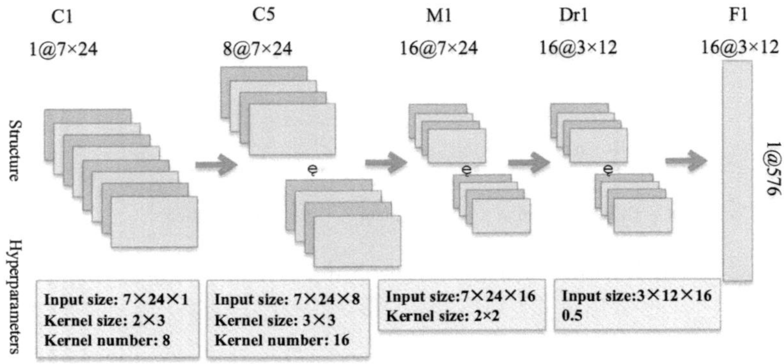

Fig. 2. M1-CNN module structure diagram.

The input of the M1-CNN module is preprocessed P24 image data. These image data contain various types of sociodemographic information related to residents, such as behavioural patterns and living habits, presented in pixel form. Local features in the images are automatically extracted through two convolutional operations. Each convolutional layer uses multiple convolutional kernels to slide over the input data, performing weighted summation and activation (ReLU activation function) to generate new feature maps. The size of the feature maps is reduced through pooling operations, lowering the computational load while extracting higher-level feature representations and enhancing the robustness of the model. During training, a portion of the neuron outputs are randomly dropped to prevent overfitting and improve the generalization ability of the model. The flatten layer converts the multidimensional feature maps into one-dimensional vectors, preparing them for the fully connected layer.

2.3 SC Module

The SC module includes two parts: the spatial reconstruction unit (SRU) and the channel reconstruction unit (CRU), which are used to replace standard convolutional operations. The SRU is introduced to leverage spatial redundancy in features. The separation operation separates feature maps with rich information from those with less information corresponding to spatial content. The reconstruction operation adds feature maps with rich information to those with less information to generate more informative feature maps and save space. The separation-reconstruction process is illustrated by Formula (1).

$$X_w = \text{Reconstruct}\ (\text{Gate}\ (\text{Sigmoid}\ (W_\gamma(\text{GN}(X)))) \tag{1}$$

In Formula (1), GN represents group normalization, which calculates normalized related weights. Reconstruct denotes the reconstruction operation, Gate represents a threshold gating function, and denotes the concatenation operation. Through the separation-reconstruction process, the SRU effectively utilizes spatial redundancy to enhance feature expression and reduce unnecessary information.

CRU employs a split-transform-fuse strategy to further reduce redundancy in spatially refined feature maps along the channel dimension.

In the split stage, the spatial feature $X^{\omega} \in R^{c\times h\times w}$ is divided into two parts, X_{up} and X_{low}, along the channel dimension according to a split ratio $\alpha(0 \leq \alpha \leq 1)$, and the channel numbers are compressed via 1×1 convolutions with a compression ratio r (commonly set to 2 in experiments) to improve computational efficiency. The transformation stage is described by Formulas 2 and 3:

$$Y_1 = M^G X_{up} + M^{P1} X_{up} \tag{2}$$

$$Y = \beta_1 Y_1 + \beta_2 Y_2 \tag{3}$$

where $Y = \beta_1 Y_1 + \beta_2 Y_2$, $Y = \beta_1 Y_1 + \beta_2 Y_2$ and M^{P2} are the weights of the groupwise convolution (GWC), upper pointwise convolution (PWC) and lower PWC, respectively, and $\cup$ is a concatenation. In the fusion stage, the SKNet scheme [18] is adopted to adaptively merge the output features $Y = \beta_1 Y_1 + \beta_2 Y_2$ and $Y = \beta_1 Y_1 + \beta_2 Y_2$ under the guidance of a feature importance vector β_1 and β_2, forming an output $Y = \beta_1 Y_1 + \beta_2 Y_2$ adapted to channel importance, as shown in Formula (4).

$$Y = \beta_1 Y_1 + \beta_2 Y_2 \tag{4}$$

CRU strategically minimizes feature redundancy in spatially refined mappings, enhances feature importance, and controls computational efficiency.

Pairing the SRU with the CRU to replace conventional convolutional operations yields more efficient performance and efficiency.

2.4 MP-CNN Module

This paper proposes a padding method for CNN models, which involves padding meaningful sociodemographic data around the feature data to be trained or after a certain convolution, thereby expanding the data. The module implementing this method is the meaningful CNN padding module, abbreviated as MP-CNN.

The MP-CNN module consists of the initial data input, feature padding processing, the first convolutional layer, the subsequent padding operation, the second convolutional layer, a max pooling layer, a dropout layer, and a flattening layer, which serves as the preparation for the fully connected layer. The detailed network architecture of each layer, together with the corresponding hyperparameters, is illustrated in Fig. 3.

Consider an initial feature map $X^{\omega} \in R^{n\times c\times h\times w}$, where N denotes the batch dimension, H and W correspond to the spatial dimensions of height and width, and C denotes the channel dimension. Then, attributes P are used to pad the input features X. The specific process is shown in Formula 5.

$$X_{out\_p} = Mapping(X, P, 1) \tag{5}$$

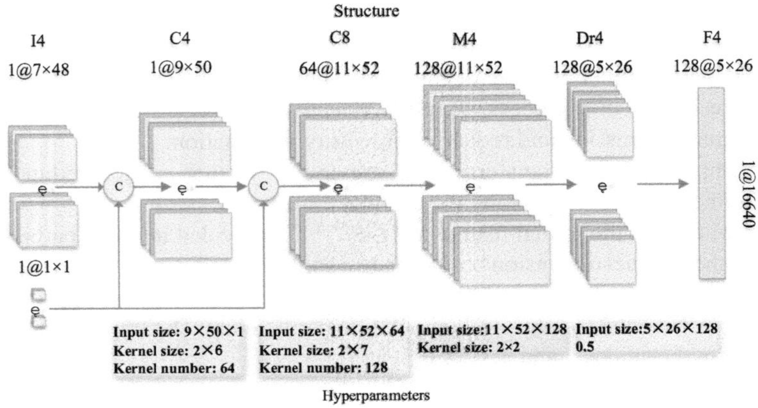

Fig. 3. MP-CNN unit structure diagram.

Among them, 1 represents the dimension that is padded or expanded, and $X_{out\_p} \in R^{N \times (H+2) \times (W+2) \times C}$ denotes the computed intermediate features. These features are then input into a new convolutional neural network for identifying the classification of other features. In other words, padding can be applied to either initial features or intermediate features to accommodate different model requirements and architectural designs. The process is shown in Eq. (6).

$$X_{other\_p} = CNN_{mp}(X_{out\_p}) \tag{6}$$

This generates new features that include other social attributes. These new features, concatenated with the features obtained from other channels, are jointly used as the input to a fully connected network for classification tasks.

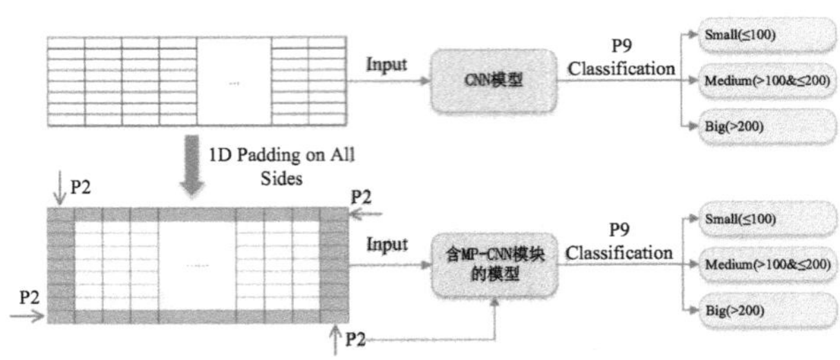

Fig. 4. MP-CNN schematic diagram with P2 filling classification P9 attributes as an example.

The schematic diagram in Fig. 4 illustrates the MP-CNN (the CNN Model Module of a Meaning Padding) process for classifying P9 attributes via P2 filling. This diagram clearly visualizes the key components and data flows involved in the classification task.

In this schematic, the original pixel data, represented as a P48 image, are first constructed into a 9 × 50 image data format. This transformation is part of a feature enhancement or feature fusion strategy aimed at improving the accuracy and robustness of the classification. The process involves filling the surrounding area of the initial or intermediate feature data with meaningful sociological attribute data.

In summary, Fig. 4 provides a comprehensive overview of the MP-CNN process for classifying P9 attributes via P2 filling. This schematic diagram highlights the key steps and components involved in transforming the original electrical consumption data into a format suitable for classification and demonstrates how additional sociological attributes can be used to enhance the feature set and improve classification accuracy.

3 Experimental Results and Analysis

This section introduces the dataset, experimental environment, and experimental results, followed by a comprehensive analysis of the experimental outcomes.

3.1 CER Dataset

This paper utilizes the user behavior trial dataset obtained from an 18-month smart metering power user behavior trial conducted by the Irish Commission for Energy Regulation (CER) in 2009 and 2010 [16], referred to as the CER dataset. In addition to time series power data, the CER dataset includes various sociological attribute data of users. This paper selects nine user sociological attributes (referred to as P1 to P9) for identification research, including age, retirement status, social type, presence of children, residence type, house age, number of bedrooms, cooking facility type, and occupied area.

The user's power consumption time series data undergo data cleaning, anomaly handling, and validation. Pixel-based data representation has multiple advantages in user behavior recognition, capturing complex patterns in time and magnitude more efficiently than raw sequence representations and spectrogram representations do, providing a more intuitive data representation, and being able to be naturally integrated with CNN models. Power consumption data for one day are used to construct images with different pixel resolutions: grayscale and color images with a resolution of 7 × 24 (hereinafter referred to as P24) and grayscale and color images with a resolution of 7 × 48 (hereinafter referred to as P48), as illustrated in Fig. 5.

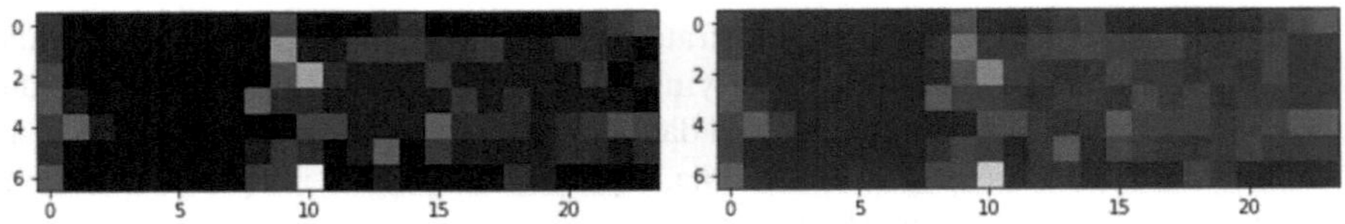

A schematic diagram illustrating the daily electricity consumption data used to construct P24

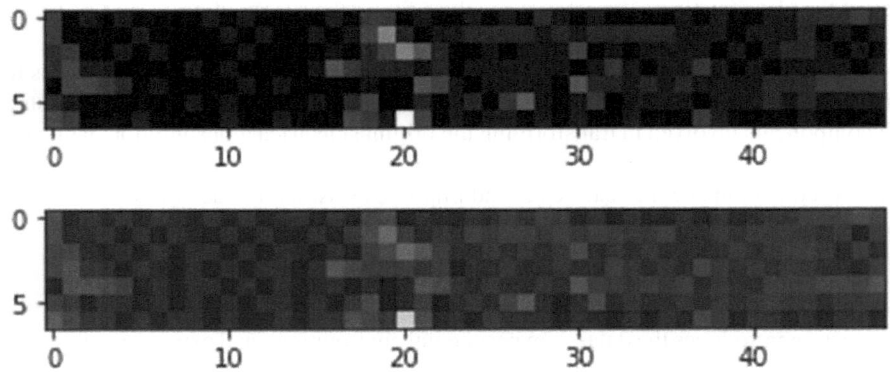

Schematic diagram illustrating the daily electricity consumption data used to construct P48

Fig. 5. A schematic diagram constructed from one-day power consumption data.

3.2 Experimental Environment and Evaluation Metrics

This paper adopts the TensorFlow Keras library and conducts programming and model validation in a Python 3 environment. The experiments are conducted on a server equipped with an Intel(R) Xeon(R) Silver 4208 CPU (2.10 GHz, 8 cores) and 96 GB of RAM.

This paper uses accuracy to evaluate the classification accuracy of the model and F1 to measure the model's performance in maintaining a balance between precision and recall. The definitions of these two indicators are as follows: true positive (TP), false positive (FP), false negative (FN), and true negative (TN).

$$Accuracy = \frac{TP+TN}{TP+FP+FN+TN} \tag{7}$$

$$F_1 = \frac{2 \cdot TP}{2 \cdot TP + FP + FN} \tag{8}$$

3.3 Analysis of the Experimental Results

Three sets of experiments are conducted in this paper to verify the effectiveness of the attribute-filling method and the model.

Verification of the MP-CNN Module Filling Effect

This experiment compares the performance of the CNN, MP-CNN, and MP-CNN (BN) models in identifying resident user sociological attributes on the basis of P24 and

P48 pixel data. The MP-CNN (BN) model refers to the CNN model with an added batch normalization (BN) layer combined with the MP-CNN module.

Figures 6 and 7 show the experimental results of identifying the P2 attribute via image data filled with P1, P3 and P9 attributes for P24 and P48 pixels in the CNN, MP-CNN, and MP-CNN (BN) models, respectively.

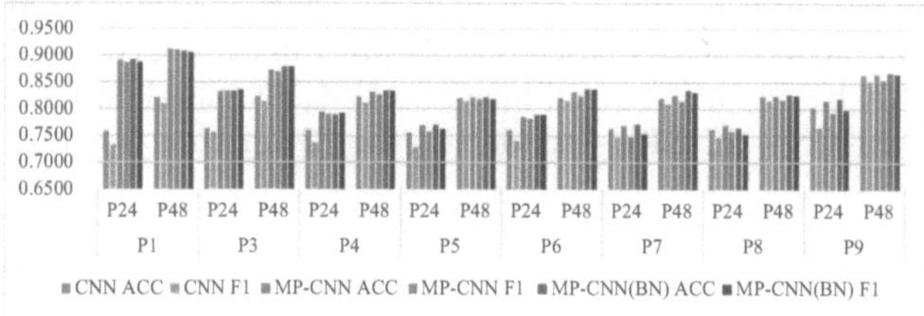

Fig. 6. P2 attribute classification performance under different numbers of pixels after P1, P3 to P9 filling.

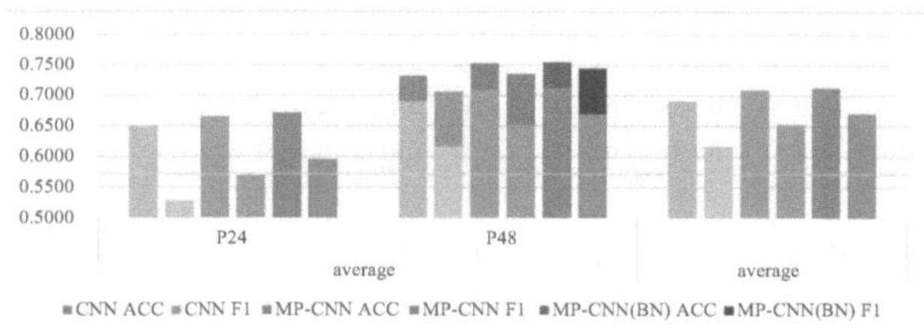

Fig. 7. P1, P3 to P9 attributes after the P2 attributes are filled to classify the average performance changes compared.

The experimental results indicate that the models constructed with P48 pixels significantly outperform those with P24 pixels in terms of performance. This suggests that higher pixel resolution retains more image details and temporal information, thereby enhancing the model's classification capabilities.

The MP-CNN and MP-CNN(BN) models achieve significant improvements in both accuracy and F1 values. This finding indicates that meaningful filling not only increases the data richness but also more effectively captures complex features in the images, whereas batch normalization enhances the model's stability and generalization ability. Specific attribute combinations (such as P1 with P2, P2 with P3, etc.) exhibit greater improvements in classification performance, reflecting a potential causal relationship between the filled attributes and the target attribute. This relationship enables the model to more effectively utilize filling information, thereby improving classification performance. Compared with the original CNN model, the MP-CNN and its variants exhibit greater stability and consistency while maintaining high performance, which is significant for the practical deployment and maintenance of the model.

The experiment revealed that using P9 (house size) as the filling attribute significantly boosts the classification performance of all other attributes. This suggests that in New Zealand society, "house size" may be an important comprehensive manifestation of a person's sociological attributes, containing information about personal economic status, social status, and other aspects.

Verification of the 3 M-CNN-SC-SRU Model Effectiveness

This experiment validated the effectiveness of the 3 M-CNN-SC-SRU model by using six different models—CNN based on P24 pixel images, CNN-SC based on P24 pixel images, CNN based on P48 pixel images, MP-CNN based on P48 pixel images, 3 M-CNN, and 3 M-CNN-SC-SRU—to identify the P2 attribute. The results are shown in Table 1.

Table 1. Experimental results of P2 attribute classification under six models.

MODEL PADDING	CNN(P24)		CNN-SC(P24)		CNN(P48)		MP-CNN (P48)		3 M-CNN		3 M-CNN-SC-SRU	
	Acc	F1	Acc	F1	Acc	F1	Acc	F1	Acc	F1	Acc	F1
P1	0.7623	0.7468	0.7732	0.7550	0.8166	0.8076	0.9137	0.9121	0.9271	0.9252	0.9402	0.9397
P3	0.7677	0.7498	0.7767	0.7736	0.8235	0.8180	0.8832	0.8818	0.9063	0.9058	0.9147	0.9141
P4	0.7608	0.7324	0.7743	0.7597	0.8249	0.8202	0.8328	0.8270	0.8688	0.8683	0.8614	0.8573
P5	0.7554	0.7231	0.7723	0.7557	0.8208	0.8122	0.8273	0.8210	0.8493	0.8435	0.8562	0.8529
P6	0.7613	0.7416	0.7703	0.7537	0.8200	0.8106	0.8355	0.8285	0.8684	0.8668	0.8767	0.8744
P7	0.7604	0.7520	0.7676	0.7433	0.8212	0.8088	0.8316	0.8243	0.8584	0.8509	0.8635	0.8571
P8	0.7623	0.7457	0.7737	0.7629	0.8187	0.8062	0.8260	0.8185	0.8564	0.8514	0.8613	0.8568
P9	0.8045	0.7690	0.8215	0.8006	0.8621	0.8529	0.8732	0.8662	0.8943	0.8871	0.9018	0.8967

The experimental results demonstrate that the performance of CNN-SC (P24) is superior to that of the CNN (P24). Specifically, the average increases in accuracy and F1 score are 0.0119 and 0.0180, respectively. This indicates that the SC module plays a significant role in enhancing classification performance. Furthermore, the MP-CNN(P48), 3 M-CNN, and 3 M-CNN-SC-SRU exhibit better performance than the CNN(P48), highlighting the substantial improvement in classification capabilities brought about by the introduction of the MP-CNN module. In particular, the 3 M-CNN achieves average increases in accuracy and F1 score of 0.0257 and 0.0274, respectively, compared with those of the MP-CNN (P48), demonstrating a significant increase in classification performance due to multifeature fusion.

Compared with the CNN (P48), the 3 M-CNN-SC-SRU model performs the best, with relative improvements of 7.08% and 7.84% in accuracy and F1 score, respectively. Compared with those of CNN (P24), the improvements are even more pronounced, reaching 15.34% and 18.26% for accuracy and F1 score, respectively. The 3 M-CNN-SC-SRU model integrates time, frequency, and difference features, forming a multifeature fusion model architecture that overcomes the limitations of single-feature recognition. This allows the model to capture complex information in the data more comprehensively.

Additionally, the model incorporates three scales of information: 7×24, 7×48, and 6×48. By inputting data with different resolutions, the model not only expands its receptive field but also maintains precision. This multiscale design helps the model remain efficient and accurate when processing data of different granularities.

Each residential user is a complex individual, and their electricity consumption data contain various latent features. Through the design of multifeature, multiscale models, these features can be more effectively mined, thereby achieving high-quality identification of residents' social attributes.

Verification of the Effects of the SC Module and the SRU Module

The performance of the 3 M-CNN, 3 M-CNN-SRU, 3 M-CNN-SC, and 3 M-CNN-SC-SRU on images with P24 pixels was tested, and the results are shown in Table 2.

Table 2. Performance of the 3 M-CNN, 3 M-CNN-SRU, 3 M-CNN-SC and 3 M-CNN-SC-SRU models.

MODEL PADDING	3 M-CNN		3 M-CNN-SRU		3 M-CNN-SC		3 M-CNN-SC-SRU		Classified Attribute
	Acc	F1	Acc	F1	Acc	F1	Acc	F1	
P6	0.8028	0.7874	0.8255	0.8190	0.8128	0.8062	0.8358	0.8302	P9
	0.8299	0.8201	0.8186	0.8067	0.8209	0.8117	0.8396	0.8346	P8
	0.7399	0.7378	0.7581	0.7558	0.7195	0.7148	0.7511	0.7492	P7
P7	0.7965	0.7965	0.8068	0.8068	0.7942	0.7942	0.8034	0.8034	P6
P8	0.7624	0.7622	0.7742	0.7742	0.7835	0.7835	0.7852	0.7852	
P9	0.8270	0.8221	0.8167	0.8145	0.8174	0.8154	0.8331	0.8307	

The experiments revealed that neither the SRU nor the SC module could comprehensively enhance the performance of the 3 M-CNN model. However, when combined, the SRU and SC modules could enhance the performance of the 3 M-CNN model.

4 Conclusion

This paper focuses primarily on identifying users' sociodemographic attributes on the basis of their electricity consumption data. By utilizing other attribute values as filling values in the CNN model and incorporating multiscale and multifeature fusion, the proposed 3 M-CNN-SC-SRU model significantly enhances the performance of user sociodemographic attribute classification. The results indicate that for the classification of P2 attributes, the proposed model achieves an average accuracy improvement of 15.34% and an average F1 score improvement of 18.26% over those of existing models. Moreover, for the classification of user sociodemographic attributes in general, the proposed model achieves average accuracy and F1 score improvements of 24.84% and 33.22%, respectively. The introduction of this model is highly important for improving the performance of classification tasks and enhancing the understanding of feature relationships. One of the potential directions for our future work is to extend the dataset to support the study of more complex scenarios so that the proposed paradigm can be applied to a wider range of real-world scenarios.

5 Disclosure of Interests.

The authors have no competing interests to declare that are relevant to the content of this article.

Acknowledgments. This work is supported by the National Natural Science Foundation of China [under Grant No. 62171060].

References

1. Wang, Y., Zhang, N., Kang, C., et al.: Power user behavior model: basic concepts and research framework. Trans. China Electrotech. Soc. **34**, 2056–2068 (2019). (in Chinese)
2. Zhu, T.,Ai, Q., He, X., et al.: Review of data-driven electricity behavior analysis methods and applications. Grid Technol. **44**(9), 11 (2020). https://doi.org/10.13335/j.1000-3673.pst.2020.0226a
3. Wang, Y., Chen, Q., Hong, T., et al.: Review of smart meter data analytics: applications, methodologies,and challenges. IEEE Trans. Smart Grid **10**(3), 3125–3148 (2018)
4. Beckel, C.,Sadamori, L., Santini, S.: Towards automatic classification of private households using electricity consumption data. In: Pappas, G.J. (ed.) Proceedings of the Fourth ACM Workshop on Embedded Sensing Systems for Energy-Efficiency in Buildings, pp. 169–176. ACM, Toronto (2012)
5. Beckel, C., Sadamori, L., Staake, T., Santini, S.: Revealing household characteristics from smart meter data. Energy **78**, 397–410 (2014)
6. Yan, S., et al.: Time–frequency feature combination based household characteristic identification approach using smart meter data. IEEE Trans. Ind. Appl. **56**(3), 2251–2262 (2020)
7. Beckel, C., Sadamori, L., Staake, T., Santini, S.: Revealing household characteristics from smart meter data. Energy **78**, 397–410 (2014)
8. Hopf, K., Sodenkamp, M., Kozlovkiy, I., Staake, T.: Feature extraction and filtering for household classification based on smart electricity meter data. Comput. Sci.-Res. Dev. **31**(3), 141–148 (2016)
9. Sodenkamp, M., Kozlovskiy, I., Staake, T.: Supervised classification with interdependent variables to support targeted energy efficiency measures in the residential sector. Decis. Anal. **3**(1), 1 (2016)
10. Sun, G., Cong, Y., Hou, D., et al.: Joint household characteristic prediction via smart meter data. IEEE Trans. Smart Grid **10**(2), 1834–1844 (2017)
11. Wang, Y., Bennani, I.L., Liu, X., Sun, M., Zhou, Y.: Electricity consumer characteristics identification: a federated learning approach. IEEE Trans. Smart Grid **12**(4), 3637–3647 (2021)
12. Gajowniczek, K., Ząbkowski, T., Sodenkamp, M.: Revealing household characteristics from electricity meter data with grade analysis and machine learning algorithms. Appl. Sci. **8**(9), 1654 (2018)
13. Fahim, M., Sillitti, A.: Analyzing load profiles of energy consumption to infer household characteristics using smart meters. Energies **12**(5), 773 (2019)
14. Wang, Y., Chen, Q., Gan, D., et al.: Deep learning-based sociodemographic information identification from smart meter data. IEEE Trans. Smart Grid **10**(3), 2593–2602 (2018)
15. Cui, Y., Yan, R., Sharma, R., et al.: Realizing multifractality of smart meter data for household characteristic prediction. Int. J. Electr. Power Energy Syst. **139**, 108003 (2022)
16. Irish Social Science Data Archive. Commission for Energy Regulation (CER) Smart Metering Project (2012). http://www.ucd.ie/issda/data/commissionforenergyregulationcer/

17. Li, J.,Wen, Y., He, L.: Scconv: spatial and channel reconstruction convolution for feature redundancy. In: Proceedings of the IEEE/CVF Conference on Computer Vision and Pattern Recognition, pp. 6153–6162 (2023)
18. Li, X., Wang, W., Hu, X., Yang, J.: Selective kernel networks. In: Proceedings of the IEEE/CVF Conference on Computer Vision and Pattern Recognition, pp. 510–519 (2019)

Identification of Parkinson's Disease Subtypes via a Multimodal Data Fusion Method

Hongbao Lu(✉) and Xiangyu Zhao

School of Information Technology Engineering, Tianjin University of Technology and Education, Tianjin, China
1694646321@qq.com

Abstract. Parkinson's disease (PD), a progressive neurological disorder, manifests substantial clinical heterogeneity. Its phenotypic spectrum encompasses diverse motor and nonmotor manifestations, indicating probable discrete disease subtypes. Accurate identification of these subtypes is essential for guiding clinical management and personalized interventions. Currently, the classification of PD subtypes primarily depends on clinical behavioral assessments. Nevertheless, additional data modalities contribute significantly to Parkinson's disease subtyping. This investigation implemented similarity network fusion (SNF) to integrate multimodal datasets—encompassing functional MRI, DNA methylation profiles, and clinical behavioral metrics—obtained from the Parkinson's Progression Markers Initiative (PPMI) repository. Initially, we constructed individual patient similarity networks for each data modality, which were subsequently fused via the SNF method to generate a more comprehensive patient similarity network. Therefore, we applied the spectral clustering technique and successfully categorized PD patients into two distinct subtypes. Notably, significant differences in symptom severity and functional connectivity strength were observed between the two subtypes. Subtype 1 demonstrated markedly greater motor and nonmotor impairments, accompanied by a general reduction in functional connectivity levels relative to healthy controls. In contrast, compared with the control group, the Subtype 2 group presented with less severe symptoms and increased functional connectivity strength. The present work thus introduces an innovative analytical framework for the use of functional magnetic resonance imaging (fMRI) data to facilitate the clinical subtyping of Parkinson's disease.

Keywords: Parkinson's disease · fMRI · Clinical behavioral assessments · Similarity network fusion · Spectral clustering

1 Introduction

The chronic neurological condition known as Parkinson's disease (PD) primarily affects middle-aged and older people. It has a complex etiology that includes environmental, genetic, and aging-related variables. This reduction in striatal dopamine neurotransmitter levels subsequently triggers a spectrum of both motor and nonmotor manifestations [1]. Parkinson's disease motor presentations display marked heterogeneity, with core

L. Huang et al. (Eds.): CCF NCCA 2025, CCIS 2774, pp. 642–657, 2026.
https://doi.org/10.1007/978-981-95-5294-8_42

features manifesting as rigidity, bradykinesia, and tremor [2]. These clinical signs are directly attributed to the degeneration and depletion of dopaminergic neurons within the substantia nigra pars compacta of the midbrain and its striatal projections, coupled with the formation of Lewy bodies. Notably, motor symptomatology extends beyond these fundamental characteristics to encompass additional disturbances, such as abnormal posture and gait. Postural instability and gait disturbances represent cardinal manifestations of Parkinson's disease (PD), predisposing patients to falls and restricted mobility. In addition to motor impairments, PD encompasses nonmotor features that substantially compromise quality of life, including cognitive deficits, depression, and sleep disorders [3]. Evidence suggests that depressive symptoms and sleep disturbances may serve as indicators predicting a deteriorating cognitive course in PD patients [4]. Concurrently, depression alongside nonmotor manifestations—notably sleep disturbances—constitutes highly prevalent and incapacitating characteristics of Parkinson's disease. These features substantially compromise patients' quality of life, diminish their cognitive capacity, and increase mortality risk within affected populations. Critically, such nonmotor complications not only impede functional independence in daily activities but also potentially accelerate the deterioration of motor capabilities. The documented constellation of motor and nonmotor manifestations points to underlying heterogeneity in the pathogenesis of Parkinson's disease (PD), suggesting the existence of phenotypically distinct subtypes. Accurate characterization of these PD variants is pivotal for informing stratified therapeutic interventions and guiding personalized care pathways.

Presently, the classification of Parkinson's disease (PD) into distinct subtypes predominantly relies on clinical behavioral evaluations alone [5]. Clinical presentations, for example, allow categorization into tremor-dominant, rigidity-dominant, and mixed forms. However, classification approaches extend beyond clinical observation, incorporating significant contributions from diverse data sources [6]. Noninvasive neuroimaging modalities, exemplified by functional magnetic resonance imaging (fMRI), which monitors neural activation patterns, yield critical perspectives. Work by Amée F. Wolters et al. revealed disruptions within the resting-state functional connectivity networks of PD patients [7]. Furthermore, Ross D. Markello and colleagues demonstrated the utility of MRI data in distinguishing patient subgroups with divergent PD pathophysiologies, suggesting its potential as a biomarker for precision medicine [8]. At the molecular level, DNA methylation—a vital epigenetic process involving the attachment of methyl (-CH_3) groups to specific DNA bases [9]—offers another dimension. Research employing methylation-specific polymerase chain reaction (MSP) by LIN Qing-ling et al. indicated that abnormal expression of clock genes in PD may be linked to methylation changes in their promoter regions [10]. The research group led by Juan I. Young revealed distinct DNA methylation variations in Parkinson's disease patients compared with controls, with pronounced alterations observed specifically within the dorsal motor nucleus of the vagus (DMV), a neuroanatomical locus known to be involved in early disease pathogenesis. The predominant methylation alterations observed in the DMV reinforce the hypothesis that vagal nerve dysfunction is involved in PD pathogenesis. These findings also independently support the involvement of Wnt signalling pathway genes as pivotal factors in PD neurodegeneration. The evident importance of diverse data modalities in identifying PD subtypes underscores that integrative, multimodal approaches offer a more comprehensive and accurate classification strategy [11]. Nevertheless, prevalent methods for multimodal disease recognition often limit themselves to simple concatenation of features from different sources, failing to integrate these attributes deeply. A critical issue in this concatenation process is the tendency for high-dimensional data (e.g., MRI, with its vast feature sets) to overshadow low-dimensional clinical data. This imbalance can lead to the modality possessing the most features disproportionately influencing classification outcomes, consequently neglecting the inherent interrelationships and balance among the multimodal datasets [8]. Therefore, effectively integrating multimodal data from PD patients and profoundly leveraging the synergistic relationships between modalities have become paramount challenges demanding prompt attention within the field of multimodal data-driven subtype identification [12].

Similarity network fusion (SNF) is a sophisticated methodology for integrating insights from heterogeneous data sources or networks. It aims to increase the robustness and accuracy of data analysis by leveraging the synergistic relationships of similarity measures across distinct networks. SNF has been effectively applied to disease subtype categorization [13]. To illustrate its utility, Yang Kuan et al. optimized the SNF algorithm by incorporating the normalized Euclidean distance. This optimization reduced analytical noise within similarity networks and fostered complementarity among diverse network datasets, achieving greater precision in cancer subtype classification [14]. In Parkinson's disease research, Ross D. Markello et al. applied SNF to integrate structural T1-weighted MRI, clinical behavioral assessments, cerebrospinal fluid analyses, and SPECT data. Their findings revealed SNF's capacity to uncover subtle interdependencies among data patterns that might otherwise remain undetected, emphasizing its utility for multimodal data integration across diverse patient cohorts [8]. To further demonstrate its applicability, Mei Yang investigated a cohort spanning the cognitive aging spectrum via SNFs. This study revealed that multimodal subtypes derived via SNF distinctly characterize the nuances of cognitive variability, surpassing the capabilities of single-modality subtypes [15]. Taken together, these studies demonstrate that applying similarity network fusion to integrate patients' multimodal data yields superior discriminative performance in disease subtyping.

This study synthesized integrated multimodal datasets derived from the Parkinson's Progression Markers Initiative (PPMI) database, incorporating functional magnetic resonance imaging (fMRI), DNA methylation profiles, and clinical behavioral evaluations. Following the generation of distinct patient similarity networks for each modality, we utilized similarity network fusion (SNF) to integrate these networks, thereby constructing an encompassing patient similarity network. Building upon this robust framework, we applied spectral clustering to delineate subtypes within the Parkinson's disease patient population. In conclusion, we examined the distinct characteristics of these subtypes across different modalities, presenting a new endeavor in the classification of Parkinson's disease subtypes. The comprehensive flow of the experimental process is illustrated in Fig. 1.

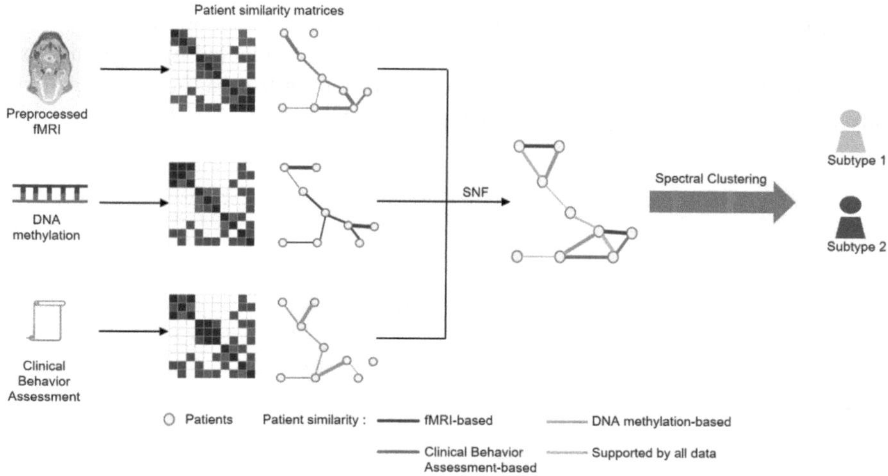

Fig. 1. This study's overall workflow involves constructing patient similarity networks for three modalities: preprocessed functional magnetic resonance imaging (fMRI), DNA methylation, and clinical behavioral assessments. These networks are subsequently fused via similarity network fusion (SNF), yielding a comprehensive patient similarity network that encapsulates information from all three modalities. Spectral clustering is then applied to this integrated network, resulting in the identification of two Parkinson's disease subtypes, which are subsequently analysed for their distinct characteristics.

2 Materials and Methodology

2.1 Data Acquisition and Cleansing

The complete dataset was procured through the Parkinson's Progression Markers Initiative (PPMI), a multi-institutional consortium aggregating participant-derived inputs from nine transatlantic research centers, with full open-access availability.

The dataset, which was collected between 2012 and 2019, comprises 384 resting-state functional MRI scans, DNA methylation cell composition data, and clinical behavioral assessment scores from 76 participants. During the MRI data procurement phase, the acquired data were retrieved in ZIP file format, thereafter uncompressed and transformed into the NIfTI format, adhering to the Brain Imaging Data Structure (BIDS) standardization criteria [16]. We employed the fMRIPrep pipeline (version 23.2.1) to subject the blood-oxygen-level-dependent (BOLD) cerebral blood flow data to a series of preprocessing steps [17]. In particular, the preprocessing sequence involved the estimation of head motion, slice timing correction, organization of the postprocessed BOLD data within the original space, alignment of EPI images with T1W images, resampling of BOLD data into the standardized space, estimation of confounding variables, and aggregation of T2*-weighted echo sequences. To delve deeper into the analysis, we leveraged the Harvard–Oxford brain atlas to delineate 48 regions of interest (ROIs) spanning the entire cerebrum [18] and subsequently utilized NiLearn software to derive BOLD time series data from these designated ROIs [19]. When analysing the DNA methylation cell composition data and patient motion assessment data, we excluded a number of patients (n = 8) with incomplete or aberrant data entries. For a minor cohort of patients (n = 1) missing clinical behavior assessment scores, we conducted mean estimations. The resulting dataset incorporated three-modal information from a total of 67 patients. Figure 2 depicts the sex distribution among patients stratified by age group.

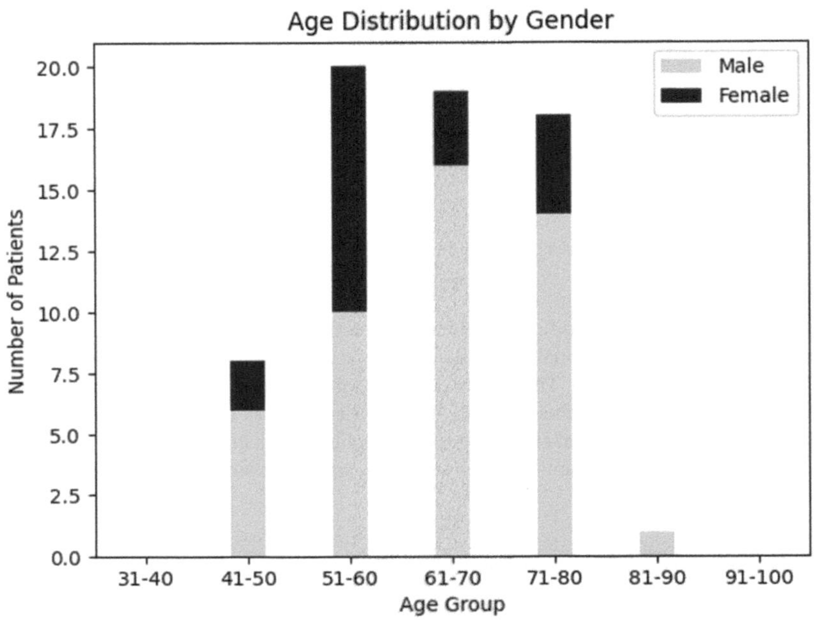

Fig. 2. Age and sex distributions of the patients

2.2 Similar Network Fusion

Similarity network fusion (SNF) is a sophisticated approach in which separate similarity networks are established for each exploitable data type within a sample set, followed by seamless integration into a singular network that encapsulates the entirety of the data [13]. For example, in constructing an encompassing perspective of a particular group of Parkinson's disease patients, the SNF method calculates and amalgamates patient similarity networks derived from each distinct data type, thereby harnessing the complementary aspects of the various datasets. This integrated approach substantially outperforms the analysis of single datasets. We utilized SNF to consolidate neuroimaging data, DNA methylation cell composition data, and clinical behavior assessment data from Parkinson's disease patient datasets.

Exemplifying the patient network, we formalize the interpatient similarity structure as a graph G(V, E). Within this network representation, vertex set V comprises the clinical cohort, whereas edge set E encapsulates dyadic similarity measures between subjects. The edge weight distribution is encoded in an n-dimensional similarity matrix W. Transformation from the distance space to the similarity space employs a scaled exponential kernel function:

$$W(i,j) = e^{\left[-\frac{p^2(x_i,x_j)}{\mu\varepsilon_{i,j}}\right]} \tag{1}$$

Here, the Euclidean distance between patients xi and xj is represented by the function p(xi, xj). The scaling parameter μ serves as an adjustable hyperparameter, whereas ε i,j mitigates scaling variance through the following calculation:

$$\varepsilon_{i,j} = \frac{\overline{p}(x_i, N_i) + \overline{p}(x_j, N_j) + \overline{p}(x_i, y_j)}{3}$$

where $\overline{\mathrm{p}}(x_i, N_i)$ represents the distance between xi and its neighborhood points.

For multimodal data integration, the primary procedure entails normalizing all the constituent similarity networks. Standard normalization methods, however, are vulnerable to numerical instability due to diagonal self-similarity entries. The enhanced normalization scheme operates as follows:

$$\mathrm{p}(i,j) = \begin{cases} \frac{\mathrm{W}(i,j)}{2\sum_{k\neq i} W(i,k)}, j \neq i \\ 1/2, j = i \end{cases} \tag{2}$$

We propose that proximity-driven similarity metrics are more reliable than their distance-based counterparts are. By implementing graph diffusion processes, similarity measures propagate to nonadjacent nodes, thereby establishing an affinity matrix manifesting enhanced sparsity characteristics:

$$\mathrm{S}(i,j) = \begin{cases} \frac{W(i,j)}{\sum_{k\in N_i} W(i,k)}, & j \in N_i \\ 0\ otherwise \end{cases} \tag{3}$$

Consequently, matrix P comprehensively encodes global affinity relationships across all patient pairs, whereas matrix S specifically captures localized neighborhood similarities restricted to K-nearest neighbors.

Subsequent similarity network integration proceeds iteratively until convergence. The core iterative procedure of SNF propagates updated similarity matrices across modalities through cross-network diffusion:

$$P_{t+1}^{(1)} = S^{(1)} \times P_t^{(2)} \times \left(S^{(1)}\right)^T \tag{4}$$

$$\mathrm{P}_{t+1}^{(2)} = S^{(2)} \times P_t^{(1)} \times \left(S^{(2)}\right)^T \tag{5}$$

Among these, the state matrix of the first data type after t iterations is given by $P_{t+1}^{(1)}$, and the similarity matrix corresponding to the second data type is denoted by $P_{t+1}^{(2)}$. Each state matrix update iteration initiates bidirectional cross-diffusion processes. Upon completing t iterations, the fused state matrix is derived via:

$$P^{(c)} = \frac{P_t^{(1)} + P_t^{(2)}}{2}$$

When m > 2, the following equation is obtained from Eqs. (4) and (5):

$$P^{(v)} = S^{(v)} \times \left(\frac{\sum_{k \neq v} P^{(k)}}{m - 1} \right) \times \left(S^{(v)} \right)^T, v = 1, 2, 3, \cdots, m \tag{6}$$

Every iterative sequence requires renormalizing the updated matrix via the previously stated equation. Algorithm termination triggers when either matrix convergence is achieved or the iteration count attains a preset threshold, typically set at 20 by empirical convention.

2.3 Spectral Clustering

Cluster identification is achieved by exploiting the intrinsic affinities between data points, necessitating solely a similarity matrix for data representation. This approach demonstrates exceptional efficacy when clustering sparse datasets, rendering it particularly applicable to the current research framework. RatioCut—a graph-theoretic partitioning algorithm—operates by segmenting graphs into discrete components through the minimization of intercomponent edge weights. As a graph-based clustering methodology, it is particularly appropriate for analysing relationally structured data. The algorithm's core objective is the minimization of the RatioCut value.

3 Results

3.1 The Choice of the Hyperparameter K and the Number of Clusters

Within the SNF framework, the hyperparameter K serves as a pivotal determinant in configuring the similarity matrix architecture. This parameter delineates the number of nearest neighbors each node connects to, thereby defining the extent of patient-centric neighborhoods during network construction. Optimal K-selectin is critical: undersized values induce excessive sparsity that may neglect significant associations, whereas oversized values produce overconnected networks capable of masking meaningful relationships. Achieving appropriate parameter calibration ensures topological fidelity in node connectivity, consequently enhancing the discriminative capacity of the integrated network. Minimalist sparse connectivity emerges with lower K values, whereas higher values foster denser interconnectivity fabrics. To determine the optimal value, we compare the results of the Kalinsky–Habbash index and the Davis–Balding index in different cases [20]. The influence of different K values on the clustering evaluation is shown in Fig. 3, where subplots A and B show the results of the Kalinsky–Habbash index and Davis–Balding index, respectively, which indicate that the best k value to choose is 5.

The second objective involves assessing how cluster count affects clustering outcomes; subgraphs C and D in Fig. 3 show the clustering evaluation results with the number of clusters ranging from 2–5. According to the results of the Kalinsky–Habbash index and Davis–Balding index, the optimal number of clusters is 2.

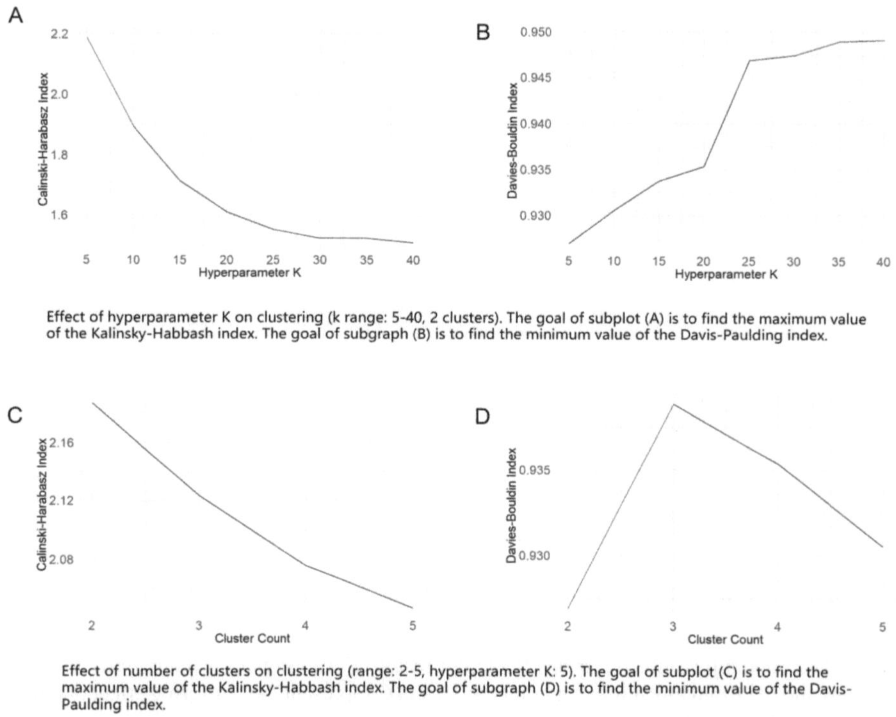

Fig. 3. Clustering performance indicators for different k values and numbers of clusters. Subplots A and B depict the Calinski–Harabasz index and Davies–Bouldin index, respectively, across k values, whereas subplots C and D display both metrics for different numbers of clusters.

3.2 Cluster Effect Diagram

The effects of spectral clustering on the respective patient similarity networks constructed by the three modalities are shown in Fig. 4, where subfigures A, B and C represent the clustering effects of the three modalities of fMRI, DNA methylation and clinical behavior assessment, respectively. Figure D shows the clustering effect diagram after fusing the patient similarity networks of the three modalities, and there are obvious differences between the clustering effects of each modality [21].

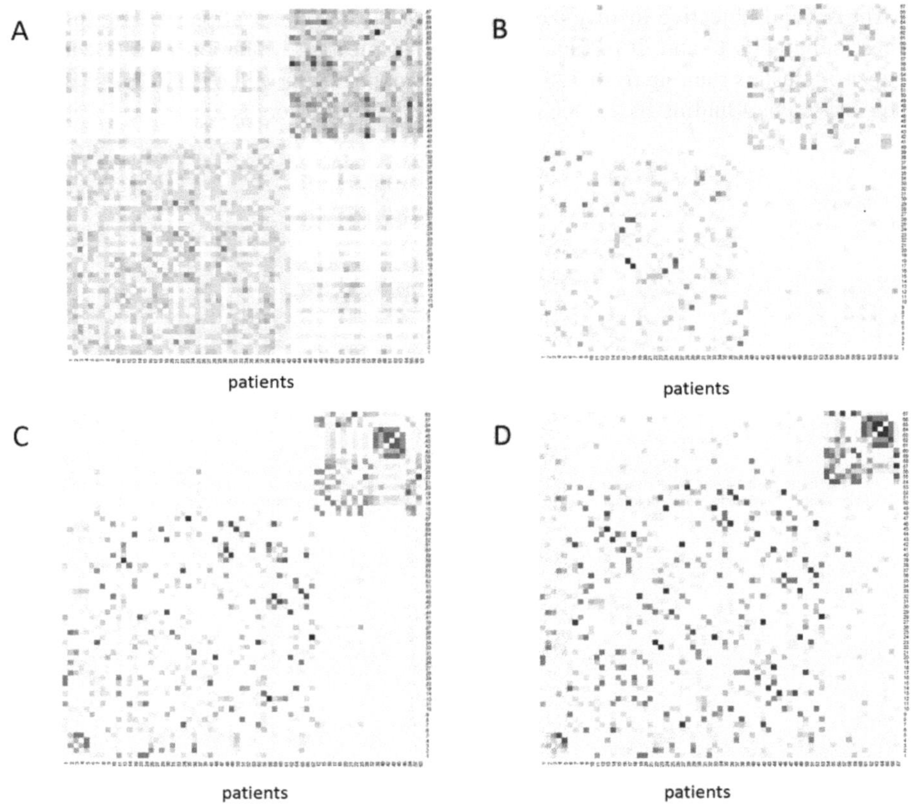

Fig. 4. Three modalities and clustering effect after fusion. Clustering effect diagrams of the three modalities and the fused modalities. Subplots A, B, and C represent the clustering effect diagrams of the three modalities (fMRI, DNA methylation, and Clinical Behavior Assessment), respectively. Subplot D is the clustering effect diagram after fusing the patient similarity networks of the three modalities.

3.3 Statistical Analysis of PD Subtypes and Differences

By adjusting the hyperparameter k and the number of clusters, two different clusters were obtained: subtype 1, 53 cases (79.1%), and subtype 2, 14 cases (20.9%).

Clinical Features. By adjusting the hyperparameter k and the number of clusters, two different clusters were obtained: subtype 1, with 53 cases (79.1%), and subtype 2, with 14 cases (20.9%). We used statistical tests to determine the differences between the two subtypes. The UPDRS is currently a widely used scale internationally and is divided into four main parts: the MDS-UPDRS comprises four distinct parts: Part I evaluates nonmotor experiences of daily living; Part II focuses on motor aspects of daily living; Part III involves the objective assessment of motor functions; and Part IV records complications of therapy. Individual items are rated on a 5-point ordinal scale (0–4), with higher scores indicating greater symptom severity in Parkinson's disease patients. [21].

As shown in Table 1 above, the two subtypes demonstrated comparable mean ages, sex distributions, and overall MDS-UPDRS I scores, with no statistically significant differences observed; however, numerically, subtype 1 was more severe, and subtype 2 was less severe (1.23 ± 1.34 vs. 0.86 ± 1.17, $p = 0.256$). In terms of the overall MDS-UPDRS-II score (5.58 ± 3.73 vs. 1.50 ± 3.81, $p < 0.001$), a significantly higher score was observed in subtype 1 than in subtype 2. In terms of the overall MDS-UPDRS III score (20.06 ± 8.81 vs. 2.86 ± 5.71, $p < 0.001$), Subtype 1 patients had significantly higher scores than did Subtype 2 patients. Furthermore, subtype 1 had higher scores specifically on items assessing posture, gait and balance, and tremor. However, the overall MDS-UPDRS IV scores were comparable between the two subtypes (1.58 ± 8.74 vs. 2.14 ± 3.76, $p = 0.386$). Therefore, according to the MDS-UPDRS scale score, the PD symptoms of subtype 1 patients were more severe, whereas those of subtype 2 patients were milder.

Table 1. Demographic differences between subtypes 1 and 2, as well as variations in the UPDRS score

Subtype	Subtype 1(n = 53)	Subtype 2(n = 14)	P
Gender			
Male	15(28.30%)	4(28.57%)	
Female	38(71.70%)	10(71.43%)	
Average Age	61.30 ± 10.17	60.41 ± 11.45	0.779
Nonmotor Symptoms			
UPDRSINP1RTOT	1.23 ± 1.34	0.86 ± 1.17	0.256
Motor Symptoms			
UPDRSIINP2PTOT	5.58 ± 3.73	1.50 ± 3.81	< 0.001
UPDRSIIINP3RISNG	0.11 ± 0.32	0.00 ± 0.00	0.19
UPDRSIIINP3GAIT	0.38 ± 0.53	0.07 ± 0.27	0.038
UPDRSIIINP3FRZGT	0.02 ± 0.14	0.00 ± 0.00	0.607
UPDRSIIINP3PSTBL	0.04 ± 0.19	0.00 ± 0.00	0.464
UPDRSIIINP3POSTR	0.42 ± 0.53	0.07 ± 0.27	0.022
UPDRSIIINP3PTRMR	0.26 ± 0.56	0.14 ± 0.36	0.484
UPDRSIIINP3PTRML	0.43 ± 0.64	0.00 ± 0.00	0.009
UPDRSIIINP3KTRMR	0.28 ± 0.45	0.07 ± 0.27	0.101
UPDRSIIINP3KTRML	0.42 ± 0.63	0.00 ± 0.00	0.012
UPDRSIIINP3RTARU	0.60 ± 0.91	0.00 ± 0.00	0.01
UPDRSIIINP3RTALU	0.72 ± 1.03	0.00 ± 0.00	0.009
UPDRSIIINP3RTARL	0.17 ± 0.47	0.00 ± 0.00	0.154
UPDRSIIINP3RTALL	0.15 ± 0.46	0.00 ± 0.00	0.191
UPDRSIIINP3RTALJ	0.11 ± 0.38	0.00 ± 0.00	0.236
UPDRSIIINP3RTCON	1.43 ± 1.28	0.07 ± 0.27	< 0.001
UPDRSIIINP3TOT	20.06 ± 8.81	2.86 ± 5.71	< 0.001
UPDRSIVNP4TOT	1.58 ± 8.74	2.14 ± 3.76	0.368

Neuroimaging Features. In the fMRI modality, we summarized the graphical indices of the functional connectivity matrices for the subtype 1, subtype 2, and healthy control groups, as shown in Table 2. Among them, the clustering coefficient, K-core, and global efficiency of subtype 1 were all lower than those of subtype 2 and the healthy control group; however, subtype 1 exhibited superior local efficiency and modularity compared with both subtype 2 and the healthy controls. Its network homophily was marginally greater than that of the control group but remained lower than that observed in subtype 2. Notably, subtype 2 displayed a substantially higher K-core value than did subtype 1, indicative of an overall stronger functional connectivity matrix in the former group. This observation is visually confirmed in Fig. 5, which illustrates the generally enhanced connectivity strength in subtype 2 relative to subtype 1. By subtracting the average functional connectivity matrix of a normal person from the average functional connectivity matrix of subtype 1, the matrix graph is mostly blue, indicating that the functional connectivity strength of subtype 1 is generally lower than that of a normal person, whereas the matrix graph of subtype 2 minus a normal person is mostly red, indicating that the functional connectivity strength of subtype 2 is generally greater than that of a normal person.

Table 2. Graphic indices of sFC for subtype 1, subtype 2, and healthy controls

	subtype 1	subtype 2	healthy controls
Clustering coefficient	0.737	0.819	0.794
k core	10.97	19.375	16.604
Degree of modulus	0.332	0.181	0.248
Network assortativity	0.202	0.768	0.193
Global efficiency	0.641	0.906	0.737
Local efficiency	0.864	0.318	0.896

DNA Methylation Features. For the DNA methylation cell composition data, we employed nonparametric statistical analysis (based on the Mann–Whitney U test) to test the six sets of cell composition data [22] to determine whether there were significant differences in the DNA methylation data between subtypes 1 and 2. The results are presented in Table 3. As shown in the table, for the Bcell and Gran data, the p values of subtype 1 and subtype 2 are less than 0.05, indicating a significant difference. Subtype 1 presented significantly lower mean B-cell counts than did subtype 2, whereas markedly elevated Gran values were observed in this subtype compared with those in subtype 2.

We also utilized violin plots to reflect the differences between the two subtypes more visually (Fig. 6). The violin plots show that the distribution ranges of the various data within the six types of methylation data for the two subtypes are distinct.

Table 3. Mann–Whitney U test results for the DNA methylation data of subtypes 1 and 2

	subtype 1	subtype 2	u	p
CD8T	0.083	0.103	239	0.0508
CD4T	0.105	0.133	243	0.0587
NK	0.047	0.050	375.5	0.8630
Bcell	0.049	0.064	180	0.0039(<0.05)
Mono	0.097	0.109	261	0.1078
Gran	0.637	0.562	577	0.0008(<0.05)

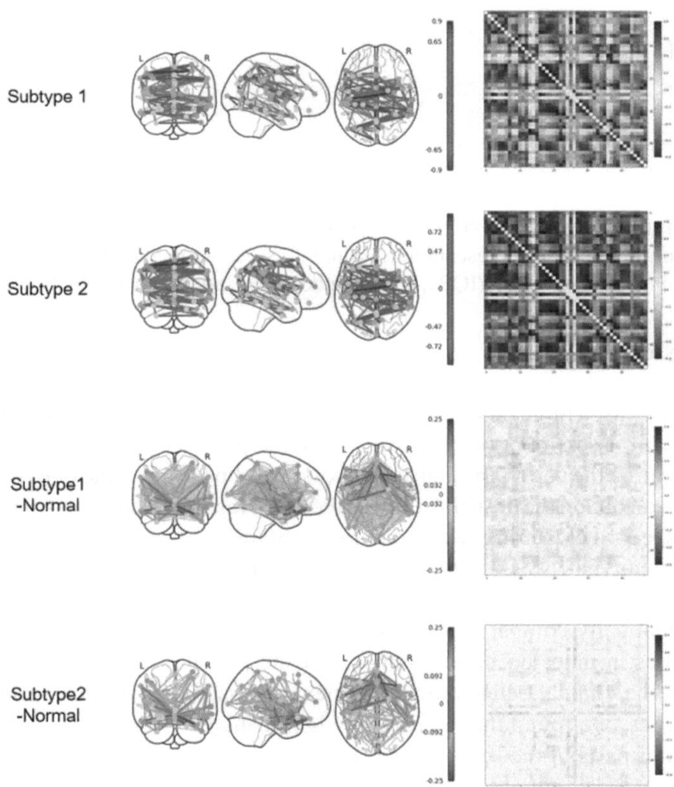

Fig. 5. The average functional connectivity matrix and brain connectivity map of subtypes 1 and 2 and the functional connectivity matrix and brain connectivity map obtained by subtracting the functional connectivity matrix of normal subjects from subtypes 1 and 2. Only the top 5% of connections with strength were plotted in the brain connectivity map. (Blue edges indicate decreased connectivity, and red edges indicate increased connectivity.)

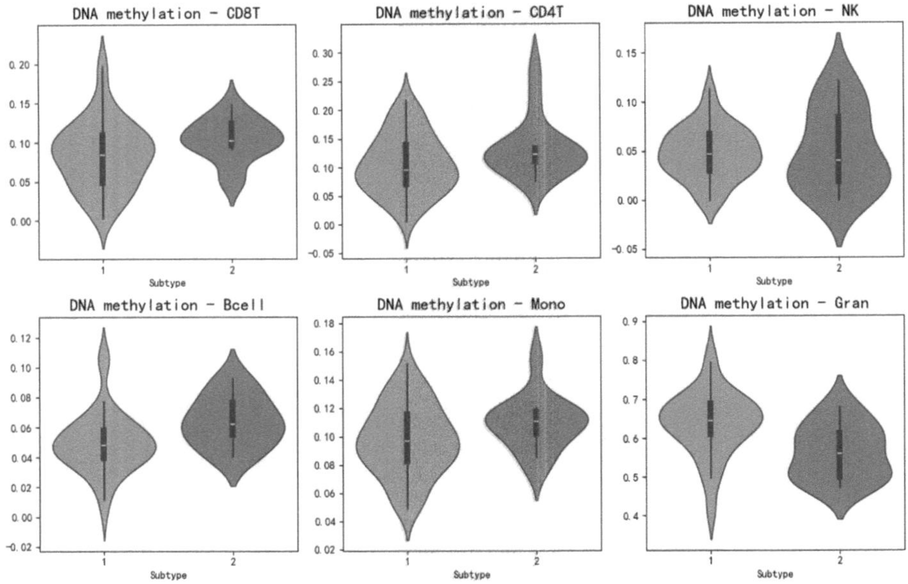

Fig. 6. Kernel density estimation of the distribution of DNA methylation features displayed by violin plots. In all graphical representations, white dots denote median values, box boundaries demarcate the interquartile range (IQR), and extended lines indicate values at 1.5 × IQR.

4 Discussion

In this study, we fused three modalities of data from patient fMRI, DNA methylation, and clinical behavior assessment of Parkinson's disease patients through the SNF method. When constructing the patient similarity network, we choose the value of hyperparameter K as 5, which can obtain the best clustering effect. By clustering the final patient similarity network via spectral clustering, we obtain two subtypes of Parkinson's disease patients. Among them, subtype 1 was severe, and the condition was more severe. Subtype 2 refers to mild patients with mild symptoms.

By systematically comparing the differences between subtypes 1 and 2 across the three modalities, a more accurate diagnosis of the patient's current condition can be achieved. For the clinical behavioral assessment, we used the UPDRS scale to analyse the two subtypes. We found that there were significant differences between subtypes 1 and 2 in all four parts of the scale. In particular, in the MDS-UPDRS III table, subtype 1 patients presented significantly elevated scores relative to those of subtype 2 patients across the majority of the assessed items. Therefore, it can be concluded that the symptoms of subtype 1 patients are more severe than those of subtype 2 patients. Through the analysis of fMRI data, various graph theory metrics of the functional connectivity matrix, analysis of network topological metrics—clustering coefficient, k-core, modularity, network homogeneity, global efficiency, and local efficiency—indicated reduced functional connectivity in subtype 1, contrasting with enhanced connectivity in subtype 2. Relative to healthy controls, subtype 1 patients consistently displayed diminished functional connectivity strength, whereas subtype 2 patients presented elevated levels. These results provide valuable insights for leveraging MRI data in the diagnosis of Parkinson's disease subtypes. For patient DNA methylation data, Mann–Whitney U nonparametric tests combined with violin plots revealed pronounced disparities between the subtypes, with marked variations specifically observed in the B-cell and granulocyte profiles. However, these differences are insufficient for definitive diagnostic assistance in determining the disease status of Parkinson's disease patients. Therefore, further enhancements in the analysis of DNA methylation data are warranted.

However, the present study still has notable limitations. First, in the assessment of clinical information, only the data from the UPDRS were utilized, while many other clinical data were not considered. Furthermore, during the subtype analysis, no clinical domain expertise was incorporated; instead, the approach relied exclusively on data-driven methods. Integrating both clinical knowledge and data-driven techniques would likely yield a more precise subtype classification. Additionally, this study involved a limited sample size (67 participants). Therefore, future investigations with larger cohorts are necessary to validate the applicability of this classification approach across more diverse Parkinson's disease (PD) patient populations.

5 Conclusion

This investigation applied the similarity network fusion (SNF) approach to integrate functional magnetic resonance imaging (fMRI), DNA methylation, and clinical behavioral assessments of Parkinson's disease patients, thereby generating a comprehensive Parkinson's patient similarity network that encompasses information from these three modalities. Spectral clustering was implemented on the patient similarity network, leading to the discovery of two discrete clinical Parkinson's disease (PD) subtypes. Significant distinctions characterized these subtypes: Subtype 1 displayed heightened severity in both motor and nonmotor deficits alongside generally reduced functional connectivity relative to healthy controls. Conversely, Subtype 2 was associated with attenuated symptoms and increased functional connectivity strength compared with healthy controls. Collectively, these results provide valuable evidence supporting the utility of MRI data for diagnosing PD subtypes.

To summarize, this study integrated three Parkinson's disease (PD) patient multimodal datasets via similarity network fusion (SNF), classified patients into two distinct subtypes via spectral clustering, and contrasted these subtypes against a healthy control cohort. The analysis revealed substantial intersubtype differences, establishing a new methodological framework for identifying PD subtypes.

References

1. Raju, V.R.: Analysis of Parkinson's disease. IP Indian J. Neurosci. **7**(2), 148–155 (2021). https://doi.org/10.18231/j.ijn.2021.025
2. Kandharkar, S.S., Kashid,N.A., Kanase, S.B., Mane, D.: Parkinson s disease, PDQ - 39 questionnaire, quality of life (QOL), motor and non-motor symptoms. Activ. Daily Living 20490 (2021)

3. Heimrich, K.G., Schönenberg, A., Santos-García, D., Mir, P., Prell, T.: The impact of nonmotor symptoms on health-related quality of life in parkinson's disease: a network analysis approach. J. Clin. Med. **12**(7), 2573 (2023). https://doi.org/10.3390/jcm12072573
4. Deuschl, G., Krack, P.: Intrepidly studying deep brain stimulation in patients with Parkinson's disease. Lancet Neurol. **19**(6), 472–473 (2020). https://doi.org/10.1016/s1474-4422(20)30138-1
5. Storey, K., Foster, E., Kalishman, S., Doty, T.: Cognitive stimulation therapy for people with parkinson's disease dementia. Arch. Phys. Med. Rehabil. **102**(10), e37–e38 (2021). https://doi.org/10.1016/j.apmr.2021.07.570
6. Perez-Lloret, S., et al.: The genetic basis of probable REM sleep behavior disorder in Parkinson's disease. Brain Sci. **13**(8), 1146 (2023). https://doi.org/10.20944/preprints202306.2047.v1
7. Wolters, A.F., et al.: Brain network characteristics and cognitive performance in motor subtypes of Parkinson's disease: a resting state fMRI study. Parkinsonism Relat. Disord. **105**, 32–38 (2022). https://doi.org/10.1016/j.parkreldis.2022.10.027
8. Markello, R.D., Shafiei, G., Tremblay, C., Postuma, R.B., Dagher,A., Mišić, B.: Multimodal phenotypic axes of Parkinson's disease. npj Parkinson's Dis. **7**(1) (2021). https://doi.org/10.1038/s41531-020-00144-9
9. Yakovenko, E.V., Elena, V.Y., Fedotova, E.Yu., Fedotova, E.Yu., Иллариошкин, С.Н., Sergey, N.I.: DNA methylation in Parkinson disease. Ann. Clin. Exp. Neurol. **14**(4) (2020). https://doi.org/10.25692/acen.2020.4.10
10. Lin, Q., et al.: Promoter methylation analysis of seven clock genes in Parkinson's disease. Neurosci. Lett. **507**(2), 147–150 (2012). https://doi.org/10.1016/j.neulet.2011.12.007
11. Shah, A., et al.: Combining multimodal biomarkers to guide deep brain stimulation programming in parkinson disease. Neuromodul. Technol. Neural Interface **26**(2), 320–332 (2023). https://doi.org/10.1016/j.neurom.2022.01.017
12. Acosta, J., Falcone, G.J., Rajpurkar, P., Topol, E.J.: Multimodal biomedical AI. Nat. Med. **28**(9), 1773–1784 (2022). https://doi.org/10.1038/s41591-022-01981-2
13. Wang, B., et al.: Similarity network fusion for aggregating data types on a genomic scale. Nat. Methods **11**(3), 333–337 (2014). https://doi.org/10.1038/nmeth.2810
14. Yang, K.: Clustering of cancer subtypes based on similarity network fusion of normalized euclidean distance and spectral neighbor. Pure Math. **09**(10), 1115–1122 (2019). https://doi.org/10.12677/pm.2019.910137
15. Yang, M., Matan-Lithwick, S., Wang, Y., De Jager, P.L., Bennett, D.A., Felsky, D.: Multi-omics integration via similarity network fusion to detect subtypes of aging. Alzheimers Dement. **18**(S4), e069431 (2022). https://doi.org/10.1002/alz.069431
16. Bulcke, C.V., et al.: BMAT: an open-source BIDS managing and analysis tool. NeuroImage: Clin. **36**, 103252 (2022). https://doi.org/10.1016/j.nicl.2022.103252
17. Estéban, O., et al.: FMRIPrep: a robust preprocessing pipeline for functional MRI. bioRxiv (Cold Spring Harbor Laboratory) (2018). https://doi.org/10.1101/306951
18. Guimaraes, A.A.R., Abdou, T., Başar, A.: Deep learning brain regions in Harvard-Oxford atlas using SegNet model (2019)
19. Estéban, O., et al.: FMRIprep: a robust preprocessing pipeline for task-based and resting-state fMRI data. F1000Research **6** (2017). https://doi.org/10.7490/f1000research.1114420.1
20. Yue, G., Deng, A., Qu, Y., Cui, H., Wang, X.: Stratified multi-density spectral clustering using Gaussian mixture model. Inf. Sci. **633**, 182–203 (2023). https://doi.org/10.1016/j.ins.2023.03.067

21. Brendel, M., Su, C., Hou, Y., Henchcliffe, C., Wang, F.: Comprehensive subtyping of Parkinson's disease patients with similarity fusion: a case study with BioFIND data. npj Parkinson's Dis. **7**(1) (2021). https://doi.org/10.1038/s41531-021-00228-0
22. Mann–Whitney U test. the SAGE encyclopedia of research design (2022). https://doi.org/10.4135/9781071812082.n316

Anomaly Prediction in Power Data Platforms Using A Transformer-BiLSTM Model

Ruixuan Lu, Zhaohong Jia(✉), Shichuan Ding , Jin Zhang , Zhen Yao, and Zhenhua Huang

Anhui University, State Grid Anhui Electricpowerco.Ltd. Informaton and Telecommunication branch, Hefei, China
zhjia@mail.ustc.edu.cn

Abstract. With the ongoing digitalization and intelligent transformation of power systems, the grid data platform, as a critical data infrastructure, has a direct impact on the operational continuity and regulation security of power systems. However, traditional anomaly detection methods struggle to address the high-dimensional, multi-source heterogeneous, and dynamically evolving characteristics of platform operational data, necessitating intelligent prediction approaches with global modeling and temporal perception capabilities. To address this challenge, this study proposes an anomaly trend prediction model integrating a Transformer encoder and BiLSTM decoder architecture. The model combines multi-head attention mechanisms with bidirectional sequence modeling to achieve high-precision trend forecasting for multi-indicator system states. Additionally, we develop a comprehensive system implementation framework supporting batch modeling and scheduled training for multiple subsystem tasks, demonstrating strong engineering adaptability and scalability. Experimental results show that the proposed model achieves a coefficient of determination (R^2) of 0.9003 on the test set, exhibiting superior early-warning capability and cross-task adaptability. The model has been deployed on a provincial-level power data platform, enabling joint prediction of key indicators across subsystems including OGG, DataWorks, and MaxCompute. This work provides an efficient solution for intelligent monitoring and operation management of power system platforms, effectively balancing global attention modeling and temporal continuity.

Keywords: Power data platform · Time series forecasting · Anomaly detection · Transformer · BiLSTM · Multi-task modeling

1 Introduction

With the rapid development of next-generation power systems and smart energy ecosystems, the grid data platform has emerged as a core infrastructure for

L. Huang et al. (Eds.): CCF NCCA 2025, CCIS 2774, pp. 658–671, 2026.
https://doi.org/10.1007/978-981-95-5294-8_43

power enterprise data governance, undertaking critical functions such as data acquisition, resource scheduling, task orchestration, and business support. By integrating multi-source heterogeneous data from dispatch systems, production monitoring systems, marketing platforms, and integrated energy management platforms, the data platform facilitates data-driven decision-making and intelligent operations for power systems. However, the platform's complex architecture, tightly coupled modules, and lengthy dataflow pipelines frequently lead to operational anomalies such as task delays, interruptions, and resource exhaustion, posing significant risks to system stability and reliability.

Traditional anomaly detection methods, predominantly reliant on threshold rules, static configurations, or manual scripting, exhibit notable limitations when handling modern multi-dimensional, dynamic, and highly nonlinear time-series indicators . With advancements in deep learning, data-driven models have increasingly been adopted for anomaly prediction and modeling. Sequence models—including recurrent neural networks (RNNs), long short-term memory networks (LSTMs) [1], and gated recurrent units (GRUs) [2]—have demonstrated effectiveness in industrial monitoring, financial risk control, and medical early-warning systems [3,4]. These models capture temporal dependencies within sequences to identify potential risks proactively.

Nevertheless, conventional RNN-based models suffer from information loss and dependency decay when processing long-range or multi-variable coupled features. Furthermore, their limited capacity to model cross-dimensional interactions among high-dimensional indicators hinders precise identification of anomaly evolution patterns. Recent breakthroughs in Transformer architectures, powered by their attention-based representation strength, have achieved state-of-the-art performance in natural language processing [5], time-series modeling [6], and multimodal data analysis [7]. By leveraging multi-head self-attention mechanisms, Transformers model global dependencies across arbitrary time steps, overcoming the local perception constraints of RNNs.

As the grid data platform plays a pivotal role in dispatching, monitoring, and data aggregation, its operational status profoundly impacts the efficiency and security of grid-wide regulation. With expanding system scales and increasing indicator complexity, traditional anomaly detection strategies struggle to address hybrid anomalies involving abrupt and gradual evolution patterns. Intelligent methods with robust temporal modeling and trend prediction capabilities are urgently needed. Early identification of "leading signals"—such as CPU surges, memory accumulation, and disk congestion—before resource loads reach critical thresholds is vital for timely operational responses.

In practice, the platform's subsystems (e.g., OGG, DataWorks, MaxCompute) present challenges such as heterogeneous feature dimensions, varying data update frequencies, and diverse anomaly patterns. Existing models lack unified adaptability and deployable frameworks for cross-system prediction. To address these issues, this paper proposes a Transformer-BiLSTM hybrid model for anomaly trend forecasting. The model extracts global dynamic features from platform monitoring data via a Transformer encoder and performs temporal modeling through a bidirectional LSTM (BiLSTM) decoder, enhancing sensitivity to resource trends and responsiveness to sudden anomalies. By integrating global dependency modeling with sequential continuity, our approach employs multilayer perceptrons (MLPs) for feature fusion, enabling high-quality input representations across diverse platform scenarios.

The primary contributions of this work are as follows:

We propose a Transformer-BiLSTM fusion architecture for anomaly trend prediction, effectively combining global attention mechanisms with bidirectional temporal modeling to address high-dimensional monitoring indicators in power data platforms.We develop a modular multi-system training and scheduling framework that supports batch modeling and deployment for OGG, DataWorks, MaxCompute, and other subsystems, ensuring strong engineering deployability and task scalability.We conduct extensive multi-indicator, multi-task experiments on real-world power system datasets, validating the model's superiority in anomaly warning accuracy, early detection capability, and stability.

2 Related Work

2.1 Time Series Forecasting Models

Time series forecasting is a core task in power system operational state analysis. Early research primarily relied on statistical methods such as ARIMA and autoregressive moving average (ARMA) models [8], which performed well for stationary sequences but exhibited limitations in handling nonlinear, multivariate, and high-volatility data. With the rise of deep learning, recurrent neural networks (RNNs) and their variants—including long short-term memory (LSTM) networks [3] and gated recurrent units (GRUs)—have been widely adopted in industrial monitoring, financial risk management, and traffic prediction, demonstrating robust temporal modeling capabilities.

However, LSTM and GRU still face challenges such as gradient vanishing and information decay when modeling long-span sequences. To mitigate these issues, researchers introduced attention mechanisms to enhance the perception of long-range dependencies. For example, the Seq2Seq+Attention architecture [9] laid the foundation for subsequent Transformer models.

2.2 Transformer for Temporal Modeling

The Transformer model was initially proposed for natural language processing [5], employing self-attention mechanisms to model dependencies between arbitrary positions in input sequences. This innovation overcame the recursive constraints of traditional RNN-based models, offering superior parallelism and modeling capacity.

In recent years, Transformers have been extensively adapted for time series forecasting. Notable examples include:

Informer [10], which introduced sparse self-attention for long-sequence prediction; Autoformer [11], enhancing interpretability through trend-seasonality decomposition; Temporal Fusion Transformer (TFT) [6], achieving explainable multi-source time-series predictions via gating mechanisms and contextual fusion. These advancements provide critical insights for the design of our Transformer-BiLSTM hybrid architecture.

2.3 Anomaly Detection and Prediction in Power Systems

Anomaly detection and prediction are pivotal for ensuring stable power system operations and have become a research focus in recent years. Traditional approaches, such as rule-based expert systems, threshold alert mechanisms, and principal component analysis (PCA), struggle to adapt to high-dimensional, dynamically evolving data environments .

Deep learning has revolutionized anomaly modeling. For instance, Wang et al. [12] developed an LSTM-based data-driven model for distribution network node load forecasting, while Zhang et al. [13] employed graph neural networks to analyze anomaly propagation paths in distribution systems. In contrast, the proposed Transformer-BiLSTM architecture better integrates global and local sequential information, significantly improving the accuracy and lead time of anomaly trend identification.

3 Transformer-BiLSTM Model

This section details the proposed Transformer-BiLSTM hybrid architecture for anomaly trend prediction. The model comprises four core modules: a multilayer perceptron (MLP) feature extractor, a multi-layer Transformer encoder, a BiLSTM decoder, and a prediction output layer. The overall framework is illustrated in Fig. 1:

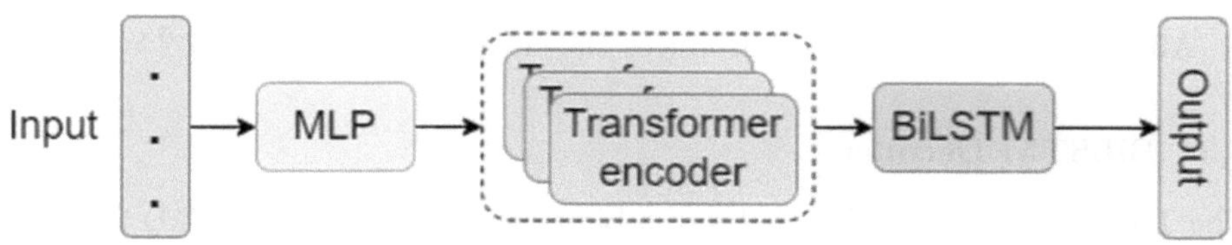

Fig. 1. Transformer-BiLSTM hybrid architecture.

3.1 Feature Modeling and Input Encoding

The model input consists of multi-dimensional resource monitoring sequences, including CPU utilization, memory usage, disk I/O, network bandwidth, and business request volume. Input data is preprocessed using sliding windows to generate fixed-length time segments, followed by normalization. Each sample has the shape:

$$X \in \mathbb{R}^{B \times T \times d} \tag{1}$$

where B is the batch size, T is the sequence length, and d is the feature dimension. A two-layer MLP performs nonlinear mapping:

$$h = ReLU(W_1 x + b_1), \quad x' = LayerNorm(W_2 h + b_2 + h) \tag{2}$$

Here, W1,W2 are weight matrices, b1,b2 are bias terms, and residual connections are applied. The MLP provides high-level abstract representations while unifying feature dimensions.

3.2 Transformer Encoder Module

The Transformer encoder contains L stacked layers, each consisting of: Multi-head self-attention mechanism Point-wise feed-forward network (FFN) Residual connections and layer normalization

The computation for each layer is:

$$Attention(\mathbf{Q}, \mathbf{K}, \mathbf{V}) = softmax\left(\frac{\mathbf{Q}\mathbf{K}^\top}{\sqrt{d_k}}\right)\mathbf{V} \tag{3}$$

where$\mathbf{Q}, \mathbf{K}, \mathbf{V}$ are obtained via linear projections:

$$\mathbf{Q} = \mathbf{H}W_Q, \quad \mathbf{K} = \mathbf{H}W_K, \quad \mathbf{V} = \mathbf{H}W_V \tag{4}$$

Each attention head outputs features of dimension H/h, where h is the number of heads. Concatenated head outputs are processed by the FFN:

$$FFN(\mathbf{z}) = ReLU(\mathbf{z}W_1 + \mathbf{b}_1)W_2 + \mathbf{b}_2 \tag{5}$$

After L layers, the encoder produces the encoded sequence$\mathbf{Z} \in \mathbb{R}^{B \times T \times H}$ 。

3.3 BiLSTM Decoder

The Transformer output $\mathbf{Z}$ is fed into a bidirectional LSTM (BiLSTM) for temporal modeling. The forward and backward hidden states at each time step are concatenated:

$$\mathbf{h}_f, \mathbf{h}_b = BiLSTM(\mathbf{Z}), \mathbf{H}_t^{bilstm} = [\mathbf{h}_f; \mathbf{h}_b] \in \mathbb{R}^{2H} \tag{6}$$

A linear projection with residual connection compresses the dimension:

$$\hat{\mathbf{z}}_t = Dropout(W \cdot \mathbf{H}_t^{bilstm} + \mathbf{b}) + \mathbf{z}_t \tag{7}$$

This structure enhances local trend modeling and compensates for the Transformer's limitations in temporal continuity.

3.4 Trend Prediction Output Layer

The final output layer maps decoded features to anomaly trend scores over P future steps using a three-layer fully connected network:

$$y = W_3 \cdot ReLU(W_2 \cdot ReLU(W_1 x + b_1) + b_2) + b_3 \tag{8}$$

The output $y \in \mathbb{R}^{B \times P}$ serves for regression prediction or subsequent anomaly classification, where B is the batch size, P is the prediction horizon, and Wi,bi are layer parameters.

3.5 Training Strategy

The model is trained via supervised learning to minimize prediction errors. Key strategies include:

Loss Function: Mean squared error (MSE) is used:

$$\mathcal{L}_{MSE} = \frac{1}{BP} \sum_{i=1}^{B} \sum_{t=1}^{P} (\hat{y}_{i,t} - y_{i,t})^2 \tag{9}$$

where B is batch size,P is prediction steps,$\hat{y}_{i,t}$ and $y_{i,t}$ denote predicted and true values.

Optimizer: Adam with initial learning rate 0.001, decayed every 1,000 steps. Batch size = 64, epochs = 7,000.

Regularization: Dropout (rate = 0.1) in Transformer and BiLSTM layers.

Gradient Clipping: Global gradient norm clipped to 1.0.

Early Stopping: Training halts if validation loss plateaus for 10 consecutive epochs.

The model demonstrates stable validation performance and achieves superior prediction accuracy on the test set.

4 System Implementation

To deploy the proposed Transformer-BiLSTM anomaly trend prediction model in power data platforms, we design and implement a modular, extensible, and batch-schedulable modeling system. Developed in Python with the PyTorch deep learning framework, the system ensures efficient training and robust engineering compatibility.

4.1 System Architecture

The system consists of seven functionally decoupled modules:

Data Processing Module: Reads raw CSV data, performs standardization, constructs sliding window samples, splits datasets into training/validation/test sets, and caches results in PKL format.

Model Definition Module: Encapsulates the Transformer-BiLSTM hybrid architecture, including MLP pre-encoding, multi-layer Transformer blocks, BiLSTM decoder, and output layers, with automatic dimension adaptation for different subsystems.

Training Control Module: Manages single-model training workflows, including loss tracking, validation performance evaluation, optimal model weight saving, and logging.

Batch Scheduling Module: Implements batch training scheduling for subsystem tasks (e.g., OGG, DataWorks, MaxCompute) and centralizes result management.

Prediction Evaluation Module: Loads trained models for test set predictions, computes metrics (MSE, MAE, R^2), and generates structured outputs.

Visualization Module: Plots training/validation loss curves and prediction vs. ground truth comparisons, saving images for model analysis.

Parameter Configuration Module: Centralizes global parameters (sequence length, prediction horizon, hyperparameters, data paths, task labels) to decouple code from configurations.

4.2 Multi-task Training Scheduling

To accommodate multi-table/multi-task prediction requirements, the system employs a task scheduling mechanism. The scheduler automatically traverses task directories (e.g., `ogg/cpu/`, `dataworks/mem/`) and independently trains models for each subtask, enabling efficient batch deployment. Training logs and visualizations are stored in hierarchical paths for post-hoc analysis and model comparison.

4.3 Model Persistence and Evaluation Interface

The system saves optimal models (based on validation loss) for each task. During testing, it loads these models to generate: Prediction vs. actual value plots Quantitative metrics (MSE, MAE, R^2) Model comparison curves This visualization-quantification dual mechanism supports model selection and iterative improvement.

4.4 Visualization and Engineering Adaptability

The visualization module automatically generates and categorizes high-quality images (e.g., loss curves, prediction plots) across resource metrics. The modular design ensures extensibility to future deep architectures like graph neural networks or hybrid models, maintaining engineering maintainability and upgrade potential.

5 Experiments and Evaluation

To validate the effectiveness of the proposed Transformer-BiLSTM hybrid model in anomaly trend prediction for power data platforms, we conduct systematic experiments focusing on dataset construction and evaluation metrics. The experiments utilize multi-source datasets derived from operational logs of core subsystems (OGG, DataWorks, MaxCompute) and employ MSE, MAE, and R^2 as quantitative evaluation criteria.

5.1 Dataset Composition

The experimental data originates from operational logs of a provincial power data platform, covering key metrics from multiple subsystems. Tasks are categorized into subtables based on subsystem types, each forming an independent prediction subtask. The input feature dimensions for each subsystem are as follows:

OGG System: Input dimension $d = 5$, including CPU, memory, disk, and network metrics.

DataWorks System: Input dimension $d = 7$, augmented with scheduling traffic and task queue features.

MaxCompute System: Input dimension $d = 9$, comprehensively covering resource utilization and job status indicators.

The model input is a multivariate time series with sequence length $T = 10$ and prediction horizon $P = 1$, aiming to predict numerical trends of target metrics at the next timestep. Data is split chronologically into training (60%), validation (20%), and test sets (20%) to prevent future information leakage.

To further investigate prediction capabilities under multi-resource anomaly scenarios, we construct a multi-label anomaly subset focusing on four resource metrics: CPU utilization, Memory occupancy rate, Disk I/O write rate, Slot utilization.

This subset contains 16,000 samples, partitioned into 60%-20%-20% training/validation/test splits, with consistent standardization and sliding window preprocessing to ensure input uniformity (Fig. 2).

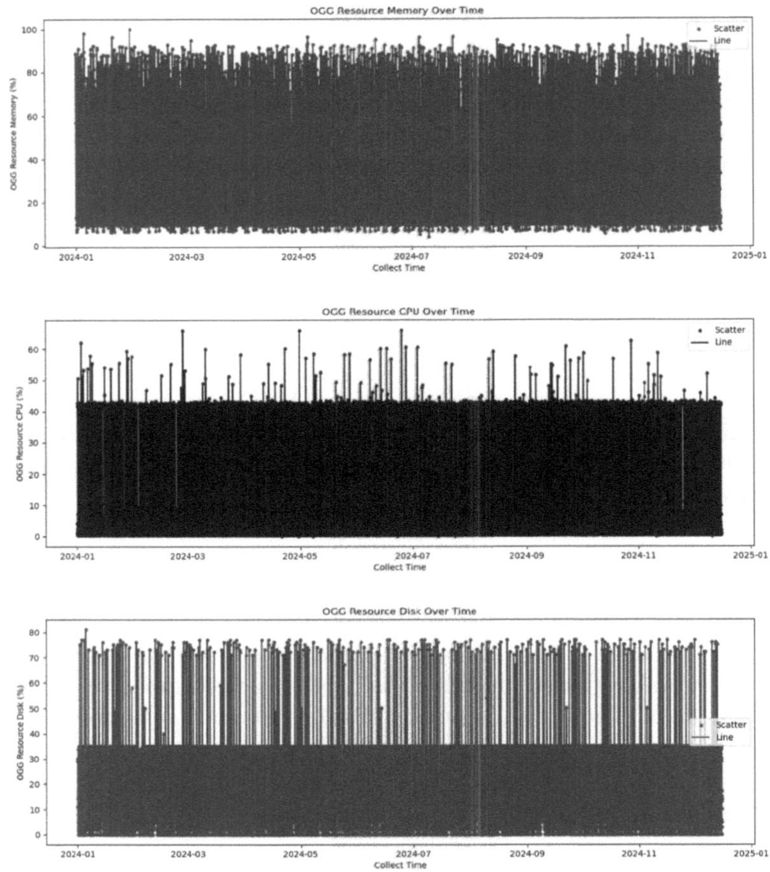

Fig. 2. Visualization of dataset generation.

5.2 Evaluation Metrics

Three standard metrics are adopted for performance evaluation:

Mean Squared Error (MSE):

$$MSE = \frac{1}{n}\sum_{i=1}^{n}(y_i - \hat{y}_i)^2 \tag{10}$$

Mean Absolute Error (MAE):

$$MAE = \frac{1}{n}\sum_{i=1}^{n}|y_i - \hat{y}_i| \tag{11}$$

Coefficient of Determination (R^2):

$$R^2 = 1 - \frac{\sum_i (y_i - \hat{y}_i)^2}{\sum_i (y_i - \bar{y})^2} \tag{12}$$

Here, y_i denotes ground truth values,$\hat{y}_i$ represents predicted values, and$\bar{y}$ is the sample mean.R^2 values closer to 1 indicate better model fitting.

6 Experimental Results and Analysis

This section presents experimental evaluations of the proposed Transformer-BiLSTM anomaly prediction model, validating its generalization capability and accuracy across diverse system metrics. To further assess prediction performance on heterogeneous indicators, we visualize prediction-ground truth comparisons for five critical subtasks selected from the test set:

DataWorks System: Task queue traffic prediction.

MaxCompute System: Job scheduling status prediction.

OGG System: CPU utilization prediction, Disk read/write prediction, Memory usage prediction.

Figures 3, 4, 5, 6 and 7 illustrate the comparative results:

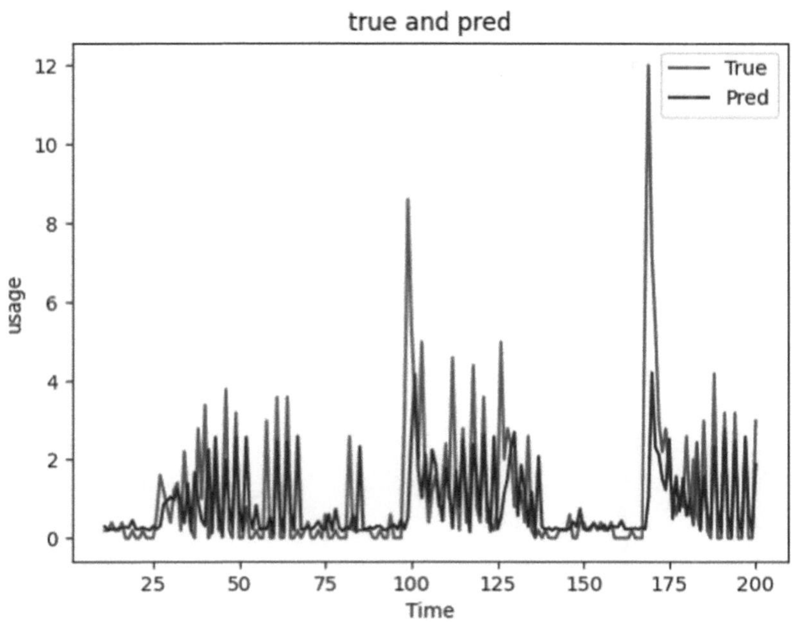

Fig. 3. DataWorks System Prediction Results.

Figure 3 illustrates the comparative analysis between predicted values and ground truth values for the DataWorks system. The model demonstrates high consistency in capturing periodic fluctuations and abrupt changes in task queue traffic, with prediction curves closely tracking actual trends. Notably, the hybrid architecture exhibits enhanced sensitivity to early-stage resource contention signals (e.g., sudden spikes in scheduling latency), enabling timely anomaly warnings. Figure 4 demonstrates the prediction performance for job scheduling status in the MaxCompute system. The model accurately forecasts sudden job queue surges and resource allocation conflicts, with predicted values aligning closely with ground truth trends. Notably, the hybrid architecture shows robust adaptability to heterogeneous job types (e.g., batch processing vs. real-time tasks), enabling early identification of scheduling bottlenecks.

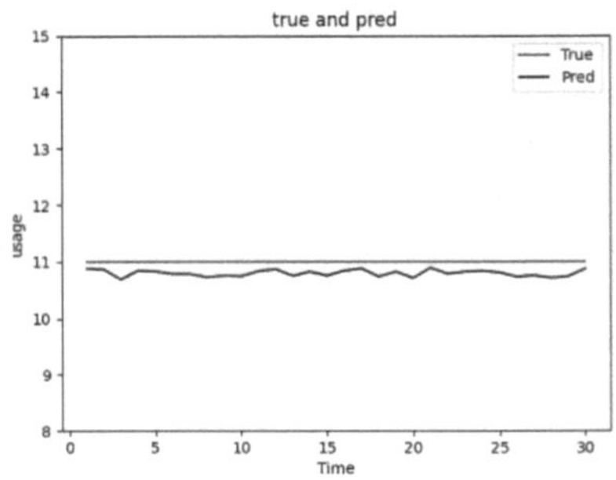

(a) Disk usage

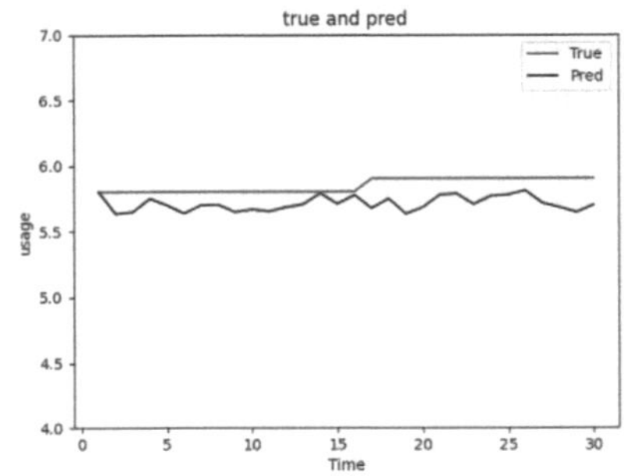

(b) Memory usage

Fig. 4. Visualization of system resource usage.

Figure 5 illustrates CPU utilization prediction results for the OGG system. The model effectively captures periodic workload fluctuations and transient spikes caused by data synchronization tasks. Predictions exhibit minimal lag during abrupt load changes, demonstrating its value in early anomaly warnings (e.g., preempting CPU saturation risks).

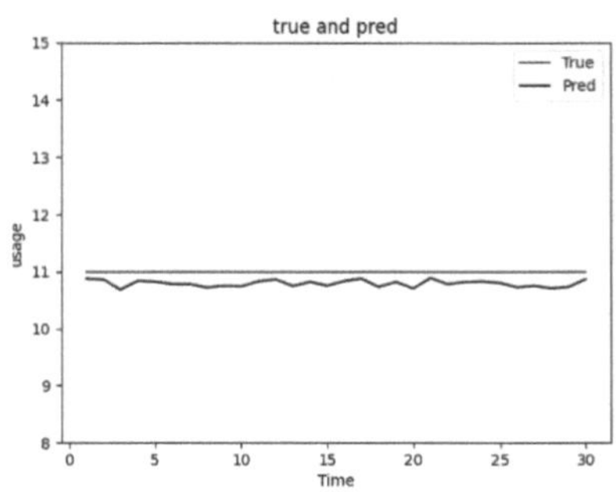

Fig. 5. OGG System - Disk Read/ Write Prediction.

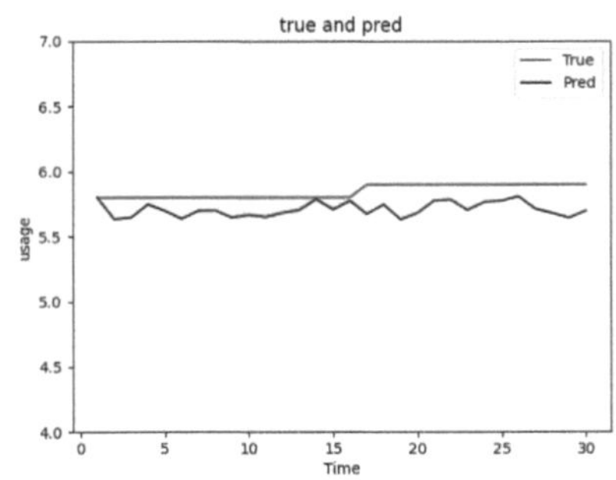

Fig. 6. OGG System - Memory Usage Prediction.

Figures 6, 7 presents disk I/O rate and memory usage prediction results for the OGG system. For disk I/O (Fig. 6a), the hybrid model precisely tracks both steady-state patterns and irregular spikes (e.g., bulk data migration events), with its bidirectional temporal modeling enhancing responsiveness to write-intensive workloads. Simultaneously, for memory utilization (Fig. 6b), the model reliably identifies gradual accumulation trends (e.g., cache leaks) and sharp increases during parallel task execution. Prediction curves consistently align with ground truth values across both metrics, demonstrating its dual capability in disk health monitoring (throughput optimization) and early-stage memory anomaly detection (proactive resource scaling).

Aggregated across all experimental trials, the model achieves a global MSE of 33.7006, MAE of 3.9604, and (R^2) of 0.9003, demonstrating prediction robustness independent of specific workload phases. From the five representative prediction visualizations in Figs. 3, 4, 5, 6 and 7, it is evident that the model's prediction curves closely track the variation trends of ground truth values, demonstrating stable forecasting capabilities under both periodic fluctuations and abrupt disturbance scenarios.

Key observations include:

Consistent performance across load patterns: The model achieves robust fitting accuracy under both steady-state and transient load conditions. Transformer's superiority in periodic metrics: For CPU and disk I/O indicators with inherent periodic patterns, the Transformer encoder effectively captures medium-term periodic fluctuations. BiLSTM's responsiveness to abrupt changes: In modeling MaxCompute job state transitions, the BiLSTM decoder significantly enhances sensitivity to abrupt state transitions (e.g., job queue saturation). Early anomaly indications: During initial stages of system congestion (e.g., gradual memory usage escalation), the prediction curves exhibit anticipatory inflection points, validating its early-warning capability.

As shown in the figures, the proposed Transformer-BiLSTM model reliably approximates complex temporal patterns across subsystems. Notably, under scenarios of drastic load fluctuations, the prediction curves remain tightly aligned with ground truth trajectories, underscoring the model's strong robustness and temporal pattern capture capabilities.

7 Conclusions and Future Work

This study addresses the challenge of multidimensional resource anomaly trend prediction in power data platforms by proposing a deep learning architecture that integrates a Transformer encoder and a BiLSTM decoder. The model effectively combines the global attention modeling capability of Transformer with the temporal dependency awareness of BiLSTM, enabling robust prediction and trend analysis for continuous multi-indicator sequential data in power systems.

Furthermore, we establish a comprehensive system implementation framework encompassing data loading, model training, result evaluation, visualization, and multi-task scheduling mechanisms, validating the efficacy of the proposed method on real-world grid platform datasets. The model demonstrates strong performance on conventional metrics (MSE, MAE, R^2) and exhibits engineering advantages in prediction response time, anomaly trend smoothness, and deployment flexibility. Currently deployed in an operational power data platform, the model supports multi-system operational monitoring and anomaly early warning, laying the foundation for future city-level smart grid regulation systems.

Future research directions include:

Incorporating Graph Structural Information: Leveraging graph neural networks to enhance feature representation by modeling physical/operational dependencies among heterogeneous indicators.

Multi-task Learning Framework: Developing cross-system joint prediction frameworks with parameter sharing and task-specific optimization.

Anomaly Classification and Explainability: Integrating anomaly detection modules with interpretability techniques to improve prediction transparency and controllability.

References

1. Sherstinsky, A.: Fundamentals of recurrent neural network (RNN) and long short-term memory (LSTM) network. Physica D **404**, 132306 (2020)
2. Dey, R., Salem, F.M.: Gate-variants of gated recurrent unit (GRU) neural networks. In: 2017 IEEE 60th International Midwest Symposium on Circuits and Systems (MWSCAS), pp. 1597–1600. IEEE (2017)
3. Graves, A.: Long short-term memory. Supervised Seq. Label. Recurrent Neural Netw. 37–45 (2012)
4. Malhotra, P., Vig, L., Shroff, G., et al.: Long short term memory networks for anomaly detection in time series. In: Proceedings 2015, vol. 89, no. 9, p. 94 (2015)
5. Vaswani, A., Shazeer, N., Parmar, N., et al.: Attention is all you need. Adv. Neural Inf. Process. Syst. **30** (2017)
6. Lim, B., Arik, S.Ö., Loeff, N., et al.: Temporal fusion transformers for interpretable multi-horizon time series forecasting. Int. J. Forecast. **37**(4), 1748–1764 (2021)
7. Dosovitskiy, A., Beyer, L., Kolesnikov, A., et al.: An image is worth 16x16 words: transformers for image recognition at scale. arXiv preprint arXiv:2010.11929 (2020)
8. Box, G.E.P., Pierce, D.A.: Distribution of residual autocorrelations in autoregressive-integrated moving average time series models. J. Am. Stat. Assoc. **65**(332), 1509–1526 (1970)
9. Bahdanau, D., Cho, K., Bengio, Y.: Neural machine translation by jointly learning to align and translate. arXiv preprint arXiv:1409.0473 (2014)
10. Zhou, H., Zhang, S., Peng, J., et al.: Informer: beyond efficient transformer for long sequence time-series forecasting. In: Proceedings of the AAAI Conference on Artificial Intelligence, vol. 35, no. 12, pp. 11106–11115 (2021)
11. Wu, H., Xu, J., Wang, J., et al.: Autoformer: decomposition transformers with auto-correlation for long-term series forecasting. Adv. Neural. Inf. Process. Syst. **34**, 22419–22430 (2021)

12. Inteha, A., Hussain, F., Khan, I.A.: A data driven approach for day ahead short term load forecasting. IEEE Access **10**, 84227–84243 (2022)
13. Pazho, A.D., Noghre, G.A., Purkayastha, A.A., et al.: A survey of graph-based deep learning for anomaly detection in distributed systems. IEEE Trans. Knowl. Data Eng. **36**(1), 1–20 (2023)

Correction to: Research on the Precision Prediction of Geospatial Choices for Postgraduate Graduation Destinations via Machine Learning Algorithms

Ling Quan, Diying Li, Sijia Xie, Ting Ye, and Tianyue Tang

Correction to: Chapter 31 in: L. Huang et al. (Eds.): *Computer Applications*, CCIS 2774, https://doi.org/10.1007/978-981-95-5294-8_31

In the originally published version of chapter 31, in reference citation "Error! Reference source not found" content was present in reference 8 and 9. This error has been removed and correct citation updated.

The updated version of this chapter can be found at
https://doi.org/10.1007/978-981-95-5294-8_31

L. Huang et al. (Eds.): CCF NCCA 2025, CCIS 2774, p. C1, 2026.
https://doi.org/10.1007/978-981-95-5294-8_44

Author Index

L. Huang et al. (Eds.): CCF NCCA 2025, CCIS 2774, pp. 673–675, 2026.
https://doi.org/10.1007/978-981-95-5294-8

MIX
Papier aus verantwortungsvollen Quellen
Paper from responsible sources
FSC® C105338

If you have any concerns about our products,
you can contact us on
ProductSafety@springernature.com

In case Publisher is established outside the EU,
the EU authorized representative is:
Springer Nature Customer Service Center GmbH
Europaplatz 3, 69115 Heidelberg, Germany

Printed by Libri Plureos GmbH
in Hamburg, Germany